Springer Texts in Business and Economics

Springer Texts in Business and Economics (STBE) delivers high-quality instructional content for undergraduates and graduates in all areas of Business/Management Science and Economics. The series is comprised of self-contained books with a broad and comprehensive coverage that are suitable for class as well as for individual self-study. All texts are authored by established experts in their fields and offer a solid methodological background, often accompanied by problems and exercises.

Pak-Sing Choi • Eric Dunaway • Felix Muñoz-Garcia

Industrial Organization

Practice Exercises with Answer Keys

Second Edition

Springer

Pak-Sing Choi
Graduate Institute of Industrial Economics
National Central University
Taoyuan City, Taiwan

Eric Dunaway
Department of Economics
Wabash College
Crawfordsville, IN, USA

Felix Muñoz-Garcia
School of Economic Sciences
Washington State University
Pullman, WA, USA

ISSN 2192-4333 ISSN 2192-4341 (electronic)
Springer Texts in Business and Economics
ISBN 978-3-031-38634-3 ISBN 978-3-031-38635-0 (eBook)
https://doi.org/10.1007/978-3-031-38635-0

This Springer imprint is published by the registered company Springer Nature Switzerland AG
The registered company address is: Gewerbestrasse 11, 6330 Cham, Switzerland

Paper in this product is recyclable.

Preface

This textbook presents exercises on Industrial Organization with detailed answer keys. Our presentation is relatively easy to read and targets undergraduate and master's level students in Economics and Finance, although some of the more challenging questions can be used as introductory material for Ph.D. students in these fields.

There are, currently, several textbooks on Industrial Organization at this level, such as Shy (1996), Carlton and Perloff (2004), Motta (2004), Pepall et al. (2008), the more recent contributions by Belleflamme and Peitz (2015) and Cabral (2017), and the more advanced textbooks by Tirole (1988) and Martin (2003). However, our presentation differs from these textbooks along several dimensions:

1. *Worked-out exercises*. We provide 122 step-by-step exercises with detailed answer keys, so readers can understand how to solve similar questions on their own. We also present the intuition behind each mathematical assumption and result, covering the typical material required in Industrial Organization courses at the undergraduate level and by most courses at the master's level.

   In contrast, the above textbooks on Industrial Organization tend to focus on theoretical tools and the main empirical findings in the literature, but rarely offer step-by-step answers to the exercises. In our experience, this presentation style hinders students' ability to apply the theoretical concepts to other fields or his or her own research projects. Our manuscript, then, seeks to fill this void by complementing the theoretical concepts in current textbooks, offering practice exercises that predict firm behavior and regulation in different industries.
2. *Tools*. We emphasize the game-theoretic tools used in each type of exercise, so readers easily perceive that these tools can be systematically applied to other markets, forms of competition, or information environments where firms, consumers, and regulating agencies interact.
3. *Algebra support and step-by-step calculations*. We assume a little mathematical background in algebra and calculus, walking the reader through most algebraic steps, which motivate students to reproduce each result on their own. From our recent experiences, students' calculus for this course is appropriate, but their algebra is often rusty. Hence, we detail the algebraic simplifications, making sure that students can more easily follow every step.
4. *Exercises based on journal articles*. Several exercises are based on published articles. However, we reduce the model to its main ingredients, providing tractable functional forms for demand and costs and dividing the question into several easy-to-answer parts. This would help students to reproduce the main findings of the article on their own. This teaching tool, which introduces them to published articles in a more friendly way, is highly effective for students in their senior year and for graduate students in seeing the main steps in the authors' analysis which students can subsequently use in their own research.

5. *Ranked exercises*. Finally, we rank exercises according to their difficulty (either because of its more general analysis or its algebra steps): with a letter A next to its title to indicate easy exercises, a B to indicate intermediate difficulty, and a C for relatively more challenging exercises. This ranking should help undergraduate (master's) students start reviewing exercises with an A (B) and, once they feel comfortable with this type of questions, move to exercises marked with a letter B (C, respectively).

Organization of the Book

We first examine industries with a single firm (monopolist), allowing for different cost functions and demands, and then extend our analysis to settings where the monopolist serves more than one market, operates more than one plant, faces entry threats, practices price discrimination, or invests in advertising before selling its product. In Chap. 2, we consider the canonical Cournot model of simultaneous quantity competition, first between two firms and then between more than two firms. We then study firms facing the same production costs or different ones and analyze how entry, equity shares, the presence of a publicly owned firm, or incomplete information can affect these results.

Chapter 3 follows a similar approach but considers the Bertrand model of simultaneous price competition. We also examine how equilibrium prices are affected when firms sell homogeneous or heterogeneous products, when their costs are symmetric or asymmetric, and when they are perfectly informed about each other's costs or not. We then explore pricing decisions in settings where goods are horizontally differentiated, vertically differentiated, or both and finish the chapter describing the role of capacity constraints to reconcile the radically different equilibrium results in the Cournot and Bertrand models.

While previous chapters assumed firms competed simultaneously in either quantities or prices, Chap. 4 considers that firms compete sequentially and shows how equilibrium behavior is affected. For presentation purposes, we start with two or more symmetric firms and then extend our analysis to asymmetric firms, incomplete information settings, sequential price competition, and models of strategic pre-commitment based on Fudenberg and Tirole (1984). Chapter 5 applies the main results in Chaps. 2–4 to study optimal regulation in imperfectly competitive industries, such as monopolies and oligopolies, where we also allow for the possibility that firms' production generates pollution, thus entailing an additional market failure that regulators seek to correct.

In Chap. 6, we explore firms' incentives to invest in research and development (R&D), identifying that these incentives depend on the market where firms interact (monopolies or oligopolies, quantity or price competition). We also study how R&D incentives are affected by the presence of research spillovers and how "research joint ventures" between firms during the R&D stage can help mitigate some inefficiencies. We finish the chapter by analyzing R&D investments in polluting industries (often known as "green innovation") and how regulators can design the optimal length of a patent.

In Chap. 7, we switch gears toward mergers, first between two firms, then among more companies, analyze conditions for a merger to be sustainable in equilibrium, and how these incentives are affected when the merging firms benefit from cost efficiencies (synergies). Chapter 8 turn to models analyzing collusion, between two or more firms, selling homogenous or heterogeneous goods, when antitrust authorities cannot detect collusive practices and when they can, and whether mergers can facilitate subsequent collusive practices. Chapter 9 examines settings where a monopolist selling two products can choose to sell each at a different price or offer a bundle, where we allow for different contexts depending on the correlation between the valuation that the consumer assigns to each product, and we finish this chapter by considering tied sales where firms offer a product for free and charge for its usage.

Chapter 10 considers again some of the models analyzed in previous chapters, but within a context of incomplete information, where at least one firm privately observes a piece of information that other firms cannot observe. In this chapter, we study strategies such as offering sales, as in Varian (1980); practicing limit pricing, as in Milgrom and Roberts (1982); selling damaged goods, as in Deneckere and McAfee (1996); allowing customers to return the goods if they are unsatisfied with its quality, as in Che (1998); offering incentives to customers so they switch brands or they stay with their current brand, as in Shaffer and Zhang (2000); or investing in product quality as a signal, as in Calveras and Ganuza (2018). Finally, Chap. 11 studies markets with network effects, where firms simultaneously or sequentially choose their strategies, as in Farrell and Saloner (1985); the role of switching costs at limiting competition, as in Klemperer (1988); buyer power coordination, as in Fumagalli and Motta (2008); and retail price maintenance, as in Winter (1993).

Changes to the Second Edition

The second edition expands Chap. 7, which is now separated into two chapters (Chapter 7, only dedicated to mergers; and Chap. 8, which focuses on collusion). The book now provides 52 new exercises in all chapters, with over 200 new pages; mostly in Chaps. 1–4, 7, and 9; and introduces all the errata identified by the authors and by several readers who generously pointed them to us, particularly Justin Leroux (HEC Montreal). Thanks again for your helpful comments!

How to Use This Book

Instructors may cover specific exercises in class to illustrate how to extend typical models to other settings, such as industries with more firms, selling differentiated goods, interacting sequentially, or under incomplete information contexts. Alternatively, instructors may assign some exercises as a required reading before class—helping students prepare for the models and applications covered in class—or after class, so students can use the exercises as a guideline for their own homework assignments. Finally, instructors at the graduate level may assign exercises based on published articles as they can facilitate students' understanding of the theoretical approach, steps, and results in the article but using a more tractable setting that students can easily reproduce on their own.

Since exercises are ranked according to their difficulty, instructors at the undergraduate level can assign the reading of A-type exercises, spending class time on more challenging topics. Similarly, instructors at the graduate level can assign the reading of B-type exercises, letting them focus on more difficult extensions of the basic models covered in class.

Table 1 includes a list of suggested exercises for each chapter, for instructors teaching undergraduate and master's courses, such as Industrial Organization or senior-level topics in Microeconomic Theory. This list is, of course, flexible, as it allows instructors teaching an undergraduate course to pick individual exercises recommended at the master's level if their course is intended for seniors or students with a strong mathematical background.

Table 1 Suggested exercises per chapter at the undergraduate and master's level

Chapter	Undergraduate level	Master's level
Chapter 1	1.1, 1.2, 1.5, 1.8, 1.10, 1.11, 1.13	1.3, 1.4, 1.6, 1.7, 1.9, 1.12, 1.15, 1.16, 1.17
Chapter 2	2.1, 2.3, 2.6, 2.9, 2.15	2.2, 2.4, 2.5, 2.8, 2.10, 2.11, 2.12, 2.13, 2.15
Chapter 3	3.1, 3.2, 3.4, 3.10	3.3, 3.5, 3.7, 3.11, 3.13
Chapter 4	4.1, 4.9, 4.10	4.2, 4.3, 4.5, 4.6, 4.10, 4.11, 4.12, 4.13
Chapter 5	5.1, 5.3, 5.6	5.2, 5.4, 5.7, 5.8
Chapter 6	6.1, 6.3	6.2, 6.3, 6.6, 6.7, 6.8, 6.12
Chapter 7	7.1, 7.3	7.2, 7.4, 7.7, 7.8
Chapter 8	8.1, 8.3, 8.7	8.2, 8.4, 8.8, 8.10, 8/19
Chapter 9	9.1, 9.2, 9.3, 9.6	9.4, 9.5, 9.7
Chapter 10	10.7	10.1, 10.2, 10.7
Chapter 11	11.1	11.1, 11.2

Table 2 Suggested exercises per chapter at the Ph.D. level

Chapter	Ph.D. level
Chapter 1	1.3, 1.4, 1.6, 1.14, 1.17, 1.18
Chapter 2	2.5, 2.7, 2.8, 2.14, 2.16, 2.17
Chapter 3	3.6, 3.8, 3.9, 3.12, 3.14
Chapter 4	4.4, 4.7, 4.8, 4.13
Chapter 5	5.5, 5.9
Chapter 6	6.4, 6.5, 6.7, 6.9, 6.10, 6.11
Chapter 7	7.6, 7.9, 7.10, 7.12
Chapter 8	8.4, 8.9, 8.11, 8.12, 8.15, 8.20
Chapter 9	9.5, 9.7
Chapter 10	10.3, 10.4, 10.5, 10.6, 10.8
Chapter 11	11.3, 11.4, 11.5, 11.6

Similarly, Table 2 provides a similar list of suggested exercises for PhD-level courses, such as in Industrial Organization or Topics in Microeconomics.

Taoyuan City, Taiwan
Crawfordsville, IN, USA
Pullman, WA, USA

Pak-Sing Choi
Eric Dunaway
Felix Muñoz-Garcia

Acknowledgements

The exercises in this book were inspired by the courses we all taught at Washington State University, the courses that Pak-Sing teaches at the Graduate Institute of Industrial Economics, National Central University (Taiwan), and the courses that Eric teaches at Wabash College.

We are extremely grateful to faculty members and students at Washington State University, for their constant feedback and suggestions, and to our teaching assistants in several courses who, along the years, helped improve the presentation and clarity of many of the exercises in this volume. We particularly thank Ana Espinola-Arredondo, Kennedy Odongo, Jikhan Jeong, Xueying Ma, Joseph Navelski, GC Apar, Dindu Lama, Mianfeng Liu, Kiriti Kanjilal, Tongzhe Li, and Azzar Uddin. We are also thankful to Lorraine Klimowich, Sudha Kannan, and all the Springer-Nature team, who helped us from the beginning of this project. Last but not least, we thank our families and friends for their constant support and encouragement.

Contents

Monopoly

1

Introduction

This chapter explores monopoly models, where a single firm operates in an industry. We first study the output decisions of this type of firm in a simplified setting with linear demand and constant marginal costs. Exercise 1.2 then extends our analysis to a context where the monopolist faces a convex cost function (i.e., increasing marginal costs) which may occur when, intuitively, producing further units becomes increasingly expensive. Exercises 1.3 also examine more general environments where the firm faces a generic inverse demand function and a generic cost function, while Exercise 1.4 focuses on settings where the monopolist faces a convex, concave, or linear demand.

Exercise 1.5 studies the monopolist's output decisions when it maximizes revenue or profits and comparing these output levels to perfectly competitive markets. Exercise 1.6 considers a richer environment, where the monopolist benefits from "learning-by-doing," namely, its marginal costs decrease in the output produced in previous periods. We allow for commitment and no commitment in its output decisions, showing output levels are unaffected.

In Exercises 1.7–1.13, we consider monopolists producing more than one good, those producing a single good but serving two separated markets, and those producing a good in more than one plant (multiplant monopolies). We also elaborate on the monopolist's ability to practice third-degree price discrimination. (In Chap. 9, after introducing incomplete information games, we consider a monopolist practicing second-degree price discrimination to distinguish consumers with different willingness to pay for an object.) Exercise 1.14 considers another price practice where the monopolist charges a two-part tariff (i.e., a fee and a price per unit) which is often used by firms like Costco or amusement parks.

Exercise 1.15 analyzes a monopolist facing entry threats and how this firm can profitably prevent entry. Exercise 1.16 considers a monopolist producing more than one good, as in Exercise 1.7, but assuming that it benefits from economies of scope, that is, its overall costs from producing both goods are lower than the sum of its costs when producing each product independently.

Exercise 1.17 considers a setting where vertical product differentiation may lead to a natural monopoly even in the absence of decreasing average costs. Specifically, the firm that consumers regard as superior may strategically choose its price to prevent competitors from staying in the industry. As a consequence, the market becomes a natural monopoly since a single firm profitably operates and other companies have no incentives to enter despite the absence of entry barriers. Exercises 1.18 and 1.19 study two forms of advertising by a monopolist, persuasive and informative, and identify the

P.-S. Choi et al., *Industrial Organization*, Springer Texts in Business and Economics,
https://doi.org/10.1007/978-3-031-38635-0_1

profit-maximizing investments in advertising in each setting showing, as expected, that the monopolist invests more in advertising when it is more effective in boosting demand and enhancing profits. Finally, Exercise 1.20 examines behavior-based price discrimination, where a monopolist selling a non-durable good in two periods can set a different price to customers who bought the good in previous periods than to those who did not and how this price pattern is affected when the firm can commit to a specific future price.

Exercise #1.1: Monopoly with Linear Costs^A

1.1 Consider a monopolist facing a linear inverse demand curve $p(q) = a - bq$, where q denotes units of output. This firm faces cost function $C(q) = F + cq$, where F denotes its fixed costs and c represents the monopolist's (constant) marginal cost of production, and $a > c \geq 0$.

(a) Find the monopolist's profit-maximizing output q^m.

- The monopolist chooses the output level q that solves

$$\max_{q \geq 0} \pi = (a - bq)q - (F + cq),$$

where $(a - bq)q$ denotes total revenue (price times the number of units sold), while $F + cq$ represents the monopolist's total cost.

Differentiating with respect to q, we obtain

$$\frac{d\pi}{dq} = a - 2bq - c = 0,$$

which simplifies to

$$a - 2bq = c.$$

The left side represents the monopolist's marginal revenue from increasing its production by one unit, MR, while the right side denotes its marginal costs, MC, as depicted in Fig. 1.1. Solving for q yields a profit-maximizing output of

$$q^m = \frac{a - c}{2b},$$

Fig. 1.1 Monopoly output level

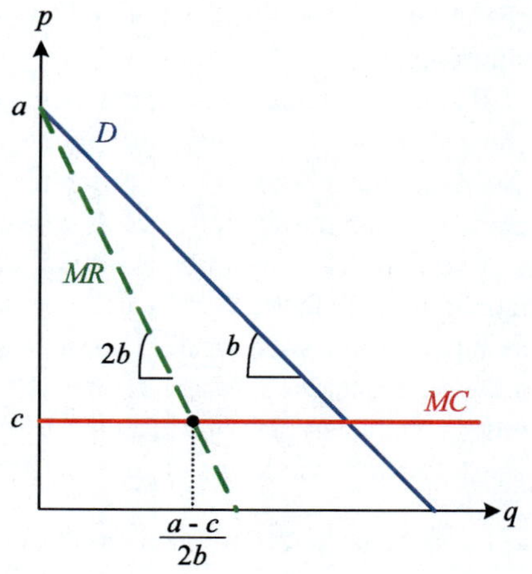

which is illustrated in Fig. 1.1 as the point where the marginal revenue curve crosses the marginal cost.

- Intuitively, the monopolist does not have incentive to produce an output level q such that $q < q^m$ (to the left of q^m in Fig. 1.1) since its marginal revenue from selling one more unit offsets its associated marginal cost. Similarly, this firm does not have incentives to produce an output level q such that $q > q^m$ (to the right of q^m in Fig. 1.1) because, by decreasing its output, it would decrease its costs more rapidly than the decrease in revenues, ultimately experiencing an increase in profits.

(b) What is the market price and the profit level?

- Plugging output level q^m into the inverse demand function, $p(q) = a - bq$, yields a monopoly price of

$$\begin{aligned} p^m = p(q^m) &= a - b\overbrace{\left(\frac{a-c}{2b}\right)}^{q^m} \\ &= \frac{2a - a + c}{2} \\ &= \frac{a+c}{2}. \end{aligned}$$

Profits are calculated as

$$\begin{aligned} \pi^m &= p^m q^m - (F + cq^m) \\ &= \left(\frac{a+c}{2}\right)\left(\frac{a-c}{2b}\right) - \left(F + c\left(\frac{a-c}{2b}\right)\right) \\ &= \left(\frac{a+c}{2} - c\right)\left(\frac{a-c}{2b}\right) - F \\ &= \frac{(a-c)^2}{4b} - F, \end{aligned}$$

which are positive if fixed costs F satisfy $F < \overline{F} \equiv \frac{(a-c)^2}{4b}$.

(c) Find the socially optimal output level, q^*. Is it larger or smaller than the profit-maximizing output, q^m, that you found in part (a)?

- At the social optimum, we have that the inverse demand crosses the marginal cost function, $p(q) = C'(q)$, as this point maximizes total welfare or eliminates deadweight loss. In this context, it means

$$a - bq = c.$$

Solving for q yields a socially optimal output q^* of

$$q^* = \frac{a-c}{b},$$

which is larger than the monopolist's profit-maximizing output q^m, as illustrated in Fig. 1.2. In particular, for this parametric specification with a linear demand and constant

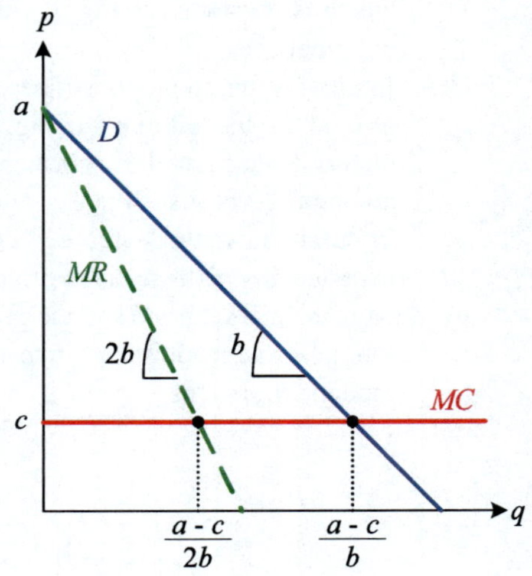

Fig. 1.2 Monopoly and socially optimal output levels

marginal costs, the socially optimal output q^* satisfies

$$q^* = 2\frac{a-c}{2b} = 2q^m,$$

thus being twice as large as the monopoly output.

(d) Does the monopolist earn positive profits when producing q^* units? Discuss.

- When the monopolist produces q^* units of output, its profits are

$$\begin{aligned}\pi^* &= (a - bq^*)q^* - (F + cq^*)\\ &= \left(a - b\overbrace{\left(\frac{a-c}{b}\right)}^{q^*}\right)\overbrace{\left(\frac{a-c}{b}\right)}^{q^*} - \left(F + c\overbrace{\left(\frac{a-c}{b}\right)}^{q^*}\right)\\ &= (a - (a-c) - c)\left(\frac{a-c}{b}\right) - F\\ &= -F.\end{aligned}$$

Our results then depend on whether fixed cost F is sunk:

- If the fixed cost F is sunk, the monopolist makes zero profits when producing q^* units of output.
- However, if the fixed cost is recoverable (i.e., non-sunk), the monopolist does not invest in the fixed cost F so that consumers cannot buy any units of output.

(e) *Numerical example.* Evaluate the monopolist's price, output, and profits assuming parameter values $a = 1, b = 1/2$, and $c = 1/4$, and compare them to socially optimal levels. How much fixed costs will the monopolist stop production?

- The monopolist produces

$$q^m = \frac{1 - \frac{1}{4}}{2 \times \frac{1}{2}} = \frac{3}{4} \text{ units}$$

and charges a price of

$$p^m = \frac{1 + \frac{1}{4}}{2} = \$\frac{5}{8}$$

to earn profits of

$$\pi^m = \frac{\left(1 - \frac{1}{4}\right)^2}{4 \times \frac{1}{2}} - F = \frac{9}{32} - F.$$

In this context, if $F > \overline{F} = \frac{9}{32}$, the monopolist will earn negative profits and choose not to produce any output.

- The socially optimal output is

$$q^* = \frac{1 - \frac{1}{4}}{\frac{1}{2}} = \frac{3}{2} \text{ units,}$$

which is twice as much as the monopolist's profit-maximizing output, but at the marginal cost of $c = \frac{1}{4}$ the monopolist earns negative profits of $-F$ from producing q^* units and must be subsidized if a government agency asks this firm to produce this output level.

Exercise #1.2: Monopoly with Convex Costs^A

1.2 Consider a monopolist facing a linear inverse demand curve $p(q) = a - bq$, where q denotes units of output, Exercise 1.2 and with cost function $C(q) = cq^2$, where $a > c \geq 0$.

(a) Find the monopolist's profit-maximizing output q^m.

- The monopolist chooses the output level q that solves

$$\max_{q \geq 0} \pi = (a - bq)\, q - cq^2.$$

Differentiating with respect to q, we obtain

$$\frac{d\pi}{dq} = a - 2bq - 2cq = 0,$$

which we can rearrange to

$$a - 2bq = 2cq.$$

Like in Exercise 1.1, the left-hand side represents the monopolist's marginal revenue from increasing its production in one unit, MR, while the right-hand side denotes its marginal costs, MC, as depicted in Fig. 1.3. Solving for q yields a profit-maximizing output of

$$q^m = \frac{a}{2\,(b + c)}.$$

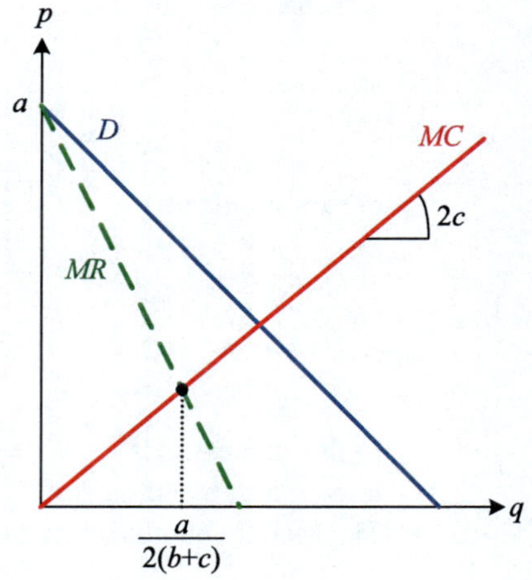

Fig. 1.3 Equilibrium condition

- *Comparative statics of monopoly output q^m:*
  - The monopoly output is increasing in a because as a increases, the demand for the monopolist's product also increases, which leads to an increase in output for any given price.
  - It is decreasing in b because as the quantity demanded becomes more sensitive to price, the monopolist scales back its production due to lower demand.
  - It is decreasing in cost parameter c. Intuitively, the monopolist produces less because it becomes increasing costly to produce every extra unit of output (convex costs).

(b) What are the market price and profit level?

- *Equilibrium price.* Plugging this output level q^m into the inverse demand function yields a monopoly price of

$$\begin{aligned} p^m &= a - bq^m \\ &= a - b\left(\frac{a}{2(b+c)}\right) \\ &= \frac{a(b+2c)}{2(b+c)}. \end{aligned}$$

- *Equilibrium profits.* Substituting price p^m into the profit function, we find that profits become

$$\begin{aligned} \pi^m &= p^m q^m - C(q^m) \\ &= \left(\frac{a(b+2c)}{2(b+c)}\right)\left(\frac{a}{2(b+c)}\right) - c\left(\frac{a}{2(b+c)}\right)^2 \\ &= \frac{a^2(b+2c-c)}{4(b+c)^2} \\ &= \frac{a^2}{4(b+c)}, \end{aligned}$$

which are increasing in demand, a, but decreasing in the demand's sensitivity to price, b, and in cost convexity, c.

(c) Under what conditions does a monopolist charge a higher price with a convex cost function than when the monopolist faces a linear cost function, $C(q) = cq$?

- We first compare equilibrium prices (see Exercise 1.1 for the price under a linear cost function), finding that

$$p^m_{\text{linear}} = \frac{a+c}{2} < \frac{a(b+2c)}{2(b+c)} = p^m_{\text{convex}}$$

holds as long as

$$(a+c)(b+c) < a(b+2c)$$
$$ab + ac + bc + c^2 < ab + 2ac$$
$$a > b + c.$$

Thus, the monopolist charges a higher price under a convex cost function if the demand is sufficiently large, i.e., $a > b + c$.

- At the same time, we should expect a lower price with linear costs to produce a higher output level due to the law of demand. In this setting,

$$q^m_{\text{linear}} = \frac{a-c}{2b} > \frac{a}{2(b+c)} = q^m_{\text{convex}}$$

We rearrange, obtaining the same condition as before,

$$(a-c)(b+c) > ab$$
$$ac > bc + c^2$$
$$a > b + c,$$

which coincides with the condition on equilibrium demand, $a > b + c$.

- Intuitively, if the demand is strong enough (in particular, $a > b + c$), the monopolist produces more units of output when its cost does not increase sharply (as in the linear cost function, in contrast with the convex cost function). Since the demand function is downward-sloping, a higher output necessarily implies a lower price.

(d) Find the socially optimal output level, q^*. Is it larger or smaller than the profit-maximizing output, q^m, that you found in part (a)? How is your answer compared to the context in which the monopolist faces a linear cost function $C(q) = cq$? Explain.

- At the social optimum, we have that the inverse demand crosses the marginal cost function, $p(q) = C'(q)$, which in this context means

$$a - bq = 2cq.$$

Solving for q yields a socially optimal output q^* of

$$q^* = \frac{a}{b+2c},$$

which is larger than the monopolist's profit-maximizing output, q^m.

- When comparing socially optimal output between the two cost functions, we observe that

$$q^*_{\text{linear}} = \frac{a-c}{b} > \frac{a}{b+2c} = q^*_{\text{convex}}$$

$$(a-c)(b+2c) > ab$$

$$2a > b+2c,$$

which holds if $a > b + c$ since $a > c$ by definition. This means that if the monopolist operates in a market with sufficiently strong demand ($a > b+c$), then the socially optimal output is higher when the monopolist faces a linear than a quadratic cost function.

(e) *Numerical example.* Evaluate the monopolist's price, output, and profits assuming parameter values $a = 1$, $b = 1/2$, and $c = 1/4$, and compare it to socially optimal output. How are your results compared to that when the monopolist's cost is linear?

- The monopolist produces

$$q^m = \frac{1}{2\left(\frac{1}{2}+\frac{1}{4}\right)} = \frac{2}{3} \text{ units}$$

and charges a price of

$$p^m = \frac{1\left(\frac{1}{2}+2\times\frac{1}{4}\right)}{2\left(\frac{1}{2}+\frac{1}{4}\right)} = \$\frac{2}{3}$$

to earn profits of

$$\pi^m = \frac{1^2}{4\left(\frac{1}{2}+\frac{1}{4}\right)} = \frac{1}{3}.$$

When the monopolist faces convex costs, it produces less output, charges a higher price, and earns higher profits in comparison to that when the monopolist's cost is linear.

- The socially optimal output is

$$q^* = \frac{1}{\frac{1}{2}+2\times\frac{1}{4}} = 1 \text{ unit},$$

which is 1.5 times of the monopolist's profit-maximizing output in the context of convex cost, but falls below the socially optimal output of $\frac{3}{2}$ when the monopolist's cost is linear.

Exercise #1.3: Monopolist with Linear Inverse Demand and Generic Cost Function^B

1.3 Consider a monopolist with inverse demand $p(q)$, which is strictly decreasing and linear in output q. The monopolist faces cost function $C(q)$, which is strictly increasing and weakly convex in q. In addition, assume that the inverse demand function originates above the marginal cost function, that is, $p(0) > C'(0)$.

(a) Set up the monopolist profit maximization problem, and find the implicit function that characterizes the monopolist's output. Interpret.

- The monopolist solves the following profit maximization problem:

$$\max_{q\geq 0} \pi(q) = p(q)q - C(q).$$

Differentiating with respect to q yields

$$p'(q)q + p(q) - C'(q) \leq 0.$$

Since $p(0) > C'(0)$ by definition, the problem has an interior solution, i.e., $q > 0$, and the first-order condition holds with equality. We can then rearrange the above expression as follows:

$$p'(q)q + p(q) = C'(q),$$

where the term on the left side represents the marginal revenue from increasing output, q, and the term on the right side reflects the marginal cost of increasing production. In addition, note that the marginal revenue, $MR \equiv p'(q)q + p(q)$, lies below the inverse demand function $p(q)$ since $p'(q) < 0$ by definition. The value of q^* that solves this equation is our implicit solution.

(b) Assume now that the monopolist's costs increase, so it faces cost function $C_1(q)$, where $C_1(q) > C(q)$ and $C_1'(q) > C'(q)$ for all q. Cost function $C_1(q)$ is still strictly increasing and weakly convex in q and $p(0) > C_1'(0)$. How are the monopolist's equilibrium output and price affected?

- The monopolist's profit-maximizing condition now becomes

$$p'(q)q + p(q) = C_1'(q).$$

Since $C_1'(q) > C'(q)$, the right-hand side of this expression has increased, which means the left-hand side must also increase for the equilibrium condition to hold with equality. If we differentiate the left-hand side of our equilibrium condition,

$$\begin{aligned}\frac{\partial MR}{\partial q} &= p''(q)q + p'(q) + p'(q) \\ &= \underbrace{p''(q)q}_{=0} + 2\underbrace{p'(q)}_{-} \\ &= 2p'(q) < 0,\end{aligned}$$

where $p''(q) = 0$ because the inverse demand function is linear in q. Therefore, the MR curve decreases in q, implying that the monopolist decreases its output to equate its marginal revenue with its new marginal cost and increases its price.

Exercise #1.4: Convex, Concave, and Linear Demand in Monopoly^C

1.4 Consider a monopolist with cost function $C(q) = cq$, where $c > 0$ denotes its constant marginal production cost. The monopolist faces an inverse demand curve $p(q)$, which is strictly decreasing in output, $p'(q) < 0$, whose vertical intercept $p(0)$ satisfies $p(0) > c$.

(a) Set up the monopolist's profit maximization problem and derive its first-order condition. Interpret.

- The monopolist solves the following profit maximization problem:

$$\max_{q \geq 0} \pi(q) = p(q)q - cq.$$

Differentiating with respect to q yields

$$p'(q)q + p(q) - c \leq 0.$$

Since $p(0) > c$, the problem has an interior solution, i.e., $q > 0$ and the first-order condition holds with equality. Hence, the above expression can be rearranged in terms of the standard marginality condition

$$p'(q)q + p(q) = c, \tag{1.1}$$

where the term on the left side represents the marginal revenue from increasing output, q, and the term on the right side reflects marginal cost of increasing production. In addition, note that the marginal revenue, $MR \equiv p'(q)q + p(q)$ lies below the inverse demand function $p(q)$ since $p'(q) < 0$ by definition.

(b) How is the monopoly price affected by a marginal increase in marginal cost c when the inverse demand function is (i) strictly concave; (ii) strictly convex; and (iii) linear?

- *Roadmap of the proof:* We first analyze how an increase in the constant marginal cost c affects equilibrium output for each of the three cases mentioned above and then describe how such variation in output affects the equilibrium price.
- If marginal cost c increases marginally, the left-hand side of expression (1.1), marginal revenue, must also increase to guarantee that the equality in (1.1) holds. Differentiating the marginal revenue expression, we obtain

$$p''(q)q + p'(q) + p'(q) = p''(q)q + 2p'(q), \tag{1.2}$$

which we now separately evaluate in three types of inverse demand functions: (i) strictly concave; (ii) linear; and (iii) strictly convex.

- *Case 1. Concave Demand Curve.* If the inverse demand function $p(q)$ is strictly concave, $p''(q) < 0$, expression (1.2) becomes

$$\underbrace{p''(q)q}_{-} + \underbrace{2p'(q)}_{-} < 0.$$

Therefore, expression (1.2) is unambiguously negative, so marginal revenue is strictly decreasing. As a consequence, when the marginal cost c increases, the monopolist responds decreasing its output q, ultimately leading to an increase in monopoly prices.

- *Parametric example.* Consider inverse demand function $p(q) = a - bq^2$, where $a, b > 0$, which is decreasing in output since $p'(q) = -2bq < 0$ and strictly concave given that $p''(q) = -2b < 0$. Therefore, evaluating expression (1.2) in this setting, we find

$$(-2b)(q) + 2(-2bq) = -6bq < 0,$$

thus confirming our above result about marginal revenue being decreasing in output q when the inverse demand curve is strictly concave.

- *Case 2. Linear Demand Curve.* If the inverse demand function $p(q)$ is linear, $p''(q) = 0$, expression (1.2) simplifies to

$$\underbrace{0}_{0} + \underbrace{2p'(q)}_{-} = \underbrace{2p'(q)}_{-} < 0,$$

yielding the same results as in Case 1: Expression (1.2) is unambiguously negative, implying that the marginal revenue curve is strictly decreasing. As a result, when the marginal cost c increases, the monopolist responds decreasing its output q, ultimately leading to an increase in monopoly prices.

- *Parametric example.* Consider, for instance, the inverse demand function $p(q) = a - bq$, where $a, b > 0$, which is linear in output since $p'(q) = -b$ and $p''(q) = 0$. We can then evaluate expression (1.2) in this setting to obtain

$$0q + 2(-b) = -2b < 0,$$

thus confirming our above result about marginal revenue being decreasing in output q when the inverse demand curve is linear.

- *Case 3. Convex Demand Curve.* If the inverse demand function $p(q)$ is strictly convex, $p''(q) > 0$, expression (1.2) becomes

$$\underbrace{p''(q)q}_{+} + \underbrace{2p'(q)}_{-}.$$

This result gives rise to two cases:

- (a) $|2p'(q)| > |p''(q)q|$, which implies that marginal revenue curve is still downward sloping; and
- (b) $|2p'(q)| < |p''(q)q|$, which entails that the marginal revenue curve is positively sloped.

The last case is, however, incompatible with the finding that the marginal revenue curve lies below the demand curve. For this to occur, the marginal revenue curve must have a negative slope and be steeper than the demand curve.

Therefore, only case (a) can be sustained, which yields the same results as in Cases 1–2: the monopolist's marginal revenue curve is still downward sloping, implying that a marginal increase in c leads the monopolist to reduce its output level, ultimately increasing monopoly prices.

- *Parametric example.* Consider, for instance, the inverse demand function $p(q) = a - b \ln q$, where $a, b > 0$, which is decreasing in output since $p'(q) = -\frac{b}{q} < 0$ and strictly convex given that $p''(q) = \frac{b}{q^2} > 0$. We can then evaluate expression (1.2) in this setting to obtain

$$\left(\frac{b}{q^2}\right)q + 2\left(-\frac{b}{q}\right) = -\frac{b}{q},$$

thus confirming our above result about marginal revenue being decreasing in output q when the inverse demand curve is convex.

Exercise #1.5: Maximizing Revenue or Maximizing Profit?^{A}

1.5 Consider a monopolist with inverse demand function $p(q)$, which is decreasing in output, $p'(q) < 0$, and exhibits a negatively sloped marginal revenue function, that is, $2p'(q) + p''(q)q < 0$. Intuitively, this allows for the inverse demand function to be concave ($p''(q) < 0$) or convex ($p''(q) > 0$), as long as its convexity is not too severe. The monopolist faces total cost function $C(q)$, where $C'(q) > 0$ and $C''(q) \geq 0$. In addition, assume that $p(0) > C'(0)$.

(a) *Revenue-maximizing output.* If the monopolist seeks to maximize total revenue, rather than profits, which output does the monopolist choose?

- In this setting, the monopolist solves

$$\max_{q \geq 0} p(q)q.$$

Differentiating with respect to q, we obtain

$$MR(q) \equiv p(q) + p'(q)q = 0,$$

which denotes marginal revenue. Graphically, the monopolist, when maximizing revenue alone, chooses the output level q^R where the marginal revenue curve crosses the horizontal axis, that is,

$$MR(q) = 0.$$

(b) *Profit-maximizing output.* If the monopolist seeks to maximize total profit, which output level does it choose?

- In this setting, the monopolist solves

$$\max_{q \geq 0} p(q)q - C(q)$$

Differentiating with respect to q, we obtain

$$p(q) + p'(q)q - C'(q) = 0$$

or, after rearranging,

$$p(q) + p'(q)q = C'(q),$$

which indicates that the monopolist increases its output q^M until the point where marginal revenue and marginal cost cross each other, that is,

$$MR(q) = MC(q).$$

(c) *Output comparison.* Show that the monopolist chooses a larger output when maximizing revenue than when maximizing total profits. Justify.

- The marginal revenue function $MR(q)$ is decreasing if $\frac{\partial MR}{\partial q} < 0$, or

$$p'(q) + p''(q)q + p'(q) = 2p'(q) + p''(q)q < 0,$$

which holds because the inverse demand function satisfies $2p'(q) + p''(q)q < 0$ by assumption. The marginal cost function $MC(q) \equiv C'(q)$ is weakly increasing since $C'(q) > 0$ and $C''(q) \geq 0$.

Therefore, $MC(q)$ crosses $MR(q)$ at a strictly interior point, entailing that $MC(q)$ crosses $MR(q)$ before $MR(q)$ crosses the horizontal axis. As a consequence, the output level solving $MR(q) = MC(q)$ is smaller than that solving $MR(q) = 0$; that is, $q^M < q^R$.

- **Intuition:** When the monopolist ignores total costs, it produces a larger output than when it takes these costs into account. Figure 1.4 illustrates the output level found in part (a) where $MR(q)$ crosses the horizontal axis, q^R, that from part (b) where $MR(q)$ crosses $C'(q)$, q^M.

(d) *Perfectly competitive output.* If the monopolist was forced by law to produce the perfectly competitive output, would this firm produce a larger output than the profit-maximizing monopolist of part (b)?

- The perfectly competitive output, q^{PC}, satisfies

$$p(q) = C'(q),$$

graphically depicted by the point where the inverse demand curve crosses the marginal cost curve. Since $p'(q) < 0$, the marginal revenue curve, $MR(q) \equiv p(q) + p'(q)q$, lies below the inverse demand curve, $p(q)$, implying that $q^{PC} > q^M$.

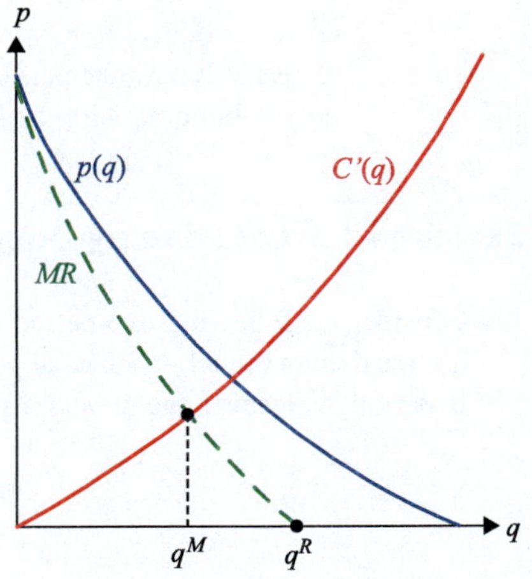

Fig. 1.4 Profit versus revenue maximization

(e) *Parametric example*. Assume that the inverse demand function is $p(q) = 1 - q$ and the total cost function is $C(q) = cq$, where $c \in [0, 1]$. Find the revenue-maximizing output, the profit-maximizing output, and the perfectly competitive output. Compare them.

- The marginal revenue curve in this setting is $MR(q) = 1 - 2q$, entailing that the revenue-maximizing output is that solving $MR(q) = 0$, or

$$1 - 2q = 0 \Longleftrightarrow q^R = \frac{1}{2}.$$

However, when the monopolist seeks to maximize total profits, the firm chooses the output level that solves $MR(q) = MC(q)$, which in this case is

$$1 - 2q = c \Longleftrightarrow q^M = \frac{1-c}{2}.$$

- Finally, the perfectly competitive output satisfies $p(q) = C'(q)$, which in this parametric example entails $1 - q = c$, or

$$q^{PC} = 1 - c.$$

- *Comparisons.*
  - Comparing output levels q^R and q^M, we see that $q^R \geq q^M$ since

$$\frac{1}{2} \geq \frac{1-c}{2}$$

simplifies to $c \leq 1$, which holds by definition.
  - Comparing q^{PC} against q^M, we see that $q^{PC} > q^M$ since $1 - c > \frac{1-c}{2}$ for all values of marginal cost c.
  - Finally, comparing q^{PC} against q^R, we find that $q^{PC} > q^R$ if and only if

$$1 - c > \frac{1}{2},$$

or $c < \frac{1}{2}$. Intuitively, the perfectly competitive output is decreasing in the marginal cost of production, c, while the revenue-maximizing output ignores this cost. When c is relatively low, the perfectly competitive output is larger than the revenue-maximizing output, but otherwise the former is lower than the latter.

Exercise #1.6: Learning-by-Doing and Commitment in Monopoly^B

1.6 Consider the following two-period monopoly model: a firm is a monopolist in a market with an inverse demand function (in each period) of $p(q) = a - bq$. The cost per unit in period 1 is c_1. In period 2, however, the monopolist has "learned by doing" so that its marginal costs decrease to

$$c_2 = c_1 - mq_1,$$

where q_1 is the monopolist's period 1 output level and m measures the learning effects. Assume that parameters satisfy $a > c_1 > 0$ and $b > m > 0$. Also assume that the monopolist does not discount future earnings (i.e., the discount factor δ is $\delta = 1$).

(a) *No commitment.* Find the monopolist's time-consistent production plan by first identifying the output level it chooses in the second period as a function of a generic output from the first period, i.e., $q_2(q_1)$, and then finding the first-period output level q_1^*.

- *Second period.* In the second period, the monopolist solves the following profit maximization problem:

$$\begin{aligned}\max_{q_2 \geq 0} \pi_2(q_2) &= (a - bq_2)\,q_2 - c_2 q_2 \\ &= (a - bq_2 - c_1 + mq_1)\,q_2.\end{aligned}$$

Differentiating with respect to q_2, and assuming interior solutions, we find

$$a - 2bq_2 - c_1 + mq_1 = 0,$$

which, after rearranging, we obtain the best response function of the monopolist

$$q_2(q_1) = \frac{a - (c_1 - mq_1)}{2b}.$$

This best response function is increasing in market size a, decreasing in the slope of the inverse demand curve b, and decreasing in the monopolist's second-period "net" cost, $c_1 - mq_1$. Intuitively, the monopolist becomes more efficient as the cost-reducing effect, m, increases and as its first-period output, q_1, increases.

Therefore, second-period profits evaluated at $q_2(q_1)$ are

$$\begin{aligned}\pi_2(q_1) &= (a - bq_2(q_1))\,q_2(q_1) - (c_1 - mq_1)\,q_2(q_1) \\ &= (a - c_1 + mq_1 - bq_2(q_1))q_2(q_1) \\ &= \left(a - c_1 + mq_1 - b\left(\frac{a - c_1 + mq_1}{2b}\right)\right)\left(\frac{a - c_1 + mq_1}{2b}\right) \\ &= \frac{(a - c_1 + mq_1)^2}{4b}.\end{aligned}$$

- *First period.* In the first period, the monopolist solves the following profit maximization problem:

$$\begin{aligned}\max_{q_1 \geq 0} \pi_1(q_1) &= \underbrace{(a - bq_1)\,q_1 - c_1 q_1}_{\text{First period}} + \underbrace{\pi_2(q_1)}_{\text{Second period}} \\ &= (a - bq_1 - c_1)\,q_1 + \frac{(a - c_1 + mq_1)^2}{4b}\end{aligned}$$

Differentiating with respect to q_1, and assuming interior solutions, we obtain

$$a - 2bq_1 - c_1 + \frac{m\,(a - c_1 + mq_1)}{2b} = 0,$$

which, after rearranging,

$$2bq_1 - \frac{m^2 q_1}{2b} = a - c_1 + \frac{m(a - c_1)}{2b}$$

$$q_1(4b^2 - m^2) = (a - c_1)(2b + m).$$

Recall that $4b^2 - m^2$ can be factored into $(2b + m)(2b - m)$. Lastly, we solve for q_1 to obtain the monopolist's optimal output in period 1

$$q_1^* = \frac{a - c_1}{2b - m},$$

which is positive by assumptions $a > c_1 > 0$ and $b > m > 0$. This output is increasing in market size a and in the cost-reduction effect of first-period output, m, but decreasing in its unit cost c_1 and the sensitivity of demand to price b. We can then summarize the subgame perfect equilibrium of the game as follows:

$$\left\{q_1^*, q_2\left(q_1\right)\right\} = \left\{\frac{a - c_1}{2b - m}, \frac{a - (c_1 - mq_1)}{2b}\right\},$$

where we write the monopolist's second-period output decision as $q_2\left(q_1\right)$, which allows for both in-equilibrium and off-the-equilibrium behavior.

- Finally, substituting the first-period output into the second-period response function,

$$\begin{aligned} q_2^* &= \frac{a - \left(c_1 - mq_1^*\right)}{2b} \\ &= \frac{a - c_1}{2b} + \frac{m}{2b} q_1^* \\ &= \frac{a - c_1}{2b} + \frac{m}{2b} \times \frac{a - c_1}{2b - m} \\ &= \frac{a - c_1}{2b - m}, \end{aligned}$$

which coincides with its first-period output q_1^* because the monopolist balances the marginal cost of output expansion in period 1, $\frac{m(a-c_1+mq_1)}{2b}$, with the marginal benefit of cost reduction in period 2, mq_1, to maximize profits across both periods. Note that if learning effects are absent, $m = 0$, the above output levels simplify to the standard monopoly output $q_1^* = q_2^* = \frac{a-c_1}{2b}$ in both periods. However, when learning effects become more significant (i.e., m increases), the monopolist's production increases in both periods.

(b) *Commitment.* Assume that the monopolist can commit to both q_1 and q_2 in the first period. Find the output levels that the firm chooses in this context.

- The monopolist's intertemporal problem is that of selecting q_1 and q_2 that solve

$$\max_{q_1, q_2 \geq 0} \underbrace{(a - bq_1)q_1 - c_1 q_1}_{\text{First-period profits}} + \underbrace{(a - bq_2)q_2 - (c_1 - mq_1)\, q_2}_{\text{Second-period profits}}.$$

Differentiating with respect to q_1 yields:

$$a - 2bq_1^m - c_1 + mq_2^m = 0,$$

and differentiating with respect to q_2, we obtain

$$a - 2bq_2^m - c_1 + mq_1^m = 0,$$

implying that the first-order conditions are symmetric. In a symmetric equilibrium, $q_1^m = q_2^m = q^m$, which helps us rewrite any of the above first-order conditions as

$$a - 2bq^m - c_1 + mq^m = 0,$$

or $a - c_1 = (2b - m)\, q^m$. Solving for q^m, we obtain output levels

$$q_1^m = q_2^m = \frac{a - c_1}{2b - m},$$

which is positive by the exercise's assumptions, i.e., $a > c_1 > 0$ and $b > m > 0$.

- When learning effects are absent, i.e., $m = 0$, the monopolist produces the standard monopoly output $\frac{a-c_1}{2b}$ in both periods. However, as learning effects become more significant (i.e., m increases), the monopolist production increases.

(c) Compare your results in parts (a) and (b). Interpret.

- The monopolist's sequential output decisions in part (a) is *time consistent* with its commitment to output levels in part (b). This is true because when the monopolist operates by backward induction in part (a), it considers the effect of first-period output level on second-period profits, which objective is aligned with its choice of profit-maximizing output levels across both periods in part (b).

(d) Consider now that the monopolist discounts future profits with a discount factor $\delta \in (0, 1)$. Would this firm produce the same output level with and without commitment?

- *Without commitment.* We first reconsider part (a), where the monopolist chooses its output level at the beginning of each stage. In period 2, it chooses the same output level as in part (a), i.e., $q_2(q_1) = \frac{a-(c_1-mq_1)}{2b}$ yielding second-period profits $\pi_2(q_1) = \frac{(a-c_1+mq_1)^2}{4b}$. However, in the first period, the monopolist now solves

$$\max_{q_1 \geq 0} \pi_1(q_1) = \underbrace{(a - bq_1)\, q_1 - c_1 q_1}_{\text{First period}} + \underbrace{\delta \pi_2(q_1)}_{\text{Second period}}$$

$$= (a - bq_1 - c_1)\, q_1 + \delta \frac{(a - c_1 + mq_1)^2}{4b}.$$

Differentiating with respect to q_1, and assuming interior solutions, we obtain

$$a - 2bq_1 - c_1 + \delta \frac{m\,(a - c_1 + mq_1)}{2b} = 0,$$

which, after rearranging,

$$2bq_1 - \delta\frac{m^2 q_1}{2b} = a - c_1 + \delta\frac{m(a-c_1)}{2b}$$

$$q_1(4b^2 - \delta m^2) = (a - c_1)(\delta m + 2b)$$

and solving for q_1 yields the monopolist's optimal output in period 1

$$q_1^* = \frac{(a-c_1)(\delta m + 2b)}{4b^2 - \delta m^2},$$

which is positive by the assumption $a > c_1 > 0$.

This output is increasing in market size a, discount rate δ, and the cost-reduction effect of first-period output, m, but decreasing in the slope of the inverse demand curve b and its per unit cost c_1, because

$$\frac{\partial q_1^*}{\partial a} = \frac{2b + \delta m}{4b^2 - \delta m^2} > 0$$

$$\frac{\partial q_1^*}{\partial b} = -\frac{(a-c_1)\left(\delta m^2 + 4b\delta m + 4b^2\right)}{\left(4b^2 - \delta m^2\right)^2} < 0$$

$$\frac{\partial q_1^*}{\partial c_1} = -\frac{2b + \delta m}{4b^2 - \delta m^2} < 0$$

$$\frac{\partial q_1^*}{\partial m} = \frac{\delta\,(a-c_1)\left(\delta m^2 + 2b\delta m + 4b^2\right)}{\left(4b^2 - \delta m^2\right)^2} > 0$$

$$\frac{\partial q_1^*}{\partial \delta} = \frac{2mb\,(a-c_1)\,(2b+m)}{\left(4b^2 - \delta m^2\right)^2} > 0.$$

We can then summarize the subgame perfect equilibrium of the game as follows:

$$\left\{q_1^*, q_2\left(q_1\right)\right\} = \left\{\frac{(a-c_1)(\delta m + 2b)}{4b^2 - \delta m^2}, \frac{a - (c_1 - mq_1)}{2b}\right\}.$$

- Finally, substituting the first-period output into the second-period response function, we find the second-period output in equilibrium as follows:

$$q_2^* = \frac{a - \left(c_1 - mq_1^*\right)}{2b}$$

$$= \frac{a - c_1}{2b} + \frac{m}{2b} \times \frac{(a-c_1)(\delta m + 2b)}{4b^2 - \delta m^2}$$

$$= \frac{(a-c_1)\,(2b+m)}{4b^2 - \delta m^2},$$

which is larger than its first-period output, that is, $q_2^* \geq q_1^*$, if

$$\frac{(a-c_1)\,(2b+m)}{4b^2 - \delta m^2} \geq \frac{(a-c_1)(\delta m + 2b)}{4b^2 - \delta m^2},$$

which rearranges to

$$(1-\delta)\, m \geq 0.$$

Therefore, when the monopolist does not discount future profits, $\delta = 1$, it chooses to produce the same output in both periods, i.e., $q_2^* = q_1^*$. However, when its values present profits more significantly than future profits, $\delta < 1$, the monopolist produces more units in the second than in the first period.

- We can also check that second-period equilibrium output, q_2^*, is increasing in the discount factor δ since

$$\frac{\partial q_2^*}{\partial \delta} = \frac{m^2\,(a-c_1)\,(2b+m)}{\left(4b^2-\delta m^2\right)^2} > 0.$$

Intuitively, as the discount factor δ increases, the monopolist puts a stronger emphasis on second-period profits. In this case, the firm increases first-period output q_1^* to generate stronger cost-reduction effects that enables the monopolist to increase its second-period profits. When δ increases to $\delta = 1$, the monopolist increases q_1^* enough to make it coincide with q_2^*.

- *With commitment.* The monopolist in this setting selects q_1 and q_2 that solve

$$\max_{q_1, q_2 \geq 0} \underbrace{(a-bq_1)q_1 - c_1 q_1}_{\text{First-period profits}} + \delta \underbrace{\left[(a-bq_2)q_2 - (c_1 - mq_1)\, q_2\right]}_{\text{Second-period profits}}.$$

Differentiating with respect to q_1 yields

$$a - 2bq_1^m - c_1 + \delta m q_2^m = 0$$

and differentiating with respect to q_2, we obtain

$$a - 2bq_2^m - c_1 + mq_1^m = 0.$$

Simultaneously solving for q_1^m and q_2^m in the above equalities yields

$$q_1^m = \frac{(a-c_1)\,(2b+\delta m)}{4b^2-\delta m^2}$$

$$q_2^m = \frac{(a-c_1)\,(2b+m)}{4b^2-\delta m^2},$$

which is straightforward to show that $q_2^m \geq q_1^m$ for all discount factors $\delta \in [0, 1]$.

- *Comparison.* Comparing the output levels we obtained with and without commitment, we find that in both periods the monopolist produces the same output since

$$q_1^m = q_1^* = \frac{(a-c_1)\,(2b+\delta m)}{4b^2-\delta m^2}, \text{ and}$$

$$q_2^m = q_2^* = \frac{(a-c_1)\,(2b+m)}{4b^2-\delta m^2}.$$

(e) *Numerical example.* Find the monopolist's output assuming parameter values $a = b = 1$, $c_1 = 1/2$, and $m = 1/4$. How much will the monopolist produce when (i) it does not discount future payoff, that is, $\delta = 1$, and (ii) it has a discount factor of $\delta = 1/2$?

- The monopolist produces

$$q_1^m = \frac{\left(1 - \frac{1}{2}\right)\left(2 + \frac{\delta}{4}\right)}{4 - \frac{\delta}{4^2}} = \frac{2\,(8 + \delta)}{64 - \delta} \text{ units}$$

in period 1, and

$$q_2^m = \frac{\left(1 - \frac{1}{2}\right)\left(2 + \frac{1}{4}\right)}{4 - \frac{\delta}{4^2}} = \frac{18}{64 - \delta} \text{ units}$$

in period 2.

- When the monopolist does not discount future profits, $\delta = 1$, it produces the same output across both periods, where $q_1^m = q_2^m = 2/7 \simeq 0.286$.
- However, when the monopolist has a discount factor of $\delta = \frac{1}{2}$, it produces $q_1^m = 34/127 \simeq 0.268$ units in period 1 which falls below $q_2^m = 36/127 \simeq 0.283$ units in period 2, both of which fall below the output level when the monopolist does not discount future profits (0.286).

Exercise #1.7: Monopolist Serving Two Interdependent Markets^B

1.7 Consider a monopolist producing two goods, 1 and 2, at a marginal cost $c > 0$. The demand function for product i is

$$q_i(p_i, p_j) = a - bp_i + gp_j \text{ where } i = \{1, 2\}$$

and parameters satisfy $a > c\,(b - g)$, $b, c > 0$, and $|b| > |g|$, entailing that own-price effects dominate cross-price effects. For generality, we allow for $g > 0$ and $g < 0$.

(a) Assume that the monopolist sets the price of good i separate to that of good j. Find the equilibrium price pair (p_i, p_j).

- The monopolist sets the price of good i by solving the following profit maximization problem:

$$\max_{p_i > 0} \pi_i\,(p_i) = \left(a - bp_i + gp_j\right)(p_i \geq 0 - c)\,.$$

Differentiating with respect to p_i, we obtain

$$a - bp_i + gp_j - b\,(p_i - c) = 0,$$

and solving for p_i, we find

$$p_i\left(p_j\right) = \frac{a + bc + gp_j}{2b},$$

which is analogous to a best response function in a standard duopoly Bertrand game of price competition where firms sell heterogeneous products. Graphically, this best response function originates at $\frac{a+bc}{2b}$ and increases at a slope of $\frac{g}{2b}$ if goods are substitutes ($g > 0$) but decreases at that slope if goods are complements ($g < 0$, respectively).

Following similar steps, the monopolist sets the price of good j by solving its own maximization problem and obtains the semetric best response function:

$$p_j(p_i) = \frac{a + bc + gp_i}{2b},$$

which has the same interpretation for the price of good i.

In a symmetric equilibrium, the monopolist sets the same price for both goods, i.e., $p_1 = p_2 = p_c$, implying that

$$p_c = \frac{a + bc + gp_c}{2b}$$

and solving for p_c yields an equilibrium price of

$$p_c = \frac{a + bc}{2b - g}.$$

(b) Suppose now that the monopolist sets the prices of goods i and j simultaneously. Find the equilibrium price pair (p_i, p_j).

- In this setting, the monopolist sets the prices of goods i and j by maximizing the joint profits across both goods,

$$\max_{p_i, p_j > 0} \quad \pi_i(p_i) + \pi_j(p_j) = \left(a - bp_i + gp_j\right)(p_i - c) + \left(a - bp_j + gp_i\right)\left(p_j - c\right).$$

Differentiating with respect to p_i and p_j, we obtain

$$a - bp_i + gp_j - b(p_i - c) + g(p_j - c) = 0$$
$$a - bp_j + gp_i - b(p_j - c) + g(p_i - c) = 0,$$

and solving for p_i and p_j, we obtain

$$p_i = p_j = \frac{a + c(b - g)}{2(b - g)}.$$

(c) Under what condition will the distinct monopolist charge a higher (lower) price than a multiproduct monopolist?

- As shown in the previous section, the multiproduct monopolist will charge a price of $p_m = \frac{a+c(b-g)}{2(b-g)}$. Comparing this price against $p_c = \frac{a+bc}{2b-g}$, we find that

$$\frac{a + c\,(b - g)}{2\,(b - g)} > \frac{a + bc}{2b - g}$$

and rearranging yields

$$[a + c(b - g)](2b - g) > 2(a + bc)(b - g)$$
$$2ab - ag + (b - g)(2bc - cg) > (2a + 2bc)(b - g)$$
$$2ab - ag > (2a + cg)(b - g)$$
$$2ab - ag > 2ab - 2ag + cg(b - g)$$
$$ag > cg(b - g)$$
$$g(a - c(b - g)) > 0.$$

Note that we do not cancel the parameter g out since it could be either positive or negative, which would influence the directionality of the inequality. Since $a > c(b - g)$ holds by assumption, the above inequality, $p_m > p_c$, holds (does not hold) if and only if $g > 0$ ($g < 0$).

Therefore, the multiproduct monopolist charges a higher (lower) price than distinct monopolists if goods are substitutes (complements). Intuitively, when goods are substitutes, setting a higher price for good i decreases the demand for good j. This negative effect is ignored by the monopolist when setting prices as independent units, but internalized by the multiproduct monopolist, such that independent units will set a lower price for good i. The opposite argument applies when good i and j are complements and a monopolist with independent units would set a higher price for each good than the multiproduct monopolist.

Exercise #1.8: Multiplant Monopolist-I^A

1.8 Consider a monopolist facing an inverse demand function of $p(q) = a - bq$ and allocating production between two production facilities. In plant 1, production cost is linear at $C_1(q_1) = c_1 q_1$, and in plant 2, production cost is convex at $C_2(q_2) = \frac{1}{2}c_2 q_2^2$, where $a > c_1, c_2 > 0$, and $b > 0$.

(a) Find the monopolist's output allocation plan. Does the firm produce in both, or only one plant? Explain your results.

- The monopolist chooses an output at where the marginal revenue equals the minimum marginal cost between the two production facilities, that is,

$$MR = a - 2b(q_1 + q_2) = \min\{c_1, c_2 q_2\},$$

which means that the monopolist concentrates its production at the plant where the last unit of output is the least costly to produce.

- *Producing in plant 2 alone.* The monopolist only produces at plant 2 if the ranking of marginal costs satisfies $c_1 > c_2 q_2$, which after solving for q_2 yields $q_2 < \frac{c_1}{c_2}$. Figure 1.5 illustrates the monopolist's inverse demand function, its marginal revenue, and the marginal cost when only plant 2 produces. In this context, we have

$$a - 2b(0 + q_2) = c_2 q_2$$

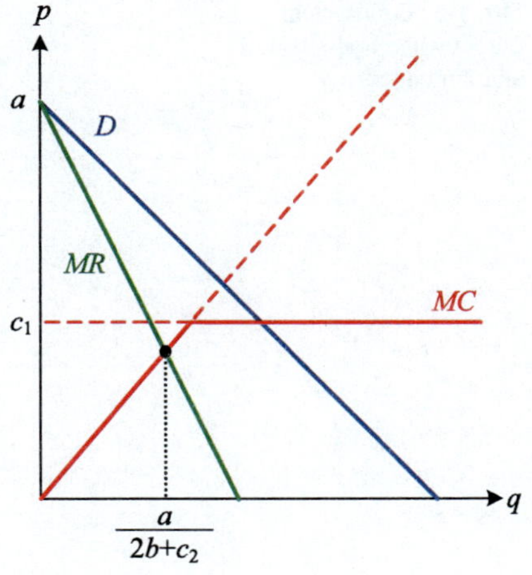

Fig. 1.5 Equilibrium output when only plant 2 produces

entailing an output level of

$$q_2 = \frac{a}{2b + c_2}.$$

We now must check that this output level satisfies the condition $q_2 < \frac{c_1}{c_2}$ for the monopolist to only produce in plant 2; that is,

$$\frac{a}{2b + c_2} < \frac{c_1}{c_2}$$

which, solving for c_2, yields

$$\begin{aligned} ac_2 &< (2b + c_2)c_1 \\ c_2(a - c_1) &< 2bc_1 \\ c_2 &< \frac{2bc_1}{a - c_1}. \end{aligned}$$

Intuitively, the monopolist produces in plant 2 alone when its costs are significantly lower than those in plant 1.

- *Producing in plant 1 alone.* The monopolist only produces at plant 1 if the ranking of marginal costs satisfies $c_1 < c_2 q_2$, which after solving for q_2 yields $q_2 > \frac{c_1}{c_2}$. In this scenario, we have

$$a - 2b(q_1 + 0) = c_1$$

entailing an output level of

$$q_1 = \frac{a - c_1}{2b}.$$

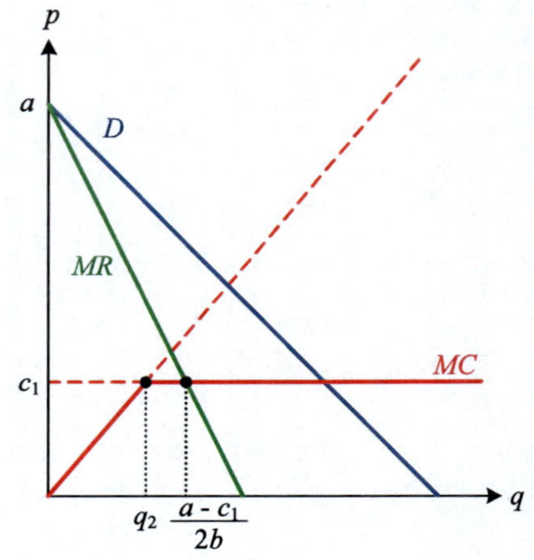

Fig. 1.6 Equilibrium output when both plants 1 and 2 produce

We now must check that the monopolist producing at this output level will not produce at plant 2 because this entails a higher marginal cost, by checking that

$$MC_2 = c_2 q_2 = c_2(0) > c_1 = MC_1,$$

which cannot hold for any value of c_1 or c_2. Thus the monopolist never produces at plant 1 alone.

- *Producing in both plants*. Finally, when $c_2 > \frac{2bc_1}{a-c_1}$ the monopolist optimizes its production by producing the first q_2 units at plant 2 (with its initially lower marginal cost of production); then the remaining $q^* - q_2$ units at plant 1 as the marginal cost of plant 2 now exceeds that of plant 1. This situation is represented visually in Fig. 1.6. Mathematically, we calculate q^* the same way we calculate q_1 when plant 1 produces alone, obtaining

$$q^* = \frac{a - c_1}{2b}.$$

From here, we know that the monopolist produces at plant 2 until the marginal cost of production at plant 2 is equal to the marginal cost of production at plant 1, $c_1 = c_2 q_2$. Solving this expression for q_2 yields

$$q_2 = \frac{c_1}{c_2}.$$

Thus, the first $\frac{c_1}{c_2}$ units are products at plant 2, with the remaining units produced at plant 1, providing an equilibrium quantity for plant 1 of

$$q_1 = q^* - q_2 = \frac{a - c_1}{2b} - \frac{c_1}{c_2}.$$

- Summarizing, we have the monopolist's efficient output allocation plan, (q_1^*, q_2^*), as follows:

$$(q_1^*, q_2^*) = \begin{cases} \left(0, \frac{a}{2b+c_2}\right) & \text{if } c_2 \leq \frac{2bc_1}{a-c_1}, \text{ and} \\ \left(\frac{a-c_1}{2b} - \frac{c_1}{c_2}, \frac{c_1}{c_2}\right) & \text{if } c_2 > \frac{2bc_1}{a-c_1}. \end{cases}$$

(b) *Numerical example.* Find the monopolist's output assuming parameter values $a = b = 1$, $c_1 = 1/2$, and $c_2 = 1/4$. How will your results be changed if $c_1 = 1/12$?

- When $c_1 = \frac{1}{2}$, the cutoff becomes

$$\frac{1}{4} < 2 = \frac{2 \times \frac{1}{2}}{1 - \frac{1}{2}}$$

so that the monopolist produces zero units in plant 1 and $q_2^* = \frac{1}{2+\frac{1}{4}} = \frac{4}{9}$ units in plant 2.

- When $c_1 = \frac{1}{12}$, the cutoff becomes

$$\frac{1}{4} > \frac{2}{11} = \frac{2 \times \frac{1}{12}}{1 - \frac{1}{12}},$$

implying that the monopolist produces $q_1^m = \frac{1-\frac{1}{12}}{2} - \frac{\frac{1}{12}}{\frac{1}{4}} = \frac{1}{8}$ units in plant 1 and $q_2^m = \frac{\frac{1}{12}}{\frac{1}{4}} = \frac{1}{3}$ units in plant 2.

Exercise #1.9: Multiplant Monopolist-II^B

1.9 Consider a monopolist with two plants, A and B, with total cost function

$$TC_i(q_i) = c_i q_i + d_i (q_i)^2$$

in plant $i = \{A, B\}$, where $c_i, d_i \geq 0$. Therefore, each plant exhibits a different cost function if $c_A \neq c_B$, if $d_A \neq d_B$, or both. Finally, the monopolist faces inverse demand function $p(Q) = 1 - Q$, where $Q = q_A + q_B$ represents aggregate output across both plants.

(a) Find the optimal output that the monopolist produces at each plant and the price it will charge.

- The monopolist maximizes its joint profit in both plants:

$$\begin{aligned}\max_{q_A, q_B \geq 0} \quad \pi = \pi_A + \pi_B =& (1 - q_A - q_B)q_A - (c_A q_A + d_A (q_A)^2) \\ &+ (1 - q_A - q_B)q_B - (c_B q_B + d_B (q_B)^2).\end{aligned}$$

Differentiating with respect to q_A, we obtain

$$1 - 2q_A - q_B - (c_A + 2d_A q_A) - q_B = 0,$$

which we can rearrange as follows:

$$1 - c_A - 2(1 + d_A)q_A - 2q_B = 0.$$

Solving for q_A, we have that

$$q_A(q_B) = \frac{1 - c_A}{2(1 + d_A)} - \frac{1}{1 + d_A} q_B.$$

Differentiating with respect to q_B yields a symmetric expression

$$1 - c_B - 2(1 + d_B)q_B - 2q_A = 0$$

and, solving for q_B, we find

$$q_B(q_A) = \frac{1 - c_B}{2(1 + d_B)} - \frac{1}{1 + d_B} q_A.$$

Inserting this result into the one we obtained after differentiating with respect to q_A yields

$$q_B = \frac{1 - c_B}{2(1 + d_B)} - \frac{1}{1 + d_B} \overbrace{\left(\frac{1 - c_A}{2(1 + d_A)} - \frac{1}{1 + d_A} q_B \right)}^{q_A},$$

which, after rearranging, entails

$$q_B = \frac{c_A - c_B + d_A(1 - c_B) + 2q_B}{2(1 + d_A + d_B + d_A d_B)}$$

and, solving for q_B, we find the equilibrium output in plant B, that is,

$$q_B = \frac{c_A + d_A - c_B(1 + d_A)}{2(d_A + d_B + d_A d_B)}.$$

Inserting this expression into $q_A(q_B) = \frac{1-c_A-2q_B}{2(1+d_A)}$, we obtain the equilibrium output in plant A as follows:

$$q_A = \frac{1 - c_A - 2\overbrace{\left(\frac{c_A + d_A - c_B(1 + d_A)}{2(d_A + d_B + d_A d_B)} \right)}^{q_B}}{2(1 + d_A)}$$

$$= \frac{c_B + d_B - c_A(1 + d_B)}{2(d_A + d_B + d_A d_B)}.$$

(b) Under which parameter conditions does the monopolist produce in plant A alone, in plant B alone, in none, or both plants?

- Equilibrium output in plant A, $q_A = \frac{c_B+d_B-c_A(1+d_B)}{2(d_A+d_B+d_A d_B)}$, is positive if and only if

$$c_A < \frac{c_B + d_B}{1 + d_B}.$$

Figure 1.7 depicts this cutoff with c_A in the vertical axis and c_B in the horizontal axis. Cutoff $c_A = \frac{c_B+d_B}{1+d_B}$ originates at a vertical intercept of $c_A = \frac{d_B}{1+d_B}$, crosses the 45-degree line at $c_B = 1$ (since $\frac{c_B+d_B}{1+d_B} = 1$ at $c_B = 1$), and then continues under the 45-degree line. Intuitively, when plant A is relatively efficient (cost pairs below cutoff $c_A = \frac{c_B+d_B}{1+d_B}$), this plant produces a positive output.

Fig. 1.7 Production cutoff for plant A

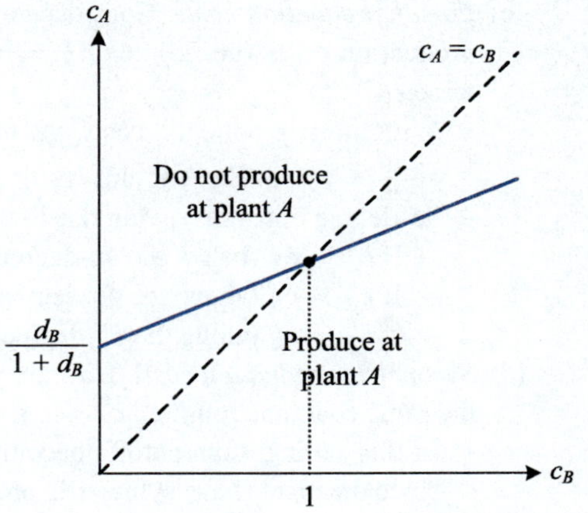

Fig. 1.8 Production cutoff for both plants

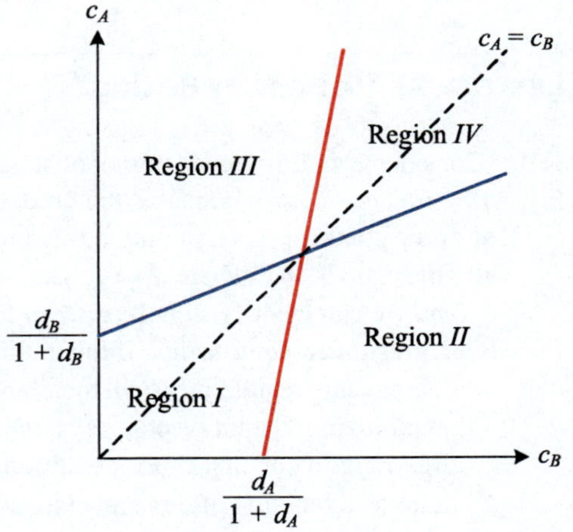

- Similarly, equilibrium output in plant B, $q_B = \frac{c_A+d_A-c_B(1+d_A)}{2(d_A+d_B+d_A d_B)}$, is positive if and only if

$$c_B < \frac{c_A + d_A}{1 + d_A}.$$

Figure 1.8 depicts this cutoff (superimposing it to $c_A = \frac{c_B+d_B}{1+d_B}$), which originates at $c_B = \frac{d_A}{1+d_A}$ and crosses the 45-degree line at $c_A = 1$. (To understand this cutoff graphically, one can rotate the figure counterclockwise.)

- Therefore, the following four regions arise:
  - *Region I.* If $c_A < \frac{c_B+d_B}{1+d_B}$ and $c_B < \frac{c_A+d_A}{1+d_A}$, in the southwest of Fig. 1.8, both plants are relatively efficient and, hence, both are active.
  - *Region II.* If $c_A < \frac{c_B+d_B}{1+d_B}$ and $c_B > \frac{c_A+d_A}{1+d_A}$, in the southeast of the figure, plant A is efficient relative to B, being the only active plant.
  - *Region III.* If $c_A > \frac{c_B+d_B}{1+d_B}$ and $c_B < \frac{c_A+d_A}{1+d_A}$, in the northwest of the figure, plant B is efficient relative to A, being the only active plant.
  - *Region IV.* If $c_A > \frac{c_B+d_B}{1+d_B}$ and $c_B > \frac{c_A+d_A}{1+d_A}$, in the northeast of the figure, both plants are relatively inefficient and, hence, both are inactive.

(c) *Linear production costs.* Consider now the special case in which both plants exhibit linear production costs (i.e., $d_A = d_B = 0$ but $c_A, c_B > 0$). How are your results in part (b) affected?

- With linear production costs, the above two cutoffs become $c_A < \frac{c_B+0}{1+0} = c_B$ and $c_B < \frac{c_A+0}{1+0} = c_A$. In Fig. 1.8, this result entails that the cutoff of both plants coincides with the 45-degree line, thus giving rise to three possible regions of production profiles:
  - If $c_A < c_B$ (below the 45-degree line), only plant A is active.
  - If $c_A > c_B$ (above the 45-degree line), only plant B is active.
  - If $c_A = c_B$ (along the 45-degree line), both plants are active.

(d) *Symmetric production costs.* How are your results from part (b) affected if both plants exhibit the same cost function (i.e., $c_A = c_B = c > 0$ and $d_A = d_B = d > 0$).

- In this setting, both cutoffs become $c < \frac{c+d}{1+d}$, which is simplified to $c < 1$. Intuitively, since the plants have symmetric production costs, they make the same decision whether to produce or not. As long as costs are relatively low ($c < 1$), both plants produce.

Exercise #1.10: Ramsey Pricing^B

1.10 Consider a multiproduct monopolist selling $N \geq 2$ goods with cost function $C(q)$, where $q = (q_1, q_2, \ldots, q_N)$ denotes the output vector. For simplicity, assume that the direct demand of every good i, $q_i(p_i)$, is only a function of good i's price, p_i, but unaffected by the prices of all other goods, p_j, where $j \neq i$. The inverse demand function $p_i(q_i)$ is strictly decreasing in q_i, and we can identify a choke price $\overline{p}$ for which $q_i(p_i) = 0$ for all $p_i \geq \overline{p}$.

(a) *Unregulated equilibrium.* If the multiproduct monopolist seeks to maximize profits, without facing any regulation, find the implicit equation characterizing its (unregulated) profit-maximizing output vector, q^U, where the superscript U denotes "unregulated." [*Hint*: Rearrange your first-order conditions as in the Lerner index, to express the price markup ratio as a function of demand elasticity.]

- The monopolist solves

$$\max_{q \geq 0} \pi(q) = \sum_{i=1}^{N} p_i(q_i) q_i - C(q).$$

Differentiating with respect to q_i, and assuming an interior solution, yields

$$\frac{\partial \pi(q)}{\partial q_i} = p_i + q_i \frac{\partial p_i(q_i)}{\partial q_i} - MC_i = 0$$

where the first two terms denote the marginal revenue from selling one more unit of good i, while the third term represents the marginal cost of producing this unit, $MC_i \equiv \frac{\partial C(q)}{\partial q_i}$. Interestingly, our setting is general enough to allow for the cost function $C(q)$ to be non-additively separable in each good. For instance, producing more units of good i increases the marginal cost of other goods, but these effects are fully internalized by the multiproduct monopolist.

Rearranging, we obtain $p_i - MC_i = -\frac{\partial p_i(q_i)}{\partial q_i} q_i$. Dividing both sides by p_i (to express the right side as a function of price elasticity), we have

$$1 - \frac{MC_i}{p_i} = -\frac{\partial p_i(q_i)}{\partial q_i} \frac{q_i}{p_i}$$

which we can rearrange to yield

$$\frac{p_i - MC_i}{p_i} = \frac{1}{\varepsilon_i}$$

where $\varepsilon_i \equiv -\frac{\partial q_i}{\partial p_i(q_i)} \frac{p_i}{q_i}$ is the price elasticity of good i's demand, so its inverse is $\frac{1}{\varepsilon_i} = -\frac{\partial p_i(q_i)}{\partial q_i} \frac{q_i}{p_i}$.

(b) *Regulated equilibrium.* Assume now that a regulator seeks to identify an output vector q^R, where the superscript R denotes "regulated," to maximize social welfare, subject to the monopolist earning at least profit $\widehat{\pi}$, where $\widehat{\pi} \geq 0$ [*Hint*: Rearrange your first-order conditions as in the Lerner index, to express the price markup ratio as a function of demand elasticity.]

- The regulator solves the following constrained welfare maximization problem:

$$\max_{q \geq 0} \quad W(q) = \sum_{i=1}^{N} \left(\int_0^{q_i} p_i(\widetilde{q}_i)\, d\widetilde{q}_i \right) - C(q)$$

$$\text{subject to} \quad \pi(q) \geq \widehat{\pi}$$

The Lagrangian associated with the above problem is

$$\mathcal{L} = \sum_{i=1}^{N} \left(\int_0^{q_i} p_i(\widetilde{q}_i)\, d\widetilde{q}_i \right) - C(q) + \lambda \left[\pi(q) - \widehat{\pi} \right]$$

where $\lambda \geq 0$ is the Lagrange multiplier.

Differentiating with respect to q_i, and assuming an interior solution, yields

$$\frac{\partial \mathcal{L}}{\partial q_i} = p_i - MC_i + \lambda \frac{\partial \pi(q)}{\partial q_i} = 0$$

which we rearrange to yield

$$p_i - MC_i = -\lambda \left(p_i + q_i \frac{\partial p_i(q_i)}{\partial q_i} - MC_i \right)$$

that further simplifies to

$$(1 + \lambda)(p_i - MC_i) = -\lambda q_i \frac{\partial p_i(q_i)}{\partial q_i}$$

Dividing both sides by p_i, we have

$$\frac{p_i - MC_i}{p_i} = \frac{\kappa}{\varepsilon_i}$$

where, for compactness, we define $\kappa \equiv \frac{\lambda}{1+\lambda}$.

(c) *Output comparison across regulatory settings.* Compare your results in parts (a) and (b). Under which setting the price markup is larger? Interpret.

- Comparing the price markups we found in parts (a) and (b), we obtain that the price markup is lower under regulation since

$$\frac{\kappa}{\varepsilon_i} < \frac{1}{\varepsilon_i}$$

holds because $\lambda \geq 0$ entails $\frac{\lambda}{1+\lambda} < 1$, or $\kappa < 1$. Intuitively, the regulator seeks a larger output level than the unregulated monopolist to maximize social welfare rather than producer surplus alone, which leads to lower prices and markup.

(d) *Effect of price elasticity.* How are your results in parts (a) and (b) affected by an increase in price elasticity?

- Differentiating $\frac{p_i - MC_i}{p_i} = \frac{1}{\varepsilon_i}$ in part (a) with respect to ε_i, we obtain that the price-markup ratio decreases by $\frac{1}{\varepsilon_i^2}$. However, in part (b), the price-markup ratio decreases by $\frac{\kappa}{\varepsilon_i^2}$. Comparing these effects, we find that

$$\frac{1}{\varepsilon_i^2} > \frac{\kappa}{\varepsilon_i^2}$$

simplifies to $\kappa < 1$, which holds given that $\lambda \geq 0$ entails that $\frac{\lambda}{1+\lambda} < 1$, or $\kappa < 1$. Intuitively, when the price elasticity increases (e.g., consumers have closer substitutes to good *i*), the price-markup decreases, with and without regulation. Under regulation, however, this markup decreases less significantly (informally, $\frac{\kappa}{\varepsilon_i^2}$ is closer to zero than $\frac{1}{\varepsilon_i^2}$), because the markup is lower under regulation.

(e) Discuss the effect of an increase in profit $\widehat{\pi}$ on your results in part (b).

- The price-markup condition we found in part (b), $\frac{p_i - MC_i}{p_i} = \frac{\kappa}{\varepsilon_i}$, seems unaffected by profit $\widehat{\pi}$, but it is indirectly a function of $\widehat{\pi}$ via λ. To see this point, first note that, from the Lagrangian in part (b), an increase in $\widehat{\pi}$ makes the profit constraint $\pi(q) \geq \widehat{\pi}$ more difficult to hold. To account for the increased $\widehat{\pi}$, the regulator needs to increase λ, which, in turn, increases κ since $\frac{\partial \kappa}{\partial \lambda} = \frac{1}{(1+\lambda)^2} > 0$, and, thus, the price markup. Indeed, an increase in λ induces the monopolist to reduce its output level in the regulated setting, increasing its profits, which approach to those under no regulation in part (a).
- Finally, in the extreme case where profit $\widehat{\pi}$ increases enough to coincide with the monopolist's profits under no regulation, $\widehat{\pi} = \pi^U$, the Lagrange multiplier $\lambda \to +\infty$ since $\lim_{\lambda \to +\infty} \kappa = \frac{1}{1 + \lim_{\lambda \to +\infty} \frac{1}{\lambda}} = 1$, yielding $\frac{p_i - MC_i}{p_i} = \frac{1}{\varepsilon_i}$ as in part (a).

(f) *Linear demand.* Evaluate your results in parts (a) and (b) at $p_i(q_i) = 1 - q_i$ for every good *i*, and $C(q) = \sum_{i=1}^{N} c_i q_i$.

- First, note that since $\frac{\partial p_i(q_i)}{\partial q_i} = -1$, the price elasticity of demand becomes

$$\varepsilon_i = -\frac{\partial q_i}{\partial p_i(q_i)} \frac{p_i}{q_i}$$
$$= \frac{p_i}{q_i}$$

so its inverse is

$$\frac{1}{\varepsilon_i} = \frac{q_i}{p_i} = \frac{q_i}{1-q_i}$$

given that $p_i\left(q_i\right) = 1 - q_i$.

- *Unregulated monopolist.* Substituting the above result into the monopolist's Lerner index, $\frac{p_i - c_i}{p_i} = \frac{1}{\varepsilon_i}$, we find that

$$\frac{(1-q_i) - c_i}{1-q_i} = \frac{q_i}{1-q_i}$$

where $p_i\left(q_i\right) = 1 - q_i$ and $MC_i = \frac{\partial C(q)}{\partial q_i} = c_i$. Rearranging and solving for q_i yields the monopolist's profit-maximizing output level,

$$q_i^U = \frac{1-c_i}{2}$$

- *Regulated monopolist.* Substituting the above result into the regulator's Lerner index expression in this case, $\frac{p_i - c_i}{p_i} = \frac{\kappa}{\varepsilon_i}$, we obtain

$$\frac{(1-q_i) - c_i}{1-q_i} = \frac{\kappa q_i}{1-q_i}$$

since $\frac{\kappa}{\varepsilon_i} = \frac{\kappa q_i}{p_i} = \frac{\kappa q_i}{1-q_i}$. Rearranging and solving for q_i yields the regulated output level:

$$q_i^R = \frac{1-c_i}{1+\kappa}$$

- Comparing q_i^U and q_i^R, it is obvious that $q_i^U < q_i^R$ since

$$\frac{1-c_i}{2} < \frac{1-c_i}{1+\kappa}$$

simplifies to $2 > 1 + \kappa$, or to $\kappa < 1$, which holds according to part (c).

Exercise #1.11: Monopolist Serving Two Separated Markets-I^A

1.11 Consider a monopolist serving two geographically separated markets, denoted as A and B, with demand functions $q_A = 1 - p_A$ and $q_B = \frac{1}{2} - p_B$, respectively. For simplicity, assume that production and transportation costs are zero.

(a) *Uniform pricing.* Assume that the firm chooses to set a uniform price across both markets. What is the profit-maximizing uniform price, p^u? What are the quantities sold on the two markets at this price, q_A^u and q_B^u?

- Total demand is found by horizontally summing individual demands, that is, when $p \geq 1/2$, aggregate demand coincides with that in market A, $q = q_A$; but when $p < 1/2$, aggregate demand is

$$\begin{aligned} q &= q_A + q_B \\ &= (1 - p_A) + \left(\frac{1}{2} - p_B\right) \\ &= \frac{3}{2} - (p_A + p_B) \\ &= \frac{3}{2} - 2p, \end{aligned}$$

where we set $p = p_A = p_B$ since the monopolist charges the same price in both markets. Solving for p, we find the inverse demand function

$$p(q) = \frac{3}{4} - \frac{1}{2}q.$$

- The profit maximization problem of the monopolist becomes

$$\max_{q \geq 0} \pi(q) = \left(\frac{3}{4} - \frac{1}{2}q\right) q$$

since it faces no production or transportation costs. Differentiating with respect to q yields

$$\frac{3}{4} - q = 0$$

and assuming interior solutions, we find the equilibrium output under uniform pricing

$$q^u = \frac{3}{4}.$$

Substituting $q^u = \frac{3}{4}$ into the above inverse demand function,

$$\begin{aligned} p^u &= \frac{3}{4} - \frac{1}{2}q^u \\ &= \frac{3}{4} - \frac{1}{2} \times \frac{3}{4} \\ &= \frac{3}{8}, \end{aligned}$$

which satisfies $p < 1/2$, thus being in the segment of the aggregate demand function analyzed above. Plugging the uniform price p^u into the individual demand functions, we obtain

$$\begin{aligned} q_A^u &= 1 - \frac{3}{8} = \frac{5}{8} \\ q_B^u &= \frac{1}{2} - \frac{3}{8} = \frac{1}{8}. \end{aligned}$$

Therefore, the monopolist sets a uniform price of $\frac{3}{8}$, selling $\frac{5}{8}$ units in market A and $\frac{1}{8}$ units in market B, for a total of $\frac{5}{8}+\frac{1}{8}=\frac{3}{4}$ units.

(b) *Third-degree price discrimination.* Assume now that the monopolist practices third-degree price discrimination. What are the profit maximizing prices and quantities in the two markets?

- *Market A*. In this setting, the monopolist treats each market as a separate monopoly. Therefore, in market A, the monopolist solves the following profit maximization problem:

$$\max_{q_A \geq 0} \pi\left(q_A\right)=\left(1-q_A\right) q_A .$$

Differentiating with respect to q_A, and assuming interior solutions, yields

$$q_A^*=\frac{1}{2}$$

Substituting $q_A^*=\frac{1}{2}$ into the inverse demand function, we find

$$p_A^*=1-q_A^*=\frac{1}{2} .$$

Therefore, in market A, the monopolist sells $\frac{1}{2}$ units at a price of $\frac{1}{2}$.

- *Market B*. In this market, the monopolist solves the following profit maximization problem:

$$\max_{q_B \geq 0} \pi\left(q_B\right)=\left(\frac{1}{2}-q_B\right) q_B .$$

Differentiating with respect to q_B, and assuming interior solutions, yields

$$q_B^*=\frac{1}{4}$$

Substituting $q_B^*=\frac{1}{4}$ into the inverse demand function, we find

$$p_B^*=\frac{1}{2}-q_B^*=\frac{1}{4} .$$

Therefore, in market B, the monopolist sells $\frac{1}{4}$ units at a price of $\frac{1}{4}$. Intuitively, the demand in market B is relatively more elastic, inducing the monopolist to set a lower price in this market than in market A, $p_B^*<p_A^*$.

(c) *Comparison.* Evaluate consumer surplus, profit, and total welfare, under a uniform price and under third-price discrimination. Compare both settings and interpret your results.

- Under uniform pricing, profit, consumer surplus, and social welfare are calculated using triangle formulas, as depicted in Fig. 1.9:

$$\pi^u=p^u q^u=\frac{3}{8} \times \frac{3}{4}=\frac{9}{32}$$

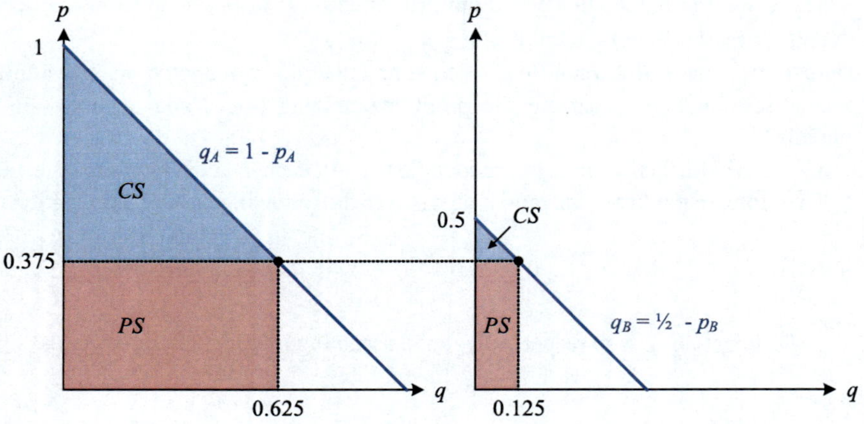

Fig. 1.9 Social welfare under uniform pricing

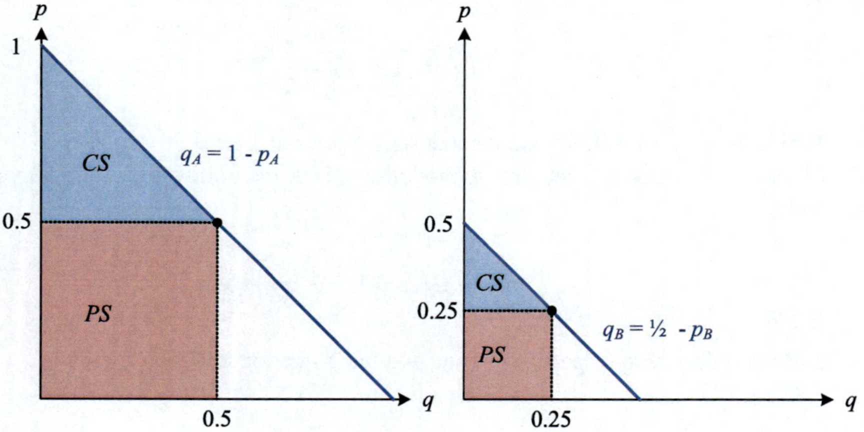

Fig. 1.10 Social welfare under third-degree price discrimination

$$CS^u = \frac{1}{2}\left(\frac{5}{8}\right)\left(\frac{5}{8}\right) + \frac{1}{2}\left(\frac{1}{8}\right)\left(\frac{1}{8}\right) = \frac{25}{128} + \frac{1}{128} = \frac{13}{64}$$

$$SW^u = \pi^u + CS^u = \frac{9}{32} + \frac{13}{64} = \frac{31}{64}.$$

- Under third-degree price discrimination, profit, consumer surplus, and social welfare are depicted in Fig. 1.10:

$$\pi_A = p_A^* q_A^* = \frac{1}{2} \times \frac{1}{2} = \frac{1}{4}$$

$$CS_A = \frac{1}{2}\left(\frac{1}{2}\right)\left(\frac{1}{2}\right) = \frac{1}{8}$$

$$SW_A = \pi_A + CS_A = \frac{1}{4} + \frac{1}{8} = \frac{3}{8}$$

$$\pi_B = p_B^* q_B^* = \frac{1}{4} \times \frac{1}{4} = \frac{1}{16}$$

$$CS_B = \frac{1}{2}\left(\frac{1}{4}\right)\left(\frac{1}{4}\right) = \frac{1}{32}$$

$$SW_B = \pi_B + CS_B = \frac{1}{16} + \frac{1}{32} = \frac{3}{32}$$

$$\pi^{3rd} = \pi_A + \pi_B = \frac{5}{16}$$

$$SW^{3rd} = SW_A + SW_B = \frac{15}{32}$$

$$CS^{3rd} = CS_A + CS_B = \frac{5}{32}.$$

- Since $\pi^{3rd} > \pi^u$, the monopolist obtains larger profits from price discrimination. However, since $p_A > p^u > p_B$, consumers in market *A* (market *B*) pay a higher (lower) price than a uniform price, and are thus worse off (better off, respectively) under price discrimination than uniform pricing. Overall speaking, $SW^u > SW^{3rd}$ indicates that the monopolist practicing third-degree price discrimination reduces social welfare, because its profit gain is more than offset by surplus loss of the consumers.

Exercise #1.12: Monopolist Serving Two Separated Markets-II^A

1.12 Cougar movie house is a local monopoly selling movie tickets in a college town. The demand for adult and student tickets are $q_A(p_A) = 130 - p_A$ and $q_S(p_S) = 30 - p_S$, respectively. Tickets are non-transferable and are printed at zero marginal costs.

(a) *Uniform pricing*. Assume that Cougar movie house chooses to set a uniform price across both markets. What is the profit-maximizing uniform price, p^u? What are the quantities sold on the two markets at this price, q_A^u and q_S^u?

- Total demand is found by horizontally summing individual demands. That is, when $p \geq 30$, aggregate demand coincides with that from adults, $q = q_A$; but when $p < 30$, aggregate demand is

$$\begin{aligned} q &= q_A + q_S \\ &= (130 - p_A) + (30 - p_S) \\ &= 160 - (p_A + p_S) \\ &= 160 - 2p \end{aligned}$$

where we set $p = p_A = p_S$ since the monopolist charges the same price in both markets. Solving for p, we find the inverse demand function

$$p(q) = 80 - \frac{1}{2}q$$

- The profit maximization problem of the monopolist becomes

$$\max_{q \geq 0} \pi(q) = \left(80 - \frac{1}{2}q\right) q$$

since it faces no production costs. Differentiating with respect to q yields

$$80 - q = 0$$

In this context, equilibrium output under uniform pricing is

$$q^u = 80.$$

Substituting $q^u = 80$ into the above inverse demand function,

$$\begin{aligned} p^u &= 80 - \frac{1}{2}q^u \\ &= 80 - \frac{1}{2} \times 80 \\ &= 40 \end{aligned}$$

which does not satisfy $p < 30$. In this setting, the monopolist will not sell to the student market and, instead, target solely the adult market. As such, we must recalculate our equilibrium values with the demand function $q = q_A$. The profit maximization problem of the monopolist now becomes

$$\max_{q \geq 0} \pi(q) = (130 - q)\, q$$

Differentiating with respect to q, yields

$$130 - 2q = 0$$

In this context, equilibrium output under uniform pricing is

$$q^u = 65.$$

Substituting $q^u = 65$ into the above inverse demand function,

$$\begin{aligned} p^u &= 130 - q^u \\ &= 130 - 65 \\ &= 65 \end{aligned}$$

which satisfies $p \geq 30$. Therefore, the monopolist sells $q_A^u = 65$ tickets to the adults at a price of $p^u = 65$ and $q_S^u = 0$ tickets to the students.

(b) *Third-degree price discrimination.* Assume now that the monopolist practices third-degree price discrimination. What are the profit-maximizing prices and quantities for the adults and students?

- In this setting, the monopolist treats each market as a separate monopoly.
- *Adults*. The results are unchanged from part (a), where the monopolist sells 65 tickets to the adults at a price of \$65.
- *Students*. In the market for students, the monopolist solves the following profit maximization problem:

$$\max_{q_S \geq 0} \pi (q_S) = (30 - q_S)\, q_S$$

Differentiating with respect to q_S yields

$$q_S^* = 15.$$

Substituting $q_S^* = 15$ into the inverse demand function, we find

$$p_A^* = 30 - q_A^* = 15$$

Therefore, the monopolist sells 15 tickets to the students at a price of \$15.

(c) *Comparison.* Evaluate consumer surplus, profits, and total welfare, under a uniform price in part (a) and third-price discrimination in part (b). Compare your results in both settings and interpret.

- Under uniform pricing, profits, consumer surplus, and social welfare are

$$\begin{aligned}
\pi^u &= p^u q^u = 65 \times 65 = 4{,}225 \\
CS^u &= \frac{1}{2}\left(130 - p^u\right) q_S^u \\
&= \frac{1}{2}(130 - 65) \times 65 \\
&= 2{,}112.5 \\
SW^u &= \pi^u + CS^u = 6{,}337.5
\end{aligned}$$

- Under third-degree price discrimination, profits, consumer surplus, and social welfare are

$$\begin{aligned}
\pi_S &= p_S^* q_S^* = 15 \times 15 = 225 \\
CS_S &= \frac{1}{2}\left(30 - p_S^*\right) q_S^* \\
&= \frac{1}{2}(30 - 15) \times 15 \\
&= 112.5 \\
SW_S &= \pi_S + CS_S = 225 + 112.5 = 337.5 \\
\pi_A &= p_A^* q_A^* = 65 \times 65 = 4{,}225
\end{aligned}$$

$$
\begin{aligned}
CS_A &= \frac{1}{2}\left(130 - p_A^*\right) q_S^* \\
&= \frac{1}{2}\left(130 - 65\right) \times 65 \\
&= 2{,}112.5 \\
SW_A &= \pi_A + CS_A \\
&= 4{,}225 + 2{,}112.5 = 6{,}337.5 \\
\pi^{3rd} &= \pi_A + \pi_S = 4{,}450 \\
CS^{3rd} &= CS_A + CS_S = 2{,}225 \\
SW^{3rd} &= SW_A + SW_S = 6{,}675.
\end{aligned}
$$

Since students are left out of this market under uniform pricing, there are no consumer surplus, profits, or total welfare generated by that group (in fact, they are all deadweight loss), but by implementing third-degree price discrimination, the monopolist includes this group in the market without harming their profits from the adult group. This leads to a special case in third-degree price discrimination where profits, consumer surplus, and total welfare all increase when a monopolist price-discriminates.

Exercise #1.13: Geographical Price Discrimination^B

1.13 Consider a metropolitan city which can be separated into two distinct markets, area 1 (e.g., suburb area) and area 2 (e.g., urban area). Crimson company has established its presence in both areas and is an exclusive dealer in area 1 with demand $q_{C1} = 600 - p_{C1}$. Gray company is active in area 2 only and is competing à la Bertrand with Crimson company on differentiated goods, which demands are

$$
\begin{aligned}
q_{C2} &= 1800 - 3p_{C2} + 2p_{G2} \\
q_{G2} &= 1800 - 3p_{G2} + 2p_{C2},
\end{aligned}
$$

where subscripts $C1$ and $C2$ denote Crimson sales in area 1 and 2, respectively. Similarly, subscript $G2$ represent Gray sales in area 2. For simplicity, products of firm i, where $i \in \{C, G\}$, are produced at zero marginal costs.

(a) *Third-degree price discrimination.* Assume that Crimson company adapts its price to the local conditions. Find the profit-maximizing prices and quantities of each firm.
   - *Area 1.* Crimson company solves the following profit maximization problem:

$$\max_{p_{C1} \geq 0} \pi_1 (p_{C1}) = p_{C1} (600 - p_{C1}) .$$

Differentiating with respect to p_{C1}, we obtain

$$600 - 2p_{C1} = 0.$$

Rearranging, optimal price in area 1 becomes

$$p_{C1}^* = 300.$$

Substituting $p_{C1}^* = 300$ into the profit function, equilibrium profit in area 1 becomes

$$\pi_1 = p_{C1}^* \left(600 - p_{C1}^*\right) = 300\,(600 - 300) = 90{,}000.$$

- *Area 2.* Crimson company solves the following profit maximization problem:

$$\max_{p_{C2} \geq 0} \pi_2\,(p_{C2}) = p_{C2}\,(1800 - 3p_{C2} + 2p_{G2})\,.$$

Differentiating with respect to p_{C2}, we obtain

$$1800 - 6p_{C2} + 2p_{G2} = 0.$$

Rearranging, the best response function of Crimson company becomes

$$p_{C2}\,(p_{G2}) = 300 + \frac{1}{3}p_{G2}.$$

Symmetrically, the best response function of Gray company is

$$p_{G2}\,(p_{C2}) = 300 + \frac{1}{3}p_{C2}.$$

Intersecting the firms' best response functions, equilibrium price in area 2 solves

$$p_2^* = 300 + \frac{1}{3}p_2^*,$$

which simplifies to

$$p_2^* = 450.$$

Substituting $p_2^* = 450$ into the profit function, equilibrium profit in area 2 becomes

$$\pi_2 = p_2^* \left(1800 - p_2^*\right) = 450\,(1800 - 450) = 607{,}500.$$

- Therefore, under price discrimination, total profits of each firm become

$$\pi_C^* = \pi_1 + \pi_2 = 697{,}500$$
$$\pi_G^* = \pi_2 = 607{,}500.$$

(b) *Uniform pricing.* Assume that Crimson company commits to one price in both markets. Find the equilibrium prices and resulting profits of each firm.

- In this situation, Crimson company faces an aggregate demand curve of $q_C = q_{C1} + q_{C2} = 600 - p_{C1} + 1800 - 3p_{C2} + 2p_{G2} = 2400 - 4p_C + 2p_G$ when $p_C \leq 600$. Crimson company then solves the following profit maximization problem:

$$\max_{p_C \geq 0} \pi_C^u (p_C) = p_C (2400 - 4p_C + 2p_G).$$

Differentiating with respect to p_C, we obtain

$$2400 - 8p_C + 2p_G = 0.$$

Rearranging, the best response function of Crimson company is

$$p_C (p_G) = 300 + \frac{1}{4} p_G.$$

The best response function of Gray company is the same as part (a) and is given by

$$p_G (p_C) = 300 + \frac{1}{3} p_C.$$

Intersecting the best response functions, equilibrium prices are

$$p_C^u = \frac{4500}{11} \approx 409.09$$

$$p_G^u = \frac{4800}{11} \approx 436.36.$$

- Therefore, under uniform pricing, total profits of each firm become

$$\pi_C^u = \frac{4500}{11}\left(600 - \frac{4500}{11}\right) + \frac{4500}{11}\left(1800 - 3 \times \frac{4500}{11} + 2 \times \frac{4800}{11}\right)$$

$$= \frac{81{,}000{,}000}{121} \simeq 669{,}421.49$$

$$\pi_G^u = \frac{4800}{11}\left(1800 - 3 \times \frac{4800}{11} + 2 \times \frac{4500}{11}\right)$$

$$= \frac{6{,}9120{,}000}{121} \simeq 571{,}239.67.$$

(c) Compare your results in parts (a) and (b), and interpret.

- Comparing equilibrium prices under different pricing strategies, we obtain that

$$p_{C1}^* < p_C^u < p_G^u < p_2^*.$$

Under uniform pricing, Crimson company aggregates the demand over both areas and offers a lower price than its rival and that of price discrimination. However, consumers in area 1 pay higher prices if they are not price discriminated.

- Comparing the firms' total profits under different pricing strategies, we find that

$$\pi_C^* > \pi_C^u$$
$$\pi_G^* > \pi_G^u.$$

Since price discrimination softens competition, both firms obtain higher profits if they practice third-degree price discrimination than uniform pricing.

(d) Suppose now that the demand for area 1 shrinks to $q_{C1} = 200 - p_{C1}$. Will Crimson company still serve area 1 if it practices uniform pricing?

- Now Crimson company faces an aggregate demand curve of $q_C = q_{C1} + q_{C2} = 200 - p_{C1} + 1800 - 3p_{C2} + 2p_{G2} = 2000 - 4p_C + 2p_G$ when $p_C \leq 200$. Crimson company solves the following profit maximization problem:

$$\max_{p_C \geq 0} \pi_C^u (p_C) = p_C (2000 - 4p_C + 2p_G).$$

Differentiating with respect to p_C, we obtain

$$2000 - 8p_C + 2p_G = 0.$$

Rearranging, we find the best response function of this firm,

$$p_C (p_G) = 250 + \frac{1}{4} p_G.$$

The best response function for the Gray company is the same as that of part (a):

$$p_G (p_C) = 300 + \frac{1}{3} p_C.$$

Intersecting the best response functions, equilibrium prices become

$$p_C^u = \frac{3900}{11} \simeq 354.54$$
$$p_G^u = \frac{4600}{11} \simeq 418.18.$$

Since $p_C^u > 200$ in this situation, Crimson company does not serve area 1 (since quantity cannot be negative). Intuitively, since the market for area 1 is now much smaller than that of area 2, Crimson company does not serve area 1 and focuses solely on area 2 if it practices uniform pricing.

Exercise #1.14: Two-Part Pricing[A]

1.14 Monopolists use two-part pricing to price discriminate and increase their profit. Examples include pricing of amusement parks, where a consumer often pays an entry fee and a separate price for some rides. These are also common in bars and in wholesale clubs like Costco. Let us investigate how a firm sets the optimal two-part tariff by assuming we have $N \geq 2$ consumers each with individual (not to be confused with market) demand for rides of $p = 1 - q$, and the costs of running the amusement park are $C(Q) = cq$, where c denotes the marginal cost and $1 > c \geq 0$.

(a) *Uniform pricing.* If the park acts as a monopoly, setting a single price, what is its profit-maximizing price, quantity of rides (per person and aggregate), and profit?

- Setting up the monopolist's profit maximization problem,

$$\max_{q \geq 0} \; N\left[(1-q)q - cq\right]$$

with its corresponding first-order condition,

$$\frac{\partial \Pi^U}{\partial q} = N\left(1 - 2q - c\right) = 0$$

Rearranging, we find our standard result that marginal revenue equals marginal cost

$$\underbrace{1 - 2q}_{MR(q)} = \underbrace{c}_{MC(q)} .$$

which simplifies to $2q = 1 - c$. Solving for q, we obtain

$$q = \frac{1-c}{2} \text{ rides per person}$$

that yields the total number of rides of

$$Q = Nq = \frac{N(1-c)}{2}.$$

In this context, the price that the monopolist charges per ride is

$$p = 1 - q = 1 - \frac{1-c}{2} = \frac{1+c}{2},$$

leading to total profits of

$$\begin{aligned}
\Pi^U &= N\left[(1-q)q - cq\right] \\
&= N\left(\frac{1+c}{2} - c\right)\left(\frac{1-c}{2}\right) \\
&= N\left(\frac{1-c}{2}\right)^2 .
\end{aligned}$$

(b) *Marginal cost pricing.* Suppose instead that the monopolists were forced to compete as if they were in a perfectly competitive market. What is the number of rides it will sell (per person and aggregate) and the associated consumer surplus?

- If the park sets the price equal to its marginal cost, it will charge c per ride since $MC(q) = c$. At this price, each consumer will purchase

$$c = 1 - q,$$

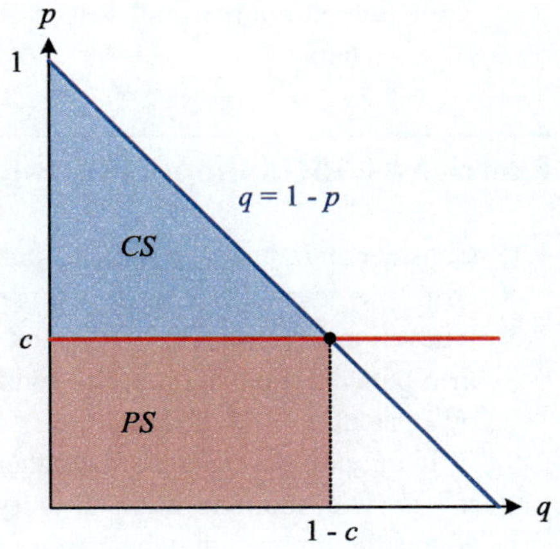

Fig. 1.11 Consumer surplus

or $q = 1 - c$ rides per customer. This means that the amusement park will produce a total of $N(1-c)$ rides with N consumers. Each consumer enjoys a consumer surplus of

$$CS = \frac{1}{2}(1-c)(1-c) = \frac{(1-c)^2}{2}.$$

Recall that the consumer surplus is the area below the inverse demand curve $p(q) = 1-q$ and above the equilibrium price, which, in this setting, is $p = c$. This relationship is depicted in Fig. 1.11. Aggregate consumer surplus for N consumers is then $\frac{N(1-c)^2}{2}$.

(c) *Two-part pricing.* Suppose the monopolist implemented a two-part pricing scheme (*Hint*: To do this, the monopolist charges an access fee equal to the consumer surplus under perfect competition and a per-unit price equal to the price under perfect competition). What is the monopolist's total profit?

- From part (b), we calculated the consumer surplus for each individual as $\frac{(1-c)^2}{2}$ and set the price equal to marginal cost, c. Furthermore, under perfect competition, each consumer purchases $q = 1 - p = 1 - c$ rides from the monopolist. Thus, the monopolist's total profits under a two-part pricing scheme are

$$\begin{aligned}\Pi^T &= N\left[\text{Access Fee} + (1-q)q + cq\right] \\ &= N\left[\frac{(1-c)^2}{2} + c(1-c) - c(1-c)\right] \\ &= N\left[\frac{(1-c)^2}{2}\right].\end{aligned}$$

This profit is larger than the profit that the park would earn if it uses uniform pricing, that is, $\Pi^T > \Pi^U$, since

$$\begin{aligned} N\left[\frac{(1-c)^2}{2}\right] &> N\left[\left(\frac{1-c}{2}\right)^2\right] \\ \frac{(1-c)^2}{2} &> \frac{(1-c)^2}{4},\end{aligned}$$

which holds for all values of c, so that the park benefits from using a two-part pricing scheme.

Exercise #1.15: Monopoly Facing Entry Threats – Based on Tirole (1988)^C

1.15 Consider an industry with an incumbent and a potential entrant, both facing marginal production cost $1 > c > 0$. A third firm cannot produce in this output market but has generated an innovation which can lower the unit production cost from c to c' (process innovation). This third firm puts the innovation up for bidding between firms 1 and 2, and the innovation is protected by a patent of unlimited duration.

If no entry occurs, the incumbent's monopoly profit is $\pi^m(c)$ without the innovation and $\pi^m(c')$ with the innovation. If entry occurs, let $\pi_i^d(c, c)$ denote the duopoly profits if no firm adopts the innovation, where firm $i = \{I, E\}$, I represents the incumbent, and E is the entrant. In this setting, profit $\pi_i^d(c', c)$ denotes firm i's duopoly profits when only firm i adopts the innovation, so $\pi_i^d(c', c) > \pi_j^d(c, c')$ where $j \neq i$. Specifically, the first argument in the profit function describes the cost of firm i, and the second argument describes its rival's cost. For simplicity, we assume that the entrant does not profitably enter the industry when the incumbent wins the bidding competition for the innovation.

(a) Find the incumbent's profit gain from winning the innovation, V_I, and the entrant's profit gain, V_E.

- This profit gain is often known as the "value of the innovation," since it represents the difference in profits when the firm wins the innovation and when it does not.
- *Incumbent.* The incumbent's value of the innovation is

$$V_I = \pi^m(c') - \pi_I^d(c, c')$$

since the entrant remains outside when the incumbent's costs are low, c', letting the incumbent earn monopoly profit $\pi^m(c')$. When the entrant wins the bidding competition for the innovation, the incumbent's cost is c while that of the entrant is c', yielding duopoly profit $\pi_I^d(c, c')$ for the incumbent.

- *Entrant.* The entrant's value of the innovation is

$$V_E = \pi_E^d(c', c) - 0 = \pi_E^d(c', c)$$

since the entrant's duopoly profit when winning the bidding competition for the innovation is $\pi_E^d(c', c)$. When losing the bidding competition, the entrant remains outside the industry and earns zero profits.

(b) Show that $\pi^m(c') \geq \pi_I^d(c, c') + \pi_E^d(c', c)$ is a sufficient condition for $V_I \geq V_E$. Interpret.

- Condition $V_I \geq V_E$ entails

$$\pi^m(c') - \pi_I^d(c, c') \geq \pi_E^d(c', c)$$

or

$$\pi^m(c') \geq \pi_I^d(c, c') + \pi_E^d(c', c).$$

This inequality says that the profit gain from winning the innovation (the value of the innovation) is larger for the incumbent than for the potential entrant if aggregate profits are larger in monopoly than under duopoly. Intuitively, this property suggests that we should expect monopolies to persist in contexts where the monopoly is more efficient in adopting the process innovation than the potential entrants.

(c) *Parametric example*. Consider an industry with inverse demand $p(Q) = 1 - Q$. Evaluate the condition on profits in part (b), and solve for cost c'. Interpret.

- Monopoly profit is $\pi^m(c') = \frac{(1-c')^2}{4}$, the incumbent's duopoly profit when its marginal cost is c while that of the entrant decreases to c' is $\pi_I^d(c, c') = \frac{(1-2c+c')^2}{9}$, and the entrant's duopoly profit if it wins the innovation is $\pi_E^d(c', c) = \frac{(1-2c'+c)^2}{9}$. Therefore, the condition on profits we found in part (b) can be written as

$$\frac{(1-c')^2}{4} \geq \frac{(1-2c+c')^2}{9} + \frac{(1-2c'+c)^2}{9},$$

which simplifies to

$$5\left(1-c'\right)^2 \geq 4\left(1-c\right)^2 + 16\left(c-c'\right)^2.$$

Rearranging, we obtain

$$(1+10c-11c')(2c-c'-1) \leq 0.$$

Solving for c', we find two roots $c'_1 \equiv 2c-1$ and $c'_2 \equiv \frac{1+10c}{11}$.

First, note that $c'_2 > c$ since

$$\frac{1+10c}{11} > c$$

simplifies to $c < 1$, which holds by definition.

Therefore, the condition on profits from part (b) holds as long as c' satisfies

$$c'_1 \leq c' \leq c.$$

For example, when $c' = c$, condition $V_I \geq V_E$ becomes

$$\frac{(1-c)^2}{4} \geq \frac{2\,(1-c)^2}{9},$$

which simplifies to $\frac{1}{4} > \frac{2}{9}$ that is satisfied.

However, when the innovation is highly efficient, where $c' < c_1$, the incumbent will produce no output if the entrant wins the innovation since $q_I\left(c, c'\right) = \frac{1-2c+c'}{3} < \frac{1-2c+2c-1}{3} = 0$, which happens in situations where the innovation makes the entrant way more efficient than the incumbent, driving the incumbent out of the industry. Thus, if the innovation is "non-drastic," in the sense of not driving its rival out of the market,[1] the monopoly will have incentives to bid for the innovation even if the cost-reduction effect, $c' - c$, is minimal.

[1] Generally, an innovation is defined as non-drastic if, in the case that only one firm innovates, its monopoly price after the innovation, $p(c')$, lies above its rival's marginal cost before the innovation, c. If this holds, the innovator sets a price slightly below the marginal cost of the non-innovating firm, $p = c - \varepsilon$ where $\varepsilon \to 0$, driving it out of the market.

- For instance, when the initial marginal cost is $c = \frac{2}{3}$, condition $2c - 1 \leq c' \leq c$ becomes $\frac{1}{3} \leq c' \leq \frac{2}{3}$. Intuitively, the incumbent benefits from the process innovation when $c' < c = 2/3$ and stays in the market even if the entrant wins the innovation when $c' > c'_1 = 1/3$. In contrast, when $c' < 1/3$, the incumbent will leave the industry if the entrant wins the innovation.

Exercise #1.16: Multiproduct Monopoly with Economies of Scope^B

1.16 Consider Ferdinand's food company, a monopolist producing two goods, ice cream (good 1) and cheese (good 2), which are regarded as substitutes for consumers. The inverse demand function of good i is

$$p_i\left(q_i, q_j\right) = a - bq_i - gq_j,$$

where $a, b, g > 0$ and $|b| > |g|$, entailing that own-price effects dominate cross-price effects.

In addition, Ferdinand's cost function is

$$C\left(q_1, q_2\right) = \frac{c}{2}\left(q_1^2 + q_2^2\right) - \beta q_1 q_2$$

where $c > 0$ and $\beta > 0$ indicates that the marginal cost of producing one good decreases in the output of another good, i.e., there are cost complementarities in production often referred as "economies of scope." When $\beta = 0$, the cost of one output is independent of the other.

(a) Find the profit-maximizing output and associated profits of Ferdinand's food company.

- Ferdinand's food company chooses q_1 and q_2 to solve

$$\max_{q_1, q_2 \geq 0} \pi\left(q_1, q_2\right) = \left(a - bq_1 - gq_2\right) q_1 + \left(a - bq_2 - gq_1\right) q_2 - \frac{c}{2}\left(q_1^2 + q_2^2\right) + \beta q_1 q_2$$

Differentiating with respect to q_1 and q_2, and assuming interior solutions, we find

$$a - 2bq_1 - 2gq_2 - cq_1 + \beta q_2 = 0, \text{ and}$$
$$a - 2bq_2 - 2gq_1 - cq_2 + \beta q_1 = 0.$$

Invoking symmetry, where $q^* = q_1 = q_2$, we obtain

$$a - 2bq^* - 2gq^* - cq^* + \beta q^* = 0.$$

Rearranging, equilibrium output becomes

$$q^* = \frac{a}{2b + 2g + c - \beta}.$$

- Substituting equilibrium output into Ferdinand's profit function, we have

$$\begin{aligned}
\pi^* &= 2\left(a - (b + g)\, q^*\right) q^* + (\beta - c)\left(q^*\right)^2 \\
&= \frac{2a^2}{2b + 2g + c - \beta} - (2b + 2g + c - \beta)\left(\frac{a}{2b + 2g + c - \beta}\right)^2 \\
&= \frac{a^2}{2b + 2g + c - \beta}.
\end{aligned}$$

(b) How does equilibrium output change in parameters β and g? Interpret your results.
- Differentiating the equilibrium output, q^*, with respect to β, we obtain

$$\frac{\partial q^*}{\partial \beta} = \frac{a}{(2b + 2g + c - \beta)^2} > 0,$$

so that as cost complementarity, β, increases, the monopolist increases the output of both goods. Intuitively, producing more units of one good makes the other good less costly to produce, increasing the incentive to produce further units of both goods.
- Differentiating the equilibrium output, q^*, with respect to g, we find

$$\frac{\partial q^*}{\partial g} = -\frac{2a}{(2b + 2g + c - \beta)^2} < 0.$$

Therefore, as the cross-price effect is strengthened, goods are more easily substitutable (more homogeneous). In this context, the monopolist reduces output of both goods.

(c) *Numerical example.* Evaluate your results in parts (a) and (b) assuming parameter values $a = 1$, $\beta = 1/2$, $c = 1/3$, and $g = 1/4$. How do they change with b? Interpret.
- Substituting $a = 1$, $\beta = 1/2$, $c = 1/3$, and $g = 1/4$ into equilibrium output, we find

$$q^* = \frac{1}{2b + 2 \times \frac{1}{4} + \frac{1}{3} - \frac{1}{2}} = \frac{3}{1 + 6b}.$$

- Similarly, substituting the above parameter values into equilibrium profit, we have

$$\pi^* = \frac{1^2}{2b + 2 \times \frac{1}{4} + \frac{1}{3} - \frac{1}{2}} = \frac{3}{1 + 6b}.$$

Therefore, both equilibrium output and profit decrease in own price effect b. Intuitively, the more sensitive is own price to the quantity demanded of one good, the fewer units will this good be produced in equilibrium.

Exercise #1.17: Vertical Differentiation and Natural Monopoly^B

1.17 Consider the following model of vertically differentiated products with two firms. In the first stage, every firm i chooses its quality, s_i, in the interval $[0, 1]$, where $i = \{1, 2\}$. In the second stage, every firm i, observing the quality pair (s_1, s_2) from the first stage, responds setting a price p_i. In the third stage, given firms' quality and prices, consumers buy one unit of the good from either firm 1 or 2. Every consumer when buying from firm i enjoys utility

$$r - p_i + \theta s_i,$$

where parameter r denotes the reservation utility that the consumer enjoys from the good and $\theta \sim U[\underline{\theta}, \overline{\theta}]$ denotes how much this consumer cares about quality. Intuitively, a consumer with $\theta = \underline{\theta}$ assigns minimal concern to quality, while $\theta = \overline{\theta}$ assigns the maximal importance to quality. Assume that, for simplicity, both firms' marginal production cost is $c = 0$. Consider that qualities are given, where $s_2 > s_1$, but firms compete in prices.

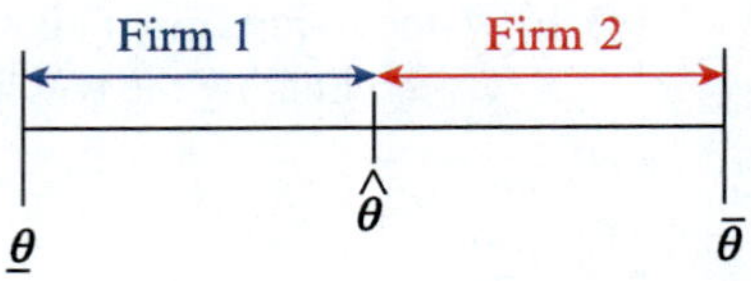

Fig. 1.12 Firm demand

(a) *Third stage – finding demand.* For given locations and prices, find the demand that each firm faces in the third stage of the game.
   - If a consumer purchases from firm 1, his utility is $r - p_1 + \theta s_1$, while purchasing from firm 2 yields $r - p_2 + \theta s_2$. Therefore, the indifferent consumer $\widehat{\theta}$ solves

$$r - p_1 + \widehat{\theta} s_1 = r - p_2 + \widehat{\theta} s_2.$$

   Solving this expression for $\hat{\theta}$ yields

$$\widehat{\theta} = \frac{p_2 - p_1}{s_2 - s_1}.$$

   Therefore, firm 1's demand is $\widehat{\theta} - \underline{\theta}$, while firm 2's demand is $\overline{\theta} - \widehat{\theta}$, as depicted in Fig. 1.12. Graphically, cutoff $\widehat{\theta}$ shifts rightward along the interval $[\underline{\theta}, \overline{\theta}]$ when the price differential $p_2 - p_1$ increases (meaning that firm 2 sets higher prices than firm 1), expanding the demand for firm 1 while shrinking that of firm 2. In contrast, cutoff $\widehat{\theta}$ shifts leftward when the quality differential $s_2 - s_1$ increases (i.e., firm 2 offers a higher quality than firm 1), shrinking the demand of firm 1 and expanding that of firm 2.

(b) *Second stage – prices.* For given locations, find the price that each firm charges in the second stage.
   - *Firm 1's best response function.* Firm 1 chooses the price p_1 that solves

$$\max_{p_1 \geq 0} \; p_1 \underbrace{\left(\frac{p_2 - p_1}{s_2 - s_1} - \underline{\theta} \right)}_{\text{Demand, } \widehat{\theta} - \underline{\theta}}.$$

   Differentiating with respect to p_1, we obtain

$$\frac{\partial \pi_1}{\partial p_1} = \frac{p_2 - p_1}{s_2 - s_1} - \underline{\theta} - \frac{p_1}{s_2 - s_1} = 0$$

   Solving for p_1, we find firm 1's best response function

$$p_1(p_2) = \frac{\underline{\theta}(s_1 - s_2)}{2} + \frac{1}{2} p_2$$

   with vertical intercept at $\frac{\underline{\theta}(s_1 - s_2)}{2}$ and slope $\frac{1}{2}$ (note that since $s_2 > s_1$, this vertical intercept is negative). Intuitively, when firm 2 increases its price by \$1, firm 1 responds increasing its own by \$0.5.

- *Firm 2's best response function.* Operating similarly for firm 2, we have that this firm chooses price p_2 to solve

$$\max_{p_2 \geq 0} \; p_2 \underbrace{\left(\overline{\theta} - \frac{p_2 - p_1}{s_2 - s_1}\right)}_{\text{Demand, } \overline{\theta} - \widehat{\theta}}.$$

Differentiating with respect to p_2, we obtain

$$\begin{aligned}\frac{\partial \pi_2}{\partial p_2} &= \overline{\theta} - \frac{p_2 - p_1}{s_2 - s_1} - p_2\left(\frac{1}{s_2 - s_1}\right) = 0 \\ &= \frac{p_1 - 2p_2}{s_2 - s_1} + \overline{\theta} = 0.\end{aligned}$$

Solving for p_2, we find firm 2's best response function

$$p_2(p_1) = \frac{\overline{\theta}(s_2 - s_1)}{2} + \frac{1}{2}p_1$$

with vertical intercept at $\frac{\overline{\theta}(s_2 - s_1)}{2}$ and slope $\frac{1}{2}$. Intuitively, when firm 1 increases its price by \$1, firm 2 responds increasing its own by \$0.5. In contrast, increasing the quality differential, $s_2 - s_1$, shifts the best response function upward in a parallel fashion, since firm 2 can command a higher price to further differentiate itself from firm 1.

- *Finding equilibrium prices.* Simultaneously solving for p_1 and p_2 in the above best response functions, we find equilibrium prices

$$p_1^*(s_1, s_2) = \frac{\left(\overline{\theta} - 2\underline{\theta}\right)(s_2 - s_1)}{3}, \text{ and}$$

$$p_2^*(s_1, s_2) = \frac{\left(2\overline{\theta} - \underline{\theta}\right)(s_2 - s_1)}{3}.$$

Since quality is given and satisfies $s_2 > s_1$ by definition, we can claim that $p_1^*(s_1, s_2) > 0$ if and only if $\overline{\theta} > 2\underline{\theta}$. Similarly, $p_2^*(s_1, s_2) > 0$ if and only if $2\overline{\theta} > \underline{\theta}$, which holds given that $\overline{\theta} > \underline{\theta}$ by definition.

Thus, we can identify two regions based on whether parameter $\overline{\theta}$ lies above or below $2\underline{\theta}$, as depicted in Fig. 1.13. For illustration purposes, Fig. 1.13 plots the 45^0-line, where $\overline{\theta} = \underline{\theta}$, and shades the admissible $\left(\overline{\theta}, \underline{\theta}\right)$-pairs where $\overline{\theta} > \underline{\theta}$, above the 45^0-line. The line $\overline{\theta} = 2\underline{\theta}$ then divides the shaded area into two regions, which we separately describe next:

- *Region I.* When $\overline{\theta} > 2\underline{\theta}$, both firms set positive prices. Intuitively, individuals with the highest concern for quality (those with $\theta = \overline{\theta}$) have such a high concern, relative to individuals with the lowest concern (those with $\theta = \underline{\theta}$), that both firms can sell positive units and make a profit.
- *Region II.* When $\overline{\theta} \leq 2\underline{\theta}$, the price of firm 1 is zero while that of firm 2 is positive (i.e., only firm 2 is active). In this case, the "quality concern differential" between individuals with the highest and lowest quality concern, $\overline{\theta} - \underline{\theta}$, is lower, implying that only firm 2 (the firm with the highest quality) can make a profit.

Therefore, when parameter $\overline{\theta}$ takes relatively lower values, $\overline{\theta} \leq 2\underline{\theta}$, we can then claim that firm 2 is a "natural monopoly" since firm 1 voluntarily exits the market.

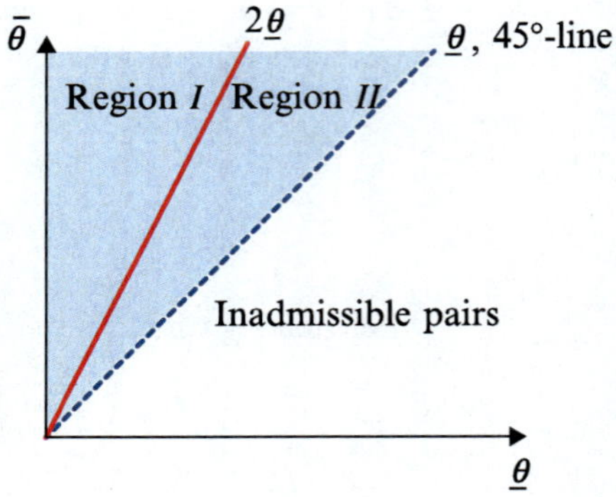

Fig. 1.13 Regions of equilibrium prices

Exercise #1.18: Persuasive Advertising in Monopoly^B

1.18 Consider a monopolist facing the demand function

$$Q(p, A) = a - pA^{-\frac{1}{2}}$$

where a is the market size and the firm spends advertising dollars A in promoting its products. For simplicity, assume that production cost is zero, and consider that $a, A > 0$.

(a) Find the price elasticity of demand $\varepsilon_{Q,P}$. Does advertising make demand more inelastic?

- The price elasticity of demand is

$$\varepsilon_{Q,P} = -\frac{\partial Q(p, A)}{\partial p}\frac{p}{Q(p, A)} = \frac{p}{A^{\frac{1}{2}}\left(a - pA^{-\frac{1}{2}}\right)} = \frac{p}{a\sqrt{A} - p}$$

Differentiating $\varepsilon_{Q,P}$ with respect to A yields

$$\frac{\partial \varepsilon_{Q,P}}{\partial A} = -\frac{ap}{2\sqrt{A}\left(a\sqrt{A} - p\right)^2},$$

which is unambiguously negative. Therefore, advertising expenditure A makes demand more inelastic, and consumers are less likely to switch from the good sold by this firm to other goods regarded as close substitutes. The advertising that makes demand more inelastic is known as "persuasive."

(b) Consider that the firm simultaneously chooses its price p and advertising expenditure, A. Setup the firm's profit maximization problem and solve for the firm's equilibrium price p^*, advertising A^*, and the resulting equilibrium profits π^*. Then, evaluate your results assuming $a = 2$.

- The firm chooses p and A to solve the following profit maximization problem (recall that, for simplicity, the model assumes no production costs):

$$\max_{p,A\geq 0} \pi(p,A) = p\underbrace{\left(a - pA^{-\frac{1}{2}}\right)}_{Q(p,A)} - A$$

Differentiating with respect to p, we obtain

$$a - \frac{2p}{\sqrt{A}} = 0$$

which we rearrange to yield

$$p(A) = \frac{a\sqrt{A}}{2}.$$

Similarly, differentiating with respect to A, we find that

$$\frac{p^2}{2A^{\frac{3}{2}}} - 1 = 0$$

which we rearrange to yield

$$A(p) = \left(\frac{p^2}{2}\right)^{\frac{2}{3}}.$$

Substituting $A(p)$ into $p(A)$, we find

$$p(A) = \frac{a\sqrt{\left(\frac{p^2}{2}\right)^{\frac{2}{3}}}}{2}$$

which simplifies to

$$p = \frac{a}{2}\sqrt[3]{\frac{p^2}{2}}$$

Rearranging, we obtain the equilibrium price, as follows:

$$p^* = \frac{a^3}{16}$$

Inserting $p^* = \frac{a^3}{16}$ into $A(p)$ yields equilibrium advertising, $A^* = A(p^*)$

$$A^* = \left(\frac{\left(\frac{a^3}{16}\right)^2}{2}\right)^{\frac{2}{3}} = \frac{a^4}{64}$$

where both equilibrium price and advertising expenditure are increasing in market demand, a.

- Plugging $p^* = \frac{a^3}{16}$ and $A^* = \frac{a^4}{64}$ into the profit function, equilibrium profits become

$$\pi^* = p^* \left[a - p^* \left(A^* \right)^{-\frac{1}{2}} \right] - A^*$$

$$= \frac{a^3}{16} \left[a - \frac{\frac{a^3}{16}}{\sqrt{\frac{a^4}{64}}} \right] - \frac{a^4}{64}$$

$$= \frac{a^4}{64}$$

which are increasing in market demand, a.

- Finally, substituting $a = 2$ into the above results yields

$$p^* = \frac{2^3}{16} = \frac{1}{2},$$

$$A^* = \frac{2^4}{64} = \frac{1}{4}, \quad \text{and}$$

$$\pi^* = \frac{2^4}{64} = \frac{1}{4}.$$

Exercise #1.19: Informative Advertising in Monopoly^B

1.19 Consider a monopolist facing the demand function

$$Q(p, A) = A^k (1 - p)$$

where $k \in (0, 1)$. For simplicity, assume that production cost is zero.

(a) Show that advertising does not affect the price elasticity of demand $\varepsilon_{Q,P}$.

- The price elasticity of demand is

$$\varepsilon_{Q,P} = -\frac{\partial Q(p, A)}{\partial p} \frac{p}{Q(p, A)} = \frac{A^k p}{A^k (1 - p)} = \frac{p}{1 - p},$$

which is independent of A, so advertising does not affect price elasticity of demand.

- In this case, advertising only shifts the demand function outward, indicating that more consumers know about the product. Advertising, however, does not change price elasticity, $\varepsilon_{Q,P}$, implying that it does not affect consumers' willingness to switch between products. This advertising is known as "informative advertising."

(b) Consider that the firm simultaneously chooses its price p and advertising expenditure, A. Setup the firm's profit maximization problem and solve for the firm's equilibrium price p^*, advertising A^*, and equilibrium profits π^*.

- The firm chooses p and A to solve the following profit maximization problem:

$$\max_{p,A\geq 0} \ \pi(p,A) = p\underbrace{A^k(1-p)}_{Q(p,A)} - A$$

Differentiating with respect to p, we obtain

$$1 - 2p = 0$$

which yields equilibrium price of

$$p^* = \frac{1}{2}.$$

Similarly, differentiating with respect to A, we have

$$kp(1-p)A^{k-1} - 1 = 0$$

Substituting $p^* = \frac{1}{2}$ into the above expression, we find

$$\frac{k}{4}A^{k-1} = 1$$

Rearranging and solving for A results in

$$A^* = \left(\frac{k}{4}\right)^{\frac{1}{1-k}}$$

- Inserting $p^* = \frac{1}{2}$ and $A^* = \left(\frac{k}{4}\right)^{\frac{1}{1-k}}$ into the profit function, equilibrium profits are

$$\begin{aligned}\pi^* &= \left(A^*\right)^k p^*\left(1-p^*\right) - A^* \\ &= \frac{1}{4}\left(\frac{k}{4}\right)^{\frac{k}{1-k}} - \left(\frac{k}{4}\right)^{\frac{1}{1-k}} \\ &= \frac{1-k}{k}\left(\frac{k}{4}\right)^{\frac{1}{1-k}}.\end{aligned}$$

(c) How do equilibrium profits change in k? Interpret.

- Differentiating π^* with respect to k, we find

$$\begin{aligned}\frac{\partial\pi^*}{\partial k} &= -\frac{1}{k^2}\left(\frac{k}{4}\right)^{\frac{1}{1-k}} + \frac{1-k}{k}\left[\frac{1}{4(1-k)}\left(\frac{k}{4}\right)^{\frac{1}{1-k}}\frac{4}{k} + \frac{1}{(1-k)^2}\log\left(\frac{k}{4}\right)\right] \\ &= \frac{1}{k(1-k)}\log\left(\frac{k}{4}\right)\end{aligned}$$

which is negative since $k \in (0, 1)$. Intuitively, as demand becomes less responsive to advertising (smaller shift out in the demand function, which occurs when k increases), the monopolist's profits decrease.

(d) Find the equilibrium output Q^* and social welfare SW^* (sum of consumer and producer surplus).

- Equilibrium output in this context becomes

$$Q^* = (A^*)^k (1 - p^*)$$
$$= \frac{1}{2}\left(\frac{k}{4}\right)^{\frac{k}{1-k}}$$

- Social welfare is the sum of consumer and producer surplus, as follows:

$$SW^* = \frac{1}{2}(Q^*)^2 + \pi^*$$
$$= \frac{1}{8}\left(\frac{k}{4}\right)^{\frac{2k}{1-k}} + \frac{1-k}{k}\left(\frac{k}{4}\right)^{\frac{1}{1-k}}$$
$$= \frac{1}{k}\left(\frac{k}{4}\right)^{\frac{1}{1-k}}\left[\frac{1}{2}\left(\frac{k}{4}\right)^{\frac{k}{1-k}} + 1 - k\right]$$

(e) Evaluate your results in parts (b) and (c) assuming $k = 1/2$.

- Substituting $k = 1/2$ into our results in part (b), we obtain that equilibrium prices, advertising, and profits are, respectively,

$$p^* = \frac{1}{2},$$
$$A^* = \left(\frac{\frac{1}{2}}{4}\right)^{\frac{1}{1-\frac{1}{2}}} = \frac{1}{64},$$
$$\pi^* = \frac{1-\frac{1}{2}}{\frac{1}{2}}\left(\frac{\frac{1}{2}}{4}\right)^{\frac{1}{1-\frac{1}{2}}} = \frac{1}{64}.$$

- Similarly, inserting $k = 1/2$ into our results in part (c), we find that aggregate output and social welfare in equilibrium are, respectively,

$$Q^* = \frac{1}{2}\left(\frac{\frac{1}{2}}{4}\right)^{\frac{\frac{1}{2}}{1-\frac{1}{2}}} = \frac{1}{16},$$
$$SW^* = \frac{1}{\frac{1}{2}}\left(\frac{\frac{1}{2}}{4}\right)^{\frac{1}{1-\frac{1}{2}}}\left[\frac{1}{2}\left(\frac{\frac{1}{2}}{4}\right)^{\frac{\frac{1}{2}}{1-\frac{1}{2}}} + 1 - \frac{1}{2}\right] = \frac{9}{512}.$$

Exercise #1.20: Behavior-Based Price Discrimination in Monopoly – Based on Fudenberg (2007) and Armstrong (2007)[B]

1.20 Consider a monopolist selling a non-durable good so that consumption occurs in the period of purchase. Suppose there are two periods and consumers can buy at one, both, or no period. A consumer's willingness to pay in each period is given by r, which is uniformly distributed in $[0, 1]$, and all players have the same discount factor δ. For simplicity, assume that the firm's marginal cost of production c is normalized to zero, $c = 0$. The firm cannot commit to future prices but it can make the second-period price conditional on whether the consumer has purchased in the first period, that is, the firm sets prices

$$(p^1, p^2(p^1), \widetilde{p}^2(p^1)),$$

where p^1 denotes first-period price; $p^2(p^1)$ is the second-period price to consumers who bought the good in the first period, thus being a function of the price in previous period; and $\widetilde{p}^2(p^1)$ represents the second-period price to consumers who did not buy the good in the first period, which can also be a function of p^1.

We next solve the game by backward induction, starting from the second period.

(a) *Second period.* Find the second-period price that the monopolist sets on consumers who purchased the product in the first period, p^2, and the price to consumers who didn't purchase the good, $\widetilde{p}^2$. [*Hint*: These prices should be a function of the first-period price p^1.]

- The consumer's willingness to pay, r, in the unit interval can be divided into the group of consumers who purchased the good in the first period, those with $r \geq \widehat{r}$, and those who did not, with $r < \widehat{r}$. The indifferent consumer, with willingness to pay $r = \widehat{r}$, is indifferent between buying in the first period at p^1 and in the second period at p^2 and buying only in the second period at $\widetilde{p}^2$, that is,

$$\underbrace{\left(\widehat{r} - p^1\right) + \delta(\widehat{r} - p^2)}_{\text{Buying in the 1st and 2nd period}} = \underbrace{\delta(\widehat{r} - \widetilde{p}^2)}_{\text{Buying only in the 2nd period}}$$

Solving for $\widehat{r}$ yields

$$\widehat{r} = p^1 + \delta(p^2 - \widetilde{p}^2)$$

Now, assuming $\widehat{r} \geq \frac{1}{2}$, the profit-maximizing price for consumers who bought in period 1 is $p^2 = \widehat{r}$, since this is the reservation price for consumers with relatively high valuation, where $r \geq \widehat{r}$, to make repeated purchases. Whereas, for consumers who did not buy in period 1, the monopolist chooses $\widetilde{p}^2$ to solve

$$\max_{\widetilde{p}^2 \geq 0} \widetilde{p}^2(\widehat{r} - \widetilde{p}^2)$$

which, differentiating with respect to $\widetilde{p}^2$, yields $2\widetilde{p}^2 - r = 0$. Solving for $\widetilde{p}^2$, the monopolist sets a price

$$\widetilde{p}^2 = \frac{\widehat{r}}{2}.$$

Plugging $p^2 = \widehat{r}$ and $\widetilde{p}^2 = \frac{\widehat{r}}{2}$ into the expression of $\widehat{r}$, we obtain

$$\widehat{r} = p^1 + \delta(p^2 - \widetilde{p}^2)$$
$$= p^1 + \delta\left(\widehat{r} - \frac{\widehat{r}}{2}\right)$$

Rearranging and solving for $\widehat{r}$ yields

$$\widehat{r} = \frac{2}{2-\delta}p^1.$$

Therefore, prices p^2 and $\widetilde{p}^2$ become

$$p^2 = \widehat{r} = \frac{2}{2-\delta}p^1$$
$$\widetilde{p}^2 = \frac{\widehat{r}}{2} = \frac{1}{2-\delta}p^1.$$

Intuitively, second-period prices increase in the first-period price p^1 since a higher first-period price leaves more customers available in the second period.

(b) *First period.* Write down the monopolist profit maximization problem in the first period, which includes its profits in that period and its expected discounted profits in the second period after inserting the prices p^2 and $\widetilde{p}^2$ that you found in part (a). Find the equilibrium first-period price p_1 that the monopolist sets.

- In the first period, the monopolist chooses p^1 to maximize its profit, as follows:

$$\max_{p^1 \geq 0} \ \pi(p^1) = p^1(1-\widehat{r}) + \delta\left[p^2(1-\widehat{r}) + \widetilde{p}^2(\widehat{r} - \widetilde{p}^2)\right]$$

where the first term denotes first-period profits, since all customers with $r > \widehat{r}$ purchase at a price p^1, yielding sales of $1 - Prob\,(r \leq \widehat{r}) = 1 - \widehat{r}$. The second term represents second-period profits, which originate from the same group of customers who purchased in period 1, $1 - \widehat{r}$, each paying a second-period price p^2, or from new customers (who did not buy the good in the first period), each paying $\widetilde{p}^2$, and yielding sales of $\widehat{r} - \widetilde{p}^2$.

From part (a), second-period prices are $p^2 = \widehat{r}$ and $\widetilde{p}^2 = \frac{\widehat{r}}{2}$. In addition, we use the result that $\widehat{r} = \frac{2}{2-\delta}p^1$ which, after solving for p^1, yields first-period price $p_1 = \frac{2-\delta}{2}\widehat{r}$, entailing that profits become

$$\max_{p^1 \geq 0} \ \pi(p^1) = \underbrace{\left(\frac{2-\delta}{2}\widehat{r}\right)}_{p^1}(1-\widehat{r})$$
$$+\delta\left[\underbrace{\widehat{r}}_{p^2}(1-\widehat{r}) + \underbrace{\frac{\widehat{r}}{2}}_{\widetilde{p}^2}\left(\widehat{r} - \underbrace{\frac{\widehat{r}}{2}}_{\widetilde{p}^2}\right)\right]$$

which is only a function of cutoff $\widehat{r}$. Rearranging and simplifying the above expression, we find

$$\max_{\widehat{r}\geq 0} \pi(r) = \frac{2+\delta}{2}\widehat{r}(1-\widehat{r}) + \frac{\delta}{4}\widehat{r}^2$$
$$= \frac{2(2+\delta)\widehat{r} - (4+\delta)\widehat{r}^2}{4}$$

Differentiating with respect to $\widehat{r}$ yields

$$2(2+\delta) - 2(4+\delta)\widehat{r} = 0$$

Solving for $\widehat{r}$, we have

$$\widehat{r} = \frac{2+\delta}{4+\delta},$$

which satisfies $\frac{2+\delta}{4+\delta} > \frac{1}{2}$, as we initially assumed ($\widehat{r} > \frac{1}{2}$), since this inequality simplifies to $4 + 2\delta > 4 + \delta$, and further reduces to $2\delta > \delta$, which holds for all values of $\delta \in [0, 1]$.

- Substituting $\widehat{r} = \frac{2+\delta}{4+\delta}$ into the prices we found in part (a) yields

$$p^1 = \frac{2-\delta}{2}\frac{2+\delta}{4+\delta} = \frac{4-\delta^2}{2(4+\delta)}$$
$$p^2 = \widehat{r} = \frac{2+\delta}{4+\delta}$$
$$\widetilde{p}^2 = \frac{\widehat{r}}{2} = \frac{2+\delta}{2(4+\delta)}$$

Finally, the monopolist's equilibrium profits are

$$\pi^* = \frac{2+\delta}{2}\widehat{r}(1-\widehat{r}) + \frac{\delta}{4}\widehat{r}^2$$
$$= \frac{2+\delta}{2}\underbrace{\frac{2+\delta}{4+\delta}}_{\widehat{r}}\left(1 - \underbrace{\frac{2+\delta}{4+\delta}}_{\widehat{r}}\right) + \frac{\delta}{4}\left(\underbrace{\frac{2+\delta}{4+\delta}}_{\widehat{r}}\right)^2$$
$$= \frac{(2+\delta)^2}{4(4+\delta)}.$$

(c) How do equilibrium prices and profits you found in part (b) change in δ? Interpret.

- Differentiating p^1 with respect to δ, we obtain

$$\frac{\partial p^1}{\partial \delta} = \frac{-2\delta(4+\delta) - (4-\delta^2)}{2(4+\delta)^2}$$
$$= -\frac{\delta^2 + 8\delta + 4}{2(4+\delta)^2} < 0.$$

- Differentiating p^2 with respect to δ, we find

$$\frac{\partial p^2}{\partial \delta} = \frac{2}{(4+\delta)^2} > 0.$$

- Differentiating $\widetilde{p}^2$ with respect to δ, we have

$$\frac{\partial \widetilde{p}^2}{\partial \delta} = \frac{1}{(4+\delta)^2} > 0.$$

- Differentiating π^* with respect to δ yields

$$\begin{aligned}\frac{\partial \pi^*}{\partial \delta} &= \frac{2\,(2+\delta)\,(4+\delta) - (2+\delta)^2}{4\,(4+\delta)^2} \\ &= \frac{(2+\delta)\,(6+\delta)}{4\,(4+\delta)^2} > 0,\end{aligned}$$

implying that the monopolist charges a lower (higher) price in the first (second) period to earn more profits when it becomes more patient (higher δ).

(d) *Price comparison.* Compare equilibrium prices p^1, p^2, and $\widetilde{p}^2$. Interpret your results.

- From part (b), we found that $p^1 = \frac{4-\delta^2}{2(4+\delta)}$, and from part (a), we know that

$$p^2 = \frac{2}{2-\delta} p^1 \text{ and } \widetilde{p}^2 = \frac{1}{2-\delta} p^1.$$

Therefore, since $\delta \in [0, 1]$, we have that

$$p^2 > p^1 > \widetilde{p}^2,$$

as depicted in Fig. 1.14. That is, consumers who bought in period 1 are offered a higher price in period 2 than in period 1, $p^2 > p^1$; but consumers who did not buy in period 1 are offered a lower price than in period 1, $p^1 > \widetilde{p}^2$. Intuitively, consumers who bought in the first period are recognized as having a relatively high willingness to pay for the product, r, and the firm charges an even higher price in the second period. In contrast, consumers who did not buy in the first period are recognized as having a relatively low willingness to pay, r, and the firm charges an even lower price to attract them in the second period.

(e) *Commitment.* How would your equilibrium results change if the monopolist could, in the first period, commit to future prices?

- If the firm was able to commit to future prices, it would *not* be in its interest to make prices conditional upon consumers' purchasing history. Indeed, with commitment, it is profit-maximizing to set the price equal to $\frac{1}{2}$ in each period. This yields a profit of $\frac{1}{4}+\delta\frac{1}{4} = \frac{1+\delta}{4}$, which is higher than that without commitment, because

$$\frac{1+\delta}{4} > \frac{(2+\delta)^2}{4\,(4+\delta)}$$

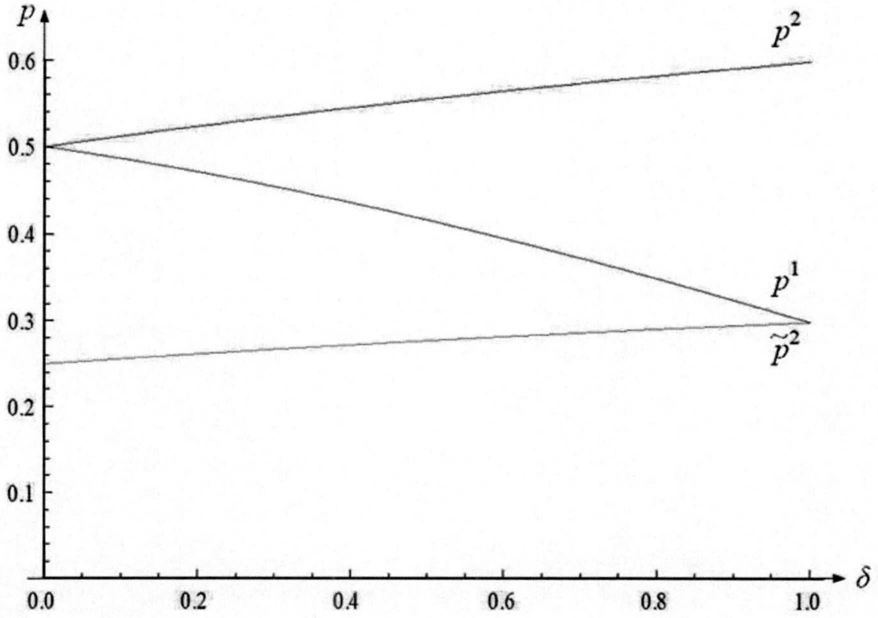

Fig. 1.14 Price comparison

simplifies to

$$4 + 5\delta + \delta^2 > 4 + 4\delta + \delta^2$$

that yields $\delta > 0$, which holds since $\delta \in [0, 1]$ by definition. The monopolist's challenge is, of course, to credibly commit to a price of $\frac{1}{2}$ in both periods.

Simultaneous Quantity Competition

2

Introduction

After studying markets with only one firm in Chap. 1, we now turn to industries with two or more firms (oligopolies) either compete in quantities (Chap. 2) or in prices (Chap. 3). In this chapter, we assume that every firm chooses independently and simultaneously its output level, yielding an equilibrium output for each firm and an equilibrium aggregate output for the industry. Given this aggregate output, the equilibrium price is determined by the demand function entailing equilibrium profits for each firm.

To provide a clear understanding of firm incentives in this type of industry, Exercise 2.1 considers a simplified setting with only two firms facing the same marginal cost of production. Exercise 2.2 then considers a more generic context, providing necessary and sufficient conditions for firms to choose profit-maximizing output levels. Exercise 2.3 (2.4) extends the setting of Exercise 2.2 but allowing for three ($N \geq 2$) firms, still facing a common marginal cost of production. Exercise 2.5 then compares the equilibrium and socially optimal output levels, showing that firms produce too few units, relative to those that maximize social welfare. Exercises 2.6–2.8 explore different pricing formulas in the Cournot context.

Exercises 2.9–2.11 allow for firms with asymmetric costs. Specifically, Exercise 2.9 considers a setting with only two firms, Exercise 2.10 extends this setting to an industry with $N \geq 2$ firms where some have a high marginal cost, while all others benefit from a low marginal cost, and Exercise 2.11 assumes that only one firm benefits from a cost advantage. Exercise 2.12 examines social welfare in these situations.

Exercises 2.13–2.16 explore situations of Cournot oligopolies that experience simple product differentiation, where the latter two allow for mergers between firms. Exercise 2.17 follows up by examining social welfare in these situations.

Exercises 2.18–2.20 continue our study of Cournot oligopolies with asymmetric firms, analyzing how entry of more firms, rather than being beneficial for consumers, can in certain industries reduce aggregate output, increasing equilibrium prices, and hurt consumers. Exercise 2.19 also examines a setting with asymmetric firms, but now assuming that they differ in their fixed, rather than marginal, costs.

The remaining exercises of this chapter consider extensions of the standard Cournot oligopoly model. First, Exercise 2.21 analyzes an industry where firms hold equity shares on each other's profits, showing that every firm has incentives to produce fewer units of output. In contrast, Exercises 2.22 and 2.23 study a market with a private firm (which maximizes profits) and a publicly owned firm

P.-S. Choi et al., *Industrial Organization*, Springer Texts in Business and Economics,
https://doi.org/10.1007/978-3-031-38635-0_2

(which maximizes a combination of profits and social welfare), demonstrating that firms produce more units in this context than when both firms are private, as the standard Cournot duopoly model assumes. Exercise 2.24 follows a similar approach but now considers that the firm manager receives a share of profits as part of his compensation, identifying how his output decisions are affected. Exercises 2.25–2.28 introduce incomplete information in the standard Cournot duopoly, where one firm cannot accurately observe its rival's production cost, which may represent a new firm that just entered an industry, thus having to maximize its expected profits.

Finally, Exercise 2.29 studies the use of two-part tariffs by firms competing in an oligopoly, showing that this pricing practice generates higher profits and welfare, but a lower consumer surplus, than uniform pricing regardless of the number of firms in the market.

Exercise #2.1: Cournot Duopoly with Symmetric Costs^A

2.1 Consider a market with two firms, 1 and 2, selling a homogeneous good (i.e., undifferentiated product). Firms face an inverse demand function $p(Q) = a - bQ$, where $Q = q_1 + q_2$ denotes aggregate output, and $a, b > 0$. Graphically, parameter $a > 0$ represents the vertical intercept of this inverse demand function and thus indicates the strength of demand. Parameter $b > 0$ is the slope of the inverse demand, measuring how much consumers reduce their willingness to pay for the product if aggregate output Q increases. Every firm faces marginal cost c, where $a > c > 0$. Every firm $i = \{1, 2\}$ simultaneously and independently selects its output level q_i, thus competing in quantities.

(a) Set up firm i's profit maximization problem, and find its best response function, $q_i(q_j)$. Interpret.

- Every firm i chooses q_i to maximize its profit, π_i, as follows:

$$\max_{q_i \geq 0} \pi_i = \underbrace{\overbrace{(a - b(q_i + q_j))}^{p(Q)} q_i}_{\text{Total revenue}} - \underbrace{cq_i}_{\text{Total cost}}.$$

Differentiating with respect to q_i, we obtain

$$a - 2bq_i - bq_j - c = 0$$

and solving for q_i, we find

$$q_i = \frac{a-c}{2b} - \frac{q_j}{2}.$$

To derive firm i's best response function, recall that for a linear equation in the form of $y = ax + b$, we must identify the vertical intercept b , which is found by setting $q_i = 0$ that yields $\frac{a-c}{2b} - \frac{q_j}{2} = 0$, which is rearranged into $q_j = \frac{a-c}{b}$. Therefore, firm i's best response function is depicted in Fig. 2.1 and expressed as follows:

$$q_i(q_j) = \begin{cases} \frac{a-c}{2b} - \frac{q_j}{2} \text{ if } q_j < \frac{a-c}{b} \\ 0 \text{ otherwise.} \end{cases}$$

Intuitively, firm i's production originates at a vertical intercept of $q_i = \frac{a-c}{2b}$, which describes its output when its rival, firm j, is inactive ($q_j = 0$), decreases with its rivals

Fig. 2.1 Firm 1's best response function

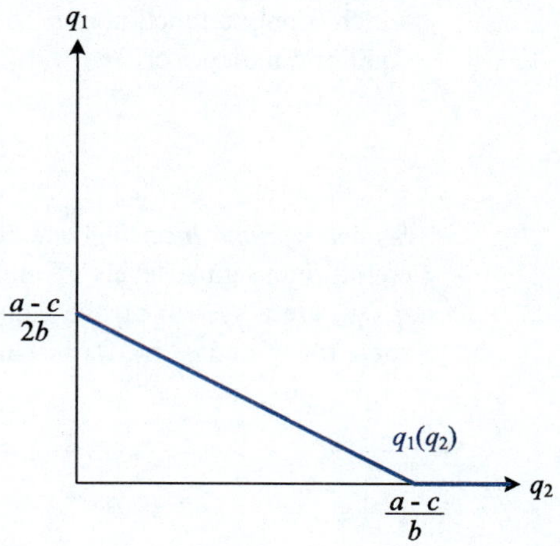

Fig. 2.2 Firm 2's best response function

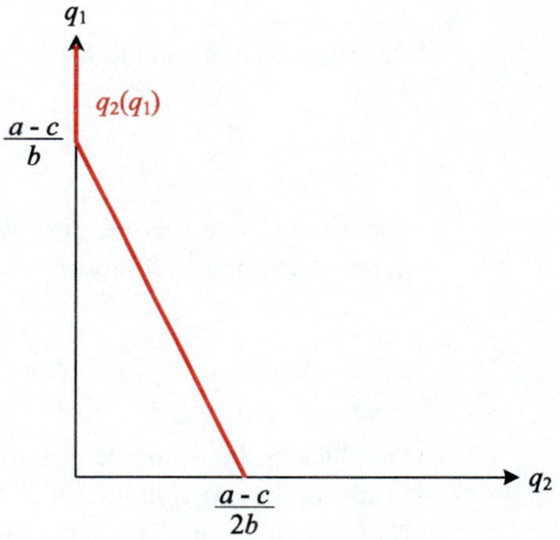

production q_j, and reaches zero when $q_j = \frac{a-c}{b}$. For all other output levels, $q_j > \frac{a-c}{b}$, firm i chooses to optimally remain inactive. Intuitively, this says that, if firm j were to produce $q_j = \frac{a-c}{b}$ (the perfectly competitive output level), firm i earns a higher profit staying inactive (zero profits) than producing a positive amount since in that case its profits would be negative. Firm j has a symmetric best response function, that is,

$$q_j(q_i) = \begin{cases} \frac{a-c}{2b} - \frac{q_i}{2} \text{ if } q_i < \frac{a-c}{b} \\ 0 \text{ otherwise,} \end{cases}$$

as depicted in Fig. 2.2.

(b) Find the Nash equilibrium in this Cournot duopoly, i.e., equilibrium output levels q_i^* and q_j^*.

- *First option: Invoking symmetry.* In a symmetric Nash equilibrium, both firms choose the same output level, so that $q_i^* = q_j^* = q^*$. Inserting this property in the best response function, we found in part (a) yields

$$q^* = \frac{a-c}{2b} - \frac{q^*}{2}$$

which is only a function of q^* on both sides of the equality. Solving for q^*, we obtain an equilibrium output of

$$q^* = \frac{a-c}{3b}.$$

- *Second option: Inserting one BRF into another.* An alternative approach to solve for equilibrium output levels q_i^* and q_j^* is to notice that best response functions $q_i(q_j)$ and $q_j(q_i)$ are a system of two equations with two unknowns. We can then simultaneously solve for q_i^* and q_j^* by, for instance, inserting $q_j(q_i)$ into $q_i(q_j)$, as follows:

$$\begin{aligned} q_i &= \frac{a-c}{2b} - \frac{1}{2}\left(\frac{a-c}{2b} - \frac{q_i}{2}\right) \\ &= \frac{a-c}{4b} + \frac{q_i}{4}. \end{aligned}$$

Rearranging and solving for q_i to obtain the equilibrium output for firm i as follows:

$$q_i^* = \frac{a-c}{3b}.$$

Inserting q_i^* into the best response function of firm j, $q_j(q_i)$, we obtain the equilibrium output for firm j as follows:

$$q_j^* = q_j(q_i^*) = \frac{a-c}{2b} - \frac{1}{2}\left(\frac{a-c}{3b}\right) = \frac{a-c}{3b}.$$

Graphically, both approaches look for the equilibrium output pair (q_i^*, q_j^*) where both firms are playing a mutual best response to each other's output, which then constitutes a Nash equilibrium of the Cournot game of simultaneous quantity competition. Figure 2.3 depicts the best response functions of both firms in the same figure and their crossing point at the equilibrium output $q_i^* = q_j^* = \frac{a-c}{3b}$.

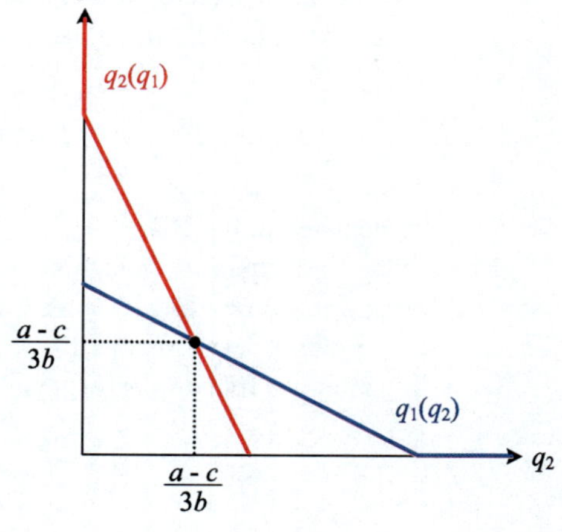

Fig. 2.3 Cournot equilibrium

(c) Is equilibrium output increasing in parameters a, b, and c? Interpret your results.

- Differentiating q^* with respect to a, we obtain

$$\frac{\partial q^*}{\partial a} = \frac{1}{3b} > 0$$

which is positive given that $b > 0$ by assumption. Intuitively, a stronger demand (higher a) induces the firm to produce more units of output.

- Differentiating q^* with respect to b, we find

$$\frac{\partial q^*}{\partial b} = -\frac{a-c}{3b^2} < 0$$

which is negative since $a > c$ by assumption. Intuitively, the firm produces fewer units when the willingness to pay for additional units decreases at a higher rate (higher b).

- We finally differentiate q^* with respect to c, to obtain

$$\frac{\partial q^*}{\partial c} = -\frac{1}{3b}$$

which is negative given that $b > 0$ by definition. Intuitively, firm i produces fewer units when its marginal cost of production increases.

(d) What is the market price in equilibrium? What are the profits each firm earns in equilibrium?

- Equilibrium price is

$$\begin{aligned} p^* &= a - b\left(q^* + q^*\right) \\ &= a - 2b\left(\frac{a-c}{3b}\right) \\ &= \frac{a+2c}{3}. \end{aligned}$$

Price is increasing in both a and c. Intuitively, a stronger demand (higher a) increases equilibrium price. A higher marginal cost induces the firm to charge a higher price.

- Equilibrium profits are

$$\pi^* = p^*q^* - cq^* = \frac{(a-c)^2}{9b}$$

which we can more compactly express as $\pi^* = (q^*)^2$.

The firm's profits increase with a, i.e., a stronger demand in its products increases the profitability of the firm. Intuitively, a higher marginal cost eats into the firm's profits. Profits decrease in b because when the inverse demand function becomes steeper (higher b), the firm produces fewer units, ultimately leading to lower profits.

(e) *Numerical example.* Assume that $a = b = 1$ and $c = 1/2$. Evaluate equilibrium output, price, and profits.

- Equilibrium output in this context becomes $q_i^* = \frac{1-\frac{1}{2}}{3} = \frac{1}{6}$, equilibrium price is $p^* = \frac{1+2\frac{1}{2}}{3} = \frac{2}{3}$, and equilibrium profits are $\pi^* = \frac{\left(1-\frac{1}{2}\right)^2}{9} = \frac{1}{36}$.

Exercise #2.2: Cournot Duopoly – Necessary and Sufficient Conditions^B

2.2 Consider a Cournot duopoly game with two firms facing an inverse demand function $p(Q) = a - bQ$, where $Q = q_1 + q_2$ denotes aggregate output, and $a, b > 0$. Every firm i's cost function is $c_i(q_i) = cq_i$, where c satisfies $a > c > 0$.

(a) Set up firm i's profit maximization problem and derive firm i's first-order condition.

- Every firm i chooses its output level q_i to solve

$$\max_{q_i \geq 0} \pi_i = (a - bq_i - bq_j)q_i - cq_i.$$

Differentiating with respect to q_i, we find the first-order condition

$$\frac{\partial \pi_i}{\partial q_i} = a - 2bq_i - bq_j - c \leq 0.$$

(b) Assuming interior solutions, find firm i's best response function $q_i(q_j)$. Interpret your result.

- In an interior solution where $q_i > 0$, the above first-order condition holds with equality, so that

$$a - 2bq_i - bq_j - c = 0.$$

Solving for q_i, we obtain

$$q_i = \frac{a-c}{2b} - \frac{1}{2}q_j$$

which is positive as long as $\frac{a-c}{2b} - \frac{1}{2}q_j > 0$, or solving for q_j, $q_j < \frac{a-c}{b}$. Therefore, we can summarize firm i's best response function as follows:

$$q_i(q_j) = \begin{cases} \frac{a-c}{2b} - \frac{1}{2}q_j \text{ if } q_j < \frac{a-c}{b}, \text{ and} \\ 0 \text{ otherwise.} \end{cases}$$

- *Vertical intercept.* Graphically, $q_i = \frac{a-c}{2b}$ represents the vertical intercept of firm i's best response function; that is, the output that firm i produces when firm j is inactive (i.e., $q_i = \frac{a-c}{2b}$ when $q_j = 0$). This output coincides with that under monopoly, which does not come at a surprise: when firm j is inactive, firm i maximizes its own profits by producing the monopoly output.
- *Slope.* In addition, the best response function has a slope of $-1/2$, indicating that, when firm j produces one more unit, firm i responds reducing its output by half a unit since in this context, firm i and firm j's output are substitutes to each other, and this is referred to as "strategic substitutes " in the industrial organization literature (alternatively, when firm j increases its output by 2 units, firm i reduces its own by 1 unit).
- *Horizontal intercept.* Finally, the best response function hits the horizontal axis at $q_j = \frac{a-c}{b}$. When firm j produces $q_j = \frac{a-c}{b}$ or more units, firm i chooses to remain inactive (i.e., $q_i = 0$) since the price it would obtain for the good would be lower than the marginal cost of production.

(c) Using the first-order conditions you found in part (a), identify the boundary condition guaranteeing that every firm i produces a weakly positive output, $q_i \geq 0$, and guaranteeing that the equilibrium price is not negative, i.e., $Q \leq a/b$.

- *Positive output.* Evaluating the above first-order condition at $q_i = 0$, we obtain that firm i has incentives to produce the first unit if and only if

$$\left.\frac{\partial \pi_i}{\partial q_i}\right|_{q_i=0} = a - 2b\underbrace{0}_{q_i=0} - bq_j - c$$
$$= a - bq_j - c \geq 0$$

which holds as long as $q_j \leq \frac{a-c}{b}$. Intuitively, this says that if firm j produces the perfectly competitive output or less (where $Q = \frac{a-c}{b}$), firm i has incentives to produce its first unit.

- *Positive price.* Evaluating the above first-order condition at $Q = q_i + q_j = \frac{a}{b}$, or $q_i = \frac{a}{b} - q_j$, we obtain that firm i has incentives to reduce its output level when the price is zero if and only if

$$\left.\frac{\partial \pi_i}{\partial q_i}\right|_{q_i=\frac{a}{b}-q_j} = a - 2b\underbrace{\left(\frac{a}{b} - q_j\right)}_{q_i=\frac{a}{b}-q_j} - bq_j - c$$
$$= -a - c + bq_j \leq 0$$

or, solving for q_j, we obtain $q_j \geq \frac{a+c}{b}$.

(d) Find the sufficient conditions for a maximum.

- Differentiating the above first-order condition with respect to q_i again, we obtain

$$\frac{\partial^2 \pi_i}{\partial q^2} = \begin{bmatrix} \frac{\partial^2 \pi_i}{\partial q_i^2} & \frac{\partial^2 \pi_i}{\partial q_i \partial q_j} \\ \frac{\partial^2 \pi_i}{\partial q_j \partial q_i} & \frac{\partial^2 \pi_i}{\partial q_j^2} \end{bmatrix}$$
$$= \begin{bmatrix} -2b & -b \\ -b & 0 \end{bmatrix}$$
$$= -2b \times 0 - (-b)(-b)$$
$$= -b^2$$

which is clearly negative given that $b > 0$ by assumption. Therefore, firm i's profit function is strictly concave, implying that the best response function found in part (b) identifies an output level that maximizes firm i's profit.

Exercise #2.3: Cournot Oligopoly with Three Symmetric Firms[A]

2.3 Consider a market with three firms producing a homogeneous good and facing a linear demand function $p(Q) = 1 - Q$, where $Q \equiv q_1 + q_2 + q_3$ denotes aggregate output. All firms face a constant marginal cost of production given by c.

(a) Set up firm 1's profit maximization problem, and obtain this firm's best response function. [*Hint*: It should be a function of firm 2's and 3's output, q_2 and q_3.]
- Firm 1's profit maximization problem is

$$\max_{q_1 \geq 0} \ \pi_1 = (1 - q_1 - q_2 - q_3)q_1 - cq_1.$$

Differentiating with respect to q_1, we obtain

$$\frac{\partial \pi_1}{\partial q_1} = 1 - 2q_1 - q_2 - q_3 - c = 0$$

rearranging, we find $2q_1 = 1 - q_2 - q_3 - c$, and solving for q_1, we obtain firm 1's best response function

$$q_1(q_2, q_3) = \frac{1-c}{2} - \frac{1}{2}(q_2 + q_3)$$

which originates at a vertical intercept of $\frac{1-c}{2}$ and decreases at a rate of $\frac{1}{2}$ when either firm 2 or 3 marginally increases its output.

(b) Repeat the process for firms 2 and 3 to obtain their best response functions. [*Hint*: You should find that all firms have symmetric best response functions.]
- *Firm 2*. Firm 2's profit maximization problem is

$$\max_{q_2 \geq 0} \ \pi_2 = (1 - q_1 - q_2 - q_3)q_2 - cq_2.$$

Differentiating with respect to q_2, we obtain

$$\frac{\partial \pi_2}{\partial q_2} = 1 - q_1 - 2q_2 - q_3 - c = 0$$

rearranging, we find $2q_2 = 1 - q_1 - q_3 - c$, and solving for q_2, we obtain firm 2's best response function

$$q_2(q_1, q_3) = \frac{1-c}{2} - \frac{1}{2}(q_1 + q_3)$$

which is symmetric to firm 1's best response function. This comes at no surprise since all firms face the same inverse demand function and total cost function.
- *Firm 3*. Firm 3's profit maximization problem is

$$\max_{q_3 \geq 0} \ \pi_3 = (1 - q_1 - q_2 - q_3)q_3 - cq_3.$$

Differentiating with respect to q_3, we obtain

$$\frac{\partial \pi_3}{\partial q_3} = 1 - q_1 - q_2 - 2q_3 - c = 0$$

rearranging, we find $2q_3 = 1 - q_1 - q_2 - c$, and solving for q_3, we obtain firm 3's best response function

$$q_3(q_1, q_2) = \frac{1-c}{2} - \frac{1}{2}(q_1 + q_2)$$

which is symmetric to firm 1's best response function. Again, this is not surprising since all firms face the same inverse demand function and total cost function.

(c) Interpret firm 1's best response function: if firm 2 where to marginally increase its output, does firm 1 increase or decrease its own output? By how much?

- To find this, we differentiate firm 1's best response function with respect to firm 2's output q_2:

$$\frac{\partial q_1(q_2, q_3)}{\partial q_2} = -\frac{1}{2}.$$

For each unit increase in firm 2's output, firm 1 decreases their output by half unit.

(d) Using the above three best response functions, find the point where they cross, the triplet (q_1^*, q_2^*, q_3^*) that characterizes the Nash equilibrium of this Cournot game.

- In a symmetric equilibrium, all firms produce the same output level, $q_1^* = q_2^* = q_3^* = q^*$. We can insert this property into firm 2's best response function to obtain

$$q^* = \frac{1-c}{2} - \frac{1}{2}\left(q^* + q^*\right)$$

which is equivalent to dropping the subscripts of all output levels. Simplifying, we find

$$2q^* = 1 - c - 2q^*,$$

or $4q^* = 1 - c$. Solving for q^*, we find that every firm produces an equilibrium output of

$$q^* = \frac{1-c}{4}.$$

(e) Is the equilibrium output that you found in part (d) increasing/decreasing in marginal cost c?

- Differentiating the equilibrium output with respect to c, we obtain

$$\frac{\partial q^*}{\partial c} = -\frac{1}{4} < 0.$$

Therefore, as marginal cost increases, the equilibrium output for each firm decreases.

(f) Find the price that emerges in equilibrium, along with the profits that every firm earns.

- The price each firm faces is

$$\begin{aligned} p^* &= 1 - q_1 - q_2 - q_3 \\ &= 1 - 3q^* \\ &= 1 - 3\frac{(1-c)}{4} \\ &= \frac{1+3c}{4}. \end{aligned}$$

- Each firm earns the same profit in equilibrium, that is,

$$\pi = pq - cq$$

or, plugging in for p^* and q^*,

$$\pi^* = \left(\frac{1+3c}{4}\right)\frac{1-c}{4} - c\frac{1-c}{4}$$
$$= \frac{(1-c)^2}{16}$$

which coincides with the square of the individual output in equilibrium, $\pi = (q^*)^2$.

(g) *Numerical example.* Assume that $a = b = 1$ and $c = 1/2$. Evaluate equilibrium output, price, and profits.

- Equilibrium output in this context becomes $q_i^* = \frac{1-\frac{1}{2}}{4} = \frac{1}{8}$, equilibrium price is $p^* = \frac{1+3\frac{1}{2}}{4} = \frac{5}{8}$, and equilibrium profits are $\pi^* = \frac{\left(1-\frac{1}{2}\right)^2}{16} = \frac{1}{64}$.

Exercise #2.4: Cournot Oligopoly with $N \geq 2$ Symmetric Firms[B]

2.4 Consider again a market where firms face inverse demand function $p(Q) = a - bQ$, but assume now that $N \geq 2$ firms compete in quantities, so aggregate output becomes $Q = \Sigma_{i=1}^{N} q_i$. For simplicity, assume that all firms face the same marginal cost c, where $a > c > 0$.

(a) Set up firm i's profit maximization problem and find its best response function, $q_i(q_j)$. Interpret. (Hint: For compactness, you may use Q_{-i} to denote the aggregate output that firm i's rivals produce, that is, $Q_{-i} = \sum_{j \neq i} q_j$.)

- Every firm i chooses q_i to maximize its profit, π_i, as follows:

$$\max_{q_i \geq 0} \pi_i = \underbrace{\overbrace{(a - b\,(q_i + Q_{-i}))}^{p(Q)} q_i}_{\text{Total revenue}} - \underbrace{cq_i}_{\text{Total cost}}.$$

Differentiating with respect to q_i, we obtain

$$a - 2bq_i - bQ_{-i} - c = 0$$

and solving for q_i, we find firm i's best response function

$$q_i(Q_{-i}) = \frac{a-c}{2b} - \frac{Q_{-i}}{2}.$$

- Intuitively, firm i's output originates at an intercept of $\frac{a-c}{2b}$ and decreases in its $N-1$'s rivals output Q_{-i} at a rate of $\frac{1}{2}$.

(b) Find the Nash equilibrium output, q_i^*.

- In a symmetric Nash equilibrium, both firms choose the same output level, so that $q_i^* = q_j^* = q^*$ for any two firms i and $j \neq i$. Therefore, $Q_{-i} = (N-1)q^*$. Inserting this result in the best response function we found in part (a) yields

$$q^* = \frac{a-c}{2b} - \frac{(N-1)q^*}{2}$$

which is only a function of q^* on both sides of the equality. Solving for q^*, we obtain an equilibrium output of

$$q^* = \frac{a-c}{b(N+1)}.$$

(c) Is equilibrium output increasing in parameters a, b, c, and N? Interpret your results.

- Differentiating q^* with respect to a, we obtain

$$\frac{\partial q^*}{\partial a} = \frac{1}{b(N+1)}$$

which is positive given that $b > 0$ by assumption. Intuitively, a stronger demand (higher a) induces the firm to produce more units of output.

- Differentiating q^* with respect to b, we find

$$\frac{\partial q^*}{\partial b} = -\frac{a-c}{b^2(N+1)}$$

which is negative since $a > c$ by assumption. Intuitively, the firm produces fewer units when, for a given value of a, the inverse demand function becomes steeper (higher b).

- Next, differentiating q^* with respect to c, to obtain

$$\frac{\partial q^*}{\partial c} = \frac{-1}{b(N+1)}$$

which is negative. The firm's equilibrium production decreases in its own marginal cost.

- We can finally differentiate q^* with respect to N, to obtain

$$\frac{\partial q^*}{\partial N} = -\frac{a-c}{b(N+1)^2}$$

which is negative. As the number of firms in the industry increases, each firm decreases its own equilibrium output.

(d) Which is the aggregate output in equilibrium, Q^*? What about the market price in equilibrium? Which are the profits each firm earns in equilibrium?

- Aggregate output is

$$Q^* = Nq^* = \frac{N(a-c)}{b(N+1)}.$$

- Equilibrium price is

$$p^* = a - b\left(\frac{N(a-c)}{b(N+1)}\right) = \frac{a+Nc}{N+1}.$$

Q^* increases in a and N but decreases in b and c. Intuitively, aggregate industry output increases in the number of firms in the industry and decreases in the marginal cost experienced by each firm.

- Equilibrium profits are

$$\pi^* = p^*q^* - cq^* = \frac{(a-c)^2}{b(N+1)^2}$$

which are decreasing in N, b, and c. However, when the market size a increases, equilibrium profits increase. A stronger demand induces the firm to produce more units of output leading to higher profits.

(e) How do individual output, q^*; aggregate output, Q^*; price, p^*; and profits, π^*, change as the number of firms, N, increases? Evaluate these equilibrium results in the special case in which only one firm operates, $N = 1$, and when infinitely many firms compete, $N \to +\infty$.

- When only one firm operates in the industry, $N = 1$, we obtain

$$q^* = \frac{a-c}{2b},\ Q^* = \frac{a-c}{2b},\ p^* = \frac{a+c}{2b},\ \text{and}\ \pi^* = \frac{(a-c)^2}{4b}$$

which coincides with our standard results in a monopolist facing linear inverse demand $p(Q) = a - bQ$ and constant marginal cost c.

- In contrast, when infinitely many firms compete, $N \to +\infty$, we find

$$q^* = \frac{a-c}{b},\ Q^* = \frac{a-c}{b},\ p^* = c,\ \text{and}\ \pi^* = 0$$

which coincides with our results of a perfectly competitive industry facing linear inverse demand $p(Q) = a - bQ$ with firms having symmetric marginal cost c.

(f) *Numerical example.* Assume that $a = b = 1$ and $c = 1/2$. Evaluate equilibrium output, price, and profits.

- Equilibrium output in this context becomes

$$q_i^* = \frac{1-\frac{1}{2}}{N+1} = \frac{1}{2(N+1)},$$

equilibrium price is $p^* = \frac{1+N\frac{1}{2}}{N+1} = \frac{N+2}{2(N+1)}$, and equilibrium profits are

$$\pi^* = \frac{\left(1-\frac{1}{2}\right)^2}{(N+1)^2} = \frac{1}{4(N+1)^2}.$$

Exercise #2.5: Comparing Equilibrium and Socially Optimal Outputs Under Cournot Competition^B

2.5 Consider an industry with $N \geq 2$ firms competing ? la Cournot, facing inverse demand function $p(Q)$, where $Q = \sum_{i=1}^{N} q_i$ denotes aggregate output, which is strictly decreasing, i.e., $p'(Q) < 0$, and exhibits a negatively sloped marginal revenue function. Intuitively, this allows for the inverse demand function to be concave ($p''(q) < 0$) or convex ($p''(q) > 0$), as long as its convexity is not too severe. Firm i faces cost function $c_i(q_i)$, which satisfies $c_i(0) = 0$ and is strictly increasing and strictly convex, i.e., $c_i'(q_i) > 0$ and $c_i''(q_i) > 0$.

(a) Find the implicit function describing the equilibrium output of every firm i, q_i^*.

- Every firm i solves

$$\max_{q_i \geq 0} p(Q)q_i - c_i(q_i).$$

Differentiating with respect to q_i yields

$$p(Q^*) + p'(Q^*)q_i^* = c_i'(q_i^*),$$

where $Q^* = q_1^* + q_2^* + \cdots + q_N^*$ denotes aggregate equilibrium output. Figure 2.4 separately depicts the left- and right-hand side of this expression as a function of q_i^*. The figure assumes that costs are convex (increasing marginal costs) but a similar figure applies if marginal costs are constant in q_i^*.

(b) Consider now a regulator seeking to maximize social welfare, defined as

$$SW = \int_0^Q p(Q)dQ - \sum_{i=1}^{N} c_i(q_i).$$

Find the implicit function describing socially optimal output of every firm i, q_i^{SO}.

Fig. 2.4 Cournot equilibrium

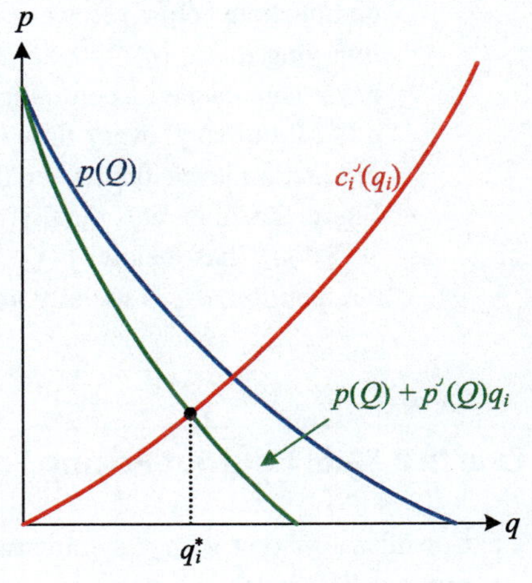

- Differentiating with respect to every firm i's output in the above social welfare function SW, we obtain

$$p(Q^{SO}) = c_i'(q_i^{SO}),$$

where $Q^{SO} = q_1^{SO} + q_2^{SO} + \cdots + q_N^{SO}$ denotes aggregate socially optimal output.

(c) Show that $q_i^* < q_i^{SO}$.

- We now compare the first-order condition we found in part (a) of the exercise against that from part (b). Since inverse demand has a negative slope, the second term in the first-order condition of part (a) satisfies $p'(Q^*)q_i^* < 0$. Therefore,

$$p(Q^*) - c_i'(q_i^*) = -p'(Q^*)q_i^* > 0 = p(Q^{SO}) - c_i'(q_i^{SO})$$

which can be more compactly expressed as

$$p(Q^*) - c_i'(q_i^*) > p(Q^{SO}) - c_i'(q_i^{SO})$$

or, after rearranging,

$$c_i'(q_i^{SO}) - c_i'(q_i^*) > p(Q^{SO}) - p(Q^*).$$

Operating by contradiction, let us assume that output levels satisfy the opposite ranking of what we are supposed to show, $q_i^* \geq q_i^{SO}$. Then, we have that $p(Q^{SO}) > p(Q^*)$, implying that the right side of the above inequality must be positive; that is, $p(Q^{SO}) - p(Q^*) > 0$, entailing that we need the left side to satisfy

$$c_i'(q_i^{SO}) - c_i'(q_i^*) > 0.$$

(The above steps actually mean that the difference $c_i'(q_i^{SO}) - c_i'(q_i^*)$ is not only positive, but larger than the positive number $p(Q^{SO}) - p(Q^*)$, but for our proof we only need to show that the less demanding condition $c_i'(q_i^{SO}) - c_i'(q_i^*) > 0$ cannot hold in our setting, as we demonstrate below.) However, marginal cost is, by definition, increasing in firm i's output, implying that $c_i'(q_i^{SO}) > c_i'(q_i^*)$ is incompatible with the output ranking $q_i^* \geq q_i^{SO}$. We have then reached a contradiction, and, as a consequence, output levels must satisfy $q_i^* < q_i^{SO}$. Intuitively, every firm i produces a larger output when the social planner chooses production levels than when the firm selects its own production.

- Figure 2.5 illustrates the first-order conditions of parts (a) and (b), showing that $p(Q^*) + p'(Q^*)q_i^*$ lies below $p(Q^{SO})$, implying that the equilibrium output under Cournot competition, q_i^*, is socially insufficient.

Exercise #2.6: Cournot Pricing Formula – Derivation^A

2.6 Consider a market with n symmetric firms competing ? la Cournot and selling a homogeneous product. The inverse demand function is $P(Q)$, where $Q = \sum_{i=1}^{n} q_i$ denotes aggregate output.

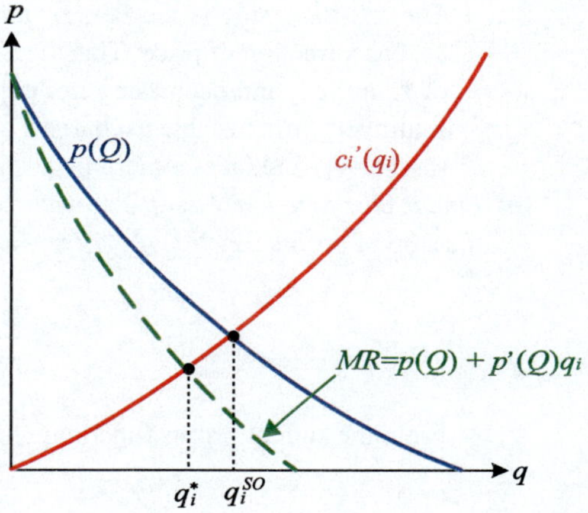

Fig. 2.5 Socially optimal equilibrium

Every firm i faces a cost function $C_i(q_i)$ that is increasing and convex in output, q_i, that is, $C_i(0) = 0$, $C_i'(q_i) > 0$, and $C_i''(q_i) > 0$.

(a) Set up firm i's profit maximization problem, and show that

$$\frac{P(Q) - C_i'(q_i)}{P(Q)} = \frac{\alpha_i}{\eta}$$

where $\alpha_i \equiv \frac{q_i}{Q}$ represents firm i's market share and $\eta = -\frac{P(Q)}{Q}\left(\frac{\partial P(Q)}{\partial Q}\right)^{-1}$ denotes the price elasticity of demand.

- Every firm i chooses q_i to solve the profit maximization problem:

$$\max_{q_i \geq 0} \pi_i(q_i) = P(Q)\, q_i - C_i(q_i)$$

Differentiating with respect to q_i, and assuming interior solutions, yields

$$P(Q) + P'(Q)\, q_i - C_i'(q_i) = 0$$

Rearranging and dividing both sides by $P(Q)$, we obtain

$$\frac{P(Q) - C_i'(q_i)}{P(Q)} = -P'(Q)\frac{q_i}{P(Q)}$$

Multiplying the right-hand side by $\frac{Q}{Q}$, we find

$$\frac{P(Q) - C_i'(q_i)}{P(Q)} = \underbrace{-P'(Q)\frac{Q}{P(Q)}}_{1/\eta}\underbrace{\frac{q_i}{Q}}_{\alpha_i}$$

where the inverse of price elasticity of demand is $\frac{1}{\eta} = -P'(Q)\frac{Q}{P(Q)}$. Simplifying yields the Cournot pricing formula, as follows:

$$\frac{P(Q) - C_i'(q_i)}{P(Q)} = \frac{\alpha_i}{\eta}$$

The left-hand side is the *Lerner index*, which measures the price markup over marginal cost as a fraction of price. This measure of market power is increasing in the market share of firm i, α_i, and decreasing in the price elasticity of demand, η, on the right-hand side. Intuitively, firm i is able to charge a higher price markup if it captures a larger market share (higher α_i) and faces a more inelastic demand (lower η).

(b) *Linear cost function.* Assume a linear cost function, $C_i(q_i) = c_i q_i$, and let $p \equiv P(Q)$. Use the sum of profits function, $\sum_{i=1}^{n} \pi_i = \sum_{i=1}^{n} (p - c_i)\alpha_i Q$, to show that

$$\frac{p - \sum_{i=1}^{N} \alpha_i c_i}{p} = \frac{\sum_{i=1}^{N} \alpha_i^2}{\eta}$$

- From the sum of profits function, we obtain

$$\begin{aligned}\sum_{i=1}^{n} \pi_i &= \sum_{i=1}^{n} \alpha_i p Q - \sum_{i=1}^{n} c_i \alpha_i Q \\ &= \left(p - \sum_{i=1}^{n} c_i \alpha_i\right) Q\end{aligned}$$

since $\sum_{i=1}^{n} \alpha_i = \sum_{i=1}^{n} \frac{q_i}{Q} = \frac{Q}{Q} = 1$.

From the Cournot pricing formula in part (a), we have that

$$\frac{p - c_i}{p} = \frac{\alpha_i}{\eta}$$

which we rearranged to yield

$$c_i = p\left(1 - \frac{\alpha_i}{\eta}\right).$$

Substituting the above results into the sum of profits function, we find

$$\begin{aligned}\left(p - \sum_{i=1}^{n} c_i \alpha_i\right) Q &= \left(p - \sum_{i=1}^{n} \overbrace{p\left(1 - \frac{\alpha_i}{\eta}\right)}^{c_i} \alpha_i\right) Q \\ &= \left(1 - \sum_{i=1}^{n} \left(1 - \frac{\alpha_i}{\eta}\right) \alpha_i\right) pQ \\ &= \left(1 - \sum_{i=1}^{n} \alpha_i + \sum_{i=1}^{n} \frac{\alpha_i^2}{\eta}\right) pQ \\ &= \frac{\sum_{i=1}^{n} \alpha_i^2}{\eta} pQ\end{aligned}$$

Finally, dividing both sides by pQ results in

$$\frac{p - \sum_{i=1}^{n} c_i \alpha_i}{p} = \frac{\sum_{i=1}^{n} \alpha_i^2}{\eta}$$

where the left-hand side is the price markup over average cost as a fraction of price. This measure of market power is increasing in the Herfindahl index, $\sum_{i=1}^{n} \alpha_i^2$, and decreasing in the price elasticity of demand η.

(c) Show that the above formula reduces to the inverse elasticity pricing rule (IEPR) for a monopolist who is subject to a constant marginal cost c.

- When the market is monopolized by one firm, this firm owns 100% of the market share, so $\sum_{i=1}^{n} \alpha_i^2 = 1$. In this context, the average market cost is just the cost of the monopolist itself, implying $\sum_{i=1}^{n} c_i \alpha_i = c$. Substituting these results into the formula we identified in part (b) yields

$$\frac{p - c}{p} = \frac{1}{\eta}$$

which is exactly the inverse elasticity pricing rule (IEPR) for the monopolist.

Exercise #2.7: Cournot Pricing Formula – Application^A

2.7 Consider the Cournot pricing formula

$$\frac{p - c}{p} = \frac{\sum_{i=1}^{n} \alpha_i^2}{\eta}$$

where η represents price elasticity of demand, p is the market price, and α_i denotes firm i's market share, where $i \in \{1, \ldots, n\}$. Firms share a common marginal cost c.

(a) Assume that the market has unitary price elasticity of demand, where a 1% increase in price induces a 1% decrease in demand. Find the Lerner index, defined as $L(n) \equiv \frac{p-c}{p}$, in the setting where every firm i has an equal market share.

- Since price elasticity of demand is unitary, $\eta = 1$, and every firm i has equal market share, $\alpha_i = \frac{1}{n}$, the Lerner index becomes

$$L(n) = \frac{p - c}{p} = \frac{\sum_{i=1}^{n} \alpha_i^2}{\eta}$$

$$= \frac{\sum_{i=1}^{n} \left(\frac{1}{n}\right)^2}{1} = n \left(\frac{1}{n}\right)^2 = \frac{1}{n}.$$

Intuitively, the price markup over marginal cost (as captured by Lerner index) is decreasing in the number of firms competing in this industry, n.

(b) Continue with the setting where n firms equally split the market. Now consider that two (out of n firms) merge into one firm. Compare the Lerner index before and after the merger.

- When two out of n firms merge, we have $n-1$ firms in the market, so that the Lerner index after the merger becomes

$$L(n-1) = \frac{1}{n-1}.$$

The market is more concentrated after the merger, and Lerner index increases since

$$L(n-1) = \frac{1}{n-1} > \frac{1}{n} = L(n)$$

which simplifies to $n > n-1$ that holds for all $n \geq 2$.

(c) The US Department of Justice and the Federal Trade Commission consider that markets with a Herfindahl-Hirschman index (HHI) exceeding 2500/10,000 to be highly concentrated. From part (a), what is the number of firms that fit into this criteria?

- The HHI is the sum of firms' squared market shares, which in this setting is

$$\underbrace{\left(\frac{1}{n}\right)^2 + \left(\frac{1}{n}\right)^2 + \cdots + \left(\frac{1}{n}\right)^2}_{n \text{ times}} = \frac{n}{n^2} = \frac{1}{n}.$$

Therefore, a highly concentrated market must satisfy

$$HHI = \frac{1}{n} \geq \frac{2500}{10{,}000} = \frac{1}{4}.$$

Simplifying, we obtain that $n \leq 4$, meaning that the market is highly concentrated if there are 4 or fewer firms.

Exercise #2.8: Cournot Pricing Formula – Parametric Example^A

2.8 Consider n symmetric firms competing à la Cournot, selling a homogeneous good, with demand function $q(p) = Ap^{-b}$ and a common marginal cost c, where $A, b, c > 0$.

(a) Find the price elasticity of demand, $\varepsilon_d \equiv \frac{\partial q(p)}{\partial p}\frac{p}{q}$.

- Differentiating the demand function $q(p)$ with respect to p, we obtain

$$\frac{\partial q(p)}{\partial p} = -bAp^{-b-1}$$

Therefore, the price elasticity of demand is given by

$$\varepsilon_d \equiv \frac{\partial q(p)}{\partial p}\frac{p}{q} = -bAp^{-b-1}\frac{p}{Ap^{-b}} = -b.$$

(b) Derive the inverse demand function, $p(q)$.

- From the demand function $q(p) = Ap^{-b}$, we obtain that

$$\frac{A}{q} = p^b$$

Rearranging and solving for p, the inverse demand function is

$$p(q) = \left(\frac{A}{q}\right)^{\frac{1}{b}}$$

(c) Derive the Cournot pricing rule. (*Hint*: Every firm i has an equal market share.)

- Since at the profit-maximizing output, it must be that $MR = MC$, we have that

$$p'(q)\, q_i + p(q) = c$$

where the left-hand (right-hand) side represents the marginal revenue (marginal cost, respectively).

Differentiating the inverse demand function $p(q)$ with respect to q, we find

$$\frac{\partial p(q)}{\partial q} = -\frac{1}{b} A^{\frac{1}{b}} q^{-\frac{1}{b}-1}$$

Together with the fact that every firm has equal market share, $q_i = \frac{q}{n}$, yields

$$\underbrace{\left(-\frac{1}{b} A^{\frac{1}{b}} q^{-\frac{1}{b}-1}\right)}_{\frac{\partial p(q)}{\partial q}} \frac{q}{n} + \underbrace{\left(\frac{A}{q}\right)^{\frac{1}{b}}}_{p(q)} = c.$$

Simplifying the above expression, we obtain

$$\left(\frac{A}{q}\right)^{\frac{1}{b}} \left(1 - \frac{1}{bn}\right) = c$$

Rearranging, we find the Cournot pricing rule, as follows:

$$\frac{p - c}{p} = \frac{1}{bn}$$

where the left-hand side measures the price markup over marginal cost as a proportion of market price (also known as the *Lerner index*) and the right-hand side is the inverse of the price elasticity of demand, b, times the number of firms n.

(d) Evaluate the price markup when the demand is perfectly elastic, and show that its effect is equivalent to that of a perfectly competitive market.

- When the demand is perfectly elastic, $b \to +\infty$ so that $\frac{p-c}{p} = 0$ yielding $p = c$. This is exactly the condition of a perfectly competitive market when $n \to +\infty$.

Exercise #2.9: Cournot Duopoly with Asymmetric Marginal Costs^A

2.9 Consider the Cournot duopoly in the previous exercise. Assume that firm 1 faces marginal cost c_1, while firm 2's is c_2, where $c_1 < c_2$ (so firm 1 enjoys a cost advantage relative to firm 2) and $a > c_2$.

(a) Find the best response function of firm 1 and of firm 2. Compare them.

- *Firm 1's best response function.* In this setting, firm 1 chooses its output level q_1 to solve the profit maximization problem

$$\max_{q_1 \geq 0} \; \pi_1 = (a - bq_1 - bq_2)q_1 - c_1 q_1.$$

Differentiating with respect to q_1, we obtain

$$\frac{\partial \pi_1}{\partial q_1} = a - 2bq_1 - bq_2 - c_1 = 0$$

rearranging, $a - bq_2 - c_1 = 2bq_1$, and solving for q_1 yields

$$q_1(q_2) = \frac{a - c_1}{2b} - \frac{1}{2}q_2$$

which is firm 1's best response function.

- *Firm 2's best response function.* In this setting, firm 2 chooses its output level q_2 to solve

$$\max_{q_2 \geq 0} \; \pi_2 = (a - bq_1 - bq_2)q_2 - c_2 q_2.$$

Differentiating with respect to q_2, we obtain

$$\frac{\partial \pi_2}{\partial q_2} = a - 2bq_2 - bq_1 - c_2 = 0$$

rearranging, $a - bq_1 - c_2 = 2bq_2$, and solving for q_2 yields

$$q_2(q_1) = \frac{a - c_2}{2b} - \frac{1}{2}q_1$$

which is firm 2's best response function.

- Relative to firm 1's best response function, firm 2's originates at a lower vertical intercept since $\frac{a-c_2}{2b} < \frac{a-c_1}{2b}$ given that $c_1 < c_2$ by assumption. Intuitively, the firm benefiting from a cost advantage produces more units for a given output level of its rival. Both best response functions, however, have the same slope.

(b) Insert firm 2's best response function into firm 1's to find the output that each firm produces in the Nash equilibrium of the Cournot game of quantity competition. Which firm produces a larger output?

- In this exercise, we cannot invoke symmetry in equilibrium output since firms are asymmetric (they face different marginal costs). We need to insert firm 2's best response function into firm 1's, finding

$$q_1 = \frac{a - c_1}{2b} - \frac{1}{2}\underbrace{\left(\frac{a - c_2}{2b} - \frac{1}{2}q_1\right)}_{q_2(q_1)}$$

simplifying, we obtain

$$q_1 = \frac{a - c_1}{2b} - \frac{a - c_2}{4b} + \frac{1}{4}q_1$$

further rearranging

$$\frac{3}{4}q_1 = \frac{2a - 2c_1 - a + c_2}{4b}$$

and solving for q_1, we find the equilibrium output for firm 1

$$q_1 = \frac{a - 2c_1 + c_2}{3b}.$$

- We can now insert this output into firm 2's best response function to find their equilibrium output:

$$q_2 = \frac{a - c_2}{2b} - \frac{1}{2}\underbrace{\left(\frac{a - 2c_1 + c_2}{3b}\right)}_{q_1}$$

$$q_2 = \frac{a - c_2}{2b} - \frac{a - 2c_1 + c_2}{6b}$$

$$q_2 = \frac{3a - 3c_2 - a + 2c_1 - c_2}{6b}$$

$$q_2 = \frac{2a - 4c_2 + 2c_1}{6b}$$

$$q_2 = \frac{a - 2c_2 + c_1}{3b}$$

which is firm 2's equilibrium output.

- *Comparing output levels.* We can directly compare q_1 and q_2 to find that the firm with the lowest marginal cost will produce more output in equilibrium, that is, $q_1 > q_2$

$$\frac{a - 2c_1 + c_2}{3b} > \frac{a - 2c_2 + c_1}{3b}$$

$$a - 2c_1 + c_2 > a - 2c_2 + c_1$$

$$3c_2 > 3c_1$$

$$c_2 > c_1$$

which is true by assumption. Therefore, our first inequality, $q_1 > q_2$, holds.

- *Corner solution.* There is a situation where firm 1 is the only active firm in the market, i.e., $q_2 = 0$. To see when this situation is possible, we set firm 2's equilibrium output equal to (or less than) zero:

$$q_2 = \frac{a - 2c_2 + c_1}{3b} \leq 0.$$

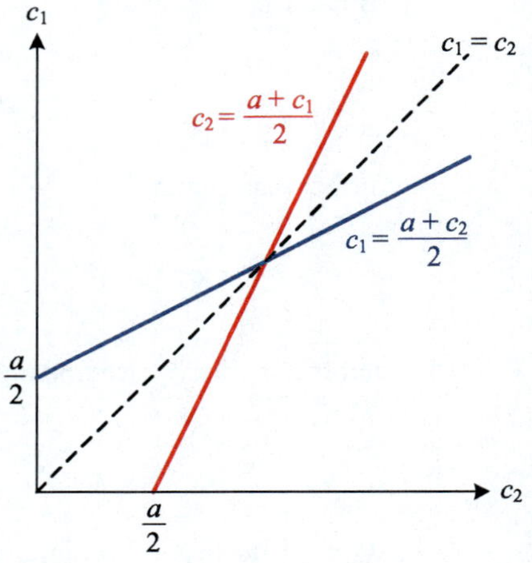

Fig. 2.6 Production decisions

First, multiply each side by $3b$ to simplify the inequality to

$$a - 2c_2 + c_1 \leq 0$$

rearranging, we obtain $2c_2 \geq a + c_1$, and solving for c_2, we find that the high-cost firm 2 will not produce when c_2 satisfies

$$c_2 \geq \frac{a + c_1}{2}.$$

This condition means that firm 2's costs are so high, relative to firm 1's, that it cannot profitably supply output to the market. As the cost differential between the two firms grows (as c_2 increases), the low-cost firm's output increases and the high-cost firm's output decreases. Once the cost discrepancy is large enough, the high-cost firm does not compete in the market. Figure 2.6 depicts cutoff $c_2 = \frac{a+c_1}{2}$ in the (c_2, c_1)-quadrant, so for all cost pairs above cutoff $c_2 = \frac{a+c_1}{2}$ firm 2 remains inactive. The figure also includes a symmetric cutoff for firm 1, $c_1 = \frac{a+c_2}{2}$, so for points below it firm 1 remains inactive. In summary, only firm 2 is active in the northwest region, as it is relatively more efficient than its rival; only firm 1 is active in the southeast region given its cost advantage; both firms are active in the southwest region as they are both relatively efficient and symmetric; and no firm is active in the northeast region because both are relatively inefficient.

(c) Find equilibrium price and equilibrium profits for each firm. Which firm earns a larger profit?

- *Price.* Each firm faces the same price, that is,

$$\begin{aligned}
p &= a - b(q_1 + q_2) \\
&= a - b\left(\frac{a - 2c_1 + c_2}{3b} + \frac{a - 2c_2 + c_1}{3b}\right) \\
&= a - \frac{a - 2c_1 + c_2 + a - 2c_2 + c_1}{3} \\
&= a - \frac{2a - c_1 - c_2}{3}
\end{aligned}$$

$$= \frac{3a - 2a + c_1 + c_2}{3}$$
$$= \frac{a + c_1 + c_2}{3}.$$

- *Firm 1's profit.* In equilibrium, firm 1 earns profit

$$\pi_1 = pq_1 - c_1q_1 = (p - c_1)q_1$$

or, substituting the equilibrium price and output,

$$\pi_1 = \left(\frac{a + c_1 + c_2}{3} - c_1\right)\frac{a - 2c_1 + c_2}{3b}$$
$$= \left(\frac{a + c_1 + c_2 - 3c_1}{3}\right)\frac{a - 2c_1 + c_2}{3b}$$
$$= \left(\frac{a - 2c_1 + c_2}{3}\right)\frac{a - 2c_1 + c_2}{3b}$$
$$= \frac{(a - 2c_1 + c_2)^2}{9b}.$$

- *Firm 2's profit.* In equilibrium, firm 2 earns profit

$$\pi_2 = pq_2 - c_2q_2 = (p - c_2)q_2$$

or, substituting the equilibrium price and output,

$$\pi_2 = \left(\frac{a + c_1 + c_2}{3} - c_2\right)\frac{a - 2c_2 + c_1}{3b}$$
$$= \left(\frac{a + c_1 + c_2 - 3c_2}{3}\right)\frac{a - 2c_2 + c_1}{3b}$$
$$= \left(\frac{a + c_1 - 2c_2}{3}\right)\frac{a - 2c_2 + c_1}{3b}$$
$$= \frac{(a - 2c_2 + c_1)^2}{9b}.$$

- *Comparing profits.* We can show that firm 1 will earn a higher profit by showing that $\pi_1 > \pi_2$:

$$\frac{(a - 2c_1 + c_2)^2}{9b} > \frac{(a - 2c_2 + c_1)^2}{9b}$$
$$(a - 2c_1 + c_2)^2 > (a - 2c_2 + c_1)^2$$
$$a - 2c_1 + c_2 > a - 2c_2 + c_1$$
$$3c_2 > 3c_1$$
$$c_2 > c_1$$

which holds by assumption. Therefore, our first inequality, $\pi_1 > \pi_2$, holds. Intuitively, the firm benefiting from a cost advantage (firm 1) earns a higher profit.

(d) *Symmetric firms.* Assume that both firms now become cost symmetric, so that $c_1 = c_2 = c$. Evaluate your results from parts (b) and (c) at $c_1 = c_2 = c$, showing that you obtain the same results as in the previous exercise.

- Evaluating each result at $c_1 = c_2 = c$, we obtain

$$q_1 = \frac{a - 2c + c}{3b} = \frac{a - c}{3b}$$

$$q_2 = \frac{a - 2c + c}{3b} = \frac{a - c}{3b}$$

$$p = \frac{a + c + c}{3} = \frac{a + 2c}{3}$$

$$\pi_1 = \frac{(a - 2c + c)^2}{9b} = \frac{(a - c)^2}{9b}$$

$$\pi_2 = \frac{(a - 2c + c)^2}{9b} = \frac{(a - c)^2}{9b}$$

which all line up with the results of a Cournot duopoly model with symmetric firms we studied in Exercise 2.1.

Exercise #2.10: Cournot Competition with n Firms Facing Asymmetric Costs^C

2.10 Consider an industry of $n \geq 2$ firms competing ? la Cournot. Firms face an inverse demand curve $p(Q) = a - Q$, where $Q \geq 0$ denotes aggregate output. Every firm i has a marginal cost of production c_i, where $a > c_i \geq 0$.

(a) Set up firm i's profit maximization problem and find its first-order condition.

- Every firm i chooses its output q_i to solve

$$\max_{q_i \geq 0} \; [a - (q_i + Q_{-i})]\, q_i - c_i q_i,$$

where Q_{-i} denotes the aggregate output of firm i's rivals. Differentiating with respect to q_i yields

$$a - 2q_i - Q_{-i} - c_i = 0$$

which we can rearrange as

$$a - c_i = 2q_i + Q_{-i}.$$

(b) Find equilibrium output. [*Hint*: Find the aggregate output Q by summing over n firms from the first-order condition that you found in part (a).]

- From the first-order condition found in part (a), we have that

$$a - c_i = 2q_i + Q_{-i},$$

which can be rewritten as

$$a - c_i = q_i + Q$$

since $Q = q_i + Q_{-i}$ by definition. Summing over n firms, we obtain

$$na - \sum_{i=1}^{n} c_i = \sum_{i=1}^{n} q_i + nQ$$

which simplifies to

$$na - \sum_{i=1}^{n} c_i = (1+n)Q$$

since $Q = \sum_{i=1}^{n} q_i$ by definition. Denoting, for compactness, $C = \sum_{i=1}^{n} c_i$ for the aggregate costs, and solving for Q, we find an expression for aggregate output in equilibrium as follows:

$$Q = \frac{na - C}{1+n}.$$

Returning to firm i's first-order condition again, $a - c_i = q_i + Q$, we can rewrite it as $q_i = a - c_i - Q$. Inserting the above expression of aggregate output in equilibrium yields the equilibrium output of firm i as follows:

$$\begin{aligned} q_i^* &= a - c_i - \overbrace{\frac{na - C}{1+n}}^{Q} \\ &= \frac{a - (1+n)c_i + C}{1+n}, \end{aligned}$$

which can also be expressed, alternatively, as

$$q_i^* = \frac{a - nc_i + \sum_{j \neq i} c_j}{1+n}.$$

(c) *First example.* Consider a setting with $n = 2$ firms (firm 1 and 2) facing inverse demand function $p(Q) = 1 - Q$, and marginal production costs c_1 and c_2, where $1 > c_i \geq 0$ for every firm $i = \{1, 2\}$. Evaluate your results from part (b) to find the equilibrium output for each firm, aggregate output, and profits. Then evaluate your results in the case that marginal production costs coincide, $c_1 = c_2 = c$, where $1 > c \geq 0$.

- *Asymmetric costs,* $c_1 \neq c_2$. In this context, the sum of marginal costs is $C = c_1 + c_2$, and demand parameters are $a = b = 1$. Therefore, aggregate output becomes

$$Q = \frac{2 - (c_1 + c_2)}{2+1} = \frac{2 - (c_1 + c_2)}{3}$$

individual output is

$$q_i = \frac{1 - 2c_i + c_j}{3}$$

and profits become

$$\pi_i = \left(1 - \frac{1 - 2c_i + c_j}{3} - \frac{1 - 2c_j + c_i}{3} - c_i\right)\frac{1 - 2c_i + c_j}{3}$$
$$= \frac{\left(1 - 2c_i + c_j\right)^2}{9}.$$

- *Symmetric costs,* $c_1 = c_2 = c$. In this setting, the above results become

$$Q = \frac{2(1 - c)}{3}$$

individual output is

$$q_i = \frac{Q}{2} = \frac{1 - c}{3}$$

and profits become

$$\pi_i = \frac{(1 - c)^2}{9}.$$

(d) *Second example.* Consider a setting with $n \geq 2$ firms facing inverse demand function $p(Q) = 1 - Q$, and symmetric marginal production cost c, where $1 > c \geq 0$. Assuming that k firms merge, benefiting from a lower marginal cost $c - x$, while the $n - k$ unmerged firms still face marginal cost c. Find the aggregate output in equilibrium when k firms merge, and compare it against aggregate output before the merger. For which parameter values the merger produces an increase in aggregate output?

- *Before the merger.* With n firms in the industry, all facing marginal cost c, the sum of marginal costs is $C = nc$. Therefore, expression $Q = \frac{na - C}{n+1}$, we can then write aggregate output in this setting as

$$Q^{NM} = \frac{n - nc}{n + 1} = \frac{n(1 - c)}{n + 1}$$

since $a = 1$, where superscript NM denotes "no merger."

- *After the merger.* If k out of n firms merge, leaving $n - k$ firms unmerged, then there are $(n - k) + 1$ firms in the industry. In this context, the sum of marginal costs is

$$C = \underbrace{(c - x)}_{\text{Merged firm}} + \underbrace{(n - k)c}_{\text{Unmerged firms}} = (n - k + 1)c - x.$$

Using expression $Q = \frac{na-C}{n+1}$, we can then write aggregate output in this setting as

$$Q^M = \frac{[(n-k)+1] - [(n-k+1)c - x]}{[(n-k)+1]+1}$$
$$= \frac{(n-k+1)(1-c)+x}{n-k+2}$$

since $a = 1$, where superscript M denotes "merger."

- *Output comparison.* Aggregate output after the merger increases if $Q^M \geq Q^{NM}$, which entails

$$\frac{(n-k+1)(1-c)+x}{n-k+2} \geq \frac{n(1-c)}{n+1}.$$

Rearranging, we obtain

$$\theta \equiv \frac{x}{1-c} \geq \frac{k-1}{n+1}.$$

Intuitively, the merger increases aggregate output (and, as a consequence, consumer surplus) if the cost-reduction effect relative to firms' margin (left-hand side, θ) is sufficiently large.

As an illustration, we can fix the total number of firms at $n = 10$ and evaluate cutoff $\frac{k-1}{n+1}$ at $k = 2$, obtaining that

$$\frac{2-1}{10+1} = 0.09.$$

Intuitively, the cost-reduction effect, relative to per-unit margins (as measured by θ), must be larger than 9% for the merger to increase consumer surplus. Mergers between more firms (higher k) produce an even larger ratio $\frac{k-1}{n+1}$, thus increasing the minimum cost-reduction effect, θ, required for the merger to increase consumer surplus.

Exercise #2.11: Cournot with Only One Firm Benefiting from a Cost Advantage^B

2.11 Consider an industry with n firms selling a homogeneous good, competing à la Cournot, with inverse demand function $p(Q) = 1 - Q$, where Q denotes aggregate output. One of the firms (firm 1) faces marginal cost $c \in (0, 1)$, while its $n - 1$ rivals face a higher marginal cost $c' \geq c$. This can be explained because firm 1 had more experience in the industry, or operating in similar markets, providing it with a cost advantage.

(a) Find firm 1's best response function and the best response function of each of its rivals.

- Firm 1 chooses its output, q_1, to solve

$$\max_{q_1 \geq 0} (1 - q_1 - Q_{-1})q_1 - cq_1,$$

where Q_{-1} denotes the aggregate output by firm 1's rivals. Differentiating with respect to q_1, we obtain

$$1 - 2q_1 - Q_{-1} - c = 0$$

and solving for q_1 yields best response function

$$q_1(Q_{-1}) = \frac{1-c}{2} - \frac{1}{2}Q_{-1}$$

which is decreasing in the aggregate output of firm 1's rivals at a rate of $\frac{1}{2}$.

- Every firm $i \neq 1$ chooses its output q_i to solve

$$\max_{q_i \geq 0} \,(1 - q_i - q_1 - Q_{-i})q_i - c'q_i,$$

where Q_{-i} denotes the aggregate output by firm i's rivals (without firm 1's output, q_1). Differentiating with respect to q_i, we obtain

$$1 - 2q_i - q_1 - Q_{-i} - c' = 0$$

and solving for q_i yields best response function

$$q_i(q_1, Q_{-i}) = \frac{1-c'}{2} - \frac{1}{2}q_1 - \frac{1}{2}Q_{-i}.$$

(b) Find equilibrium output levels for each firm.

- Every firm other than firm 1 chooses the same output, so $q_i = q_j$ for all $i \neq j \neq 1$. Therefore, $Q_{-i} = (n-2)q_i$. Inserting this property in firm i's best response function, we find that

$$q_i = \frac{1-c'}{2} - \frac{1}{2}q_1 - \frac{1}{2}(n-2)q_i.$$

Solving for q_i yields

$$q_i = \frac{1-c'-q_1}{n}$$

implying that the aggregate output of firm 1's rivals is

$$Q_{-1} = \frac{(n-2)(1-c'-q_1)}{n}.$$

We can now insert this expression into firm 1's best response function to obtain

$$q_1 = \frac{1-c}{2} - \frac{1}{2}\underbrace{\frac{(n-2)(1-c'-q_1)}{n}}_{Q_{-1}}.$$

Rearranging and solving for q_1, we find firm 1's equilibrium output

$$q_1^* = \frac{n(1-c) - (n-1)\left(1-c'\right)}{n+1}.$$

which is positive since $n > n-1$ and $1-c \geq 1-c'$. We can finally insert this output level into firm i's best response function to find its equilibrium output, that is,

$$\begin{aligned} q_i^* &= \frac{1-c'}{n} - \frac{1}{n}\left(\frac{n(1-c)}{n+1} - \frac{(n-1)\left(1-c'\right)}{n+1}\right) \\ &= \frac{1+c-2c'}{n+1} \end{aligned}$$

(c) Evaluate the equilibrium output for firm 1 and its $n-1$ rivals when all firms are symmetric, that is, $c = c'$.

- Evaluating firm 1's equilibrium output at $c = c'$, we obtain

$$\begin{aligned} q_1^* &= \frac{n(1-c) - (n-1)(1-c)}{n+1} \\ &= \frac{1-c}{n+1} \end{aligned}$$

Similarly, evaluating its rivals' equilibrium output at $c = c'$, we find

$$\begin{aligned} q_i^* &= \frac{1+c-2c}{n+1} \\ &= \frac{1-c}{n+1} \end{aligned}$$

where $q_1^* = q_i^*$ coincides with the equilibrium output in a standard Cournot model with n symmetric firms.

(d) Under which conditions on firm 1's cost, c, will all firms produce a positive output in equilibrium? Under which conditions will only firm 1 be active? Interpret.

- First, recall that, as found in part (b), firm 1 is unambiguously active. However, each of its rivals is active if and only if $q_i^* \geq 0$, or

$$\frac{1+c-2c'}{n+1} \geq 0$$

which, solving for firm i's costs, c', simplifies to

$$c' \leq \frac{1+c}{2} \equiv \overline{c}$$

Therefore, when $c' \leq \overline{c}$, all firms are active. Otherwise, only firm 1 is active, as it benefits from a sufficiently strong cost advantage.

(e) *Symmetric firms.* Evaluate your results in part (d) in the case that all firms are symmetric, that is, $c = c'$.

- When $c = c'$, the inequality in part (d) becomes

$$c \leq \frac{1+c}{2}$$

that simplifies to $c \leq 1$, a condition that holds by definition, so that all firms produce positive units when they are cost symmetric (as in the standard Cournot model with n symmetric firms).

Exercise #2.12: Social Welfare with Cost-Symmetric Firms Selling Heterogeneous Goods and Competing in Quantities^{A}

2.12 Consider two firms competing à la Cournot selling a differentiated good at a common marginal cost c, where $0 \leq c < 1$. Every firm i has an inverse demand function

$$p_i(q_i, q_j) = 1 - q_i - \gamma q_j,$$

where $\gamma \in [0, 1]$ denotes the degree of product differentiation. When $\gamma = 0$, products are completely differentiated, but when $\gamma = 1$ they are homogeneous.

(a) Find firm i's best response function. How is it affected by a marginal increase in parameter γ?

- Every firm i chooses q_i to solve the profit maximization problem, as follows:

$$\max_{q_i \geq 0} \; \pi(q_i) = \left(1 - q_i - \gamma q_j\right) q_i - c q_i$$

Taking the first-order condition with respect to q_i, and assuming interior solutions,

$$1 - 2q_i - \gamma q_j - c = 0$$

Solving for q_i, we find that firm i's best response function is

$$q_i\left(q_j\right) = \frac{1-c}{2} - \frac{\gamma}{2} q_j$$

meaning that the best response function originates at a vertical intercept of $\frac{1-c}{2}$ and decreases in q_j at a rate $\frac{\gamma}{2}$. Graphically, an increase in parameter γ (goods are more homogeneous) produces an inward rotation of the best response function, leaving its vertical intercept unchanged at $\frac{1-c}{2}$ but making this function steeper. Intuitively, that means that, as firms sell more homogeneous products, a given increase in firm j's output yields a larger reduction in firm i's output level, so that goods become stronger strategic substitutes.

- *Extreme cases.*
  - When goods are completely differentiated ($\gamma = 0$), this best response function simplifies to

$$q_i = \frac{1-c}{2},$$

thus coinciding with the monopoly output under linear demand.

– In contrast, when goods are homogeneous ($\gamma = 1$), this best response function becomes

$$q_i\left(q_j\right) = \frac{1-c}{2} - \frac{1}{2}q_j,$$

as in standard Cournot models without product differentiation.

(b) Find the equilibrium output of every firm i.

- In a symmetric equilibrium, every firm i produces the same output level, $q_i = q_j$. Inserting this property in the above best response function, we obtain that

$$q_i = \frac{1-c}{2} - \frac{\gamma}{2}q_i.$$

Rearranging and solving for q_i yields an equilibrium output of

$$q_i^* = \frac{1-c}{2+\gamma}.$$

This output is decreasing in γ indicating that, as firms' output become more homogeneous (higher γ), each firm produces fewer units. As expected, equilibrium output is also decreasing in firms' common marginal cost, c.

(c) Find the equilibrium profits of every firm i.

- Inserting the equilibrium output, $q_i^* = \frac{1-c}{2+\gamma}$, into firm i's objective function, we find that its equilibrium profits are

$$\begin{aligned}\pi_i^* &= \left(1 - \frac{1-c}{2+\gamma} - \frac{\gamma(1-c)}{2+\gamma} - c\right)\frac{1-c}{2+\gamma}\\ &= \left(\frac{(2+\gamma)(1-c)-(1+\gamma)(1-c)}{2+\gamma}\right)\frac{1-c}{2+\gamma}\\ &= \left(\frac{1-c}{2+\gamma}\right)^2\end{aligned}$$

which is decreasing in the parameter measuring product homogeneity, γ, and in marginal cost, c.

(d) If a social planner could determine individual output level for each firm, find the socially optimal levels that he would choose.

- Social welfare is the sum of consumer and producer surplus, as follows:

$$\begin{aligned}SW\left(q_i, q_j\right) &= \underbrace{\frac{q_i^2 + 2\gamma q_i q_j + q_j^2}{2}}_{CS} + \underbrace{\left[p_i(q_i, q_j) - c\right]q_i + \left[p_j(q_i, q_j) - c\right]q_j}_{PS}\\ &= \left(1 - \frac{1}{2}q_i - c\right)q_i + \left(1 - \frac{1}{2}q_j - c\right)q_j - \gamma q_i q_j\end{aligned}$$

Differentiating with respect to q_i, and assuming interior solutions, we find that

$$1 - q_i - \gamma q_j - c = 0$$

Rearranging and solving for q_i, socially optimal output becomes

$$q_i^{SO} = \frac{1-c}{1+\gamma}$$

which is also decreasing in the parameter measuring product homogeneity, γ, and in marginal cost, c.

(e) Evaluate the difference in output levels you found in parts (b) and (d), and conduct comparative statics with respect to parameter γ. Interpret your results.

- Comparing the socially optimal and equilibrium output levels, we see that

$$q_i^{SO} = \frac{1-c}{1+\gamma} > \frac{1-c}{2+\gamma} = q_i^*$$

which, intuitively, means that firms produce a socially insufficient output. Specifically, this output difference is

$$q_i^{SO} - q_i^* = \frac{1-c}{1+\gamma} - \frac{1-c}{2+\gamma} = \frac{1-c}{2+3\gamma+\gamma^2}$$

which is decreasing as products become more homogeneous (higher γ) since

$$\frac{\partial \left(q_i^{SO} - q_i^*\right)}{\partial \gamma} = -\frac{(3+2\gamma)(1-c)}{\left(2+3\gamma+\gamma^2\right)^2}$$

is negative given that c satisfies $0 \leq c < 1$ by definition. Intuitively, firms engage in more intensive competition when products become more homogeneous, thus reducing the output difference between the competitive and socially optimal levels.

(f) Evaluate the social welfare at the equilibrium output level found in part (b).

- Substituting $q_i^* = \frac{1-c}{2+\gamma}$ into the social welfare function in part (d), we find that

$$\begin{aligned} SW^* &= 2\left(1 - \frac{1-c}{2(2+\gamma)} - c\right)\frac{1-c}{2+\gamma} - \gamma\left(\frac{1-c}{2+\gamma}\right)^2 \\ &= 2\left(\frac{[2(2+\gamma)-1](1-c)}{2(2+\gamma)}\right)\frac{1-c}{2+\gamma} - \gamma\left(\frac{1-c}{2+\gamma}\right)^2 \\ &= \frac{(3+\gamma)(1-c)^2}{(2+\gamma)^2}. \end{aligned}$$

(g) Evaluate the social welfare at the socially optimal output level found in part (d).

- Substituting $q_i^{SO} = \frac{1-c}{1+\gamma}$ into the social welfare function in part (d), we find that

$$SW^{SO} = 2\left(1 - \frac{1-c}{2(1+\gamma)} - c\right)\frac{1-c}{1+\gamma} - \gamma\left(\frac{1-c}{1+\gamma}\right)^2$$

$$= 2\left(\frac{[2(1+\gamma)-1](1-c)}{2(1+\gamma)}\right)\frac{1-c}{1+\gamma} - \gamma\left(\frac{1-c}{1+\gamma}\right)^2$$

$$= \frac{(1-c)^2}{1+\gamma}$$

(h) Compare social welfare in the equilibrium, from part (f), and that at the social optimum, from part (g). How is this difference affected by parameter γ? Explain.

- Social welfare in the equilibrium from part (f) unambiguously falls below that evaluated at the socially optimal level from part (g) since

$$SW^{SO} - SW^* = \frac{(1-c)^2}{1+\gamma} - \frac{(3+\gamma)(1-c)^2}{(2+\gamma)^2}$$

$$= \frac{(1-c)^2}{(1+\gamma)(2+\gamma)^2}$$

which is unambiguously positive for all values of $0 \leq c < 1$, so that the social planner chooses an output level for every firm i that yields a welfare level higher than what the firms will choose when they compete ? la Cournot.

- In addition, differentiating SW^* with respect to γ, we find

$$\frac{\partial SW^*}{\partial \gamma} = \frac{\left[(2+\gamma)^2 - 2(2+\gamma)(3+\gamma)\right](1-c)^2}{(2+\gamma)^4}$$

$$= -\frac{(4+\gamma)(1-c)^2}{(2+\gamma)^3}$$

- Similarly, differentiating SW^{SO} with respect to γ, we have

$$\frac{\partial SW^{SO}}{\partial \gamma} = -\frac{(1-c)^2}{(1+\gamma)^2}$$

- When output becomes more homogeneous (higher γ), social welfare decreases since the fall in producer surplus due to more intensive output competition more than offsets the increase in consumer surplus. Furthermore, the decrease in social welfare is faster when firms produce socially optimal output since

$$\frac{\partial SW^{SO}}{\partial \gamma} < \frac{\partial SW^*}{\partial \gamma}$$

$$-\frac{(1-c)^2}{(1+\gamma)^2} < -\frac{(4+\gamma)(1-c)^2}{(2+\gamma)^3}$$

$$\gamma^3 + 6\gamma^2 + 12\gamma + 8 > \gamma^3 + 6\gamma^2 + 9\gamma + 4$$

$$3\gamma + 4 > 0$$

which holds for all values of $\gamma \in [0, 1]$. Therefore, the difference in social welfare, $(SW^{SO} - SW^*)$, narrows down when goods become more homogeneous.

Exercise #2.13: Cournot Competition with Heterogeneous Goods^A

2.13 Consider a Cournot duopoly where every firm i faces the following inverse demand function:

$$p_i(q_i, q_j) = 1 - q_i - \gamma q_j$$

where $\gamma \in (-1, 1]$ measures cross-price effects, that is, how sensitive the demand for good i is affected by the number of units that firm $j \neq i$ offers. When $\gamma = 0$, every firm i is, essentially, unaffected by the its rival's sales. In contrast, when $\gamma = 1$, every unit that firm j produces affects firm i's sales as much as every unit that firm i produces or, more compactly, own- and cross-price effects are approximately identical. Lastly, when $\gamma \to -1$, goods are perfect complements. More generally, an increase in γ indicates that the goods of firms i and j become more homogeneous. For simplicity, assume that every firm i faces a common marginal cost c, where $1 > c \geq 0$.

(a) Find firm i's best response function, $q_i(q_j)$. How is it affected by an increase in parameter γ? Evaluate its limit at $\gamma = 0$ and at $\gamma = 1$. Interpret your results.

- Every firm i chooses its output to solve

$$\max_{q_i \geq 0} \left(1 - q_i - \gamma q_j\right) q_i - c q_i$$

Differentiating with respect to q_i and assuming an interior solution yields

$$1 - 2q_i - \gamma q_j - c = 0.$$

Solving for q_i, we find firm i's best response function, as follows:

$$q_i(q_j) = \frac{1-c}{2} - \frac{\gamma}{2} q_j$$

which originates at a vertical intercept of $\frac{1-c}{2}$. When goods are substitutes (complements), where $\gamma \in (0, 1)$ ($\gamma \in (-1, 0)$), the slope of the best response function, $\frac{\gamma}{2}$, satisfies $\frac{\gamma}{2} > 0$ ($\frac{\gamma}{2} < 0$), implying that the best response function is downward (upward) sloping. In this context, the higher the absolute value of γ, the steeper is the best response function, and the more responsive is firm i's output to firm j's output.[1]

- Evaluating the best response function, $q_i(q_j)$, at $\gamma = 0$, we obtain that $q_i = \frac{1-c}{2}$ which reconciles with the monopolist's profit-maximizing output. Whereas, evaluating at $\gamma = 1$, we find that $q_i = \frac{1-c}{2} - \frac{1}{2} q_j$ which resembles the best response function of homogeneous firms.

(b) Use your results from part (a) to find the equilibrium output, q^*.

- In a symmetric equilibrium, both firms produce the same output level, $q_i = q_j = q$. Inserting this property in the best response function of part (a), we obtain that

$$2q = 1 - c - \gamma q$$

[1] This allows for γ to be positive and high (e.g., $\gamma \to 1$), producing a best response function that decreases significantly in q_j, but also allows for γ to be negative and large in absolute value (e.g., $\gamma \to -1$), entailing a positively sloped best response function that increases substantially as a response to q_j.

Solving for q yields an equilibrium output of

$$q^* = \frac{1-c}{2+\gamma}$$

(c) How is the equilibrium output of part (b), q^*, affected by a marginal increase in parameter γ? Evaluate equilibrium output q^* at $\gamma = 0$, $\gamma = 1$, and $\gamma \to -1$. Interpret your results.

- Differentiating q^* with respect to γ yields

$$\frac{\partial q^*}{\partial \gamma} = -\frac{1-c}{(2+\gamma)^2}$$

which is unambiguously negative, meaning that output decreases with γ. Intuitively, when goods are weaker complements or stronger substitutes (higher γ), competition intensifies, thereby reducing the equilibrium output of every firm.

- In this setting, evaluating q^* at $\gamma = 0$, we obtain that

$$q^* = \frac{1-c}{2+0} = \frac{1-c}{2}$$

which coincides with the profit-maximizing output of the monopolist.

- Next, evaluating q^* at $\gamma = 1$, we find that

$$q^* = \frac{1-c}{2+1} = \frac{1-c}{3}$$

which coincides with the equilibrium output of homogeneous duopoly.

- Finally, evaluating q^* at $\gamma \to -1$, we find that

$$q^* = \frac{1-c}{2-1} = 1-c$$

which coincides with the socially optimal output that maximizes welfare.

(d) Find equilibrium profits, π^*. How are they affected by a marginal increase in parameter γ? Evaluate π^* at $\gamma = 0$, $\gamma = 1$, and $\gamma \to -1$. Interpret your results.

- We first find the equilibrium profits, π^*, by inserting the equilibrium output found in part (b), q^*, in firm i's objective function, that is,

$$\begin{aligned}\pi^* &= \left(1 - q^* - \gamma q^* - c\right) q^* \\ &= \left(1 - \frac{1-c}{2+\gamma} - \frac{\gamma(1-c)}{2+\gamma} - c\right)\frac{1-c}{2+\gamma} \\ &= \left(\frac{1-c}{2+\gamma}\right)^2\end{aligned}$$

- Next, differentiating the above equilibrium profits, π^*, with respect to γ yields

$$\frac{\partial \pi^*}{\partial \gamma} = -\frac{2(1-c)^2}{(2+\gamma)^3}$$

which is unambiguously negative, entailing that fiercer competition (higher γ) reduces equilibrium profits.

- In this setting, evaluating π^* at $\gamma = 0$, we obtain that

$$q^* = \left(\frac{1-c}{2+0}\right)^2 = \frac{(1-c)^2}{4}$$

which coincides with the equilibrium profits of the monopolist.

- Next, evaluating π^* at $\gamma = 1$, we find that

$$\pi^* = \left(\frac{1-c}{2+1}\right)^2 = \frac{(1-c)^2}{9}$$

which coincides with the equilibrium profits of homogeneous duopoly.

- Finally, evaluating π^* at $\gamma \to -1$, we find that

$$\pi^* = \left(\frac{1-c}{2-1}\right)^2 = (1-c)^2$$

which are the largest when goods are perfect complements.

Exercise #2.14: Cournot Competition with $N \geq 2$ Asymmetric Firms Allowing for Product Differentiation^B

2.14 Consider an industry of $N \geq 2$ firms competing ? la Cournot. Every firm i faces an inverse demand function

$$p_i(q_i) = a - q_i - \gamma \sum_{j \neq i} q_j,$$

where $q_i \geq 0$ denotes firm i's output. Parameter $\gamma \in [0, 1]$ captures the degree of product differentiation between firm i and its rivals, so that when $\gamma = 0$, all firms sell completely differentiated products while when $\gamma = 1$ they sell completely homogeneous goods. Every firm i has a constant marginal cost of production c_i, where $a > c_i \geq 0$.

(a) Set up firm i's profit maximization problem and find its best response function.

- Every firm i chooses output q_i to solve

$$\max_{q_i \geq 0} \left[a - q_i - \gamma \sum_{j \neq i} q_j - c_i\right] q_i$$

Differentiating with respect to q_i, and assuming interior solutions, yields

$$a - 2q_i - \gamma \sum_{j \neq i} q_j - c_i = 0$$

Solving for q_i, the best response function of firm i becomes

$$q_i(q_j) = \frac{a - c_i}{2} - \frac{\gamma}{2} \sum_{j \neq i} q_j$$

which originates at $\frac{a-c_i}{2}$ and decreases in the output of every rival firm at a rate of $\frac{\gamma}{2}$. When goods become more homogeneous (γ increases), the best response function becomes steeper, indicating that firm i's output becomes more responsive to the output of all its rivals.

(b) Find the equilibrium output, price, and associated profits of every firm i.

- Summing up the first-order functions found in part (a), we have

$$Na - 2\sum_{i=1}^{N} q_i - \gamma \sum_{i=1}^{N} \sum_{j \neq i} q_j - \sum_{i=1}^{N} c_i = 0$$

Defining $Q = \sum_{i=1}^{N} q_i$, the above expression becomes

$$Na - 2Q - \gamma (N-1) Q - \sum_{i=1}^{N} c_i = 0$$

which is rearranged to yield

$$Q = \frac{Na - \sum_{i=1}^{N} c_i}{2 + \gamma (N-1)}$$

Substituting the above result back into the best response function, we find

$$q_i = \frac{a - c_i}{2} - \frac{\gamma}{2} (Q - q_i)$$

that can be rearranged as

$$(2-\gamma) q_i = a - c_i - \gamma \overbrace{\frac{Na - \sum_{i=1}^{N} c_i}{2 + \gamma (N-1)}}^{Q}$$

Simplifying, the equilibrium output of firm i becomes

$$q_i^* = \frac{(2-\gamma) a - [2 + \gamma (N-2)] c_i + \gamma \sum_{j \neq i} c_j}{(2-\gamma) [2 + \gamma (N-1)]}$$

Substituting the equilibrium output into the inverse demand function, we obtain

$$p_i^* = a - \overbrace{\frac{(2-\gamma) a - [2 + \gamma (N-2)] c_i + \gamma \sum_{j \neq i} c_j}{(2-\gamma) [2 + \gamma (N-1)]}}^{q_i^*}$$

$$- \gamma \sum_{j \neq i} \overbrace{\frac{(2-\gamma) a - [2 + \gamma (N-2)] c_j + \gamma \sum_{k \neq j} c_k}{(2-\gamma) [2 + \gamma (N-1)]}}^{q_j^*}$$

$$= \frac{(2-\gamma) a + \left[2 + \gamma (N-2) - \gamma^2 (N-1)\right] c_i + \gamma \sum_{j \neq i} c_j}{(2-\gamma) [2 + \gamma (N-1)]}$$

Finally, substituting p_i^* and q_i^* into the profit function, equilibrium profits are

$$\pi_i^* = \left(\frac{(2-\gamma)\,a - [2+\gamma\,(N-2)]\,c_i + \gamma \sum_{j \neq i} c_j}{(2-\gamma)\,[2+\gamma\,(N-1)]} \right)^2$$

(c) *Duopoly.* Consider a setting with $N = 2$ firms (firm 1 and 2) facing marginal costs c_1 and c_2, respectively, where $a > c_1 > c_2 \geq 0$. Evaluate your results in part (b) to find the equilibrium output, price, and profits of every firm, and compare. Then evaluate your results in the case of symmetric marginal costs, where $c = c_1 = c_2 > 0$.

- *Asymmetric costs.* When $c_1 > c_2$, equilibrium results in part (b) become

$$q_i^* = \frac{(2-\gamma)\,a - 2c_i + \gamma c_j}{(2-\gamma)\,(2+\gamma)}$$

$$p_i^* = \frac{(2-\gamma)\,a + \left(2-\gamma^2\right) c_i + \gamma c_j}{(2-\gamma)\,(2+\gamma)}$$

$$\pi_i^* = \left(\frac{(2-\gamma)\,a - 2c_i + \gamma c_j}{(2-\gamma)\,(2+\gamma)} \right)^2$$

When firm 2 has a lower marginal cost than firm 1, we show that $q_2^* > q_1^*$ since

$$q_2^* = \frac{(2-\gamma)\,a - 2c_2 + \gamma c_1}{(2-\gamma)\,(2+\gamma)} > \frac{(2-\gamma)\,a - 2c_1 + \gamma c_2}{(2-\gamma)\,(2+\gamma)} = q_1^*$$

simplifies to $(2+\gamma)\,c_1 > (2+\gamma)\,c_2$, which holds because $c_1 > c_2$ by definition.

In this context, we show that firm 1 charges a higher price than firm 2 since

$$p_1^* = \frac{(2-\gamma)\,a + \left(2-\gamma^2\right) c_1 + \gamma c_2}{(2-\gamma)\,(2+\gamma)} > \frac{(2-\gamma)\,a + \left(2-\gamma^2\right) c_2 + \gamma c_1}{(2-\gamma)\,(2+\gamma)} = p_2^*$$

simplifies to $\left(2-\gamma-\gamma^2\right) c_1 > \left(2-\gamma-\gamma^2\right) c_2$, which holds given that $\gamma \leq 1$ and $c_1 > c_2$ by assumption.

Finally, since equilibrium profits are square of output, it follows that $\pi_2^* > \pi_1^*$.

- *Symmetric costs.* In this setting where $c = c_1 = c_2$, the above results simplify to

$$q_i^* = \frac{a-c}{2+\gamma}$$

$$p_i^* = \frac{a + (1+\gamma)\,c}{2+\gamma}$$

$$\pi_i^* = \left(\frac{a-c}{2+\gamma} \right)^2$$

Exercise #2.15: Cournot Competition with $N \geq 2$ Asymmetric Firms Allowing for Product Differentiation: Merger^B

2.15 Continue with the setting in Exercise 2.14, but consider that all firms are merged into a single entity.

(a) Set up the merging entity's profit maximization problem and derive the first-order condition.

- The merging entity chooses output q_i for every merged firm i to solve

$$\max_{\{q_i\}_{i=1}^N} \sum_{i=1}^{N} \left[a - q_i - \gamma \sum_{j \neq i} q_j - c_i \right] q_i$$

Differentiating with respect to q_i, and assuming interior solutions, we obtain

$$a - 2q_i - 2\gamma \sum_{j \neq i} q_j - c_i = 0$$

(b) Find the equilibrium output, price, and associated profits of every firm i.

- Summing up the first-order conditions found in part (a), we have

$$Na - 2\sum_{i=1}^{N} q_i - 2\gamma \sum_{i=1}^{N} \sum_{j \neq i} q_j - \sum_{i=1}^{N} c_i = 0$$

Defining $Q = \sum_{i=1}^{N} q_i$, the above expression becomes

$$Na - 2Q - 2\gamma (N-1) Q - \sum_{i=1}^{N} c_i = 0$$

which is rearranged to yield

$$Q = \frac{Na - \sum_{i=1}^{N} c_i}{2[1 + \gamma (N-1)]}$$

Substituting the above result back into the first-order condition, we find

$$a - 2q_i - 2\gamma (Q - q_i) - c_i = 0$$

that can be rearranged as

$$a - c_i - 2\gamma \overbrace{\frac{Na - \sum_{i=1}^{N} c_i}{2[1 + \gamma (N-1)]}}^{Q} = 2(1-\gamma) q_i$$

Simplifying, the equilibrium output of every firm i becomes

$$q_i^M = \frac{(1-\gamma)\,a - [1+\gamma\,(N-2)]\,c_i + \gamma \sum_{j\neq i} c_j}{2\,(1-\gamma)\,[1+\gamma\,(N-1)]}$$

where the superscript M stands for merger.

(c) Find the equilibrium price and profit of every firm i.

- Substituting the equilibrium output into the inverse demand function, we obtain

$$\begin{aligned} p_i^M &= a - \overbrace{\frac{(1-\gamma)\,a - [1+\gamma\,(N-2)]\,c_i + \gamma \sum_{j\neq i} c_j}{2\,(1-\gamma)\,[1+\gamma\,(N-1)]}}^{q_i^M} \\ &\quad - \gamma \sum_{j\neq i} \overbrace{\frac{(1-\gamma)\,a - [1+\gamma\,(N-2)]\,c_j + \gamma \sum_{k\neq j} c_k}{2\,(1-\gamma)\,[1+\gamma\,(N-1)]}}^{q_j^M} \\ &= \frac{(1-\gamma)\,[1+\gamma\,(N-1)]\,a + \left[1+\gamma\,(N-2) - \gamma^2\,(N-1)\right] c_i}{2\,(1-\gamma)\,[1+\gamma\,(N-1)]} \end{aligned}$$

so that every merged firm i earns a profit margin of

$$\begin{aligned} p_i^M - c_i &= \frac{(1-\gamma)\,[1+\gamma\,(N-1)]\,a + \left[1+\gamma\,(N-2) - \gamma^2\,(N-1)\right] c_i}{2\,(1-\gamma)\,[1+\gamma\,(N-1)]} - c_i \\ &= \frac{a-c_i}{2} \end{aligned}$$

which coincides with that obtained from the multiproduct monopolist.

- Finally, inserting p_i^M and q_i^M into the profit function, equilibrium profits of every firm i become

$$\begin{aligned} \pi_i^M &= \left(p_i^M - c_i\right) q_i^M \\ &= \frac{(a-c_i)\left[(1-\gamma)\,a - (1+\gamma\,(N-2))\,c_i + \gamma \sum_{j\neq i} c_j\right]}{4\,(1-\gamma)\,[1+\gamma\,(N-1)]} \end{aligned}$$

Exercise #2.16: Cournot Competition with $N \geq 2$ Asymmetric Firms Allowing for Product Differentiation: Merger Incentives^B

2.16 Let us compare the results in Exercises 2.14 and 2.15 to understand firms' incentives to merge in this type of industry.

(a) Find the condition on c_i that firm i has incentives to join the merger.

- Firm i has incentives to join the merger if and only if $\pi_i^M \geq \pi_i^*$, where

$$\frac{(a-c_i)\left[(1-\gamma)a-(1+\gamma(N-2))c_i+\gamma\sum_{j\neq i}c_j\right]}{4(1-\gamma)[1+\gamma(N-1)]} \geq \left(\frac{(2-\gamma)a-[2+\gamma(N-2)]c_i+\gamma\sum_{j\neq i}c_j}{(2-\gamma)[2+\gamma(N-1)]}\right)^2$$

simplifies to

$$\left[4(1+\gamma(N-1))+(2-\gamma)^2(N-1)^2\right](a-c_i)^2 - \gamma\left[4(1+\gamma(N-1))+(2-\gamma)^2(N-1)^2\right](a-c_i)\left(a+(N-2)c_i-\sum_{j\neq i}c_j\right) - 4(1-\gamma)(1+\gamma(N-1))\left(a+(N-2)c_i-\sum_{j\neq i}c_j\right)^2 \geq 0$$

Since $a - c_i > 0$, or else firm i does not produce any units even if it is a monopoly, we can rearrange the above expression that c_i solves the following quadratic inequality:

$$4(1-\gamma)(1+\gamma(N-1))x^2+\gamma\left[4(1+\gamma(N-1))+(2-\gamma)^2(N-1)^2\right]x - \left[4(1+\gamma(N-1))+(2-\gamma)^2(N-1)^2\right] \leq 0$$

where $x \equiv \frac{a+(N-2)c_i-\sum_{j\neq i}c_j}{a-c_i}$ that can be rearranged as $c_i = \frac{(x-1)a+\sum_{j\neq i}c_j}{N-2+x}$.

(b) *Homogeneous products.* Evaluate your results in part (a) in the case that products are homogeneous, $\gamma = 1$.

- Substituting $\gamma = 1$ into the inequality found in part (a), we obtain

$$\left[4N+(N-1)^2\right](a-c_i)^2-\left[4N+(N-1)^2\right](a-c_i)\left(a+(N-2)c_i-\sum_{j\neq i}c_j\right) \geq 0$$

which simplifies into

$$a-c_i-a-(N-2)c_i+\sum_{j\neq i}c_j \geq 0$$

and is rearranged to yield

$$c_i \leq \frac{1}{N-1}\sum_{j\neq i}c_j$$

Intuitively, firm i must have a below-average cost to profit from the merger.

(c) *Duopoly*. Evaluate your results in part (a) in the case that only two firms compete, $N = 2$. Then, evaluate your results assuming that $a = 1$ and $\gamma = \frac{1}{2}$, and interpret.

- Substituting $N = 2$ into the inequality that we found in part (a), we find

$$\left(8+\gamma^2\right)(a-c_i)^2 - \gamma\left(8+\gamma^2\right)(a-c_i)\left(a-c_j\right) - 4\left(1-\gamma^2\right)\left(a-c_j\right)^2 \geq 0$$

Solving for c_i, and dismissing the negative root, we obtain

$$\frac{a-c_i}{a-c_j} \geq \frac{\gamma\left(8+\gamma^2\right)+\left(4-\gamma^2\right)\sqrt{8+\gamma^2}}{2\left(8+\gamma^2\right)}$$

Defining $y \equiv \frac{\gamma(8+\gamma^2)+(4-\gamma^2)\sqrt{8+\gamma^2}}{2(8+\gamma^2)}$ yields

$$c_i \leq \overline{c}_i \equiv (1-y)\,a + yc_j$$

where a higher market size, a, or a more costly rival, c_j, induces firm i to merge even when it is less efficient than its rival, as long as this inefficiency is not extreme, that is, $c_i \leq \overline{c}_i$.

- Substituting $\gamma = \frac{1}{2}$ into the expression of y, we have that

$$y = \frac{\frac{1}{2}\left(8+\frac{1}{4}\right)+\left(4-\frac{1}{4}\right)\sqrt{8+\frac{1}{4}}}{2\left(8+\frac{1}{4}\right)} = \frac{33+15\sqrt{33}}{132} \simeq 0.903$$

Finally, substituting $a = 1$ into the cutoff $\overline{c}_i$, we obtain that

$$c_i \leq 0.097 + 0903c_j$$

so that when $c_j = 0$, firm i cannot be too inefficient to find the merger profitable, i.e., $c_i \leq 0.097$. However, when firm j is inefficient, say, $c_j = 1$, then firm i has incentives to acquire its relatively inefficient rival under all parameter conditions, i.e., $0 \leq c_i \leq 1$.

Exercise #2.17: Social Welfare with Cost-Asymmetric Firms Selling Heterogeneous Goods and Competing in Quantities^A

2.17 Consider the setting in Exercise 2.12, but assume now that firm i's marginal cost is c_i, where $0 \leq c_i < 1$.

(a) Find the best response function of firm i, $q_i\left(q_j\right)$.

- Every firm i chooses q_i to solve the profit maximization problem, as follows:

$$\max_{q_i \geq 0} \pi\left(q_i\right) = \left(1 - q_i - \gamma q_j\right) q_i - c_i q_i$$

Taking the first-order condition with respect to q_i, and assuming interior solutions,

$$1 - 2q_i - \gamma q_j - c_i = 0$$

such that the best response function of firm i, in response to firm j's output, becomes

$$q_i\left(q_j\right) = \frac{1-c_i}{2} - \frac{\gamma}{2}q_j$$

meaning that the best response function originates at a vertical intercept of $\frac{1-c_i}{2}$ and decreases in q_j at a rate $\frac{\gamma}{2}$. Graphically, an increase in parameter γ (goods are more homogeneous) produces an inward rotation of the best response function, leaving its vertical intercept unchanged at $\frac{1-c_i}{2}$ but making this function steeper. Intuitively, that means that, as firms sell more homogeneous products, a given increase in firm j's output yields a larger reduction in firm i's output level.

- *Extreme cases.*
  - When goods are completely differentiated ($\gamma = 0$), this best response function simplifies to

$$q_i = \frac{1-c_i}{2},$$

thus coinciding with the monopoly output under linear demand.

  - In contrast, when goods are homogeneous ($\gamma = 1$), this best response function becomes

$$q_i\left(q_j\right) = \frac{1-c_i}{2} - \frac{1}{2}q_j,$$

as in standard Cournot models without product differentiation.

(b) Find the equilibrium output of every firm i.

- Inserting firm j's best response function, $q_j(q_i)$, into firm i's, $q_i\left(q_j\right)$, we obtain

$$q_i = \frac{1-c_i}{2} - \frac{\gamma}{2}\underbrace{\left(\frac{1-c_j}{2} - \frac{\gamma}{2}q_i\right)}_{q_j(q_i)}$$

which simplifies to

$$q_i = \frac{2\left(1-c_i\right) - \gamma\left(1-c_j\right)}{4} + \frac{\gamma^2}{4}q_i$$

Solving for q_i, we find that firm i's equilibrium output is

$$q_i^* = \frac{2\left(1-c_i\right) - \gamma\left(1-c_j\right)}{4-\gamma^2}$$

Firm j's equilibrium output is, then,

$$q_j^* = \frac{1-c_j}{2} - \frac{\gamma}{2}\underbrace{\left(\frac{2\left(1-c_i\right) - \gamma\left(1-c_j\right)}{4-\gamma^2}\right)}_{q_i^*}$$

$$= \frac{2\left(1-c_j\right) - \gamma\left(1-c_i\right)}{4-\gamma^2}.$$

(c) For which parameter values does the equilibrium output of part (b) entails that (i) only firm i produces output, (ii) both firms produce output, and (iii) only firm j produces output? Interpret your results.

- Firm i produces positive units, $q_i^* \geq 0$, if and only if

$$\frac{2(1-c_i) - \gamma(1-c_j)}{4-\gamma^2} \geq 0$$

or

$$2(1-c_i) \geq \gamma(1-c_j)$$

which we can rearrange as

$$\frac{1-c_i}{1-c_j} \geq \frac{\gamma}{2}$$

Defining, for compactness, $m_i \equiv \frac{1-c_i}{1-c_j}$, which measures the efficiency of firm i relative to firm j. To understand this efficiency measure, note that:
  - When $c_i < c_j$, firm i is relatively more efficient than firm j, so $1-c_i > 1-c_j$ entails that $m_i > 1$.
  - The opposite happens when firm i is, relatively speaking, less efficient than firm j, where $c_i > c_j$ entails $1-c_i < 1-c_j$, or $m_i < 1$.
  We can, then, say that firm i produces positive units if and only if

$$m_i \geq \frac{\gamma}{2}$$

which means that firm i is active when its relative efficiency, as captured by $m_i \equiv \frac{1-c_i}{1-c_j}$, is not too low.

- Otherwise, suppose firm i is relatively inefficient, where $m_i < \frac{\gamma}{2}$, firm j is the only active firm in the industry, solving

$$\max_{q_j \geq 0} \pi(q_j) = (1-q_j)q_j - c_j q_j$$

  Taking the first-order condition with respect to q_j, and assuming interior solutions,

$$1 - 2q_j - c_j = 0$$

  Solving for the output of firm j, we obtain

$$q_j^* = \frac{1-c_j}{2}$$

- Summarizing the above results, we obtain equilibrium output pair, as follows:

$$\left(q_i^*, q_j^*\right) = \begin{cases} 0, \frac{1-c_j}{2} & \text{if } m_i < \frac{\gamma}{2} \\ \frac{2(1-c_i)-\gamma(1-c_j)}{4-\gamma^2}, \frac{2(1-c_j)-\gamma(1-c_i)}{4-\gamma^2} & \text{if } \frac{\gamma}{2} \leq m_i < \frac{2}{\gamma} \\ \frac{1-c_i}{2}, 0 & \text{if } m_i \geq \frac{2}{\gamma} \end{cases}$$

(d) Find the equilibrium profits of every firm i.

- When both firms are active, every firm i earns equilibrium profits of

$$\pi_i^* = \left(1 - \underbrace{\frac{2(1-c_i) - \gamma(1-c_j)}{4-\gamma^2}}_{q_i^*} - \gamma\underbrace{\frac{2(1-c_j) - \gamma(1-c_i)}{4-\gamma^2}}_{q_j^*}\right)$$

$$\times \underbrace{\frac{2(1-c_i) - \gamma(1-c_j)}{4-\gamma^2}}_{q_i^*} - c_i\underbrace{\frac{2(1-c_i) - \gamma(1-c_j)}{4-\gamma^2}}_{q_i^*}$$

$$= \frac{(4-\gamma^2-2+\gamma^2)(1-c_i) - (2\gamma-\gamma)(1-c_j)}{4-\gamma^2} \times \frac{2(1-c_i)-\gamma(1-c_j)}{4-\gamma^2}$$

$$= \left(\frac{2(1-c_i)-\gamma(1-c_j)}{4-\gamma^2}\right)^2$$

- Whereas, when only firm i is active, this firm obtain equilibrium profits of

$$\pi_i^* = \left(1 - \frac{1-c_i}{2} - \gamma 0\right)\frac{1-c_i}{2} - c_i\frac{1-c_i}{2}$$

$$= \frac{(1-c_i)^2}{4}.$$

- Summarizing, equilibrium profits are

$$\left(\pi_i^*, \pi_j^*\right) = \begin{cases} 0, \frac{(1-c_j)^2}{4} & \text{if } m_i < \frac{\gamma}{2} \\ \left(\frac{2(1-c_i)-\gamma(1-c_j)}{4-\gamma^2}\right)^2, \left(\frac{2(1-c_j)-\gamma(1-c_i)}{4-\gamma^2}\right)^2 & \text{if } \frac{\gamma}{2} \le m_i < \frac{2}{\gamma} \\ \frac{(1-c_i)^2}{4}, 0 & \text{if } m_i \ge \frac{2}{\gamma} \end{cases}$$

(e) If a social planner could determine individual output level for each firm, find the socially optimal levels that he would choose.

- According to part (d), Exercise 2.12, social welfare is the sum of consumer and producer surplus, as follows:

$$SW(q_i, q_j) = \underbrace{\frac{q_i^2 + 2\gamma q_i q_j + q_j^2}{2}}_{CS} + \underbrace{\left[p_i(q_i, q_j) - c_i\right]q_i + \left[p_j(q_i, q_j) - c_j\right]q_j}_{PS}$$

$$= \left(1 - \frac{1}{2}q_i - c_i\right)q_i + \left(1 - \frac{1}{2}q_j - c_j\right)q_j - \gamma q_i q_j$$

Differentiating with respect to q_i, and assuming interior solutions, we find that

$$1 - c_i - q_i - \gamma q_j = 0$$

Differentiating with respect to q_j, and assuming interior solutions, we find that

$$1 - c_j - q_j - \gamma q_i = 0$$

which is rearranged to yield

$$q_j = 1 - c_j - \gamma q_i$$

Substituting this into the first order condition for q_i, we obtain

$$q_i = 1 - c_i - \gamma \underbrace{\left(1 - c_j - \gamma q_i\right)}_{q_j}$$

Rearranging and solving for q_i, socially optimal output becomes

$$q_i^{SO} = \frac{(1 - c_i) - \gamma\left(1 - c_j\right)}{1 - \gamma^2}$$

which is positive if and only if

$$1 - c_i \geq \gamma\left(1 - c_j\right)$$

or $\frac{1-c_i}{1-c_j} \geq \gamma$, which we compactly express as $m_i \geq \gamma$ since $m_i \equiv \frac{1-c_i}{1-c_j}$.

- Otherwise, when $m_i < \gamma$, the social planner solves

$$SW\left(0, q_j\right) = \left(1 - \frac{1}{2}q_j - c_j\right)q_j$$

Differentiating with respect to q_j, and assuming interior solutions, we find that

$$1 - q_j - c_j = 0$$

Rearranging and solving for q_j, socially optimal output becomes

$$q_j^{SO} = 1 - c_j,$$

- Summarizing, socially optimal output levels are

$$\left(q_i^{SO}, q_j^{SO}\right) = \begin{cases} 0, 1 - c_j & \text{if } m_i < \gamma \\ \frac{(1-c_i)-\gamma(1-c_j)}{1-\gamma^2}, \frac{(1-c_j)-\gamma(1-c_i)}{1-\gamma^2} & \text{if } \gamma \leq m_i < \frac{1}{\gamma} \\ 1 - c_i, 0 & \text{if } m_i \geq \frac{1}{\gamma} \end{cases}$$

(f) Compare the competitive output level in part (c) to the socially optimal level in part (e).

- When $m_i < \frac{\gamma}{2}$,

$$q_i^* = q_i^{SO} = 0,$$

so the relatively inefficient firm i does not produce any output whether it competes with firm j or under the social planner's directive.

- When $\frac{\gamma}{2} \leq m_i < \gamma$, output satisfies

$$q_i^{SO} = 0 < \frac{2(1-c_i) - \gamma(1-c_j)}{4-\gamma^2} = q_i^*,$$

so while firm i produces output under Cournot competition, the social planner shuts down this firm.

- When $\gamma \leq m_i < \frac{1}{\gamma}$, the output condition

$$q_i^{SO} = \frac{(1-c_i) - \gamma(1-c_j)}{1-\gamma^2} > \frac{2(1-c_i) - \gamma(1-c_j)}{4-\gamma^2} = q_i^*$$

reduces to $(2+\gamma^2)(1-c_i) - 3\gamma(1-c_j) > 0$, or more compactly,

$$m_i > \widehat{m}_i \equiv \frac{3\gamma}{2+\gamma^2}.$$

It is straightforward to verify that $\gamma < \widehat{m}_i < \frac{1}{\gamma}$, indicating that when firm i's relative efficiency parameter m_i satisfies $\gamma \leq m_i \leq \widehat{m}_i$ ($\widehat{m}_i < m_i < \frac{1}{\gamma}$), the social planner reduces (increases) this firm's output relative to the competitive level.

- When $\frac{1}{\gamma} \leq m_i < \frac{2}{\gamma}$, output ranking satisfies

$$q_i^{SO} = 1 - c_i > \frac{2(1-c_i) - \gamma(1-c_j)}{4-\gamma^2} = q_i^*$$

since this inequality reduces to $(2-\gamma^2)(1-c_i) + \gamma(1-c_j) > 0$ which holds for all values of $\gamma \in [0, 1]$, so that the social planner allows firm i to dominate the market and produce socially optimal output that is above the competitive level.

- When $m_i \geq \frac{2}{\gamma}$, output satisfies

$$q_i^{SO} = 1 - c_i > \frac{1-c_i}{2} = q_i^*.$$

Intuitively, firm i dominates the market whether under the social planner's directive or not, but the social planner seeks more output from this firm until its marginal cost coincides with marginal benefit.

(g) Evaluate social welfare at the competitive output levels found in part (c).

- We analyze welfare in three cases: (i) $m_i < \frac{\gamma}{2}$, (ii) $\frac{\gamma}{2} \leq m_i < \frac{2}{\gamma}$, and (iii) $m_i \geq \frac{2}{\gamma}$.
  (i) When $m_i < \frac{\gamma}{2}$, substituting equilibrium output levels $q_i^* = 0$ and $q_j^* = \frac{1-c_j}{2}$ into the social welfare function in part (e), we find

$$SW^* = \left(1 - \frac{1}{2} \times \underbrace{0}_{q_i^*} - c_i\right) \times \underbrace{0}_{q_i^*} + \left(1 - \frac{1}{2}\underbrace{\frac{1-c_j}{2}}_{q_j^*} - c_j\right)\underbrace{\frac{1-c_j}{2}}_{q_j^*} - \gamma$$

$$\times \underbrace{0}_{q_i^*} \underbrace{\frac{1-c_j}{2}}_{q_j^*}$$

$$= \frac{(4-1)\left(1-c_j\right)}{4} \frac{1-c_j}{2}$$

$$= \frac{3\left(1-c_j\right)^2}{8}$$

(ii) When $\frac{\gamma}{2} \leq m_i < \frac{2}{\gamma}$, inserting output levels $q_i^* = \frac{2(1-c_i)-\gamma(1-c_j)}{4-\gamma^2}$ and $q_j^* = \frac{2(1-c_j)-\gamma(1-c_i)}{4-\gamma^2}$ into the social welfare function in part (e), we have

$$SW^* = \left(1 - \frac{1}{2} \underbrace{\frac{2(1-c_i) - \gamma\left(1-c_j\right)}{4-\gamma^2}}_{q_i^*} - c_i\right) \underbrace{\frac{2(1-c_i) - \gamma\left(1-c_j\right)}{4-\gamma^2}}_{q_i^*}$$

$$+ \left(1 - \frac{1}{2} \underbrace{\frac{2\left(1-c_j\right) - \gamma(1-c_i)}{4-\gamma^2}}_{q_j^*} - c_j\right) \underbrace{\frac{2\left(1-c_j\right) - \gamma(1-c_i)}{4-\gamma^2}}_{q_j^*}$$

$$- \gamma \underbrace{\frac{2(1-c_i) - \gamma\left(1-c_j\right)}{4-\gamma^2}}_{q_i^*} \underbrace{\frac{2\left(1-c_j\right) - \gamma(1-c_i)}{4-\gamma^2}}_{q_j^*}$$

$$= \frac{4\left(3-\gamma^2\right)(1-c_i)^2 - 2\gamma\left(2-\gamma^2\right)(1-c_i)\left(1-c_j\right) - \gamma^2\left(1-c_j\right)^2}{2\left(4-\gamma^2\right)^2}$$

$$+ \frac{4\left(3-2\gamma^2\right)\left(1-c_j\right)^2 - 2\gamma\left(2-\gamma^2\right)(1-c_i)\left(1-c_j\right) - \gamma^2(1-c_i)^2}{2\left(4-\gamma^2\right)^2}$$

$$+ \frac{2\gamma\left[2\gamma(1-c_i)^2 - \left(4+\gamma^2\right)(1-c_i)\left(1-c_j\right) + 2\gamma\left(1-c_j\right)^2\right]}{2\left(4-\gamma^2\right)^2}$$

$$= \frac{\left(12-\gamma^2\right)(1-c_i)^2 - 2\gamma\left(8-\gamma^2\right)(1-c_i)\left(1-c_j\right) + \left(12-\gamma^2\right)\left(1-c_j\right)^2}{2\left(4-\gamma^2\right)^2}$$

(iii) When $m_i \geq \frac{2}{\gamma}$, substituting $q_i^* = \frac{1-c_i}{2}$ and $q_j^* = 0$ into the social welfare function in part (e), we obtain a result reciprocal to that when $m_i < \frac{\gamma}{2}$, where

$$SW^* = \frac{3(1-c_i)^2}{8}$$

- Summarizing, social welfare evaluated at the competitive output levels is

$$SW^* = \begin{cases} \frac{3(1-c_j)^2}{8} & \text{if } m_i < \frac{\gamma}{2} \\ \frac{(12-\gamma^2)(1-c_i)^2 - 2\gamma(8-\gamma^2)(1-c_i)(1-c_j) + (12-\gamma^2)(1-c_j)^2}{2(4-\gamma^2)^2} & \text{if } \frac{\gamma}{2} \le m_i < \frac{2}{\gamma} \\ \frac{3(1-c_i)^2}{8} & \text{if } m_i \ge \frac{2}{\gamma} \end{cases}$$

(h) Evaluate social welfare at the socially optimal output levels found in part (e).

- We analyze welfare in three cases: (i) $m_i < \gamma$, (ii) $\gamma \le m_i < \frac{1}{\gamma}$, and (iii) $m_i \ge \frac{1}{\gamma}$.
  (i) When $m_i < \gamma$, substituting socially optimal output levels $q_i^{SO} = 0$ and $q_j^{SO} = 1 - c_j$ into the social welfare function in part (e), we find

$$SW^{SO} = \left(1 - \frac{1}{2} \times \underbrace{0}_{q_i^{SO}} - c_i\right) \underbrace{0}_{q_i^{SO}} + \left(1 - \frac{1}{2} \underbrace{(1-c_j)}_{q_j^{SO}} - c_j\right) \underbrace{(1-c_j)}_{q_j^{SO}}$$
$$- \gamma \times \underbrace{0}_{q_i^{SO}} \underbrace{(1-c_j)}_{q_j^{SO}}$$
$$= \frac{(2-1)(1-c_j)}{2}(1-c_j)$$
$$= \frac{(1-c_j)^2}{2}$$

  (ii) When $\gamma \le m_i < \frac{1}{\gamma}$, inserting the socially optimal output levels $q_i^{SO} = \frac{(1-c_i)-\gamma(1-c_j)}{1-\gamma^2}$ and $q_j^{SO} = \frac{(1-c_j)-\gamma(1-c_i)}{1-\gamma^2}$ into the social welfare function in part (e), we have

$$SW^{SO} = \left(1 - \frac{1}{2} \underbrace{\frac{(1-c_i)-\gamma(1-c_j)}{1-\gamma^2}}_{q_i^{SO}} - c_i\right) \underbrace{\frac{(1-c_i)-\gamma(1-c_j)}{1-\gamma^2}}_{q_i^{SO}}$$
$$+ \left(1 - \frac{1}{2} \underbrace{\frac{(1-c_j)-\gamma(1-c_i)}{1-\gamma^2}}_{q_j^{SO}} - c_j\right) \underbrace{\frac{(1-c_j)-\gamma(1-c_i)}{1-\gamma^2}}_{q_j^{SO}}$$
$$- \gamma \underbrace{\frac{(1-c_i)-\gamma(1-c_j)}{1-\gamma^2}}_{q_i^{SO}} \underbrace{\frac{(1-c_j)-\gamma(1-c_i)}{1-\gamma^2}}_{q_j^{SO}}$$

$$= \frac{(1-2\gamma^2)(1-c_i)^2 + 2\gamma^3(1-c_i)(1-c_j) - \gamma^2(1-c_j)^2}{2(1-\gamma^2)^2}$$

$$+ \frac{(1-2\gamma^2)(1-c_j)^2 + 2\gamma^3(1-c_i)(1-c_j) - \gamma^2(1-c_i)^2}{2(1-\gamma^2)^2}$$

$$+ \frac{2\gamma\left[\gamma(1-c_i)^2 - (1+\gamma^2)(1-c_i)(1-c_j) + \gamma(1-c_j)^2\right]}{2(1-\gamma^2)^2}$$

$$= \frac{(1-c_i)^2 - 2\gamma(1-c_i)(1-c_j) + (1-c_j)^2}{2(1-\gamma^2)}$$

(iii) When $m_i \geq \frac{1}{\gamma}$, substituting $q_i^{SO} = 1 - c_i$ and $q_j^{SO} = 0$ into the social welfare function in part (e), we obtain a result reciprocal to that when $m_i < \gamma$, where

$$SW^{SO} = \frac{(1-c_i)^2}{2}$$

- Summarizing, social welfare evaluated at the socially optimal output levels is

$$SW^{SO} = \begin{cases} \frac{(1-c_j)^2}{2} & \text{if } m_i < \gamma \\ \frac{(1-c_i)^2 - 2\gamma(1-c_i)(1-c_j) + (1-c_j)^2}{2(1-\gamma^2)} & \text{if } \gamma \leq m_i < \frac{1}{\gamma} \\ \frac{(1-c_i)^2}{2} & \text{if } m_i \geq \frac{1}{\gamma} \end{cases}$$

(i) Compare social welfare in the competitive equilibrium, from part (g), and that evaluated at the social optimum, from part (h). Interpret your results.

- When $m_i < \frac{\gamma}{2}$, social welfare satisfies

$$SW^{SO} = \frac{(1-c_j)^2}{2} > \frac{3(1-c_j)^2}{8} = SW^*$$

which reduces to $\frac{1}{2} > \frac{3}{8}$ that is satisfied for all values of $0 \leq c_j < 1$.

- When $\frac{\gamma}{2} \leq m_i < \gamma$, which is equivalent to $\frac{1}{\gamma} < m_j \leq \frac{2}{\gamma}$, we have that

$$SW^{SO} = \frac{(1-c_j)^2}{2}$$

$$> \frac{(12-\gamma^2)(1-c_i)^2 - 2\gamma(8-\gamma^2)(1-c_i)(1-c_j) + (12-\gamma^2)(1-c_j)^2}{2(4-\gamma^2)^2}$$

$$= SW^*$$

which reduces to

$$f(\gamma, m_j) \equiv \left(4 - 7\gamma^2 + \gamma^4\right)m_j^2 + 2\gamma\left(8-\gamma^2\right)m_j - \left(12-\gamma^2\right) > 0$$

Here, we want to find the minimum value of $f(\gamma, m_j)$ to show that this function is positive for all parameter values. First, differentiating with respect to m_j yields

$$f_2(\gamma, m_j) = 2\left[\left(4 - 7\gamma^2 + \gamma^4\right) m_j + \gamma\left(8 - \gamma^2\right)\right]$$

Depending on the value of γ, the function $f(\gamma, m_j)$ can be increasing or decreasing in m_j. In this context, we separately identify two cases: (i) $0 \le \gamma < \sqrt{\frac{7-\sqrt{33}}{2}}$, where $4 - 7\gamma^2 + \gamma^4 > 0$, and (ii) $\sqrt{\frac{7-\sqrt{33}}{2}} < \gamma \le 1$, where $4 - 7\gamma^2 + \gamma^4 < 0$.

- In case (i) where $0 \le \gamma < \sqrt{\frac{7-\sqrt{33}}{2}}$, we have that $f_2(\gamma, m_j) > 0$ when $(4 - 7\gamma^2 + \gamma^4) m_j + \gamma(8 - \gamma^2) > 0$, which is rearranged to yield $m_j > -\frac{\gamma(8-\gamma^2)}{4-7\gamma^2+\gamma^4}$. This condition is satisfied since the left-hand side of this inequality, m_j, must be positive, while the fraction on the right-hand side, $-\frac{\gamma(8-\gamma^2)}{4-7\gamma^2+\gamma^4}$, is negative.
- In case (ii) where $\sqrt{\frac{7-\sqrt{33}}{2}} < \gamma \le 1$, we find that $f_2(\gamma, m_j) > 0$ when $(4 - 7\gamma^2 + \gamma^4) m_j + \gamma(8 - \gamma^2) > 0$, which is rearranged to yield $m_j < -\frac{\gamma(8-\gamma^2)}{4-7\gamma^2+\gamma^4}$. This condition holds since the maximum value of m_j, $\frac{2}{\gamma}$, satisfies $\frac{2}{\gamma} < -\frac{\gamma(8-\gamma^2)}{4-7\gamma^2+\gamma^4}$, which is equivalent to $8 - 6\gamma^2 + \gamma^4 > 0$ that holds for all values of γ.

Combining the above two cases, since $f_2(\gamma, m_j) > 0$ holds for all values of $0 \le \gamma \le 1$, we can plug in the minimum value of m_j, $\frac{1}{\gamma}$, into $f(\gamma, m_j)$ to obtain

$$f\left(\gamma, \frac{1}{\gamma}\right) = \frac{4 - 7\gamma^2 + \gamma^4}{\gamma^2} + \frac{2\gamma(8 - \gamma^2)}{\gamma} - \left(12 - \gamma^2\right) = \frac{4 - 3\gamma^2}{\gamma^2}$$

which is positive, so that $SW^{SO} > SW^*$ holds for all $\frac{\gamma}{2} \le m_i < \gamma$ and $0 \le \gamma \le 1$.

- When $\gamma \le m_i < \frac{1}{\gamma}$, the condition

$$\begin{aligned} SW^{SO} &= \frac{(1-c_i)^2 - 2\gamma(1-c_i)(1-c_j) + (1-c_j)^2}{2(1-\gamma^2)} \\ &> \frac{(12-\gamma^2)(1-c_i)^2 - 2\gamma(8-\gamma^2)(1-c_i)(1-c_j) + (12-\gamma^2)(1-c_j)^2}{2(4-\gamma^2)^2} \\ &= SW^* \end{aligned}$$

reduces to

$$\left(4 + 5\gamma^2\right) m_i^2 - 2\gamma\left(8 + \gamma^2\right) m_i + \left(4 + 5\gamma^2\right) > 0.$$

Since the discriminant of this function is

$$\begin{aligned} \Delta &= 4\gamma^2\left(8 + \gamma^2\right)^2 - 4\left(4 + 5\gamma^2\right)^2 \\ &= -4\left(1 - \gamma^2\right)(2-\gamma)^2(2+\gamma)^2 < 0, \end{aligned}$$

and, in addition, $4 + 5\gamma^2 > 0$, we obtain that $SW^{SO} > SW^*$ for all $\gamma \leq m_i < \frac{1}{\gamma}$ and $0 \leq \gamma \leq 1$.

- Similarly, since $\frac{1}{\gamma} \leq m_i < \frac{2}{\gamma}$ $\left(m_i \geq \frac{2}{\gamma}\right)$ is reciprocal to the case of $\frac{\gamma}{2} \leq m_i < \gamma$ $\left(m_i < \frac{\gamma}{2}\right)$, the condition $SW^{SO} > SW^*$ also holds for the respective range of γ.
- In summary, $SW^{SO} > SW^*$ holds for all values of $\gamma \in [0, 1]$ and $0 \leq c_i, c_j < 1$. Intuitively, the social planner maximizes both consumer and producer surplus, unlike firms which focus on producer surplus alone, so that welfare at the socially optimal output levels is unambiguously above that at the competitive output levels.

Exercise #2.18: Entry that Reduces Aggregate Output^A

2.18 Consider an industry with $N \geq 2$ firms competing à la Cournot, facing inverse demand function $p(Q) = a - bQ$, and a symmetric cost function $C(q_i) = cq_i - \frac{d}{2}q_i^2$, where $a > c$ and $2b > d > 0$. As a remark, note that marginal costs in this setting are $C'(q_i) = c - dq_i$, which is decreasing in firm i's output, thus indicating the presence of economies of scale.

(a) Find every firm i's equilibrium output, q_i^*, and profits, π_i^*.

- Every firm i chooses q_i to solve the following profit maximization problem,

$$\max_{q_i \geq 0} \; [a - b(q_i + Q_{-i})]q_i - \left(cq_i - \frac{d}{2}q_i^2\right),$$

where $Q_{-i} = \sum_{j \neq i} q_j$ denotes the aggregate output of firm i's rivals.
Differentiating with respect to q_i, and assuming interior solution, yields

$$a - 2bq_i - bQ_{-i} = c - dq_i.$$

Solving for q_i, we find firm i's best response function

$$q_i(Q_{-i}) = \frac{a-c}{2b-d} - \frac{b}{2b-d}Q_{-i}.$$

In a symmetric equilibrium, every firm produces the same output level $q_i^* = q_j^*$ for every firm $j \neq i$, so that $Q_{-i} = (N-1)q_i^*$. Inserting this property into the above best response function, we obtain

$$q_i^* = \frac{a-c}{2b-d} - \frac{b}{2b-d}(N-1)q_i^*$$

which yields the equilibrium output level for every firm in this Cournot game,

$$q_i^* = \frac{a-c}{b(N+1)-d}.$$

This equilibrium output is positive if $b(N+1) > d$, which holds since $2b > d$ by assumption. In addition, equilibrium output is increasing in the market size (as captured by $a - c$), decreasing in the number of firms, N, and increasing in the economies of scale (higher d).

- Substituting q_i^* into the profit function of firm i, equilibrium profits become

$$\pi_i^* = \left[a - Nbq_i^* - c + \frac{d}{2}q_i^*\right] q_i^*$$
$$= \frac{(a-c)^2(2b-d)}{2[b(N+1)-d]^2}.$$

which is positive since $2b > d$ by assumption.

(b) Find the aggregate output in equilibrium, Q^*. Under which parameter conditions does it decrease with entry? Interpret.

- The aggregate output is

$$Q^* = Nq_i^* = \frac{N(a-c)}{b(N+1)-d}$$

which is positive since $2b > d$ by assumption.

- Differentiating Q^* with respect to N, we find

$$\frac{\partial Q^*}{\partial N} = \frac{(a-c)(b-d)}{[b(N+1)-d]^2}$$

which is negative if and only if $d > b$. Therefore, when economies of scale (as captured by parameter d) are strong enough, aggregate output decreases when more firms enter into the market. Intuitively, entry induces every incumbent firm to reduce its output, hindering its ability to benefit from economies of scale. As a result, every firm produces at a higher cost after entry, ultimately leading to a lower aggregate output than before entry. Alternatively, the output that the entrant contributes to the industry is offset by the reduction in output from all incumbent firms, which all suffer from higher production costs.

(c) Evaluate the aggregate output when $N = 1$ and when $N \to +\infty$. Compare.

- Substituting $N = 1$ into the aggregate output in part (b), we obtain

$$Q^* = \frac{a-c}{2b-d}$$

Substituting $N \to +\infty$ into the aggregate output in part (b), we find

$$\lim_{N\to+\infty} Q^* = \lim_{N\to+\infty} \frac{a-c}{b\left(1+\frac{1}{N}\right)-\frac{d}{N}} = \frac{a-c}{b}$$

Therefore, aggregate output under monopoly ($N = 1$) exceeds that under competitive market ($N \to +\infty$) when

$$\frac{a-c}{2b-d} > \frac{a-c}{b}$$

which reduces to

$$d > b$$

that happens when the economies of scale is relatively strong.

(d) Show that π_i^* decreases with entry. Explain.

- Differentiating π_i^* with respect to N, we have

$$\frac{\partial \pi_i^*}{\partial N} = -\frac{b\,(a-c)^2\,(2b-d)}{[b(N+1)-d]^3}$$

which is negative since $2b > d$ holds by definition. Therefore, we obtain two cases, depending on the relative size of the economies of scale (parameter d):
  - When d is relatively high, $d > b$, entry decreases aggregate output and profits.
  - When d is relatively low, $d < b$, entry increases aggregate output but still decreases profits.

(e) Evaluate consumer surplus, producer surplus, and welfare, in equilibrium. Then, identify the number of firms, N, that maximizes each of them.

- Consumer surplus is one-half of the square of aggregate output, where

$$CS^* = \frac{(Q^*)^2}{2} = \frac{N^2\,(a-c)^2}{2\,[b\,(N+1)-d]^2}$$

- Aggregate profits sum up the profits of every firm i, given by

$$\Pi^* = N\pi_i^* = \frac{N\,(a-c)^2\,(2b-d)}{2\,[b(N+1)-d]^2}$$

- Social welfare is the sum of consumer and producer surplus, that is

$$SW^* = CS^* + \Pi^* = \frac{N\,(a-c)^2\,(N+2b-d)}{2\,[b(N+1)-d]^2}$$

which derivative with respect to N is

$$\frac{\partial SW^*}{\partial N} = -\frac{(a-c)^2\,[N\,(2\,(d-b)+b\,(2b-d))+(d-b)\,(2b-d)]}{2\,[b(N+1)-d]^3}$$

which is unambiguously negative when $d > b$ since $2b > d$ by assumption. Therefore, we find two cases:
  - When firms exhibit significant economies of scale ($d > b$), a monopoly ($N^* = 1$) maximizes social welfare. This explains why economies of scale are used to justify natural monopolies such as public utilities (gas, electricity, and water supplies).
  - In contrast, when economies of scale are less significant $d < b$, the number of firms that maximizes social welfare is

$$N^* = \frac{(b-d)\,(2b-d)}{b\,(2b-d)-2\,(b-d)}$$

which is positive since $b > d$ and $2b > d$.

Exercise #2.19: Cournot with Asymmetric Fixed Costs^B

2.19 Consider a Cournot duopoly, allowing firms to face fixed costs. In particular, assume that firm 1 faces a total cost function $TC_1(q_1) = F_1 + cq_1$, where $F_1 > 0$ denotes its fixed cost and $1 > c > 0$ represents its marginal cost. Firm 2's total cost function is similar, $TC_2(q_2) = F_2 + cq_2$, where $F_2 > 0$ denotes its fixed cost, and satisfies $F_2 > F_1$, and $c > 0$ is the same marginal cost as firm 1's. Firms face a linear inverse demand function $p(Q) = 1 - Q$, where Q denotes aggregate output.

(a) Find the best response functions of each firm and the equilibrium output.

- *Firm 1's best response function.* In this setting, firm 1 chooses its output level q_1 to solve

$$\max_{q_1 \geq 0} \pi_1 = (1 - q_1 - q_2)q_1 - cq_1 - F_1.$$

Differentiating with respect to q_1, we obtain

$$\frac{\partial \pi_1}{\partial q_1} = 1 - 2q_1 - q_2 - c = 0$$

rearranging, $1 - q_2 - c = 2q_1$, and solving for q_1 yields

$$q_1(q_2) = \frac{1-c}{2} - \frac{1}{2}q_2$$

which is firm 1's best response function. The best response function coincides with that in a Cournot duopoly where firms face no fixed costs. Intuitively, when choosing its optimal output, firm 1 only considers its marginal revenues and costs, but ignores its fixed cost. The fixed cost should only impact its profits and, as a consequence, the conditions under which this firm produces a positive output.

- *Firm 2's best response function.* In this setting, firm 2 chooses its output level q_2 to solve

$$\max_{q_2 \geq 0} \pi_2 = (1 - q_1 - q_2)q_2 - cq_2 - F_2.$$

Differentiating with respect to q_2, we obtain

$$\frac{\partial \pi_2}{\partial q_2} = 1 - 2q_2 - q_1 - c = 0$$

rearranging, $1 - q_1 - c = 2q_2$, and solving for q_2 yields

$$q_2(q_1) = \frac{1-c}{2} - \frac{1}{2}q_1$$

which is firm 2's best response function. This best response function is symmetric to firm 1's.

- The best response functions of firm 1 and firm 2 do not differ from the Cournot duopoly with symmetric marginal costs. Therefore, the equilibrium quantity each firm produces is the same as in that situation, that is,

$$q_1^* = q_2^* = \frac{1-c}{3}.$$

(b) How are the equilibrium results affected? Interpret.

- Given that the best response functions and output do not change relative to the case of symmetric firms, the price in the market will not change, i.e., $p = \frac{1+2c}{3}$. Profits for the two firms will be altered slightly by the fixed cost:

$$\pi_1 = \frac{(1-c)^2}{9} - F_1$$

$$\pi_2 = \frac{(1-c)^2}{9} - F_2.$$

For firm 1, if its fixed cost F_1 satisfies $F_1 > \frac{(1-c)^2}{9}$, the firm chooses to remain inactive as its overall profits from participating in this market would be negative. A similar argument applies for firm 2 if its fixed cost F_2 satisfies $F_2 > \frac{(1-c)^2}{9}$. Therefore, four cases arise, each of them depicted in one of the regions of Fig. 2.7, with F_1 in the vertical axis and F_2 in the horizontal axis:

- *Both firms active.* If $F_1, F_2 < \frac{(1-c)^2}{9}$, both firms produce the equilibrium output found in part (a) and make positive profits.
- *Only firm 1 is active.* If only firm 1 faces a sufficiently low fixed cost, that is, $F_1 < \frac{(1-c)^2}{9}$ but $F_2 > \frac{(1-c)^2}{9}$, then firm 1 produces the monopoly output $\frac{1-c}{2}$, earning monopoly profit $\frac{(1-c)^2}{4}$, while its rival remains inactive.
- *Only firm 2 is active.* If only firm 2 faces a sufficiently low fixed cost, that is, $F_2 < \frac{(1-c)^2}{9}$ but $F_1 > \frac{(1-c)^2}{9}$, then firm 2 produces the monopoly output $\frac{a-c}{2}$, earning monopoly profit $\frac{(1-c)^2}{4}$, while its rival remains inactive.
- *No active firms.* If both firms have relatively high fixed costs, that is, $F_1, F_2 > \frac{(1-c)^2}{9}$, then neither firm is active in the market.

(c) *Numerical example.* Evaluate your equilibrium results in part (b) at $c = 1/2$. What happens if c decreases to $c = 1/10$? Interpret.

- When $c = 1/2$, the cutoffs of F_1 and F_2 in Fig. 2.7 become

$$\frac{(1-\frac{1}{2})^2}{9} = \frac{1}{36} \simeq 0.03.$$

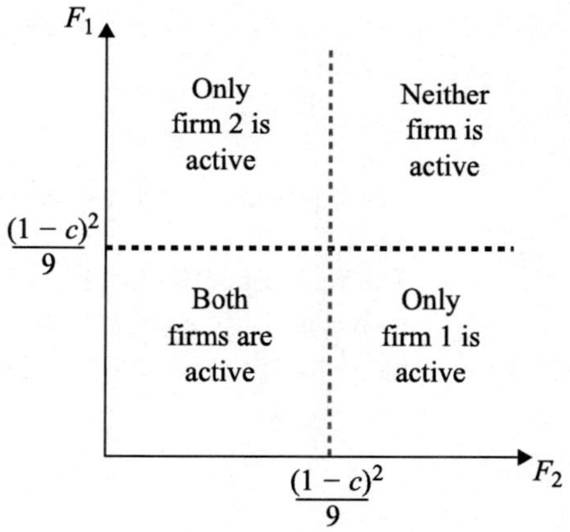

Fig. 2.7 Production decisions

When marginal costs decrease to $c = 1/10$, these cutoffs increase to $\frac{(1-\frac{1}{10})^2}{9} = \frac{9}{100} = 0.09$. Graphically, the horizontal cutoff at the vertical axis shifts upward, and the vertical cutoff in the horizontal axis shifts rightward as marginal costs decrease, intuitively describing that the region where both firms are active expands (in the southwest of Fig. 2.7) and that the regions where only one firm is active also expand (at the northwest or southeast of the figure). In contrast, the region of (F_1, F_2)-pairs where neither firm is active (at the northeast of Fig. 2.7) shrinks.

Exercise #2.20: Can Fewer Firms Decrease Prices?^{B}

2.20 Consider an industry with $n \geq 2$ firms competing ? la Cournot, facing an inverse demand function $p(Q) = 1 - Q$, where $Q \geq 0$ denotes aggregate output. Firms in this industry are asymmetric in their marginal costs. Specifically, a share $\alpha \in [0, 1]$ of them have marginal cost c_H, which we regard as "inefficient," and the remaining share, $1 - \alpha$, are firms with marginal cost c_L, which we refer as "efficient," where $a > c_H > c_L$.

(a) Find each type of firm's best response function. Interpret.

- The demand function is

$$p = 1 - Q = 1 - \sum_{i \in L} q_i - \sum_{j \in H} q_j,$$

where the last two terms represent the total production from low- and high-cost firms, respectively. The profit function of every high-cost firm $j \in H$ is

$$\pi_j = (p(Q) - c_H)q_j.$$

Differentiating with respect to q_j, we find

$$1 - \left(q_j + \sum_{i \in L} q_i + \sum_{j \in H} q_j + c_H \right) = 0.$$

Similarly, the profit function of every low-cost firm $i \in L$ is

$$\pi_i = (p(Q) - c_L)q_i \; .$$

Differentiating with respect to q_i, we find

$$1 - \left(q_i + \sum_{i \in L} q_i + \sum_{j \in H} q_j + c_L \right) = 0.$$

In a strategy profile where all low-cost firms produce the same output level q_i, we must have

$$\sum_{i \in L} q_i = (1 - \alpha)nq_i$$

since there are $(1-\alpha)n$ firms with low costs in this industry. Similarly, in a strategy profile where all high-cost firms produce the same output q_j, we must have that

$$\sum_{j\in H} q_j = \alpha n q_j$$

since there are αn firms with high costs. Plugging these results into the above first-order conditions, we obtain

$$1-\left(q_j+(1-\alpha)nq_i+\alpha nq_j+c_H\right)=0, \text{ and}$$
$$1-\left(q_i+(1-\alpha)nq_i+\alpha nq_j+c_L\right)=0.$$

Solving for q_i in the last expression, we find the best response function of every low-cost firm $i \in L$ as follows:

$$q_L(q_H)=\frac{1-c_L}{1+(1-\alpha)n}-\frac{\alpha n}{1+(1-\alpha)n}q_H$$

which, as expected, decreases in the production of each high-cost rival, q_H. Similarly, we can solve for q_j in the first expression above, obtaining the best response function of every high-cost firm $j \in H$ as follows:

$$q_H(q_L)=\frac{1-c_H}{1+\alpha n}-\frac{(1-\alpha)n}{1+\alpha n}q_L$$

which is decreasing in the production of each low-cost rival, q_L.

(b) Find equilibrium output for every high-cost firm and every low-cost firm.

- *Equilibrium output.* Inserting best response function $q_H(q_L)$ into $q_L(q_H)$ yields

$$q_L=\frac{1-c_L}{1+(1-\alpha)n}-\frac{\alpha n}{1+(1-\alpha)n}\underbrace{\left(\frac{1-c_H}{1+\alpha n}-\frac{(1-\alpha)n}{1+\alpha n}q_L\right)}_{q_H}.$$

Solving for q_L yields an equilibrium output of

$$q_L^*=\frac{1-c_L+n\alpha(c_H-c_L)}{n+1}.$$

Therefore, the equilibrium output of every high-cost firm is

$$q_H^*=\frac{1-c_H}{1+\alpha n}-\frac{(1-\alpha)n}{1+\alpha n}\underbrace{\frac{1-c_L+n\alpha(c_H-c_L)}{n+1}}_{q_L^*}$$
$$=\frac{1-c_H-n(1-\alpha)(c_H-c_L)}{n+1}.$$

- *Equilibrium price.* Inserting the above equilibrium outputs, q_L^* and q_H^*, into the inverse demand function, we obtain an equilibrium price

$$
\begin{aligned}
p^* &= 1 - (1-\alpha)nq_L^* - \alpha n q_H^* \\
&= 1 - \frac{(1-\alpha)n(1-c_L+n\alpha(c_H-c_L))}{n+1} - \frac{\alpha n(1-c_H-n(1-\alpha)(c_H-c_L))}{n+1} \\
&= \frac{1+n\alpha c_H + n(1-\alpha)c_L}{1+n}.
\end{aligned}
$$

(c) Find under which parameter conditions do high-cost firms produce a positive output level. Examine how this parameter condition is affected by the number of firms in the industry, n, and by the proportion of high-cost firms, α.

- The denominator of equilibrium output q_H^* is positive given that $1+n>0$ by definition. Therefore, q_H^* is positive as long as the numerator is positive, $1-c_H-n(1-\alpha)(c_H-c_L) > 0$ which, solving for c_H, yields

$$
c_H \leq \frac{1+n(1-\alpha)c_L}{1+n(1-\alpha)} \equiv \overline{c}_H.
$$

Intuitively, condition $c_H \leq \overline{c}_H$ says that every high-cost firm produces a positive output only when its cost c_H is not too high, relative to its low-cost rivals. Figure 2.8 illustrates cutoff $\overline{c}_H$ in the (c_H, c_L)-quadrant. Cutoff $\overline{c}_H$ originates at $c_H = \frac{1}{1+n(1-\alpha)}$ when $c_L = 0$ and reaches a height of $c_H = 1$ when $c_L = 1$, thus lying above the 45°-line where $c_H = c_L$. Therefore, among all admissible (c_H, c_L)-pairs (those above the 45°-line so $c_H > c_L$), only those below cutoff $\overline{c}_H$ induce every high-cost firm to produce a positive output.

- *Comparative statics.*
  - We can now check if cutoff $\overline{c}_H$ increases or decreases as more firms compete in this industry. In particular,

$$
\frac{\partial \overline{c}_H}{\partial n} = -\frac{(1-\alpha)(1-c_L)}{[1+n(1-\alpha)]^2} < 0, \text{ where } 1 > c_L
$$

implying that, as more firms compete, the condition on high-cost firms to remain active, $c_H \leq \overline{c}_H$, becomes more stringent. Intuitively, as n increases, the number of low-cost firms, $(1-\alpha)n$, also increases, making it harder for every high-cost firm

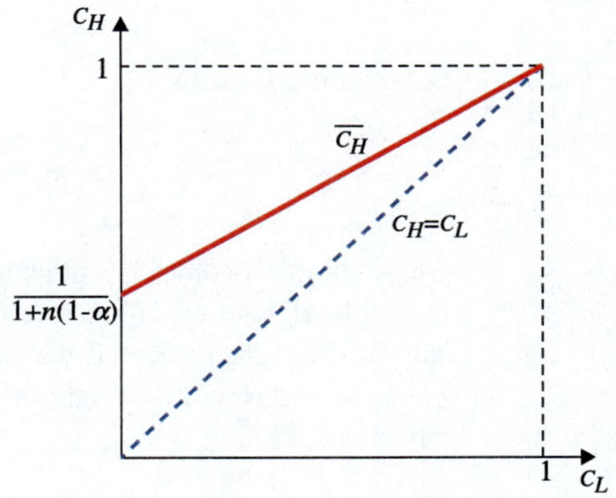

Fig. 2.8 Cutoff $\overline{c}_H$

to compete. When n becomes sufficiently high, condition $c_H \leq \overline{c}_H$ does not hold, inducing all high-cost firms to exit the industry.

Graphically, an increase in n produces a downward shift in the vertical intercept of cutoff $\overline{c}_H$ in Fig. 2.8, shrinking the region of (c_H, c_L)-pairs for which every high-cost firm produces a positive output.

– We now check whether cutoff $\overline{c}_H$ increases or decreases in the proportion of high-cost firms in the industry, α, as follows:

$$\frac{\partial \overline{c}_H}{\partial \alpha} = \frac{1 - c_L}{[1 + n(1-\alpha)]^2} > 0.$$

Intuitively, this result indicates that, as high-cost firms become a larger share of the industry, the condition on high-cost firms to remain active, $c_H \leq \overline{c}_H$, becomes less stringent.

(d) Find equilibrium output and prices if all high-cost firms exit the industry. Are consumers better off when all high-cost firms remain active or when they exit?

- If all high-cost firms exit, $(1-\alpha)n$ low-cost types of firms will remain in the market. The equilibrium quantity in that setting becomes

$$q_L^{**} = \frac{1 - c_L}{1 + n(1-\alpha)}$$

which does not depend on high-cost firm's marginal cost, c_H, since these firms are not active. The equilibrium price in this setting is

$$\begin{aligned} p^{**} &= 1 - (1-\alpha)nq_L^* = \\ &= 1 - (1-\alpha)n \underbrace{\frac{1 - c_L}{1 + n(1-\alpha)}}_{q_L^{**}} \\ &= \frac{1 + n(1-\alpha)c_L}{1 + n(1-\alpha)}. \end{aligned}$$

Comparing equilibrium prices when all high-cost firms are active, p^*, and when they exit the industry, p^{**}, we obtain that $p^* > p^{**}$ holds if

$$\frac{1 + n\alpha c_H + n(1-\alpha)c_L}{1+n} > \frac{1 + n(1-\alpha)c_L}{1 + n(1-\alpha)}.$$

Solving for c_H yields

$$c_H > \frac{1 + n(1-\alpha)c_L}{1 + n(1-\alpha)} \equiv \overline{c}_H,$$

where cutoff $\overline{c}_H$ coincides with the one we found above. Therefore, when condition $c_H > \overline{c}_H$ holds, all high-cost firms exit the industry, and prices are lower than when these firms stay in the industry. Recall that Fig. 2.8 illustrates that this cutoff originates at $c_H = \frac{1}{1+n(1-\alpha)}$ and reaches a height of $c_H = 1$ when $c_L = 1$, thus lying above the 45°-line where $c_H = c_L$.

- *Comparative statics.* The vertical intercept of cutoff $\overline{c}_H$ (see Fig. 2.8) is decreasing in the number of firms competing in the industry, n, but increasing in the share of firms with high marginal costs, α. Intuitively, as the industry becomes less competitive (low n) and a larger share of firms have a high marginal cost (high α), cutoff $\overline{c}_H$ shifts upward, expanding the region of (c_H, c_L)-pairs for which prices are lower after the exist of all high-cost firms.
- This example illustrates that, if an industry has a subset of inefficient firms with sufficiently high costs, it may actually be better for consumers that these firms exit the industry, leaving the market with fewer firms, than if these inefficient firms stay.

Exercise #2.21: Cournot with Equity Shares – Based on Reynolds and Snapp (1986)^B

2.21 Consider a Cournot duopoly with linear inverse demand curve $p(Q) = 1 - Q$, where Q denotes aggregate output. Both firms have a common constant marginal cost $c > 0$, where $1 > c$. Assume that firms own an equity share of α, i.e., every firm i receives a share $0 < \alpha \leq \frac{1}{2}$ in its rival's profit.

(a) Find firm i's best response function. How is it affected by equity shares, α?

- Firm i's profit maximization problem (PMP) is given by

$$\max_{q_i \geq 0} \underbrace{(1-\alpha)(1-q_i-q_j-c)q_i}_{\text{Firm } i\text{'s profit}} + \underbrace{\alpha(1-q_i-q_j-c)q_j}_{\text{Firm } j\text{'s profit}}.$$

Taking first-order conditions with respect to q_i yields

$$(1-\alpha)(1-q_i-q_j-c) - (1-\alpha)q_i - \alpha q_j = 0.$$

Solving for q_i, we find firm i's best response function

$$q_i(q_j) = \frac{1-c}{2} - \frac{1}{2(1-\alpha)}q_j.$$

Figure 2.9 depicts this best response function, showing that its vertical intercept is unaffected by α, but it becomes steeper as α increases. Intuitively, as a firm receives a higher share of its rival's profit, it decreases its own output. For instance, when $\alpha = 0$, the above best response function becomes $q_i(q_j) = \frac{1-c}{2} - \frac{1}{2}q_j$, coinciding with that in standard Cournot duopolies with linear marginal costs (see Fig. 2.9). However, when firms equally share their profits, $\alpha = \frac{1}{2}$, the above best response function simplifies to $q_i(q_j) = \frac{1-c}{2} - q_j$, thus being steeper (see Fig. 2.10).

(b) Use your results from part (a) to find the Cournot equilibrium output.

- In a symmetric equilibrium $q_i = q_j = q^*$. Solving for q^* yields

$$q^* = \frac{(1-\alpha)(1-c)}{3-2\alpha}.$$

(c) Evaluate equilibrium output q_i^* at $\alpha = 0$ and $\alpha = \frac{1}{2}$. Interpret.

Fig. 2.9 BRF with $\alpha = 0$

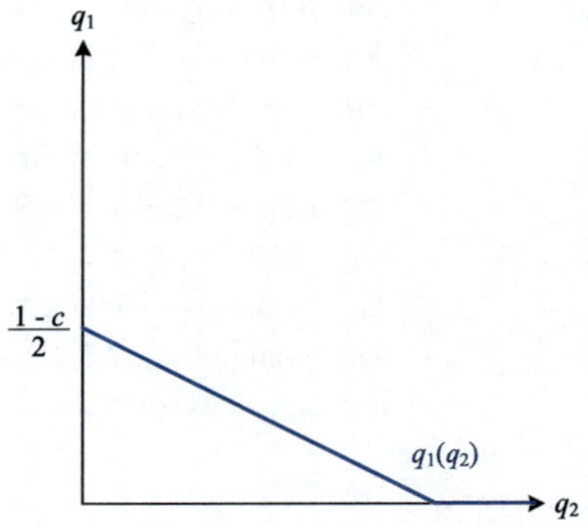

Fig. 2.10 BRF with $\alpha = 0.5$

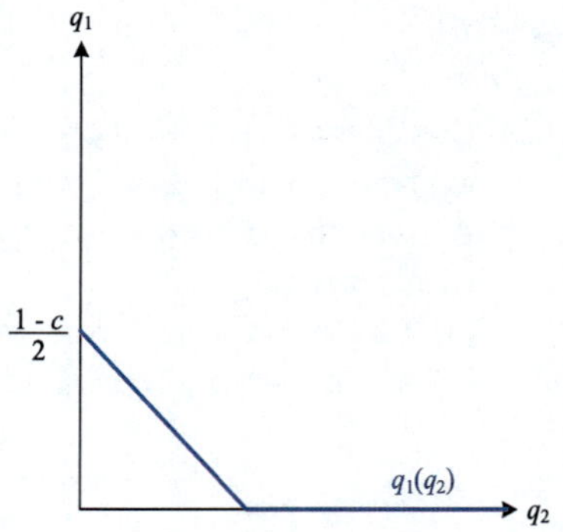

- When firms do not benefit from each other's profits, $\alpha = 0$, equilibrium output q^C becomes $\frac{1-c}{3}$, thus coinciding with that under the standard Cournot model with linear inverse demand curve $p(q) = 1 - q$.
- In contrast, when firms fully share their profits, $\alpha = 1/2$, equilibrium output q^C becomes $\frac{1-c}{4}$, thus coinciding with half of monopoly output (or cartel output).

(d) Does equilibrium output q_i^* increase or decrease in α? Explain.

- Differentiating q^* with respect to share α yields

$$\frac{\partial q^*}{\partial \alpha} = -\frac{1-c}{3-2\alpha} + \frac{2(1-\alpha)(1-c)}{(3-2\alpha)^2}$$
$$= -\frac{1-c}{(3-2\alpha)^2}.$$

Intuitively, as firms share more of each other's profits, their individual PMP resembles the joint PMP in a cartel, leading each of them to reduce its production.

Exercise #2.22: Cournot Competition Between a Private and a Public Firm^B

2.22 Consider an industry where two firms are producing a homogeneous good, and facing a linear inverse demand function $p(Q) = 1 - Q$, where Q denotes aggregate output. All firms face a constant marginal cost of production c, where $1 > c > 0$. Firms compete à la Cournot. Firm 1 (a private firm) seeks to maximize its profits, while firm 2 (a public firm) maximizes a combination of welfare and profits as follows:

$$V = \alpha W + (1-\alpha)\pi_2,$$

where parameter $\alpha \in [0, 1]$ indicates that weight that the manager of the public firm assigns to welfare W and $W = CS + PS$ denotes the sum of consumer and producer surplus. Furthermore, $1-\alpha$ represents the weight he assigns to profit π_2, where $\pi_i = (1 - q_i - q_j)q_i - cq_i$ for every firm $i = \{1, 2\}$.

(a) Find the firm 1's and 2's best response functions. How is firm 2's best response function affected by an increase in parameter α?

- *Private firm.* Firm 1 chooses its output q_1 to solve

$$\max_{q_1 \geq 0} \pi_1 = (1 - q_1 - q_2)q_1 - cq_1.$$

Differentiating with respect to q_1 yields

$$1 - 2q_1 - q_2 - c = 0$$

and solving for q_1, we find firm 1's best response function

$$q_1(q_2) = \frac{1-c}{2} - \frac{1}{2}q_2$$

which originates at $\frac{1-c}{2}$ and decreases in its rival's output, q_2, at a rate of 1/2.

- *Public firm.* Firm 2 chooses its output q_2 to solve

$$\max_{q_2 \geq 0} \alpha \underbrace{\left[\int_0^{q_1+q_2} (1-x)dx - c(q_1+q_2)\right]}_{W} + (1-\alpha)\underbrace{[(1-q_1-q_2)q_2 - cq_2]}_{\pi_2}.$$

Differentiating with respect to q_2 yields

$$\alpha\,[1 - q_1 - q_2 - c] + (1-\alpha)\,[1 - 2q_2 - q_1 - c] = 0$$

and solving for q_2, we find

$$q_2(q_1) = \frac{1-c}{2-\alpha} - \frac{1}{2-\alpha}q_1.$$

- *Comparative statics.* The best response function of the private firm is, of course, unaffected by the weight that the public firm assigns to welfare, α. However, the public firm's best response function is affected. In particular, an increase in α increases its vertical intercept and slope, leaving its horizontal intercept unaffected. Intuitively, as the public firm assigns a larger weight on welfare, it produces a larger output q_2 for a given output level of the private firm, q_1. For instance, when $\alpha = 1$, the above best response function simplifies to

$$q_2(q_1) = 1 - c - q_1$$

while when $\alpha = 0$, the best response function becomes

$$q_2(q_1) = \frac{1-c}{2} - \frac{1}{2}q_1$$

thus coinciding with that under the standard Cournot duopoly.

(b) Find equilibrium output levels for each firm.

- Inserting $q_2(q_1)$ into $q_1(q_2)$, we find

$$q_1 = \frac{1-c}{2} - \frac{1}{2}\underbrace{\left(\frac{1-c}{2-\alpha} - \frac{1}{2-\alpha}q_1\right)}_{q_2(q_1)}.$$

Rearranging, and solving for q_1, we obtain the equilibrium output of firm 1 (the private firm):

$$q_1^* = \frac{(1-\alpha)(1-c)}{3-2\alpha}.$$

We can now insert this outcome into the best response function of the public firm, which yields

$$\begin{aligned} q_2^* &= \frac{1-c}{2-\alpha} - \frac{1}{2-\alpha}\underbrace{\frac{(1-\alpha)(1-c)}{3-2\alpha}}_{q_1^*} \\ &= \frac{1-c}{3-2\alpha}. \end{aligned}$$

(c) How are equilibrium output levels affected by an increase in parameter α? Evaluate these output levels in the extreme cases where $\alpha = 1$ and $\alpha = 0$. Interpret.

- Differentiating firm 1's equilibrium output with respect to α, we find

$$\frac{\partial q_1^*}{\partial \alpha} = -\frac{1-c}{(3-2\alpha)^2} < 0$$

while differentiating firm 2's equilibrium output with respect to α, we obtain

$$\frac{\partial q_2^*}{\partial \alpha} = \frac{2(1-c)}{(3-2\alpha)^2} > 0.$$

In summary, when the public firm assigns a larger weight to welfare (higher α), it produces a larger output, leading the private firm to produce fewer units (as both types of firms regard output levels as strategic substitutes).

- *Extreme cases:*
  - When the public firm only considers welfare in its objective function ($\alpha = 1$), it produces the perfectly competitive output, $q_2^* = 1 - c$, while the private firm produces zero units. Intuitively, the public firm operates in this setting as a regulator seeking to produce the output level that maximizes social welfare, which coincides with the perfectly competitive output at the aggregate level since $q_1^* + q_2^* = 1 - c$.

- When, instead, the public firm only considers profits in its objective function ($\alpha = 0$), both firms behave as in standard Cournot duopoly models, producing $q_1^* = q_2^* = \frac{1-c}{3}$ units. In this context, both firms face the same profit maximization problem as Cournot duopolists, thus yielding the same equilibrium outcomes.

(d) *Numerical example.* Evaluate equilibrium output at $c = 1/2$. How are they affected by an increase in α? Then evaluate your results at $\alpha = 1$.

- *Private firm.* Equilibrium output for the private firm evaluated at $c = 1/2$ is

$$q_1^* = \frac{(1-\alpha)\left(1-\frac{1}{2}\right)}{3-2\alpha} = \frac{1-\alpha}{2(3-2\alpha)}$$

which is decreasing in α given that

$$\frac{\partial q_1^*}{\partial \alpha} = -\frac{1}{2\,(3-2\alpha)^2} < 0.$$

When $\alpha = 0$, this firm produces $q_1^* = \frac{1-0}{2(3-0)} = \frac{1}{6}$ units; and when $\alpha = 1$, this firm produces $q_1^* = \frac{1-1}{2(3-2)} = 0$ units.

- *Public firm.* Equilibrium output of the public firm in this context is

$$q_2^* = \frac{1-\frac{1}{2}}{3-2\alpha} = \frac{1}{2(3-2\alpha)}$$

which is increasing in α because

$$\frac{\partial q_2^*}{\partial \alpha} = \frac{1}{(3-2\alpha)^2} > 0.$$

When $\alpha = 0$, this firm produces $q_2^* = \frac{1}{2(3-0)} = \frac{1}{6}$ units; and when $\alpha = 1$, this firm produces $q_2^* = \frac{1}{2(3-2)} = \frac{1}{2}$ units.

(e) Evaluate consumer surplus, producer surplus, and total welfare as a function of c and α. Find the value of α that maximizes social welfare.

- To calculate our consumer and producer surpluses, we must first obtain our total output, Q^*,

$$Q^* = \frac{(1-\alpha)(1-c)}{3-2\alpha} + \frac{1-c}{3-2\alpha} = \frac{(2-\alpha)(1-c)}{3-2\alpha}$$

- *Consumer surplus.* Consumer surplus is

$$CS = \frac{1}{2}(Q^*)^2 = \frac{(2-\alpha)^2(1-c)^2}{2(3-2\alpha)^2}$$

- *Producer surplus.* Producer surplus in this context is

$$\begin{aligned} PS &= p^* Q^* - cQ^* \\ &= (p^* - c)Q^* \\ &= (1 - Q^* - c)Q^* \end{aligned}$$

Simplifying,

$$
\begin{aligned}
PS &= \left(1 - \frac{(2-\alpha)(1-c)}{3-2\alpha} - c\right)\left(\frac{(2-\alpha)(1-c)}{3-2\alpha}\right) \\
&= \left(\frac{(3-2\alpha)(1-c)-(2-\alpha)(1-c)}{3-2\alpha}\right)\left(\frac{(2-\alpha)(1-c)}{3-2\alpha}\right) \\
&= \left(\frac{(1-\alpha)(1-c)}{3-2\alpha}\right)\left(\frac{(2-\alpha)(1-c)}{3-2\alpha}\right) \\
&= \frac{(1-\alpha)(2-\alpha)(1-c)^2}{(3-2\alpha)^2}
\end{aligned}
$$

- *Welfare*. Total welfare in this context is

$$
\begin{aligned}
W &= CS + PS \\
&= \frac{(2-\alpha)^2(1-c)^2}{2(3-2\alpha)^2} + \frac{(1-\alpha)(2-\alpha)(1-c)^2}{(3-2\alpha)^2} \\
&= \frac{(4-3\alpha)(2-\alpha)(1-c)^2}{2(3-2\alpha)^2}
\end{aligned}
$$

- *Maximizing social welfare*. Differentiating our total welfare with respect to α, and setting it equal to zero,

$$
\begin{aligned}
\frac{\partial W}{\partial \alpha} &= \frac{(1-c)^2}{2} \\
&\quad \times \left[\frac{(-3(2-\alpha)-1(4-3\alpha))(3-2\alpha)^2-(4-3\alpha)(2-\alpha)(2(3-2\alpha)(-2))}{(3-2\alpha)^4}\right] \\
&= 0
\end{aligned}
$$

which is equal to zero when the numerator is zero, that is,

$$4(2-\alpha)(4-3\alpha) = 2(5-3\alpha)(3-2\alpha)$$

which simplifies to

$$16 - 20\alpha + 6\alpha^2 = 15 - 19\alpha + 6\alpha^2,$$

and this holds if and only if $\alpha^* = 1$. Thus, social welfare is maximized when the public firm's objective function assigns full weight on social welfare and no weight to profits. Intuitively, when $\alpha^* = 1$, the public firm sets the perfectly competitive quantity whereas the private firm does not produce. This leads to the perfectly competitive price and no deadweight loss, thus maximizing efficiency. This result goes in line with that in Matsumura (1998) (see Proposition 2), who allows for more general demand and cost functions.

Exercise #2.23: Cournot Competition Between *N* Private Firms and a Public Firm^B

2.23 Consider the setting in Exercise 2.22, but let us now allow for $N \geq 1$ private firms. In this context, aggregate output is $Q \equiv q_0 + \sum_{i=1}^{N} q_i$, where $i = 1, 2, \ldots, N$ denotes private firms. Every private firm still maximizes its profits

$$\pi_i = p(X)q_i - cq_i$$

while the public firm maximizes a combination of social welfare and profits

$$V_0 = \alpha W + (1-\alpha)\pi_0$$

where social welfare is given by $W = \int_0^Q p(y)dy - cQ$ and its profits are $\pi_0 = p(Q)q_0 - cq_0$. Intuitively, parameter α represents the weight that the manager of the public firm assigns to social welfare, while $1-\alpha$ is the weight that he assigns to profits. All firms simultaneously and independently choose their output levels.

(a) Find the best response function of every private firm, $q_i(q_0)$. How are your results affected by a marginal increase in N? Interpret.

- Each private firm i maximizes

$$\max_{q_i \geq 0} \pi_i(q_i) = \left[1 - \left(q_0 + q_i + \sum_{j \neq i}^{N} q_j\right)\right] q_i - cq_i$$

Differentiating with respect to q_i yields

$$1 - q_0 - 2q_i - \sum_{j \neq i}^{N} q_j - c = 0$$

In a symmetric equilibrium, $q_i = q_j$, implying that

$$\sum_{j \neq i}^{N} q_j = (N-1)q_i,$$

so we can rewrite the above first-order condition as follows:

$$1 - q_0 - (N+1)q_i - c = 0$$

Solving for q_i, we find the best response function of every private firm i

$$q_i(q_0) = \frac{1-c}{N+1} - \frac{1}{N+1}q_0.$$

- *Comparative statics.*
  - As expected, this best response function is unaffected by the weight that the public firm assigns on social welfare, α.

- However, an increase in the number of private firms (higher N) produces a downward shift in the vertical intercept, $\frac{1-c}{N+1}$, and in the slope, $\frac{1}{N+1}$, entailing a lower output level for every private firm. In the special case where a single private firm competes against a public firm, $N = 1$, this best response function simplifies to

$$q_i(q_0) = \frac{1-c}{2} - \frac{1}{2}q_0,$$

as in standard Cournot duopolies.

(b) Find the best response function of the public firm, $q_0(q_i)$. How are your results affected by a marginal increase in N? And how are they affected by a marginal increase in α? Interpret.

- The public firm maximizes

$$\begin{aligned}\max_{q_0 \geq 0} \; V_0 &= \alpha W + (1-\alpha)\pi_0 \\ &= \alpha \left[\int_0^{q_0+\sum_{i=1}^N q_i} p(y)dy - c\left(q_0 + \sum_{i=1}^N q_i\right)\right] \\ &\quad +(1-\alpha)\left(1 - q_0 - \sum_{i=1}^N q_i - c\right) q_0 \\ &= \alpha \left[\int_0^{q_0+Nq_i} (1-y)dy - c\,(q_0 + Nq_i)\right] + (1-\alpha)\,(1 - q_0 - Nq_i - c)\, q_0\end{aligned}$$

Differentiating with respect to x_0, and using Leibniz's rule, we obtain

$$\frac{\partial V}{\partial q_0} = \alpha\,[1 - q_0 - Nq_i - c] + (1-\alpha)\,(1 - 2q_0 - Nq_i - c) = 0$$

Solving for q_0, we find the public firm's best response function

$$q_0(q_i) = \frac{1-c}{2-\alpha} - \frac{N}{2-\alpha}q_i$$

- *Comparative statics.*
  - An increase in the weight that the public firm assigns to welfare, α, produces an outward shift in the best response function, as described in part (a) of Exercise 2.13 in the case of $N = 1$. Intuitively, as the public firm assigns more weight on welfare, it produces a larger output q_0 for a given aggregate output of its private rivals, $Q = Nq_i$.
  - An increase in the number of private firms, N, does not affect the vertical intercept of the public firm's best response function, $\frac{1-c}{2-\alpha}$, but makes it steeper, since $\frac{N}{2-\alpha}$ increases. Intuitively, as the public firm faces more private rivals, it decreases its output more significantly for a given increase in q_i, since the simultaneous output competition of more firms would bring the total output closer to the perfectly competitive level (and reaches that level when $N \to \infty$). Needless to say, when $N = 1$, the above best response function for the public firm, $q_0(q_i)$, coincides with that in part (a) of Exercise 2.13, $q_0(q_i) = \frac{1-c}{2-\alpha} - \frac{1}{2-\alpha}q_i$.

(c) Evaluate the above best response functions at $\alpha = 0$, $\alpha = 1/2$, and $\alpha = 1$. Interpret.

- When $\alpha = 0$, the best response function of the public firm, $q_0(q_i)$, is

$$q_0(q_i) = \frac{1-c}{2} - \frac{N}{2}q_i$$

which becomes steeper in the number of private firms N and coincides with the best response function in standard Cournot oligopoly models where all firms are private ($\alpha = 0$).

- When $\alpha = 1/2$, the best response function of the public firm, $q_0\,(q_i)$, is

$$\begin{aligned} q_0\,(q_i) &= \frac{2\,(1-c)}{3} - \frac{2N}{3}q_i \\ &= \frac{2\,[1-c-Nq_i]}{3} \end{aligned}$$

which lies above the best response function evaluated at $\alpha = 0$.

- When $\alpha = 1$, the best response function of the public firm, $q_0(q_i)$, becomes

$$q_0(q_i) = 1 - c - Nq_i$$

which is larger than the best response function evaluated at $\alpha = 1/2$.

- In summary, when α increases from 0 to 1, the best response function of the public firm pivots outward, becoming steeper, with its horizontal intercept being unaffected.[2]

(d) Calculate the equilibrium quantities for the private and public firms. Find the aggregate output in equilibrium. How are your results affected by a marginal increase in N? Interpret.

- *Individual output.* The equilibrium quantities solve the system of two equations:

$$\begin{aligned} q_i\,(q_0) &= \frac{1-c}{N+1} - \frac{1}{N+1}q_0 \quad \text{and} \\ q_0(q_i) &= \frac{1-c}{2-\alpha} - \frac{N}{2-\alpha}q_i \end{aligned}$$

Simultaneously solving for q_0 and q_i, we find the individual output levels:

$$\begin{aligned} q_0 &= \frac{1-c}{N(1-\alpha)-\alpha+2} \quad \text{and} \\ q_i &= \frac{(1-c)(1-\alpha)}{N(1-\alpha)-\alpha+2} \end{aligned}$$

Both q_0 and q_i are decreasing in N. Intuitively, as the number of private firms in the market increases, the public and every private firm i would lower their respective output

[2] Recall that, to find the horizontal intercept, we set the equation of the best response function equal to zero, $\frac{1-c}{2-\alpha} - \frac{N}{2-\alpha}q_i = 0$. Rearranging, we obtain $1 - c = Nq_i$, which solving for q_i yields $q_i = \frac{1-c}{N}$ that is independent of α.

levels. Differentiating with respect to α, we obtain that

$$\frac{\partial q_0}{\partial \alpha} = \frac{(1-c)\,(N+1)}{[N(1-\alpha)-\alpha+2]^2} > 0 \text{ and}$$

$$\frac{\partial q_i}{\partial \alpha} = -\frac{1-c}{[N(1-\alpha)-\alpha+2]^2} < 0$$

implying that, as the public firm assigns a larger weight to welfare, it produces more units while every private firm responds producing fewer units – referred to as the "crowding out" effect in public economics.

- *Aggregate output.* Summing the above output levels, we find that the aggregate output is

$$\begin{aligned} Q &= q_0 + \sum_{i=1}^{N} q_i \\ &= q_0 + N q_i \\ &= \frac{1-c}{N(1-\alpha)-\alpha+2} + N\frac{(1-c)(1-\alpha)}{N(1-\alpha)-\alpha+2} \\ &= \frac{(1-c)\,[N(1-\alpha)+1]}{N(1-\alpha)-\alpha+2}. \end{aligned}$$

 which is increasing in the number of private firms, N, since

$$\frac{\partial Q}{\partial N} = \frac{(1-c)(1-\alpha)^2}{[N(1-\alpha)-\alpha+2]^2} > 0$$

and in the weight that the public firm assigns to welfare, α, because

$$\frac{\partial Q}{\partial \alpha} = \frac{1-c}{[N(1-\alpha)-\alpha+2]^2} > 0.$$

Intuitively, aggregate output increases in the total number of private firms N and in the weight that the public firm assigns to welfare, α.

(e) Calculate the socially optimal output level and compare it with the equilibrium outcome you obtained in part (d). How are your results affected by a marginal increase in N? Interpret.

- We know the social welfare is given by

$$W = \int_0^Q p(y)dy - cQ.$$

Differentiating with respect to Q yields

$$\frac{\partial W}{\partial Q} = p(Q) - c = 0$$

Solving for Q, we obtain

$$p\,(Q) = c.$$

- Hence, the socially optimal output level happens where the inverse demand function crosses the marginal cost curve, $p(Q) = c$, which in this context implies $1 - Q = c$, i.e., $Q = 1-c$. The aggregate equilibrium output Q that we found in part (d), $\frac{(1-c)[N(1-\alpha)+1]}{N(1-\alpha)-\alpha+2}$, lies below the socially optimal level, $1 - c$, because the output difference

$$(1-c) - \frac{(1-c)\,[N(1-\alpha)+1]}{N(1-\alpha)-\alpha+2} = \frac{(1-c)(1-\alpha)}{N(1-\alpha)-\alpha+2}$$

is weakly positive given that $c, \alpha \in [0, 1]$ and $N \geq 1$ by assumption.

However, this output difference, $\frac{(1-c)(1-\alpha)}{N(1-\alpha)-\alpha+2}$, is decreasing in N, approaching zero when $N \to +\infty$ (perfectly competitive private market). Similarly, this output difference is decreasing in α, becoming zero at $\alpha = 1$ (when the public firm assigns complete weight to consumer surplus and producing $q_0 = 1 - c$ units).

Exercise #2.24: Managerial Incentives in Cournot – Based on Fershtman and Judd (1987)^C

2.24 Consider an industry with two firms competing à la Cournot, facing a marginal production cost $c = 1/2$, and an inverse demand function $p(Q) = a - Q$, where Q denotes aggregate output, and $a > 1/2$.

We seek to analyze equilibrium behavior in the following sequential-move game, which describes the incentives between the firm owner and the firm manager:

(i) In stage 1, the owner of firm i simultaneously and independently chooses the profit share α_i that the manager receives in the second stage of the game. The owner seeks to maximize expected profits.

(ii) In stage 2, the realization of parameter a is observed only by the firm manager (not by the owner, who only knows that $a = 1$ with probability β and $a = 2$ with probability $1 - \beta$).

Observing the realization of a, the manager chooses firm i's output, q_i, to maximize his objective function

$$M_i = \alpha_i \pi_i + (1-\alpha_i) S_i,$$

where π_i denotes firm i's profit $\pi_i = (a - Q)q_i - \frac{1}{2}q_i$ and $S_i = (a - Q)q_i$ represents this firm's sales. For simplicity, let us assume that $\alpha_i \in [0, 1]$.

(a) *Second stage.* Find firm i's best response function in the second stage, $q_i(q_j, \alpha_i)$. How is it affected by α_i? Interpret.

- The manager of every firm i takes the profit shares α_i and α_j as given and solves

$$\max_{q_i \geq 0} M_i = \alpha_i \underbrace{\left[(a-Q)q_i - \frac{1}{2}q_i\right]}_{\pi_i} + (1-\alpha_i)\underbrace{[(a-Q)q_i]}_{S_i}.$$

Differentiating with respect to q_i yields

$$\alpha_i \left[a - 2q_i - q_j - \frac{1}{2}\right] + (1-\alpha_i)\left[a - 2q_i - q_j\right] = 0$$

which simplifies to

$$a - \frac{1}{2}\alpha_i - 2q_i - q_j = 0.$$

Solving for q_i, we find manager i's best response function in the second stage of the game:

$$q_i(q_j, \alpha_i) = \frac{a - \frac{1}{2}\alpha_i}{2} - \frac{1}{2}q_j.$$

Graphically, this best response function shifts downward when α_i increases, without altering its slope. Intuitively, when the manager receives a larger share of profits, he decreases firm i's output. However, when a larger share of his incentives comes from sales, he increases output.

(b) Find equilibrium output in the second stage, $q_i^*(\alpha_i, \alpha_j)$. Is it increasing or decreasing in α_i? What about in α_j?

- Manager j's best response function is symmetric, that is,

$$q_j(q_i, \alpha_j) = \frac{a - \frac{1}{2}\alpha_j}{2} - \frac{1}{2}q_i.$$

Inserting this best response function into $q_i(q_j, \alpha_i)$ yields

$$q_i = \frac{a - \frac{1}{2}\alpha_i}{2} - \frac{1}{2}\left(\frac{a - \frac{1}{2}\alpha_j}{2} - \frac{1}{2}q_i\right)$$

which is a function of q_i alone. Rearranging, we find

$$3q_i = a - \alpha_i + \frac{1}{2}\alpha_j$$

and solving for q_i, we obtain the equilibrium output in the second stage

$$q_i^* = \frac{2a - 2\alpha_i + \alpha_j}{6}.$$

(c) Find equilibrium price and profits in the second stage.

- Equilibrium price is

$$p^* = a - q_i^* - q_j^* = \frac{2a + \alpha_i + \alpha_j}{6}$$

and equilibrium profits are

$$\pi_i^* = p^* q_i^* - \frac{1}{2}q_i^* = \frac{(2a + \alpha_i + \alpha_j - 3)(2a - 2\alpha_i + \alpha_j)}{36}.$$

(d) *First stage.* Find firm i's best response function in the first stage, $\alpha_i(\alpha_j)$. Is it negatively sloped? Interpret.

- In the first stage, the owner, anticipating equilibrium output and profits in the second and third stages, respectively, chooses the profit share he provides to the manager, α_i, as follows:

$$\begin{aligned}\max_{\alpha_i} E[\pi_i^*] &= \beta\pi_i^*(a=1) + (1-\beta)\pi_i^*(a=2)\\ &= \beta\frac{(2+\alpha_i+\alpha_j-3)(2-2\alpha_i+\alpha_j)}{36}\\ &\quad+(1-\beta)\frac{(4+\alpha_i+\alpha_j-3)(4-2\alpha_i+\alpha_j)}{36}.\end{aligned}$$

Differentiating with respect to α_i yields

$$\frac{\beta(2-2\alpha_i+\alpha_j)-2\beta(\alpha_i+\alpha_j-1)}{36}$$

$$+\frac{(1-\beta)(4-2\alpha_i+\alpha_j)-2(1-\beta)(\alpha_i+\alpha_j+1)}{36}=0.$$

Rearranging, we obtain

$$2\beta-4\alpha_i-\alpha_j+2=0.$$

Solving for α_i, we find owner i's best response function in the first stage of the game:

$$\alpha_i(\alpha_j)=\frac{2\beta-\alpha_j+2}{4}.$$

Given that

$$\frac{\partial\alpha_i(\alpha_j)}{\partial\alpha_j}=-\frac{1}{4}<0,$$

this best response function is decreasing in α_j.

- Intuitively, when the owner of firm j increases the profit share he provides to his manager, the manager responds decreasing firm j's output in the second stage of the game. Anticipating this lower output from its rival, firm i's owner seeks to induce his manager to produce a larger output, which is achieved by reducing the profit share (or, alternatively, increasing the sales incentive). In short, profit shares are strategic substitutes.

(e) Find the equilibrium profit shares that the owner provides to the manager, α_i^*.

- The owner of firm j has a symmetric best response function $\alpha_j(\alpha_i)$. Inserting it into $\alpha_i(\alpha_j)$, we find

$$\alpha_i=\frac{2\beta-\overbrace{\left(\frac{2\beta-\alpha_i+2}{4}\right)}^{\alpha_j(\alpha_i)}+2}{4}$$

which depends on α_i alone. Rearranging and solving for α_i, we obtain the equilibrium profit share

$$\alpha_i^*=\frac{2\beta+2}{5}.$$

Inserting this result in the equilibrium output, we obtain that

$$q_i^*(\alpha_i^*) = \frac{2a - 2\frac{2\beta+2}{5} + \frac{2\beta+2}{5}}{6}$$
$$= \frac{5a - \beta - 1}{15}.$$

Therefore, equilibrium output is increasing in demand realization, a, and decreasing in the probability of a low demand, β.

(f) *Numerical example.* Evaluate your equilibrium results at $a = 1$ and $\beta = 1/2$.

- At these parameter values, we obtain that the equilibrium profit share is $\alpha_i^* = \frac{2\frac{1}{2}+2}{5} = \frac{3}{5}$ (60% of profits) and equilibrium output becomes $q_i^*(\alpha_i^*) = \frac{5-\frac{1}{2}-1}{15} = \frac{7}{30}$.

Exercise #2.25: Cournot Competition Under Incomplete Information-I^B

2.25 Consider an industry with two firms competing à la Cournot and facing inverse demand function $p(Q) = 1 - Q$, where Q denotes aggregate output. Firms interact in an incomplete information setting where firm 2's marginal costs are either c_H or c_L, where $c_H > c_L \geq 0$, occurring with probability p and $1 - p$, respectively. For simplicity, assume that firm 1's marginal costs are c_L, which is common knowledge among all players.

(a) Find the best response function for every firm i, $q_i^k(q_j^L, q_j^H)$, where $k = \{L, H\}$ denotes firm i's marginal cost (high or low).

- *Firm 1.* Since firm 1 has low costs by definition, it chooses q_1 that solves the following expected profit maximization problem:

$$\max_{q_1 \geq 0} \pi_1(q_1) = \underbrace{(1-p)\left(1 - q_1 - q_2^L\right) q_1}_{\text{When firm 2 has low costs}} + \underbrace{p\left(1 - q_1 - q_2^H\right) q_1}_{\text{When firm 2 has high costs}} - c_L q_1$$
$$= \left(1 - c_L - q_1 - (1-p)\, q_2^L - p q_2^H\right) q_1.$$

Assuming interior solutions, that is, $q_1 > 0$, the first-order condition satisfies

$$\frac{\partial \pi_1(q_1)}{\partial q_1} = 1 - c_L - 2q_1 - (1-p)\, q_2^L - p q_2^H = 0$$

such that the best response function of firm 1 becomes

$$q_1\left(q_2^L, q_2^H\right) = \frac{1 - c_L}{2} - \frac{p q_2^H + (1-p)\, q_2^L}{2}$$

which originates at $\frac{1-c_L}{2}$ and decreases in its rival's expected production, $p q_2^H + (1-p)\, q_2^L$.

In addition, the best response function becomes steeper when the "expected output" that firm 1 assesses from firm 2, $p q_2^H + (1-p)\, q_2^L$, increases. Intuitively, this occurs when probability p increases as long as $q_2^L > q_2^H$.

- *Firm 2*. This firm is privately informed about its production costs. When it has low costs, it chooses q_2^L to solve

$$\max_{q_2^L \geq 0} \pi_2^L\left(q_2^L\right) = \left(1 - q_1 - q_2^L\right) q_2^L - c_L q_2^L$$
$$= \left(1 - c_L - q_1 - q_2^L\right) q_2^L.$$

Assuming interior solutions, that is, $q_2^L > 0$, the first-order condition satisfies

$$\frac{\partial \pi_2^L\left(q_2^L\right)}{\partial q_2^L} = 1 - c_L - q_1 - 2q_2^L = 0$$

such that the best response function of firm 2 when its costs are low becomes

$$q_2^L\left(q_1\right) = \frac{1 - c_L}{2} - \frac{1}{2} q_1$$

which originates at $\frac{1-c_L}{2}$ and decreases in its rival's output, q_1, at a rate of $1/2$.

- When firm 2 has high costs, it chooses q_2^H that solves the following expected profit maximization problem:

$$\max_{q_2^H \geq 0} \pi_2^H\left(q_2^H\right) = \left(1 - q_1 - q_2^H\right) q_2^H - c_H q_2^H$$
$$= \left(1 - c_H - q_1 - q_2^H\right) q_2^H.$$

Assuming interior solutions, that is, $q_2^H > 0$, the first-order condition satisfies

$$\frac{\partial \pi_2^H\left(q_2^H\right)}{\partial q_2^H} = 1 - c_H - q_1 - 2q_2^H = 0$$

such that the best response function of firm 2 when its costs are high becomes

$$q_2^H\left(q_1\right) = \frac{1 - c_H}{2} - \frac{1}{2} q_1$$

which originates at $\frac{1-c_H}{2}$ and decreases in its rival's output, q_1, at a rate of $1/2$.

(b) Use your results from part (a) to find the Bayesian Nash equilibrium (BNE) of the game.

- Substituting firm 2's best response functions into the best response function of firm 1 yields

$$q_1 = \frac{1 - c_L}{2} - \frac{(1-p)\overbrace{\left(\frac{1 - c_L}{2} - \frac{1}{2} q_1\right)}^{q_2^L(q_1)} + p\overbrace{\left(\frac{1 - c_H}{2} - \frac{1}{2} q_1\right)}^{q_2^H(q_1)}}{2}$$

which simplifies to

$$4q_1 = 2 - 2c_L - 1 + q_1 + (1-p)\, c_L + p c_H$$

and solving for output q_1, we obtain firm 1's equilibrium output

$$q_1^* = \frac{1-(1+p)\,c_L + pc_H}{3}.$$

- Substituting the above results into the best response functions of firm 2,

$$q_2^{L*} = \frac{1-c_L}{2} - \frac{1}{2}\left(\frac{1-(1+p)\,c_L+pc_H}{3}\right) = \frac{2-(2-p)\,c_L - pc_H}{6}$$

$$q_2^{H*} = \frac{1-c_H}{2} - \frac{1}{2}\left(\frac{1-(1+p)\,c_L+pc_H}{3}\right) = \frac{2+(1+p)\,c_L-(3+p)\,c_H}{6}.$$

(c) *Comparative statics.* How are the equilibrium output levels you found in part (b) affected by changes in c_H, c_L, and p? Interpret.

- Differentiating equilibrium output with respect to c_H, we find

$$\frac{\partial q_1^*}{\partial c_H} = \frac{p}{3} > 0$$

$$\frac{\partial q_2^{L*}}{\partial c_H} = -\frac{p}{6} < 0$$

$$\frac{\partial q_2^{H*}}{\partial c_H} = -\frac{3+p}{6} < 0.$$

In other words, as it becomes more costly for the high-cost type to produce the good, output of firm 1 increases and that of firm 2 decreases.

- Differentiating equilibrium output with respect to c_L,

$$\frac{\partial q_1^*}{\partial c_L} = -\frac{1+p}{3} < 0$$

$$\frac{\partial q_2^{L*}}{\partial c_L} = -\frac{2-p}{6} < 0$$

$$\frac{\partial q_2^{H*}}{\partial c_L} = \frac{1+p}{6} > 0$$

such that, as it becomes more costly for the low-cost type to produce the good, output of firm 1 decreases while that of low-cost (high-cost) firm 2 decreases (increases, respectively).

- Differentiating equilibrium output with respect to probability p,

$$\frac{\partial q_1^*}{\partial p} = \frac{c_H - c_L}{3} > 0$$

$$\frac{\partial q_2^{L*}}{\partial p} = -\frac{c_H - c_L}{6} < 0$$

$$\frac{\partial q_2^{H*}}{\partial p} = -\frac{c_H - c_L}{6} < 0.$$

Intuitively, as firms are more likely to face high costs, output of firm 1 (2) increases (decreases, respectively).

(d) *Numerical example.* Evaluate equilibrium output at $c_H = 1/2$, $c_L = 1/4$, and $p = 1/3$.
- At these parameter values, we obtain that firm 1's output is

$$q_1^* = \frac{1 - \left(1 + \frac{1}{3}\right)\frac{1}{4} + \frac{1}{3}\frac{1}{2}}{3} = \frac{5}{18} \simeq 0.28$$

and firm 2's output is

$$q_2^{L*} = \frac{2 - \left(2 - \frac{1}{3}\right)\frac{1}{4} - \frac{1}{3}\frac{1}{2}}{6} = \frac{17}{72} \simeq 0.24, \text{ and}$$

$$q_2^{H*} = \frac{2 + \left(1 + \frac{1}{3}\right)\frac{1}{4} - \left(3 + \frac{1}{3}\right)\frac{1}{2}}{6} = \frac{1}{9} \simeq 0.11.$$

Exercise #2.26: Cournot Competition Under Incomplete Information-II^C

2.26 Consider an industry with two firms competing à la Cournot and facing inverse demand function $p(Q) = 1 - Q$, where $Q = q_1 + q_2$ denotes aggregate output. Every firm i is privately informed about its marginal cost, high or low, denoted as c_H and c_L, respectively, where $1 > c_H > c_L = 0$. Finally, consider that, while firm j cannot observe the realization of firm i's marginal cost (c_H or c_L), firm j knows that both types are equally likely. Firms then interact in a simultaneous-move game of incomplete information, and in this exercise we seek to find the Bayesian Nash equilibrium of the game.

(a) Find the best response function for every firm i when its marginal costs are low, $q_i^L(q_j^H, q_j^L)$.
- When firm i has low costs, it chooses q_i^L to solve the following expected profit maximization problem:

$$\max_{q_i^L \geq 0} \pi_i^L\left(q_i^L\right) = \overbrace{\frac{1}{2}\left(1 - q_i^L - q_j^L\right)q_i^L}^{\text{Profits if } j \text{ is low cost}} + \overbrace{\frac{1}{2}\left(1 - q_i^L - q_j^H\right)q_i^L}^{\text{Profits if } j \text{ is high cost}}$$

$$= \left(1 - q_i^L - \frac{q_j^L + q_j^H}{2}\right)q_i^L$$

which does not include the production cost of firm i because $c_L = 0$.

Differentiating with respect to q_i^L and assuming an interior solution yield

$$\frac{\partial \pi_i^L\left(q_i^L\right)}{\partial q_i^L} = 1 - 2q_i^L - \frac{q_j^L + q_j^H}{2} = 0.$$

Solving for q_i^L, we find the best response function of firm i when its costs are low as follows:

$$q_i^L\left(q_j^L, q_j^H\right) = \frac{1}{2} - \frac{q_j^L + q_j^H}{4}$$

which originates at $1/2$ and decreases in its rival's output at a rate of $1/4$ (both when its rival has low and high costs).

(b) Find the best response function for every firm i when its marginal costs are high, $q_i^H(q_j^H, q_j^L)$.

- When firm i has high costs, it chooses q_i^H to solve the following expected profit maximization problem:

$$\max_{q_i^H \geq 0} \pi_i^H\left(q_i^H\right) = \overbrace{\frac{1}{2}\left(1 - q_i^H - q_j^L\right) q_i^H}^{\text{Profits if } j \text{ is low cost}} + \overbrace{\frac{1}{2}\left(1 - q_i^H - q_j^H\right) q_i^H}^{\text{Profits if } j \text{ is high cost}} - c_H q_i^H$$

$$= \left(1 - c_H - q_i^H - \frac{q_j^L + q_j^H}{2}\right) q_i^H.$$

Assuming interior solutions, that is, $q_i^H > 0$, the first-order condition satisfies

$$\frac{\partial \pi_i^H\left(q_i^H\right)}{\partial q_i^H} = 1 - c_H - 2q_i^H - \frac{q_j^L + q_j^H}{2} = 0$$

such that the best response function of firm i when its costs are high becomes

$$q_i^H\left(q_j^L, q_j^H\right) = \frac{1 - c_H}{2} - \frac{q_j^L + q_j^H}{4}$$

which originates at $\frac{1-c_H}{2}$, but decreases in its rival's output at a rate of $1/4$ (both when its rival has low and high costs).

- Comparing it with firm i's best response function when its costs are low, $q_i^L\left(q_j^L, q_j^H\right)$, we can see that, for a given profile of firm j's output, $\left(q_j^L, q_j^H\right)$, firm i responds producing a larger output when its own costs are low than when they are high if

$$\frac{1}{2} > \frac{1 - c_H}{2}$$

which holds since $c_H > 0$ by definition.

(c) Use your results from parts (a) and (b) to find the Bayesian Nash equilibrium (BNE) of the game.

- Since firms i and j are symmetric, we impose symmetry on the equilibrium output that

$$q^L = q_i^L = q_j^L$$
$$q^H = q_i^H = q_j^H.$$

Substituting the above results into the best response functions we found in parts (a) and (b) yields

$$q^L = \frac{1}{2} - \frac{q^L + q^H}{4} \quad \text{from part (a), and}$$

$$q^H = \frac{1-c_H}{2} - \frac{q^L+q^H}{4} \text{ from part (b).}$$

Simultaneously solving for q^L and q^H, we find that the equilibrium output levels are

$$q^{L*} = \frac{4+c_H}{12} \text{ and } q^{H*} = \frac{4-5c_H}{12},$$

where, as expected, every firm i produces more output when its marginal cost is low than when it is high, $q^{L*} > q^{H*}$.

- Therefore, the BNE is

$$\left(q_i^{L*}, q_i^{H*}\right) = \left(\frac{4+c_H}{12}, \frac{4-5c_H}{12}\right)$$

for every firm i. For instance, if $c_H = \frac{1}{2}$, we obtain equilibrium output pair of

$$\left(q_i^{L*}, q_i^{H*}\right) = \left(\frac{3}{8}, \frac{1}{8}\right).$$

Exercise #2.27: Cournot Competition Under Incomplete Information-III^C

2.27 Consider the industry in Exercise 2.26, but assume that the marginal production cost of firm 1 (2) is high, $c_H > 0$, with probability p (q, respectively), where $p, q \in [0, 1]$. Similarly, the marginal cost of firm 1 (2) is low, $c_L = 0$, with probability $1-p$ ($1-q$, respectively).

(a) Find firm 1's best response function when its marginal cost is low, $q_1^L(q_2^H, q_2^L)$.

- When firm 1 has low marginal cost, it chooses q_1^L to solve the following expected profit maximization problem:

$$\max_{q_1^L \geq 0} \pi_1^L\left(q_1^L\right) = \overbrace{(1-q)\left(1-q_1^L-q_2^L\right)q_1^L}^{\text{Profits if firm 2 is of low cost}} + \overbrace{q\left(1-q_1^L-q_2^H\right)q_1^L}^{\text{Profits if firm 2 is of high cost}}$$

$$= \left(1-q_1^L-(1-q)q_2^L-qq_2^H\right)q_1^L$$

which does not include the production cost of firm 1 because $c_L = 0$.

Differentiating with respect to q_1^L, and assuming an interior solution, yields

$$\frac{\partial \pi_1^L\left(q_1^L\right)}{\partial q_1^L} = 1 - 2q_1^L - (1-q)q_2^L - qq_2^H = 0.$$

Solving for q_1^L, we find the best response function of firm 1 when its cost is low, as follows:

$$q_1^L\left(q_2^L, q_2^H\right) = \frac{1}{2} - \frac{(1-q)q_2^L + qq_2^H}{2}$$

which originates at $1/2$ and decreases in its rival's output at a rate of $\frac{1-q}{2}$ ($\frac{q}{2}$) when its rival has low (high) marginal cost. Alternatively, firm 1's output decreases in its rival's expected output at a rate of $1/2$, as captured by the term $(1-q)q_2^L + qq_2^H$.

(b) Find firm 2's best response function when its marginal cost is low, $q_2^L(q_1^H, q_1^L)$.

- When firm 2 has low marginal cost, it chooses q_2^L to solve the following expected profit maximization problem:

$$\max_{q_2^L \geq 0} \pi_2^L\left(q_2^L\right) = \overbrace{(1-p)\left(1-q_2^L-q_1^L\right)q_2^L}^{\text{Profits if firm 1 is of low cost}} + \overbrace{p\left(1-q_2^L-q_1^H\right)q_2^L}^{\text{Profits if firm 1 is of high cost}}$$

$$= \left(1-q_2^L-(1-p)q_1^L-pq_1^H\right)q_2^L$$

which does not include the production cost of firm 2 because $c_L = 0$.
Differentiating with respect to q_2^L, and assuming an interior solution, yields

$$\frac{\partial \pi_2^L\left(q_2^L\right)}{\partial q_2^L} = 1-2q_2^L-(1-p)q_1^L-pq_1^H = 0.$$

Solving for q_2^L, we find the best response function of firm 2 when its cost is low, as follows:

$$q_2^L\left(q_1^L, q_1^H\right) = \frac{1}{2} - \frac{pq_1^H+(1-p)q_1^L}{2}$$

which originates at $1/2$ and decreases in its rival's output at a rate of $\frac{1-p}{2}$ ($\frac{p}{2}$) when its rival has low (high) marginal cost. Alternatively, firm 2's output decreases in its rival's expected output at a rate of $1/2$, as captured by the term $pq_1^H + (1-p)q_1^L$.

(c) Find firm 1's best response function when its marginal cost is high, $q_1^H(q_2^H, q_2^L)$.

- When firm 1 has high cost, it chooses q_1^H to solve the following expected profit maximization problem:

$$\max_{q_1^H \geq 0} \pi_1^H\left(q_1^H\right) = \overbrace{(1-q)\left(1-q_1^H-q_2^L\right)q_1^H}^{\text{Profits if firm 2 is of low cost}} + \overbrace{q\left(1-q_1^H-q_2^H\right)q_1^H}^{\text{Profits if firm 2 is of high cost}} - c_H q_1^H$$

$$= \left(1-c_H-q_1^H-(1-q)q_2^L-qq_2^H\right)q_1^H$$

Assuming an interior solution, that is, $q_1^H > 0$, the first-order condition satisfies

$$\frac{\partial \pi_1^H\left(q_1^H\right)}{\partial q_1^H} = 1-c_H-2q_1^H-(1-q)q_2^L-qq_2^H = 0$$

such that the best response function of firm 1 when its cost is high becomes

$$q_1^H\left(q_2^L, q_2^H\right) = \frac{1-c_H}{2} - \frac{(1-q)q_2^L+qq_2^H}{2}$$

which originates at $\frac{1-c_H}{2}$, but decreases in its rival's output at a rate of $\frac{1-q}{2}$ ($\frac{q}{2}$) when its rival has low (high) marginal cost. Alternatively, firm 1's output decreases in its rival's expected output at a rate of $1/2$, as captured by the term $(1-q)q_2^L + qq_2^H$.

- Comparing it with firm i's best response function when its cost is low, $q_i^L\left(q_j^L, q_j^H\right)$, we can see that, for a given profile of firm j's output, $\left(q_j^L, q_j^H\right)$, firm i responds producing a larger output when its own cost is low than when it is high. Graphically, $q_i^L\left(q_j^L, q_j^H\right)$ originates at $\frac{1}{2}$ while $q_i^H\left(q_j^L, q_j^H\right)$ originates at $\frac{1-c_H}{2}$, and both best response functions have the same slope, thus being parallel to each other since $c_H > 0$ by definition.

(d) Find firm 2's best response function when its marginal cost is high, $q_2^H(q_1^H, q_1^L)$.

- When firm 2 has high cost, it chooses q_2^H to solve the following expected profit maximization problem:

$$\max_{q_2^H \geq 0} \pi_2^H\left(q_2^H\right) = \overbrace{(1-p)\left(1-q_2^H-q_1^L\right)q_2^H}^{\text{Profits if firm 1 is of low cost}} + \overbrace{p\left(1-q_2^H-q_1^H\right)q_2^H}^{\text{Profits if firm 1 is of high cost}} - c_H q_2^H$$

$$= \left(1-c_H-q_2^H-(1-p)q_1^L-pq_1^H\right)q_2^H$$

Assuming an interior solution, that is, $q_2^H > 0$, the first order condition satisfies

$$\frac{\partial \pi_2^H\left(q_2^H\right)}{\partial q_2^H} = 1-c_H-2q_2^H-(1-p)q_1^L-pq_1^H = 0$$

such that the best response function of firm 2 when its cost is high becomes

$$q_2^H\left(q_1^L, q_1^H\right) = \frac{1-c_H}{2} - \frac{(1-p)q_1^L+pq_1^H}{2}$$

which originates at $\frac{1-c_H}{2}$, but decreases in its rival's output at a rate of $\frac{1-q}{2}$($\frac{q}{2}$) when its rival has low (high) marginal cost. Alternatively, firm 2's output decreases in its rival's expected output at a rate of $1/2$, as captured by the term $(1-p)q_1^L + pq_1^H$.

- The comparison between $q_2^H\left(q_j^L, q_j^H\right)$ and $q_2^L\left(q_j^L, q_j^H\right)$ is analogous to that in part (c) for firm 1.

(e) Use your results from parts (a) to (d) to find the BNE of the game. [*Hint*: You cannot invoke symmetry since the best response functions are not symmetric in this setting.]

- Substituting $q_2^H\left(q_1^L, q_1^H\right)$ and $q_2^L\left(q_1^L, q_1^H\right)$ into $q_1^H\left(q_2^L, q_2^H\right)$ and $q_1^L\left(q_2^L, q_2^H\right)$, we obtain

$$q_1^H = \frac{1-c_H}{2} - \frac{(1-q)}{2}\overbrace{\left[\frac{1}{2} - \frac{(1-p)q_1^L+pq_1^H}{2}\right]}^{q_2^L}$$

$$-\frac{q}{2}\overbrace{\left[\frac{1-c_H}{2} - \frac{(1-p)q_1^L+pq_1^H}{2}\right]}^{q_2^H}$$

$$= \frac{1-(2-q)\,c_H + pq_1^H + (1-p)\,q_1^L}{4}$$

$$q_1^L = \frac{1}{2} - \frac{(1-q)}{2}\overbrace{\left[\frac{1}{2} - \frac{(1-p)q_1^L + pq_1^H}{2}\right]}^{q_2^L} - \frac{q}{2}\overbrace{\left[\frac{1-c_H}{2} - \frac{(1-p)q_1^L + pq_1^H}{2}\right]}^{q_2^H}$$

$$= \frac{1+qc_H + pq_1^H + (1-p)\,q_1^L}{4}$$

Rearranging the above set of linear equations, we obtain

$$(4-p)\,q_1^H - (1-p)\,q_1^L = 1-(2-q)\,c_H$$
$$-pq_1^H + (3+p)\,q_1^L = 1+qc_H$$

Solving for the system of equations simultaneously, we obtain

$$q_1^{H*} = \frac{2-(3-2q+p)c_H}{6}$$
$$q_1^{L*} = \frac{2+(2q-p)c_H}{6}$$

Substituting q_1^{H*} and q_1^{L*} into firm 2's best response functions, we have

$$q_2^{H*} = \frac{1-c_H}{2} - \frac{(1-p)}{2}\overbrace{\left[\frac{2+(2q-p)c_H}{6}\right]}^{q_1^L} - \frac{p}{2}\overbrace{\left[\frac{2-(3-2q+p)c_H}{6}\right]}^{q_1^H}$$

$$= \frac{2-(3-2p+q)c_H}{6}$$

$$q_2^{L*} = \frac{1}{2} - \frac{(1-p)}{2}\overbrace{\left[\frac{2+(2q-p)c_H}{6}\right]}^{q_1^L} - \frac{p}{2}\overbrace{\left[\frac{2+(2q-p-3)c_H}{6}\right]}^{q_1^H}$$

$$= \frac{2+(2p-q)c_H}{6}$$

Therefore, equilibrium output levels are

$$q_1^{L*} = \frac{2+(2q-p)c_H}{6}$$
$$q_1^{H*} = \frac{2-(3-2q+p)c_H}{6}$$
$$q_2^{L*} = \frac{2+(2p-q)c_H}{6}$$
$$q_2^{H*} = \frac{2-(3-2p+q)c_H}{6}$$

(f) *Comparative statics.* How are the equilibrium output levels $(q_1^H, q_1^L, q_2^H, q_2^L)$ affected by a marginal increase in p? And how about a marginal increase in q? Interpret.

- Differentiating equilibrium output levels with respect to p, we obtain

$$\frac{\partial q_1^{L*}}{\partial p} = -\frac{c_H}{6} < 0$$

$$\frac{\partial q_1^{H*}}{\partial p} = -\frac{c_H}{6} < 0$$

$$\frac{\partial q_2^{L*}}{\partial p} = \frac{c_H}{3} > 0$$

$$\frac{\partial q_2^{H*}}{\partial p} = \frac{c_H}{3} > 0$$

As firm 1's marginal cost is more likely to be high (p increases), firm 1 decreases its output while firm 2 increases its own, which happens regardless of firm 1's true marginal cost.

- Differentiating equilibrium output levels with respect to q, we find that

$$\frac{\partial q_1^{L*}}{\partial q} = \frac{c_H}{3} > 0$$

$$\frac{\partial q_1^{H*}}{\partial q} = \frac{c_H}{3} > 0$$

$$\frac{\partial q_2^{L*}}{\partial q} = -\frac{c_H}{6} < 0$$

$$\frac{\partial q_2^{H*}}{\partial q} = -\frac{c_H}{6} < 0$$

As firm 2's marginal cost is more likely to be high (q increases), firm 2 decreases its output while firm 1 increases its own, which happens regardless of firm 2's true marginal cost.

(g) *Symmetric probabilities.* Evaluate your equilibrium results in the special case where both firms' costs occur with the same probability, $p = q$. What if, in addition, these probabilities are both 1/2?

- When $p = q$, firms are symmetric, equilibrium output levels become

$$q_1^{L*} = q_2^{L*} = q^{L*} = \frac{2 + pc_H}{6} \quad \text{and}$$

$$q_1^{H*} = q_2^{H*} = q^{H*} = \frac{2 - (3 - p)\,c_H}{6}$$

- In addition, if $p = q = \frac{1}{2}$, equilibrium output levels become

$$q_1^{L*} = q_2^{L*} = q^{L*} = \frac{4 + c_H}{12} \quad \text{and}$$

$$q_1^{H*} = q_2^{H*} = q^{H*} = \frac{4 - 5c_H}{12}$$

(h) *Special cases*. Evaluate your equilibrium results in the special case where both firms' types are certain, as under complete information: (1) $p = q = 1$, (2) $p = 1$ and $q = 0$, (3) $p = 0$ and $q = 1$, and (4) $p = q = 0$. Interpret.

- When $p = q = 1$, both firms have high marginal costs, the equilibrium output levels become

$$q_1^{H*} = q_2^{H*} = q^{H*} = \frac{1 - c_H}{3}$$

which coincide with the results in the standard Cournot game under complete information when both firms face the same marginal cost, c_H.

- When $p = 1$ and $q = 0$, firm 1 has high marginal cost while firm 2 has low marginal cost, the equilibrium output levels are

$$q_1^{H*} = \frac{1 - 2c_H}{3}$$

$$q_2^{L*} = \frac{1 + c_H}{3}$$

As expected, the firm with low marginal cost produces more units than the firm with high marginal cost.

- When $p = 0$ and $q = 1$, firm 2 has high marginal cost and firm 1 has low marginal cost, the equilibrium output levels are

$$q_1^{L*} = \frac{1 + c_H}{3}$$

$$q_2^{H*} = \frac{1 - 2c_H}{3}$$

As expected, the firm with low marginal cost produces more units than the firm with high marginal cost.

- When $p = q = 0$, both firms have low marginal costs, the equilibrium output levels become

$$q_1^{L*} = q_2^{L*} = q^{L*} = \frac{1}{3}$$

which coincide with the results in the standard Cournot game under complete information when both firms face zero marginal costs.

Exercise #2.28: Cournot Competition When Firms Are Asymmetrically Informed of Product Differentiation^B

2.28 Consider two firms competing à la Cournot on differentiated goods facing an inverse demand function

$$p(q_1, q_2) = 1 - q_1 - \gamma q_2,$$

where parameter γ measures the degree of product differentiation. When $\gamma = 0$, products are completely differentiated, and when $\gamma = 1$, products are completely homogeneous. Assume that firm 1 has complete knowledge on the realization of γ, but firm 2 only knows that this parameter is uniformly distributed in the interval $[0, 1]$. For simplicity, assume that every firm faces a common marginal cost c, where $c \geq 0$.

(a) *Informed firm's best response function.* Set up firm 1's profit maximization problem and derive its best response function.

- Firm 1 chooses q_1 to solve the following profit maximization problem:

$$\max_{q_1 \geq 0} \pi_1(q_1) = (1 - q_1 - \gamma q_2) q_1 - c q_1$$

Taking the first-order condition with respect to q_1, and assuming interior solutions,

$$1 - 2q_1 - \gamma q_2 - c = 0$$

Rearranging and solving for q_1, the best response function of firm 1 becomes

$$q_1(q_2) = \frac{1-c}{2} - \frac{\gamma}{2} q_2$$

which originates at a vertical intercept of $\frac{1-c}{2}$ and decreases in the output of firm 2 at a rate of $\frac{\gamma}{2}$. For example, when firm 2 produces zero units, firm 1 produces $\frac{1-c}{2}$ units that resembles the monopolist's profit-maximizing output. However, when firm 2 produces $\frac{1-c}{\gamma}$ units, $q_1 = 0$ so that firm 1 is driven out of the market.

(b) *Uninformed firm's best response function.* Set up firm 2's expected profit maximization problem, and find its best response function.

- Firm 2 chooses q_2 to solve the expected profit maximization problem:

$$\max_{q_2 \geq 0} E[\pi_2(q_2)] = \int_0^1 [(1 - q_2 - \gamma q_1) q_2 - c q_2] f(\gamma) d\gamma$$

Since $\gamma \sim U[0, 1]$, its cumulative distribution function is $F(\gamma) = \gamma$ with density function of $f(\gamma) = F'(\gamma) = 1$. Substituting $f(\gamma) = 1$ into the expected profit maximization problem, we obtain

$$\max_{q_2 \geq 0} E[\pi_2(q_2)] = \int_0^1 (1 - q_2 - \gamma q_1 - c) q_2 d\gamma$$

$$= \left[(1 - q_2)\gamma - \frac{\gamma^2}{2} q_1 - c\gamma \right]_0^1 q_2$$

$$= \left[(1 - q_2) - \frac{1}{2} q_1 - c \right] q_2$$

Taking the first-order condition with respect to q_2, and assuming interior solutions,

$$1 - 2q_2 - \frac{1}{2} q_1 - c = 0$$

Rearranging and solving for q_2, the best response function of firm 2 becomes

$$q_2(q_1) = \frac{1-c}{2} - \frac{1}{4}q_1$$

which originates at a vertical intercept of $\frac{1-c}{2}$ and decreases in the output of firm 1 at a rate of $\frac{1}{4}$. This happens because firm 2 takes the expected value of $E(\gamma) = \frac{1}{2}$. Intuitively, when $\gamma < \frac{1}{2}$ $(\gamma \geq \frac{1}{2})$, firm 2's expectation of γ is higher (lower) than its actual value, that is, $\gamma < E(\gamma)$ $(\gamma \geq E(\gamma))$, so firm 2's output decreases at a faster (slower) rate in firm 1's output than firm 1's output decreases in firm 2's, where the slope of its best response function is steeper (less steep) than its rival's.

(c) *Equilibrium outcome.* Find the equilibrium output and profits of every firm.

- Substituting the best response function of firm 2 into that of firm 1, we find

$$q_1 = \frac{1-c}{2} - \frac{\gamma}{2}\underbrace{\left(\frac{1-c}{2} - \frac{1}{4}q_1\right)}_{q_2}$$

which we can rearrange to yield

$$\left(1 - \frac{\gamma}{8}\right)q_1 = \frac{2-\gamma}{2}\frac{1-c}{2}$$

Solving for q_2, equilibrium output of firm 1 becomes

$$q_1^* = \frac{2(2-\gamma)(1-c)}{8-\gamma}$$

Plugging into the best response function of firm 2, its equilibrium output becomes

$$q_2^* = \frac{1-c}{2} - \frac{1}{4}\underbrace{\frac{2(2-\gamma)(1-c)}{8-\gamma}}_{q_1^*}$$

$$= \frac{3(1-c)}{8-\gamma}$$

- Finally, inserting equilibrium output into the profit function of firm 1 yields

$$\pi_1^* = \left(1 - q_1^* - \gamma q_2^* - c\right)q_1^*$$
$$= \left(\frac{2(2-\gamma)(1-c)}{8-\gamma}\right)^2$$

- Similarly, substituting equilibrium output into the profit function of firm 2 yields

$$\pi_2^* = \left(1 - q_2^* - \gamma q_1^* - c\right)q_2^*$$
$$= \frac{3\left(5 - 5\gamma + 2\gamma^2\right)(1-c)^2}{(8-\gamma)^2}$$

(d) When will firm 1 produce more units of output than firm 2? Explain.

- Firm 1 produces more output than firm 2 when

$$\frac{2(2-\gamma)(1-c)}{8-\gamma} > \frac{3(1-c)}{8-\gamma}$$

which simplifies to

$$2-\gamma > \frac{3}{2}$$

that yields $\gamma < \frac{1}{2}$. Intuitively, when goods are less homogeneous than what firm 2 expects, firm 2 produces less output, which, in turn, induces firm 1 to produce more units.

(e) *Comparative statics.* How do equilibrium output and profits change with γ? Explain.

- Differentiating equilibrium output of firm 1 and 2 with respect to γ, we find that

$$\frac{\partial q_1^*}{\partial \gamma} = -\frac{12(1-c)}{(8-\gamma)^2} < 0$$

$$\frac{\partial q_2^*}{\partial \gamma} = \frac{3(1-c)}{(8-\gamma)^2} > 0$$

When firm 1 observes that goods become more homogeneous (higher γ), it expects a more intensive competition and, thus, reduces output. Firm 2, however, only observes that firm 1 reduces output but does not know that its product becomes more homogeneous with firm 1, so firm 2 increases output as firms regard each other's output as strategic substitutes.

- Differentiating equilibrium profits of firm 1 and 2 with respect to γ, we obtain that

$$\frac{\partial \pi_1^*}{\partial \gamma} = 2\left(\frac{2(2-\gamma)(1-c)}{8-\gamma}\right) \times \underbrace{\left[\frac{-12(1-c)}{(8-\gamma)^2}\right]}_{\frac{\partial q_1^*}{\partial \gamma}}$$

$$= -\frac{48(2-\gamma)(1-c)^2}{(8-\gamma)^3} < 0$$

and

$$\frac{\partial \pi_2^*}{\partial \gamma} = -\frac{3\left[(5-4\gamma)(8-\gamma) - 2\left(5-5\gamma+2\gamma^2\right)\right](1-c)^2}{(8-\gamma)^3}$$

$$= -\frac{9(10-9\gamma)(1-c)^2}{(8-\gamma)^3} < 0$$

Intuitively, when competition becomes more intensive, every firm obtains lower profits.

(f) *Numerical example.* Evaluate your results in part (c) assuming $\gamma = \frac{1}{2}$. Interpret.

- Substituting $\gamma = \frac{1}{2}$ into the equilibrium output in part (c), we find that

$$q_1^* = \frac{2\left(2 - \frac{1}{2}\right)(1-c)}{8 - \frac{1}{2}} = \frac{2(1-c)}{5}$$

$$q_2^* = \frac{3(1-c)}{8 - \frac{1}{2}} = \frac{2(1-c)}{5}$$

- Inserting $\gamma = \frac{1}{2}$ into the equilibrium profits in part (c), we obtain that

$$\pi_1^* = \left(\frac{2\left(2 - \frac{1}{2}\right)(1-c)}{8 - \frac{1}{2}}\right)^2 = \frac{4(1-c)^2}{25}$$

$$\pi_2^* = \frac{3\left(5 - 5 \times \frac{1}{2} + 2 \times \frac{1}{2^2}\right)(1-c)^2}{\left(8 - \frac{1}{2}\right)^2} = \frac{4(1-c)^2}{25}$$

Since firm 2's expected value of γ coincides with its actual value, that is, $E(\gamma) = \gamma = \frac{1}{2}$, equilibrium output, as well as profits, of firm 1 and 2 coincide.

Exercise #2.29: Nonlinear Pricing in Oligopoly – Based on Harrison and Kline (2001)^C

2.29 Consider an industry with $n \geq 2$ firms competing ? la Cournot, facing an inverse demand function $p(Q) = 1 - Q$, where $Q \geq 0$ denotes aggregate output. Every firm i faces a total cost function $C(q) = F + cq$, where $F \geq 0$ denotes fixed costs while $c > 0$ represents the marginal production cost. In the first stage, every firm i simultaneously and independently offers a contract (q_i, A_i), specifying an output level, q_i, and an access fee, A_i, to consumers. In the second stage, consumers respond choosing a contract from one of the firms, (q_i, A_i), or not accepting any contract at all (no purchase).

Harrison and Kline (2001) show that, in equilibrium, every firm i chooses an output level $q_i = q^c$, which coincides with that under marginal cost pricing. The equilibrium access fee is

$$A_i = \min\left\{CS(q^c), \frac{cq^c}{|\varepsilon(q^c)|(n-1)}\right\},$$

where $CS(q^c)$ denotes the consumer surplus evaluated at q^c and $\varepsilon(q^c)$ represents the price elasticity of demand evaluated at q^c. (For a formal proof, see Harrison and Kline, 2001, pp. 368–371.)

(a) Find the equilibrium access fee A_i in the above setting.

- Consumer surplus in this setting is

$$CS(q^c) = \frac{1}{2}\left(q^c\right)^2$$

$$= \frac{1}{2}(1-c)^2$$
$$= \frac{(1-c)^2}{2}$$

while price elasticity of demand evaluated at q^c is

$$\varepsilon(q^c) = \frac{dQ}{dP} \times \frac{P(q^c)}{q^c} = -\frac{c}{1-c}$$

entailing that the second term in the min operator of the access fee becomes

$$\frac{cq^c}{|\varepsilon(q^c)|\,(n-1)} = \frac{c(1-c)}{\frac{c}{1-c}(n-1)}$$
$$= \frac{(1-c)^2}{n-1}.$$

Therefore, $CS(q^c) \leq \frac{cq^c}{|\varepsilon(q^c)|(n-1)}$ when $\frac{1}{2} \leq \frac{1}{n-1}$, which occurs if and only if $n \leq 3$. In summary, we can say that

$$A_i = \begin{cases} \frac{(1-c)^2}{2} & \text{if } n \leq 3, \\ \frac{(1-c)^2}{n-1} & \text{otherwise.} \end{cases}$$

(b) Evaluate the equilibrium profits, consumer surplus, and total welfare when the firm offers the above nonlinear contract.

- Equilibrium profits for each individual firm under the nonlinear contract are

$$\pi^{NL} = A_i + (c-c)q^c - F = A_i - F$$

since firms charge an access fee A_i to consumers but then sell q^c units at marginal cost pricing, so $p = c$. Profits then originate exclusively from the access fee A_i, which gives rise to two cases:
  - $A_i = CS(q^c) = \frac{(1-c)^2}{2}$ if $n \leq 3$, and
  - $A_i = \frac{(1-c)^2}{n-1}$ otherwise.
- Consumer surplus is then zero, $CS^{NL} = 0$, if $n \leq 3$, since in this case the consumer pays an access fee A_i that coincides with his surplus. Otherwise, the consumer surplus becomes

$$CS^{NL} = CS(q^c) - A_i$$
$$= \frac{(1-c)^2}{2} - \frac{(1-c)^2}{n-1}$$
$$= \frac{(n-3)(1-c)^2}{2\,(n-1)}.$$

- Finally, total welfare when $n \leq 3$ is the sum of consumer surplus (zero in this case) and aggregate profits, that is,

$$W^{NL} = \frac{n(1-c)^2}{2} - nF.$$

Otherwise, total welfare becomes

$$W^{NL} = \frac{(n-3)(1-c)^2}{2\,(n-1)} - nF + n\frac{(1-c)^2}{n-1}$$
$$= \frac{3(1-c)^2}{2} - nF.$$

(c) Consider now a setting where firms cannot offer a contract, but, instead, sell all its output at the same price p without an access fee. Find the equilibrium price that firms offer, their equilibrium profits, consumer surplus, and total welfare.

- In this context, every firm solves

$$\max_{q_i \geq 0} \pi_i = [1 - q_i - (n-1)q_{-i}]q_i - (F + cq_i)$$

yielding best response function

$$q_i(q_{-i}) = \frac{1-c}{2} - \frac{n-1}{2}q_{-i}.$$

In a symmetric equilibrium $q_i = q$ for every firm i, we obtain

$$q = \frac{1-c}{2} - \frac{n-1}{2}q$$

which, solving for q, yields an equilibrium output

$$q = \frac{1-c}{n+1}$$

with equilibrium price

$$p = 1 - n\underbrace{\frac{1-c}{n+1}}_{q} = \frac{n+1-n(1-c)}{n+1} = \frac{1+nc}{n+1}.$$

- Equilibrium profits for each individual firm are then

$$\pi = \frac{1+nc}{n+1}\frac{1-c}{n+1} - \left(F + c\frac{1-c}{n+1}\right)$$
$$= \frac{(1-c)^2}{(n+1)^2} - F$$

and consumer surplus is

$$CS = \frac{1}{2}\left(n\frac{1-c}{n+1}\right)^2 = \frac{n^2(1-c)^2}{2(n+1)^2}$$

implying that total welfare is

$$W = \frac{n^2(1-c)^2}{2(n+1)^2} + n\left[\frac{(1-c)^2}{(n+1)^2} - F\right]$$
$$= \frac{n(n+2)(1-c)^2}{2(n+1)^2} - nF$$

(d) Compare equilibrium profits, consumer surplus, and total welfare with and without access fees.

- Equilibrium profits satisfy
  - Case 1, where $n \leq 3$:

$$\pi^{NL} = \frac{(1-c)^2}{2} - F > \frac{(1-c)^2}{(n+1)^2} - F = \pi$$

which simplifies to

$$\frac{[(n+1)^2 - 2](1-c)^2}{2(n+1)^2} > 0$$

and further rearranging yields

$$\frac{(n^2 + 2n - 1)(1-c)^2}{2(n+1)^2} > 0.$$

Since $(1-c)^2 > 0$ and $(n+1)^2 > 0$, we need to check if $n^2 + 2n - 1 > 0$. Solving for n, we find that this inequality holds if $n > \sqrt{2} - 1 \simeq 0.41$, which holds since $n \geq 2$ by definition. Therefore, profits are higher when the firm offers a nonlinear contract, $\pi^{NL} > \pi$.

  - Case 2, where $n > 3$

$$\pi^{NL} = \frac{(1-c)^2}{n-1} - F > \frac{(1-c)^2}{(n+1)^2} - F = \pi$$

which simplifies to

$$\frac{[(n+1)^2 - (n-1)](1-c)^2}{(n-1)(n+1)^2} > 0$$

and further rearranging yields

$$\frac{(n^2 + n + 2)(1-c)^2}{(n-1)(n+1)^2} > 0.$$

Since every term on the left side of the above inequality is positive, profits in this case are also higher when the firm offers a nonlinear contract than otherwise, $\pi^{NL} > \pi$.

 - In summary, profits are larger under a nonlinear contract regardless of the number of firms competing in the industry, i.e., both in Case 1 and in Case 2.
- Consumer surplus satisfies
 - Case 1, where $n \leq 3$

$$CS^{NL} = 0 < \frac{n^2(1-c)^2}{2(n+1)^2} = CS$$

 implying that consumers derive lower surplus when the firm offers a nonlinear contract, $CS^{NL} < CS$.
 - Case 2, where $n > 3$

$$CS^{NL} = \frac{(n-3)(1-c)^2}{2(n-1)} < \frac{n^2(1-c)^2}{2(n+1)^2} = CS$$

 which simplifies to

$$5n + 3 > 0$$

 so that consumer surplus is smaller when the firm offers a nonlinear contract, $CS^{NL} < CS$.
 - In summary, consumer surplus is smaller under a nonlinear contract regardless of the number of firms competing in the industry, i.e., both in Case 1 and in Case 2.
- Finally, total welfare satisfies
 - Case 1, where $n \leq 3$

$$W^{NL} = \frac{n(1-c)^2}{2} - nF > \frac{n(n+2)(1-c)^2}{2(n+1)^2} - nF = W$$

 which simplifies to

$$\frac{n\left[(n+1)^2 - n - 2\right](1-c)^2}{2(n+1)^2} > 0$$

 and further rearranging yields

$$\frac{n(n^2+n-1)(1-c)^2}{2(n+1)^2} > 0$$

 which is positive if $n^2 + n - 1 > 0$ holds. Solving for n, we find that this inequality is satisfied if $n > \frac{\sqrt{5}-1}{2} \simeq 0.62$ which holds since $n \geq 2$ by definition. Hence, welfare is larger when the firm offers a nonlinear contract than otherwise, where $W^{NL} > W$.
 - Case 2, where $n > 3$

$$W^{NL} = \frac{3(1-c)^2}{2} - nF > \frac{n(n+2)(1-c)^2}{2(n+1)^2} - nF = W$$

which simplifies to

$$\frac{\left[3(n+1)^2 - n(n+2)\right](1-c)^2}{2(n+1)^2} > 0$$

and further rearranging yields

$$\frac{(2n^2+4n+3)(1-c)^2}{2(n+1)^2} > 0.$$

Since $n \geq 3$, every term in the left side of the above inequality is positive, so that this inequality holds. Therefore, welfare is also is larger when the firm offers a nonlinear contract than otherwise, $W^{NL} > W$.

– In summary, welfare is larger under a nonlinear contract regardless of the number of firms competing in the industry, i.e., both in Case 1 and in Case 2.

Simultaneous Price Competition 3

Introduction

This chapter considers similar industries as those we analyzed in Chap. 2, but assuming that firms compete in prices. In this setting, every firm simultaneously and independently chooses the price for its product. When firms sell a homogeneous good (such as the same cereal variety, cement, or other minerals), the firm setting the lowest price captures all sales while all other firms sell zero units. When firms sell heterogeneous goods (such as clothing), the firm setting the lowest price may attract more, but not all customers.

Exercise 3.1 considers an industry where firms sell undifferentiated goods and face the same marginal costs of production. In this context, price competition is so intense that all firms set their price equal to their common marginal cost, thus yielding the same equilibrium result as a perfectly competitive market with infinitely many firms which take prices as given. In Exercise 3.2, we show that this extreme competition is attenuated when firms face different marginal costs of production and, similarly, Exercise 3.3 demonstrates that price competition is also attenuated when firms offer price-matching guarantees. While at first glance these guarantees seem to strengthen price competition, we show that they actually reduce it, helping firms earn larger profits in equilibrium.

Exercises 3.4 and 3.5 analyze price competition in settings where firms sell differentiated (i.e., heterogeneous) goods, assuming companies with symmetric costs in Exercise 3.4 and with asymmetric costs in Exercise 3.5. Similarly, Exercises 3.6–3.7 consider price competition in industries where one firm privately observes its marginal cost of production while its rival does not (e.g., a newcomer in the market), first in the context of homogeneous goods (Exercise 3.6) and then in a setting with heterogeneous goods (Exercise 3.7).

Exercise 3.8 then examines how an incumbent firm can use prices to deter entry by potential competitors in the market, while Exercise 3.9 studies how to reconcile the extremely different equilibrium predictions in the Bertrand model of price competition (as presented in Exercise 3.1) and in the Cournot model of quantity competition (as described in Exercise 2.1).

The remaining exercises consider models of horizontal and vertical product differentiation. First, Exercises 3.11 and 3.12 present the Hotelling model where two firms simultaneously and independently choose their product attributes (graphically represented as the firm's position in a line describing customers' ideal points). In models of horizontal product differentiation, consumers may buy from the firm offering the product closest to their ideal (in terms of attributes) if both firms set the same prices. Exercise 3.13 further explores the socially optimal location for each firm, while

P.-S. Choi et al., *Industrial Organization*, Springer Texts in Business and Economics,
https://doi.org/10.1007/978-3-031-38635-0_3

Exercise 3.14 relaxes the uniform distribution assumption. Exercise 3.15 examines a model of spatial competition where $N \geq 2$ firms compete in the so-called Salop circle.

Exercise 3.16 extends the Hotelling model allowing for firms to horizontally differentiate their products along two dimensions, showing that firms have incentives to seek maximum differentiation in one dimension but do not differentiate their products in the other dimension. In contrast, Exercise 3.17 considers vertical differentiation, where the customers regard the product of one firm as technologically superior to that of its rival, implying that all customers buy the superior good if both had the same price. Finally Exercise 3.18 investigates horizontal and vertical differentiation, showing that a firm captures a larger demand if its quality premium offsets cost asymmetry.

Exercise #3.1: Price Competition with Homogeneous Products and Symmetric Costs[A]

3.1 Consider two firms competing in prices (à Bertrand) on homogeneous goods. The demand function of firm i, where $i, j \in \{1, 2\}$, is

$$q_i\left(p_i, p_j\right) = \begin{cases} 1 - p_i & \text{if } p_i < p_j \\ \frac{1-p_i}{2} & \text{if } p_i = p_j \\ 0 & \text{if } p_i > p_j. \end{cases}$$

For simplicity, assume that both face a marginal production cost c, where $1 > c \geq 0$.

(a) Find the best response function $p_i\left(p_j\right)$ for every firm i.

- When firms compete in price for homogeneous goods, firm i's demand is

$$q_i\left(p_i, p_j\right) = \begin{cases} 1 - p_i & \text{if } p_i < p_j \\ \frac{1-p_i}{2} & \text{if } p_i = p_j \\ 0 & \text{if } p_i > p_j. \end{cases}$$

  Intuitively, firm i captures all the market when its price is lower than firm j's, and it splits the market evenly with firm j when both companies set the same price for its products and has no sales when its price is strictly higher than that of its rival.

- In this context, every firm i's best response function is depicted in Fig. 3.1 and can be formally expressed as

$$p_i\left(p_j\right) = \begin{cases} p^m & \text{if } p_j > p^m \\ p_j - \varepsilon & \text{if } c < p_j \leq p^m \\ c & \text{if } p_j \leq c, \end{cases}$$

  where p^m denotes the monopoly price, and it is defined as the solution to the monopolist's profit maximization problem as follows:

$$\max_{p_i \geq 0} \; (p_i - c)(1 - p_i).$$

  Differentiating with respect to p_i yields $1 - 2p_i + c = 0$, and solving for p_i, we obtain the monopoly price $p^m = \frac{1+c}{2}$. We can interpret this best response function as follows:

  – First, when firm j sets a price above the monopoly price p^m, firm i can set a price $p_i = p^m$ since, by doing that, it captures all sales and maximizes its profits. This is

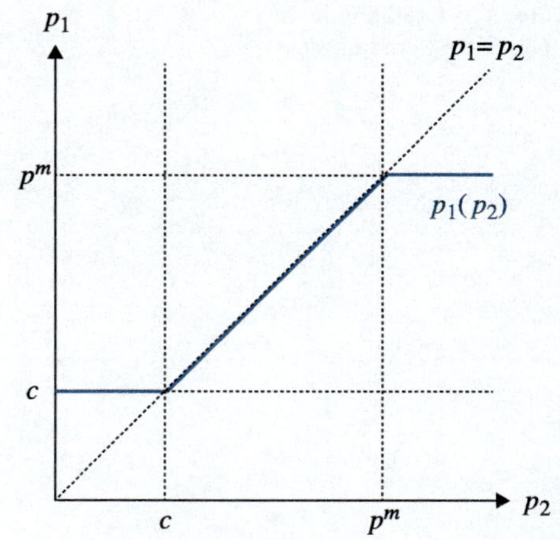

Fig. 3.1 Firm 1's best response function in Bertrand

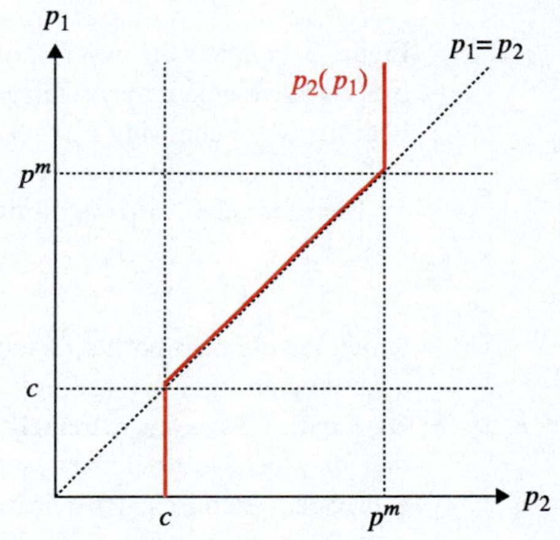

Fig. 3.2 Firm 2's best response function in Bertrand

illustrated in the horizontal segment of the best response function, at the northeast of Fig. 3.1.

- Second, when firm j sets a price between p^m and c, firm i optimally responds by slightly undercut its rival's price. This is illustrated in the figure by the segment slightly below the 45-degree line, so that $p_i = p_j - \varepsilon$.
- Finally, when firm j sets a price below the common marginal cost for both firms, c, firm i responds by setting a price $p_i = c$. This price guarantees no sales since $p_i = c > p_j$ but at least generates no losses. This is illustrated in Fig. 3.1 by the horizontal segment in the southwest of the figure.

• Firm j's best response function is symmetric, that is,

$$p_j(p_i) = \begin{cases} p^m & \text{if } p_i > p^m \\ p_i - \varepsilon & \text{if } c < p_i \leq p^m \\ c & \text{if } p_i \leq c \end{cases}$$

as depicted in Fig. 3.2.

(b) Find the equilibrium prices and profits.

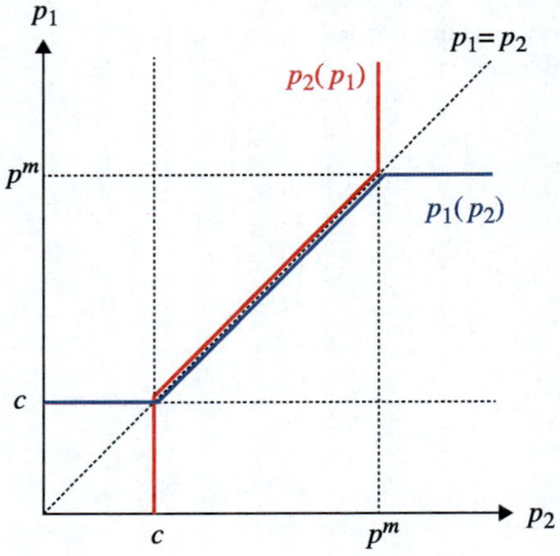

Fig. 3.3 Equilibrium price pair in the Bertrand game

- Figure 3.3 depicts the best response functions of firms i and j, superimposing them, so we can visualize where they intersect. Their point of intersection identifies a price pair where both firms are choosing best responses to each other's strategies, i.e., a Nash equilibrium of the price competition game.

  Therefore, the Nash equilibrium price pair is

$$p_i^* = p_j^* = c,$$

  which entails zero profits for both firms, $\pi_i^* = \pi_j^* = 0$. In other words, firms make no economic profits when selling a homogeneous product.
- **Remark:** This is an extremely competitive result since, with only two firms, we find that they behave as if they operated in a perfectly competitive industry. As we see in subsequent exercises, when firms sell heterogeneous products, or when they suffer capacity constraints, their competition is ameliorated.

Exercise #3.2: Price Competition with Homogeneous Products and Asymmetric Costs^A

3.2 Consider two firms competing à Bertrand selling homogeneous goods. Firm i's demand, where $i \in \{1, 2\}$, is

$$Q_i(p_i) = \begin{cases} Q(p_i) & \text{if } p_i < p_j \\ \alpha_i Q(p_i) & \text{if } p_i = p_j \\ 0 & \text{if } p_i > p_j, \end{cases}$$

where $\alpha_i \in [0, 1]$ stands for the market share of firm i when firms set the same price. Most economic applications assume that, when both firms set the same price, $p_i = p_j$, they evenly split the market, entailing that $\alpha_i = \frac{1}{2}$, but here we allow for any market share $\alpha_i \in [0, 1]$. As we show below, this more general assumption does not impact the equilibrium results. Without loss of generality, firm 1 has a lower marginal cost of production than firm 2, that is, $c_1 < c_2$.

(a) Assume that $c_2 \leq p^m(c_1)$, where $p^m(c_1)$ denotes the price that firm 1 would set under monopoly. Find the equilibrium price pair in this game and the equilibrium profits each firm earns.

- Condition $c_2 \leq p^m(c_1)$ says that, if the most competitive player (firm 1) were to set a price equal to its monopoly price $p^m(c_1)$, its rival (firm 2) could set a price between its own marginal cost and firm 1's price, $p^m(c_1) > p_2 > c_2$, capture all the market, and earn a strictly positive profit. Firm 1 would, of course, have incentives to deviate from such pricing profile to a price p_1 below $p^m(c_1)$. More generally, firm 1 has incentives to set a price p_1 such that $p_1 \leq c_2$ to captures all sales. We describe this price in more detail below.
- *Most competitive firm.* Firm 1 (the one with the lowest marginal cost of production) can set a price

$$p_1 = c_2 - \varepsilon, \text{ where } \varepsilon \to 0.$$

Intuitively, firm 1 sets a price extremely close to the marginal cost of its (less competitive) rival, c_2. By doing so, firm 1 captures all the market and makes the highest possible per-unit margin. Specifically, any price p_1 below c_2, $p_1 < c_2$, helps firm 1 capture all the sales and $p_1 = c_2 - \varepsilon$, where $\varepsilon \to 0$ yields the highest per-unit margin.
- *Least competitive firm.* In this setting, firm 2 sets a price

$$p_2 = c_2,$$

captures no sales whatsoever, and earns zero profits. Note that any price $p_2 \geq c_2$ would yield the same result and firm 2 does not have incentives to set its price below its own marginal cost, $p_2 < c_2$, as that would lead to losses for any sales that this firm achieves. Figure 3.4 summarizes our results.
- *Equilibrium profits.* Therefore, firm 1 earns equilibrium profits

$$\pi_1^* = (p_1 - c_1)Q(p_1) = (c_2 - \varepsilon - c_1)\,Q(c_2 - \varepsilon)$$

since $p_1 = c_2 - \varepsilon$. If firm 1 makes $\varepsilon \to 0$, this profit simplifies to

$$\pi_1^* = (c_2 - c_1)\,Q(c_2),$$

which is positive since marginal costs satisfy $c_2 > c_1$ by assumption, and the demand at a price $p_1 = c_2$ is positive, $Q(c_2) > 0$. In contrast, firm 2 earns zero profits in equilibrium.

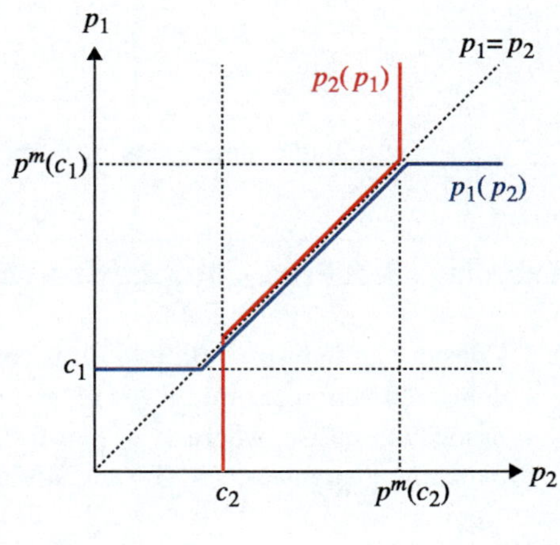

Fig. 3.4 Equilibrium price pair when $c_2 \leq p^m(c_1)$

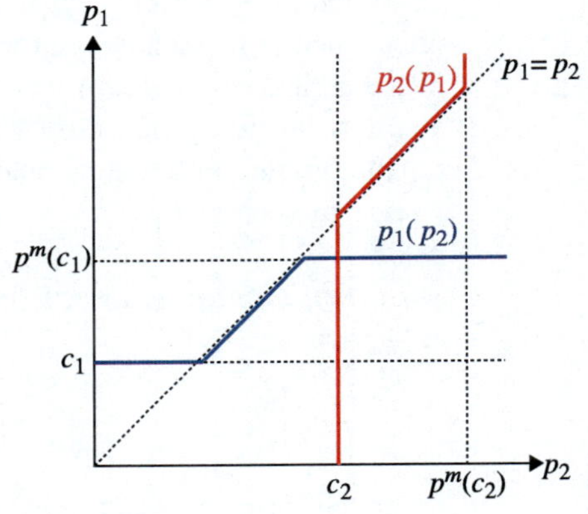

Fig. 3.5 Equilibrium price pair when $c_2 > p^m(c_1)$

(b) Assume now that $c_2 > p^m(c_1)$. Find the equilibrium price pair in this game and the equilibrium profits each firm earns.

- Condition $c_2 > p^m(c_1)$ means that, if the most competitive player (firm 1) were to set a price equal to its monopoly price $p^m(c_1)$, it would capture all sales since firm 2 would have no incentives to set a price $p_2 \leq p^m(c_1)$.
- Therefore, firm 1's equilibrium price in this context is its monopoly price

$$p_1 = p^m(c_1),$$

while firm 2 still sets $p_2 = c_2$, making no sales. Figure 3.5 summarizes our results.

- Equilibrium profits for firm 1 then become

$$\pi_1^* = (p_1 - c_1)Q(p_1) = \left(p^m(c_1) - c_1\right) Q(p^m(c_1))$$

since $p_1 = p^m(c_1)$, while those of firm 2 are still zero.

(c) *Numerical example.* Consider a demand function $Q(p) = 1 - p$, $c_1 = 1/4$, and $c_2 = 1/2$. Evaluate your equilibrium results in parts (a) and (b).

- Since $c_1 = \frac{1}{4} < \frac{1}{2} = c_2$, firm 1 sets $p_1^m = \frac{1}{2}$ and earns a profit of

$$\begin{aligned}\pi_1^* &= \left(p_1^m - c_1\right) \times q_1^m \\ &= \left(\frac{1}{2} - \frac{1}{4}\right) \times \left(1 - \frac{1}{2}\right) \\ &= \frac{1}{4} \times \frac{1}{2} = \frac{1}{8}.\end{aligned}$$

- Firm 2, in contrast, sets $p_2 = c_2 = \frac{1}{2}$ and earns $\pi_2^* = 0$.

Exercise #3.3: Price Competition with Price-Matching Guarantees^B

3.3 Consider an industry with two firms competing in prices and selling a homogeneous good. The demand function of firm i is $q_i\left(p_i, p_j\right) = 1 - p_i$, and we assume that both face the same marginal production cost c, where $1 > c \geq 0$. In this exercise, we examine if the introduction of price-matching guarantees (PMGs) actually helps firms soften their price competition in equilibrium.

PMGs are common in several stores and entail that, for a given price profile (p_i, p_j), firm i receives a price p_i when it is lower than its rival's, $p_i \leq p_j$ but receives a price p_j when $p_i > p_j$. In this case, the PMG "kicks in" because firm i is charging a higher price than firm j's, implying that a customer can buy the product from firm i but pay only p_j. In other words, firm i sells its product but receives the lowest of the two prices, i.e., $p_i = \min\{p_i, p_j\}$.

(a) If firm i operated as a monopolist, find its monopoly price, p^m, output, q^m, and profit.

- Firm i maximizes its monopoly profit as follows:

$$\max_{p \geq 0} p(1-p) - c(1-p) = (1-p)(p-c).$$

Differentiating with respect to p, we obtain

$$1 - 2p + c = 0,$$

and solving for p, we find the monopoly price

$$p^m = \frac{1+c}{2},$$

which entails a monopoly output of

$$q^m = 1 - p^m = 1 - \frac{1+c}{2} = \frac{1-c}{2}$$

and monopoly profits of

$$\begin{aligned} \pi^m &= p^m(1-p^m) - c(1-p^m) \\ &= \frac{1+c}{2}\frac{1-c}{2} - c\frac{1-c}{2} \\ &= \frac{(1-c)^2}{4}. \end{aligned}$$

(b) Assume now that both firms offer a PMG. Show that a symmetric Nash equilibrium can be sustained where both firms set the same price $p^* = p^m$. [*Hint*: You only need to check that every firm i has no incentive to unilaterally deviate from p^* by setting a price p_i lower or higher than p^*.]

- *Equilibrium profits*. Under the PMG, when both firms set a common price p^*, firm i's profits become

$$\pi_i = \min\{p_i, p_j\}\underbrace{\frac{1}{2}(1 - \min\{p_i, p_j\})}_{\text{Firm } i\text{'s sales}} - c\underbrace{\frac{1}{2}(1 - \min\{p_i, p_j\})}_{\text{Firm } i\text{'s sales}}.$$

This profit is, essentially, identical to $p_i(1-p_i) - c(1-p_i)$, but evaluated at $p_i = \min\{p_i, p_j\}$ since firm i only receives the lowest of the two prices, and includes $\frac{1}{2}$ in the sales of firm i given that both firms equally share the market when selling their goods at the same price.

Because $p^* = p_i = p_j$, then $\min\{p_i, p_j\} = p^*$, which helps us simplify the above profit to

$$\pi_i = p^*\frac{1}{2}(1-p^*) - c\frac{1}{2}(1-p^*)$$
$$= \frac{1-p^*}{2}(p^*-c).$$

For a given price, this profit is exactly half of the monopoly profits (see the profit maximization problem at the beginning of part a). Evaluating this profit at $p^m = \frac{1+c}{2}$, we obtain that the profit under the PMG is

$$\pi_i^{PMG}(p^m) = \frac{1}{2}\frac{(1-c)^2}{4} = \frac{(1-c)^2}{8}.$$

We will use this profit as a benchmark next, to see if firm i has unilateral incentives to reduce its price below p^* or to increase it above p^*.

- *Downward deviations.* When firm i undercuts its rival's price, $p_i < p^*$, firm i sells its product at price p_i given that $\min\{p_i, p^*\} = p_i$. Its profits, hence, become

$$\pi_i(p_i) = \min\{p_i, p^*\}\frac{1}{2}(1-\min\{p_i, p^*\}) - c\frac{1}{2}(1-\min\{p_i, p^*\})$$
$$= \frac{1-p_i}{2}(p_i-c),$$

which is equal to half of the monopoly profits (see the profit maximization problem at the beginning of part a). Therefore, we must have that

$$\pi_i(p_i) = \frac{1-p_i}{2}(p_i-c) < \frac{(1-c)^2}{8} = \pi_i^{PMG}(p^m),$$

implying that firm i does not have incentives to undercut price p^m when both firms set a price p^m for their product and offer a PMG.

- *Upward deviations.* When firm i sets its price above its rival's, $p_i > p^*$, firm i must still sell its product for the lowest to these two prices, p^*, due to the PMG, i.e., $\min\{p_i, p^*\} = p^*$. Its profits, then, become

$$\pi_i(p_i) = \min\{p_i, p^*\}\frac{1}{2}(1-\min\{p_i, p^*\}) - c\frac{1}{2}(1-\min\{p_i, p^*\})$$
$$= \frac{1-p^*}{2}(p^*-c).$$

Given that in this exercise we are checking if $p^* = p^m$ can be sustained as a Nash equilibrium, where $p^m = \frac{1+c}{2}$, this deviating profit becomes

$$\pi_i(p_i) = \frac{(1-c)^2}{8},$$

which coincides with the profit that firm i earns when choosing price p^* (see our discussion above).

- We can then conclude that firm i does not have strict incentives to deviate from the common price p^m to a price $p_i \neq p^m$, since it would not (strictly) increase its profits. This applies to both downward deviations, where price p_i satisfies $p_i < p^*$, and to upward deviations where $p_i > p^*$.
- **Intuition:** The introduction of PMGs, rather than strengthening competition, actually softens price competition between firms. The reason for this result to emerge is straightforward: by undercutting firm j's price, firm i is not really stealing customers. Instead, firm i keeps selling the same units but at a lower price. Firm j's customers still purchase from firm j because the PMG allows them to benefit from firm i's lower price when purchasing from firm j.

(c) *Numerical example.* Assume a marginal cost $c = 2/5$. Evaluate equilibrium prices and profits in this context when firms can offer PMGs.

- Firms offering PMGs set the same price as what the monopolist charges, where

$$p^* = \frac{1+c}{2} = \frac{1+\frac{2}{5}}{2} = \frac{7}{10} = 0.7.$$

- In this context, every firm i that practices PMGs obtains a profit of

$$\pi_i^{PMG} = \frac{1}{8}\left(1 - \frac{2}{5}\right)^2 = \frac{9}{200} = 0.045.$$

Exercise #3.4: Price Competition with Heterogeneous Goods and Symmetric Costs^A

3.4 Consider two firms competing à Bertrand selling heterogeneous goods. The demand function of firm i, where $i, j \in \{1, 2\}$, is

$$q_i\left(p_i, p_j\right) = 1 - \gamma p_i + p_j,$$

where $\gamma \geq 1$ represents the degree of product differentiation (homogeneous when $\gamma = 1$ but differentiated when $\gamma > 1$). Every firm i faces a constant marginal cost of c, where $1 > c > 0$, in producing every unit of the good.

(a) Characterize the firms' best response functions and graphically illustrate your results.

- Firm i, where $i \in \{1, 2\}$, chooses p_i to solve the following profit maximization problem:

$$\max_{p_i > 0} \pi_i\left(p_i\right) = \left(p_i - c\right)\left(1 - \gamma p_i + p_j\right).$$

Differentiating with respect to p_i and assuming interior solutions, we obtain

$$\frac{\partial \pi_i\left(p_i\right)}{\partial p_i} = 1 - 2\gamma p_i + p_j + \gamma c = 0.$$

The best response function of firm i becomes

$$p_i\left(p_j\right) = \frac{1+\gamma c}{2\gamma} + \frac{1}{2\gamma} p_j,$$

which originates at $\frac{1+\gamma c}{2\gamma}$ and increases in p_j at a rate of $\frac{1}{2\gamma}$, as depicted in Fig. 3.6.

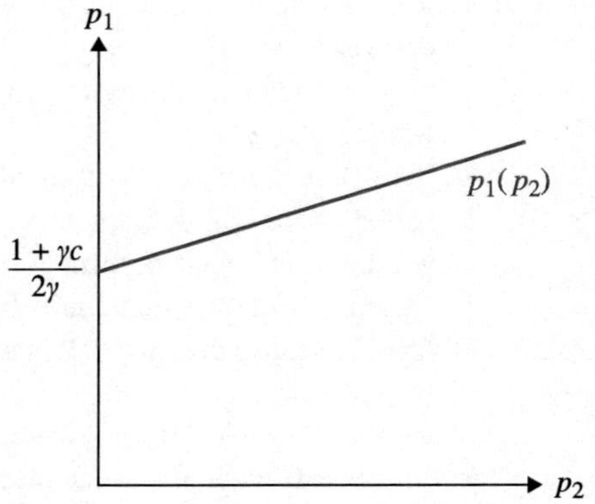

Fig. 3.6 Firm i's best response function

- When we differentiate with respect to γ, we find that

$$\frac{\partial}{\partial \gamma}\left[\frac{1+\gamma c}{2\gamma}\right] = \frac{2\gamma c - 2 - 2\gamma c}{4\gamma^2} = -\frac{1}{2\gamma^2} < 0$$

$$\frac{\partial}{\partial \gamma}\left[\frac{1}{2\gamma}\right] = -\frac{1}{4\gamma^2} < 0$$

so that both the intercept and the slope of the best response function decrease as goods become more differentiated, suggesting that competition becomes less intense.

(b) What are the equilibrium price, output, and profit of each firm? Find the condition on γ in which both firms produce output and the output level if every firm sets its price at the marginal cost.

- Invoking symmetry, where $p = p_1 = p_2$, we obtain

$$p = \frac{1+\gamma c + p}{2\gamma},$$

which yields equilibrium price for every firm i, where $i \in \{1, 2\}$ as follows:

$$p^* = \frac{1+\gamma c}{2\gamma - 1}.$$

Figure 3.7 depicts this equilibrium price pair. Differentiating the equilibrium price p^* with respect to γ, we obtain

$$\frac{\partial p^*}{\partial \gamma} = \frac{(2\gamma - 1)\,c - 2\,(1+\gamma c)}{(2\gamma-1)^2}$$

$$= -\frac{2+c}{(2\gamma-1)^2} < 0$$

so that as goods become more differentiated, equilibrium price decreases.

Fig. 3.7 Equilibrium price pair

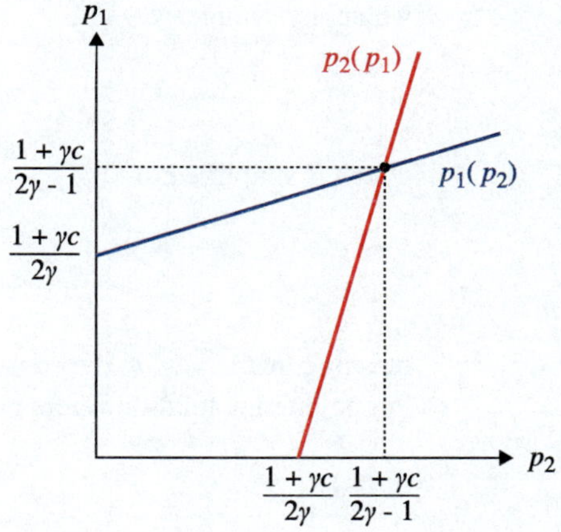

- Substituting p^* into the demand function, equilibrium output becomes

$$\begin{aligned} q^* &= 1 - \gamma p_1^* + p_2^* \\ &= 1 - (\gamma - 1)\frac{1 + \gamma c}{2\gamma - 1} \\ &= \frac{\gamma\left[1 - (\gamma - 1)c\right]}{2\gamma - 1}. \end{aligned}$$

Differentiating the equilibrium output q^* with respect to γ, we obtain

$$\begin{aligned} \frac{\partial q^*}{\partial \gamma} &= \frac{(1 + (1 - 2\gamma)c)(2\gamma - 1) - 2\gamma\left[1 - (\gamma - 1)c\right]}{(2\gamma - 1)^2} \\ &= -\frac{\left[2\gamma(\gamma - 1) + 1\right]c + 1}{(2\gamma - 1)^2} \end{aligned}$$

so that equilibrium output decreases when goods become more differentiated.

- Substituting q_1^* into the profit function, equilibrium profit becomes

$$\begin{aligned} \pi^* &= \left(p^* - c^*\right) q^* \\ &= \left(\frac{1 + \gamma c}{2\gamma - 1} - c\right)\frac{\gamma\left[1 - (\gamma - 1)c\right]}{2\gamma - 1} \\ &= \gamma\left(\frac{1 - (\gamma - 1)c}{2\gamma - 1}\right)^2. \end{aligned}$$

- Next, let us also find the reservation price of the consumer (that is, the "choke price" where the inverse demand function crosses the vertical axis) by setting the inverse demand functions equal to zero as follows:

$$\begin{aligned} 0 &= 1 - \gamma p_i + p_j \\ 0 &= 1 - \gamma p_j + p_i, \end{aligned}$$

which, by symmetry, yields

$$\gamma p = 1 + p,$$

which is simplified to

$$\overline{p} = \frac{1}{\gamma - 1},$$

meaning that if $p \geq \overline{p}$, the consumer will consume zero unit of the good.

- For an interior solution where both firms produce output, we need

$$p^* < \overline{p},$$

which can be expressed as

$$\frac{1 + \gamma c}{2\gamma - 1} < \frac{1}{\gamma - 1},$$

which is rearranged to yield

$$\gamma - 1 + \gamma c\,(\gamma - 1) < 2\gamma - 1$$

so that the condition for interior solution becomes

$$c < \frac{1}{\gamma - 1} = \overline{p}.$$

Figure 3.8 depicts this cutoff, with $c \in [0, 1]$ in the vertical axis and $\gamma \geq 1$ in the horizontal axis. As goods become more differentiated (γ increases), every firm i needs to be more efficient (lower c) in order to mitigate the stronger price effect on consumers' willingness to pay for its output. Otherwise, no firm will produce output in the market when the marginal cost is above the consumer's reservation price.

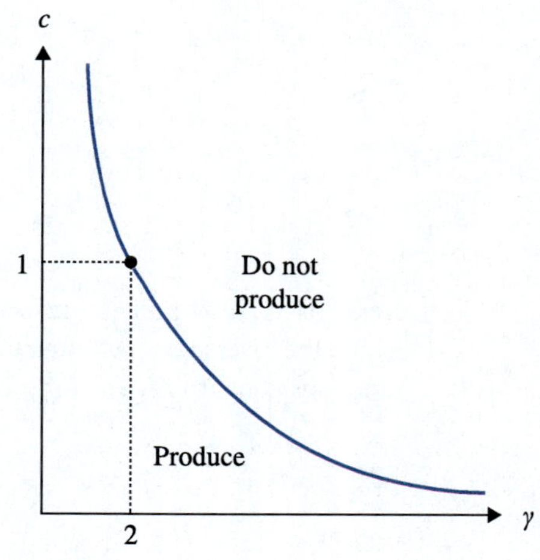

Fig. 3.8 Production decisions as a function of γ and c

- When every firm sets its price at the marginal cost, the output becomes

$$q_1 = 1 - \gamma c + c$$
$$q_2 = 1 - \gamma c + c,$$

which yields

$$q(c) = 1 - (\gamma - 1)c$$

so that $c < \overline{p}$ is the necessary condition for $q(c) \geq 0$.

(c) *Numerical example.* Evaluate the equilibrium price, output, and profit under $\gamma = 5/4$.
- Equilibrium outcomes are

$$p^* = \frac{1 + \frac{5c}{4}}{\frac{10}{4} - 1} = \frac{4 + 5c}{6}$$

$$q^* = \frac{\frac{5}{4}\left[1 - \left(\frac{5}{4} - 1\right)c\right]}{\frac{10}{4} - 1} = \frac{5(4 - c)}{24}$$

$$\pi^* = \frac{5}{4}\left(\frac{1 - \left(\frac{5}{4} - 1\right)c}{\frac{10}{4} - 1}\right)^2 = \frac{5(4 - c)^2}{144}.$$

Exercise #3.5: Price Competition with Heterogeneous Goods and Asymmetric Costs^B

3.5 Consider two firms competing à Bertrand selling heterogeneous goods. The demand function of firm i, where $i, j \in \{1, 2\}$, is

$$q_i(p_i, p_j) = 1 - \gamma p_i + p_j,$$

where $\gamma \geq 1$ represents the degree of product differentiation (homogeneous when $\gamma = 1$ but differentiated when $\gamma > 1$). Without loss of generality, assume that firm 1 has a lower marginal cost than firm 2 in producing every unit of the good, that is, $0 < c_1 < c_2 < 1$.

(a) Characterize the firms' best response functions and graphically illustrate your results.
- Firm i, where $i \in \{1, 2\}$, chooses p_i to solve the following profit maximization problem:

$$\max_{p_i > 0} \pi_i(p_i) = (p_i - c_i)(1 - \gamma p_i + p_j).$$

Differentiating with respect to p_i and assuming interior solutions, we obtain

$$\frac{\partial \pi_i(p_i)}{\partial p_i} = 1 - 2\gamma p_i + p_j + \gamma c_i = 0.$$

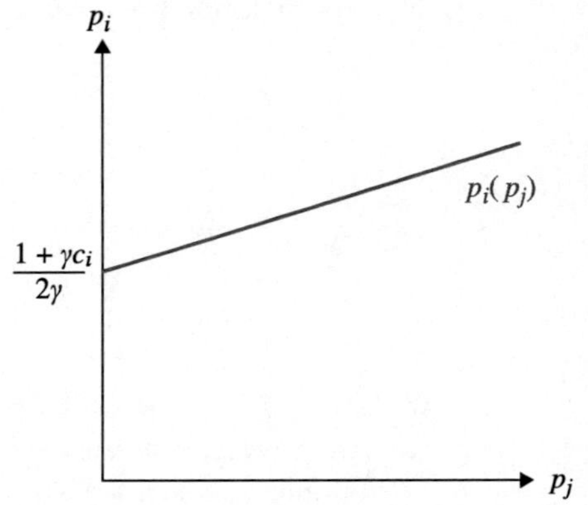

Fig. 3.9 Firm i's best response function

The best response function of firm i becomes

$$p_i\left(p_j\right) = \frac{1+\gamma c_i}{2\gamma} + \frac{1}{2\gamma} p_j,$$

which originates at $\frac{1+\gamma c_i}{2\gamma}$ and increases in p_j at a rate of $\frac{1}{2\gamma}$, as depicted in Fig. 3.9.

- *Comparative statics.* When we differentiate with respect to γ, we check that

$$\frac{\partial}{\partial \gamma}\left[\frac{1+\gamma c_i}{2\gamma}\right] = \frac{2\gamma c_i - 2 - 2\gamma c_i}{4\gamma^2} = -\frac{1}{2\gamma^2} < 0$$

$$\frac{\partial}{\partial \gamma}\left[\frac{1}{2\gamma}\right] = -\frac{1}{4\gamma^2} < 0$$

so that both the intercept and the slope of the best response function decrease as goods become more differentiated, indicating that competition becomes less intense.

(b) What are the equilibrium price, output, and profit of each firm? Find the *sufficient* condition on γ in which both firms produce output and the output level if every firm sets its price at the marginal cost.

- Rearranging the best response functions, we obtain

$$2\gamma p_1 - p_2 = 1 + \gamma c_1$$
$$2\gamma p_2 - p_1 = 1 + \gamma c_2$$

represented in matrix form as follows:

$$\begin{bmatrix} 2\gamma & -1 \\ -1 & 2\gamma \end{bmatrix} \begin{bmatrix} p_1 \\ p_2 \end{bmatrix} = \begin{bmatrix} 1+\gamma c_1 \\ 1+\gamma c_2 \end{bmatrix}.$$

Solving by matrix inverse, we obtain

$$\begin{bmatrix} p_1 \\ p_2 \end{bmatrix} = \frac{1}{4\gamma^2 - 1} \begin{bmatrix} 2\gamma & 1 \\ 1 & 2\gamma \end{bmatrix} \begin{bmatrix} 1+\gamma c_1 \\ 1+\gamma c_2 \end{bmatrix} = \frac{1}{4\gamma^2 - 1} \begin{bmatrix} 2\gamma^2 c_1 + \gamma c_2 + 2\gamma + 1 \\ 2\gamma^2 c_2 + \gamma c_1 + 2\gamma + 1, \end{bmatrix}$$

which yields equilibrium price for firm i, where $i \in \{1, 2\}$ as follows:

$$p_1^* = \frac{2\gamma^2 c_1 + \gamma c_2 + 2\gamma + 1}{4\gamma^2 - 1}$$

$$p_2^* = \frac{2\gamma^2 c_2 + \gamma c_1 + 2\gamma + 1}{4\gamma^2 - 1}.$$

Differentiating the equilibrium price p_1^* with respect to γ, we obtain

$$\frac{\partial p_1^*}{\partial \gamma} = \frac{(4\gamma c_1 + c_2 + 2)\left(4\gamma^2 - 1\right) - 8\gamma\left(2\gamma^2 c_1 + \gamma c_2 + 2\gamma + 1\right)}{\left(4\gamma^2 - 1\right)^2}$$

$$= -\frac{4\gamma c_1 + \left(4\gamma^2 + 1\right) c_2 + 2\left(4\gamma^2 + 1\right) + 8\gamma}{\left(4\gamma^2 - 1\right)^2} < 0$$

so that as goods become more differentiated, the equilibrium price of firm 1 decreases. A symmetric result is obtained if we differentiate p_2^* with respect to γ.

- Substituting p_1^* and p_2^* into the demand function, the equilibrium output becomes

$$q_1^* = 1 - \gamma p_1^* + p_2^*$$

$$= 1 - \gamma\left(\frac{2\gamma^2 c_1 + \gamma c_2 + 2\gamma + 1}{4\gamma^2 - 1}\right) + \left(\frac{2\gamma^2 c_2 + \gamma c_1 + 2\gamma + 1}{4\gamma^2 - 1}\right)$$

$$= \frac{\gamma}{4\gamma^2 - 1}\left[\gamma c_2 - \left(2\gamma^2 - 1\right) c_1 + 2\gamma + 1\right]$$

$$q_2^* = 1 - \gamma p_2^* + p_1^*$$

$$= 1 - \gamma\left(\frac{2\gamma^2 c_2 + \gamma c_1 + 2\gamma + 1}{4\gamma^2 - 1}\right) + \left(\frac{2\gamma^2 c_1 + \gamma c_2 + 2\gamma + 1}{4\gamma^2 - 1}\right)$$

$$= \frac{\gamma}{4\gamma^2 - 1}\left[\gamma c_1 - \left(2\gamma^2 - 1\right) c_2 + 2\gamma + 1\right].$$

Differentiating the equilibrium output q_1^* with respect to γ, we obtain

$$\frac{\partial q_1^*}{\partial \gamma} = \frac{\left[2\gamma c_2 - \left(6\gamma^2 - 1\right) c_1 + 4\gamma + 1\right]\left(4\gamma^2 - 1\right) - 8\gamma^2\left[\gamma c_2 - \left(2\gamma^2 - 1\right) c_1 + 2\gamma + 1\right]}{\left(4\gamma^2 - 1\right)^2}$$

$$= -\frac{2\gamma c_2 + \left(8\gamma^4 - 2\gamma^2 + 1\right) c_1 + \left(4\gamma^2 + 4\gamma - 1\right)}{\left(4\gamma^2 - 1\right)^2} < 0$$

so that as goods become more differentiated, equilibrium output of firm 1 decreases. A symmetric result is obtained if we differentiate q_2^* with respect to γ.

- Substituting q_1^* and q_2^* into the profit function, equilibrium profit becomes

$$\pi_1^* = \left(p_1^* - c_1\right) q_1^*$$

$$= \frac{\gamma}{4\gamma^2 - 1}\left(\frac{2\gamma^2 c_1 + \gamma c_2 + 2\gamma + 1}{4\gamma^2 - 1} - c_1\right)\left[\gamma c_2 - \left(2\gamma^2 - 1\right) c_1 + 2\gamma + 1\right]$$

$$= \gamma \left(\frac{\gamma c_2 - (2\gamma^2 - 1)\, c_1 + 2\gamma + 1}{4\gamma^2 - 1} \right)^2$$

$$\pi_2^* = (p_2^* - c_2)\, q_2^*$$

$$= \frac{\gamma}{4\gamma^2 - 1} \left(\frac{2\gamma^2 c_2 + \gamma c_1 + 2\gamma + 1}{4\gamma^2 - 1} - c_2 \right) \left[\gamma c_1 - \left(2\gamma^2 - 1\right) c_2 + 2\gamma + 1 \right]$$

$$= \gamma \left(\frac{\gamma c_1 - (2\gamma^2 - 1)\, c_2 + 2\gamma + 1}{4\gamma^2 - 1} \right)^2 .$$

- Next, let us also find the reservation price of the consumer (that is, the "choke price" where the inverse demand function crosses the vertical axis) by setting the inverse demand functions equal to zero as follows:

$$0 = 1 - \gamma p_1 + p_2$$
$$0 = 1 - \gamma p_2 + p_1,$$

which, by symmetry, yields,

$$\gamma p = 1 + p$$

which is simplified to

$$\overline{p} = \frac{1}{\gamma - 1},$$

meaning that if $p \geq \overline{p}$, the consumer will consume zero unit of the output.
- For an interior solution where both firms produce output, we need

$$p_1^* < \overline{p}$$
$$p_2^* < \overline{p},$$

which can be expressed as

$$\frac{2\gamma^2 c_1 + \gamma c_2 + 2\gamma + 1}{4\gamma^2 - 1} < \frac{1}{\gamma - 1}$$
$$\frac{2\gamma^2 c_2 + \gamma c_1 + 2\gamma + 1}{4\gamma^2 - 1} < \frac{1}{\gamma - 1},$$

which is rearranged to yield

$$(2\gamma c_1 + c_2)(\gamma - 1) < (2\gamma + 1)$$
$$(2\gamma c_2 + c_1)(\gamma - 1) < (2\gamma + 1)$$

so that the *sufficient* condition by combining the above inequalities becomes

$$(\gamma - 1)(2\gamma + 1) \max\{c_1, c_2\} < (2\gamma + 1)$$
$$\Rightarrow \max\{c_1, c_2\} < \frac{1}{\gamma - 1} = \overline{p}.$$

As goods become more differentiated (γ increases), every firm i needs to be more efficient (lower c) in order to mitigate the stronger price effect on consumers' willingness to pay for its output. Otherwise, the less efficient firm may exit the market if $c_i \geq \overline{p}$ when the marginal cost is above the consumer's reservation price.

- When every firm sets its price at the marginal cost, the output becomes

$$q_1(c_1, c_2) = 1 - \gamma c_1 + c_2$$

$$q_2(c_1, c_2) = 1 - \gamma c_2 + c_1$$

so that $c < \overline{p}$ is the *necessary* condition for $q(c) \geq 0$.

(c) *Numerical example.* Evaluate equilibrium outcomes under $c_1 = 1/4$ and $c_2 = 1/2$ as a function of γ. Under what conditions of γ will both firms produce a positive output?

- Equilibrium outcomes are

$$p_1^* = \frac{2\gamma^2\frac{1}{4} + \gamma\frac{1}{2} + 2\gamma + 1}{4\gamma^2 - 1} = \frac{\gamma^2 + 5\gamma + 2}{2(4\gamma^2 - 1)}$$

$$p_2^* = \frac{2\gamma^2\frac{1}{2} + \gamma\frac{1}{4} + 2\gamma + 1}{4\gamma^2 - 1} = \frac{4\gamma^2 + 9\gamma + 4}{4(4\gamma^2 - 1)}$$

$$q_1^* = \frac{\gamma}{4\gamma^2 - 1}\left[\gamma\frac{1}{2} - \left(2\gamma^2 - 1\right)\frac{1}{4} + 2\gamma + 1\right] = \frac{\gamma(-2\gamma^2 + 10\gamma + 5)}{4(4\gamma^2 - 1)}$$

$$q_2^* = \frac{\gamma}{4\gamma^2 - 1}\left[\gamma\frac{1}{4} - \left(2\gamma^2 - 1\right)\frac{1}{2} + 2\gamma + 1\right] = \frac{\gamma(-4\gamma^2 + 9\gamma + 6)}{4(4\gamma^2 - 1)}$$

$$\pi_1^* = \gamma\left(\frac{\gamma\frac{1}{2} - (2\gamma^2 - 1)\frac{1}{4} + 2\gamma + 1}{4\gamma^2 - 1}\right)^2 = \frac{\gamma(-2\gamma^2 + 10\gamma + 5)^2}{16(4\gamma^2 - 1)^2}$$

$$\pi_2^* = \gamma\left(\frac{\gamma\frac{1}{4} - (2\gamma^2 - 1)\frac{1}{2} + 2\gamma + 1}{4\gamma^2 - 1}\right)^2 = \frac{\gamma(-4\gamma^2 + 9\gamma + 6)^2}{16(4\gamma^2 - 1)^2}.$$

Firms will produce positive units of output if $q_1^* > 0$ and $q_2^* > 0$, where

$$2\gamma^2 - 10\gamma - 5 < 0$$

$$4\gamma^2 - 9\gamma - 6 < 0.$$

Solving for γ, we obtain

$$\frac{5 - \sqrt{35}}{2} < \gamma < \frac{5 + \sqrt{35}}{2}$$

$$\frac{9 - \sqrt{177}}{8} < \gamma < \frac{9 + \sqrt{177}}{8}.$$

Since γ, it suffices to say that both firms produce output if

$$1 < \gamma < 2.79.$$

Exercise #3.6: Price Competition with Homogeneous Goods and Uncertain Costs^C

3.6 Consider an industry with $n \geq 2$ firms selling a homogeneous good. Firms face linear demand $Q(p) = 1 - p$. Every firm i privately observes its marginal cost c_i, independently drawn from a uniform distribution $c_i \sim U[0, 1]$. The lowest price charged by the competitors of firm i is denoted as

$$\widehat{p}_{-i} = \min\{p_1, \ .., \ p_{i-1}, \ p_{i+1}, .., \ p_n\}.$$

(a) *Demand.* Describe firm i's demand function.

- We need to separately consider three cases:
  - When firm i sets a price p_i strictly below the lowest competing price, $\widehat{p}_{-i} = \min\{p_1, .., \ p_{i-1}, \ p_{i+1}, .., \ p_n\}$, firm i captures all the market, which means that its demand is given by $q_i = 1 - p_i$.
  - When firm i sets a price p_i strictly above the lowest competing price, $\widehat{p}_{-i}$, firm i makes no sales, entailing a demand $q_i = 0$.
  - Finally, when firm i's price coincides with the lowest competing price, $\widehat{p}_{-i}$, market demand is evenly shared among the m firms charging the lowest price, that is, $q_i = \frac{1-p_i}{m}$.
- In summary, firm i's demand is

$$q_i(p_i, \widehat{p}_{-i}) = \begin{cases} 1 - p_i & \text{if } p_i < \widehat{p}_{-i}, \\ \frac{1-p_i}{m} & \text{if } p_i = \widehat{p}_{-i}, \text{ and} \\ 0 & \text{otherwise.} \end{cases}$$

(b) *Profit Maximization.* We seek to find Bayesian Nash Equilibria (BNE) where every firm i uses a price function $p^*(c_i)$ mapping its marginal cost c_i into a price p_i. Write firm i's expected profit maximization problem.

- Every firm i solves

$$\max_{p_i \geq 0} \ \underbrace{(p_i - c_i)}_{\text{Margin per unit}} \ \underbrace{(1 - p_i)}_{\text{Demand}} \Pr(p_i < \widehat{p}_{-i}), \tag{3.1}$$

where the last term represents the probability that firm i charges a price strictly below the lowest competing price. This occurs, in particular, when firm i's price lies below each of its competitors' prices, which we write as follows:

$$\begin{aligned} \Pr\left(p_i < \widehat{p}_{-i}\right) = {} & \Pr\left(p_i < p^*(c_1)\right) \times \cdots \times \Pr(p_i < p^*(c_{i-1})) \\ & \times \Pr(p_i < p^*(c_{i+1})) \times \cdots \times \Pr(p_i < p^*(c_n)). \end{aligned} \tag{3.2}$$

To express each of these $n-1$ probabilities, we now claim that the price function is strictly increasing (we confirm this point below). Therefore, the probability that firm i's price lies strictly below that of firm j is

$$\Pr\left(p_i < p^*(c_j)\right) = \Pr\left(p^{*-1}(p_i) < c_j\right)$$
$$= 1 - p^{*-1}(p_i),$$

where in the first equality, we applied the inverse on both sides of the inequality, and in the second equality, we use the property that costs are uniformly distributed, so $p^{*-1}(p_i) < c_j$ is represented by $1 - p^{*-1}(p_i)$.

Intuitively, the inverse $p^{*-1}(p_j)$ denotes the marginal cost of firm j when setting price p_j and following the equilibrium price function $p^*(c_j) = p_j$, so that $c_j = p^{*-1}(p_j)$.

- Therefore, probability in (expression (3.2)) can be rewritten as

$$\Pr(p_i < \widehat{p}_{-i}) = \left[1 - p^{*-1}(p_i)\right] \times \cdots \times \left[1 - p^{*-1}(p_i)\right]$$
$$\times \left[1 - p^{*-1}(p_i)\right] \times \cdots \times \left[1 - p^{*-1}(p_i)\right].$$

We are then ready to rewrite the expected profit maximization problem in (1) as follows:

$$\max_{p_i \geq 0} (p_i - c_i)(1 - p_i)[1 - p^{*-1}(p_i)]^{n-1}. \tag{3.3}$$

(c) Differentiate the expected profit maximization problem with respect to price p_i.

- Differentiating expression (3.3) with respect to price p_i, we obtain

$$(1 + c_i - 2p_i)[1 - p^{*-1}(p_i)]^{n-1} - (p_i - c_i)(1 - p_i)(n-1)$$
$$[1 - p^{*-1}(p_i)]^{n-2} \frac{\partial p^{*-1}(p_i)}{\partial p_i} = 0. \tag{3.4}$$

To simplify this first-order condition, recall that, in a symmetric equilibrium, $p^*(c_i) = p_i$, and that the derivative of the inverse price-setting function is the inverse of the derivative. Therefore,

$$p'^*(c_i)(1 + c_i - 2p^*(c_i))[1 - c_i]^{n-1} - (p^*(c_i) - c_i)(1 - p^*(c_i))(n-1)[1 - c_i]^{n-2} = 0. \tag{3.5}$$

Dividing Eq. (3.5) by $[1 - c_i]^{n-2}$ yields

$$p'^*(c_i) = \frac{(n-1)(p^*(c_i) - c_i)(1 - p^*(c_i))}{(1 - c_i)(1 + c_i - 2p^*(c_i))}. \tag{3.6}$$

(d) *Linear price function.* Suppose that this price function is linear, that is, $p(c_i) = a + bc_i$, where a and b are parameters we need to find. In this context, $p'^*(c_i) = b$. Assume also that $p^*(1) = 1$, which you should confirm at the end of the exercise.

- Assumption $p^*(1) = 1$ is relatively reasonable, as it means that the firm with the highest marginal cost $c_i = 1$ charges a price equal to this marginal cost. A lower price would entail losses per unit and a higher price will clearly yield no sales since the probability of another firm having the highest marginal cost too is negligible. Using assumption $p^*(1) = 1$ in the linear price function $p(c_i) = a + bc_i$, we obtain

$$a + b1 = 1,$$

or $b = 1 - a$. Therefore, the price function $p(c_i) = a + bc_i$ can be rewritten as $p(c_i) = a + (1 - a)c_i$, entailing that

$$p - c_i = a(1 - c_i).$$

Our above results also imply that

$$1 - p = (1 - a)(1 - c_i)$$

and that

$$1 + c_i - 2p = (1 - 2a)(1 - c_i).$$

Each of the above results is one of the terms in expression (3.6), so we can now insert them, where appropriate, to obtain

$$\underbrace{1 - a}_{b} = \frac{(n-1)\overbrace{a(1 - c_i)}^{p^*(c_i)-c_i}\overbrace{(1 - a)(1 - c_i)}^{1-p^*(c_i)}}{(1 - c_i)\underbrace{(1 - 2a)(1 - c_i)}_{1+c_i-2p^*(c_i)}}.$$

This expression is only a function of one unknown, a. Solving for a yields $a = \frac{1}{n+1}$. Therefore, parameter b is

$$b = 1 - a = 1 - \frac{1}{n+1} = \frac{n}{n+1}$$

yielding a price function

$$p(c_i) = \underbrace{\frac{1}{n+1}}_{a} + \underbrace{\frac{n}{n+1}}_{b}c_i.$$

When $c_i = 1$, the price becomes \$1, as predicted, since

$$p(1) = \frac{1}{n+1} + \frac{n}{n+1}1 = \frac{1+n}{n+1} = \$1.$$

(e) Does firm i charge a price above its marginal cost c_i in the price function found in part (d)? Interpret.

- Figure 3.10 depicts this price function $p(c_i)$, which originates at $a = \frac{1}{n+1}$ when firm i's marginal cost is the lowest, $c_i = 0$, and increases as the firm's cost increases (as it becomes less efficient), at a rate of $\frac{n}{n+1}$.
- In other words, the price function $p(c_i)$ lies above the 45-degree line, especially when the firm is relatively efficient (low c_i). In particular, every firm i sets a price strictly above its marginal cost c_i for all $0 \leq c_i < 1$ and a price that coincides with its marginal cost when $c_i = 1$.

(f) How is the price function that you found in part (d) affected by an increase in the number of firms, n? For illustration purposes, evaluate the price function that you found in part (d) at $n = 1$, $n = 2$, and $n = 10$?

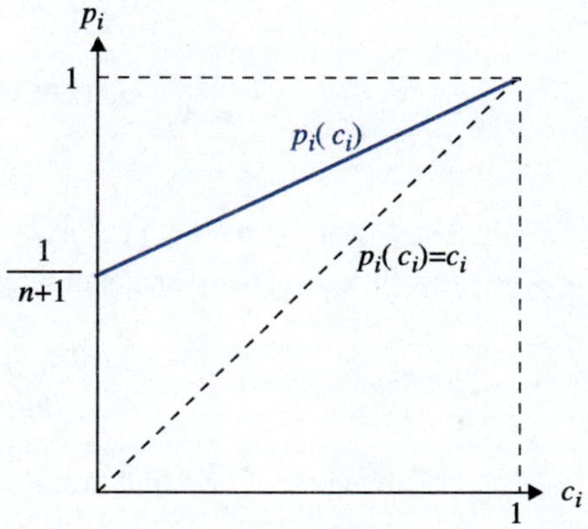

Fig. 3.10 Price function $p(c_i)$

- In Fig. 3.10, an increase in the number of firms, n, produces a downward shift in the vertical intercept $\frac{1}{n+1}$ (without affecting $p(1) = 1$ at the right-hand side of the figure). Therefore, every firm i sets a lower price as the market becomes more competitive.
- Substituting $n = 1$, $n = 2$, and $n = 10$ into the price function, we obtain

$$p(1) = \frac{1}{1+1} + \frac{1}{1+1}c_i = \frac{1+c_i}{2}$$
$$p(2) = \frac{1}{2+1} + \frac{2}{2+1}c_i = \frac{1+2c_i}{3}$$
$$p(10) = \frac{1}{10+1} + \frac{10}{10+1}c_i = \frac{1+10c_i}{11},$$

which illustrates the above downward shift in the price function.

Exercise #3.7: Price Competition with Heterogeneous Goods and Uncertain Costs^B

3.7 Consider two firms competing in prices à Bertrand and selling heterogeneous goods. The demand function of every firm i is

$$q_i\left(p_i, p_j\right) = 1 - \gamma p_i + p_j,$$

where $\gamma \geq 1$ denotes the degree of product differentiation (i.e., homogeneous goods when $\gamma = 1$ but differentiated when $\gamma > 1$). Every firm i faces a constant marginal cost of c_H with probability β and a marginal cost c_L with the remaining probability $1 - \beta$, where $1 > c_H > c_L \geq 0$. Every firm i privately observes its own marginal cost but does not observe the marginal cost of its rival. The probability distribution over costs c_H and c_L is common knowledge among firms.

(a) Find every firm i's best response function when its marginal cost is high, c_H, and its best response function when its marginal cost is low, c_L.
   - Firm i, after observing its marginal cost realization c_k, where $k = \{H, L\}$, chooses its price p_i to solve the following profit maximization problem:

$$\max_{p_i>0} \pi_i(p_i) = (p_i - c_k)\left(1 - \gamma p_i + \underbrace{\beta p_j^H + (1-\beta)p_j^L}_{\text{Expected } p_j}\right),$$

where $\overline{p}_j \equiv \beta p_j^H + (1-\beta)p_j^L$ denotes the expected price of firm i's rival. Differentiating with respect to p_i and assuming interior solutions, we obtain

$$\frac{\partial \pi_i(p_i)}{\partial p_i} = 1 - 2\gamma p_i + \overline{p}_j + \gamma c_k = 0.$$

Solving for p_i, we find firm i's best response function

$$p_i^k(\overline{p}_j) = \frac{1+\gamma c_k}{2\gamma} + \frac{1}{2\gamma}\overline{p}_j \text{ for every cost } k,$$

which originates at $\frac{1+\gamma c_k}{2\gamma}$ and increases in its rival's expected price, p_j, at a rate of $\frac{1}{2\gamma}$.

- Graphically, as firm i observes a lower realization of its marginal cost (c_L rather than c_H), the vertical intercept of its best response function, $\frac{1+\gamma c_k}{2\gamma}$, decreases, shifting the best response function downward, thus indicating that firm i sets a lower price on its product, for a given expected price of its rival.
- We can now separately evaluate this best response function at each of its possible cost realizations, c_H and c_L, and use the definition of the expected price $\overline{p}_j \equiv \beta p_j^H + (1-\beta)p_j^L$, to obtain

$$p_i^H\left(p_j^H, p_j^L\right) = \frac{1+\gamma c_H}{2\gamma} + \frac{1}{2\gamma}\left(\beta p_j^H + (1-\beta)p_j^L\right) \text{ when firm } i\text{'s costs are high}$$

$$p_i^L\left(p_j^H, p_j^L\right) = \frac{1+\gamma c_L}{2\gamma} + \frac{1}{2\gamma}\left(\beta p_j^H + (1-\beta)p_j^L\right) \text{ when firm } i\text{'s costs are low.}$$

(b) What are the equilibrium prices?

- In a symmetric equilibrium, both firms set the same price when their marginal costs coincide, that is, $p^k = p_1^k = p_2^k$ for every $k = \{H, L\}$. Using this property in the above system of four equations (two for each firm), we obtain only two equations

$$p^H = \frac{1+\gamma c_H}{2\gamma} + \frac{1}{2\gamma}\left(\beta p^H + (1-\beta)p^L\right) \text{ and}$$
$$p^L = \frac{1+\gamma c_L}{2\gamma} + \frac{1}{2\gamma}\left(\beta p^H + (1-\beta)p^L\right)$$

which simplify into

$$2\gamma p^H = 1 + \gamma c_H + \beta p^H + (1-\beta)p^L \text{ and}$$
$$2\gamma p^L = 1 + \gamma c_L + \beta p^H + (1-\beta)p^L.$$

Simultaneously solving for p^H and p^L, we find

$$p^H = \frac{2 + c_L(1-\beta) + c_H(\beta + 2\gamma - 1)}{2(2\gamma - 1)}$$

and

$$p^L = \frac{2(1+\gamma c_L) + \beta(c_H - c_L)}{2(2\gamma - 1)}.$$

(c) How are equilibrium prices affected by changes in parameters γ and β?

- Differentiating the equilibrium prices with respect to γ, we obtain

$$\frac{\partial p^H}{\partial \gamma} = \frac{\partial p^L}{\partial \gamma} = -\frac{2 + c_L(1-\beta) + \beta c_H}{(2\gamma - 1)^2},$$

which is unambiguously negative. Therefore, as goods become more differentiated (higher γ), equilibrium price decreases, both when firm i observes a high marginal cost and when it observes a low marginal cost.

- Differentiating the equilibrium prices with respect to β, we find

$$\frac{\partial p^H}{\partial \beta} = \frac{\partial p^L}{\partial \beta} = \frac{c_H - c_L}{2(2\gamma - 1)},$$

which is clearly positive. Intuitively, as the high marginal cost becomes relatively more likely (higher β), equilibrium price increases.

(d) *Numerical example.* Assume that $\gamma = 3/2$, $c_L = 1/4$, and $c_H = 1/2$. Find the equilibrium prices p^H and p^L, and confirm that they increase in β. Then, evaluate the equilibrium prices at $\beta = 0$ and at $\beta = 1$. Interpret.

- Substituting $\gamma = 3/2$, $c_L = 1/4$, and $c_H = 1/2$ into the equilibrium prices, we obtain

$$p^L(\beta) = \frac{2\left(1 + \frac{3}{2} \times \frac{1}{4}\right) + \beta\left(\frac{1}{2} - \frac{1}{4}\right)}{2\left(2 \times \frac{3}{2} - 1\right)} = \frac{11 + \beta}{16}$$

$$p^H(\beta) = \frac{2 + \frac{1}{4}(1-\beta) + \frac{1}{2}\left(\beta + 2 \times \frac{3}{2} - 1\right)}{2\left(2 \times \frac{3}{2} - 1\right)} = \frac{13 + \beta}{16},$$

and both are monotonically increasing in β. In addition,

$$p^L(0) = \frac{11}{16}$$

$$p^H(0) = \frac{13}{16}$$

$$p^L(1) = \frac{3}{4}$$

$$p^H(1) = \frac{7}{8}$$

so that firms set higher equilibrium prices when they have higher probability to realize a high marginal cost.

Exercise #3.8: Entry-Deterring Prices^C

3.8 Consider the following entry deterrence game. In the first stage, an incumbent monopolist, facing constant marginal cost c_1, sets its price p_1. In the second stage, a potential entrant observes the incumbent's price p_1 and responds choosing whether or not to join the market, at a cost F. The entrant's constant marginal cost is c_2, where $c_2 > c_1$, which may be explained because the incumbent has more experience in the industry and thus benefits from a cost advantage. If entry occurs, the entrant sets its price p_2 as a function of p_1, which we can express as the best response function $p_2(p_1)$. If entry does not occur, the entrant's profit is normalized to zero. The demand function is

$$q_i(p_i, p_j) = 1 - \gamma p_i + p_j,$$

where $\gamma > 1$ indicates that own-price effects dominate cross-price effects.

(a) *Second stage.* If the entrant joins the industry, find its best response function $p_2(p_1)$ and its associated profits, $\pi_{ent}(p_1)$. How does the entrant's profit vary γ?

- Upon entry, the entrant takes the incumbent's price p_1 as given and solves

$$\begin{aligned}\max_{p_2 \geq 0} \pi_2(p_2) &= p_2(1 - \gamma p_2 + p_1) - c_2(1 - \gamma p_2 + p_1) \\ &= (p_2 - c_2)(1 - \gamma p_2 + p_1).\end{aligned}$$

Differentiating with respect to p_2 yields

$$\frac{d\pi_2(p_2)}{dp_2} = 1 + \gamma c_2 + p_1 - 2\gamma p_2 = 0.$$

Rearranging, we obtain

$$1 + \gamma c_2 + p_1 = 2\gamma p_2,$$

and solving for p_2, we find the entrant's best response function

$$p_2(p_1) = \frac{1 + \gamma c_2}{2\gamma} + \frac{1}{2\gamma} p_1,$$

which originates at $p_2 = \frac{1+\gamma c_2}{2\gamma}$ and increases in the incumbent's price, p_1, at a rate of $\frac{1}{2\gamma}$.

- The entrant's profits are then

$$\begin{aligned}\pi_{ent}(p_1) &= \left[\left(\frac{1 + \gamma c_2}{2\gamma} + \frac{1}{2\gamma} p_1\right) - c_2\right]\left(1 - \gamma\left(\frac{1 + \gamma c_2}{2\gamma} + \frac{1}{2\gamma} p_1\right) + p_1\right) \\ &= \frac{1 + \gamma c_2 + p_1 - 2\gamma c_2}{2\gamma} \times \frac{2 - 1 - \gamma c_2 - p_1 + 2p_1}{2} \\ &= \frac{(1 - \gamma c_2 + p_1)^2}{4\gamma},\end{aligned}$$

which is increasing in its rival's price p_1 and decreasing in own-price effects γ as

$$\frac{\partial \pi_{ent}(p_1)}{\partial \gamma} = \frac{-2\gamma c_2 (1-\gamma c_2 + p_1) - (1-\gamma c_2 + p_1)^2}{4\gamma^2}$$
$$= -\frac{(1-\gamma c_2 + p_1)(1+\gamma c_2 + p_1)}{4\gamma^2},$$

which is negative as long as $c_2 < \frac{1+p_1}{\gamma}$, as the next section shows.

(b) *First stage, entry deterrence.* Find the minimal price that the incumbent must set to deter entry, p_1^{DE}, where superscript DE denotes "deter entry." How is this price affected by F and γ? Interpret.

- The entry-deterring price p_1^{DE} makes the potential entrant indifferent between entering, earning a profit of $\pi_{ent}(p_1) - F$ and not entering, earning a zero profit. Therefore, the entry-deterring price p_1^{DE} solves $\pi_{ent}(p_1) - F = 0$, which entails

$$\frac{(1-\gamma c_2 + p_1)^2}{4\gamma} = F.$$

Rearranging, we obtain

$$1 - \gamma c_2 + p_1 = 2\sqrt{\gamma F},$$

which, solving for price p_1, yields an entry-deterring price

$$p_1^{DE} = \gamma c_2 + 2\sqrt{\gamma F} - 1.$$

Intuitively, the incumbent can set a higher price (so entry deterrence becomes easier for this firm) when the entrant's marginal cost of production, c_2, is relatively high (as the entrant is not so competitive), when entry becomes more costly (higher F), and when products are more differentiated (higher γ, thus making entry less threatening).

(c) *First stage, accommodate.* Among all prices that do not deter entry, $p_1 < p_1^{DE}$, find the one that maximizes the incumbent's profits, p_1^A, where superscript A denotes "accommodation."

- Conditional on accommodation, the incumbent solves

$$\max_{p_1 < p_1^{DE}} \pi_1 = p_1 \left(1 - \gamma p_1 + \overbrace{\frac{1+\gamma c_2}{2\gamma} + \frac{1}{2\gamma} p_1}^{p_2(p_1)}\right)$$
$$- c_1 \left(1 - \gamma p_1 + \underbrace{\frac{1+\gamma c_2}{2\gamma} + \frac{1}{2\gamma} p_1}_{p_2(p_1)}\right),$$

which is evaluated at the entrant's best response function, $p_2(p_1)$, since the incumbent anticipates this response. The above maximization problem simplifies to

$$\max_{p_1 < p_1^{DE}} \pi_1(p_1) = \frac{(p_1 - c_1)\left(1 + 2\gamma + \gamma c_2 - \left(2\gamma^2 - 1\right) p_1\right)}{2\gamma},$$

which is a function of the incumbent's price, p_1, alone.

Differentiating with respect to p_1, we find

$$\frac{d\pi_1(p_1)}{dp_1} = \frac{1 + 2\gamma + \gamma c_2 - \left(2\gamma^2 - 1\right)(2p_1 - c_1)}{2\gamma} = 0.$$

We are ignoring the constraint $p_1 < p_1^{DE}$, but we check below that our result satisfies this constraint. Rearranging, we obtain

$$2\left(2\gamma^2 - 1\right) p_1 = 1 + 2\gamma + \gamma c_2 + \left(2\gamma^2 - 1\right) c_1,$$

and solving for p_1 yields the incumbent's profit maximizing price conditional on accommodation

$$p_1^A = \frac{1 + 2\gamma + \gamma c_2 + \left(2\gamma^2 - 1\right) c_1}{2\left(2\gamma^2 - 1\right)}.$$

- Finally, we confirm that p_1^A satisfies the constraint $p_1^A < p_1^{DE}$ when

$$\frac{1 + 2\gamma + \gamma c_2 + \left(2\gamma^2 - 1\right) c_1}{2\left(2\gamma^2 - 1\right)} < \gamma c_2 + 2\sqrt{\gamma F} - 1$$

$$1 + 2\gamma + 2\left(2\gamma^2 - 1\right) + \left(2\gamma^2 - 1\right) c_1 + \gamma\left(1 - 2\left(2\gamma^2 - 1\right)\right) c_2 < 4\sqrt{\gamma F}\left(2\gamma^2 - 1\right)$$

$$4\gamma^2 + 2\gamma - 1 - \left(4\gamma^3 - 2\gamma^2 - 3\gamma + 1\right) c_2 < 4\sqrt{\gamma F}\left(2\gamma^2 - 1\right),$$

which, solving for F, yields

$$F > \underline{F} \equiv \frac{1}{16\gamma}\left(\frac{4\gamma^2 + 2\gamma - 1 - \left(4\gamma^3 - 2\gamma^2 - 3\gamma + 1\right) c_2}{2\gamma^2 - 1}\right)^2,$$

and the second last line is from $c_2 > c_1$ and $4\gamma^3 - 2\gamma^2 - 3\gamma + 1 > 0$ for all $\gamma > 1$ that give the sufficient condition for $\underline{F}$. Intuitively, when entry costs are sufficiently high, the incumbent must set a higher price when it seeks to deter entry than when it accommodates entry.

(d) In this part of the exercise, we seek to identify under which condition on the entry cost F the incumbent earns a higher profit deterring than accommodating entry. For simplicity, find this condition on F under symmetric production costs between incumbent and entrant, i.e., $c_1 = c_2 = \frac{1}{2}$.

- If the incumbent practices entry deterrence, charging price $p_1^{DE} = \gamma c_2 + 2\sqrt{\gamma F} - 1$, its profit becomes

$$\pi_1^{DE} = \frac{\left(p_1^{DE} - c_1\right)\left(1 + 2\gamma + \gamma c_2 - \left(2\gamma^2 - 1\right) p_1^{DE}\right)}{2\gamma}$$
$$= \frac{\left(\gamma c_2 + 2\sqrt{\gamma F} - 1 - c_1\right)\left(1 + 2\gamma + \gamma c_2 - \left(2\gamma^2 - 1\right)\left(\gamma c_2 + 2\sqrt{\gamma F} - 1\right)\right)}{2\gamma}$$
$$= \frac{\left(2\sqrt{\gamma F} - 1 - c_1 + \gamma c_2\right)\left(\gamma\left(\gamma + 1\right) - \gamma\left(\gamma^2 - 1\right) c_2 - \left(2\gamma^2 - 1\right)\sqrt{\gamma F}\right)}{\gamma},$$

whereas if this firm accommodates, it sets a price $p_1^A = \frac{1+2\gamma+\gamma c_2+(2\gamma^2-1)c_1}{2(2\gamma^2-1)}$, and its profit becomes

$$\pi_1^A = \frac{\left(p_1^A - c_1\right)\left(1 + 2\gamma + \gamma c_2 - \left(2\gamma^2 - 1\right) p_1^A\right)}{2\gamma}$$
$$= \frac{1}{8\gamma\left(2\gamma^2 - 1\right)^2} \times \left(1 + 2\gamma + \gamma c_2 + \left(2\gamma^2 - 1\right) c_1 - 2\left(2\gamma^2 - 1\right) c_1\right) \times$$
$$\left[2\left(2\gamma^2 - 1\right)\left(1 + 2\gamma + \gamma c_2\right) - \left(2\gamma^2 - 1\right)\left[1 + 2\gamma + \gamma c_2 + \left(2\gamma^2 - 1\right) c_1\right]\right]$$
$$= \frac{\left(1 + 2\gamma + \gamma c_2 - \left(2\gamma^2 - 1\right) c_1\right)^2}{8\gamma\left(2\gamma^2 - 1\right)}.$$

- *Symmetric production costs.* We first evaluate the incumbent's entry deterrence profit, π_1^{DE}, at $c_1 = c_2 = c$, to obtain

$$\pi_1^{DE} = \frac{\left[2\sqrt{\gamma F} - \left(1 - \left(\gamma - 1\right) c\right)\right]\left[\gamma\left(\gamma + 1\right)\left(1 - \left(\gamma - 1\right) c\right) - \left(2\gamma^2 - 1\right)\sqrt{\gamma F}\right]}{\gamma}$$

and the profit from accommodating at $c_1 = c_2 = c$ to find

$$\pi_1^A = \frac{\left(1 + 2\gamma - \left(2\gamma^2 - \gamma - 1\right) c\right)^2}{8\gamma\left(2\gamma^2 - 1\right)}.$$

Therefore, the incumbent prefers to deter entry if and only if $\pi_1^{DE} > \pi_1^A$, which is simplified to the following quadratic inequality:

$$\underbrace{16\left(2\gamma^2 - 1\right)^2}_{A}\gamma F - \underbrace{8\left(2\gamma^2 - 1\right)\left(4\gamma^2 + 2\gamma - 1\right)\left(1 - \left(\gamma - 1\right) c\right)}_{B}\sqrt{\gamma F}$$
$$\underbrace{+\, 8\gamma\left(\gamma + 1\right)\left(2\gamma^2 - 1\right)\left(1 - \left(\gamma - 1\right) c\right)^2 + \left(1 + 2\gamma - \left(2\gamma^2 - \gamma - 1\right) c\right)^2}_{C} < 0.$$

Since F cannot be negative by definition, solving for F, we obtain the following, more compact inequality, where we replaced different terms for A, B, and C as follows:

$$\sqrt{\gamma F} < \frac{-B + \sqrt{B^2 - 4AC}}{2A},$$

which is rearranged to yield

$$F < \overline{F} \equiv \frac{\left(-B + \sqrt{B^2 - 4AC}\right)^2}{4\gamma A^2}.$$

Therefore, we can conclude that

- *Low entry costs.* When $F < \underline{F}$, the incumbent prices satisfy $p_1^A > p_1^{DE}$ and the incumbent accommodates entry.
- *Intermediate entry costs.* When $\underline{F} \leq F < \overline{F}$, the incumbent prices satisfy $p_1^A < p_1^{DE}$ while its profits are $\pi_1^{DE} > \pi_1^A$ so that the incumbent can deter entry by setting a high deterring price to continue dominating the industry.
- *High entry costs.* When $F \geq \overline{F}$, the incumbent prices satisfy $p_1^A < p_1^{DE}$ and its profits are $\pi_1^{DE} < \pi_1^A$. In this setting, the entry-deterring price would be relatively high, making entry deterrence unprofitable for the incumbent, so that the incumbent would accommodate entry.

(e) *Numerical example.* Assume $\gamma = 2$ and symmetric production costs $c = c_1 = c_2 = 1/2$. Evaluate your results in part (d) and discuss for which values of entry cost F the incumbent has incentives to deter entry.

- Substituting $\gamma = 2$ and $c = 1/3$ into equilibrium prices in part (d), we obtain

$$\pi_1^{DE} = \frac{\left[2\sqrt{2F} - \left(1 - \frac{2-1}{2}\right)\right]\left[2\,(2+1)\left(1 - \frac{2-1}{2}\right) - \left(2 \times 2^2 - 1\right)\sqrt{2F}\right]}{2}$$

$$= \frac{\left(4\sqrt{2F} - 1\right)\left(3 - 7\sqrt{2F}\right)}{4}, \text{ and}$$

$$\pi_1^A = \frac{\left(1 + 2 \times 2 - \frac{2 \times 2^2 - 2 - 1}{2}\right)^2}{8 \times 2\left(2 \times 2^2 - 1\right)} = \frac{25}{448}.$$

- In this context, the incumbent has incentives to practice entry deterrence if

$$\frac{\left(4\sqrt{2F} - 1\right)\left(3 - 7\sqrt{2F}\right)}{4} > \frac{25}{448},$$

which is rearranged to yield

$$56F - 19\sqrt{2F} + 3\frac{25}{112} < 0,$$

which does not hold for all F, so that the incumbent will accommodate entry.

Exercise #3.9: Using Capacity Constraints to Reconcile Bertrand and Cournot Models^C

3.9 In the Bertrand model of price competition firms face no capacity constraints. Intuitively, this means that if one of the firms sets the lowest price in the market, it attracts all customers and can serve them positive units, regardless of how large the demand is. This assumption may describe some industries, like smartphone applications, where millions of customers can download a new game into their phone simultaneously but may not be reasonable for industries where firms face capacity constraints, such as airlines or appliances. Kreps and Scheinkman (1983) presented a model that reconciles Bertrand and Cournot models of oligopoly behavior by adding a previous stage to the standard Bertrand model of price competition where firms choose a capacity level.

Consider a market with two firms. In the first stage, every firm i chooses a production capacity $\bar{q}_i$ at a cost of c per unit of capacity where $0 \leq \bar{q}_i \leq 1$, where $c > \frac{3}{4}$. In the second stage, firms observe each others' capacity and respond competing in prices. Once capacity $\bar{q}_i$ is decided, firms can produce up to that capacity with zero marginal cost.

Each firm faces demand $p = 1 - Q$ and chooses prices simultaneously in the second stage, and sales are distributed as in the Bertrand model of price competition.

(a) *Second stage.* Show that both firms set a common price

$$p_1 = p_2 = p^* = 1 - \bar{q}_1 - \bar{q}_2$$

in the second stage.

- Let us start by assuming that firm 1 chooses a price such that $p_1 = p^*$ and show that firm 2 does not have incentives to deviate from that price.
  - If firm 2 does not deviate, it sells all of its capacity $\bar{q}_2$.
  - If it lowers its price, the firm still sells all of its capacity but will earn less profit, so this is not a profitable deviation.
  - If firm 2 increases its price such that $p^* < p_2$, its revenue is its price times the residual demand after all of firm 1's units are sold, or $\hat{Q} = 1 - p_2 - \bar{q}_1$.

$$p_2\hat{Q} = p_2(1 - p_2 - \bar{q}_1).$$

(Recall that the exercise assumes no production costs, so we only need to consider revenue.) To find the maximum of this revenue, first differentiate with respect to p_2 to obtain

$$1 - 2p_2 - \bar{q}_1 = 0.$$

Solving for p_2, we obtain the revenue maximizing price deviation of

$$p_2 = \frac{1 - \bar{q}_1}{2}.$$

However, this price is larger than firm 1's price of $p^* = 1 - \bar{q}_1 - \bar{q}_2$ if

$$\frac{1 - \bar{q}_1}{2} > 1 - \bar{q}_1 - \bar{q}_2$$

$$1 - \bar{q}_1 > 2 - 2\bar{q}_1 - 2\bar{q}_2$$

$$1 < \bar{q}_1 + 2\bar{q}_2.$$

We know that neither firm can invest in a capacity larger than $\frac{1}{8}$ (which is its profit maximizing output as if it was a monopoly in this market), so this condition cannot hold. Since $c > \frac{3}{4}$ by assumption, the maximum capacity (i.e., the monopoly quantity) is now $\frac{1}{8}$, which is why we can conclude that the condition cannot hold. This means that price $p_2 = \frac{1-\bar{q}_1}{2}$ satisfies $p_2 < p^*$, and firm 2 will earn a lower revenue at the deviating price p_2 than at the equilibrium price p^*.
Hence, each firm has no incentive to deviate from price p^*, and both set a price

$$p_1 = p_2 = p^* = 1 - \bar{q}_1 - \bar{q}_2.$$

(b) *First stage.* In the first stage, every firm i simultaneously and independently chooses its capacity $\bar{q}_i$. How much capacity does each firm invest in?

- In the first stage, firms anticipate equilibrium prices $p^* = 1 - \bar{q}_1 - \bar{q}_2$ in the second stage and simultaneously choose capacities $\bar{q}_1$ and $\bar{q}_2$. Firm 1's profit maximization problem is

$$\max_{\bar{q}_1 \geq 0} \; p^*\bar{q}_1 - c\bar{q}_1,$$

where the second term represents the cost of investing in $\bar{q}_1$ units of capacity. Substituting for p^* yields

$$\max_{\bar{q}_1 \geq 0} \; \underbrace{(1 - \bar{q}_1 - \bar{q}_2)}_{p^* \geq 0}\bar{q}_1 - c\bar{q}_1.$$

Differentiating with respect to capacity $\bar{q}_1$,

$$1 - 2\bar{q}_1 - \bar{q}_2 - c = 0,$$

and rearranging and solving for $\bar{q}_1$, we obtain firm 1's best response function

$$\bar{q}_1(\bar{q}_2) = \frac{1-c}{2} - \frac{1}{2}\bar{q}_2,$$

which originates at a capacity of $\frac{1-c}{2}$ when firm 2 does not invest in capacity and decreases at a rate of $\frac{1}{2}$ for every additional unit of capacity that firm 2 invests. Firm 2 has a symmetric best response function, that is,

$$\bar{q}_2(\bar{q}_1) = \frac{1-c}{2} - \frac{1}{2}\bar{q}_1.$$

- In a symmetric equilibrium, both firms invest the same amount in capacity, that is, $\bar{q}_1 = \bar{q}_2 = \bar{q}$. Substituting this condition into either firm's best response function, we obtain

$$\bar{q} = \frac{1-c}{2} - \frac{1}{2}\bar{q},$$

and rearranging $\frac{3}{2}\bar{q} = \frac{1-c}{2}$ and solving for $\bar{q}$, we find the equilibrium capacity each firm chooses

$$\bar{q}^* = \frac{1-c}{3}.$$

- Plugging this result into the pricing decision from the second stage, we find the equilibrium price to be

$$p^* = 1 - \bar{q}_1^* - \bar{q}_2^* = 1 - \frac{1-c}{3} - \frac{1-c}{3} = \frac{1+2c}{3}.$$

(c) How do your results compare to the standard Cournot model with two firms competing in quantities, facing inverse demand function $p(Q) = 1 - Q$, and marginal cost $c > \frac{3}{4}$?
- The standard Cournot model finds that equilibrium quantity for each firm is

$$q^* = \frac{a-c}{3b} = \frac{1-c}{3}.$$

This is the same quantity each firm chose in our game above, which means that the Bertrand game with capacity constraints yields the same results as the Cournot model.

Exercise #3.10: Generalized Hotelling Model^A

3.10 Consider a unit mass of identical consumers, uniformly distributed on the unit line, $x \in U\,[0,1]$. Each consumer buys one unit of the good from either firm 1 or firm 2 located at the ends on the unit line, with per-unit travel cost of τ_1 and τ_2, respectively, where $\tau_1, \tau_2 > 0$. The indirect utility of consumer x is

$$r_1 - \tau_1 x - p_1 \quad \text{when buying from firm 1}$$
$$r_2 - \tau_2 (1-x) - p_2 \quad \text{when buying from firm 2,}$$

where $r_1 \neq r_2$ means that consumers perceive good 1 as heterogenous to good 2. That is, if $r_1 > r_2$ consumers regard good 1 as superior, while if $r_1 < r_2$ they regard this good as inferior.

(a) *Finding demands functions.* Identify the indifferent consumer $\hat{x}$, and solve for the demands of firm 1 and firm 2.
- The indifferent consumer, $\hat{x}$, satisfies

$$r_1 - \tau_1\hat{x} - p_1 = r_2 - \tau_2\left(1-\hat{x}\right) - p_2,$$

which, after rearranging, and solving for $\hat{x}$, yields

$$\hat{x} = \frac{\tau_2}{\tau_1+\tau_2} + \frac{p_2-p_1}{\tau_1+\tau_2} - \frac{r_2-r_1}{\tau_1+\tau_2}.$$

- Therefore, demand of firm 1 is

$$D_1\left(p_1, p_2\right) = \hat{x} = \frac{\tau_2}{\tau_1+\tau_2} + \frac{p_2-p_1}{\tau_1+\tau_2} - \frac{r_2-r_1}{\tau_1+\tau_2}$$

and that of firm 2 is

$$D_2(p_1, p_2) = 1 - \hat{x} = \frac{\tau_1}{\tau_1 + \tau_2} - \frac{p_2 - p_1}{\tau_1 + \tau_2} + \frac{r_2 - r_1}{\tau_1 + \tau_2}.$$

(b) *Finding best response functions.* Suppose the cost of firm 1 is asymmetric to that of firm 2, that is, $c_1 \neq c_2$. Set up the profit maximization problem and derive the best response function of every firm.

- Firm 1 chooses its price, p_1, to solve the following profit maximization problem:

$$\max_{p_1 > 0} \pi_1(p_1) = D_1(p_1, p_2)(p_1 - c_1)$$

$$= \left(\frac{\tau_2}{\tau_1 + \tau_2} + \frac{p_2 - p_1}{\tau_1 + \tau_2} - \frac{r_2 - r_1}{\tau_1 + \tau_2} \right)(p_1 - c_1).$$

Differentiating with respect to p_1 yields

$$\frac{\tau_2 + p_2 - 2p_1 - r_2 + r_1 + c_1}{\tau_1 + \tau_2} = 0.$$

Rearranging, we obtain firm 1's best response function, as follows:

$$p_1(p_2) = \left(\frac{\tau_2 + c_1}{2} + \frac{r_1 - r_2}{2} \right) + \frac{1}{2} p_2,$$

where $\frac{\tau_2 + c_1}{2} + \frac{r_1 - r_2}{2}$ represents its vertical intercept and $\frac{1}{2}$ is its positive slope, indicating that prices are strategic complements. In addition, this vertical intercept increases in the transportation cost that the consumer experiences when buying from the rival company (firm 2), τ_2, in firm 1's production cost, c_1, and in firm 1's vertical differentiation relative to firm 2, $r_1 - r_2$.

- Firm 2 chooses its price, p_2, to solve the following profit maximization problem:

$$\max_{p_2 > 0} \pi_2(p_2) = D_2(p_1, p_2)(p_2 - c_2)$$

$$= \left(\frac{\tau_1}{\tau_1 + \tau_2} - \frac{p_2 - p_1}{\tau_1 + \tau_2} + \frac{r_2 - r_1}{\tau_1 + \tau_2} \right)(p_2 - c_2).$$

Differentiating with respect to p_2 yields

$$\frac{\tau_1 + p_1 - 2p_2 + r_2 - r_1 + c_2}{\tau_1 + \tau_2} = 0.$$

Rearranging, we obtain firm 2's best response function, as follows:

$$p_2(p_1) = \left(\frac{\tau_1 + c_2}{2} + \frac{r_2 - r_1}{2} \right) + \frac{1}{2} p_1,$$

thus exhibiting a symmetric vertical intercept, $\frac{\tau_1 + c_2}{2} + \frac{r_2 - r_1}{2}$, and slope, $\frac{1}{2}$, as firm 1's best response function.

(c) *Equilibrium prices.* Find the equilibrium price of every firm and identify the indifferent consumer $\hat{x}$.

- Substituting $p_2(p_1)$ into $p_1(p_2)$, we obtain

$$p_1 = \frac{\tau_2 + c_1}{2} - \frac{r_2 - r_1}{2} + \frac{1}{2}\underbrace{\left(\frac{\tau_1 + c_2}{2} + \frac{r_2 - r_1}{2} + \frac{1}{2}p_1\right)}_{p_2(p_1)}.$$

Rearranging and solving for p_1, the equilibrium price of firm 1 becomes

$$p_1^* = \frac{2\tau_2 + \tau_1 + 2c_1 + c_2 + r_1 - r_2}{3}.$$

Next, plugging p_1^* into $p_2(p_1)$, we find

$$p_2 = \frac{\tau_1 + c_2}{2} + \frac{r_2 - r_1}{2} + \frac{1}{2}\underbrace{\left(\frac{2\tau_2 + \tau_1 + 2c_1 + c_2 + r_1 - r_2}{3}\right)}_{p_1^*}.$$

Rearranging and solving for p_2, we find the equilibrium price of firm 2

$$p_2^* = \frac{2\tau_1 + \tau_2 + 2c_2 + c_1 + r_2 - r_1}{3}.$$

(d) Find the equilibrium demands. Under which parameter conditions does firm 1 sell more units than firm 2?

- Since the difference in equilibrium prices is

$$\begin{aligned} p_2^* - p_1^* &= \frac{2\tau_1 + \tau_2 + 2c_2 + c_1 + r_2 - r_1}{3} - \frac{2\tau_2 + \tau_1 + 2c_1 + c_2 + r_1 - r_2}{3} \\ &= \frac{\tau_1 - \tau_2 + c_2 - c_1 + 2r_2 - 2r_1}{3}, \end{aligned}$$

the indifferent consumer $\hat{x}$ becomes

$$\begin{aligned} \hat{x} &= \frac{\tau_2}{\tau_1 + \tau_2} + \frac{p_2^* - p_1^*}{\tau_1 + \tau_2} - \frac{r_2 - r_1}{\tau_1 + \tau_2} \\ &= \frac{\tau_2}{\tau_1 + \tau_2} + \frac{\tau_1 - \tau_2 + c_2 - c_1 + 2r_2 - 2r_1}{3(\tau_1 + \tau_2)} - \frac{r_2 - r_1}{\tau_1 + \tau_2} \\ &= \frac{2\tau_2 + \tau_1 + c_2 - c_1 - r_2 + r_1}{3(\tau_1 + \tau_2)}. \end{aligned}$$

The equilibrium demand of firm 1 is, then,

$$D_1^* = \hat{x} = \frac{2\tau_2 + \tau_1 + c_2 - c_1 - r_2 + r_1}{3(\tau_1 + \tau_2)}.$$

- The equilibrium demand of firm 2 is

$$D_2^* = 1 - \hat{x}$$
$$= 1 - \frac{2\tau_2 + \tau_1 + c_2 - c_1 - r_2 + r_1}{3(\tau_1 + \tau_2)}$$
$$= \frac{2\tau_1 + \tau_2 - c_2 + c_1 + r_2 - r_1}{3(\tau_1 + \tau_2)}.$$

- Therefore, firm 1 sells more units than firm 2 if and only if

$$\frac{2\tau_2 + \tau_1 + c_2 - c_1 - r_2 + r_1}{3(\tau_1 + \tau_2)} > \frac{2\tau_1 + \tau_2 - c_2 + c_1 + r_2 - r_1}{3(\tau_1 + \tau_2)},$$

which is rearranged to yield

$$\frac{\tau_2 - \tau_1}{2} > (r_2 - r_1) - (c_2 - c_1).$$

(e) For compactness, define $\Delta r \equiv r_2 - r_1$, $\Delta c \equiv c_2 - c_1$, and $\Delta\tau \equiv \tau_2 - \tau_1$. Rewrite the condition for which firm 1 sells more units than firm 2 found in part (d) with $\Delta\tau$ on the left side. Interpret.

- Firm 1 sells more units than firm 2 if and only if

$$\frac{\Delta\tau}{2} > \Delta r - \Delta c.$$

The left side of the above inequality represents the transportation cost disadvantage of firm 2, $\Delta\tau \equiv \tau_2 - \tau_1$. The right side measures by how much its quality premium, $\Delta r \equiv r_2 - r_1$, overcomes its cost inefficiency, $\Delta c \equiv c_2 - c_1$. Intuitively, when it is relatively more costly to travel to firm 2 than to firm 1 and this difference exceeds its quality-cost differential, as captured by $\Delta r - \Delta c$, firm 1 sells more units than firm 2. Otherwise, firm 2 sells more units than firm 1.

(f) *Symmetric firms.* What if firms are symmetric in product quality ($\Delta r = 0$), production cost ($\Delta c = 0$), and transportation cost ($\Delta\tau = 0$)? Find the equilibrium demand in this setting, and compare your results to the standard Hotelling model.

- Substituting $r_1 = r_2$, $c_1 = c_2$, and $\tau_1 = \tau_2$ into the results in part (d), the indifferent consumer becomes

$$\hat{x} = \frac{2\tau_1 + \tau_1 + c_1 - c_1 - r_1 + r_1}{3(\tau_1 + \tau_1)} = \frac{1}{2},$$

yielding the equilibrium demand of firm 1 of

$$D_1^* = \hat{x} = \frac{1}{2}$$

and similarly the equilibrium demand of firm 2 of

$$D_2^* = 1 - \hat{x} = \frac{1}{2}$$

which coincide with the results in the standard Hotelling model, where homogeneous firms equally split the market.

(g) *Numerical example.* Evaluate equilibrium demands at parameter values $r_1 = 1$, $r_2 = 2$, $c_1 = 0$, and $c_2 = \frac{1}{2}$. When will firm 1 sell more units than firm 2? Interpret.

- Substituting $r_1 = 1, r_2 = 2, c_1 = 0$, and $c_2 = \frac{1}{2}$ into the results in part (d), the equilibrium demand of firm 1 becomes

$$D_1^* = \frac{2\tau_2 + \tau_1 + \frac{1}{2} - 0 - 2 + 1}{3(\tau_1 + \tau_2)} = \frac{2(2\tau_2 + \tau_1) - 1}{6(\tau_1 + \tau_2)}$$

and that of firm 2 is

$$D_2^* = \frac{2\tau_1 + \tau_2 - \frac{1}{2} + 0 + 2 - 1}{3(\tau_1 + \tau_2)} = \frac{2(2\tau_1 + \tau_2) + 1}{6(\tau_1 + \tau_2)}.$$

- Therefore, firm 1 sells more units than firm 2 if and only if

$$D_1^* = \frac{2(2\tau_2 + \tau_1) - 1}{6(\tau_1 + \tau_2)} > \frac{2(2\tau_1 + \tau_2) + 1}{6(\tau_1 + \tau_2)} = D_2^*,$$

which simplifies to

$$\Delta\tau = \tau_2 - \tau_1 > 1.$$

Intuitively, when traveling to firm 2 is much more costly than traveling to firm 1, where $\Delta\tau > 1$, firm 1 is able to overcome its quality disadvantage in this setting, $\Delta r = r_2 - r_1 = 2 - 1 = 1$, relative to its cost advantage, $\Delta c = c_2 - c_1 = \frac{1}{2} - 0 = \frac{1}{2}$, and ends up selling more units than firm 2.

Exercise #3.11: Hotelling Model of Horizontal Product Differentiation^A

3.11 Consider the following model of horizontally differentiated products with two firms. In the first stage, every firm i chooses its location, l_i, in the interval $[0, 1]$, where $i = \{1, 2\}$. In the second stage, every firm, observing the location pair (l_1, l_2) from the first stage, responds setting a price p_i. In the third stage, given firms' location and prices, consumers buy one unit of the good from either firm 1 or 2. Consumers are, for simplicity, uniformly distributed in the unit line. Assume that consumers suffer quadratic transportation costs, and both firms' marginal production cost is $c > 0$.

(a) *Third stage—Finding demand.* For given locations from the first stage, and given prices from the second stage, find the demand that each firm has in the third stage.

- If a consumer purchases from firm 1, his utility is $r - p_1 - t(x - l_1)^2$, while purchasing from firm 2 yields $r - p_2 - t(x - l_2)^2$. Therefore, the indifferent consumer $\widehat{x}$ solves

$$r - p_1 - t(\widehat{x} - l_1)^2 = r - p_2 - t(\widehat{x} - l_2)^2,$$

which yields

$$\widehat{x} = \frac{l_1 + l_2}{2} - \frac{p_1 - p_2}{2t(l_2 - l_1)}.$$

Firm 1's demand is $\widehat{x}$, while firm 2's demand is $1-\widehat{x}$.

- When both firms set the same prices, $p_1 = p_2$, firm 1's demand simplifies to $\widehat{x} = \frac{l_1+l_2}{2}$, while firm 2's demand becomes $1-\widehat{x} = 1-\frac{l_1+l_2}{2}$.

(b) *Second stage–Prices.* For given locations from the first stage, find the price that each firm sets in the second stage.

- *Finding firm 1's best response function.* Firm 1 chooses the price p_1 that solves

$$\max_{p_1 \geq 0} \; (p_1 - c)\underbrace{\left(\frac{l_1+l_2}{2} - \frac{p_1 - p_2}{2t(l_2 - l_1)}\right)}_{\text{Demand, } \widehat{x}}.$$

Differentiating with respect to p_1, we obtain

$$\left(\frac{l_1+l_2}{2} - \frac{p_1 - p_2}{2t(l_2 - l_1)}\right) + (p_1 - c)\left(-\frac{1}{2t(l_2 - l_1)}\right) = 0.$$

Solving for p_1, we find firm 1's best response function

$$p_1(p_2) = \frac{t(l_1+l_2)(l_2-l_1)+c}{2} + \frac{1}{2}p_2$$

with vertical intercept at $\frac{t(l_1+l_2)(l_2-l_1)+c}{2}$ and slope $\frac{1}{2}$. Intuitively, when firm 2 increases its price by \$1, firm 1 responds by increasing its own by \$0.5.

- In addition, a marginal increase in firm 1's location, l_1, or in firm 2's location, l_2, yields the following change in the above best response functions:

$$\frac{\partial p_1(p_2)}{\partial l_1} = -2tl_1 < 0 \quad \text{and} \quad \frac{\partial p_1(p_2)}{\partial l_2} = 2tl_2 > 0,$$

respectively. Therefore, when firm 1 moves its position rightward, its best response function shifts downward in a parallel fashion, indicating that the firm charges less for its product. Intuitively, its position is closer to firm 2's, attenuating product differentiation and ultimately decreasing the price that firm 1 can charge. In contrast, when firm 2 moves its position rightward, both firms move further away from each other, entailing more differentiated products. In this case, firm 1's best response function shifts upward, indicating a higher price p_1.

- *Finding firm 2's best response function.* Operating similarly for firm 2, we have that this firm chooses price p_2 to solve

$$\max_{p_2 \geq 0} \; (p_2 - c)\underbrace{\left(1 - \left(\frac{l_1+l_2}{2} - \frac{p_1 - p_2}{2t(l_2 - l_1)}\right)\right)}_{\text{Demand, } 1-\widehat{x}}.$$

Differentiating with respect to p_2, we obtain

$$\left(1 - \left(\frac{l_1+l_2}{2} - \frac{p_1 - p_2}{2t(l_2 - l_1)}\right)\right) + (p_2 - c)\left(-\frac{1}{2t(l_2 - l_1)}\right) = 0.$$

Solving for p_2, we find firm 2's best response function

$$p_2(p_1) = \frac{t(l_2 - l_1)(2 - l_1 - l_2) + c}{2} + \frac{1}{2}p_1$$

with vertical intercept at $\frac{t(l_2-l_1)(2-l_1-l_2)+c}{2}$ and slope $\frac{1}{2}$, thus exhibiting the same intuition as the positively sloped best response function of firm 1.

- Similar to the case of firm 1, a marginal increase in firm 1's location, l_1, or in firm 2's location, l_2, yields

$$\frac{\partial p_2(p_1)}{\partial l_1} = -2t(1 - l_1) < 0 \text{ and } \frac{\partial p_2(p_1)}{\partial l_2} = 2t(1 - l_2) > 0,$$

respectively. Therefore, when firm 1 moves its position rightward, it becomes closer to firm 2, attenuating product differentiation and ultimately decreasing the price that firm 2 charges (downward shift in firm 2's best response function). In contrast, when firm 2 moves its position rightward, both firms move further away from each other, entailing more differentiated products. In this case, firm 2's best response function shifts upward.

- *Finding equilibrium prices.* Simultaneously solving for p_1 and p_2 in the above best response functions, we find equilibrium prices

$$p_1^*(l_1, l_2) = c + \frac{t}{3}(l_2 - l_1)(2 + l_1 + l_2), \text{ and}$$

$$p_2^*(l_1, l_2) = c + \frac{t}{3}(l_2 - l_1)(4 - l_1 - l_2).$$

Following with our above discussion, note that when both firms are located at the same position, $l_1 = l_2$, equilibrium prices simplify to marginal cost pricing

$$p_1^*(l_1, l_2) = p_2^*(l_1, l_2) = c.$$

Our results also help us examine the case of pricing under exogenous product differentiation (e.g., $l_1 = 0$ and $l_2 = 1$). In this setting, equilibrium prices become

$$p_1^*(l_1, l_2) = c + \frac{t}{3}(1 - 0)(2 + 0 + 1) = c + t, \text{ and}$$

$$p_2^*(l_1, l_2) = c + \frac{t}{3}(1 - 0)(4 - 0 - 1) = c + t.$$

Therefore, second-stage profits are

$$\pi_1^*(l_1, l_2) = \frac{t}{18}(l_2 - l_1)(2 + l_1 + l_2)^2 \text{ and}$$

$$\pi_2^*(l_1, l_2) = \frac{t}{18}(l_2 - l_1)(4 - l_1 - l_2)^2$$

which also collapse to zero when both firms are located at the same position, $l_1 = l_2$, and to $\pi_1^*(l_1, l_2) = \pi_2^*(l_1, l_2) = \frac{t}{2}$ when firms' locations are exogenously determined at $l_1 = 0$ and $l_2 = 1$.

(c) *First stage—Equilibrium location.* Anticipating equilibrium behavior in the second and third stages, find the equilibrium location choice of each firm in the first stage of the game.

- *Finding firm 1's best response function.* In the first stage, firm 1 anticipates the equilibrium prices that firms charge in the second stage and chooses its location l_1 to solve

$$\max_{l_1} \frac{t}{18}(l_2 - l_1)(2 + l_1 + l_2)^2.$$

Differentiating with respect to l_1, we find

$$-\frac{t}{18}(2 + l_1 + l_2)^2 + \frac{t}{9}(l_2 - l_1)(2 + l_1 + l_2) = 0,$$

which, solving for l_1, yields firm 1's best response function

$$l_1(l_2) = -\frac{2}{3} + \frac{1}{3}l_2.$$

- *Finding firm 2's best response function.* Similarly, firm 2 chooses location l_2 to solve

$$\max_{l_2} \frac{t}{18}(l_2 - l_1)(4 - l_1 - l_2)^2.$$

Differentiating with respect to l_2, we find

$$\frac{t}{18}(4 - l_1 - l_2)^2 - \frac{t}{9}(l_2 - l_1)(4 - l_1 - l_2) = 0,$$

which, solving for l_2, yields firm 2's best response function

$$l_2(l_1) = \frac{4}{3} + \frac{1}{3}l_1.$$

- *Finding equilibrium location.* Simultaneously solving for l_1 and l_2 in the above best response functions yields

$$l_1 = -\frac{2}{3} + \frac{1}{3}\underbrace{\left(\frac{4}{3} + \frac{1}{3}l_1\right)}_{l_2(l_1)}.$$

Solving for l_1, we obtain

$$l_1 = -\frac{1}{4}.$$

Inserting this result into firm 2's best response function, we find that

$$l_2 = \frac{4}{3} + \frac{1}{3}\left(-\frac{1}{4}\right) = \frac{5}{4}.$$

Therefore, firms differentiate their products as much as possible. In the interval [0, 1], they locate at the endpoints of the line, $l_1 = 0$ and $l_2 = 1$. From our above discussion, we know that these positions yield equilibrium prices

$$p_1^*(0,1) = p_2^*(0,1) = c + t$$

and equilibrium profits

$$\pi_1^*(0,1) = \pi_2^*(0,1) = \frac{t}{2}.$$

Exercise #3.12: Hotelling with Asymmetric Transportation Costs^B

3.12 Consider a mass of identical consumers uniformly distributed on the unit line, $x \in U[0,1]$. Each consumer buys one unit of the good from either firm 1 or firm 2, located at two ends on the unit line, with per-unit travel cost of τ_1 and τ_2 respectively, where $\tau_1 > \tau_2 > 0$. The indirect utility that consumer x derives from buying the good is

$$\begin{aligned} r_1 - \tau_1 x - p_1 \quad & \text{when buying from firm 1} \\ r_2 - \tau_2(1-x) - p_2 \quad & \text{when buying from firm 2,} \end{aligned}$$

where $r_1 > r_2 > 0$ means that consumers perceive good 1 as superior to good 2 (vertical product differentiation).

(a) *Finding demands.* Identify the indifferent consumer $\hat{x}$, and solve for the demands of firm 1 and firm 2.

- The indifferent consumer solves

$$r_1 - \tau_1\hat{x} - p_1 = r_2 - \tau_2\left(1-\hat{x}\right) - p_2,$$

which, after rearranging, yields

$$\hat{x} = \frac{r_1 - r_2 + \tau_2}{\tau_1 + \tau_2} + \frac{p_2 - p_1}{\tau_1 + \tau_2}.$$

- The demand for firm 1 is

$$D_1(p_1, p_2) = \hat{x} = \frac{r_1 - r_2 + \tau_2}{\tau_1 + \tau_2} + \frac{p_2 - p_1}{\tau_1 + \tau_2}$$

and that for firm 2 is

$$\begin{aligned} D_2(p_1, p_2) &= 1 - \hat{x} = 1 - \left(\frac{r_1 - r_2 + \tau_2}{\tau_1 + \tau_2} + \frac{p_2 - p_1}{\tau_1 + \tau_2}\right) \\ &= \frac{r_2 - r_1 + \tau_1}{\tau_1 + \tau_2} - \frac{p_2 - p_1}{\tau_1 + \tau_2}. \end{aligned}$$

(b) *Equilibrium prices.* Suppose that firms' production costs are symmetric, where $c = c_1 = c_2$. Find the equilibrium prices.

- Firm 1 chooses p_1 to solve the following profit maximization problem:

$$\max_{p_1>0} \pi_1(p_1) = D_1(p_1, p_2)(p_1 - c)$$
$$= \left(\frac{r_1 - r_2 + \tau_2}{\tau_1 + \tau_2} + \frac{p_2 - p_1}{\tau_1 + \tau_2}\right)(p_1 - c).$$

Differentiating with respect to p_1 yields

$$r_1 - r_2 + \tau_2 + p_2 - p_1 - p_1 + c = 0.$$

Rearranging, we obtain firm 1's best response function, as follows:

$$p_1(p_2) = \frac{r_1 - r_2 + \tau_2 + c}{2} + \frac{p_2}{2},$$

which originates at $\frac{r_1-r_2+\tau_2+c}{2}$ and increases in its rival's price, p_2, at a rate of $1/2$. When consumers regard the good from firm 1 as more vertically superior than that of firm 2, term $r_1 - r_2$ increases, shifting firm 1's best response function upward. In contrast, when products exhibit a smaller vertical differentiation, term $r_1 - r_2$ decreases, shifting firm 1's best response function downward.

In addition, when the transportation cost that consumers suffer when buying from firm 2 increases (higher τ_2) or when both firms' production cost increases (higher c), firm 1's best response function shifts upward, indicating that, for a given price of its rival, p_2, firm 1 can charge a higher price p_1.

- Firm 2 chooses p_2 to solve the following profit maximization problem:

$$\max_{p_2>0} \pi_2(p_2) = D_2(p_1, p_2)(p_2 - c)$$
$$= \left(\frac{r_2 - r_1 + \tau_1}{\tau_1 + \tau_2} - \frac{p_2 - p_1}{\tau_1 + \tau_2}\right)(p_2 - c).$$

Differentiating with respect to p_2 yields

$$r_2 - r_1 + \tau_1 - p_2 + p_1 - p_2 + c = 0.$$

Rearranging, we obtain firm 2's best response function, as follows:

$$p_2(p_1) = \frac{r_2 - r_1 + \tau_1 + c}{2} + \frac{p_1}{2}.$$

The comparative statics of this best response function are symmetric to those of firm 1's and discussed above.

- Inserting best response function $p_2(p_1)$ into $p_1(p_2)$, we find that

$$p_1 = \frac{r_1 - r_2 + \tau_2 + c}{2} + \frac{1}{2}\left(\frac{r_2 - r_1 + \tau_1 + c}{2} + \frac{p_1}{2}\right).$$

Rearranging and solving for p_1, we obtain firm 1's equilibrium price

$$p_1^* = c + \frac{r_1 - r_2 + \tau_1 + 2\tau_2}{3}.$$

Inserting p_1^* into $p_2(p_1)$, we find firm 2's equilibrium price, $p_2^* = p_2(p_1^*)$, that is,

$$\begin{aligned} p_2^* &= \frac{r_2 - r_1 + \tau_1 + c}{2} + \frac{1}{2}\left(c + \frac{r_1 - r_2 + \tau_1 + 2\tau_2}{3}\right) \\ &= c + \frac{r_2 - r_1 + 2\tau_1 + \tau_2}{3}. \end{aligned}$$

(c) *Equilibrium demands.* Find the equilibrium demands. When will firm 2 sell more units than firm 1? Interpret.

- Inserting equilibrium prices p_1^* and p_2^*, we find that firm 1's equilibrium demand becomes

$$\begin{aligned} &D_1\left(p_1^*, p_2^*\right) \\ &= \frac{r_1 - r_2 + \tau_2}{\tau_1 + \tau_2} + \frac{p_2^* - p_1^*}{\tau_1 + \tau_2} \\ &= \frac{r_1 - r_2 + \tau_2}{\tau_1 + \tau_2} + \frac{1}{\tau_1 + \tau_2}\left(c + \frac{r_2 - r_1 + 2\tau_1 + \tau_2}{3} - c - \frac{r_1 - r_2 + \tau_1 + 2\tau_2}{3}\right) \\ &= \frac{r_1 - r_2 + \tau_1 + 2\tau_2}{3(\tau_1 + \tau_2)}. \end{aligned}$$

- Operating similarly, we obtain that firm 2's equilibrium demand is

$$\begin{aligned} &D_2\left(p_1^*, p_2^*\right) \\ &= \frac{r_2 - r_1 + \tau_1}{\tau_1 + \tau_2} - \frac{p_2^* - p_1^*}{\tau_1 + \tau_2} \\ &= \frac{r_2 - r_1 + \tau_1}{\tau_1 + \tau_2} - \frac{1}{\tau_1 + \tau_2}\left(c + \frac{r_2 - r_1 + 2\tau_1 + \tau_2}{3} - c - \frac{r_1 - r_2 + \tau_1 + 2\tau_2}{3}\right) \\ &= \frac{r_2 - r_1 + 2\tau_1 + \tau_2}{3(\tau_1 + \tau_2)}. \end{aligned}$$

- Therefore, firm 2 sells more units than firm 1 if $D_2\left(p_1^*, p_2^*\right) > D_1\left(p_1^*, p_2^*\right)$, that is,

$$\frac{r_2 - r_1 + 2\tau_1 + \tau_2}{3(\tau_1 + \tau_2)} > \frac{r_1 - r_2 + \tau_1 + 2\tau_2}{3(\tau_1 + \tau_2)}.$$

Rearranging, this inequality simplifies to

$$\tau_1 - \tau_2 > 2(r_1 - r_2)$$

Intuitively, firm 2 sells more units than firm 1 when the difference in transportation cost, $\tau_1 - \tau_2$, is more than double the difference in quality of the good, $2(r_1 - r_2)$.

(d) *Extreme cases*. Evaluate your results of part (c) at the case that $r_1 \to r_2$ and at the case that $\tau_1 \to \tau_2$. Interpret.

- When consumers regard both goods as almost not vertically differentiated, so that $(r_1 - r_2) \to 0$, condition

$$\tau_1 - \tau_2 > 2(r_1 - r_2)$$

holds given that $\tau_1 > \tau_2 > 0$ by assumption. Intuitively, because consumers pay a higher transportation cost when buying from firm 1 than from firm 2, and they value both goods similarly, firm 2 sells more units than firm 1.

- When consumers regard the good from firm 1 as significantly better than that of firm 2, so that r_1 is larger than r_2, but suffering similar transportation costs when buying from either firm, so that $(\tau_1 - \tau_2) \to 0$, condition

$$\tau_1 - \tau_2 > 2(r_1 - r_2)$$

cannot hold. Intuitively, consumers value firm 1 a lot while not suffering a larger transportation when buying its product, implying that firm 1 sells more units than firm 2.

(e) How will your result in part (c) be affected if $r_2 > r_1$? Explain.

- When $r_2 > r_1$, the inequality found in part (c), $\tau_1 - \tau_2 > 2(r_1 - r_2)$, is always satisfied since $\tau_1 - \tau_2$ on the left side is positive while $2(r_1 - r_2)$ on the right side is now negative. Intuitively, firm 2 benefits from a double advantage relative to firm 1: not only consumers face lower transportation costs when buying from firm 2 than from firm 1, but they also value firm 2's product more than firm 1's (i.e., firm 2's product is now regarded as vertically superior).

Exercise #3.13: Socially Optimal Location in Hotelling Model^B

3.13 Consider a mass of identical consumers uniformly distributed on the unit line, $x \in U[0, 1]$. Each consumer buys one unit of the good from firm 1 or firm 2 located at location l_1 and l_2, respectively, where $0 \le l_1 \le l_2 \le 1$. Assume that the price is regulated by the government at $\overline{p}$ and the per-unit transportation cost is $\tau > 0$.

Consumer x enjoys utility from consuming product i, where $i \in \{1, 2\}$, of

$$v_i(x) = r - \tau |x - l_i| - \overline{p},$$

where r is the reservation utility of the consumer located at l_i.

(a) *Finding demands*. Identify the indifferent consumer $\hat{x}$, and solve for the demands of firm 1 and firm 2.

- The indifferent consumer $\hat{x}$ solves $v_1(x) = v_2(x)$, which entails

$$r - \tau |x - l_1| - \overline{p} = r - \tau |x - l_2| - \overline{p}.$$

Since consumer $\hat{x}$ is to the right (left) of firm 1 (firm 2), $l_1 \le x \le 1_2$, we have

$$x - l_1 = l_2 - x.$$

Rearranging and solving for $\hat{x}$, we obtain

$$\hat{x} = \frac{l_1 + l_2}{2}.$$

- Therefore, the demand of firm 1 is

$$D_1(l_1, l_2) = \hat{x} = \frac{l_1 + l_2}{2}.$$

Similarly, the demand of firm 2 is

$$D_2(l_1, l_2) = 1 - \hat{x} = 1 - \frac{l_1 + l_2}{2} = \frac{2 - l_1 - l_2}{2}.$$

(b) *Equilibrium locations.* Write the firm's profit maximization problem and solve for the equilibrium locations.

- Firm i chooses l_i to solve

$$\pi_i\left(l_i, l_j\right) = \begin{cases} \frac{(\overline{p}-c)(l_i+l_j)}{2} & \text{if } l_i < l_j \\ \frac{\overline{p}-c}{2} & \text{if } l_i = l_j \\ \frac{(\overline{p}-c)(2-l_i-l_j)}{2} & \text{if } l_i > l_j. \end{cases}$$

- We now seek to show that location profile $l_1^* = l_2^* = 1/2$ is the only Nash equilibrium of the game. We do that by showing that other location profiles lead at least one firm to profitably deviate to another location:
  - Suppose that both firms are located to the left of the midpoint, $l_i < l_j < 1/2$. In this setting, firm i has incentives to deviate to the right of l_j, i.e., $l_i > l_j$, if its profits satisfy

$$\frac{(\overline{p} - c)\left(2 - l_i - l_j\right)}{2} > \frac{(\overline{p} - c)\left(l_i + l_j\right)}{2},$$

    which simplifies to $2 - l_i - l_j > l_i + l_j$ and further reduces to $l_i + l_j < 1$, which holds by assumption.
  - Suppose that both firms are located to the right of the midpoint, $l_i > l_j > 1/2$. In this context, firm i has incentives to deviate to the left of l_j, i.e., $l_i < l_j$, if its profits satisfy

$$\frac{(\overline{p} - c)\left(l_i + l_j\right)}{2} > \frac{(\overline{p} - c)\left(2 - l_i - l_j\right)}{2},$$

    which simplifies to $l_i + l_j > 2 - l_i - l_j$, or to $l_i + l_j > 1$, which holds by assumption.
  - Finally, suppose that one firm is located to the left and its rival to the right of the midpoint, $l_i < 1/2 < l_j$. In this case, firm i has incentives to deviate to $l_i = 1/2$ since its profits satisfy

$$\frac{(\overline{p} - c)\left(1/2 + l_j\right)}{2} > \frac{(\overline{p} - c)\left(l_i + l_j\right)}{2},$$

    which simplifies to $l_i < 1/2$. A similar argument applies to the setting where $l_j < 1/2 < l_i$.

- Therefore, in equilibrium, we must have that $l_1^* = l_2^* = 1/2$, since in this case no firm has incentives to deviate. It is straightforward to show that deviating will cause this firm to lose market share to the other firm, thus entailing profit losses.

(c) *Socially optimal locations.* Find the socially optimal location l_1^{SO} and l_2^{SO}, and compare your results to the equilibrium locations identified in part (b).

- The total transportation cost is

$$TC(l_1, l_2) = \tau \int_0^{l_1} (l_1 - x)\, dx + \tau \int_{l_1}^{\hat{x}} (x - l_1)\, dx$$
$$+\tau \int_{\hat{x}}^{l_2} (l_2 - x)\, dx + \tau \int_{l_2}^{1} (x - l_2)\, dx.$$

The first term integrates the transportation cost of consumers from 0 to where firm 1 is located, and the second term is for consumers from the location of firm 1 to the indifferent consumer $\hat{x}$. An analogous argument applies for firm 2's transportation cost in the third and fourth terms.

- Rearranging this expression, we find that the total transportation costs are

$$TC(l_1, l_2) = \tau \left| l_1 x - \frac{x^2}{2} \right|_0^{l_1} + \tau \left| \frac{x^2}{2} - l_1 x \right|_{l_1}^{\hat{x}} + \tau \left| l_2 x - \frac{x^2}{2} \right|_{\hat{x}}^{l_2} + \tau \left| \frac{x^2}{2} - l_2 x \right|_{l_2}^{1}.$$

Further rearranging yields

$$TC(l_1, l_2) = \tau \left(l_1^2 - \frac{l_1^2}{2} \right) + \tau \left(\frac{\hat{x}^2}{2} - l_1 \hat{x} \right) - \tau \left(\frac{l_1^2}{2} - l_1^2 \right) + \tau \left(l_2^2 - \frac{l_2^2}{2} \right)$$
$$- \tau \left(l_2 \hat{x} - \frac{\hat{x}^2}{2} \right) + \tau \left(\frac{1}{2} - l_2 \right) - \tau \left(\frac{l_2^2}{2} - l_2^2 \right)$$

and simplifies to

$$TC(l_1, l_2) = \tau \left[\hat{x}^2 - (l_1 + l_2)\hat{x} + l_1^2 + l_2^2 - l_2 + \frac{1}{2} \right].$$

Inserting $\hat{x} = \frac{l_1 + l_2}{2}$, as found in part (a), we finally find that the total transportation costs are

$$TC(l_1, l_2) = \frac{\tau}{4} \left[2 + 4l_1^2 + 4l_2^2 - 4l_2 - (l_1 + l_2)^2 \right].$$

Differentiating $TC(l_1, l_2)$ with respect to l_1 yields

$$8l_1 - 2(l_1 + l_2) = 0.$$

Rearranging and solving for l_1, we obtain

$$l_1(l_2) = \frac{l_2}{3}.$$

Differentiating $TC(l_1, l_2)$ with respect to l_2, we find

$$8l_2 - 4 - 2(l_1 + l_2) = 0.$$

Rearranging and solving for l_2, we have

$$l_2(l_1) = \frac{l_1 + 2}{3}.$$

Substituting $l_1(l_2) = \frac{l_2}{3}$ into the above expression, we obtain

$$l_2 = \frac{l_2}{9} + \frac{2}{3}.$$

Rearranging and solving for l_2 yields

$$l_2^{SO} = \frac{3}{4}.$$

Inserting $l_2^{SO} = \frac{3}{4}$ into $l_1(l_2) = \frac{l_2}{3}$ yields

$$l_1^{SO} = \frac{1}{4}.$$

- Comparing socially optimal locations, $l_1^{SO} = \frac{1}{4}$ and $l_3^{SO} = \frac{3}{4}$, against the equilibrium locations found in part (b), $l_1^* = l_2^* = 1/2$, we find that:
  - The socially efficient location of firm 1, $l_1^{SO} = \frac{1}{4}$, is halfway between the origin of the unit line (point 0) and its equilibrium location, $l_1^* = \frac{1}{2}$.
  - Similarly, the socially efficient location of firm 2, $l_2^{SO} = \frac{3}{4}$, is halfway between its equilibrium location, $l_2^* = \frac{1}{2}$, and the end of the unit line (point 1).

Exercise #3.14: Hotelling with Non-uniformly Distributed Consumers.^{C}

3.14 Consider a mass of identical consumers distributed on the unit line according to the cumulative distribution function

$$F(x) = x^2(\alpha - \beta x),$$

where $\alpha, \beta > 0$ and $x \in [0, 1]$. Each consumer buys one unit of the homogenous good from either firm 1 or firm 2 located at the end points on the unit line and suffers a per-unit travel cost of $\tau > 0$. The indirect utility of consumer x is

$$\begin{aligned} &r - \tau x - p_1 && \text{when buying from firm 1} \\ &r - \tau(1 - x) - p_2 && \text{when buying from firm 2,} \end{aligned}$$

where $r > 0$ measures consumers' valuation on the quality of the homogeneous product.

(a) Use the properties of symmetric distribution function, where $F\left(\frac{1}{2}\right) = \frac{1}{2}$ and $F(1) = 1$, to characterize the cumulative distribution function $F(x)$.

- Evaluating $F(x) = x^2(\alpha - \beta x)$ at $x = \frac{1}{2}$, we obtain

$$F\left(\frac{1}{2}\right) = \frac{1}{2^2}\left(\alpha - \frac{\beta}{2}\right) = \frac{1}{2},$$

since $F\left(\frac{1}{2}\right) = \frac{1}{2}$. Simplifying this expression yields

$$\beta = 2(\alpha - 2).$$

Substituting $\beta = 2(\alpha - 2)$ into $F(1) = 1$, we find that

$$1^2[\alpha - 2(\alpha - 2)] = 1$$

or

$$\alpha = 2(\alpha - 2).$$

Rearranging, we find $\alpha = 3$. Inserting this result into $\beta = 2(\alpha - 2)$ yields $\beta = 2(3 - 2) = 2$, so that the distribution function is

$$F(x) = x^2(3 - 2x).$$

(b) *Finding demands.* Using the cumulative distribution function you found in part (a), identify the indifferent consumer $\hat{x}$, and solve for the demands of firms 1 and 2.

- The indifferent consumer solves

$$r - \tau\hat{x} - p_1 = r - \tau\left(1 - \hat{x}\right) - p_2,$$

which, after rearranging and solving for $\hat{x}$, yields

$$\hat{x} = \frac{1}{2} + \frac{p_2 - p_1}{2\tau}.$$

- Therefore, the demand for firm 1 is

$$\begin{aligned} D_1(p_1, p_2) &= F\left(\hat{x}\right) = \hat{x}^2\left(3 - 2\hat{x}\right) \\ &= \left(\frac{1}{2} + \frac{p_2 - p_1}{2\tau}\right)^2\left[3 - 2\left(\frac{1}{2} + \frac{p_2 - p_1}{2\tau}\right)\right] \\ &= \frac{(\tau + p_2 - p_1)^2(2\tau - p_2 + p_1)}{4\tau^3} \end{aligned}$$

and that for firm 2 is

$$D_2(p_1, p_2) = 1 - F(\hat{x})$$
$$= 1 - \frac{(\tau + p_2 - p_1)^2 (2\tau - p_2 + p_1)}{4\tau^3}$$
$$= \frac{4\tau^3 - (\tau + p_2 - p_1)^2 (2\tau - p_2 + p_1)}{4\tau^3}.$$

(c) *Equilibrium prices.* Suppose that production costs are normalized to zero. Set up firm i's profit maximization problem, derive the first-order conditions, and then invoke symmetry to find equilibrium prices.

- Firm 1 chooses p_1 to solve the following profit maximization problem:

$$\max_{p_1>0} \pi_1(p_1) = p_1 D_1(p_1, p_2)$$
$$= \frac{p_1 (\tau + p_2 - p_1)^2 (2\tau - p_2 + p_1)}{4\tau^3}.$$

Differentiating with respect to p_1 yields

$$\frac{(\tau + p_2 - p_1)\left[(\tau + p_2 - p_1)(2\tau - p_2 + 2p_1) - 2p_1 (2\tau - p_2 + p_1)\right]}{4\tau^3} = 0,$$

which best response function, $p_1(p_2)$, implicitly solves

$$4p_1^2 + (4\tau - 5p_2) p_1 - (\tau + p_2)(2\tau - p_2) = 0.$$

- Firm 2 chooses p_2 to solve the following profit maximization problem:

$$\max_{p_2>0} \pi_2(p_2) = p_2 D_2(p_1, p_2)$$
$$= \frac{p_2\left[4\tau^3 - (\tau + p_2 - p_1)^2 (2\tau - p_2 + p_1)\right]}{4\tau^3}.$$

Taking the first order condition with respect to p_2 yields

$$\frac{4\tau^3 - (\tau + p_2 - p_1)\left[(\tau + p_2 - p_1)(2\tau - 2p_2 + p_1) + 2p_2 (2\tau - p_2 + p_1)\right]}{4\tau^3} = 0,$$

which best response function, $p_2(p_1)$, implicitly solves

$$4p_2^3 - 9p_1 p_2^2 - 6(\tau - p_1)(\tau + p_1) p_2 - (\tau - p_1)^2 (2\tau + p_1) + 4\tau^3 = 0.$$

- In a symmetric equilibrium, $p^* = p_1^* = p_2^*$, so that the implicit function for $p_1(p_2)$ becomes

$$4p^2 + (4\tau - 5p) p - (\tau + p)(2\tau - p) = 0,$$

which simplifies to

$$\tau\,(3p - 2\tau) = 0.$$

Solving for p^*, we find an equilibrium price of

$$p^* = \frac{2\tau}{3}.$$

(d) Find the equilibrium demand and profits of every firm.

- Inserting $p^* = p_1^* = p_2^* = \frac{2\tau}{3}$ into $D_1\,(p_1, p_2)$, firm 1's equilibrium demand becomes

$$D_1^* = \frac{\left(\tau + p_2^* - p_1^*\right)^2 \left(2\tau - p_2^* + p_1^*\right)}{4\tau^3}$$
$$= \frac{(\tau + 0)^2\,(2\tau - 0)}{4\tau^3} = \frac{2\tau^3}{4\tau^3}$$
$$= \frac{1}{2}$$

with associated profits of

$$\pi_1^* = \frac{2\tau}{3}\frac{1}{2} = \frac{\tau}{3}.$$

- Substituting $p^* = \frac{2\tau}{3}$ into $D_2\,(p_1, p_2)$, firm 2's equilibrium demand becomes

$$D_2^* = \frac{4\tau^3 - \left(\tau + p_2^* - p_1^*\right)^2 \left(2\tau - p_2^* + p_1^*\right)}{4\tau^3}$$
$$= \frac{4\tau^3 - (\tau + 0)^2\,(2\tau - 0)}{4\tau^3} = \frac{4\tau^3 - 2\tau^3}{4\tau^3}$$
$$= \frac{1}{2}$$

with associated profits of

$$\pi_2^* = \frac{2\tau}{3}\frac{1}{2} = \frac{\tau}{3}.$$

(e) *Comparison with uniformly distributed consumers.* Compare your results (i.e., firms' equilibrium prices, demand, and profits) to the setting where the consumers are uniformly distributed in [0, 1]. Explain.

- When consumers are uniformly distributed, recall from previous exercises that

$$p^U = \tau$$
$$D^U = \frac{1}{2}$$
$$\pi^U = \frac{\tau}{2},$$

where the superscript U stands for the setting of uniform distribution.

- Comparing to the setting where consumers are distributed according to $F(x)$, it is straightforward to verify that every firm charges a lower price under $F(x)$, that is,

$$p^* = \frac{2\tau}{3} < \tau = p^U$$

in order to attract the mass of consumers concentrated around the mean $1/2$.

Furthermore, since firms equally split the consumers, where $D_1^* = D_2^* = D^U = 1/2$, but charging a lower price, every firm generates lower profits than when consumers are uniformly distributed, that is,

$$\pi_1^* = \pi_2^* = \frac{\tau}{3} < \frac{\tau}{2} = \pi^U.$$

(f) *Socially optimal location.* Assume that the social planner decides the firms' locations to minimize the total transportation costs of all consumers that are distributed according to $F(x)$. Find the socially optimal location of firm 1, l_1, and of firm 2, l_2. Then, compare your results to those in a setting of uniformly distributed customers. Interpret.

- The total transportation cost in this setting is given by

$$TC(l_1, l_2) = \underbrace{\overbrace{\tau \int_0^{l_1} (l_1 - x)\, dF(x)}^{\text{Consumers to the left of } l_1} + \overbrace{\tau \int_{l_1}^{\hat{x}} (x - l_1)\, dF(x)}^{\text{Consumers to the right of } l_1}}_{\text{Transportation cost of consumers buying from firm 1}}$$

$$+ \underbrace{\overbrace{\tau \int_{\hat{x}}^{l_2} (l_2 - x)\, dF(x)}^{\text{Consumers to the left of } l_2} + \overbrace{\tau \int_{l_2}^{1} (x - l_2)\, dF(x)}^{\text{Consumers to the right of } l_2}}_{\text{Transportation cost of consumers buying from firm 2}}.$$

The first term integrates the transportation cost of consumers from 0 to firm 1, which is located at l_1, and the second term represents the transportation cost of consumers located between firm 1 and the indifferent consumer located at $\hat{x}$. The third and fourth terms are defined analogously for firm 2, which is located at l_2.

- Since $F(x) = x^2(3 - 2x)$, we obtain that its density is $f(x) = 6x(1 - x)$, yielding

$$\begin{aligned} TC(l_1, l_2) = {} & 6\tau \int_0^{l_1} (l_1 - x)\, x(1 - x)\, dx + 6\tau \int_{l_1}^{\hat{x}} (x - l_1)\, x(1 - x)\, dx \\ & + 6\tau \int_{\hat{x}}^{l_2} (l_2 - x)\, x(1 - x)\, dx + 6\tau \int_{l_2}^{1} (x - l_2)\, x(1 - x)\, dx, \end{aligned}$$

which entails

$$\begin{aligned} TC(l_1, l_2) = {} & 6\tau \left[\frac{l_1}{2}x^2 - \frac{1 + l_1}{3}x^3 + \frac{1}{4}x^4\right]_0^{l_1} - 6\tau \left[\frac{l_1}{2}x^2 - \frac{1 + l_1}{3}x^3 + \frac{1}{4}x^4\right]_{l_1}^{\hat{x}} \\ & + 6\tau \left[\frac{l_2}{2}x^2 - \frac{1 + l_2}{3}x^3 + \frac{1}{4}x^4\right]_{\hat{x}}^{l_2} - 6\tau \left[\frac{l_2}{2}x^2 - \frac{1 + l_2}{3}x^3 + \frac{1}{4}x^4\right]_{l_2}^{1}, \end{aligned}$$

- Furthermore, since the distribution is symmetric, we have that the indifferent consumer is located at $\hat{x} = \frac{l_1+l_2}{2}$, simplifying the above expression to

$$TC(l_1, l_2) = \frac{\tau}{16}\Big[16l_1^3(2 - l_1) + 16l_2^3(2 - l_2) - 8(2l_2 - 1)$$
$$-(l_1 + l_2)^2(4l_1 - 2l_1l_2 + 4l_2 - l_1^2 - l_2^2)\Big].$$

Differentiating the total cost function with respect to l_1, we obtain

$$30l_1^3 - 42l_1^2 - 6l_1^2l_2 + 12l_1l_2 - 6l_1l_2^2 + 6l_2^2 - 2l_2^3 = 0.$$

Similarly, differentiating the total cost function with respect to l_2 yields

$$30l_2^3 - 42l_2^2 - 6l_2^2l_1 + 12l_1l_2 - 6l_1^2l_2 + 6l_1^2 - 2l_1^3 + 8 = 0.$$

Simultaneously solving for l_1 and l_2 in the two first-order conditions above, we find that the socially optimal locations are

$$l_1^{SO,F} = 0.326 \quad \text{and} \quad l_2^{SO,F} = 0.674,$$

where the superscript *SO,F* stands for socially optimal location under $F(x)$.

- *Comparison with uniformly distributed consumers.* When consumers are uniformly distributed, it is socially optimal for firm to be equidistant, that is, $l_1^{SO,U} = 0.25$ and $l_2^{SO,U} = 0.75$, where *SO,U* denotes socially optimal location under uniform distribution.
  - Comparing our above results, we see that for firm 1

$$l_1^{SO,U} = 0.25 < 0.326 = l_1^{SO,F}.$$

  - In contrast, for firm 2 we find that

$$l_2^{SO,U} = 0.75 > 0.674 = l_2^{SO,F}.$$

  - To understand this result, first note that both $F(x)$ and $U[0, 1]$ have the same mean of $1/2$, where

$$E_F[x] = 6\int_0^1 x^2(1-x)\,dx$$
$$= 6\left[\frac{x^3}{3}\right]_0^1 - 6\left[\frac{x^4}{4}\right]_0^1$$
$$= \frac{1}{2}, \text{ and}$$
$$E_U[x] = \int_0^1 x\,dx$$
$$= \left[\frac{x^2}{2}\right]_0^1 = \frac{1}{2},$$

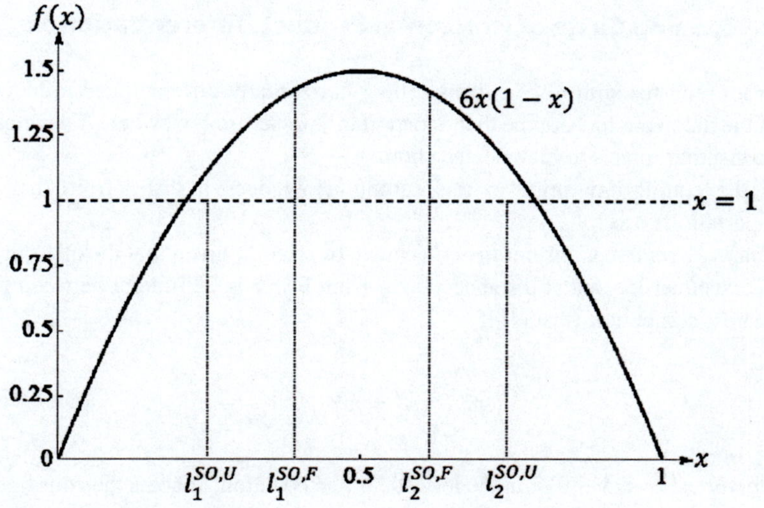

Fig. 3.11 Firm's socially optimal location under $U[0, 1]$ and $F(x)$.

but consumers are less dispersed under $F(x)$ than under $U[0, 1]$, since

$$
\begin{aligned}
E_F\left[x^2\right] &= 6\int_0^1 x^3(1-x)\,dx \\
&= 6\left[\frac{x^4}{4}\right]_0^1 - 6\left[\frac{x^5}{5}\right]_0^1 \\
&= \frac{3}{10}, \text{ and} \\
E_U\left[x^2\right] &= \int_0^1 x^2 dx \\
&= \left[\frac{x^3}{3}\right]_0^1 = \frac{1}{3},
\end{aligned}
$$

yielding the following variances for each distribution function

$$
\begin{aligned}
Var_F(x) &= E_F\left[x^2\right] - (E_F[x])^2 = \frac{3}{10} - \frac{1}{4} = \frac{1}{20} = 0.05 \\
Var_U(x) &= E_U\left[x^2\right] - (E_U[x])^2 = \frac{1}{3} - \frac{1}{4} = \frac{1}{12} \simeq 0.08.
\end{aligned}
$$

Since $Var_F(x) < Var_U(x)$, consumers are more concentrated around the mean under $F(x)$ than under $U[0, 1]$, which induces firms to locate closer to the mean under $F(x)$ than under $U[0, 1]$, that is,

$$l_1^{SO,U} < l_1^{SO,F} < \frac{1}{2} < l_2^{SO,F} < l_2^{SO,U},$$

as Fig. 3.11 illustrates. For easier reference, the figure also plots the density function of $F(x) = x^2(3-2x)$, $f(x) = 6x(1-x)$ and that of the uniform $F(x) = x$, $f(x) = 1$.

Exercise #3.15: Salop Circle of Horizontal Product Differentiation^B

3.15 Consider an industry with $n \geq 2$ firms selling horizontally differentiated products, which we model as the Salop circle. Assume that transportation costs are τd^2, where d denotes the distance that the consumer travels to his selected shop.

(a) Find the equilibrium price in this setting. How does it differ from that under linear transportation costs?

- *Finding demand.* Consider firm i's choice of price p_i given that the other firms charge p. A consumer located at distance $x < \frac{1}{n}$ from firm i is indifferent between firm i and the nearest competitor (firm j) if

$$p_i + \tau x^2 = p_j + \tau \left(\frac{1}{n} - x\right)^2,$$

where $\tau x^2 = \tau(x - 0)^2$, in the left side of the equation, denotes the consumer's quadratic transportation cost of moving from x to 0, buying from firm i, while $\tau \left(\frac{1}{n} - x\right)^2$ in the right side measures this quadratic transportation cost when moving from x to $\frac{1}{n}$, buying from firm j.

Solving for x, we obtain

$$x = \frac{n(p_j - p_i)}{2\tau} + \frac{1}{2n},$$

which implies that firm i's demand is twice this amount (consumers to the left and right of the indifferent consumer), that is,

$$Q_i(p_i, p) = 2x = \frac{1}{n} - \frac{n(p_i - p_j)}{\tau}.$$

- *Finding equilibrium prices.* Anticipating this demand, the firm solves the following profit maximization problem:

$$\max_{p_i \geq 0} \; (p_i - c)Q_i(p_i, p).$$

Differentiating with respect to p_i yields

$$\frac{1}{n} - \frac{n(2p_i - p_j - c)}{\tau} = 0.$$

In a symmetric equilibrium, all firms set the same prices, $p_i = p_j = p$ for any two firms i and $j \neq i$. Inserting this property in the above first-order condition yields

$$\frac{1}{n} = \frac{n(p - c)}{\tau}.$$

Solving for p, we obtain the equilibrium price

$$p^* = c + \frac{\tau}{n^2},$$

which is increasing in the marginal cost of production, c, in the unit transportation cost, τ, but decreases in the number of competing firms in the industry, n. Interestingly, this price collapses to the perfectly competitive price $p^* = c$ when an infinite number of firms compete, $n \to +\infty$.

(b) Find the equilibrium number of firms entering the industry, n^e, when the entry cost is $e > 0$.

- Firm profits in equilibrium are

$$\begin{aligned}\pi^*(e) &= (p^* - c)Q_i(p^*, p^*) - e \\ &= \left(c + \frac{\tau}{n^2} - c\right)\left(\frac{1}{n} - \frac{n(p_i - p_j)}{\tau}\right) - e \\ &= \frac{\tau}{n^3} - e.\end{aligned}$$

Using the zero-profit condition, $\pi^*(e) = 0$, so no more firms have incentives to join the industry, and solving for e, we obtain the equilibrium number of firms entering this market as follows:

$$n^e = \left(\frac{\tau}{e}\right)^{1/3},$$

which entails an equilibrium price of

$$p^* = c + \frac{\tau}{(n^e)^2} = c + \frac{\tau}{\left[\left(\frac{\tau}{e}\right)^{1/3}\right]^2} = c + \tau^{1/3}e^{2/3},$$

which is increasing in the marginal cost of production, the unit transportation cost, and the entry cost.

(c) Find the socially optimal number of firms entering the industry, n^{SO}, when the entry cost is $e > 0$.

- At the social optimum, all consumers purchase one unit, implying that the price they pay goes to the firm in the form of total revenue. However, each consumer incurs a transportation cost and, in addition, every firm that enters the industry incurs an entry cost. The sum of total transportation costs and total entry costs gives us, then, the welfare to consider at the social optimum.

  Therefore, we only need to minimize the sum of total entry costs and total transportation costs as follows:

$$\min_n ne + 2n\tau \int_0^{\frac{1}{2n}} x^2 dx = \min_n ne + \frac{\tau}{12n^2}.$$

Differentiating with respect to n yields

$$e - \frac{\tau}{6n^3} = 0,$$

which, solving for n, yields the socially optimal number of firms entering the industry

$$n^{SO} = \left(\frac{\tau}{6e}\right)^{1/3} = \frac{1}{6^{1/3}}\underbrace{\left(\frac{\tau}{e}\right)^{1/3}}_{n^e} = \frac{1}{6^{1/3}}n^e \simeq 0.55n^e.$$

(d) Compare your results in parts (b) and (c). Interpret.
- The socially optimal number of firms is lower than the equilibrium number of firms, $n^{SO} < n^e$, since every firm does not internalize the business-stealing effect that its entry imposes on other firms. The regulator internalizes this external effect, seeking less entry.

(e) Compare your result in part (d) against the case in which transportation costs are linear. Interpret.
- Equilibrium entry with linear transportation costs is $n^e = \left(\frac{\tau}{e}\right)^{1/2}$, thus being larger than under quadratic transportation costs. We can also measure the excessive entry in each context as the difference between the equilibrium and the socially optimal number of firms in the industry, $n^e - n^{SO}$.
  – Under linear transportation costs, excessive entry is given by

$$\text{Excessive Entry} = \left(\frac{\tau}{e}\right)^{1/2} - \frac{1}{2}\left(\frac{\tau}{e}\right)^{1/2} = 0.5\left(\frac{\tau}{e}\right)^{1/2}.$$

  – Under quadratic transportation costs, we have

$$\text{Excessive Entry} = \left(\frac{\tau}{e}\right)^{1/3} - \left(\frac{\tau}{6e}\right)^{1/3} \simeq 0.45\left(\frac{\tau}{e}\right)^{1/3}.$$

(f) *Numerical example.* Assume a unit transportation cost $\tau = 1/12$ and an entry cost $e = 1/4$. Evaluate equilibrium entry, n^e, and socially optimal entry, n^{SO}, in the case of linear transportation costs and quadratic transportation costs.
- Under linear transportation costs, we obtain

$$n^e = \sqrt{\frac{\frac{1}{12}}{\frac{1}{4}}} = \frac{1}{\sqrt{3}} \approx 0.58$$

$$n^{SO} = \frac{1}{2}\sqrt{\frac{\frac{1}{12}}{\frac{1}{4}}} = \frac{1}{2\sqrt{3}} \approx 0.29.$$

- Under quadratic transportation costs, we have

$$n^e = \sqrt[3]{\frac{\frac{1}{12}}{\frac{1}{4}}} = \frac{1}{\sqrt[3]{3}} \approx 0.69$$

$$n^{SO} = \sqrt[3]{\frac{\frac{1}{12}}{6 \times \frac{1}{4}}} = \frac{1}{\sqrt[3]{18}} \approx 0.38.$$

Exercise #3.16: Horizontal Differentiation in Two Dimensions, Based on Irmean and Thisse (1998)[C]

3.16 Consider a model of horizontally differentiated products with two firms. In the first stage, firm A (B) chooses its location in the two-dimensional space (a_1, a_2) ((b_1, b_2)), where $0 \leq a_1 \leq b_1 \leq 1$ and $0 \leq a_2 \leq b_2 \leq 1$. Intuitively, firms differentiate along two dimensions (e.g., sweetness and color), implying that consumer preferences in this setting are uniformly distributed in a unit square (i.e., a square of side one). Let r be the reservation utility of the consumer and the per-unit disutility from purchasing a good be t_1 (t_2) in dimension 1 (2), where $t_1(b_1 - a_1) > t_2(b_2 - a_2)$ suggests that the consumer suffers more disutility from spatial divergence along the first than the second dimension when the firm's location does not coincide with his ideal.

(a) *Third stage—Indifferent consumer.* Given locations from the first stage, (a_1, a_2) for firm A and (b_1, b_2) for firm B, and prices from the second stage (p_A, p_B), identify the location of indifferent consumer, x_1 (x_2), in the third stage.

- If the consumer purchases from firm A, his utility is $r - p_A - t_1(x_1 - a_1)^2 - t_2(x_2 - a_2)^2$, while purchasing from firm B yields $r - p_B - t_1(x_l - b_1)^2 - t_2(x_2 - b_2)^2$. Therefore, the indifferent consumer solves

$$r - p_A - t_1(x_1 - a_1)^2 - t_2(x_2 - a_2)^2 = r - p_B - t_1(x_l - b_1)^2 - t_2(x_2 - b_2)^2,$$

which we rearrange to

$$t_1(a_1^2 - 2a_1x_1 - b_1^2 + 2b_1x_1) = p_B - p_A + t_2(b_2^2 - 2b_2x_2 - a_2^2 + 2a_2x_2).$$

Solving for x_1 yields

$$x_1(x_2) = \frac{p_B - p_A + t_1(b_1^2 - a_1^2) + t_2(b_2^2 - a_2^2)}{2t_1(b_1 - a_1)} - \frac{t_2(b_2 - a_2)}{t_1(b_1 - a_1)}x_2.$$

(b) *Third stage—Finding demand.* Consider for simplicity that firms only differentiate in dimension 1. Derive the firms' demand as a function of second-stage prices.

- Since firms only differentiate in dimension 1, the indifferent consumer solves

$$x_1 = \int_0^1 \left[\frac{p_B - p_A + t_1(b_1^2 - a_1^2) + t_2(b_2^2 - a_2^2)}{2t_1(b_1 - a_1)} - \frac{t_2(b_2 - a_2)}{t_1(b_1 - a_1)}x_2\right] dF(x_2)$$

when firms do not seek spatial differentiation in dimension 2, yielding

$$\begin{aligned} x_1 &= \frac{p_B - p_A + t_1(b_1^2 - a_1^2) + t_2(b_2^2 - a_2^2)}{2t_1(b_1 - a_1)} - \frac{t_2(b_2 - a_2)}{t_1(b_1 - a_1)}\int_0^1 x_2 dx_2 \\ &= \frac{p_B - p_A + t_1(b_1^2 - a_1^2) + t_2(b_2^2 - a_2^2) - t_2(b_2 - a_2)}{2t_1(b_1 - a_1)} \end{aligned}$$

since x_2 is uniformly distributed independently from x_1 in the unit square.

- In this context, the demand of firm A becomes

$$D_A(p_A, p_B) = x_1 = \frac{p_B - p_A + t_1(b_1^2 - a_1^2) + t_2(b_2^2 - a_2^2) - t_2(b_2 - a_2)}{2t_1(b_1 - a_1)}.$$

- Whereas the demand of firm B is given by

$$\begin{aligned} D_B(p_A, p_B) &= 1 - D_A(p_A, p_B) \\ &= \frac{p_A - p_B - t_1(b_1^2 - a_1^2) - t_2(b_2^2 - a_2^2) + 2t_1(b_1 - a_1) + t_2(b_2 - a_2)}{2t_1(b_1 - a_1)}. \end{aligned}$$

(c) *Second stage—Prices.* Assume a common per-unit cost c, where $0 \leq c < 1$, for every firm $i \in \{A, B\}$. Set up the firms' profit maximization problem to find prices and profits as a function of first-period locations.

- *Finding firm A's best response function.* Firm A chooses price p_A to solve

$$\max_{p_A > 0} \pi_A(p_A) = (p_A - c)\underbrace{\frac{p_B - p_A + t_1\left(b_1^2 - a_1^2\right) + t_2\left(b_2^2 - a_2^2\right) - t_2(b_2 - a_2)}{2t_1(b_1 - a_1)}}_{D_A(p_A, p_B)}.$$

Differentiating with respect to p_A, we obtain

$$\frac{\partial \pi_A(p_A)}{\partial p_A} = \frac{p_B - 2p_A + c + t_1\left(b_1^2 - a_1^2\right) + t_2\left(b_2^2 - a_2^2\right) - t_2(b_2 - a_2)}{2t_1(b_1 - a_1)}.$$

Assuming an interior solution, $\frac{\partial \pi_A(p_A)}{\partial p_A} = 0$, firm A's best response function is

$$p_A(p_B) = \frac{c + t_1\left(b_1^2 - a_1^2\right) + t_2\left(b_2^2 - a_2^2\right) - t_2(b_2 - a_2)}{2} + \frac{p_B}{2}$$

with a vertical intercept of $\frac{c+t_1\left(b_1^2-a_1^2\right)+t_2\left(b_2^2-a_2^2\right)-t_2(b_2-a_2)}{2}$ and slope of $\frac{1}{2}$.

- *Finding firm B's best response function.* Firm B chooses price p_B to solve

$$\max_{p_B > 0} \pi_B(p_B)$$
$$= (p_B - c)\underbrace{\frac{p_A - p_B - t_1\left(b_1^2 - a_1^2\right) - t_2\left(b_2^2 - a_2^2\right) + 2t_1(b_1 - a_1) + t_2(b_2 - a_2)}{2t_1(b_1 - a_1)}}_{D_B(p_A, p_B)}.$$

Differentiating with respect to p_B, we obtain

$$\frac{\partial \pi_B(p_B)}{\partial p_B} = \frac{p_A - 2p_B + c - t_1\left(b_1^2 - a_1^2\right) - t_2\left(b_2^2 - a_2^2\right) + 2t_1(b_1 - a_1) + t_2(b_2 - a_2)}{2t_1(b_1 - a_1)}.$$

Assuming an interior solution, $\frac{\partial \pi_B(p_B)}{\partial p_B} = 0$, firm B's best response function is

$$p_B(p_A) = \frac{c - t_1\left(b_1^2 - a_1^2\right) - t_2\left(b_2^2 - a_2^2\right) + 2t_1(b_1 - a_1) + t_2(b_2 - a_2)}{2} + \frac{p_A}{2}$$

with a vertical intercept of $\frac{c-t_1\left(b_1^2-a_1^2\right)-t_2\left(b_2^2-a_2^2\right)+2t_1(b_1-a_1)+t_2(b_2-a_2)}{2}$ and slope of $\frac{1}{2}$.

- *Finding equilibrium prices.* Substituting the best response function of firm B into that of firm A, we obtain

$$p_A = \frac{c + t_1\left(b_1^2 - a_1^2\right) + t_2\left(b_2^2 - a_2^2\right) - t_2(b_2 - a_2)}{2}$$
$$+ \frac{1}{2}\left(\frac{c - t_1\left(b_1^2 - a_1^2\right) - t_2\left(b_2^2 - a_2^2\right) + 2t_1(b_1 - a_1) + t_2(b_2 - a_2)}{2} + \frac{p_A}{2}\right).$$

Rearranging, we obtain

$$p_A(a_1,a_2,b_1,b_2) = c + \frac{t_1\left(b_1^2-a_1^2\right)+t_2\left(b_2^2-a_2^2\right)+2t_1(b_1-a_1)-t_2(b_2-a_2)}{3}$$

Substituting $p_A(a_1,a_2,b_1,b_2)$ into $p_B(p_A)$, we find

$$p_B(a_1,a_2,b_1,b_2) = c - \frac{t_1\left(b_1^2-a_1^2\right)+t_2\left(b_2^2-a_2^2\right)-4t_1(b_1-a_1)-t_2(b_2-a_2)}{3}.$$

- *Finding firm A's profits.* Substituting $p_A(a_1,a_2,b_1,b_2)$ and $p_B(a_1,a_2,b_1,b_2)$ into $\pi_A(p_A)$, profits of firm A become

$$\begin{aligned}\pi_A(a_1,a_2,b_1,b_2) &= (p_A-c)\frac{p_B-p_A+t_1\left(b_1^2-a_1^2\right)+t_2\left(b_2^2-a_2^2\right)-t_2(b_2-a_2)}{2t_1(b_1-a_1)}\\ &= \frac{\left[t_1\left(b_1^2-a_1^2\right)+t_2\left(b_2^2-a_2^2\right)+2t_1(b_1-a_1)-t_2(b_2-a_2)\right]^2}{18t_1(b_1-a_1)}.\end{aligned}$$

- *Finding firm B's profits.* Substituting $p_A(a_1,a_2,b_1,b_2)$ and $p_B(a_1,a_2,b_1,b_2)$ into $\pi_B(p_B)$, profits of firm B become

$$\begin{aligned}&\pi_B(a_1,a_2,b_1,b_2)\\ &= (p_B-c)\frac{p_A-p_B-t_1\left(b_1^2-a_1^2\right)-t_2\left(b_2^2-a_2^2\right)+2t_1(b_1-a_1)+t_2(b_2-a_2)}{2t_1(b_1-a_1)}\\ &= \frac{\left[-t_1\left(b_1^2-a_1^2\right)-t_2\left(b_2^2-a_2^2\right)+4t_1(b_1-a_1)+t_2(b_2-a_2)\right]^2}{18t_1(b_1-a_1)}.\end{aligned}$$

(d) *First stage—Equilibrium location.* Anticipating equilibrium behavior in the second and third stages, find the equilibrium locations in the first stage. Interpret.

- *Firm A.* Differentiating $\pi_A(a_1,a_2,b_1,b_2)$ with respect to a_1, we obtain

$$\begin{aligned}\frac{\partial\pi_A(a_1,a_2,b_1,b_2)}{\partial a_1} &= \frac{2\alpha(b_1-a_1)(-2a_1t_1-2t_1)+\alpha^2}{18t_1(b_1-a_1)^2}\\ &= \frac{\alpha\left[\alpha-4t_1(1+a_1)(b_1-a_1)\right]}{18t_1(b_1-a_1)^2},\end{aligned}$$

where $\alpha \equiv t_1(b_1^2-a_1^2)+t_2(b_2^2-a_2^2)+2t_1(b_1-a_1)-t_2(b_2-a_2)$, and

$$\begin{aligned}&\alpha-4t_1(1+a_1)(b_1-a_1)\\ &= t_1(b_1^2-a_1^2)+t_2(b_2^2-a_2^2)+2t_1(b_1-a_1)-t_2(b_2-a_2)-4t_1(1+a_1)(b_1-a_1)\\ &= t_1(b_1-a_1)(b_1-3a_1-2)+t_2(b_2-a_2)(b_2+a_2-1).\end{aligned}$$

Since $t_1(b_1-a_1) > t_2(b_2-a_2)$, the above condition becomes

$$\begin{aligned}
&t_1(b_1-a_1)(b_1-3a_1-2)+t_2(b_2-a_2)(b_2+a_2-1)\\
&< t_1(b_1-a_1)(b_1-3a_1-2+b_2+a_2-1)\\
&= t_1(b_1-a_1)[a_2+b_1+b_2-3(1+a_1)]<0
\end{aligned}$$

because $a_2+b_1+b_2<3(1+a_1)$ entails $\frac{\partial\pi_A(a_1,a_2,b_1,b_2)}{\partial a_1}<0$ so that $a_1^*=0$.

Similarly, differentiating $\pi_A(a_1,a_2,b_1,b_2)$ with respect to a_2, we have

$$\frac{\partial\pi_A(a_1,a_2,b_1,b_2)}{\partial a_2}=\frac{t_2(1-2a_2)}{18t_1(b_1-a_1)},$$

which is zero when $a_2^*=\frac{1}{2}$.

- *Firm B.* Differentiating $\pi_B(a_1,a_2,b_1,b_2)$ with respect to b_1, we obtain

$$\begin{aligned}
\frac{\partial\pi_B(a_1,a_2,b_1,b_2)}{\partial b_1}&=\frac{2\beta(b_1-a_1)(-2b_1t_1+4t_1)-\beta^2}{18t_1(b_1-a_1)^2}\\
&=\frac{\beta[4t_1(2-b_1)(b_1-a_1)-\beta]}{18t_1(b_1-a_1)^2},
\end{aligned}$$

where $\beta\equiv -t_1\left(b_1^2-a_1^2\right)-t_2\left(b_2^2-a_2^2\right)+4t_1(b_1-a_1)+t_2(b_2-a_2)$, and

$$\begin{aligned}
&4t_1(2-b_1)(b_1-a_1)-\beta\\
&=4t_1(2-b_1)(b_1-a_1)+t_1(b_1^2-a_1^2)+t_2(b_2^2-a_2^2)-4t_1(b_1-a_1)-t_2(b_2-a_2)\\
&=t_1(b_1-a_1)(4+a_1-3b_1)+t_2(b_2-a_2)(b_2+a_2-1)\,.
\end{aligned}$$

Since $t_1(b_1-a_1)>t_2(b_2-a_2)$, the above condition becomes

$$\begin{aligned}
&t_1(b_1-a_1)(4+a_1-3b_1)+t_2(b_2-a_2)(b_2+a_2-1)\\
&> t_2(b_2-a_2)(4+a_1-3b_1+b_2+a_2-1)\\
&= t_2(b_2-a_2)[a_1+a_2+b_2+3(1-b_1)]>0
\end{aligned}$$

so that $\frac{\partial\pi_B(a_1,a_2,b_1,b_2)}{\partial b_1}>0$ implies $b_1^*=1$.

Similarly, differentiating $\pi_B(a_1,a_2,b_1,b_2)$ with respect to b_2, we have

$$\frac{\partial\pi_B(a_1,a_2,b_1,b_2)}{\partial b_2}=\frac{t_2(1-2b_2)}{18t_1(b_1-a_1)},$$

which is zero when $b_2^*=\frac{1}{2}$.

- Summarizing, equilibrium locations are

$$\left(a_1^*,a_2^*\right)=\left(0,\frac{1}{2}\right)\quad\text{for firm } A$$

$$\left(b_1^*,b_2^*\right)=\left(1,\frac{1}{2}\right)\quad\text{for firm } B$$

Intuitively, every firm seeks maximum differentiation in dimension 1, where consumers suffer the most significant disutility (per-unit travel cost), and minimum differentiation in dimension 2.

(e) *Equilibrium profits.* Evaluate firms' equilibrium profits in part (c). How are the firms' profits affected by t_1 and t_2, and which firm makes more profits? Explain.

- Inserting the equilibrium locations $(a_1^*, a_2^*, b_1^*, b_2^*) = (0, \frac{1}{2}, 1, \frac{1}{2})$ into the profit function of firm A yields

$$\pi_A\left(0, \frac{1}{2}, 1, \frac{1}{2}\right) = \frac{\left[t_1\left(1^2 - 0^2\right) + t_2\left(\frac{1}{2^2} - \frac{1}{2^2}\right) + 2t_1(1-0) - t_2\left(\frac{1}{2} - \frac{1}{2}\right)\right]^2}{18t_1(1-0)}$$
$$= \frac{t_1}{2},$$

which increases in disutility t_1 that consumers suffer from dimension 1 (where firms differentiate) but is not affected by disutility t_2 that consumers experience in dimension 2 (where firms do not differentiate).

- Substituting the equilibrium locations $(a_1^*, a_2^*, b_1^*, b_2^*) = (0, \frac{1}{2}, 1, \frac{1}{2})$ into the profit function of firm B, we find

$$\pi_B\left(0, \frac{1}{2}, 1, \frac{1}{2}\right) = \frac{\left[-t_1\left(1^2 - 0^2\right) - t_2\left(\frac{1}{2^2} - \frac{1}{2^2}\right) + 4t_1(1-0) + t_2\left(\frac{1}{2} - \frac{1}{2}\right)\right]^2}{18t_1(1-0)}$$
$$= \frac{t_1}{2},$$

which increases in disutility t_1 but does not change in disutility t_2 for the same reasons as for firm A.

- Since firms are symmetric across dimension 1 and homogeneous in dimension 2, every firm generates the same profits in equilibrium, that is, $\pi_A\left(0, \frac{1}{2}, 1, \frac{1}{2}\right) = \pi_B\left(0, \frac{1}{2}, 1, \frac{1}{2}\right) = \frac{t_i}{2}$.

(f) *Symmetric per-unit travel cost.* Evaluate the indifferent consumer, equilibrium prices, and profits when $t = t_1 = t_2$. Interpret.

- When $t = t_1 = t_2$, the price function of firm A becomes

$$p_A(a_1, a_2, b_1, b_2) = c + \frac{t\left(b_1^2 - a_1^2\right) + t\left(b_2^2 - a_2^2\right) + 2t(b_1 - a_1) - t(b_2 - a_2)}{3}$$
$$= c + \frac{t\left[(b_1 - a_1)(b_1 + a_1 + 2) + (b_2 - a_2)(b_2 + a_2 - 1)\right]}{3},$$

while the price function of firm B is given by

$$p_B(a_1, a_2, b_1, b_2) = c - \frac{t\left(b_1^2 - a_1^2\right) + t\left(b_2^2 - a_2^2\right) - 4t(b_1 - a_1) - t(b_2 - a_2)}{3}$$
$$= c - \frac{t\left[(b_1 - a_1)(b_1 + a_1 - 4) + (b_2 - a_2)(b_2 + a_2 - 1)\right]}{3},$$

yielding the indifferent consumer at

$$x_1 = \frac{p_B - p_A + t\left(b_1^2 - a_1^2\right) + t\left(b_2^2 - a_2^2\right) - t\left(b_2 - a_2\right)}{2t\left(b_1 - a_1\right)}$$
$$= \frac{\left(b_1 - a_1\right)\left(b_1 + a_1 + 2\right) + \left(b_2 - a_2\right)\left(b_2 + a_2 - 1\right)}{6\left(b_1 - a_1\right)}$$

with associated profits of

$$\pi_A\left(a_1, a_2, b_1, b_2\right) = \frac{\left[t\left(b_1^2 - a_1^2\right) + t\left(b_2^2 - a_2^2\right) + 2t\left(b_1 - a_1\right) - t\left(b_2 - a_2\right)\right]^2}{18t\left(b_1 - a_1\right)}$$
$$= \frac{t\left[\left(b_1 - a_1\right)\left(b_1 + a_1 + 2\right) + \left(b_2 - a_2\right)\left(b_2 + a_2 - 1\right)\right]^2}{18\left(b_1 - a_1\right)}$$

for firm A and

$$\pi_B\left(a_1, a_2, b_1, b_2\right) = \frac{\left[-t\left(b_1^2 - a_1^2\right) - t\left(b_2^2 - a_2^2\right) + 4t\left(b_1 - a_1\right) + t\left(b_2 - a_2\right)\right]^2}{18t\left(b_1 - a_1\right)}$$
$$= \frac{t\left[\left(b_1 - a_1\right)\left(b_1 + a_1 - 4\right) + \left(b_2 - a_2\right)\left(b_2 + a_2 - 1\right)\right]^2}{18\left(b_1 - a_1\right)}$$

for firm B.

- Substituting the equilibrium locations $(a_1^*, a_2^*, b_1^*, b_2^*) = (0, \frac{1}{2}, 1, \frac{1}{2})$ into the indifferent consumer yields

$$x_1^* = \frac{(1-0)(1+0+2) + \left(\frac{1}{2} - \frac{1}{2}\right)\left(\frac{1}{2} + \frac{1}{2} - 1\right)}{6(1-0)}$$
$$= \frac{1}{2}$$

who is exactly in the middle of a_1 and b_1, where $\frac{a_1^* + b_1^*}{2} = \frac{0+1}{2} = \frac{1}{2}$.

- Substituting now the equilibrium locations $(a_1^*, a_2^*, b_1^*, b_2^*) = (0, \frac{1}{2}, 1, \frac{1}{2})$ into the equilibrium prices yields

$$p_A^* = c + \frac{t\left[(1-0)(1+0+2) + \left(\frac{1}{2} - \frac{1}{2}\right)\left(\frac{1}{2} + \frac{1}{2} - 1\right)\right]}{3}$$
$$= c + t$$

for firm A, and similarly

$$p_B^* = c - \frac{t\left[(1-0)(1+0-4) + \left(\frac{1}{2} - \frac{1}{2}\right)\left(\frac{1}{2} + \frac{1}{2} - 1\right)\right]}{3}$$
$$= c + t$$

for firm B, which increases linearly in production cost c and disutility parameter t, as in Exercise 3.10, part (b).

- Substituting $(a_1^*, a_2^*, b_1^*, b_2^*) = (0, \frac{1}{2}, 1, \frac{1}{2})$ into the equilibrium profits,

$$\pi_A^* = \frac{t\left[(1-0)(1+0+2) + \left(\frac{1}{2}-\frac{1}{2}\right)\left(\frac{1}{2}+\frac{1}{2}-1\right)\right]^2}{18(1-0)}$$

$$= \frac{t}{2}$$

for firm A, and similarly

$$\pi_B^* = \frac{t\left[(1-0)(1+0-4) + \left(\frac{1}{2}-\frac{1}{2}\right)\left(\frac{1}{2}+\frac{1}{2}-1\right)\right]^2}{18(1-0)}$$

$$= \frac{t}{2}$$

which reconcile with our results in part (e) when $t = t_1 = t_2$ and part (c) of Exercise 3.10 where firms differentiate in one dimension.

Exercise #3.17: Vertical Differentiation, Quality Choice, and Price Competition^B

3.17 Consider the following model of vertically differentiated products with two firms. In the first stage, every firm i chooses its quality, s_i, in the interval $[0, 1]$, where $i = \{1, 2\}$. In the second stage, every firm i, observing the quality pair (s_1, s_2) from the first stage, responds setting a price p_i. In the third stage, given firms' quality and prices, consumers buy one unit of the good from either firm 1 or 2. Every consumer when buying from firm i enjoys utility

$$r - p_i + \theta s_i,$$

where parameter $\theta \sim U[0, 1]$ denotes how much this consumer cares about quality. Intuitively, a consumer with $\theta = 0$ does not assign any concern to quality, while $\theta = 1$ assigns the maximal importance to quality. Assume that, for simplicity, both firms' marginal production cost is $c > 0$.

(a) *Third stage.* For given locations and prices, find the demand that each firm faces in the third stage of the game.

- If a consumer purchases from firm 1, his utility is $r - p_1 + \theta s_1$, while purchasing from firm 2 yields $r - p_2 + \theta s_2$. Therefore, the indifferent consumer $\widehat{\theta}$ solves

$$r - p_1 + \widehat{\theta} s_1 = r - p_2 + \widehat{\theta} s_2,$$

which yields

$$\widehat{\theta} = \frac{p_2 - p_1}{s_2 - s_1}.$$

Therefore, firm 1's demand is $\widehat{\theta}$, while firm 2's demand is $1 - \widehat{\theta}$.

- Graphically, cutoff $\widehat{\theta}$ shifts rightward in the unit line when the price differential $p_2 - p_1$ increases (meaning that firm 2 sets higher prices than firm 1), expanding the demand for firm 1 while shrinking that of firm 2. In contrast, cutoff $\widehat{\theta}$ shifts leftward when the quality differential $s_2 - s_1$ increases (that is, firm 2 offers a higher quality than firm 1), shrinking the demand of firm 1 and expanding that of firm 2.

(b) *Second stage.* For given locations, find the prices that each firm charge in the second stage.

- Firm 1 chooses the price p_1 that solves

$$\max_{p_1 \geq 0} \ (p_1 - c)\underbrace{\left(\frac{p_2 - p_1}{s_2 - s_1}\right)}_{\text{Demand, } \widehat{\theta}}.$$

Differentiating with respect to p_1, we obtain

$$\frac{p_2 - 2p_1 + c}{s_2 - s_1} = 0.$$

Solving for p_1, we find firm 1's best response function

$$p_1(p_2) = \frac{c}{2} + \frac{1}{2}p_2$$

with vertical intercept at $\frac{c}{2}$ and slope $\frac{1}{2}$. Intuitively, when firm 2 increases its price by \$1, firm 1 responds increasing its own by \$0.5.

- Operating similarly for firm 2, we have that this firm chooses price p_2 to solve

$$\max_{p_2 \geq 0} \ (p_2 - c)\underbrace{\left(1 - \frac{p_2 - p_1}{s_2 - s_1}\right)}_{\text{Demand, } 1-\widehat{\theta}}.$$

Differentiating with respect to p_2, we obtain

$$\frac{s_2 - s_1 + p_1 - 2p_2 + c}{s_2 - s_1} = 0.$$

Solving for p_2, we find firm 2's best response function

$$p_2(p_1) = \frac{s_2 - s_1 + c}{2} + \frac{1}{2}p_1$$

with vertical intercept at $\frac{s_2 - s_1 + c}{2}$ and slope $\frac{1}{2}$. Intuitively, when firm 2 increases its price by \$1, firm 1 responds increasing its own by \$0.5.

- Simultaneously solving for p_1 and p_2 in the above best response functions, we find equilibrium prices

$$p_1^*(s_1, s_2) = \frac{s_2 - s_1 + 3c}{3} = \frac{s_2 - s_1}{3} + c, \text{ and}$$

$$p_2^*(s_1, s_2) = \frac{2s_2 - 2s_1 + 3c}{3} = \frac{2(s_2 - s_1)}{3} + c.$$

Therefore, second-stage profits are

$$\pi_1^*(s_1, s_2) = \frac{s_2 - s_1}{9}, \text{ and}$$

$$\pi_2^*(s_1, s_2) = \frac{4(s_2 - s_1)}{9}.$$

(c) *First stage.* Anticipating the equilibrium prices that firms charge in the second stage and the demand that entails in the third stage, find the quality level each firm chooses in the first stage.

- In the first stage, firm 1 anticipates the equilibrium prices that firms charge in the second stage and chooses its quality s_1 to solve

$$\max_{s_1} \pi_1^*(s_1, s_2) = \frac{s_2 - s_1}{9}.$$

Differentiating with respect to s_1, we find $-\frac{1}{9} < 0$, which indicates that we are in a corner solution, where firm 1 reduces s_1 as much as possible, $s_1 = 0$.

- Similarly, firm 2 chooses quality s_2 to solve

$$\max_{s_2} \pi_2^*(s_1, s_2) = \frac{4(s_2 - s_1)}{9}.$$

Differentiating with respect to s_2, we find $\frac{4}{9} > 0$, which indicates that we are in a corner solution, where firm 2 increases s_2 as much as possible, $s_2 = 1$. Therefore, firms choose completely different levels of quality.

- In this context, equilibrium prices (evaluated at the above equilibrium qualities) become

$$p_1^*(s_1^*, s_2^*) = \frac{1-0}{3} + c = \frac{1}{3} + c \text{ and}$$

$$p_2^*(s_1^*, s_2^*) = \frac{2(1-0)}{3} + c = \frac{2}{3} + c.$$

Intuitively, firm 2 offers a relatively high-quality-price pair, $s_2 = 1$ and $p_2 = \frac{2}{3} + c$, while firm 1 chooses a lower quality-price pair, $s_1 = 0$ and $p_1 = \frac{1}{3} + c$.

Exercise #3.18: Products Horizontally and Vertically Differentiated^C

3.18 In this exercise, we examine a product that consumers regard as horizontally and vertically differentiated. As in Hotelling models, firms are exogenously located at the endpoints of the unit line, and consumers are uniformly distributed. Each consumer buys at most one unit of the good from either firm 1 or 2 and incurs a travel cost of $\tau > 0$. As in models of vertically differentiated products, assume that consumers perceive firm 1's good as superior to firm 2's, that is, $r_1 > r_2$. Then the indirect utility of consumer x is

$$r_1 - \tau x - p_1 \quad \text{when buying from firm 1}$$

$$r_2 - \tau(1 - x) - p_2 \quad \text{when buying from firm 2.}$$

(a) Identify the indifferent consumer $\hat{x}$, and solve for the demands of firm 1 and firm 2.
- The indifferent consumer, located at $\hat{x}$, solves

$$r_1 - \tau\hat{x} - p_1 = r_2 - \tau\left(1 - \hat{x}\right) - p_2,$$

which, after rearranging, yields

$$\hat{x} = \frac{1}{2} + \frac{(r_1 - r_2) - (p_1 - p_2)}{2\tau}.$$

- Therefore, the demand for firm 1 is

$$D_1(p_1, p_2) = \hat{x} = \frac{1}{2} + \frac{(r_1 - r_2) - (p_1 - p_2)}{2\tau},$$

which is increasing in firm 1's quality differential $r_1 - r_2$ but decreasing in its price differential, $p_1 - p_2$, and in consumers' transportation cost, τ.
- The demand for firm 2 is

$$D_2(p_1, p_2) = 1 - \hat{x} = \frac{1}{2} - \frac{(r_1 - r_2) - (p_1 - p_2)}{2\tau},$$

which is decreasing in firm 1's quality differential $r_1 - r_2$ and in consumers' transportation cost, τ, but increasing in firm 2's price differential, $p_2 - p_1$.

(b) Suppose quality is costly to produce, that is, $c_1 > c_2$. Find the equilibrium prices that each firm charges.
- *Firm 1's best response function.* Firm 1 chooses its price p_1 to solve the following profit maximization problem:

$$\begin{aligned}\max_{p_1 > 0} \pi_1(p_1) &= D_1(p_1, p_2)(p_1 - c_1) \\ &= \left(\frac{1}{2} + \frac{(r_1 - r_2) - (p_1 - p_2)}{2\tau}\right)(p_1 - c_1).\end{aligned}$$

Differentiating with respect to p_1 yields

$$\tau + (r_1 - r_2) - (p_1 - p_2) - (p_1 - c_1) = 0.$$

Solving for p_1, we obtain firm 1's best response function

$$p_1(p_2) = \frac{p_2}{2} + \frac{r_1 - r_2 + \tau + c_1}{2},$$

which is increasing in firm 1's quality differential $r_1 - r_2$, in its rival's price, p_2, in consumer's transportation cost, τ, and in the cost of producing the good, c_1.
- *Firm 2's best response function.* Firm 2 chooses p_2 to solve the following profit maximization problem:

$$\max_{p_2>0} \pi_2(p_2) = D_2(p_1, p_2)(p_2 - c_2)$$
$$= \left(\frac{1}{2} - \frac{(r_1 - r_2) - (p_1 - p_2)}{2\tau}\right)(p_2 - c_2).$$

Differentiating with respect to p_2 yields

$$\tau - (r_1 - r_2) + (p_1 - p_2) - (p_2 - c_2) = 0.$$

Solving for p_2, we find firm 2's best response function in prices

$$p_2(p_1) = \frac{p_1}{2} + \frac{r_2 - r_1 + \tau + c_2}{2},$$

which is decreasing in firm 1's quality differential $r_1 - r_2$ but increasing in its rival's price, p_1, in consumer's transportation cost, τ, and in the cost of producing the good, c_2.

- Inserting best response function $p_2(p_1)$ into $p_1(p_2)$, we obtain

$$p_1 = \frac{\overbrace{\frac{p_1}{2} + \frac{r_2 - r_1 + \tau + c_2}{2}}^{p_2(p_1)}}{2} + \frac{r_1 - r_2 + \tau + c_1}{2},$$

which, solving for p_1, yields equilibrium price

$$p_1^* = \tau + \frac{r_1 - r_2 + 2c_1 + c_2}{3}.$$

Inserting this price into firm 2's best response function, we find this firm's equilibrium price

$$p_2^* = \frac{\overbrace{\tau + \frac{r_1 - r_2 + 2c_1 + c_2}{3}}^{p_1^*}}{2} + \frac{r_2 - r_1 + \tau + c_2}{2}$$
$$= \tau + \frac{r_2 - r_1 + c_1 + 2c_2}{3}.$$

- Firm 1's equilibrium price is increasing in its quality differential, $r_1 - r_2$, while firm 2's is decreasing in this differential. In addition, every firm i's price is increasing in consumer's transportation cost, τ, in the cost of producing its own good, c_i, and in the cost of producing the rival's good, c_j, where $j \neq i$. However, firm i's price is more sensitive to an increase in its own cost than to an increase in its rival's.

(c) Find the equilibrium demands. Under what condition firm 1 sells more units than firm 2?

- Firm 1's equilibrium demand becomes

$$D_1\left(p_1^*, p_2^*\right) = \frac{1}{2} + \frac{(r_1 - r_2) - \left(p_1^* - p_2^*\right)}{2\tau}$$
$$= \frac{1}{2} + \frac{(r_1 - r_2)}{2\tau}$$
$$- \frac{1}{2\tau}\left(\tau + \frac{r_1 - r_2 + 2c_1 + c_2}{3} - \tau - \frac{r_2 - r_1 + c_1 + 2c_2}{3}\right)$$
$$= \frac{1}{2} + \frac{(r_1 - r_2) - (c_1 - c_2)}{6\tau}.$$

- Firm 2's equilibrium demand is

$$D_2\left(p_1^*, p_2^*\right) = \frac{1}{2} - \frac{(r_1 - r_2) - \left(p_1^* - p_2^*\right)}{2\tau}$$
$$= \frac{1}{2} - \frac{(r_1 - r_2)}{2\tau}$$
$$+ \frac{1}{2\tau}\left(\tau + \frac{r_1 - r_2 + 2c_1 + c_2}{3} - \tau - \frac{r_2 - r_1 + c_1 + 2c_2}{3}\right)$$
$$= \frac{1}{2} + \frac{(r_2 - r_1) - (c_2 - c_1)}{6\tau}.$$

- Therefore, firm 1 sells more units than firm 2 if

$$D_1\left(p_1^*, p_2^*\right) > D_2\left(p_1^*, p_2^*\right)$$
$$\frac{(r_1 - r_2) - (c_1 - c_2)}{6\tau} > \frac{(r_2 - r_1) - (c_2 - c_1)}{6\tau}$$
$$r_1 - r_2 > c_1 - c_2.$$

  Intuitively, this occurs when firm 1's quality differential, $r_1 - r_2$, is larger than its cost disadvantage, $c_1 - c_2$. If firm 1 experiences a cost advantage (i.e., $c_1 - c_2 < 0$), the above condition holds for any positive quality differential.

(d) *Numerical example.* Consider that $\tau = 1/10$, $r_1 = 1/2$, $r_2 = 1/4$, $c_1 = 1/4$, and $c_2 = 1/8$. Evaluate equilibrium prices and demand for each firm, and show that firm 1 sells more units than firm 2.

- Equilibrium prices and demand are

$$p_1^* = \frac{1}{10} + \frac{\frac{1}{2} - \frac{1}{4} + \left(2 \times \frac{1}{4}\right) + \frac{1}{8}}{3} = \frac{47}{120} \approx 0.39$$
$$p_2^* = \frac{1}{10} + \frac{\frac{1}{4} - \frac{1}{2} + \frac{1}{4} + \left(2 \times \frac{1}{8}\right)}{3} = \frac{11}{60} \approx 0.18$$
$$D_1\left(p_1^*, p_2^*\right) = \frac{1}{2} + \frac{\left(\frac{1}{2} - \frac{1}{4}\right) - \left(\frac{1}{4} - \frac{1}{8}\right)}{6 \times \frac{1}{10}} = \frac{17}{24} \approx 0.71$$

$$D_2\left(p_1^*, p_2^*\right) = \frac{1}{2} + \frac{\left(\frac{1}{4} - \frac{1}{2}\right) - \left(\frac{1}{8} - \frac{1}{4}\right)}{6 \times \frac{1}{10}} = \frac{7}{24} \approx 0.29,$$

and it is straightforward to verify that firm 1 sells more units than firm 2 since

$$D_1\left(p_1^*, p_2^*\right) = \frac{17}{24} > \frac{7}{24} = D_2\left(p_1^*, p_2^*\right).$$

Sequential Competition

4

Introduction

In this chapter, we start analyzing strategic settings where firms interact sequentially, rather than simultaneously, as opposed to most of the exercises in Chaps. 2 and 3. In this type of industries, one or more firms choose their output level in a first stage (often referred to as the industry "leader") and, observing this output level/s, other firms respond by selecting their output level/s (and thus are known as the industry "followers").

We begin with the canonical Stackelberg game of quantity competition with only one leader and one follower facing the same marginal cost of production (Exercise 4.1) and then extend our analysis to a setting with three firms sequentially choosing their output levels, still facing a common marginal cost (Exercise 4.2), and to an industry with one leader and one follower but which face different marginal costs (Exercise 4.3). Exercise 4.4 then examines a more general setting without assuming specific demand and cost functions, showing how equilibrium output decisions in this sequential-move game differ from those in a simultaneous-move game (Cournot quantity competition).

Exercise 4.5 applies the tools learned in previous exercises of this chapter to different contexts, such as industries with a private firm (which maximizes profits) and a publicly owned firm (which maximizes a linear combination of profits and social welfare). We check how output levels are affected relative to the standard Stackelberg game where all firms are private. Exercise 4.6 considers a similar setting as Exercise 4.5, but assuming that the public firm is now the leader, while the private firm is the follower, seeking to compare how equilibrium results are affected.

While previous exercises assume that firms perfectly observe each other's costs, Exercise 4.7 allows for one firm to not observe its rival's cost, thus operating under incomplete information Exercise 4.8 considers a Stackelberg game with stochastic demand, where only the follower observes the actual state of demand. Exercise 4.9 follows a similar approach, but now assuming that the follower cannot observe the leader's production cost, using the leader's output to infer its type. Then, Exercise 4.10 studies a "hybrid" model of sequential competition, where the leader chooses output and the follower responds choosing its price, and vice versa, to understand in which contexts a firm may prefer to choose prices (quantities) when its rival responds with quantities (prices).

Exercise 4.11 generalizes the standard Stackelberg game of quantity competition but considering an industry with $N \geq 2$ firms, thus allowing for M leaders and $N - M$ followers and examining how equilibrium output is affected when the number of leaders or followers increases.

Exercises 4.12 and 4.13 still consider that firms interact sequentially, but now assuming that they compete in prices, first selling a homogeneous good (Exercise 4.12) and then selling heterogeneous

P.-S. Choi et al., *Industrial Organization*, Springer Texts in Business and Economics,
https://doi.org/10.1007/978-3-031-38635-0_4

goods (Exercise 4.13). Exercise 4.14 analyzes whether firms prefer to commit to price or quantity competition, as in Singh and Vives (1984), showing that their preference is unaffected by the degree of product differentiation between their products. The remainder of the chapter explores models of strategic pre-commitment, where one or both firms choose an action in the first period of the game to alter subsequent competition in quantities or prices in the second stage, as in the seminal paper by Fudenberg and Tirole (1984). Specifically, Exercise 4.15 considers that only one firm can choose a first-period action, such as investing in a cost-reducing technology that lowers its marginal costs in the second stage of the game, and that firms compete in quantities (à la Cournot) in the second stage. We elaborate on how the first-period action affects the best response functions that firms face in the second period and, as a consequence, their equilibrium output decisions. In Exercise 4.16, we consider a similar setting, but assuming that firms compete in prices in the second period, identifying how firms' incentives are affected in the first period of the game. Finally, Exercise 4.17 allows for both firms to strategically pre-commit (e.g., invest in cost-reducing technologies), showing how results differ from those settings where only one firm can invest in this technology.

Exercise #4.1: Stackelberg Competition with Two Symmetric Firms^A

4.1. Consider an industry where two firms producing a homogeneous good, and facing a linear demand function $p(Q) = 1 - Q$, where $Q \equiv q_1 + q_2$ denotes aggregate output. All firms have constant marginal cost of production c, where $1 > c > 0$. Firms interact in the following sequential-move game: In the first stage, firm 1 (the industry leader) chooses its output q_1. In the second stage, firm 2 (the industry follower) observes the leader's output q_1 and responds with its own output level q_2.

We solve the game by applying backward induction, starting with the follower's output decision in the second stage, and, then, examining the leader's output choice in the first stage.

(a) *Second stage.* Find the follower's best response function. Interpret.

- The follower's profit maximization problem is

$$\max_{q_2 \geq 0} \ \pi_2 = (1 - q_1 - q_2)q_2 - cq_2.$$

Differentiating with respect to q_2, we obtain

$$\frac{\partial \pi_2}{\partial q_2} = 1 - 2q_2 - q_1 - c = 0.$$

Rearranging, we find

$$2q_2 = 1 - c - q_1,$$

and solving for q_2, we find

$$q_2 = \frac{1-c}{2} - \frac{1}{2}q_1,$$

which helps us write firm 2's best response function as follows:

$$q_2(q_1) = \begin{cases} \frac{1-c}{2} - \frac{1}{2}q_1 & \text{if } q_1 < 1 - c \\ 0 & \text{otherwise.} \end{cases}$$

This function originates at a vertical intercept of $q_2 = \frac{1-c}{2}$, decreases in the leader's output, q_1, at a rate of 1/2, and crosses the horizontal axis at $q_1 = 1 - c$. For all output levels $q_1 > 1 - c$ from the leader, the follower responds staying inactive, $q_2 = 0$.

- This best response function coincides with that found in the Cournot game for firm 2 (see Exercise 2.1 for more details).

(b) *First stage.* Find the leader's equilibrium output.

- The leader anticipates the follower's best response function, $q_2(q_1)$, in the next stage of the game, so its profit maximization problem in the first stage is

$$\max_{q_1 \geq 0} \pi_1 = [1 - q_1 - q_2(q_1)]\, q_1 - cq_1,$$

which is evaluated at $q_2(q_1)$. Inserting firm 2's best response function, we obtain

$$\max_{q_1 \geq 0} \pi_1 = \left[1 - q_1 - \overbrace{\left(\frac{1-c}{2} - \frac{1}{2}q_1\right)}^{q_2(q_1)}\right] q_1 - cq_1.$$

Simplifying yields

$$\max_{q_1 \geq 0} \pi_1 = \left[\frac{1+c}{2} - \frac{1}{2}q_1\right] q_1 - cq_1,$$

which is only a function of the leader's choice variable (its output level q_1).

- Differentiating the leader's profit with respect to q_1, we find

$$\frac{\partial \pi_1}{\partial q_1} = \frac{1+c}{2} - q_1 - c = 0.$$

Rearranging,

$$\frac{1+c}{2} - c = q_1.$$

Solving for q_1, we obtain the leader's equilibrium output

$$q_1^* = \frac{1-c}{2}.$$

(c) Find firm 2's output level in equilibrium.

- The follower observes this quantity and inserts it into its best response function, to obtain the equilibrium output

$$\begin{aligned} q_2^* &= \frac{1-c}{2} - \frac{1}{2}\overbrace{\left(\frac{1-c}{2}\right)}^{q_1^*} \\ &= \frac{1-c}{2} - \frac{1-c}{4} \\ &= \frac{1-c}{4}. \end{aligned}$$

- **Remark:** We *do not say* that the subgame perfect equilibrium of this Stackelberg game with two symmetric firms is $(q_1^*, q_2^*) = \left(\frac{1-c}{2}, \frac{1-c}{4}\right)$. This output vector only describes firms' output along the equilibrium path. In contrast, the subgame perfect equilibrium of this game must describe what each firm does to maximize its profits, both on- and off-the-equilibrium path. Therefore, the subgame perfect equilibrium is

$$(q_1^*, q_2^*) = \left(\frac{1-c}{2}, \frac{1-c}{2} - \frac{1}{2}q_1\right),$$

where firm 2 (the follower) responds with its best response function, for any output that the leader chooses (both $q_1 = q_1^*$ and $q_1 \neq q_1^*$).

(d) Find equilibrium profits for the leader and follower:
- The leader's equilibrium profits are

$$\begin{aligned}\pi_1^* &= [1 - q_1^* - q_2^*]\, q_1^* - cq_1^* \\ &= \left[1 - \frac{1-c}{2} - \frac{1-c}{4}\right]\frac{1-c}{2} - c\frac{1-c}{2} \\ &= \frac{(1-c)^2}{8},\end{aligned}$$

while equilibrium profits for the follower are

$$\begin{aligned}\pi_2^* &= [1 - q_2^* - q_1^*]\, q_2^* - cq_2^* \\ &= \left[1 - \frac{1-c}{4} - \frac{1-c}{2}\right]\frac{1-c}{4} - c\frac{1-c}{4} \\ &= \frac{(1-c)^2}{16}.\end{aligned}$$

(e) *Numerical example.* Evaluate equilibrium output at $c = 1/2$:
- The leader produces $q_1^* = \frac{1-\frac{1}{2}}{2} = \frac{1}{4}$ units, and the follower responds producing $q_2^* = \frac{1-\frac{1}{2}}{4} = \frac{1}{8}$ units in equilibrium. More generally, the follower's best response function in this setting becomes

$$\begin{aligned}q_2(q_1) &= \frac{1-\frac{1}{2}}{2} - \frac{1}{2}q_1 \\ &= \frac{1}{4} - \frac{1}{2}q_1,\end{aligned}$$

which originates at $q_2 = \frac{1}{4}$ when the leader is inactive and decreases at a rate of $\frac{1}{2}$ for every unit that the leader produces. Equilibrium profits in this context are $\pi_1^* = \frac{\left(1-\frac{1}{2}\right)^2}{8} = \frac{1}{32}$ for the leader and $\pi_2^* = \frac{\left(1-\frac{1}{2}\right)^2}{16} = \frac{1}{64}$ for the follower.

Exercise #4.2: Stackelberg Competition with Three Symmetric Firms^B

4.2. Consider the same setting as in Exercise 4.1, but assume now that a third firm enters the industry. The time structure remains unaffected: first, firm 1 chooses output q_1; observing q_1, firm 2 responds with its output q_2; and, observing both q_1 and q_2, firm 3 responds choosing its output q_3:

(a) Find equilibrium output levels:

- We solve this game by operating by backward induction, starting from firm 3 (the last mover), then moving to firm 2, and finally examining firm 1 (the first mover).
- *Firm 3.* Starting with the last mover (firm 3), its profit maximization problem is

$$\max_{q_3 \geq 0} \ \pi_3 = (1 - q_1 - q_2 - q_3)q_3 - cq_3.$$

Differentiating with respect to q_3, we obtain

$$\frac{\partial \pi_3}{\partial q_3} = 1 - 2q_3 - q_1 - q_2 - c = 0,$$

and rearranging, we find

$$2q_3 = 1 - c - q_1 - q_2.$$

Solving for q_3, we find firm 3's best response function

$$q_3(q_1, q_2) = \frac{1-c}{2} - \frac{1}{2}(q_1 + q_2).$$

Firm 3's best response function originates at a vertical intercept of $\frac{1-c}{2}$ units (when both firms 1 and 2 are inactive) and decreases at a rate of $\frac{1}{2}$ when either firm 1 or 2 marginally increases their production levels.

- *Firm 2.* Firm 2's profit maximization problem is

$$\max_{q_2 \geq 0} \ \pi_2 = [1 - q_1 - q_2 - q_3(q_1, q_2)]\, q_2 - cq_2,$$

which is evaluated at firm 3's best response function, $q_3(q_1, q_2)$, since firm 2 can anticipate firm 3's equilibrium behavior in the last stage of the game. Inserting firm 3's best response function, $q_3(q_1, q_2)$, in the above problem, we obtain

$$\max_{q_2 \geq 0} \ \pi_2 = \left[1 - q_1 - q_2 - \overbrace{\left(\frac{1-c}{2} - \frac{1}{2}(q_1 + q_2) \right)}^{q_3(q_1,q_2)} \right] q_2 - cq_2.$$

Simplifying yields

$$\max_{q_2 \geq 0} \ \pi_2 = \left(\frac{1+c}{2} - \frac{1}{2}q_1 - \frac{1}{2}q_2 \right) q_2 - cq_2,$$

which is only a function of firm 1 and 2's output (the firms that have not yet selected their output levels). Differentiating with respect to q_2, we find

$$\frac{\partial \pi_2}{\partial q_2} = \frac{1+c}{2} - \frac{1}{2}q_1 - q_2 - c = 0.$$

Solving for q_2, we obtain firm 2's best response function

$$q_2(q_1) = \frac{1-c}{2} - \frac{1}{2}q_1.$$

In preparation for firm 1's problem (the industry leader), we insert this best response function into firm 3's best response function (found above), yielding

$$\begin{aligned} q_3(q_1) &= \frac{1-c}{2} - \frac{1}{2}\left(q_1 + \overbrace{\left(\frac{1-c}{2} - \frac{1}{2}q_1\right)}^{q_2(q_1)} \right) \\ &= \frac{1-c}{4} - \frac{1}{4}q_1, \end{aligned}$$

which helps us to write firm 3's best response function as a function of firm 1's output alone.

- *Firm 1.* The leader's profit maximization problem is

$$\max_{q_1 \geq 0} \pi_1 = [1 - q_1 - q_2(q_1) - q_3(q_1)]\, q_1 - cq_1,$$

which is evaluated at firm 2's and 3's best response functions, since firm 1 can anticipate its rivals' equilibrium behavior in all subsequent stages. Inserting firm 2 and 3's best response functions, we obtain

$$\max_{q_1 \geq 0} \pi_1 = \left[1 - q_1 - \overbrace{\left(\frac{1-c}{2} - \frac{1}{2}q_1\right)}^{q_2(q_1)} - \overbrace{\left(\frac{1-c}{4} - \frac{1}{4}q_1\right)}^{q_3(q_1)} \right] q_1 - cq_1.$$

Simplifying yields

$$\max_{q_1 \geq 0} \pi_1 = \left(\frac{1+3c}{4} - \frac{1}{4}q_1\right) q_1 - cq_1,$$

which is only a function of firm 1's choice variable (its own output q_1). Differentiating with respect to q_1, we find

$$\frac{\partial \pi_1}{\partial q_1} = \frac{1+3c}{4} - \frac{1}{2}q_1 - c = 0.$$

Rearranging yields

$$\frac{1}{2}q_1 = \frac{1-c}{4},$$

implying that firm 1's optimal output is

$$q_1^* = \frac{1-c}{2}.$$

(b) Find firm 2's and 3's output level in equilibrium.

- Plugging firm 1's equilibrium output $q_1^* = \frac{1-c}{2}$ into firm 2's best response function, we find its equilibrium output

$$q_2^* = \frac{1-c}{2} - \frac{1}{2}\underbrace{\left(\frac{1-c}{2}\right)}_{q_1}$$
$$= \frac{1-c}{4}.$$

Similarly, firm 3's equilibrium output is

$$q_3^* = \frac{1-c}{4} - \frac{1}{4}\underbrace{\left(\frac{1-c}{2}\right)}_{q_1}$$
$$= \frac{1-c}{8}.$$

(c) Find equilibrium profits for each firm.

- Firm 1's equilibrium profits are

$$\pi_1^* = \left[1 - q_1^* - q_2^* - q_3^*\right] q_1^* - c q_1^*$$
$$= \left[1 - \frac{1-c}{2} - \frac{1-c}{4} - \frac{1-c}{8}\right]\frac{1-c}{2} - c\frac{1-c}{2}$$
$$= \frac{(1-c)^2}{16},$$

while equilibrium profits for firm 2 are

$$\pi_2^* = \left[1 - q_1^* - q_2^* - q_3^*\right] q_2^* - c q_2^*$$
$$= \left[1 - \frac{1-c}{2} - \frac{1-c}{4} - \frac{1-c}{8}\right]\frac{1-c}{4} - c\frac{1-c}{4}$$
$$= \frac{(1-c)^2}{32}$$

and for firm 3 are

$$\pi_3^* = \left[1 - q_1^* - q_2^* - q_3^*\right] q_3^* - c q_3^*$$
$$= \left[1 - \frac{1-c}{2} - \frac{1-c}{4} - \frac{1-c}{8}\right]\frac{1-c}{8} - c\frac{1-c}{8}$$
$$= \frac{(1-c)^2}{64}.$$

(d) Compare output levels in the Stackelberg game with two firms (Exercise 4.1) and with three firms (this exercise).

- Relative to the setting with two firms analyzed in Exercise 4.1, the addition of a third firm does not change the output decision of the first two movers. However, the last mover (firm 3) faces a lower residual demand left from the first two firms and thus produces a smaller output than the first or second mover. A similar argument applies if we keep adding more firms (see Exercise 4.7 for a setting with m leaders and $n - m$ followers).
- While the first two movers produce the same output as when only two firms are present in the industry, the third firm's output yields a larger aggregate output, which induces a lower price and ultimately decreases the profits of both firms 1 and 2. In particular, firm 1 sees its profits decrease by half, from $\frac{(1-c)^2}{8}$ to $\frac{(1-c)^2}{16}$; and, similarly, firm 2's profits decrease from $\frac{(1-c)^2}{16}$ to $\frac{(1-c)^2}{32}$. (Compare Exercise 4.1(d) against Exercise 4.1(c).) In other words, while firms 1 and 2 produce the same units, they make less profits when firm 3 is present.

(e) *Numerical example.* Evaluate equilibrium output at $c = 1/2$.

- Firm 1 produces $q_1^* = \frac{1-\frac{1}{2}}{2} = \frac{1}{4}$ units, firm 2 responds producing $q_2^* = \frac{1-\frac{1}{2}}{4} = \frac{1}{8}$ units in equilibrium, and, finally, firm 3 responds producing $q_3^* = \frac{1-\frac{1}{2}}{8} = \frac{1}{16}$ units in equilibrium. Therefore, equilibrium profits in this context are $\pi_1^* = \frac{\left(1-\frac{1}{2}\right)^2}{16} = \frac{1}{64}$ for firm 1, $\pi_2^* = \frac{\left(1-\frac{1}{2}\right)^2}{32} = \frac{1}{128}$ for the firm 2, and $\pi_3^* = \frac{\left(1-\frac{1}{2}\right)^2}{64} = \frac{1}{256}$ for the firm 3.

Exercise #4.3: Stackelberg Competition with Two Asymmetric Firms^B

4.3. Consider an industry where two firms producing a homogeneous good, and facing a linear demand function $p(Q) = 1 - Q$, where Q denotes aggregate output.

Firms interact in the following sequential-move game: In the first stage, firm 1 (the industry leader) chooses its output q_1. In the second stage, firm 2 (the industry follower) observes the leader's output q_1 and responds with its own output level q_2.

The leader, because of its experience in the industry, faces a lower constant marginal cost c_l, while the follower's constant marginal cost is c_f, where $1 > c_f \geq c_l > 0$.

(a) Find the follower's best response function. Interpret.

- The follower's profit maximization problem is

$$\max_{q_2 \geq 0} \ \pi_2 = (1 - q_1 - q_2)q_2 - c_f q_2.$$

Differentiating with respect to q_2, we obtain

$$1 - 2q_2 - q_1 - c_f = 0,$$

rearranging, we find

$$2q_2 = 1 - c_f - q_1,$$

and solving for q_2, we find firm 2's best response function

$$q_2(q_1) = \frac{1 - c_f}{2} - \frac{1}{2}q_1,$$

which originates at a vertical intercept of $q_2 = \frac{1-c_f}{2}$ and decreases in the leader's output, q_1, at a rate of 1/2, crossing the horizontal axis at $q_1 = 1 - c$.

(b) Find the leader's equilibrium output.

- The leader anticipates the follower's best response function in the second stage, $q_2(q_1)$, so its profit maximization problem is

$$\max_{q_1 \geq 0} \pi_1 = [1 - q_1 - q_2(q_1)]\, q_1 - c_l q_1.$$

Inserting firm 2's best response function in this problem, we obtain

$$\max_{q_1 \geq 0} \pi_1 = \left[1 - q_1 - \overbrace{\left(\frac{1-c_f}{2} - \frac{1}{2}q_1\right)}^{q_2(q_1)}\right] q_1 - c_l q_1.$$

Simplifying yields

$$\max_{q_1 \geq 0} \pi_1 = \frac{1 - q_1 + c_f - 2c_l}{2} q_1,$$

which is only a function of the leader's choice variable (its output level q_1).

- Differentiating the leader's profit with respect to q_1, we find

$$1 - 2q_1 + c_f - 2c_l = 0.$$

Solving for q_1, we obtain the leader's equilibrium output

$$q_1^* = \frac{1 + c_f - 2c_l}{2},$$

which is decreasing in its own production cost but increasing in the follower's cost.

(c) Find firm 2's output level in equilibrium:

- The follower observes the leader's equilibrium output, q_1^*, and inserts it into its best response function to obtain its equilibrium output

$$\begin{aligned} q_2^* &= \frac{1-c_f}{2} - \frac{1}{2}\overbrace{\left(\frac{1+c_f-2c_l}{2}\right)}^{q_1^*} \\ &= \frac{1-c_f}{2} - \frac{1+c_f-2c_l}{4} \\ &= \frac{1-3c_f+2c_l}{4}, \end{aligned}$$

which is decreasing in its own production cost but increasing in the leader's cost.

(d) Under which conditions are both firms active in the industry? Under which conditions is only the leader active?

- *Both firms are active.* Both firms are active in the industry when both

$$q_1^* = \frac{1 + c_f - 2c_l}{2} > 0 \text{ and } q_2^* = \frac{1 - 3c_f + 2c_l}{4} > 0.$$

Solving for c_l in $q_1^* > 0$, we find that the leader is active if

$$c_l < \frac{1 + c_f}{2}.$$

Similarly, solving for c_l in $q_2^* > 0$, we find that the follower is active if

$$c_l > \frac{3c_f - 1}{2}.$$

Therefore, combining these two conditions on c_l, we obtain that both firms are active if the leader's costs are intermediate, that is,

$$\frac{1 + c_f}{2} > c_l > \frac{3c_f - 1}{2}.$$

- *Only the leader is active.* The leader is the only active firm in the industry if condition $c_l < \frac{1+c_f}{2}$ holds, but $c_l > \frac{3c_f-1}{2}$ does not. In other words, for the entrant to be inactive, we need $c_l \leq \frac{3c_f-1}{2}$. Comparing the leader and the follower's cutoffs, we find that

$$\frac{1 + c_f}{2} - \frac{3c_f - 1}{2} = 1 - c_f,$$

which is positive since $c_f < 1$ by assumption. Graphically, this result means that cutoff $\frac{1+c_f}{2}$ lies to the right-hand side of cutoff $\frac{3c_f-1}{2}$. Therefore, condition $c_l \leq \frac{3c_f-1}{2}$ implies $c_l < \frac{1+c_f}{2}$ or, formally, $c_l \leq \frac{3c_f-1}{2}$ is a sufficient condition for $c_l < \frac{1+c_f}{2}$. Then, for the leader to be the only active firm, we only need condition $c_l \leq \frac{3c_f-1}{2}$. Intuitively, the leader's cost advantage is so strong that the follower decides to be inactive.
- *Only the follower is active.* The follower is the only active firm in the industry if $c_l > \frac{3c_f-1}{2}$ and $c_l \geq \frac{1+c_f}{2}$. From our above discussion, we know that $\frac{1+c_f}{2} > \frac{3c_f-1}{2}$, implying that condition $c_l \geq \frac{1+c_f}{2}$ is sufficient for $c_l > \frac{3c_f-1}{2}$. In other words, the follower is the only active firm if $c_l \geq \frac{1+c_f}{2}$ holds.
- In summary, our above discussion identifies three regions of c_l. These regions are depicted in Fig. 4.1 with c_l in the vertical axis and c_f in the horizontal axis, along with cutoff $c_l = \frac{3c_f-1}{2}$, originating at $c_l = -\frac{1}{2}$ and reaching a maximum height of 1, and cutoff $c_l = \frac{1+c_f}{2}$, which originates at $c_l = \frac{1}{2}$ that also reaches a maximum height of 1. The three regions are the following:
  (i) When the leader is relatively efficient, $c_l \leq \frac{3c_f-1}{2}$, it is the only active firm in the industry.
  (ii) If the leader's cost advantage is moderate, $\frac{3c_f-1}{2} < c_l < \frac{1+c_f}{2}$, both firms are active.
  (iii) If the leader's cost advantage is low, $c_l \geq \frac{1+c_f}{2}$, the leader is inactive and only the follower is active.

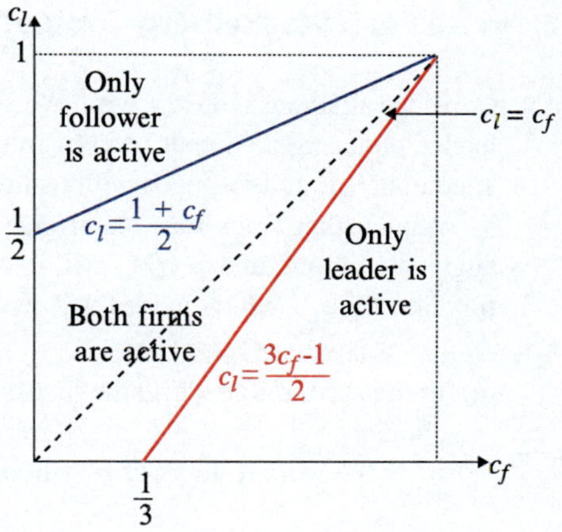

Fig. 4.1 Output profiles in Stackelberg

- **Remark:** Note that condition $c_l \geq \frac{1+c_f}{2}$ entails that the leader suffers from a cost disadvantage relative to the follower since $\frac{1+c_f}{2} > c_f$ simplifies to $1 > c_f$, which holds by assumption. Intuitively, in all cost pairs (c_l, c_f) in Region (iii), the leader suffers from a cost disadvantage, leading it to shut down, but it also suffers from a cost disadvantage (although smaller) in Region (ii), where the leader produces a positive output level.

  Similarly, it is easy to show that condition $c_l \leq \frac{3c_f-1}{2}$ entails that the leader benefits from a cost advantage relative to the follower since $\frac{3c_f-1}{2} < c_f$ simplifies to $c_f < 1$, which holds by definition. Therefore, in all cost pairs of Region (i) and in a portion of Region (ii), the leader has a cost advantage.

(e) *Symmetric firms.* Show that when the leader enjoys no cost advantage (that is, $c_l = c_f = c$), your above results coincide with those in a standard Stackelberg game with two symmetric firms.

- Evaluating the leader's equilibrium output, $q_1^* = \frac{1+c_f-2c_l}{2}$, at $c_l = c_f = c$, we obtain

$$q_1^* = \frac{1+c-2c}{2} = \frac{1-c}{2}$$

and evaluating the follower's best response function, $q_2(q_1) = \frac{1-c_f}{2} - \frac{1}{2}q_1$, at $c_l = c_f = c$, yields

$$q_2(q_1) = \frac{1-c}{2} - \frac{1}{2}q_1,$$

which implies that, in equilibrium, the follower produces $q_2^* = q_2(q_1^*)$, that is,

$$q_2^* = \frac{1-c}{2} - \frac{1}{2}\left(\frac{1-c}{2}\right)$$
$$= \frac{1-c}{4}.$$

Exercise #4.4: Stackelberg Competition, General Presentation^{C}

4.4. Consider again an industry with two firms competing in quantities, where firm 1 operates as a leader, choosing its output level q_1 in the first period. In the second period, firm 2 observes firm 1's output, q_1, and responds with its own output level q_2.

Assume that firms face an inverse demand function $p(Q)$, where $Q = q_1 + q_2$ denotes aggregate output, and $p'(Q) < 0$. In addition, firm i's total cost function is given by total cost function $C_i(q_i)$, which is weakly increasing and convex in its output level q_i, $C_i'(q_i) \geq 0$ and $C_i''(q_i) \geq 0$, where $i = \{1, 2\}$.

(a) In the second stage, find the first-order condition that implicitly characterizes firm 2's output choice.

- In the second stage, firm 2 chooses q_2 to solve

$$\max_{q_2 \geq 0} \pi_2 = p(q_1 + q_2)q_2 - C_2(q_2).$$

Differentiating with respect to q_2, we obtain

$$p(q_1 + q_2) + p'(q_1 + q_2)q_2 = C_2'(q_2),$$

and, given that aggregate output is defined as $Q = q_1 + q_2$, we can rewrite this first-order condition more compactly as

$$p(Q) + p'(Q)q_2 = C_2'(q_2).$$

Therefore, firm 2 observes firm 1's output in the first stage, q_1, and responds choosing an output level q_2 that solves the above first-order condition. Intuitively, firm 2 increases q_2 until the point where the marginal revenue of additional units of output (on the left side) coincides with its associated marginal cost (on the right side). Solving for q_2 yields firm 2's best response function $q_2(q_1)$.

(b) In the first stage, find the first-order condition that implicitly characterizes firm 1's output choice:

- In the first stage, firm 1 anticipates firm 2's best response function $q_2(q_1)$ and chooses its output q_1 to maximize its profits as follows:

$$\max_{q_1 \geq 0} \pi_1 = p(q_1 + \underbrace{q_2(q_1)}_{\text{BRF}_2})q_1 - C_1(q_1),$$

where we replaced $Q = q_1 + q_2(q_1)$. Mathematically, this profit is a function of q_1 alone, entailing that the solution to the above profit maximization problem is not a function of q_2 – as opposed to Cournot models of simultaneous quantity competition. Differentiating with respect to q_1 yields

$$p(Q) + p'(q_1 + q_2)\left(1 + \frac{\partial q_2(q_1)}{\partial q_1}\right)q_1 = C_1'(q_1)$$

or, after rearranging, and noticing that $Q = q_1 + q_2$,

$$p(Q) + p'(Q)q_1 + \underbrace{p'(Q)\frac{\partial q_2(q_1)}{\partial q_1}q_1}_{\text{New, strategic, effect}} = C_1'(q_1).$$

Therefore, the leader also increases q_1 until the point where the marginal revenue (left side) coincides with its associated marginal cost (right side). The first two terms of marginal revenue are symmetric between the leader and the follower, but the leader's includes a new, third, term that was absent from the follower's first-order condition. We describe this new effect in part (c) of the exercise.

(c) Compare your results in part (b) with those in a Cournot model of quantity competition, where both firms 1 and 2 simultaneously and independently choose output levels:

- In the Cournot model of simultaneous quantity competition, every firm i takes firm j's output, q_j, as given and, hence, faces a profit maximization problem identical to that of the follower in the Stackelberg game; that is,

$$\max_{q_i \geq 0} \ \pi_i = p(q_i + q_j)q_i - C_i(q_i).$$

Differentiating with respect to q_i yields

$$p(Q) + p'(Q)q_i = C_i'(q_i)$$

for every firm i and $j \neq i$. However, in the Stackelberg model of sequential quantity competition, the leader faces a new effect (third term in its first-order condition), $p'(Q)\frac{\partial q_2(q_1)}{\partial q_1}q_1$. In addition, this effect is positive since $p'(Q) < 0$ by definition (negatively sloped inverse demand function) and $\frac{\partial q_2(q_1)}{\partial q_1} < 0$ (firm 2 responds decreasing its output when firm 1 increases its production, i.e., firms' output levels are strategic substitutes).

- Intuitively, this means that firm 1 (the leader) has more incentives to raise its output, q_1, in the Stackelberg game than in the Cournot model with simultaneous output competition. In other words, the leader exercises its first-mover advantage, where this firm increases its output, forcing the follower to reduce its production in the subsequent stage.

Exercise #4.5: Stackelberg Competition Between a Private Leader and a Public Follower^B

4.5. Consider an industry where two firms producing a homogeneous good and facing a linear demand function $p(Q) = 1 - Q$, where Q denotes aggregate output. All firms face a constant marginal cost of production c, where $1 > c > 0$. Firms interact in the following sequential-move game: In the first stage, firm 1 (the industry leader) chooses its output q_1. In the second stage, firm 2 (the industry follower) observes the leader's output q_1 and responds with its own output level q_2.

Assume that firm 1 (a private firm) seeks to maximize its profits, while firm 2 (a public firm) maximizes a combination of welfare and profits

$$V = \alpha W + (1 - \alpha)\pi_2,$$

where parameter $\alpha \in [0, 1]$ indicates that weight that the manager of the public firm assigns to welfare W, where $W = CS + PS$ denotes the sum of consumer and producer surplus, whereas $1 - \alpha$ represents the weight he assigns to profit π_2, where $\pi_i = (1 - q_i - q_j)q_i - cq_i$ for every firm $i = \{1, 2\}$.

(a) Find the follower's best response function. Interpret:
- The follower's profit maximization problem is

$$\max_{q_2 \geq 0} \ V = \alpha W + (1-\alpha)\pi_2,$$

where W denotes social welfare

$$W = \int_0^Q p(Q)dQ - cQ = (1-c)Q - \frac{1}{2}Q^2.$$

Therefore, the above problem simplifies to

$$\max_{q_2 \geq 0} \ V = \alpha\left[(1-c)Q - \frac{1}{2}Q^2\right] + (1-\alpha)[(1-Q-c)q_2]$$
$$= \alpha\left[(1-c)(q_1+q_2) - \frac{1}{2}(q_1+q_2)^2\right] + (1-\alpha)[(1-(q_1+q_2)-c)q_2].$$

Differentiating with respect to q_2, we obtain

$$\alpha(1-c-q_1-q_2) + (1-\alpha)(1-c-q_1-2q_2) = 0.$$

Solving for q_2, we find firm 2's best response function

$$q_2(q_1) = \frac{1-c}{2-\alpha} - \frac{1}{2-\alpha}q_1,$$

which originates at $q_2 = \frac{1-c}{2-\alpha}$ and decreases in the leader's output, q_1, at a rate of $\frac{1}{2-\alpha}$. The vertical intercept shifts up as the public firm assigns a larger weight on social welfare, i.e., as α increases. However, the slope of the best response function, $\frac{1}{2-\alpha}$, also increases in α, thus making this function steeper.

(b) Find the leader's equilibrium output:
- The leader's (private firm) profit maximization problem is

$$\max_{q_1 \geq 0} \ \pi_1 = [1 - q_1 - q_2(q_1)] . q_1 - cq_1$$

Inserting firm 2's best response function, $q_2(q_1)$, we obtain

$$\max_{q_1 \geq 0} \ \pi_1 = \left[1 - q_1 - \overbrace{\left(\frac{1-c}{2-\alpha} - \frac{1}{2-\alpha}q_1\right)}^{q_2(q_1)}\right] q_1 - cq_1.$$

Simplifying yields

$$\max_{q_1 \geq 0} \ \pi_1 = \left[1 - q_1 - \frac{1-c-q_1}{2-\alpha} - c\right] q_1,$$

which is only a function of the leader's choice variable (its output level q_1).

- Differentiating the leader's profit with respect to q_1, we find

$$1 - c - 2q_1 + \frac{2q_1}{2-\alpha} - \frac{1-c}{2-\alpha} = 0.$$

Rearranging yields

$$\left(1 - \frac{1}{2-\alpha}\right) 2q_1 = 1 - c - \frac{1-c}{2-\alpha}$$

or

$$\frac{1-\alpha}{2-\alpha} 2q_1 = \frac{(1-c)(1-\alpha)}{2-\alpha}.$$

Solving for q_1, we obtain the leader's equilibrium output

$$q_1^* = \frac{1-c}{2}.$$

(c) Find firm 2's output level in equilibrium:
- The follower observes $q_1^* = \frac{1-c}{2}$ and inserts it into its best response function:

$$q_2^* = \frac{1-c}{2-\alpha} - \frac{1}{2-\alpha} \overbrace{\frac{1-c}{2}}^{q_1^*}$$
$$= \frac{1-c}{2(2-\alpha)}.$$

(d) How is each firm's output decision affected by a marginal increase in α? Interpret:
- The private firm's equilibrium output, q_1^*, is unaffected by the weight that the public firm (the follower) assigns to social welfare, α. However, the public firm's equilibrium output, q_2^*, increases in this weight since

$$\frac{\partial q_2^*}{\partial \alpha} = \frac{1-c}{2\,(2-\alpha)^2} > 0$$

given that $1 > c$ and $\alpha \in [0, 1]$ by assumption. Intuitively, the public firm increases its production, for a given output of the leader, when it assigns a larger weight on social welfare and, thus, a lower weight on profits.

(e) *Numerical example.* Evaluate equilibrium output at $c = 1/2$.
- The leader produces $q_1^* = \frac{1-\frac{1}{2}}{2} = \frac{1}{4}$ units, and the follower responds producing $q_2^* = \frac{1-\frac{1}{2}}{2(2-\alpha)} = \frac{1}{4(2-\alpha)}$ units in equilibrium. For instance, when $\alpha = 0$, the follower's output becomes $q_2^* = \frac{1}{4(2-0)} = \frac{1}{8}$ units, as in the standard Stackelberg game between two private firms analyzed in Exercise 4.1. In contrast, when $\alpha = 1$, the follower's output increases to $q_2^* = \frac{1}{4(2-1)} = \frac{1}{4}$ units.

(f) Evaluate aggregate output and social welfare in equilibrium. How do they change with α? Interpret:

- *Aggregate output.* Aggregate output is

$$Q^* = \frac{1-c}{2} + \frac{1-c}{2(2-\alpha)}$$
$$= \frac{(3-\alpha)(1-c)}{2(2-\alpha)}.$$

Differentiating Q^* with respect to α, yields

$$\frac{\partial Q^*}{\partial \alpha} = \frac{1-c}{2(2-\alpha)^2},$$

which is positive since $c, \alpha \in [0, 1]$. Intuitively, when firm 2 assigns more weight on social welfare (higher α), it increases its output more significantly than the leader decreases its own output, ultimately giving rise to an increase in aggregate output.

- *Social welfare.* Substituting the aggregate output into the welfare function, we find that equilibrium social welfare becomes

$$W^* = (1-c)\frac{(3-\alpha)(1-c)}{2(2-\alpha)} - \frac{1}{2}\left(\frac{(3-\alpha)(1-c)}{2(2-\alpha)}\right)^2$$
$$= \frac{(1-c)^2(3-\alpha)(5-3\alpha)}{8(2-\alpha)^2}.$$

- Differentiating W^* with respect to α yields

$$\frac{\partial W^*}{\partial \alpha} = \frac{(1-c)^2\left[(2-\alpha)(3\alpha-7) + \left(15 - 14\alpha + 3\alpha^2\right)\right]}{4(2-\alpha)^3}$$
$$= \frac{(1-c)^2(1-\alpha)}{4(2-\alpha)^3},$$

which is positive for all $\alpha \in [0, 1]$. Therefore, as firm 2 assigns a higher weight on social welfare, aggregate output increases, and equilibrium social welfare increases as well.

(g) Evaluate aggregate output and social welfare in equilibrium. How do they change with α? Interpret:

- *Aggregate output.* Aggregate output is

$$Q^* = \frac{1-c}{2} + \frac{1-c}{2(2-\alpha)}$$
$$= \frac{(3-\alpha)(1-c)}{2(2-\alpha)}.$$

Differentiating Q^* with respect to α, yields

$$\frac{\partial Q^*}{\partial \alpha} = \frac{1-c}{2(2-\alpha)^2},$$

which is positive since $c, \alpha \in [0, 1]$. Intuitively, when firm 2 assigns more weight on social welfare (higher α), it increases its output more significantly than the leader decreases its own output, ultimately giving rise to an increase in aggregate output.

- *Social welfare.* Substituting the aggregate output into the welfare function, we find that equilibrium social welfare becomes

$$W^* = (1-c)\frac{(3-\alpha)(1-c)}{2(2-\alpha)} - \frac{1}{2}\left(\frac{(3-\alpha)(1-c)}{2(2-\alpha)}\right)^2$$
$$= \frac{(1-c)^2(3-\alpha)(5-3\alpha)}{8(2-\alpha)^2}.$$

- Differentiating W^* with respect to α yields

$$\frac{\partial W^*}{\partial \alpha} = \frac{(1-c)^2\left[(2-\alpha)(3\alpha-7) + \left(15 - 14\alpha + 3\alpha^2\right)\right]}{4(2-\alpha)^3}$$
$$= \frac{(1-c)^2(1-\alpha)}{4(2-\alpha)^3},$$

which is positive for all $\alpha \in [0, 1]$. Therefore, as firm 2 assigns a higher weight on social welfare, aggregate output increases, and equilibrium social welfare increases as well.

Exercise #4.6: Stackelberg Competition Between a Public Leader and a Private Follower^B

4.6. Consider an industry from Exercise 4.5, but assume now that the order of play is switched, namely, the public firm is the leader and firm 2 is the follower. Firms have the same objective functions as in Exercise 4.5.

(a) Find the follower's best response function. Interpret:

- The follower's (private firm) profit maximization problem is

$$\max_{q_2 \geq 0} \pi_2 = (1 - q_1 - q_2)q_2 - cq_2.$$

Differentiating with respect to q_2, we obtain

$$1 - q_1 - 2q_2 - c = 0.$$

Solving for q_2, we find firm 2's best response function

$$q_2(q_1) = \frac{1-c}{2} - \frac{1}{2}q_1.$$

Intuitively, nothing has changed for the industry follower relative to a standard sequential competition model with two private firms.

(b) Find the leader's equilibrium output.

- The leader's (public firm) maximization problem is

$$\max_{q_1 \geq 0} V = \alpha W + (1-\alpha)\pi_1,$$

where W denotes social welfare, defined as follows:

$$W = \int_0^Q p(Q)dQ - cQ$$
$$= (1-c)Q - \frac{1}{2}Q^2.$$

Therefore, the above problem simplifies to

$$\max_{q_1 \geq 0} V = \alpha \overbrace{\left[(1-c)Q - \frac{1}{2}Q^2\right]}^{W} + (1-\alpha)\overbrace{[(1-Q)q_1 - cq_1]}^{\pi_1}$$
$$= \alpha \left[(1-c)\underbrace{(q_1+q_2)}_{Q} - \frac{1}{2}\underbrace{(q_1+q_2)^2}_{Q}\right] + (1-\alpha)\left[(1-\underbrace{(q_1+q_2)}_{Q})q_1 - cq_1\right].$$

Since the leader anticipates the follower's behavior in the subsequent stage of the game, we can insert its best response function, $q_2(q_1)$, into the above objective function, yielding

$$\max_{q_1 \geq 0} V = \alpha \left[(1-c)(q_1 + q_2(q_1)) - \frac{1}{2}(q_1 + q_2(q_1))^2\right]$$
$$+ (1-\alpha)\left[(1-(q_1+q_2(q_1)))q_1 - cq_1\right].$$

Let us simplify the expression $q_1 + q_2(q_1)$, obtaining

$$q_1 + q_2(q_1) = q_1 + \left(\frac{1-c}{2} - \frac{1}{2}q_1\right)$$
$$= \frac{1-c}{2} + \frac{1}{2}q_1.$$

Inserting this expression into firm 1's maximization problem yields

$$\max_{q_1 \geq 0} V = \alpha \left[(1-c)\left(\frac{1-c}{2} + \frac{1}{2}q_1\right) - \frac{1}{2}\left(\frac{1-c}{2} + \frac{1}{2}q_1\right)^2\right]$$
$$+ (1-\alpha)\left[(1 - \left(\frac{1-c}{2} + \frac{1}{2}q_1\right) - c)q_1\right].$$

- Differentiating with respect to q_1, we obtain

$$\alpha \left[\frac{1-c}{2} - \frac{1}{2}\left(\frac{1-c}{2} + \frac{1}{2}q_1\right)\right] + (1-\alpha)\left[1 - \frac{1-c}{2} - q_1 - c\right] = 0.$$

Rearranging yields

$$(4-3\alpha)\, q_1 = (2-\alpha)(1-c).$$

Solving for q_1, we obtain the leader's equilibrium output

$$q_1^* = \frac{(2-\alpha)(1-c)}{4-3\alpha}.$$

- This output level captures both of firm 1's extremes:
  - When $\alpha = 0$ and firm 1 is entirely interested in its own profits, firm 1's quantity reduces to $\frac{1-c}{2}$, which we would expect from a standard Stackelberg leader.
  - In contrast, when $\alpha = 1$ and firm 1 is entirely interested in maximizing social welfare, firm 1's quantity reduces to $1-c$, the perfectly competitive quantity, which necessitates an equilibrium price equal to the constant marginal cost of production, c. (The next part of the exercise confirms that, when $\alpha = 1$, the follower observes the leader producing an output of $1-c$ and responds with an output level $q_2^* = 0$.)

(c) Find firm 2's output level in equilibrium:
- The follower observes the leader's output, $q_1^* = \frac{(2-\alpha)(1-c)}{4-3\alpha}$ and inserts it into its best response function, as follows:

$$q_2^* = \frac{1-c}{2} - \frac{1}{2}\overbrace{\frac{(2-\alpha)(1-c)}{4-3\alpha}}^{q_1^*}$$
$$= \frac{(1-\alpha)(1-c)}{4-3\alpha}.$$

- As in part (b), we observe both extremes based on the value of α:
  - When $\alpha = 0$ and firm 1 is entirely interested in its own profits, firm 2's quantity reduces to $\frac{1-c}{4}$, which we would expect from a standard Stackelberg follower.
  - However, when $\alpha = 1$ and firm 1 is entirely interested in maximizing social welfare, firm 2's quantity reduces to 0, as firm 1 already produces the perfectly competitive quantity, leaving no room for firm 2 to produce and earn positive profits.

(d) How is each firm's output decision affected by a marginal increase in α? Interpret:
- Differentiating the leader's equilibrium output with respect to α yields

$$\frac{\partial q_1^*}{\partial \alpha} = \frac{[-(4-3\alpha) + 3(2-\alpha)](1-c)}{(4-3\alpha)^2}$$
$$= \frac{2(1-c)}{(4-3\alpha)^2} > 0.$$

- Differentiating the follower's equilibrium output with respect to α, we find that

$$\frac{\partial q_2^*}{\partial \alpha} = \frac{[-(4-3\alpha) + 3(1-\alpha)](1-c)}{(4-3\alpha)^2}$$
$$= -\frac{1-c}{(4-3\alpha)^2} < 0.$$

Since $c < 1$ by assumption, we find that leader increases its output in α as it assigns a higher weight to social welfare relative to its own profits. This comes at the expense of the follower, who must respond reducing its own output as a consequence. Figure 4.2

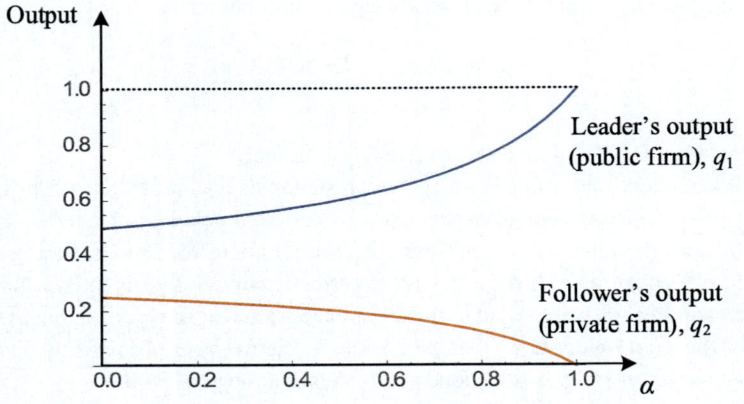

Fig. 4.2 Effect of α on equilibrium output

illustrates the effect of α on the leader's and follower's equilibrium output where, for simplicity, we assume $c = 0$.

(e) Find equilibrium prices. How are they affected by a marginal increase in α? Interpret:

- Equilibrium price is

$$\begin{aligned} p^* &= 1 - q_1^* - q_2^* \\ &= 1 - \frac{(2-\alpha)(1-c)}{4-3\alpha} - \frac{(1-\alpha)(1-c)}{4-3\alpha} \\ &= \frac{1-\alpha+(3-2\alpha)c}{4-3\alpha}. \end{aligned}$$

Differentiating the equilibrium price with respect to α,

$$\begin{aligned} \frac{\partial p^*}{\partial \alpha} &= \frac{-(1+2c)(4-3\alpha)+3\left[1-\alpha+(3-2\alpha)c\right]}{(4-3\alpha)^2} \\ &- \frac{1-c}{(4-3\alpha)^2} < 0. \end{aligned}$$

Intuitively, as α increases, the market quantity, Q, increases as the public firm places an increasing weight on social welfare, so equilibrium price decreases.

(f) Evaluate aggregate output and social welfare in equilibrium. How do they change with α? Interpret:

- *Aggregate output.* Aggregate output is

$$\begin{aligned} Q^* &= \frac{(2-\alpha)(1-c)}{4-3\alpha} + \frac{(1-\alpha)(1-c)}{4-3\alpha} \\ &= \frac{(3-2\alpha)(1-c)}{4-3\alpha}. \end{aligned}$$

Differentiating Q^* with respect to α yields

$$\frac{\partial Q^*}{\partial \alpha} = \frac{1-c}{(4-3\alpha)^2},$$

which is positive since $c, \alpha \in [0, 1]$. Intuitively, when firm 1 assigns more weight on social welfare (higher α), it increases its output more significantly than the follower decreases its own output, ultimately giving rise to an increase in aggregate output.

- *Social welfare.* Substituting the aggregate output into the welfare function, we find that equilibrium social welfare becomes

$$W^* = (1-c)\,\frac{(3-2\alpha)(1-c)}{4-3\alpha} - \frac{1}{2}\left(\frac{(3-2\alpha)(1-c)}{4-3\alpha}\right)^2$$

$$= \frac{(1-c)^2 (3-2\alpha)(5-4\alpha)}{2(4-3\alpha)^2}.$$

Differentiating W^* with respect to α yields

$$\frac{\partial W^*}{\partial \alpha} = \frac{(1-c)^2\left[(4-3\alpha)(8\alpha-11) + 3\left(15-22\alpha+8\alpha^2\right)\right]}{(4-3\alpha)^3}$$

$$= \frac{(1-c)^2(1-\alpha)}{(4-3\alpha)^3},$$

which is positive for all $\alpha \in [0, 1)$ (social welfare is maximized at $\alpha = 1$ when the public firm produces the perfectly competitive quantity). Therefore, as firm 1 assigns a higher weight on social welfare, aggregate output increases, and equilibrium social welfare increases as well.

(g) Compare the equilibrium output and welfare levels against those found in Exercise 4.5, where the public firm acted as the follower:

- Comparing our results against those in Exercise 4.5, we find that:
  - The private firm's output now decreases with α, as it no longer benefits from being a Stackelberg leader. Specifically, it produces at most half of its output as in Exercise 4.5, decreasing to 0 as α approaches $\alpha = 1$.
  - In contrast, the public firm's output increases in α but follows a similar pattern as in Exercise 4.5. When $\alpha = 0$, firm 1 doubles its output relative to the results of Exercise 4.5, but as α approaches $\alpha = 1$, firm 1 ensures in both exercises that the aggregate output approaches the perfectly competitive output.
  - Whereas, aggregate output when the public firm acts as the follower is larger than when it acts as the leader, since

$$\frac{(3-\alpha)(1-c)}{2(2-\alpha)} > \frac{(3-2\alpha)(1-c)}{4-3\alpha}$$

reduces to

$$\alpha(1-\alpha) > 0$$

that holds for all values of α. Intuitively, the public firm can produce more units after observing the private leader's behavior than committing to an output level before observing the private follower's production decisions.

– Similarly, social welfare when the public firm acts as the follower is larger than when it acts as the leader, since

$$\frac{(1-c)^2(3-\alpha)(5-3\alpha)}{8(2-\alpha)^2} > \frac{(1-c)^2(3-2\alpha)(5-4\alpha)}{2(4-3\alpha)^2}$$

simplifies to

$$(1-\alpha)^2(8-5\alpha) > 0$$

that holds for all values of $\alpha \in [0, 1]$. Since aggregate output is larger when the public firm acts as the follower than as the leader, so does equilibrium welfare of the former rise above the latter.

Exercise #4.7: Stackelberg Competition Under Incomplete Information – Uninformed Leader^B

4.7. Consider a Stackelberg game between a leader (firm 1) and a follower (firm 2). Inverse demand function is given by $p(Q) = 1 - Q$, where Q denotes aggregate output. The leader's marginal cost $MC_1 = \frac{1}{4}$ is common knowledge among the players. Intuitively, the leader is the industry incumbent, and all firms can estimate its costs with relative accuracy. However, the follower's marginal costs are either high $\left(MC_2 = \frac{1}{3}\right)$ or low $\left(MC_2 = \frac{1}{5}\right)$ with probabilities p and $1 - p$, respectively. The entrant is a newcomer, and thus the leader cannot perfectly observe the follower's costs:

(a) Find the follower's best response function:

- Let q_1, q_2^L, and q_2^H denote the output of firm 1, low- and high-cost firm 2, respectively.
- When firm 2 has low costs, it chooses the output $q_2^L \geq 0$ that solves the following expected profit maximization problem:

$$\max_{q_2^L \geq 0} \pi_2^L\left(q_2^L\right) = \left(1 - q_1 - q_2^L\right) q_2^L - \frac{1}{5} q_2^L$$
$$= \left(\frac{4}{5} - q_1 - q_2^L\right) q_2^L.$$

Assuming interior solutions, that is, $q_2^L > 0$, the first-order condition satisfies

$$\frac{\partial \pi_2^L\left(q_2^L\right)}{\partial q_2^L} = \frac{4}{5} - q_1 - 2q_2^L = 0$$

such that the best response function of firm 2 when its costs are low becomes

$$q_2^L(q_1) = \frac{2}{5} - \frac{1}{2} q_1,$$

which originates at $2/5$ and decreases in firm 1's output, q_1, at a rate of $1/2$.

- When firm 2 has high costs, it chooses the output $q_2^H \geq 0$ that solves the following expected profit maximization problem:

$$\max_{q_2^H \geq 0} \pi_2^H\left(q_2^H\right) = \left(1 - q_1 - q_2^H\right) q_2^H - \frac{1}{3} q_2^H$$
$$= \left(\frac{2}{3} - q_1 - q_2^H\right) q_2^H.$$

Assuming interior solutions, that is, $q_2^H > 0$, the first-order condition satisfies

$$\frac{\partial \pi_2^H\left(q_2^H\right)}{\partial q_2^H} = \frac{2}{3} - q_1 - 2q_2^H = 0$$

such that the best response function of firm 2 when its costs are high becomes

$$q_2^H\left(q_1\right) = \frac{1}{3} - \frac{1}{2} q_1,$$

which originates at a lower vertical intercept of $1/3$ than when its costs are low, but also decreases in q_1 at a rate of $1/2$.

(b) Find the leader's profit maximizing output, and summarize the subgame perfect Nash equilibrium of the game:

- Firm 1 chooses its output $q_1 \geq 0$ that solves the following expected profit maximization problem:

$$\max_{q_1 \geq 0} \pi_1\left(q_1\right) = (1 - p)\left[1 - q_1 - q_2^L\left(q_1\right)\right] q_1$$
$$+ p\left[1 - q_1 - q_2^H\left(q_1\right)\right] q_1 - \frac{1}{4} q_1,$$

which simplifies to

$$\max_{q_1 \geq 0} \pi_1\left(q_1\right) = \left[\frac{3}{4} - q_1 - (1 - p)\left(\frac{2}{5} - \frac{q_1}{2}\right) - p\left(\frac{1}{3} - \frac{q_1}{2}\right)\right] q_1$$
$$= \left(\frac{7}{20} + \frac{p}{15} - \frac{q_1}{2}\right) q_1.$$

Assuming interior solutions, that is, $q_1 > 0$, the first-order condition satisfies

$$\frac{\partial \pi_1\left(q_1\right)}{\partial q_1} = \frac{7}{20} + \frac{p}{15} - q_1 = 0$$

such that the output of firm 1 (leader) becomes

$$q_1^* = \frac{7}{20} + \frac{p}{15}.$$

Substituting the above into the best response function of firm 2, equilibrium output of the follower becomes

$$q_2^{L*} = \frac{2}{5} - \frac{1}{2}\left(\frac{7}{20} + \frac{p}{15}\right) = \frac{27-4p}{120}$$

$$q_2^{H*} = \frac{1}{3} - \frac{1}{2}\left(\frac{7}{20} + \frac{p}{15}\right) = \frac{19-4p}{120}.$$

- Therefore, the subgame perfect Nash equilibrium of the game is

$$\left\{q_1^*, q_2^{L*}, q_2^{H*}\right\} = \left\{\frac{21+4p}{60}, \frac{27-4p}{120}, \frac{19-4p}{120}\right\}$$

such that when firm 2 is more likely to face high costs (i.e., probability p increases), the output of firm 1 increases, but the output of both low-cost and high-cost firm 2 decreases.

(c) How would your results in parts (a) and (b) be changed if the leader could perfectly observe the follower's marginal costs when these costs are high? What if the follower's costs are low?

- If the leader could perfectly observe the follower's marginal costs, and these costs are high, we have that $p = 1$, which implies that the above output levels become

$$q_1^* = \frac{5}{12} \text{ and } q_2^{H*} = \frac{1}{8}.$$

- Similarly, if the leader could perfectly observe the follower's marginal costs, and these costs are low, we have that $p = 0$, which implies that the above output levels become

$$q_1^* = \frac{7}{20} \text{ and } q_2^{L*} = \frac{9}{40}.$$

Exercise #4.8: Stackelberg with Demand Shocks, Based on Liu (2005)[B]

4.8. Consider a Stackelberg model of sequential quantity competition between firm 1 (leader) and firm 2 (follower), with inverse demand function $p(Q) = a - Q$, where $Q = q_1 + q_2$ denotes aggregate output. For simplicity, assume symmetric marginal production cost $c \geq 0$.

Unlike in the standard Stackelberg model, let us consider that demand is uncertain at the beginning of the game. In particular, the leader chooses its output, q_1, without observing the state of market demand, where $a \in [a_L, a_H]$ with expectation $E[a] = \overline{a}$, and $a_L > c \geq 0$. In the second period, the follower observes q_1 and the realization of parameter a and responds with its output q_2.

(a) *Second stage*. Find firm 2's best response function, $q_2(q_1)$.

- The follower (firm 2) observes the realization of a, takes the leader's output, q_1, as given, and solves

$$\max_{q_2 \geq 0} \pi_2(q_2) = (a - q_1 - q_2)\, q_2 - cq_2.$$

Differentiating with respect to q_2 and assuming an interior solution yield

$$\frac{\partial \pi_2(q_2)}{\partial q_2} = a - q_1 - 2q_2 - c = 0,$$

which entails the following best response function

$$q_2(q_1) = \begin{cases} \frac{a-c}{2} - \frac{1}{2}q_1 \text{ for all } q_1 \leq a - c \\ 0 \text{ otherwise.} \end{cases}$$

As expected, this best response function of the follower originates at $\frac{a-c}{2}$ and decreases in the leader's output, q_1, at a rate of $\frac{1}{2}$.

(b) *First stage*. Identify firm 1's equilibrium output, q_1^*, and associated profits, π_1^*.

- The leader anticipates the follower's *expected* best response function,

$$E[q_2(q_1)] = E\left[\frac{a-c}{2} - \frac{1}{2}q_1\right] = \frac{\overline{a}-c}{2} - \frac{1}{2}q_1,$$

since $E[a] = \overline{a}$, and inserts it into its objective function, as follows:

$$\begin{aligned} \max_{q_1 \geq 0} \pi_1(q_1) &= [\overline{a} - q_1 - E[q_2(q_1)]] q_1 - cq_1 \\ &= \left[\overline{a} - q_1 - \frac{\overline{a}-c}{2} + \frac{1}{2}q_1\right] q_1 - cq_1 \\ &= \frac{(\overline{a} - c - q_1) q_1}{2}. \end{aligned}$$

Differentiating with respect to q_1 and assuming an interior solution yield

$$\frac{\partial \pi_1(q_1)}{\partial q_1} = \frac{\overline{a} - c - 2q_1}{2} = 0,$$

and solving for q_1, we find the leader's equilibrium output,

$$q_1^* = \frac{\overline{a}-c}{2},$$

which is positive since $a_L > c \geq 0$, implying that $\overline{a} > c \geq 0$ also holds.

- The leader's equilibrium profits are, then,

$$\begin{aligned} \pi_1^* &= \frac{\left(\overline{a} - c - \frac{\overline{a}-c}{2}\right)\frac{\overline{a}-c}{2}}{2} \\ &= \frac{(\overline{a}-c)^2}{8}. \end{aligned}$$

(c) Characterize firm 2's equilibrium output, q_2^*, and associated profits, π_2^*.

- Inserting q_1^* into the follower's best response function from part (a), the follower's equilibrium output is

$$\begin{aligned} q_2^* &= \frac{a - c - \frac{\overline{a}-c}{2}}{2} \\ &= \frac{2a - \overline{a} - c}{4}, \end{aligned}$$

which is positive if and only if $2a - c \geq \overline{a}$. Intuitively, this occurs when actual demand is relatively strong or, alternatively, when the expected demand is sufficiently low. In this context, firm 1 produces relatively few units, leaving some residual demand for firm 2 to make positive profits. Otherwise, firm 1 produces a large output, leaving firm 2 with too small of a residual demand. In this setting, firm 2 prefers to remain inactive, that is, $q_2^* = 0$.

In summary, we can report firm 2's equilibrium output as follows:

$$q_2^* = \begin{cases} \frac{2a-\overline{a}-c}{4} & \text{if } 2a - c \geq \overline{a}, \\ 0 & \text{otherwise.} \end{cases}$$

- Substituting q_1^* and q_2^* into the follower's profit function results in

$$\pi_2^* = \left(a - \frac{\overline{a} - c}{2} - \frac{2a - \overline{a} - c}{4} - c\right) \frac{2a - \overline{a} - c}{4}$$
$$= \left(\frac{2a - \overline{a} - c}{4}\right)^2,$$

which is positive if and only if firm 2 is active, that is, $2a - c \geq \overline{a}$.

(d) How are the equilibrium output levels of leader and follower affected by the expected state of demand, $\overline{a}$, and by the realized demand, a? Interpret:

- Differentiating q_1^* with respect to $\overline{a}$, we find

$$\frac{\partial q_1^*}{\partial \overline{a}} = \frac{1}{2} > 0,$$

so a higher expected demand instigates firm 1 to produce more output.

- Next, differentiating q_1^* with respect to a, we have

$$\frac{\partial q_1^*}{\partial a} = 0$$

so that firm 1, which does not know the actual demand, does not respond to it.

- Similarly, differentiating q_2^* with respect to $\overline{a}$, we obtain

$$\frac{\partial q_2^*}{\partial \overline{a}} = -\frac{1}{4} < 0.$$

Since firm 1 responds to a higher expected demand increasing its output, firm 2 reacts decreasing its own production because output levels are strategic substitutes.

- Finally, differentiating q_2^* with respect to a yields

$$\frac{\partial q_2^*}{\partial a} = \frac{1}{2} > 0$$

so firm 2 increases its output when actual demand becomes stronger.

(e) Compare the leader and follower profits from parts (b) and (c) in equilibrium. For which parameter conditions does the leader benefit from earns more profits than the follower (first-mover advantage)?

- Comparing equilibrium profits from parts (b) and (c), we obtain that

$$\pi_1^* = \left(\frac{\overline{a} - c}{2\sqrt{2}}\right)^2 \geq \left(\frac{2a - \overline{a} - c}{4}\right)^2 = \pi_2^*$$

simplifies to

$$2\overline{a} - 2c \geq 2\sqrt{2}a - \sqrt{2}\overline{a} - \sqrt{2}c$$

that further reduces to

$$\overline{a} \geq \frac{2\sqrt{2}a + \left(2 - \sqrt{2}\right)c}{2 + \sqrt{2}},$$

which happens when the expected demand is sufficiently strong.

- For instance, when $c = 0$, this condition simplifies to $\overline{a} \geq 2(\sqrt{2} - 1)a \simeq 0.82a$. Intuitively, if the state of demand that the follower observes, a, is relatively low, the leader earns a higher profit than the follower. Alternatively, if the leader assigns a relatively high probability weight on high demand states (high $\overline{a}$), but the realized demand is low (low a), the leader produces sufficiently more units than the follower and ends up earning a larger profit.

Exercise #4.9: Stackelberg Competition Under Incomplete Information – Uninformed Follower^C

4.9. Consider an industry with inverse demand function $p(Q) = 1 - Q$, where Q denotes aggregate output. A leader and a follower compete à la Stackelberg, as in previous exercises in this chapter. However, we seek to analyze how their interaction is affected when the follower is uninformed about the leader's marginal production cost. The follower's marginal cost is c_f, which is common knowledge among the firms.

(a) *Complete information.* As a benchmark, let us first assume that the leader's marginal cost is observed by the follower, c_l. Find the equilibrium output q_l of the leader and q_f of the follower in this Stackelberg game, as a function of c_l, and the equilibrium profits for each firm. Evaluate these profits when the leader's marginal cost is $c_l = c_L$ (which we interpret as a low-cost leader, since this firm is more efficient than the follower) and $c_l = c_H$ (which we interpret as a high-cost leader, as it suffers a cost disadvantage relative to the follower), where $0 \leq c_L < c_f < c_H < 1$, and $3c_f - 1 \leq 2c_l < 1 + c_f$, which guarantees an interior solution both when the leader's costs are high and low:

- The follower solves the following profit maximization problem:

$$\pi_f(q_f) = (1 - q_l - q_f)\,q_f - c_f q_f.$$

Differentiating with respect to q_f, and assuming interior solutions,

$$1 - q_l - 2q_f - c_f = 0.$$

Rearranging, we obtain the best response function of the follower:

$$q_f(q_l) = \frac{1 - c_f}{2} - \frac{1}{2}q_l.$$

- The leader solves the following profit maximization problem:

$$\pi_l(q_l) = \left(1 - q_l - q_f(q_l)\right) q_l - c_l q_l$$
$$= \frac{1 - q_l - 2c_l + c_f}{2} q_l.$$

Differentiating with respect to q_l, and assuming interior solutions,

$$1 - 2q_l - 2c_l + c_f = 0.$$

Rearranging, we obtain the equilibrium output of the leader,

$$q_l = \frac{1 - 2c_l + c_f}{2},$$

which is positive when $c_l < \frac{1+c_f}{2}$. Intuitively, the leader must be relatively efficient in order to produce a positive output.

Substituting the leader's equilibrium output q_l into the best response function of the follower, $q_f(q_l)$, we obtain the follower's equilibrium output:

$$q_f = \frac{1 - c_f}{2} - \frac{1}{2}\underbrace{\left(\frac{1 - 2c_l + c_f}{2}\right)}_{q_l}$$
$$= \frac{1 + 2c_l - 3c_f}{4},$$

which is positive when $c_f < \frac{1+2c_l}{3}$, that is, when the follower is relatively efficient compared to the leader. Solving for c_l in this condition, we obtain $c_l > \frac{3c_f - 1}{2}$, which we can now compare with the condition we found for the leader to produce a positive output, yielding that $\frac{1+c_f}{2} - \frac{3c_f-1}{2} > 0$ since $1 > c_f$ holds by assumption. We therefore identify the following results:

  - When the leader's cost is high, in particular, $c_l \geq \frac{1+c_f}{2}$, the leader does not produce any output, but the follower monopolizes the market.
  - When the leader's cost is intermediate, in particular, $\frac{3c_f-1}{2} \leq c_l < \frac{1+c_f}{2}$, both the leader and the follower produce output.
  - When the leader's cost is low, in particular, $c_l < \frac{3c_f-1}{2}$, the leader monopolizes the market and the follower does not produce any output.

- *Profits.* Substituting the equilibrium output into the follower's profit function, we find the follower's equilibrium profits as follows:

$$\pi_f = \left(1 - \frac{1 - 2c_l + c_f}{2} - \frac{1 + 2c_l - 3c_f}{4} - c_f\right)\frac{1 + 2c_l - 3c_f}{4}$$
$$= \frac{\left(1 + 2c_l - 3c_f\right)^2}{16}.$$

Substituting the equilibrium output into the leader's profit function, we obtain the leader's equilibrium profits as follows:

$$\pi_l = \left(1 - \frac{1 - 2c_l + c_f}{2} - 2c_l + c_f\right)\frac{1 - 2c_l + c_f}{4}$$

$$= \frac{\left(1 - 2c_l + c_f\right)^2}{8}.$$

(b) *Separating equilibrium.* Assume now that the leader privately observes its marginal cost. The follower does not observe this realization but knows that leader's cost is low with probability p and high with probability $1 - p$, where $p \in [0, 1]$. We first analyze whether a separating perfect Bayesian equilibrium (PBE) can be sustained when the leader chooses a type-dependent output level, that is, q_L when $c_l = c_L$ and q_H when $c_l = c_H$. For simplicity, you may focus on testing whether the separating strategy profile (q_L^{SE}, q_H^{SE}) can be sustained as a PBE, where q_L^{SE} is larger than the complete information output of the low-cost leader, q_L, but q_H^{SE} coincides with the complete information output of the high-cost leader q_H. In addition, you may assume that, upon observing an off-the-equilibrium output level $q' \neq q_L^{SE} \neq q_H^{SE}$, the follower's beliefs are $\mu\left(c_L|q'\right) = 1$ and $\mu\left(c_H|q'\right) = 0$.

- *Roadmap of the proof.* When testing if a strategy profile can be sustained as a PBE, we first update the beliefs of the player who acts as the uninformed responder (in this case, the follower) and then identify this player's optimal responses given its updated beliefs. This allows us to check if the privately informed first mover (any of the leader's types in this exercise) has incentives to behave as prescribed in the strategy profile we are testing or, instead, has incentives to deviate.
- *First step – Updated beliefs.* After observing that the leader has produced an output level q, the follower's belief of facing a low-cost leader is $\mu\left(c_L|q\right)$, while that of facing a high-cost leader is $\mu\left(c_H|q\right)$. In a separating equilibrium where the leader chooses q_L^{SE} after observing $c_l = c_L$ and q_H^{SE} after observing $c_l = c_H$, the follower's beliefs (posterior probabilities) can be updated according to Bayes' rule, as follows:

$$\mu\left(c_L|q_L^{SE}\right) = \frac{\Pr\left(q_L^{SE}|c_L\right)\Pr\left(c_L\right)}{\Pr\left(q_L^{SE}|c_L\right)\Pr\left(c_L\right) + \Pr\left(q_L^{SE}|c_H\right)\Pr\left(c_H\right)} = \frac{1p}{1p + 0(1-p)} = 1$$

$$\mu\left(c_H|q_L^{SE}\right) = \frac{\Pr\left(q_L^{SE}|c_H\right)\Pr\left(c_H\right)}{\Pr\left(q_L^{SE}|c_L\right)\Pr\left(c_L\right) + \Pr\left(q_L^{SE}|c_H\right)\Pr\left(c_H\right)} = \frac{0(1-p)}{1p + 0(1-p)} = 0$$

$$\mu\left(c_L|q_H^{SE}\right) = \frac{\Pr\left(q_H^{SE}|c_L\right)\Pr\left(c_L\right)}{\Pr\left(q_H^{SE}|c_L\right)\Pr\left(c_L\right) + \Pr\left(q_H^{SE}|c_H\right)\Pr\left(c_H\right)} = \frac{0p}{0p + 1(1-p)} = 0$$

$$\mu\left(c_H|q_H^{SE}\right) = \frac{\Pr\left(q_H^{SE}|c_H\right)\Pr\left(c_H\right)}{\Pr\left(q_H^{SE}|c_L\right)\Pr\left(c_L\right) + \Pr\left(q_H^{SE}|c_H\right)\Pr\left(c_H\right)} = \frac{1(1-p)}{0p + 1(1-p)} = 1.$$

Intuitively, these updated beliefs say that, upon observing output q_L^{SE}, the follower is convinced of dealing with a low-cost leader, i.e., $\mu\left(c_L|q_L^{SE}\right) = 1$ but not a high-cost leader, $\mu\left(c_H|q_L^{SE}\right) = 0$. In contrast, when the follower observes output q_H, it puts full probability on facing a high-cost leader, i.e., $\mu\left(c_H|q_H^{SE}\right) = 1$ but not a low-cost leader, $\mu\left(c_L|q_H^{SE}\right) = 0$.

Finally, upon observing an off-the-equilibrium output of the leader $q' \neq q_L^{SE} \neq q_H^{SE}$, Bayes' rule cannot help us update the follower's beliefs and, for generality, we leave them unrestricted between 0 and 1, i.e., $\mu\left(c_l|q'\right) \in [0, 1]$, where $l \in \{L, H\}$ denotes the leader's type. However, as suggested in the exercise, we can assume that the follower holds off-the-equilibrium beliefs $\mu\left(c_L|q'\right) = 1$ and $\mu\left(c_H|q'\right) = 0$. Intuitively, this means that the follower sustains the worst beliefs from the leader's perspective, as the follower responds producing the lowest output.

- *Second step – Follower's responses.* Using these updated beliefs, we can rewrite the follower's expected profits as follows, depending on the output level that the follower observes (q_L, q_H, or off-the-equilibrium output q').
  - When the follower observes output q_L^{SE}, its expected profit maximization problem becomes

$$\pi_f\left(q_f|q_L^{SE}\right) = \overbrace{\underbrace{\mu\left(c_L|q_L^{SE}\right)}_{=1}\left(1 - q_L^{SE} - q_f - c_f\right)q_f}^{\text{Leader's costs are low}}$$

$$+\overbrace{\underbrace{\mu\left(c_H|q_L^{SE}\right)}_{=0}\left(1 - q_L - q_f - c_f\right)q_f}^{\text{Leader's costs are high}}$$

$$= \left(1 - q_L^{SE} - q_f - c_f\right)q_f.$$

Differentiating with respect to q_f, and solving for q_f, we find that the follower's best response function upon observing q_L^{SE} is

$$q_f\left(q_L^{SE}\right) = \frac{1 - c_f}{2} - \frac{1}{2}q_L^{SE}.$$

  - When the follower observes output q_H^{SE}, its expected profit maximization problem becomes

$$\pi_f\left(q_f|q_H^{SE}\right) = \overbrace{\underbrace{\mu\left(c_L|q_H^{SE}\right)}_{=0}\left(1 - q_H - q_f - c_f\right)q_f}^{\text{Leader's costs are low}}$$

$$+\overbrace{\underbrace{\mu\left(c_H|q_H^{SE}\right)}_{=1}\left(1 - q_H^{SE} - q_f - c_f\right)q_f}^{\text{Leader's costs are high}}$$

$$= \left(1 - q_H^{SE} - q_f - c_f\right)q_f.$$

Differentiating with respect to q_f, and solving for q_f, we find that the follower's best response function upon observing q_H^{SE} is

$$q_f\left(q_H^{SE}\right) = \frac{1 - c_f}{2} - \frac{1}{2}q_H^{SE}.$$

– Finally, when the follower observes an off-the-equilibrium output q', and given the above beliefs, its expected profit maximization problem becomes identical to that after observing output q_L, that is,

$$\pi_f\left(q_f|q'\right) = \overbrace{\underbrace{\mu\left(c_L|q'\right)}_{=1}\left(1-q'-q_f-c_f\right)q_f}^{\text{Leader's costs are low}} + \overbrace{\underbrace{\mu\left(c_H|q'\right)}_{=0}\left(1-q'-q_f-c_f\right)q_f}^{\text{Leader's costs are high}}$$

$$= \left(1-q'-q_f-c_f\right)q_f.$$

Differentiating with respect to q_f, and solving for q_f, we find that the follower's best response function upon observing q' is

$$q_f\left(q'\right) = \frac{1-c_f}{2} - \frac{1}{2}q'.$$

- *Third step – Leader's output.* First, we analyze the high-cost leader produces the prescribed output in this strategy profile, q_H^{SE}, which coincides with its output under complete information, $q_H = \frac{1-2c_H+c_f}{2}$, earning profits of

$$\pi_H\left(q_H\right) = \left(1-q_H-q_f\left(q_H\right)\right)q_H - c_H q_H$$
$$= \frac{1-q_H-2c_H+c_f}{2}q_H$$
$$= \frac{\left(1-2c_H+c_f\right)^2}{8}$$

since we evaluated this profit at $q_H = \frac{1-2c_H+c_f}{2}$. Let us now check if this high-cost leader has incentives to deviate from q_H to q_L^{SE}. If it deviates to the output level of the low-cost type, q_L^{SE}, the high-cost leader earns profits of

$$\pi_H\left(q_L^{SE}\right) = \left(1-q_L^{SE}-q_f\left(q_L^{SE}\right)\right)q_L^{SE} - c_H q_L^{SE}$$
$$= \frac{1-q_L^{SE}-2c_H+c_f}{2}q_L^{SE},$$

which falls below $\pi_H\left(q_H\right)$ if

$$\frac{1-q_L^{SE}-2c_H+c_f}{2}q_L^{SE} < \frac{\left(1-2c_H+c_f\right)^2}{8}$$

that simplifies to

$$q_L^{SE} < \frac{1-2c_H+c_f}{2} = q_H.$$

Intuitively, the high-cost leader prefers to choose output level q_H than mimics the low type (choosing q_L^{SE}) as long as $q_L^{SE} < q_H$, that is, if the low-cost leader's output in the separating equilibrium is lower than the high-cost leader's output q_H.

Second, we consider low-cost leader. If it produces q_L^{SE} units, as described by this separating strategy profile, its profits are

$$\pi_L\left(q_L^{SE}\right) = \left(1 - q_L^{SE} - q_f\left(q_L^{SE}\right)\right) q_L^{SE} - c_L q_L^{SE}$$
$$= \frac{1 - q_L^{SE} - 2c_L + c_f}{2} q_L^{SE}.$$

If, instead, this firm deviates from q_L^{SE} to $q_H^{SE} = q_H$, its profits are

$$\pi_L(q_H) = (1 - q_H - q_f(q_H)) q_H - c_L q_H$$
$$= \frac{1 - q_H - 2c_L + c_f}{2} q_H,$$

and evaluating them at $q_H = \frac{1-2c_H+c_f}{2}$, we obtain

$$\pi_L(q_H) = \frac{(1 - 2c_H + c_f)(1 - 4c_L + 2c_H + c_f)}{8}.$$

Therefore, the low-cost leader prefers to choose output q_L^{SE} than q_H if

$$\frac{(1 - 2c_H + c_f)(1 - 4c_L + 2c_H + c_f)}{8} < \frac{1 - q_L^{SE} - 2c_L + c_f}{2} q_L^{SE}$$

that simplifies to

$$\frac{(1 - 2c_H + c_f - 2q_L^{SE})(1 - 4c_L + 2c_H + c_f - 2q_L^{SE})}{8} < 0.$$

Solving for q_L^{SE}, we find two roots

$$q_L^{SE} > \frac{1 - 2c_H + c_f}{2} = q_H \quad \text{and} \quad q_L^{SE} < \frac{1 + 2c_H - 4c_L + c_f}{2}.$$

Intuitively, the low-cost leader must produce more than the high-cost leader under complete information q_H, as otherwise the high-cost leader would have incentives to mimic output q_L^{SE}; but less than $q_L^{SE} = \frac{1+2c_H-4c_L+c_f}{2}$ since, otherwise, the separating effort would be too costly for the low-cost leader. In other words, $q_L^{SE} = q_H$ can be interpreted as the lower bound of the separating output to distinguish itself from the high-cost leader (and, thus, convey this information to the uninformed follower), while $q_L^{SE} = \frac{1+2c_H-4c_L+c_f}{2}$ represents, instead, the upper bound of this separating output.

- In addition, note that the profit maximizing output under complete information for the low-cost leader, $q_L = \frac{1-2c_L+c_f}{2}$, lies between the least-costly and the most-costly separating PBE, since

$$\frac{1 + 2c_H - 4c_L + c_f}{2} - \frac{1 - 2c_L + c_f}{2} = c_H - c_L, \text{ and}$$
$$\frac{1 - 2c_L + c_f}{2} - \frac{1 - 2c_H + c_f}{2} = c_H - c_L.$$

Therefore, among all separating strategy profiles in the interval

$$q_L^{SE} \in \left[\frac{1-2c_H+c_f}{2}, \frac{1+2c_H-4c_L+c_f}{2}\right],$$

the complete information output q_L yields the highest profit for the low-cost leader.

- *Intuitive Criterion.* As a remark, note that only one output level, $q_L^{SE} = q_L$, survives Cho and Kreps' (1987) Intuitive Criterion:
  - To understand this result, start from the PBE where the low-cost leader chooses output $q_L^{SE} = \frac{1+2c_H-4c_L+c_f}{2}$, in the upper bound of separating PBEs. This firm type can improve its equilibrium profit by deviating to output level q_L, but the high-cost leader does not. Therefore, upon observing the off-the-equilibrium output q_L, the follower believes that it must originate from the low-cost leader, thus responding with output $q_f(q_L)$, which indeed increases the low-cost leader's profits relative to those of choosing output q_L^{SE}. As a consequence, the PBE where the low-cost leader chooses output $q_L^{SE} = \frac{1+2c_H-4c_L+c_f}{2}$ violates the Intuitive Criterion.
  - A similar argument applies if we start from $q_L^{SE} = q_H$, in the lower bound of separating PBEs, also violating the Intuitive Criterion.
  - Finally, at output $q_L^{SE} = q_L$ (where the low-cost leader chooses the same output level as under complete information), no leader type can improve its equilibrium profit by deviating to any off-the-equilibrium output level, implying that the follower's beliefs cannot be updated and, thus, the PBE where the low-cost leader chooses $q_L^{SE} = q_L$ services the Intuitive Criterion.

(c) *Pooling equilibrium.* Can a pooling PBE be supported, where the leader chooses a type-independent output profile, that is, q both when $c_l = c_L$ and $c_l = c_H$? As in part (b), you may assume that, upon observing an off-the-equilibrium output level $q' \neq q$, the follower's beliefs are $\mu\left(c_L|q'\right) = 1$ and $\mu\left(c_H|q'\right) = 0$. In addition, you can focus on testing whether a strategy profile where both leader types choose output $q = q_L$ (i.e., an output that coincides with that of the low-cost leader under complete information) can be supported as a PBE of the game:

- We follow the same steps as in part (b) to check if type-independent output $q = q_L$ can be sustained as a PBE.
- *First step – Updated beliefs.* Suppose the leader chooses $q = q_L$ both when $c_l = c_L$ and $c_l = c_H$, and then its posterior beliefs can be updated using Bayes' rule as follows:

$$\mu(c_L|q) = \frac{\Pr(q|c_L)\Pr(c_L)}{\Pr(q|c_L)\Pr(c_L) + \Pr(q|c_H)\Pr(c_H)} = \frac{1p}{1p+1(1-p)} = p = \Pr(c_L)$$

$$\mu(c_H|q) = \frac{\Pr(q|c_H)\Pr(c_H)}{\Pr(q|c_L)\Pr(c_L) + \Pr(q|c_H)\Pr(c_H)} = \frac{1(1-p)}{1p+1(1-p)} = 1-p = \Pr(c_H).$$

Therefore, prior and posterior beliefs coincide since observing output q does not provide additional information as to whether the leader is of low-cost c_L or high-cost c_H. Whereas, upon observing off-the-equilibrium output of the leader $q' \neq q$, the follower's beliefs are $\mu\left(c_L|q'\right) = 1$ and $\mu\left(c_H|q'\right) = 0$.

- *Second step – Follower's responses.* Using these updated beliefs, we can rewrite the follower's expected profits, conditional on observing q on the equilibrium path, as follows:

$$\begin{aligned}\pi_f\left(q_f|q\right) &= \underbrace{\mu\left(c_L|q\right)}_{=p}\left(1-q-q_f\right)q_f \\ &+ \underbrace{\mu\left(c_H|q\right)}_{=1-p}\left(1-q-q_f\right)q_f - c_f q_f \\ &= \left(1-q-q_f-c_f\right)q_f\end{aligned}$$

yielding the following best response function of the follower:

$$q_f\left(q\right) = \frac{1-c_f}{2} - \frac{1}{2}q,$$

and since we are checking if $q = q_L$ can be sustained in equilibrium, and $q_L = \frac{1-2c_L+c_f}{2}$, the follower's equilibrium output $q_f\left(q_L\right)$ becomes

$$\begin{aligned}q_f\left(q_H\right) &= \frac{1-c_f}{2} - \frac{1}{2}\overbrace{\left(\frac{1-2c_L+c_f}{2}\right)}^{q_L} \\ &= \frac{1+2c_L-3c_f}{4}.\end{aligned}$$

When the follower observes off-the-equilibrium output q', its expected profits are

$$\begin{aligned}\pi_f\left(q_f|q'\right) &= \underbrace{\mu\left(c_L|q'\right)}_{=1}\left(1-q'-q_f-c_f\right)q_f + \underbrace{\mu\left(c_H|q'\right)}_{=0}\left(1-q'-q_f-c_f\right)q_f \\ &= \left(1-q'-q_f-c_f\right)q_f\end{aligned}$$

yielding the following best response function of the follower:

$$q_f\left(q'\right) = \frac{1-c_f}{2} - \frac{1}{2}q'.$$

- *Third step – Leader's output.* We first consider the high-cost leader. If it chooses the pooling output level $q = q_L$, as prescribed in this strategy profile, it earns profit of

$$\begin{aligned}\pi_H(q_L) &= p[\left(1-q_L-q_f\left(q_L\right)\right)q_L] + (1-p)[\left(1-q_L-q_f\left(q_L\right)\right)q_L] - c_H q_L \\ &= \left(1-q_L-q_f\left(q_L\right)\right)q_L - c_H q_L\end{aligned}$$

and since $q_L = \frac{1-2c_L+c_f}{2}$, and $q_f\left(q_L\right) = \frac{1+2c_L-3c_f}{4}$, this profit becomes

$$\pi_H\left(q_L\right) = \left(1-q_L-q_f\left(q_L\right)\right)q_L - c_H q_L$$

$$= \left(1 - \overbrace{\frac{1-2c_L+c_f}{2}}^{q_L} - \overbrace{\frac{1+2c_L-3c_f}{4}}^{q_f(q_L)} - c_H\right)\overbrace{\frac{1-2c_L+c_f}{2}}^{q_L}$$

$$= \frac{\left(1+2\left(c_L-2c_H\right)+c_f\right)\left(1-2c_L+c_f\right)}{8}.$$

If, instead, the high-cost leader deviates to its output level under complete information, $q_H = \frac{1-2c_H+c_f}{2}$, it will reveal its type to the follower, but avoid the "pooling effort" of having to increase its output level from q_H to q_L. If the high-cost leader chooses q_H, it earns a profit of

$$\pi_H(q_H) = \left(1 - q_H - q_f\left(q_H\right)\right) q_H - c_H q_H$$

$$= \frac{\left(1-2c_H+c_f\right)^2}{8}$$

as shown at the end of part (a). Therefore, the high-cost leader chooses the pooling output level $q = q_L$, rather than deviating to q_H if and only if

$$\frac{\left(1+2\left(c_L-2c_H\right)+c_f\right)\left(1-2c_L+c_f\right)}{8} > \frac{\left(1-2c_H+c_f\right)^2}{8},$$

which simplifies to

$$c_H^2 - 2c_H c_L + c_L^2 < 0$$

and is equivalent to $(c_H - c_L)^2 < 0$ that does not hold since $c_H > c_L$ by assumption. The high-cost leader, by choosing its complete information output level q^H, maximizes its profits and has no incentives to selecting the pooling output level q_L. Since we found one type of leader that has no incentives to select the pooling output level q_L, we do not need to analyze whether the other type (low cost) has incentives to choose q_L or not. In summary, the pooling strategy profile where both leader types choose output q_L cannot be sustained as a PBE of the game.

(d) *Numerical example*. Evaluate equilibrium output when $p = \frac{1}{2}$, $c_f = \frac{1}{2}$, $c_L = \frac{2}{5}$, and $c_H = \frac{3}{5}$.
- Under complete information, equilibrium output becomes

$$q_L = \frac{1-2c_L+c_f}{2} = \frac{1-2\times\frac{2}{5}+\frac{1}{2}}{2} = \frac{7}{20} \text{ for the low-cost leader,}$$

$$q_H = \frac{1-2c_H+c_f}{2} = \frac{1-2\times\frac{3}{5}+\frac{1}{2}}{2} = \frac{3}{20} \text{ for the high-cost leader,}$$

and equilibrium output of the follower is

$$q_f = \frac{1+2c_L-3c_f}{4} = \frac{1+2\times\frac{2}{5}-3\times\frac{1}{2}}{4} = \frac{3}{40} \text{ when facing a low-cost leader, and}$$

$$q_f = \frac{1+2c_H-3c_f}{4} = \frac{1+2\times\frac{3}{5}-3\times\frac{1}{2}}{4} = \frac{7}{40} \text{ when facing a high-cost leader.}$$

- In the separating PBE, as discussed in part (b), the low-cost leader chooses an output level

$$q_L^{SE} \in \left[\frac{1-2c_H+c_f}{2}, \frac{1+2c_H-4c_L+c_f}{2}\right],$$

which in this numerical example entails

$$q_L^{SE} \in \left[\frac{1-2\times\frac{3}{5}+\frac{1}{2}}{2}, \frac{1+2\times\frac{3}{5}-4\times\frac{2}{5}+\frac{1}{2}}{2}\right]$$

$$= \left[\frac{3}{20}, \frac{11}{20}\right].$$

Exercise #4.10: Hybrid Competition – Leader Choosing Output and Follower Choosing Price^B

4.10. Consider two firms producing differentiated goods at zero marginal costs. Every firm $i \in \{1, 2\}$ faces inverse demand function

$$p_i\left(q_i, q_j\right) = 1 - q_i - \beta q_j,$$

where parameter β satisfies $-1 < \beta < 0$ when goods are complements, $0 < \beta \leq 1$ when goods are substitutes, and $\beta = 0$ when goods are independent of each other. The game proceeds as follows: first, firm 1 (leader) chooses output q_1. Observing this output, firm 2 (follower) responds choosing its price p_2. Intuitively, this model may describe industries where the leader has a limited capacity (it can produce up to q_1 units), while the follower can serve any number of customers without facing a limiting capacity.

(a) *Second stage.* Find the follower's best response function, $p_2(q_1)$.

- The inverse demand function of firm 2 is given by

$$p_2(q_2) = 1 - q_2 - \beta q_1.$$

Rearranging and solving for q_2, we obtain the demand function, as follows:

$$q_2(p_2) = 1 - p_2 - \beta q_1.$$

- Firm 2, observing q_1, chooses p_2 to solve the following profit maximization problem:

$$\max_{p_2 \geq 0} \pi_2(p_2) = p_2\overbrace{(1 - p_2 - \beta q_1)}^{q_2(p_2)}.$$

Differentiating with respect to p_2, and assuming interior solutions,

$$1 - 2p_2 - \beta q_1 = 0.$$

Rearranging and solving for p_2, firm 2's best response function becomes

$$p_2(q_1) = \frac{1}{2} - \frac{\beta}{2} q_1,$$

which originates at an intercept of $\frac{1}{2}$. When goods are substitutes (complements), parameter β satisfies $\beta > 0$ ($\beta < 0$), implying that an increase in firm 1's output causes firm 2 to respond decreasing (increasing) its price at a rate of $\frac{\beta}{2}$. When goods are independent, $\beta = 0$, firm 2 monopolizes the market for its good so its price simplifies to $p_2 = \frac{1}{2}$.

(b) *First stage.* Before finding firm 1's profit maximizing output level in the first stage, q_1, let us find the inverse demand function that firm 1 faces. Show that this inverse demand function can be expressed in the following form:

$$p_1(q_1) = 1 - \beta(1 - p_2) - \left(1 - \beta^2\right) q_1,$$

where price p_2 is given by firm 2's best response function, as found in part (a):

- From the inverse demand function, we obtain

$$p_1 = 1 - q_1 - \beta q_2$$
$$p_2 = 1 - q_2 - \beta q_1,$$

which we can rearrange to obtain

$$q_1 + \beta q_2 = 1 - p_1$$
$$\beta q_1 + q_2 = 1 - p_2.$$

Using Cramer's rule, we find

$$q_1 = \frac{\begin{bmatrix} 1 - p_1 & \beta \\ 1 - p_2 & 1 \end{bmatrix}}{\begin{bmatrix} 1 & \beta \\ \beta & 1 \end{bmatrix}} = \frac{1 - \beta + \beta p_2 - p_1}{1 - \beta^2}$$
$$= \frac{1}{1+\beta} + \frac{\beta}{1-\beta^2} p_2 - \frac{1}{1-\beta^2} p_1.$$

Rearranging and solving for p_1 yield

$$p_1(q_1) = 1 - \beta(1 - p_2) - \left(1 - \beta^2\right) q_1.$$

Finally, inserting firm 2's best response function in $p_2(q_1) = \frac{1}{2} - \frac{\beta}{2} q_1$, the above expression simplifies to

$$p_1(q_1) = 1 - \beta\left(1 - \overbrace{\left(\frac{1}{2} - \frac{\beta}{2} q_1\right)}^{p_2(q_1)}\right) - \left(1 - \beta^2\right) q_1$$
$$= 1 - \beta + \frac{\beta}{2}(1 - \beta q_1) - \left(1 - \beta^2\right) q_1$$
$$= \frac{(2 - \beta) - (2 - \beta^2) q_1}{2}.$$

(c) Set up firm 1's profit maximization problem and find its optimal output, q_1^*.
- In this context, firm 1 chooses q_1 to solve the following profit maximization problem:

$$\max_{q_1 \geq 0} \pi_1(q_1) = \overbrace{\left[\frac{(2-\beta) - (2-\beta^2)\, q_1}{2} \right]}^{p_1(q_1)} q_1.$$

Differentiating with respect to q_1 and assuming interior solutions,

$$2 - \beta - 2\left(2 - \beta^2\right) q_1 = 0.$$

Rearranging and solving for q_1, the optimal output of firm 1 becomes

$$q_1^* = \frac{2-\beta}{2\left(2-\beta^2\right)},$$

which satisfies

$$\frac{\partial q_1^*}{\partial \beta} = -\frac{\beta^2 - 4\beta + 2}{2\left(2-\beta^2\right)^2},$$

which is positive (negative) if $-1 < \beta \leq 2 - \sqrt{2}$ ($2 - \sqrt{2} < \beta \leq 1$), meaning that if goods are close substitutes to each other (where $\beta > 2 - \sqrt{2} \simeq 0.59$), then the leader reduces output if it faces fiercer competition from its follower producing similar products.

(d) *Equilibrium outcomes.* Plug q_1^* into the best response function in part (a) to find the equilibrium price of firm 2, p_2^*. Then, evaluate the equilibrium profits of firm 1, π_1^*, and of firm 2, π_2^*.
- Inserting q_1^* into $p_2(q_1)$ in part (a), we obtain

$$\begin{aligned} p_2^* &= \frac{1}{2} - \frac{\beta}{2} \frac{2-\beta}{2\left(2-\beta^2\right)} \\ &= \frac{4 - \beta(2+\beta)}{4\left(2-\beta^2\right)}. \end{aligned}$$

- Substituting q_1^* into $\pi_1(q_1)$ in part (b), we find

$$\begin{aligned} \pi_1^* &= \frac{1}{2} \overbrace{\frac{2-\beta}{2\left(2-\beta^2\right)}}^{q_1^*} \left[(2-\beta) - \left(2-\beta^2\right) \overbrace{\frac{2-\beta}{2\left(2-\beta^2\right)}}^{q_1^*} \right] \\ &= \frac{(2-\beta)^2}{8\left(2-\beta^2\right)}. \end{aligned}$$

- Similarly, inserting q_1^* and p_2^* into $\pi_2(p_2)$, we have

$$\pi_2^* = \overbrace{\frac{4-2\beta-\beta^2}{4(2-\beta^2)}}^{p_2^*}\left(1-\overbrace{\frac{4-2\beta-\beta^2}{4(2-\beta^2)}}^{p_2^*}-\beta\overbrace{\frac{2-\beta}{2(2-\beta^2)}}^{q_1^*}\right)$$

$$= \left(\frac{4-2\beta-\beta^2}{4(2-\beta^2)}\right)^2.$$

(e) *Comparative statics.* How do equilibrium profits in part (c) change with β? Interpret:

- Differentiating π_1^* with respect to β, we obtain

$$\frac{\partial \pi_1^*}{\partial \beta} = \frac{-(2-\beta)\left[2-\beta^2-\beta(2-\beta)\right]}{4(2-\beta^2)^2}$$

$$= -\frac{(1-\beta)(2-\beta)}{2(2-\beta^2)^2},$$

which is negative for all $\beta \in (-1, 1)$. Profits are maximized when $\beta = 1$, and the goods are independent of each other.

- Differentiating π_2^* with respect to β, we find

$$\frac{\partial \pi_2^*}{\partial \beta} = -\frac{(4-2\beta-\beta^2)\left[(1+\beta)(2-\beta^2)-\beta(4-2\beta-\beta^2)\right]}{4(2-\beta^2)^3}$$

$$= -\frac{(2-2\beta+\beta^2)(4-2\beta-\beta^2)}{4(2-\beta^2)^3},$$

which is also negative for all $\beta \in (-1, 1]$.

- Intuitively, when β increases, goods become more homogeneous (when $\beta \in (0, 1]$) or less complementary (when $\beta \in (-1, 0]$), so output competition becomes more intense, reducing both the leader's and the follower's equilibrium profits.

(f) *Comparison.* Compare equilibrium profits in part (c). For what values of β does first-mover advantage exist? Explain your results:

- Firm 1 enjoys first-mover advantage if and only if $\pi_1^* > \pi_2^*$, that is,

$$\frac{(2-\beta)^2}{8(2-\beta^2)} > \left(\frac{4-2\beta-\beta^2}{4(2-\beta^2)}\right)^2,$$

which simplifies to

$$2(2-\beta)^2\left(2-\beta^2\right) > \left(4-2\beta-\beta^2\right)^2.$$

Solving for β, the above inequality holds for all

$$0 < \beta < \frac{4}{3}.$$

Since parameter β satisfies $\beta \in (-1, 1]$, we conclude that three cases can arise:

- When $\beta > 0$, we find that $\pi_1^* > \pi_2^*$. Intuitively, when goods are substitutes, firm 1 (the leader) enjoys first-mover advantage in setting the output to restrict the residual demand left over to firm 2.
- When $\beta < 0$, we obtain that $\pi_1^* < \pi_2^*$. In this case, goods are complements, and firm 2 (the follower) enjoys second-mover advantage in setting its price to soften the competition with firm 1.
- When $\beta = 0$, we have that $\pi_1^* = \pi_2^*$. In this setting, products are unrelated with each other (that is, neither substitutes nor complements), so that every firm monopolizes its market and earns the same profits regardless of the order of play.

(g) *Numerical example.* Evaluate equilibrium output of firm 1, price of firm 2, and every firm's equilibrium profits in parts (b) and (c) assuming (i) $\beta = 1$, (ii) $\beta = 0$, and (iii) $\beta = -\frac{1}{2}$. Compare:

(i) When $\beta = 1$, we obtain the following results:

- Equilibrium output of firm 1 is

$$q_1^* = \frac{2-1}{2\left(2-1^2\right)} = \frac{1}{2}$$

yielding equilibrium profits of

$$\pi_1^* = \frac{(2-1)^2}{8\left(2-1^2\right)} = \frac{1}{8}.$$

- Equilibrium price of firm 2 is

$$p_2^* = \frac{4-2\times 1-1^2}{4\left(2-1^2\right)} = \frac{1}{4}$$

yielding equilibrium profits of

$$\pi_2^* = \left(\frac{4-2\times 1-1^2}{4\left(2-1^2\right)}\right)^2 = \frac{1}{16}.$$

- When goods are completely homogeneous ($\beta = 1$), firm 1 produces twice as much output as and earns double profits than firm 2. Our results, thus, coincide with those in a standard model of Stackelberg competition, since firm 2 monopolizes the residual market and obtains the same outcomes whether it sets price or output.

(ii) When $\beta = 0$, we obtain the following results:

- Equilibrium output of firm 1 is

$$q_1^* = \frac{2-0}{2\left(2-0^2\right)} = \frac{1}{2}$$

yielding equilibrium profits of

$$\pi_1^* = \frac{(2-0)^2}{8\left(2-0^2\right)} = \frac{1}{4}.$$

- Equilibrium price of firm 2 is

$$p_2^* = \frac{4 - 2 \times 0 - 0^2}{4\left(2 - 0^2\right)} = \frac{1}{2}$$

yielding equilibrium profits of

$$\pi_2^* = \left(\frac{4 - 2 \times 0 - 0^2}{4\left(2 - 0^2\right)}\right)^2 = \frac{1}{4}.$$

- When goods are independent of each other ($\beta = 0$), every firm monopolizes its product market, producing the same units of output and earning the same profits.

(iii) When $\beta = -\frac{1}{2}$, we obtain the following results:

- Equilibrium output of firm 1 is

$$q_1^* = \frac{2 + \frac{1}{2}}{2\left(2 - \frac{1}{2^2}\right)} = \frac{5}{7}$$

yielding equilibrium profits of

$$\pi_1^* = \frac{\left(2 + \frac{1}{2}\right)^2}{8\left(2 - \frac{1}{2^2}\right)} = \frac{25}{56}.$$

- Equilibrium price of firm 2 is

$$p_2^* = \frac{4 + 2 \times \frac{1}{2} - \frac{1}{2^2}}{4\left(2 - \frac{1}{2^2}\right)} = \frac{19}{28}$$

yielding equilibrium profits of

$$\pi_2^* = \left(\frac{4 + 2 \times \frac{1}{2} - \frac{1}{2}^2}{4\left(2 - \frac{1}{2^2}\right)}\right)^2 = \frac{361}{784}.$$

- When goods are complements ($\beta = -\frac{1}{2}$), every firm faces a larger demand when the other firm expands production (graphically, demand curve experiences an upward shift in its rivals' output). This allows every firm to produce more units and earn higher profits than when goods are independent of each other.

Exercise #4.11: Stackelberg Competition with m Leaders, Based on Huck et al. (2001)^C

4.11. Consider a market with n firms, whose inverse demand function is

$$P(Q) = 1 - Q,$$

where $Q = \sum_{i=1}^{N} q_i$ represents the total output and q_i is firm i's individual output, where $i \in \{1, \ldots, N\}$. We model the game as Stackelberg, in which m firms, where $m < n$, are leaders that independently and simultaneously choose their output levels. The remaining $n - m$ firms, having observed the output levels of m leaders, are followers in making their own output decisions. For simplicity, assume zero production costs:

(a) Set up the profit maximization problem for the leader and follower.

- Let $\overline{Q}^l$ be the total output of m leaders, and Q_{-i}^f be the total output of other $n - m - 1$ followers. The follower chooses q_i^f to solve the following profit maximization problem:

$$\max_{q_i^f \geq 0} \pi^f \left(q_i^f\right) = \left(1 - \overline{Q}^l - Q_{-i}^f - q_i^f\right) q_i^f.$$

- Let Q_{-i}^l be the total output of other $m-1$ leaders, and $Q^f \left(q_i^l + Q_{-i}^l\right)$ be the total output of all $n - m$ followers. The leader chooses q_i^l to solve the following profit maximization problem:

$$\max_{q_i^l \geq 0} \pi^l \left(q_i^l\right) = \left(1 - Q_{-i}^l - Q^f \left(q_i^l + Q_{-i}^l\right) - q_i^l\right) q_i^l.$$

(b) Find the equilibrium output and associated profit of the leader and follower:

- *Second Stage – Followers.* Differentiating the follower's profit with respect to q_i^f, we find

$$\frac{\partial \pi^f \left(q_i^f\right)}{\partial q_i^f} = 1 - \overline{Q}^l - Q_{-i}^f - 2q_i^f = 0$$

that is rearranged to yield

$$q_i^f = \frac{1 - \overline{Q}^l - Q_{-i}^f}{2}.$$

Since followers are symmetric, we set the aggregate output from all i's rivals $Q_{-i}^f = (n - m - 1) q_i^f$ to obtain

$$q_i^f = \frac{1 - \overline{Q}^l - (n - m - 1) q_i^f}{2}.$$

Rearranging, we obtain the best response function of each follower:

$$q_i^f \left(\overline{Q}^l\right) = \frac{1 - \overline{Q}^l}{n - m + 1},$$

which has an intercept of $\frac{1}{n-m+1}$ and a slope of $-\frac{1}{n-m+1}$, meaning that as the total number of firms n increases or the number of leaders m decreases, firm i responds less sensitively to the total output of all leaders $\overline{Q}^l$.

- *First stage (leaders).* By backward induction, we substitute the aggregate output from all followers, $Q^f\left(q_i^l + Q_{-i}^l\right) = (n-m)\,q_i^f\left(q_i^l + Q_{-i}^l\right)$ into the leader's profit function. Therefore, the leader chooses q_i^l to solve

$$\max_{q_i^l \geq 0} \pi^l\left(q_i^l\right) = \left(1 - Q_{-i}^l - \frac{n-m}{n-m+1}\left(1 - q_i^l - Q_{-i}^l\right) - q_i^l\right) q_i^l$$

$$= \frac{1}{n-m+1}\left(1 - Q_{-i}^l - q_i^l\right) q_i^l.$$

Differentiating with respect to q_i^l, we obtain

$$\frac{\partial \pi^l\left(q_i^l\right)}{\partial q_i^l} = 1 - Q_{-i}^l - 2q_i^l = 0.$$

Since leaders are symmetric, we set $Q_{-i}^l = (m-1)\,q_i^l$ to obtain

$$q_i^l = \frac{1}{m+1},$$

which is decreasing in the number of leaders, m.

Substituting $\overline{Q}^l = mq_i^l = \frac{m}{m+1}$ into the best response function of each follower,

$$q_i^f = \frac{1 - m\overbrace{\frac{1}{m+1}}^{q_i^l}}{n-m+1} = \frac{1}{(m+1)(n-m+1)}.$$

Further differentiating q_i^f with respect to n and m, we find

$$\frac{\partial q_i^f}{\partial n} = -\frac{1}{(m+1)(n-m+1)^2} < 0$$

$$\frac{\partial q_i^f}{\partial m} = -\frac{n-m+1-m-1}{(m+1)^2(n-m+1)^2}$$

$$= \frac{2m-n}{(m+1)^2(n-m+1)^2}$$

so that when the number of followers comprises the majority of firms, where $m > \frac{n}{2}$, the follower's equilibrium output further increases in the proportion of followers in the market (holding the total number of firms n constant).

(c) Evaluate the equilibrium outputs you found in part (b) at $m = 1$ and $n = 2$, so we have only one leader and one follower. Show that you obtain the standard results in the Stackelberg model of quantity competition with only one leader and one follower (as in Exercise 4.1):

- Evaluating the leader's output at $m = 1$, we obtain $q_i^l = \frac{1}{1+1} = \frac{1}{2}$, which coincides with the standard production level under Stackelberg (with no marginal costs), $\frac{1-0}{2} = \frac{1}{2}$.

- Similarly, evaluating the follower's output at $m = 1$ and $n = 2$, we find

$$q_i^f = \frac{1}{(1+1)(2-1+1)} = \frac{1}{4},$$

which coincides with the follower's equilibrium output in a standard Stackelberg game, $\frac{1-0}{4} = \frac{1}{4}$. (See Exercise 4.1, at the beginning of this chapter, for more details.)

(d) Find the equilibrium profits for each leader and each follower:
- Substituting $q_i^l = \frac{1}{m+1}$ and $q_i^f = \frac{1}{(m+1)(n-m+1)}$ into the profit functions, we obtain

$$\pi^l(n,m) = \frac{1}{n-m+1}\left(1 - \frac{m}{m+1}\right)\frac{1}{m+1}$$
$$= \frac{1}{(m+1)^2(n-m+1)}$$
$$\pi^f(n,m) = \left(1 - \frac{m}{m+1} - \frac{n-m}{(m+1)(n-m+1)}\right)\frac{1}{(m+1)(n-m+1)}$$
$$= \frac{1}{(m+1)^2(n-m+1)^2}.$$

(e) How do profits change in the total number of firms n and the number of leaders m?
- Differentiating each leader's profit, $\pi^l(n,m)$, with respect to n and m, we find

$$\frac{\partial \pi^l(n,m)}{\partial n} = -\frac{1}{(m+1)^2(n-m+1)^2} < 0$$
$$\frac{\partial \pi^l(n,m)}{\partial m} = -\frac{2(m+1)(n-m+1) - (m+1)^2}{(m+1)^4(n-m+1)^2}$$
$$= \frac{3m-2n-1}{(m+1)^3(n-m+1)^2}.$$

And differentiating the follower's profit, $\pi^f(n,m)$, with respect to n and m, yields

$$\frac{\partial \pi^f(n,m)}{\partial n} = -\frac{2}{(m+1)^2(n-m+1)^3} < 0$$
$$\frac{\partial \pi^f(n,m)}{\partial m} = \frac{-2\left[(m+1)(n-m+1)^2 - (m+1)^2(n-m+1)\right]}{(m+1)^4(n-m+1)^4}$$
$$= \frac{2[m+1-n+m-1]}{(m+1)^3(n-m+1)^3}$$
$$= \frac{2(2m-n)}{(m+1)^3(n-m+1)^3}.$$

Intuition:
- *More firms.* Equilibrium profits of the leader and follower both unambiguously decrease in the total number of firms n, meaning that both types of firms earn lower profit if the market becomes more competitive.

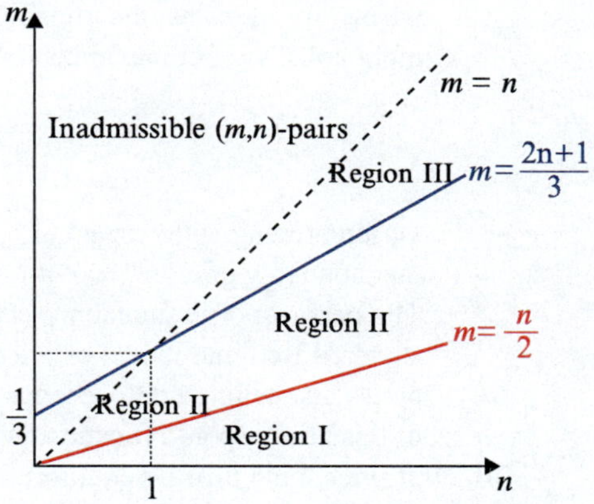

Fig. 4.3 How profits are affected by an increase in the number of followers

- *More leaders.* However, when fixing the total number of n firms, we increase the number of leaders m (which reduces the number of followers at the same time), and we obtain mixed effects on the leader's and follower's profit. For illustration purposes, Fig. 4.3 depicts these three regions in the (m, n)-quadrant where, since $m < n$ by assumption, we only focus on (m, n)-pairs below the 45^0-line:
  - *Region I.* When $m \leq \frac{n}{2}$, both leader's and follower's profit decreases in m.
  - *Region II.* When $\frac{n}{2} < m \leq \frac{2n+1}{3}$, follower's profit increases, but leader's profit decreases in m.
  - *Region III.* When $m > \frac{2n+1}{3}$, both leader's and follower's profit increases in m. Intuitively, when the proportion of followers increases, the leaders respond individually increasing (decreasing) their output, when followers represent a relatively large (small) proportion of the industry, leading to an increase (decrease) in each leader's profit.

Exercise #4.12: Sequential Price Competition with Homogeneous Goods^A

4.12. Consider an industry with two firms competing in prices and selling a homogeneous good. The demand function of every firm i is

$$q_i\left(p_i, p_j\right) = 1 - p_i + p_j.$$

Every firm i faces a constant marginal cost of c, where $1 > c > 0$. Firm 1 is the industry leader, selecting its price p_1 in the first stage of the game. In the second stage, firm 2 observes price p_1 and responds with its own price p_2.

(a) *Second stage.* Find firm 2's best response function $p_2(p_1)$.

- The follower observes the leader's price p_1, thus taking it as given, and responds choosing p_2 to maximize its profits. It is straightforward to characterize its best response function as follows:

$$p_2\left(p_1\right) = \begin{cases} p^m & \text{if } p_1 > p^m \\ p_1 - \varepsilon & \text{if } c < p_1 \leq p^m \\ c & \text{if } p_1 \leq c, \end{cases}$$

where p^m denotes the monopoly price, and it is defined as the solution to the monopolist's profit maximization problem as follows:

$$\max_{p_i > 0} \; (p_i - c)(1 - p_i).$$

Differentiating with respect to p_i yields $1 - 2p_i + c = 0$, and solving for p_i, we obtain the monopoly price $p^m = \frac{1+c}{2}$.

- The best response function $p_2(p_1)$ has the same interpretation as that of firm i in the standard Bertrand model of price competition with firms simultaneously selecting their prices and selling homogeneous goods. We refer the reader to Exercise 3.1 for a figure of this best response function and more details on its interpretation.

(b) *First stage.* Find firm 1's equilibrium price, p_1^*:

- Anticipating the follower's best response function $p_2(p_1)$, the leader chooses its price p_1. In this setting, the leader expects that its price will be:
  - Undercut by the follower when $p_1 > c$ (thus yielding no sales for the leader), or
  - Mimicked by the follower when $p_1 = c$ (thus entailing half of the market share for the leader since both firms set the same price)

  We did not mention the case when $p_1 < c$, where the follower responds setting a price $p_2 = c$ and all sales go to the leader but making a loss per unit. Therefore, the leader has (weak) incentives to set a price $p_1 = c$, implying that the subgame perfect equilibrium of the game is

$$(p_1^*, p_2^*) = (c, p_2(p_1)),$$

which yields equilibrium prices equal to c for both leader and follower, i.e., $p_1^* = p_2(p_1^*) = c$.

(c) *Comparison against simultaneous price competition.* Compare your equilibrium results with those in the Bertrand model where firms choose their prices simultaneously (Exercise 3.1):

- The equilibrium price pair in the sequential version of the game coincides with that in the simultaneous version, namely, both firms set their price equal to the marginal production cost c.

Exercise #4.13: Sequential Price Competition with Heterogeneous Goods^B

4.13. Consider an industry with two firms competing in prices and selling heterogeneous goods. The demand function of every firm i is

$$q_i(p_i, p_j) = 1 - \gamma p_i + p_j,$$

where $\gamma \geq 1$ represents the degree of product differentiation (homogeneous when $\gamma = 1$ but differentiated when $\gamma > 1$). Every firm i faces a constant marginal cost of c, where $c > 0$, in producing every unit of the good. Firm 1 is the industry leader, selecting its price p_1 in the first stage of the game. In the second stage, firm 2 observes price p_1 and responds with its own price p_2.

(a) *Second stage.* Find firm 2's best response function $p_2(p_1)$.

- The follower observes the leader's price p_1, thus taking it as given, and responds choosing p_2 to solve the following profit maximization problem:

$$\max_{p_2>0} \pi_2(p_2) = (p_2 - c)\underbrace{(1 - \gamma p_2 + p_1)}_{q_2}.$$

Differentiating with respect to p_2, and assuming interior solutions, we obtain

$$1 - 2\gamma p_2 + p_1 + \gamma c = 0.$$

Solving for p_2, we find firm 2's best response function as follows:

$$p_2(p_1) = \frac{1+\gamma c}{2\gamma} + \frac{1}{2\gamma} p_1,$$

which originates at $\frac{1+\gamma c}{2\gamma}$ and increases in the leader's price, p_1, at a rate of $\frac{1}{2\gamma}$. This best response function is identical to those we found in Chap. 3 (see Exercises 3.4 and 3.5), where firms simultaneously choose their prices, where we found that both the intercept and the slope of the best response function decrease as goods become more differentiated, indicating that competition becomes less intense.

(b) *First stage.* Find firm 1's equilibrium price, p_1^*.

- The leader anticipates the follower's best response function $p_2(p_1)$ and chooses price p_1 to solve

$$\max_{p_1>0} \pi_1(p_1) = (p_1 - c)(1 - \gamma p_1 + p_2(p_1))$$

$$= (p_1 - c)\left(1 - \gamma p_1 + \underbrace{\frac{1+\gamma c}{2\gamma} + \frac{1}{2\gamma} p_1}_{p_2(p_1)}\right)$$

$$= (p_1 - c)\frac{1 + \gamma(2 + c) + p_1(1 - 2\gamma^2)}{2\gamma}.$$

Differentiating with respect to p_1, and assuming interior solutions, we obtain

$$1 + 2\gamma - 2p_1\left(2\gamma^2 - 1\right) - c\left(1 - \gamma - 2\gamma^2\right) = 0,$$

and solving for p_1, we find the leader's equilibrium price

$$p_1^* = \frac{1 + 2\gamma - (1 - \gamma - 2\gamma^2)c}{2(2\gamma^2 - 1)}.$$

Inserting this equilibrium price into the follower's best response function, we find the equilibrium price of this firm as follows:

$$p_2\left(p_1^*\right) = \frac{1+\gamma c}{2\gamma} + \frac{1}{2\gamma}\underbrace{\left(\frac{1+2\gamma-\left(1-\gamma-2\gamma^2\right)c}{2\left(2\gamma^2-1\right)}\right)}_{p_1^*}$$

$$= \frac{4\gamma^2+2\gamma-1+\left(4\gamma^3+2\gamma^2-\gamma-1\right)c}{4\gamma\left(2\gamma^2-1\right)}.$$

(c) *Comparison against simultaneous price competition.* Compare your equilibrium results with those in the Bertrand model where firms choose their prices simultaneously (Exercise 3.4). Which firm sees its price increase in the sequential version of the game?

- In Exercise 3.3, we found that, when firms set their prices simultaneously, equilibrium prices are

$$p_i^{Sim} = \frac{1+\gamma c}{2\gamma-1} \quad \text{for every firm } i.$$

Comparing p_i^{Sim} against the equilibrium price in the sequential-move version of the game, p_1^* (which we can label as p_1^{Seq}, for clarity), we obtain that, for firm 1,

$$p_1^{Seq} - p_1^{Sim} = \frac{1+2\gamma-(1-\gamma-2\gamma^2)c}{2\left(2\gamma^2-1\right)} - \frac{1+\gamma c}{2\gamma-1}$$

$$= \frac{1-c\left(\gamma-1\right)}{2\left(2\gamma-1\right)\left(2\gamma^2-1\right)},$$

which is positive as long as $c < \frac{1}{\gamma-1}$. For firm 2, we find that

$$p_2^{Seq} - p_2^{Sim} = \frac{4\gamma^2+2\gamma-1+\left(4\gamma^3+2\gamma^2-\gamma-1\right)c}{4\gamma\left(2\gamma^2-1\right)} - \frac{1+\gamma c}{2\gamma-1}$$

$$= \frac{1-c\left(\gamma-1\right)}{4\gamma\left(2\gamma-1\right)\left(2\gamma^2-1\right)},$$

which is also positive as long as $c < \frac{1}{\gamma-1}$. Therefore, both firms increase their equilibrium prices in the sequential version of the game, relative to its simultaneous version, when cost satisfies the condition for an interior solution (see part (b) of Exercise 3.4 for details).

(d) *Numerical example.* Find the equilibrium prices and associated profits of the leader and the follower assuming parameter values $\gamma = 2$ and $c = 1/3$.

- The leader's equilibrium price in this context is

$$p_1^* = \frac{1+(2\times 2)+\frac{1}{3}\left((2\times 2^2)+2-1\right)}{2\left((2\times 2^2)-1\right)}$$

$$= \frac{4}{7} \simeq 0.571$$

with an associated profit of

$$\pi_1^* = (p_1^* - c)\left(\frac{1 + \gamma(2 + c) - p_1^*(2\gamma^2 - 1)}{2\gamma}\right)$$

$$= \left(\frac{4}{7} - \frac{1}{3}\right)\left(\frac{1 + 2\left(2 + \frac{1}{3}\right) - \frac{4}{7}(2 \times 2^2 - 1)}{2 \times 2}\right)$$

$$= \frac{25}{252} \simeq 0.099.$$

- The follower's equilibrium price in this context is

$$p_2^* = \frac{(4 \times 2^2) + (2 \times 2) - 1 + \frac{1}{3}\left((4 \times 2^3) + (2 \times 2^2) - 2 - 1\right)}{4 \times 2\left((2 \times 2^2) - 1\right)}$$

$$= \frac{47}{84} \simeq 0.560$$

with an associated profit of

$$\pi_2^* = (p_2^* - c)\left(1 - \gamma p_2^* + p_1^*\right)$$

$$= \left(\frac{47}{84} - \frac{1}{3}\right)\left(1 - 2 \times \frac{47}{84} + \frac{4}{7}\right)$$

$$= \frac{361}{3528} \simeq 0.102.$$

Exercise #4.14: Committing in Prices or Quantities, Based on Singh and Vives (1984)^{B}

4.14. Consider a duopoly, where every firm $i = \{1, 2\}$ faces inverse demand function

$$p_i(q_i, q_j) = 1 - q_i - \beta q_j,$$

where $j \neq i$, and $\beta \in (0, 1]$ denotes product differentiation. As in similar exercises, a higher value of β represents more homogeneous products. Both firms face a constant marginal production cost c, where $0 \leq c < 1$.

Firms compete in the following sequential-move game:

- In the first stage, every firm i independently and simultaneously chooses whether to commit to a price for its product (which we refer to as "commit in prices") or commit to a volume of output (which we denote as "commit in quantities").
- In the second stage, if a firm committed to price (quantity) in the first stage, it responds setting its quantity (price).

In this exercise, we seek to show that, in equilibrium, both firms find committing in quantities being a strictly dominant strategy. In addition, this result is unaffected by parameter values β and *c*. We solve the game operating by backward induction, starting from the last (second) stage:

(a) *Second stage.* Find the equilibrium behavior in the second stage of the game, for each commitment profile arising from the first stage:

- There are four commitment profiles we should consider at the beginning of the second stage of the game (i.e., four subgames): (1) both firms commit to quantities, (q_i, q_j); (2) both commit to prices, (p_i, p_j); (3) only firm 1 commits to quantities, (q_1, p_2); and (4) only firm 2 commits to quantities, (p_1, q_2). Subgames (3) and (4) can be, of course, simultaneously considered by analyzing settings where only firm i commits to quantities, (q_i, p_j) for all $j \neq i$. We separately examine each subgame below.
- *Subgame 1. Both firms commit to quantities.* In this case, every firm i's inverse demand function is $p_i(q_i, q_j) = 1 - q_i - \beta q_j$.
- *Subgame 2. Both firms commit to prices.* In this case, we first need to use the inverse demand function given at the beginning of the exercise, $p_i(q_i, q_j) = 1 - q_i - \beta q_j$, and solve for q_i to obtain

$$q_i = 1 - p_i - \beta q_j.$$

Similarly, solving for q_j in $p_j(q_i, q_j) = 1 - q_j - \beta q_i$ yields

$$q_j = 1 - p_j - \beta q_i.$$

We now seek to simultaneously solve for q_i and q_j in the above two equations. Inserting $q_j = 1 - p_j - \beta q_i$ into $q_i = 1 - p_i - \beta q_j$ yields

$$q_i = 1 - p_i - \beta\overbrace{(1 - p_j - \beta q_i)}^{q_j},$$

which, solving for q_i, entails

$$q_i(p_i, p_j) = \frac{1 - p_i - \beta(1 - p_j)}{1 - \beta^2}.$$

Similarly, we can now insert this result into $q_j = 1 - p_j - \beta q_i$, to obtain

$$\begin{aligned} q_j &= 1 - p_j - \beta\overbrace{\left(\frac{1 - p_i - \beta(1 - p_j)}{1 - \beta^2}\right)}^{q_i} \\ &= \frac{1 - p_j - \beta(1 - p_i)}{1 - \beta^2}. \end{aligned}$$

Subgames 3–4. Only one firm commits to quantity. If firm i commits to prices, p_i, while its rival commits to quantities, q_j, then firm i's inverse demand is $p_i(q_i, q_j) = 1 - q_i - \beta q_j$ or, after solving for q_i, $q_i = 1 - p_i - \beta q_j$. The demand for the firm committing to quantities (firm j) is $q_j = \frac{1-p_j-\beta(1-p_i)}{1-\beta^2}$, as shown in the previous part of this exercise. After solving for p_j, we obtain that

$$p_j = 1 - \beta(1 - p_i) - (1 - \beta^2)q_j.$$

(b) *First stage.* Identify the equilibrium behavior in the first stage of the game, and show that committing to quantities is a weakly dominant strategy for both firms:

- Conditional on firm j committing to quantities, we need to separately evaluate firm i's profits from responding committing to quantities, $\pi_i^{q,q}$, and committing to prices, $\pi_i^{p,q}$. Similarly, conditional on firm j committing to prices, we will find firm i's profits when it commits to quantities, $\pi_i^{q,p}$, and when it commits to prices, $\pi_i^{p,p}$. As a consequence, we first need to evaluate equilibrium profits in all four strategy profiles, to ultimately solve the following 2×2 matrix, where every firm simultaneously chooses whether to commit in prices or quantities in the first period.

		Firm 2	
		Quantities	Prices
Firm 1	Quantities	$\pi_1^{q,q}, \pi_2^{q,q}$	$\pi_1^{q,p}, \pi_2^{q,p}$
	Prices	$\pi_1^{p,q}, \pi_2^{p,q}$	$\pi_1^{p,p}, \pi_2^{p,p}$

- *Both firms commit to quantities.* If both firms commit to quantities, every firm i takes the inverse demand function $p_i(q_i, q_j) = 1 - q_i - \beta q_j$ as given and solves

$$\max_{q_i \geq 0} \ (1 - q_i - \beta q_j)\, q_i - c q_i.$$

Differentiating with respect to q_i, and assuming interior solution, yields

$$\frac{\partial \pi_i^{q,q}}{\partial q_i} = 1 - 2q_i - \beta q_j - c = 0,$$

and solving for q_i, we obtain the following best response function

$$q_i(q_j) = \frac{1-c}{2} - \frac{\beta}{2} q_j.$$

In a symmetric equilibrium, both firms produce the same output level, $q_i = q_j = q$, implying that firm i's best response function simplifies to $q = \frac{1-c}{2} - \frac{\beta}{2} q$. After solving for q, equilibrium output becomes

$$q^{q,q} = \frac{1-c}{2+\beta}.$$

Therefore, equilibrium profits when both firms commit to quantities are

$$\begin{aligned}
\pi_i^{q,q} &= \left(1 - q^{q,q} - \beta q^{q,q}\right) q^{q,q} - c q^{q,q} \\
&= \left(1 - \frac{1-c}{2+\beta} - \beta \frac{1-c}{2+\beta}\right) \frac{1-c}{2+\beta} - c \frac{1-c}{2+\beta} \\
&= \left(\frac{1-c}{2+\beta}\right)^2.
\end{aligned}$$

- *Both firms commit to prices.* In this case, every firm i takes the direct demand function $q_i(p_i, p_j) = \frac{1-p_i-\beta(1-p_j)}{1-\beta^2}$ found in part (a) as given and chooses the price p_i that solves

$$\max_{p_i \geq c} \ (p_i - c)\left(\frac{1 - p_i - \beta(1 - p_j)}{1 - \beta^2}\right).$$

Differentiating with respect to p_i, and assuming interior solution, yields

$$\frac{1 - 2p_i - \beta(1 - p_j) + c}{1 - \beta^2} = 0,$$

and solving for p_i, we obtain the following best response function

$$p_i(p_j) = \frac{1 + c}{2} - \frac{\beta}{2}(1 - p_j).$$

In a symmetric equilibrium, both firms set the same price, $p_i = p_j = p$, which helps us rewrite firm i's best response function as follows, $p = \frac{1+c}{2} - \frac{\beta}{2}(1 - p)$. After solving for p, equilibrium price becomes

$$p^{p,p} = \frac{1 + c - \beta}{2 - \beta}.$$

Therefore, equilibrium profits when both firms commit to prices are

$$\pi_i^{p,p} = \left(p^{p,p} - c\right)\left(\frac{1 - p^{p,p} - \beta(1 - p^{p,p})}{1 - \beta^2}\right)$$
$$= \frac{(1 - \beta)(1 - c)^2}{(2 - \beta)^2(1 + \beta)}.$$

- *Only one firm commits to prices.* If firm i commits to prices, while j commits to quantities (often known as "hybrid competition," since every firm has a different choice variable), we know from part (a) that firm i solves

$$\max_{p_i \geq c} \ (p_i - c)\,(1 - p_i - \beta q_j).$$

Differentiating with respect to p_i, and assuming interior solution, yields

$$1 - 2p_i - \beta q_j + c = 0,$$

which entails the following best response function

$$p_i(q_j) = \frac{1 + c}{2} - \frac{\beta}{2}q_j.$$

In contrast, firm j takes the inverse demand function $p_j(p_i, q_j) = 1 - \beta(1 - p_i) - (1 - \beta^2)q_j$ as given and solves

$$\max_{q_j \geq 0} \left[1 - \beta(1 - p_i) - (1 - \beta^2)q_j\right] q_j - c q_j.$$

Differentiating with respect to q_j, and assuming interior solution, yields

$$1 - \beta(1 - p_i) - 2(1 - \beta^2)q_j - c = 0,$$

which, solving for q_j, entails the following best response function

$$q_j(p_i) = \frac{1 - c - \beta}{2(1 - \beta^2)} + \frac{\beta}{2(1 - \beta^2)} p_i.$$

Inserting the best response function $p_i(q_j)$ into $q_j(p_i)$ yields

$$q_j = \frac{1 - c - \beta}{2(1 - \beta^2)} + \frac{\beta(1 + c)}{4(1 - \beta^2)} - \frac{\beta^2}{4(1 - \beta^2)} q_j,$$

which, after solving for q_j, we obtain an equilibrium output

$$q_j^{p,q} = \frac{(2 - \beta)(1 - c)}{4 - 3\beta^2}.$$

Inserting this result into the best response function $p_i(q_j)$ yields firm i's equilibrium price, as follows:

$$\begin{aligned} p_i^{p,q} &= p_i(q_j^{p,q}) \\ &= \frac{1 + c}{2} - \frac{\beta}{2} \overbrace{\left(\frac{(2 - \beta)(1 - c)}{4 - 3\beta^2} \right)}^{q_j^{p,q}} \\ &= \frac{(2 + \beta)(1 - \beta) + (2 + \beta - 2\beta^2)c}{4 - 3\beta^2}. \end{aligned}$$

We still need to find $p_j^{p,q}$ and $q_i^{p,q}$. To do that, we insert $p_i^{p,q}$ and $q_j^{p,q}$ into firm i's inverse demand function $p_i(q_i, q_j) = 1 - q_i - \beta q_j$, to obtain

$$p_i^{p,q} = 1 - q_i - \beta q_j^{p,q},$$

which entails

$$\frac{(2 + \beta)(1 - \beta) + (2 + \beta - 2\beta^2)c}{4 - 3\beta^2} = 1 - q_i - \beta \left(\frac{(2 - \beta)(1 - c)}{4 - 3\beta^2} \right).$$

Rearranging yields

$$q_i = 1 - \frac{(2 + \beta)(1 - \beta) + (2 + \beta - 2\beta^2)c}{4 - 3\beta^2} - \beta \left(\frac{(2 - \beta)(1 - c)}{4 - 3\beta^2} \right),$$

and solving for q_i, we find firm i's equilibrium output in this hybrid setting, that is,

$$q_i^{p,q} = \frac{(2 + \beta)(1 - \beta)(1 - c)}{4 - 3\beta^2}.$$

Finally, to find $p_j^{p,q}$, we just need to insert $q_j^{p,q}$ and $q_i^{p,q}$ in firm j's inverse demand function $p_j(q_i, q_j) = 1 - q_j - \beta q_i$, to obtain

$$\begin{aligned} p_j^{p,q} &= p_j(q_i^{p,q}, q_j^{p,q}) = 1 - q_j^{p,q} - \beta q_i^{p,q} \\ &= 1 - \overbrace{\left(\frac{(2-\beta)(1-c)}{4-3\beta^2}\right)}^{q_j^{p,q}} - \beta \overbrace{\left(\frac{(\beta+2)(1-\beta)(1-c)}{4-3\beta^2}\right)}^{q_i^{p,q}} \\ &= \frac{4-3\beta^2-(2+\beta-\beta^2-\beta^3)(1-c)}{4-3\beta^2}. \end{aligned}$$

Therefore, firm i's equilibrium profits in this hybrid setting are

$$\begin{aligned} \pi_i^{p,q} &= (p_i^{p,q} - c)\, q_i^{p,q} \\ &= \left(\frac{(2+\beta)(1-\beta)(1-c)}{4-3\beta^2}\right)\left(\frac{(2+\beta)(1-\beta)(1-c)}{4-3\beta^2}\right) \\ &= \left(\frac{(2+\beta)(1-\beta)(1-c)}{4-3\beta^2}\right)^2 \end{aligned}$$

recalling that firm i commits to prices, while j commits to quantities. Similarly, firm j's equilibrium profits are

$$\begin{aligned} \pi_j^{p,q} &= \left(p_j^{p,q} - c\right) q_j^{p,q} \\ &= \left(\frac{(1+\beta)(2-\beta)(1-\beta)(1-c)}{4-3\beta^2}\right)\left(\frac{(2-\beta)(1-c)}{4-3\beta^2}\right) \\ &= \frac{(1+\beta)(1-\beta)(2-\beta)^2(1-c)^2}{(4-3\beta^2)^2}. \end{aligned}$$

- We can now insert all the profits we found into the payoff matrix, $\left(\pi_i^{q,q}, \pi_j^{q,q}\right)$, $\left(\pi_i^{p,p}, \pi_j^{p,p}\right)$, $\left(\pi_i^{p,q}, \pi_j^{p,q}\right)$, and $\left(\pi_i^{q,p}, \pi_j^{q,p}\right)$, obtaining the following payoff matrix:

Firm 2

		Quantities	Prices
Firm 1	Quantities	$\left(\frac{1-c}{2+\beta}\right)^2, \left(\frac{1-c}{2+\beta}\right)^2$	$\frac{(1+\beta)(1-\beta)(2-\beta)^2(1-c)^2}{(4-3\beta^2)^2}, \left(\frac{(2+\beta)(1-\beta)(1-c)}{4-3\beta^2}\right)^2$
	Prices	$\left(\frac{(2+\beta)(1-\beta)(1-c)}{4-3\beta^2}\right)^2, \frac{(1+\beta)(1-\beta)(2-\beta)^2(1-c)^2}{(4-3\beta^2)^2}$	$\frac{(1-\beta)(1-c)^2}{(2-\beta)^2(1+\beta)}, \frac{(1-\beta)(1-c)^2}{(2-\beta)^2(1+\beta)}$

When firm 2 chooses quantities, in the left column, firm 1 earns higher profits responding with quantities since

$$\pi_1^{q,q} = \left(\frac{1-c}{2+\beta}\right)^2 > \left(\frac{(\beta+2)(1-\beta)(1-c)}{4-3\beta^2}\right)^2 = \pi_1^{p,q},$$

which, simplifying, yields

$$4 - 3\beta^2 > 4 - 3\beta^2 - \beta^3$$

or $\beta^3 > 0$, which holds for all admissible values of $\beta \in (0, 1]$. Similarly, when firm 2 chooses prices, in the right column, firm 1 earns higher profits responding with quantities because

$$\pi_1^{q,p} = \frac{(1+\beta)(1-\beta)(2-\beta)^2(1-c)^2}{(4-3\beta^2)^2} > \frac{(1-\beta)(1-c)^2}{(2-\beta)^2(1+\beta)} = \pi_1^{p,p},$$

which, simplifying (note that the $1-\beta$ term cannot be canceled out when $\beta = 1$), yields

$$(1+\beta)(1-\beta)(2-\beta)^2 > (1-\beta)(4-3\beta^2)$$

or $(1-\beta)\beta^3 > 0$, which holds for all $\beta \in (0, 1)$.

- Therefore, as long as $\beta < 1$, then regardless of what firm 2 chooses in the first stage of the game, firm 1 earns a strictly higher profit committing to quantities than to prices; and this result is unaffected by the parameter value c. In summary, this means that firm 1 finds committing in quantities to be a weakly dominant strategy (and strictly dominant if $\beta < 1$). Since profits are symmetric across firms, a similar argument applies to firm 2.

(c) *Comparative statics.* How are equilibrium profits affected by parameter β? Evaluate equilibrium profits at $\beta \to 0$ and at $\beta = 1$. Interpret:

- In equilibrium, every firm i's profits are

$$\pi_i^{q,q} = \left(\frac{1-c}{2+\beta}\right)^2,$$

which are unambiguously decreasing in β. Intuitively, as consumers regard products to be more homogeneous, output competition intensifies, and firms' profits decrease.

- At $\beta \to 0$, equilibrium profits approach $\left(\frac{1-c}{2}\right)^2$, as under monopoly, since products are completely differentiated. In contrast, at $\beta = 1$, equilibrium profits become $\left(\frac{1-c}{3}\right)^2$, as under Cournot competition with homogeneous products.

Exercise #4.15: Strategic Pre-commitment by Only One Firm Followed by Cournot Competition, Based on Fudenberg and Tirole (1984)^B

4.15. Consider an industry with two symmetric firms facing inverse demand function $p(Q) = 1 - Q$, where $Q = q_1 + q_2$ represents aggregate output. In the first stage, firm 1 chooses its investment in cost-reducing technologies, k, at a cost $\frac{1}{2}k^2$. This investment decreases its marginal production cost from c to $c - \alpha k$, where $\alpha \in [0, 1]$ denotes the effectiveness of the investment. Intuitively, when $\alpha = 0$, the investment is futile, but when $\alpha = 1$ every dollar invested reduces the initial marginal cost of the firm, c, by one dollar. Firm 2 cannot invest in cost-reducing technologies. In the second stage, both firms observe firm 1's investment k and respond competing à la Cournot:

(a) *Second stage.* Find the best response function in the second period, $q_i(q_j)$. [*Hint:* The best response function of each firm will differ.]

- In the second stage, firm 1 solves

$$\max_{q_1 \geq 0} (1 - q_1 - q_2)q_1 - (c - \alpha k)\, q_1,$$

which is evaluated at its second-period cost, $c - \alpha k$, after investing k dollars in cost-reducing technologies. Differentiating with respect to q_1, we obtain

$$1 - 2q_1 - q_2 - c + \alpha k = 0,$$

and solving for q_1, we find firm 1's best response function

$$q_1(q_2) = \frac{1 - (c - \alpha k)}{2} - \frac{q_2}{2}.$$

Graphically, firm 1's best response function shifts upward when this firm invests more in cost-reducing technologies (higher k) or when, for a given investment k, its effectiveness increases (higher α).

- Similarly, firm 2 solves

$$\max_{q_2 \geq 0} (1 - q_1 - q_2)q_2 - cq_2,$$

which is evaluated at its cost c, since this firm cannot invest in cost-reducing technologies. Differentiating with respect to q_2, we obtain

$$1 - q_1 - 2q_2 - c = 0,$$

and solving for q_2, we find firm 2's best response function

$$q_2(q_1) = \frac{1 - c}{2} - \frac{q_1}{2}.$$

(b) Find the second-period output function, $q_i(k)$.

- Inserting firm 2's best response function, $q_2(q_1)$, into firm 1's, $q_1(q_2)$, we find

$$\begin{aligned} q_1 &= \frac{1 - c + \alpha k}{2} - \frac{1}{2}\overbrace{\left(\frac{1 - c}{2} - \frac{q_1}{2}\right)}^{q_2(q_1)} \\ &= \frac{1 - c + \alpha k}{2} - \frac{(1 - c)}{4} + \frac{q_1}{4}, \end{aligned}$$

which, rearranging and solving for q_1, yields the equilibrium output function for firm 1

$$q_1(k) = \frac{1 - (c - 2\alpha k)}{3}.$$

This output level is positive since $c < 1$ by assumption and αk enters positively.

- Inserting this output into firm 2's best response function, $q_2(q_1)$, we obtain

$$q_2 = \frac{1-c}{2} - \frac{1}{2}\overbrace{\left(\frac{1-(c-2\alpha k)}{3}\right)}^{q_1(k)},$$

which, rearranging, yields the equilibrium output function for firm 2 as follows:

$$q_2(k) = \frac{1-c-\alpha k}{3}.$$

This output level is positive as long as $\alpha < \frac{1-c}{k}$. Intuitively, the effectiveness of firm 1's investment cannot be too high. Otherwise, firm 2 would face such a competitive rival (firm 1) that it would prefer to stay inactive. This allows us to express firm 2's output function as follows:

$$q_2(k) = \begin{cases} \frac{1-c-\alpha k}{3} & \text{if } \alpha < \frac{1-c}{k} \\ 0 & \text{otherwise.} \end{cases}$$

- Comparing $q_1(k)$ and $q_2(k)$, we see that $q_1(k) > q_2(k)$ since

$$\frac{1-(c-2\alpha k)}{3} > \frac{1-c-\alpha k}{3}$$

simplifies to $2\alpha k > -\alpha k$, which holds by definition. Intuitively, firm 1 produces more units of output since its second-period marginal costs are lower than firm 2's thanks to its investment in cost-reducing technologies in the first period.

(c) *First stage.* To simplify the analysis, assume that $c = \frac{1}{2}$ and $\alpha = \frac{1}{4}$. Find firm 1's equilibrium investment in cost-reducing technologies, k^*.

- We first evaluate the output functions from part (b) at parameter values $c = \frac{1}{2}$ and $\alpha = \frac{1}{4}$, obtaining

$$q_1(k) = \frac{1+k}{6} \quad \text{and}$$

$$q_2(k) = \begin{cases} \frac{2-k}{12} & \text{if } \alpha < \frac{1}{2k} \\ 0 & \text{otherwise.} \end{cases}.$$

Therefore, the second-period profits of firm 1 become

$$\pi_1^{2nd} = (1 - q_1(k) - q_2(k))q_1(k) - \left(\frac{1}{2} - \frac{1}{4}k\right) q_1(k)$$
$$= \frac{(1+k)^2}{36}.$$

As a result, firm 1 chooses its cost-reducing investment k to solve the following problem in the first-period game:

$$\max_{k \geq 0} \pi_1^{2nd} - \frac{1}{2}k^2.$$

Differentiating with respect to k, we find

$$\frac{1+k}{18} - k = 0,$$

and solving for k, we obtain firm 1's equilibrium investment in cost-reducing technologies,

$$k^* = \frac{1}{17}.$$

- *SPNE.* Therefore, the subgame perfect equilibrium of the game predicts that, in the first period, firm 1 invests $k^* = 1/17$ in cost-reducing technologies. In the second period, it responds with an output function $q_1(k) = \frac{1+k}{6}$, while its rival responds with output function $q_2(k) = \frac{2-k}{12}$. (Recall that the subgame perfect equilibrium allows for in- and off-the-equilibrium behavior after the first stage.)
- *Equilibrium path.* In equilibrium, since firm 1 invests $k^* = 1/17$ in cost-reducing technologies, firm 1 responds choosing an output level

$$q_1(k^*) = \frac{1+k^*}{6} = \frac{1+\frac{1}{17}}{6} = \frac{3}{17},$$

while its rival selects $q_2(k^*) = (2 - 1/17)/12 = 11/68$ since condition $\alpha < \frac{1}{2k}$, when evaluated at $k^* = 1/17$, becomes

$$\alpha < \frac{1}{2k^*} = \frac{1}{2 \times \frac{1}{17}} = \frac{17}{2},$$

which holds in our context since $\alpha = \frac{1}{4}$ by assumption.
- *Remark:* Firm 1's choice of k is known in the literature as a "top dog" strategy since this firm uses its investment in cost-reducing technologies during the first period to alter the subsequent Cournot competition to its advantage. This occurs when the strategy that firm 1 chooses during the first period is visible by all players and firm 1 cannot renege from it subsequent periods. In addition, an increase in its strategy must shift firm 1's best response function in the subsequent stage downward, and this best response function must be negatively sloped. For a more formal presentation of "top dog" strategies, among others, see Fudenberg and Tirole (1984).

Exercise #4.16: Strategic Pre-commitment by Only One Firm Followed by Bertrand Competition^B

4.16. Consider two firms selling heterogeneous goods. The demand function of firm i is

$$q_i\left(p_i, p_j\right) = 1 - \gamma p_i + p_j,$$

where $j \neq i$, and $\gamma \geq 1$ represents the degree of product differentiation (homogeneous when $\gamma = 1$ but differentiated when $\gamma > 1$). In the first stage, firm 1 chooses its investment in cost-reducing technologies, k, at a cost $\frac{1}{2}k^2$. This investment decreases its marginal production cost from c to $c - \alpha k$, where $\alpha \in [0, 1]$ denotes the effectiveness of the investment. Intuitively, when $\alpha = 0$, investment is futile, but when $\alpha = 1$, every dollar invested reduces the initial marginal cost of the firm, c, by one dollar. Firm 2 cannot invest in cost-reducing technologies. In the second stage, both firms observe firm 1's investment k and respond competing à la Bertrand:

(a) *Second stage.* Find the best response function in the second period, $p_i(p_j)$. [*Hint:* The best response function of each firm will differ.]

- In the second period, firm 1 solves

$$\max_{p_1>0} \pi_1(p_1) = p_1(1 - \gamma p_1 + p_2) - (c - \alpha k)(1 - \gamma p_1 + p_2).$$

Differentiating with respect to p_1, and assuming interior solutions, we obtain

$$\frac{\partial \pi_1(p_1)}{\partial p_1} = 1 - 2\gamma p_1 + p_2 + \gamma(c - \alpha k) = 0.$$

Solving for p_1, we obtain the best response function of firm 1 as follows:

$$p_1(p_2) = \frac{1 + \gamma(c - \alpha k)}{2\gamma} + \frac{1}{2\gamma} p_2,$$

which originates at $\frac{1+\gamma(c-\alpha k)}{2\gamma}$ and increases in its rival price, p_2, at a rate of $\frac{1}{2\gamma}$.

We can then anticipate that an increase in either firm 1's investment during the first period, k, or the effectiveness of this investment, α, would shift firm 1's best response function downward. Intuitively, this firm can now charge lower prices because of its lower second-period marginal cost. As we show in subsequent parts of this exercise, investing in cost-reducing technologies becomes, in this setting of price competition, unattractive for firm 1, as opposed to how attractive it is when firms compete in quantities.

- Firm 2 solves a similar profit maximization problem, but evaluated at marginal cost c, since this firm cannot invest in cost-reducing technologies during the first period.

$$\max_{p_2>0} \pi_2(p_2) = p_2(1 - \gamma p_2 + p_1) - c(1 - \gamma p_2 + p_1).$$

Differentiating with respect to p_2, and assuming interior solutions, we obtain

$$\frac{\partial \pi_2(p_2)}{\partial p_2} = 1 - 2\gamma p_2 + p_1 + \gamma c = 0.$$

Solving for p_2, we obtain firm 2's best response function as follows:

$$p_2(p_1) = \frac{1 + \gamma c}{2\gamma} + \frac{1}{2\gamma} p_1,$$

which originates at $\frac{1+\gamma c}{2\gamma}$ and increases in its rival price, p_1, at a rate of $\frac{1}{2\gamma}$. Graphically, $p_1(p_2)$ originates at a lower vertical intercept than $p_2(p_1)$ does, indicating that firm 1 can afford to set lower prices than its rival.

(b) Find the second-period price function, $p_i(k)$, and the associated profits:

- Inserting firm 2's best response function, $p_2(p_1)$, into firm 1's, $p_1(p_2)$, we find

$$p_1(p_2) = \frac{1+\gamma(c-\alpha k)}{2\gamma} + \frac{1}{2\gamma}\overbrace{\left(\frac{1+\gamma c}{2\gamma} + \frac{1}{2\gamma}p_1\right)}^{p_2(p_1)}$$
$$= \frac{(2\gamma+1)(1+\gamma c) - 2\alpha\gamma^2 k + p_1}{4\gamma^2},$$

which, rearranging and solving for p_1, yields the equilibrium price for firm 1, as a function of k,

$$p_1(k) = \frac{(2\gamma+1)(1+\gamma c) - 2\alpha\gamma^2 k}{(2\gamma+1)(2\gamma-1)},$$

which unambiguously decreases in this firm's investment in cost-reducing technologies, k, and in the effectiveness of its investment, α. This price also decreases in the degree of product differentiation, γ, if

$$\frac{\partial p_1(k)}{\partial \gamma} = \frac{4\alpha\gamma k - (2+c)(2\gamma+1)^2}{(2\gamma+1)^2(2\gamma-1)^2} < 0,$$

which holds for sufficiently differentiated products, where

$$\gamma > \frac{4\sqrt{\alpha^2 k^2 - 2(2+c)\alpha k} + \alpha k - 2 - c}{8(2+c)}.$$

- Inserting this price into firm 2's best response function, $p_2(q_1)$, we obtain

$$p_2 = \frac{1+\gamma c}{2\gamma} + \frac{1}{2\gamma}\overbrace{\frac{(2\gamma+1)(1+\gamma c) - 2\alpha\gamma^2 k}{4\gamma^2-1}}^{p_1(k)},$$

which, rearranging, yields the equilibrium price for firm 2 as follows:

$$p_2(k) = \frac{(2\gamma+1)(1+\gamma c) - \alpha\gamma k}{(2\gamma+1)(2\gamma-1)}.$$

- Comparing the numerators of prices $p_1(k)$ and $p_2(k)$, we see that $p_1(k) < p_2(k)$ since

$$-2\alpha\gamma^2 k < -\alpha\gamma k$$

simplifies to $\gamma > \frac{1}{2}$, which holds by definition, so that firm 1 sets a lower price in equilibrium than its rival given that its second-period marginal costs are lower.

- Substituting equilibrium prices into the profit functions, we obtain for firm 1

$$\pi_1(k) = [p_1(k) - c + \alpha k][1 - \gamma p_1(k) + p_2(k)]$$
$$= \frac{\gamma\left[(2\gamma+1)(1-(\gamma-1)c) + \left(2\gamma^2-1\right)\alpha k\right]^2}{(2\gamma+1)^2(2\gamma-1)^2},$$

and similarly, for firm 2,

$$\pi_2(k) = [p_2(k) - c][1 - \gamma p_2(k) + p_1(k)]$$
$$= \frac{\gamma [(2\gamma+1)(1-(\gamma-1)c) - \alpha\gamma k]^2}{(2\gamma+1)^2(2\gamma-1)^2}.$$

(c) *First stage.* To simplify our analysis, assume that $c = \frac{1}{2}$, $\alpha = \frac{1}{4}$, and $\gamma = 2$. Find firm 1's equilibrium investment in cost-reducing technologies, k^*:

- We first evaluate the price functions from part (b) at parameter values $c = \frac{1}{2}$, $\alpha = \frac{1}{4}$, and $\gamma = 2$, obtaining

$$p_1(k) = \frac{(2\times 2+1)\left(1+2\times\frac{1}{2}\right) - 2\times\frac{1}{4}\times 2^2 k}{(2\times 2+1)(2\times 2-1)} = \frac{10-2k}{15}$$

$$p_2(k) = \frac{(2\times 2+1)\left(1+2\times\frac{1}{2}\right) - \frac{1}{4}\times 2k}{(2\times 2+1)(2\times 2-1)} = \frac{20-k}{30}.$$

Therefore, the second-period profits of firm 1 become

$$\pi_1^{2nd} = \frac{2\left[(2\times 2+1)\left(1-(2-1)\frac{1}{2}\right) + \left(2\times 2^2-1\right)\frac{1}{4}k\right]^2}{(2\times 2+1)^2(2\times 2-1)^2}$$
$$= \frac{(10+7k)^2}{1800}$$

In period one, firm 1 chooses its cost-reducing investment k to solve the following problem:

$$\max_{k\geq 0} \pi_1^{2nd} - \frac{1}{2}k^2.$$

Differentiating with respect to k, we find

$$\frac{7(10+7k)}{900} - k = 0,$$

and solving for k, we obtain firm 1's equilibrium investment in cost-reducing technologies,

$$k^* = \frac{70}{851} \simeq 0.082.$$

- *SPNE.* Therefore, the subgame perfect equilibrium of the game predicts that, in the first period, firm 1 invests $k^* = 0.082$ in cost-reducing technologies. In the second period, firm 1 responds with a price function $p_1(k) = \frac{10-2k}{15}$, while firm 2 responds with a price function $p_2(k) = \frac{20-k}{30}$. (Recall that the subgame perfect equilibrium allows for in- and off-the-equilibrium behavior after the first stage.)

- *Remark:* In this setting, firm 1's choice of k corresponds to the "puppy dog ploy" strategy in Fudenberg and Tirole (1984), since an increase in k shifts firm 1's best response function in the subsequent stage downward, and this best response function is positively sloped.

Exercise #4.17: Strategic Pre-commitment by Both Firms^B

4.17. Consider the setting in Exercise 4.15, but allow now for *every* firm $i = \{1, 2\}$ to invest in cost-reducing technologies during the first stage, choosing its k_i. Therefore, in the second stage, both firms observe the investment profile (k_i, k_j) they selected in the previous stage and then compete à la Cournot:

(a) *Second stage.* Find the best response function in the second period, $q_i(q_j)$:

- In the second stage, every firm i solves

$$\max_{q_i \geq 0} (1 - q_i - q_j)q_i - (c - \alpha k_i)\, q_i,$$

which is evaluated at its second-period cost $c - \alpha k_i$, after investing k_i dollars in cost-reducing technologies. Differentiating with respect to q_i, we obtain

$$1 - 2q_i - q_j - (c - \alpha k_i) = 0$$

and solving for q_i, we find firm i's best response function

$$q_i(q_j) = \frac{1 - c + \alpha k_i}{2} - \frac{q_j}{2}.$$

This best response function is increasing in k_i, $\frac{\partial q_i}{\partial k_i} > 0$ (since firm i becomes relatively more competitive but decreasing in q_j). A symmetric expression applies for firm j's best response function, $q_j(q_i)$.

(b) Find the second-period output function, $q_i(k_i, k_j)$. Is it increasing in k_i and k_j?

- Inserting firm j's best response function, $q_j(q_i)$, into firm i's, $q_i(q_j)$, we find

$$q_i = \frac{1 - c + ak_i}{2} - \frac{(1 - c + \alpha k_j - q_i)}{4},$$

which, rearranging and solving for q_i, yields firm i's equilibrium output function

$$q_i(k_i, k_j) = \frac{1 - c + \alpha\left(2k_i - k_j\right)}{3}.$$

This output function increases in firm i's own investment in cost-reducing technologies in the previous period, k_i, but decreases in firm j's investment, k_j.

(c) *First stage.* To simplify our analysis, assume that $c = \frac{1}{2}$ and $\alpha = \frac{1}{4}$ for the remainder of the exercise. Find firm i's best response function in the first stage, $k_i(k_j)$. Is it positively or negatively sloped? Interpret:

- We first evaluate the output functions from part (b) at parameter values $c = \frac{1}{2}$ and $\alpha = \frac{1}{4}$, obtaining

$$q_i(k_i, k_j) = \frac{2 + 2k_i - k_j}{12} \quad \text{and}$$
$$q_j(k_i, k_j) = \frac{2 + 2k_j - k_i}{12}.$$

Therefore, the second-period profits of firm i become

$$\pi_i^{2nd} = (1 - q_i(k_i, k_j) - q_j(k_i, k_j))q_i(k_i, k_j) - \left(\frac{1}{2} - \frac{1}{4}k_i\right) q_i(k_i, k_j)$$
$$= \frac{(2 + 2k_i - k_j)^2}{144}.$$

As a result, every firm i chooses its cost-reducing investment k_i to solve the following problem in the first-period game:

$$\max_{k_i \geq 0} \pi_i^{2nd} - \frac{1}{2} k_i^2.$$

Differentiating with respect to k_i, we find

$$\frac{2 + 2k_i - k_j}{36} - k_i = 0,$$

and solving for k_i, we obtain firm i's best response function in the first period

$$k_i(k_j) = \frac{2 - k_j}{34}.$$

Therefore, the best response function is negatively sloped, indicating that firms regard each other's cost-reducing investments as strategic substitutes. In other words, when firm j invests one more dollar in cost-reducing technologies, k_j, firm i responds reducing its own investment, k_i.

(d) Find the equilibrium investment in cost-reducing technologies, k_i^*:

- Inserting firm j's best response function, $k_j(k_i)$, into firm i's, $k_i(k_j)$, we obtain

$$k_i = \frac{\left(2 - \frac{(2-k_i)}{34}\right)}{34},$$

which, rearranging, yields $1155k_i = 66$. Solving for k_i, we obtain firm i's equilibrium investment,

$$k_i^* = \frac{2}{35}.$$

(e) Summarize the subgame perfect equilibrium of the game:

- In the subgame perfect equilibrium of the game, every firm i invests $k_i^* = 2/35$ in cost-reducing technologies during the first period and responds with an output function

$q_i(k_i, k_j) = \frac{2+2k_i-k_j}{12}$ in the second period, yielding an equilibrium output of 6/35 units for every firm i.

(f) How would your above results change if firms coordinate their investment decisions during the first stage, while still competing à la Cournot in the second stage? Compare your results with those in part (e):

- The second-stage output function, $q_i(k_i, k_j)$ and $q_j(k_i, k_j)$, would remain unaffected in this setting. However, firms would now coordinate their investments k_i and k_j to maximize their joint profits as follows:

$$\max_{k_i,k_j\geq 0} \left(\pi_i^{2nd} - \frac{1}{2}k_i^2\right) + \left(\pi_j^{2nd} - \frac{1}{2}k_j^2\right)$$

$$= \frac{(2+2k_i-k_j)^2}{144} + \frac{(2+2k_j-k_i)^2}{144} - \frac{1}{2}k_i^2 - \frac{1}{2}k_j^2.$$

Differentiating with respect to k_i and k_j, we find

$$\frac{2-67k_i-4k_j}{72} = 0$$

$$\frac{2-67k_j-4k_i}{72} = 0.$$

Invoking symmetry, where $k = k_i = k_j$, we then find

$$2 - 67k - 4k = 0,$$

and further simplifying yields

$$k_i^{Coord} = k_j^{Coord} = \frac{2}{71}.$$

- Comparing the investment in cost-reducing technologies when firms do not coordinate their decisions, k_i^*, and when they do, k_i^{Coord}, we find that

$$k_i^* = \frac{2}{35} > \frac{2}{71} = k_i^{Coord}.$$

Intuitively, when firms independently choose their investment, they invest too much in cost-reducing technologies, compared to those when they coordinate their decisions.

Regulating Imperfectly Competitive Markets

5

Introduction

This chapter takes a regulatory approach by considering some of the models of imperfect competition analyzed in previous chapters (monopoly and oligopolies with different numbers of firms) and examines different policy tools that can induce firms to produce the output level that maximizes the social welfare (we refer to this output level as the "socially optimal output"), such as subsidies and taxes for each unit of output that the firm produces.

First, Exercise 5.1 considers a regulator facing a monopolist and who seeks to induce this firm to produce the socially optimal output. For simplicity, we consider in this setting that the regulator can observe the monopolist's production costs, and thus accurately design the per-unit subsidy that induces this firm to increase its output toward the socially optimal output. (We relax this complete information assumption in Exercises 5.5 and 5.6.) In this exercise, we assume that the regulator's welfare function considers consumer and producer surplus, thus implying that the socially optimal output exceeds the equilibrium output that the monopolist would produce in the absence of regulation.

Exercise 5.2, however, considers a more general context, where each unit of output generates environmental damages (e.g., pollution). The regulator then has one more term to consider in the welfare function, leading him to induce lower socially optimal output levels than in the absence of pollution. This is reflected in the equilibrium subsidies that he sets in this context, which are less generous than in Exercise 5.1. If the environmental damage from pollution is severe enough, the regulator may even set a tax per unit of output (e.g., emission fees) to curb pollution. Exercise 5.3 extends this setting to allow for more general objective functions by the regulator (Environmental Protection Agency), which can be utilitarian or Pigouvian as special cases, or any combination of them. Exercise 5.4 studies a natural monopoly, where the firm faces a decreasing average cost function, and discusses regulatory tools.

Exercise 5.5 considers a similar setting as in Exercise 5.1, but assuming that the regulator cannot perfectly observe the monopolist's cost. For simplicity, we consider a relatively stylized context where the monopolist has either a low or a high marginal cost of production, both equally likely, and no fixed costs, along with a simplified cost of raising public funds for the regulator. In this setting, we evaluate optimal subsidies when the regulator is perfectly informed about the monopolist's cost (as a complete information benchmark) and then examine the incomplete information game. In this case, the regulator offers a menu of lump-sum subsidies and output levels that he offers to induce each type of firm to select the subsidy–output pair meant for its type. In this environment, we identify the inefficiencies that arise relative to incomplete information, and the information rents that the regulator must offer

P.-S. Choi et al., *Industrial Organization*, Springer Texts in Business and Economics,
https://doi.org/10.1007/978-3-031-38635-0_5

to one of the firm types to induce self-selection. Exercise 5.6 then extends this setting to one where firms have fixed and variable costs, firm types are not necessarily equally likely, and the cost of raising public funds is more generic.

The remainder of the chapter considers regulation in oligopolistic markets. In Exercise 5.7, we analyze a non-polluting oligopoly with $N \geq 2$ firms and identify socially optimal output and the emission fee or subsidy that the regulator sets to induce firms to produce this output level. Exercise 5.8 extends this model to an industry where output generates environmental damages. As in the case of a polluting monopoly (Exercise 5.2), we find in which cases the regulator sets a subsidy per unit of output (i.e., when the environmental damages from pollution are minor) and the cases where, instead, the regulator sets an emission fee (i.e., when its environmental damage is severe). Exercise 5.9 allows more general objective functions by the environmental regulator, as we did in the case of monopoly in Exercise 5.3, thus considering utilitarian or Pigouvian regulators as special cases. Exercise 5.10 studies regulation of polluting duopoly when firms compete à la Stackelberg, showing that firms are subject to more stringent emission fees when they compete sequentially than simultaneously. Exercise 5.11 studies regulation of polluting duopoly when firms compete à la Stackelberg, showing that firms are subject to more stringent emission fees when they compete sequentially than simultaneously. Exercise 5.11 studies industries when entry is allowed (endogenous entry), identifying the equilibrium number of firms that enter the industry, the optimal number of firms that should enter the industry to maximize welfare, and why we may expect a socially excessive number of firms joining the market if entry is left unregulated. Finally, Exercise 5.12 examines tax incidence under monopoly allowing for different demand functions.

Exercise #5.1: Regulating a Monopolist Under Complete Information[A]

5.1. Consider a monopolist with inverse demand function $p(q) = a - q$ and that cost function $C(q) = cq$, where c denotes the constant marginal cost of production, and $a > c > 0$. In the first stage, a regulator sets a per-unit subsidy s to induce the monopolist to produce the socially optimal output q^*. In the second stage, the monopolist observes the per-unit subsidy s and responds choosing its output level q.

(a) *Second stage*. Find the output function that the monopolist chooses in the second stage as a function of s, $q(s)$:

- The monopolist chooses the output level q that solves

$$\max_{q \geq 0} \ (a-q)q - (c-s)q$$

since the marginal cost c decreases in the per-unit subsidy s, so the monopolist's marginal cost of production (net of the subsidy) is $c - s$. Differentiating with respect to q, we obtain

$$a - 2q = c - s.$$

Solving for q yields a profit maximizing output of

$$q(s) = \frac{a - (c - s)}{2},$$

which is, as expected, increasing in the per-unit subsidy s that the monopolist receives because the lower the net marginal cost $c - s$, the less costly (and hence, more profitable) the monopolist is in producing additional units of output.

(b) *First stage*. Find the optimal per-unit subsidy the regulator sets to induce the monopolist to produce the socially optimal output q^{SO}. Assume that the regulator seeks to maximize the sum of consumer and producer surplus less the cost of paying subsidy s for each of the q units produced:

- In the first stage, the regulator anticipates the output function that the monopolist has in the second stage, $q(s)$, and thus finds the subsidy s for which the monopolist's output coincides with the socially optimal output q^{SO}, that is,

$$q(s) = q^{SO}.$$

- *First step, finding the socially optimal output.* We first find the socially optimal output q^{SO}. This output solves

$$\max_{q \geq 0} \; W = \underbrace{\frac{q^2}{2}}_{CS} + \underbrace{[(a-q)q - (c-s)q]}_{\pi(s)} - sq,$$

where the first (second) term in the welfare function represents consumer (producer) surplus, and the last term denotes the total cost of paying subsidy s to q units of output. This problem simplifies to

$$\max_{q \geq 0} \; W = \frac{q^2}{2} + [(a-q)q - cq],$$

which clarifies that the subsidy is welfare neutral, namely, the monopolist receives sq to lower its production costs, but this subsidy must originate from other consumers or firms in this society. (Richer models assume that tax collection needed to raise the public funds used in the subsidy generates inefficiencies – deadweight loss from taxation plus administrative costs – so the last term in the regulator's welfare function becomes $(1 + g)sq$, rather than sq, where $g > 0$ denotes these inefficiencies. For simplicity, our model assumed that these inefficiencies are negligible, $g = 0$.)

Differentiating with respect to q yields

$$q + a - 2q - c = 0$$

and solving for q, we obtain the socially optimal output

$$q^{SO} = a - c,$$

which is increasing in the strength of demand, a, but decreasing in marginal cost c. This output level coincides, of course, with that under a perfectly competitive market where the inverse demand curve crosses the marginal cost, $p(q) = C'(q)$. In our context, this condition becomes $a - q = c$, which, solving for q, yields the socially optimal output $q^{SO} = a - c$.

- *Second step, finding the socially optimal subsidy.* We can now equate the monopolist's second-stage output to the socially optimal output, $q(s) = q^{SO}$, which entails

$$\frac{a-(c-s)}{2}=a-c.$$

Solving for the per-unit subsidy s, we obtain

$$s^*=a-c.$$

Intuitively, the regulator needs to offer a more generous subsidy when the product mark-up, $a-c$, increases.

(c) *Numerical example.* Assume an inverse demand function $p(q)=1-q$ and marginal cost $c=1/2$. Evaluate your equilibrium results in parts (a) and (b) in this setting:

- In this context, the regulator offers a per-unit subsidy $s^*=1-\frac{1}{2}=\frac{1}{2}$, and the monopolist responds producing $q^{SO}=1-\frac{1}{2}=\frac{1}{2}$ in equilibrium. More generally, for any given per-unit subsidy s, the monopolist produces according to output function $q\,(s)=\frac{1-(\frac{1}{2}-s)}{2}=\frac{1+2s}{4}$ that, as expected, increases in s.

Exercise #5.2: Regulating a Polluting Monopoly^B

5.2. Consider the industry in Exercise 5.1, but assuming that every unit of output generates one unit of pollution. The social planner considers the environmental damage of this pollution in the social welfare function, which now becomes

$$W=\left(\frac{q^2}{2}+\pi\,(s)-sq\right)-dq^2.$$

The first three terms (in brackets for clarity) denote consumer and producer surplus plus the cost of funding the subsidy program. The last term represents the environmental damage from aggregate pollution, which is increasing and convex in production (and, thus, in pollution), and parameter d satisfies $d>0$.

The timing coincides with that in Exercise 5.1, that is, in the first period, the regulator sets per-unit subsidy s; and, in the second period, the monopolist responds choosing its output q:

(a) *Second stage.* Find the monopolist's second-period output function, $q_i\,(s)$:

- The monopolist solves the same profit maximization problem as in Exercise 5.1, obtaining the same equilibrium output

$$q\,(s)=\frac{a-(c-s)}{2},$$

which has the same comparative statics and intuition as in Exercise 5.1(a).

(b) *First stage.* Find the welfare maximizing subsidy s^* that the social planner sets in the first period:

- *First step, finding the socially optimal output.* We first find the socially optimal output q^{SO}. This output solves

$$\max_{q\geq 0}\ W=\underbrace{\frac{q^2}{2}}_{CS}+\underbrace{[(a-q)q-(c-s)q]}_{\pi(s)}-sq-\underbrace{dq^2}_{\text{Env. Damage}},$$

where the first (second) term in the welfare function represents consumer (producer) surplus, the third term denotes the total cost of paying subsidy s to q units of output, and the last term is the environmental damage from pollution. This problem simplifies to

$$\max_{q \geq 0} W = \frac{q^2}{2} + [(a-q)q - cq] - dq^2.$$

Differentiating with respect to q yields

$$q + a - 2q - c - 2dq = 0.$$

After rearranging, we find

$$q + 2dq = a - c,$$

and solving for q, we obtain the socially optimal output

$$q^{SO} = \frac{a-c}{1+2d}.$$

Note that when environmental damages are absent, $d = 0$, this output level simplifies to $a - c$, thus coinciding with the socially optimal output in Exercise 5.1.

- *Second step, finding the socially optimal subsidy.* We can now equate the monopolist's second-stage output to the socially optimal output, $q(s) = q^{SO}$, which entails

$$\frac{a-(c-s)}{2} = \frac{a-c}{1+2d}.$$

Solving for the per-unit subsidy s yields

$$s^* = \frac{(a-c)(1-2d)}{1+2d},$$

which can also be expressed as $s^* = (1-2d)\, q^{SO}$.

(c) Under which conditions is s^* positive? Is the socially optimal subsidy s^* increasing or decreasing in d? Interpret:

- Subsidy s^* is positive if $1 - 2d > 0$, or $d < \frac{1}{2}$. Intuitively, the regulator faces two market failures in this market: a typical underproduction due to monopoly, which would lead the regulator to set a subsidy to induce an increase in output, and an overproduction due to pollution, which would call for negative subsidies (emission fees) to induce an output reduction:
  - When condition $d < \frac{1}{2}$ holds, indicating that pollution is not extremely damaging, the first market failure dominates, leading the regulator to set a positive subsidy, which induces the monopolist to increase its production.
  - In contrast, when $d > \frac{1}{2}$, pollution becomes more damaging, the second market failure dominates, and the regulator sets a negative subsidy (a tax), which induces the monopolist to decrease its output.
- Subsidy s^* is unambiguously decreasing in the environmental damage parameter d since

$$\frac{\partial s^*}{\partial d} = -\frac{4(a-c)}{(1+2d)^2} < 0.$$

In other words, the subsidy can be positive or negative depending on whether the environmental damage from pollution is minor or severe, respectively, but a more damaging pollution produces a clear reduction in the per-unit subsidy that the regulator offers to the monopolist.

(d) *Numerical example.* Assume an inverse demand function $p(q) = 1 - q$ and marginal cost $c = 1/2$. Evaluate the socially optimal subsidy at $d = 1/4$ and at $d = 1$. Interpret:
- The socially optimal subsidy in this context becomes

$$s^* = \frac{\left(1-\frac{1}{2}\right)(1-2d)}{1+2d} = \frac{1-2d}{2(1+2d)}.$$

When $d = 1/4$, this subsidy simplifies to $s^* = 1/6$. In contrast, when $d = 1$, we obtain that $s^* = -1/6$, indicating that the regulator sets a negative subsidy (a tax per unit of output, as in emission fees) to curb total emissions.

Exercise #5.3: Regulating a Monopolist Under Complete Information – Pigouvian EPA^A

5.3. Consider the industry in Exercise 5.2, but assuming that every unit of output generates one unit of pollution. The social planner considers environmental damages of this pollution in the social welfare function, which now becomes

$$W = (1-\beta)\left[\frac{q^2}{2} + \pi(s) - sq\right] - \beta d q^2.$$

The first three terms (in brackets for clarity) denote consumer and producer surplus ($CS + PS$) plus the cost of funding the subsidy program. The last term represents the environmental damages (ED) from aggregate pollution, which are increasing and convex in production (and, thus, in pollution), and parameter d satisfies $d > 0$.

In addition, parameter β satisfies $0 \leq \beta \leq 1$ and measures the weight that the regulator (e.g., EPA) assigns to environmental damages from output. For instance, when $\beta = 1$, the welfare function collapses to ED, indicating that the regulator only considers the external effects of pollution, setting Pigouvian emission fees to curb pollution, whereas when $\beta = 0$, the welfare function simplifies to $CS + PS$, representing that the regulator ignores the environmental damages from output. More compactly, parameter β denotes how "Pigouvian" the regulator is when setting emission fees (completely Pigouvian when $\beta = 1$, but not so Pigouvian otherwise). The timing coincides with that in Exercise 5.1, that is, in the first period, the regulator sets per-unit subsidy s; and in the second period, the monopolist responds choosing its output q.

(a) *Second stage.* Find the monopolist's second-period output function, $q_i(s)$:
- The monopolist solves the same profit maximization problem as in Exercise 5.1, obtaining the same equilibrium output

$$q(s) = \frac{a-(c-s)}{2},$$

which has the same comparative statics and intuition as in part (a) of Exercise 5.1.

(b) *First stage.* Find the welfare maximizing subsidy s^* that the social planner sets in the first period:

- *First step, finding the socially optimal output.* We first find the socially optimal output q^{SO}. This output solves

$$\max_{q \geq 0} \; W = (1-\beta) \left\{ \underbrace{\frac{q^2}{2}}_{CS} + \underbrace{[(a-q)\,q - (c-s)\,q]}_{\pi(s)} - sq \right\} - \underbrace{\beta\, dq^2}_{ED},$$

where the first (second) term in the welfare function represents consumer (producer) surplus, the third term denotes the total cost of paying subsidy s to q units of output, and the last term indicates the environmental damages from pollution. This welfare maximization problem simplifies to

$$\max_{q \geq 0} \; W = (1-\beta)\, \frac{(2a-2c-q)\,q}{2} - \beta d q^2.$$

Differentiating with respect to q yields

$$(1-\beta)\,(a-c-q) - 2\beta d q = 0.$$

After rearranging, we find

$$(1-\beta)\,q + 2\beta d q = (1-\beta)\,(a-c),$$

and solving for q, we obtain the socially optimal output

$$q^{SO} = \frac{(1-\beta)\,(a-c)}{1-\beta\,(1-2d)}.$$

To gain some intuition of this result, consider the following special cases:

  – Note that when environmental damages are absent, $d = 0$, this output level simplifies to $q^{SO} = a - c$, thus coinciding with the socially optimal output in Exercise 5.1.
  – The same result applies when the EPA assigns no weight on the environmental damage, $\beta = 0$, which simplifies the above socially optimal output to $q^{SO} = a - c$.
  – When the EPA is Pigouvian, $\beta = 1$, this socially optimal output simplifies to $q^{SO} = 0$, so the regulator seeks to reduce pollution as much as possible.
  – Finally, when the EPA assigns the same weight to $CS + PS$ than to ED, $\beta = \frac{1}{2}$, this socially optimal output simplifies to $q^{SO} = \frac{a-c}{1+2d}$, which is decreasing in the environmental damage parameter d.

- *Second step, finding the socially optimal subsidy.* We can now equate the monopolist's second-stage output to the socially optimal output, $q(s) = q^{SO}$, which entails

$$\frac{a-(c-s)}{2} = \frac{(1-\beta)\,(a-c)}{1-\beta\,(1-2d)}.$$

Solving for the per-unit subsidy s yields

$$s^* = \frac{(a-c)\left[1-\beta(1+2d)\right]}{1-\beta(1-2d)},$$

which can also be expressed as $s^* = \frac{1-\beta(1+2d)}{1-\beta} q^{SO}$.

(c) Under which conditions is s^* positive? Is the socially optimal subsidy s^* increasing or decreasing in d? Is s^* increasing or decreasing in β? Interpret:

- Subsidy s^* is positive if and only if $1-\beta(1+2d) > 0$, or $d < \frac{1-\beta}{2\beta}$. Intuitively, the regulator faces two market failures in this market: a typical underproduction due to monopoly, which would lead the regulator to set a subsidy to induce an increase in output, and an overproduction due to pollution, which would call for negative subsidies (emission fees) to induce an output reduction:
  - When condition $d < \frac{1-\beta}{2\beta}$ holds, indicating that pollution is not extremely damaging, the first market failure dominates, leading the regulator to set a positive subsidy, which induces the monopolist to increase its production.
  - In contrast, when $d > \frac{1-\beta}{2\beta}$, pollution becomes more damaging, the second market failure dominates, and the regulator sets a negative subsidy (a tax), which induces the monopolist to decrease its output.
- Subsidy s^* is unambiguously decreasing in the environmental damage parameter d since

$$\frac{\partial s^*}{\partial d} = -\frac{4\beta(1-\beta)(a-c)}{\left[1-\beta(1-2d)\right]^2} < 0.$$

  In other words, the subsidy can be positive or negative depending on whether environmental damages from pollution are minor or severe, respectively, but a more damaging pollution produces a clear reduction in the per-unit subsidy that the regulator offers to the monopolist.
- Subsidy s^* is unambiguously decreasing in the policy weight β since

$$\frac{\partial s^*}{\partial \beta} = -\frac{4(a-c)d}{\left[1-\beta(1-2d)\right]^2} < 0,$$

  entailing that, as the regulator puts more emphasis on environmental damages, he offers a lower subsidy to the monopolist.

(d) *Numerical example.* Assume an inverse demand function $p(q) = 1-q$, marginal cost $c = 1/2$, and $d = 1/4$. Evaluate the socially optimal subsidy in part (b) at $\beta = 0$, $\beta = 1/2$, and at $\beta = 1$. Interpret:

- The socially optimal subsidy in this context becomes

$$\begin{aligned} s^* &= \frac{\left(1-\frac{1}{2}\right)\left[1-\beta\left(1+2\times\frac{1}{4}\right)\right]}{1-\beta\left(1-2\times\frac{1}{4}\right)} \\ &= \frac{2-3\beta}{2(2-\beta)}. \end{aligned}$$

- When $\beta = 0$, this subsidy simplifies to

$$s^* = \frac{2-3\times 0}{2(2-0)} = \frac{1}{2}$$

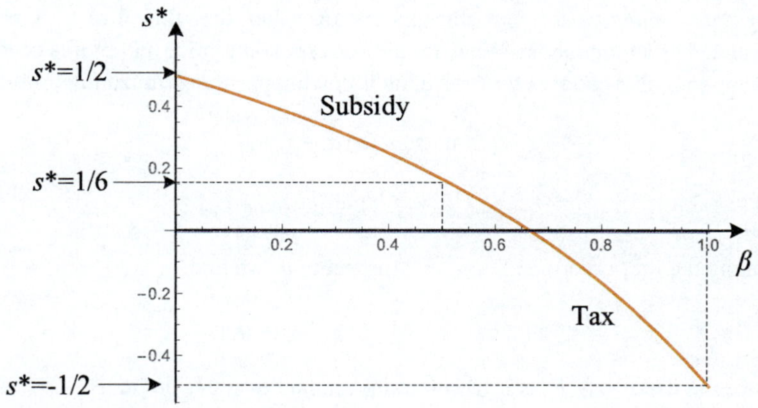

Fig. 5.1 Socially optimal subsidy as a function of β

entailing that the monopolist is subsidized to produce more units of output as the regulator completely ignores environmental damages.

- Whereas, when $\beta = 1/2$, this subsidy becomes

$$s^* = \frac{2 - 3 \times \frac{1}{2}}{2\left(2 - \frac{1}{2}\right)} = \frac{1}{6},$$

where the regulator reduces the subsidy offered to the monopolist when he considers environmental damages as equally important as consumer and producer surplus.

- Finally, when $\beta = 1$, this subsidy is

$$s^* = \frac{2 - 3 \times 1}{2\,(2 - 1)} = -\frac{1}{2},$$

which is negative, since now the regulator only considers environmental damages and taxes the monopolist until the firm stops producing polluting output.

- Figure 5.1 plots the socially optimal subsidy $s^* = \frac{2-3\beta}{2(2-\beta)}$ as a function of β, summarizing our above results. Overall, the subsidy becomes less generous as the regulator becomes more Pigouvian (higher β), since he seeks to reduce pollution more significantly.

Exercise #5.4: Regulating a Natural Monopolist^A

5.4. A water supply company provides water to a small town. The inverse demand function for water in this town is given by $p\,(q) = 1 - q$, and this company's total cost function is $C\,(q) = F + cq$, where $F \geq 0$ denotes fixed costs and c represents the marginal cost, where $1 > c \geq 0$.

(a) Why does this situation illustrate a "natural monopoly" problem?

- The average cost of production is $AC\,(q) = \frac{C(q)}{q} = \frac{F}{q} + c$, which is decreasing in output. Therefore, multiple producers are more costly than a single monopolist, that is, the sum of their average costs will be larger than the monopolist's average costs. In this context, a monopolist naturally becomes more cost-efficient than having several producers with a similar cost structure as the monopolist.

(b) *Unregulated monopolist.* Find the amount of water that this firm will produce if left unregulated as a monopolist. What are the corresponding price and profits of the firm?

- The monopolist chooses q to solve the following profit maximization problem:

$$\max_{q \geq 0} \; p(q)\, q - C(q).$$

$$= (1 - c - q)\, q - F.$$

Taking the first-order condition with respect to q, we find

$$1 - c - 2q = 0$$

since the fixed cost, F, is invariant to the amount of water produced.

Solving for q, we obtain the monopolist's profit maximizing output of

$$q^m = \frac{1-c}{2}.$$

Substituting $q^m = \frac{1-c}{2}$ into the inverse demand function, we find the price of

$$p^m = 1 - \frac{1-c}{2} = \frac{1+c}{2}.$$

This yields the monopolist's profits as follows:

$$\pi^m = \left(1 - c - \overbrace{\frac{1-c}{2}}^{q^m}\right)\overbrace{\frac{1-c}{2}}^{q^m} - F$$

$$= \left(\frac{1-c}{2}\right)^2 - F,$$

which is positive if the fixed cost is not too high, that is, $F \leq \left(\frac{1-c}{2}\right)^2$.

(c) *Marginal cost pricing.* Determine the amount of water that this firm will produce if the regulatory agency forces this firm to price according to its marginal cost (that is, to produce an amount of water q^* that solves $p(q^*) = MC(q^*)$). Find the corresponding price and profits for the firm:

- In this setting, the monopolist sets $p(q^*) = MC(q^*)$, which means

$$1 - q^* = c,$$

that is, $q^* = 1 - c$ units of water, at a price of $p^* = c$, yielding losses of

$$\pi^* = \left(1 - c - \overbrace{(1-c)}^{q^*}\right)\overbrace{(1-c)}^{q^*} - F = -F.$$

(d) *Price discrimination.* Consider now that the regulatory agency allows the monopoly to charge two different prices: a high price p_1 for the first q_1 units and a low price $p(q^*)$ for the remaining $(q^* - q_1)$ (i.e., the units from q_1 up to the output level you found in part (c), q^*). In addition, the regulatory agency imposes the condition that the firm cannot make any profits, $\pi = 0$, when charging these two prices. Find the value of q_1 and the associated price, $p(q_1)$: (Hint: Identify the condition in which the firm produces water, and then the two possible output levels.)

- First, note that the value of q_1 must satisfy the "zero profits" condition, that is,

$$\pi = p(q_1)\,q_1 + p\left(q^*\right)\left(q^* - q_1\right) - F - cq^* = 0,$$

which, after substituting $q^* = 1 - c$ from part (c), yields

$$(1 - q_1)\,q_1 + c\,(1 - c - q_1) - F - c\,(1 - c) = 0$$

that is rearranged into

$$(1 - q_1 - c)\,q_1 - F = 0$$

implying that the firm has to make zero profits on the first q_1 units of water produced, because for the additional $q^* - q_1$ units, the firm is forced to marginal cost pricing that does not generate additional revenues to cover its fixed cost.

- Rearranging the above expression, we obtain

$$q_1^2 - (1 - c)\,q_1 + F = 0.$$

Solving for q_1, we obtain two roots:

$$q_1^L = \frac{1 - c - \sqrt{(1-c)^2 - 4F}}{2} \quad \text{and}$$

$$q_1^H = \frac{1 - c + \sqrt{(1-c)^2 - 4F}}{2},$$

where superscripts L and H indicate low and high amount of water, respectively. The condition for the monopolist to produce positive amount of water is for the discriminant of the above quadratic root to be positive, where

$$(1 - c)^2 - 4F \geq 0$$

that simplifies into

$$c \leq 1 - 2\sqrt{F}$$

meaning that the marginal cost in producing every unit of water cannot be too high relative to the fixed cost of operating the water production facility.

- *Prices.* The prices associated with the two levels of output are

$$
\begin{aligned}
p_L &= 1 - q_1^L \\
&= 1 - \frac{1 - c - \sqrt{(1-c)^2 - 4F}}{2} \\
&= \frac{1 + c + \sqrt{(1-c)^2 - 4F}}{2}, \text{ and} \\
p_H &= 1 - q_1^H \\
&= 1 - \frac{1 - c + \sqrt{(1-c)^2 - 4F}}{2} \\
&= \frac{1 + c - \sqrt{(1-c)^2 - 4F}}{2}.
\end{aligned}
$$

- *Summary.* We can summarize the monopolist's regulated pricing strategy as follows:
  - *First case:* Produce q_1^L units of water and charge consumers a per-unit price of p_L; and above that amount of water, produce an additional $1 - c - q_L$ units of water that is priced at the marginal cost c.
  - *Second case:* Produce q_1^H units of water and charge consumers a per-unit price of p_H; and above that amount of water, produce an additional $1 - c - q_H$ units of water that is priced at the marginal cost c.

(e) Compare social welfare across different cases. Which policy does the regulator prefer?

- If the monopolist is unregulated, social welfare is

$$
\begin{aligned}
W^{unregulated} &= \frac{1}{2}\left(q^m\right)^2 + \pi^m \\
&= \frac{1}{2}\left(\frac{1-c}{2}\right)^2 + \left(\frac{1-c}{2}\right)^2 - F \\
&= \frac{3(1-c)^2}{8} - F.
\end{aligned}
$$

- If the monopolist practices marginal cost pricing, social welfare is

$$
\begin{aligned}
W^{mcp} &= \frac{1}{2}\left(q^*\right)^2 + \pi^* \\
&= \frac{(1-c)^2}{2} - F.
\end{aligned}
$$

- If the monopolist practices price discrimination, social welfare is

$$
W^{pd} = \frac{(1-c)^2}{2}
$$

since the total output is $q^* = 1 - c$ and the monopolist can break even (that is, $\pi = 0$).

- Comparing across different cases, social welfare is the lowest if the monopolist is unregulated, as expected. However, the monopolist practicing price discrimination in part (d) generates higher welfare than the monopolist practicing marginal cost pricing in part (c), since the former can charge a higher price on the first q_1 units to cover for the fixed costs, while the latter does not.

(f) *Numerical example.* Evaluate your results in part (d) assuming parameter values $c = \frac{1}{4}$ and $F = \frac{1}{10}$:

- In the first case identified in part (d), the monopolist produces

$$q_1^L = \frac{1 - c - \sqrt{(1-c)^2 - 4F}}{2}$$

$$= \frac{1 - \frac{1}{4} - \sqrt{\left(1 - \frac{1}{4}\right)^2 - 4 \times \frac{1}{10}}}{2} \simeq 0.17 \text{ units}$$

at a price

$$p_L = \frac{1 + c + \sqrt{(1-c)^2 - 4F}}{2}$$

$$= \frac{1 + \frac{1}{4} + \sqrt{\left(1 - \frac{1}{4}\right)^2 - 4 \times \frac{1}{10}}}{2} \simeq \$0.82.$$

In the second case found in part (d), the monopolist produces

$$q_1^H = \frac{1 - c + \sqrt{(1-c)^2 - 4F}}{2}$$

$$= \frac{1 - \frac{1}{4} + \sqrt{\left(1 - \frac{1}{4}\right)^2 - 4 \times \frac{1}{10}}}{2} \simeq 0.58 \text{ units}$$

at a price

$$p_H = \frac{1 + c - \sqrt{(1-c)^2 - 4F}}{2}$$

$$= \frac{1 + \frac{1}{4} - \sqrt{\left(1 - \frac{1}{4}\right)^2 - 4 \times \frac{1}{10}}}{2} \simeq \$0.42.$$

Exercise #5.5: Regulating a Monopolist Under Incomplete Information-I^B

5.5. Consider a monopolist that provides catering services in a small college town. Its fixed cost in setting up the venue is $\frac{1}{2}$, and marginal cost in serving every meal is constant and can either be low (c_L) or high (c_H), with equal probability, where $c_H > 0$ and $c_L = 0$. The inverse demand function is $p(q) = 1 - q$, where q represents the aggregate output (i.e., total meals served). To encourage more students to attend the welcoming party, the university president offers a subsidy S_i to the firm (which can be negative if the university sets an entry fee on this firm), where $i \in \{L, H\}$. The university's costs of paying subsidy S_i are $\frac{5}{4}S_i$, since there is an administrative cost of 25% in collecting the funds to pay the subsidy:

(a) *Complete information.* Assume that the university president has access to the cost information of the firm. Set up the university's welfare maximization problem, which takes into account firm i's participation constraint:
- Observing firm i's cost, the president solves the following welfare maximization problem (technically, one problem for each firm type):

$$\max_{q_i, S_i \geq 0} SW(q_i, S_i) = \overbrace{\frac{1}{2}q_i^2}^{CS_i} + \overbrace{(1 - q_i - c_i)\, q_i + S_i - \frac{1}{2}}^{PS_i} - \frac{5}{4}S_i$$

subject to the following participation constraint of firm i:

$$(1 - q_i - c_i)\, q_i + S_i - \frac{1}{2} \geq 0. \qquad (PC_i)$$

(b) Find the number of students served under complete information, q_i^{**}, and the subsidy, S_i^{**}, when the firm's costs are (i) low and (ii) high:
- Participation constraint PC_i must bind (otherwise the president can further reduce the subsidy and still induce the participation of firm i), so we must have that

$$S_i = \frac{1}{2} - (1 - q_i - c_i)\, q_i.$$

- Inserting this equation into the president's objective function, we can simplify the above welfare maximization problem to the following unconstrained problem:

$$\begin{aligned}
&\max_{q_i \geq 0} SW(q_i) \\
&= \frac{1}{2}q_i^2 + (1 - q_i - c_i)\, q_i - \frac{1}{2} \\
&\quad + \underbrace{\frac{1}{2} - (1 - q_i - c_i)\, q_i}_{\text{Subsidy } S_i} - \frac{5}{4}\underbrace{\left[\frac{1}{2} - (1 - q_i - c_i)\, q_i\right]}_{\text{Subsidy } S_i} \\
&= \frac{1}{2}q_i^2 + \frac{5}{4}(1 - q_i - c_i)\, q_i - \frac{5}{8}.
\end{aligned}$$

An additional advantage of this rearranged welfare maximization problem is that we have now reduced the set of choice variables from two (q_i and S_i) to one, q_i.
- Differentiating with respect to q_i, and assuming interior solutions, we find

$$q_i + \frac{5}{4}(1 - 2q_i - c_i) = 0,$$

which, after rearranging and solving for q_i, yields the equilibrium number of students being served,

$$q_i^{**} = \frac{5\,(1 - c_i)}{6}.$$

- Substituting equilibrium output q_i^{**} into the participation constraint of the monopolist, we find

$$S_i^{**} = \frac{1}{2} - \left(1 - q_i^{**} - c_i\right) q_i^{**}$$
$$= \frac{1}{2} - \left(1 - \frac{5(1-c_i)}{6} - c_i\right) \frac{5(1-c_i)}{6}$$
$$= \frac{1}{2} - \frac{5(1-c_i)^2}{36}.$$

- *Summary.* The low-cost firm receives a subsidy of $\frac{13}{36}$ to serve $\frac{5}{6}$ students. In contrast, the high-cost firm receives a subsidy of $\frac{1}{2} - \frac{5(1-c_H)^2}{36}$ to serve $\frac{5(1-c_H)}{6}$ students. It is straightforward to show that the more inefficient is the high-cost firm (c_H increases), the more subsidy this firm receives in serving fewer students.

(c) *Asymmetric information.* Assume that the university president can only observe the probability distribution of the firm's costs but does not know whether the firm's cost is low or high. Set up the university's welfare maximization problem and characterize the firm's participation and incentive compatibility constraints. Then, argue which of these constraints hold with equality, and use your results to simplify the university's welfare maximization problem:

- The university solves the following expected welfare maximization problem:

$$\max_{q_L, q_H, S_L, S_H} \frac{1}{2}\left[\overbrace{\frac{1}{2}q_L^2}^{CS_L} + \overbrace{(1-q_L)\,q_L + S_L - \frac{1}{2}}^{PS_L} - \frac{5}{4}S_L\right]$$
$$+ \frac{1}{2}\left[\overbrace{\frac{1}{2}q_H^2}^{CS_H} + \overbrace{(1-q_H-c_H)\,q_H + S_H - \frac{1}{2}}^{PS_H} - \frac{5}{4}S_H\right]$$

subject to the following participation constraints for each firm type:

$$(1-q_L)\,q_L + S_L - \frac{1}{2} \geq 0 \qquad (PC_L)$$
$$(1-q_H-c_H)\,q_H + S_H - \frac{1}{2} \geq 0, \qquad (PC_H)$$

and the following incentive compatibility constraints for each firm type,

$$(1-q_L)\,q_L + S_L - \frac{1}{2} \geq (1-q_H)\,q_H + S_H - \frac{1}{2} \qquad (IC_L)$$
$$(1-q_H-c_H)\,q_H + S_H - \frac{1}{2} \geq (1-q_L-c_H)\,q_L + S_L - \frac{1}{2}, \qquad (IC_H)$$

where constraint IC_L (IC_H) mandates that the low-(high-) cost firm has no incentives to take on the contract designated for the high-(low-) cost firm.

- We now investigate whether PC_L is slack or not. From IC_L, we have

$$(1 - q_L) q_L + S_L - \frac{1}{2} \geq (1 - q_H) q_H + S_H - \frac{1}{2}$$
$$> (1 - q_H - c_H) q_H + S_H - \frac{1}{2} \geq 0,$$

where the second inequality is due to the fact that $c_H > 0$ by definition. Using the first and last terms in the above inequality, we obtain

$$(1 - q_L) q_L + S_L - \frac{1}{2} > 0,$$

which implies that PC_L is slack (i.e., holds with strict inequality). Hence, PC_H must bind; otherwise, the university could further reduce the subsidy S_H to the high-cost firm and still induce its participation.

Substituting the binding PC_H into IC_H, the IC_H constraint becomes

$$0 \geq (1 - q_L - c_H) q_L + S_L - \frac{1}{2}$$

meaning that IC_H is slack.

Accordingly, IC_L must bind so the low-cost firm does not have incentives to mimic the high-cost firm (otherwise, the university can further reduce the subsidy S_L to the low-cost firm and still induce this firm to choose the contract meant for its type).

In this context, let us rewrite the binding PC_H and IC_L constraints as follows:

$$S_H = \frac{1}{2} - (1 - q_H - c_H) q_H$$
$$S_L = (1 - q_H) q_H + S_H - (1 - q_L) q_L$$
$$= (1 - q_H) q_H + \overbrace{\frac{1}{2} - (1 - q_H - c_H) q_H}^{=S_H} - (1 - q_L) q_L$$
$$= \frac{1}{2} + c_H q_H - (1 - q_L) q_L.$$

Substituting S_L and S_H into the university's objective function, we have

$$\max_{q_L, q_H \geq 0} \frac{1}{2} \left[\frac{1}{2} q_L^2 + (1 - q_L) q_L - \frac{1}{2} - \frac{1}{4} \overbrace{\left(\frac{1}{2} + c_H q_H - (1 - q_L) q_L \right)}^{=S_L} \right]$$
$$+ \frac{1}{2} \left[\frac{1}{2} q_H^2 + (1 - q_H - c_H) q_H - \frac{1}{2} - \frac{1}{4} \overbrace{\left(\frac{1}{2} - (1 - q_H - c_H) q_H \right)}^{=S_H} \right],$$

which reduces the number of choice variables from four to two (q_L and q_H).

(d) Find the number of students served under incomplete information and the subsidy when the firm's costs are (i) low and (ii) high:

- *Finding output levels.* Differentiating with respect to q_L and q_H, and assuming interior solutions, we find

$$q_L + \frac{5}{4}(1 - 2q_L) = 0$$

$$q_H + \frac{5}{4}(1 - 2q_H - c_H) - \frac{c_H}{4} = 0.$$

Solving the system of equations above, we obtain the equilibrium number of students being served under incomplete information:

$$q_L^* = \frac{5}{6}$$

$$q_H^* = \frac{5 - 6c_H}{6}$$

so that the high-cost firm serves fewer students than the low-cost firm.

- *Finding subsidies.* Substituting output levels q_L^* and q_H^* into the subsidy, we find that the equilibrium subsidies to each type of firm become

$$S_H^* = \frac{1}{2} - \left(1 - q_H^* - c_H\right) q_H^*$$

$$= \frac{1}{2} - \frac{5 - 6c_H}{36}$$

$$S_L^* = \frac{1}{2} + c_H q_H^* - \left(1 - q_L^*\right) q_L^*$$

$$= \frac{13}{36} + \frac{c_H(5 - 6c_H)}{6}.$$

We then compare these output levels and subsidies against those under the complete information context in part (e) of the exercise.

(e) *Comparison.* Compare your results in parts (b) and (d). Interpret:

- *Comparing output levels across information settings:*
  - The low-cost firm serves the same number of students across both information contexts, that is, $q_L^* = q_L^{**}$. In other words, asymmetric information does not cause this firm to change its output level relative to the complete information case, which is known as the "no distortion at the top" condition.
  - However, the high-cost firm serves fewer students when the university does not observe the firm's cost, since the output difference

$$q_H^* - q_H^{**} = \frac{5 - 6c_H}{6} - \frac{5(1 - c_H)}{6}$$

$$= -\frac{c_H}{6} < 0$$

is unambiguously negative.

- *Comparing subsidies across information settings:*
  - The low-cost firm receives a larger subsidy (or pays a smaller entry fee) under asymmetric information, that is, $S_L^* > S_L^{**}$, since $c_H q_H^* > 0$, which holds because $q_H^* \geq 0$ and $c_H > 0$ by assumption. Intuitively, the university needs to offer an information premium (also called the information rent) of $S_L^* - S_L^{**}$ in order to induce the low-cost firm not to mimic the high-cost firm.
  - In contrast, the high-cost firm receives a larger subsidy under asymmetric information, that is, $S_H^* > S_H^{**}$, when the following inequality holds:

$$\frac{13}{36} + \frac{c_H(5 - 6c_H)}{6} > \frac{1}{2} - \frac{5(1 - c_H)^2}{36}$$

$$5 - 10c_H + 5c_H^2 + 30c_H - 36c_H^2 > 5,$$

which simplifies to $c_H < \frac{20}{31} \simeq 0.65$. Intuitively, because the high-cost firm serves fewer students under asymmetric information, its subsidy is more generous in this setting only if its marginal cost is sufficiently low.

(f) *Numerical example.* Consider $c_H = \frac{1}{4}$. Find the firm's output and subsidies under the complete information setting in part (b) and the incomplete information context in part (d):

- *Complete information.* Substituting the values found in part (b) above, we find output levels

$$q_L^{**} = \frac{5}{6} \simeq 0.833$$

$$q_H^{**} = \frac{5\left(1 - \frac{1}{4}\right)}{6} = \frac{5}{8} \simeq 0.625.$$

In addition, equilibrium subsidies in this setting become

$$S_L^{**} = \frac{13}{36} \simeq 0.361$$

$$S_H^{**} = \frac{1}{2} - \frac{5\left(1 - \frac{1}{4}\right)^2}{36} = \frac{1}{2} - \frac{5}{64} \simeq 0.422.$$

Therefore, the university provides a lower (higher) subsidy to the low- (high-) cost firm to produce more (less) output.

- *Incomplete information.* Substituting the values found in part (b) above, we find output levels

$$q_L^* = \frac{5}{6} \simeq 0.833$$

$$q_H^* = \frac{5 - 6\frac{1}{4}}{6} = \frac{7}{12} \simeq 0.583.$$

In addition, equilibrium subsidies in this setting become

$$S_L^* = \frac{13}{36} + \frac{\frac{1}{4}\left(5-6\frac{1}{4}\right)}{6} = \frac{73}{144} \simeq 0.507$$

$$S_H^* = \frac{1}{2} - \frac{5-6\frac{1}{4}}{36} = \frac{29}{72} \simeq 0.403.$$

Therefore, the university provides a higher (lower) subsidy to the low- (high-) cost firm to produce more (less) output. In particular, a lower subsidy is offered to the high-cost firm to produce less output under incomplete than complete information.

Exercise #5.6: Regulating a Monopolist Under Incomplete Information-II^C

5.6. Consider the setting in Exercise 5.5, but allowing for a more general framework. In particular, assume now that the firm's fixed and unit costs can be low (F_L, c_L) or high (F_H, c_H), with probability p and $1-p$, respectively, where $F_H > F_L > 0$, $c_H > c_L > 0$, and $p \in [0, 1]$. The inverse demand function is $p(q) = 1 - q$, where q represents the aggregate output. The university still offers a subsidy S_i to the firm, where $i \in \{L, H\}$, and allows for the university's cost of paying subsidy S_i to be $(1+g)S_i$, where $g > 0$ denotes the administrative cost of collecting the funds to pay the subsidy:

(a) *Complete information.* Assume that the university president has access to the cost information of the firm. Set up the university's welfare maximization problem, which takes into account firm i's participation constraint:

- Observing firm i's cost, the president solves the following welfare maximization problem (technically, one problem for each firm type):

$$\max_{q_i, S_i \geq 0} SW(q_i, S_i) = \overbrace{\frac{1}{2}q_i^2}^{CS_i} + \overbrace{(1-q_i-c_i)\,q_i + S_i - F_i}^{PS_i} - (1+g)\,S_i$$

subject to the following participation constraint of firm i:

$$(1-q_i-c_i)\,q_i + S_i - F_i \geq 0. \qquad (PC_i)$$

(b) Find the number of students served under complete information, q_i^{**}, and the subsidy, S_i^{**}, when the firm's costs are (i) low and (ii) high.

- Participation constraint PC_i must bind (otherwise the president can further reduce the subsidy and still induce the participation of firm i), so we must have that

$$S_i = F_i - (1-q_i-c_i)\,q_i.$$

- Inserting this equation into the president's objective function, we can simplify the above welfare maximization problem to the following unconstrained problem:

$$\max_{q_i \geq 0} SW(q_i) = \frac{1}{2}q_i^2 + (1 - q_i - c_i)\, q_i + \underbrace{[F_i - (1 - q_i - c_i)\, q_i]}_{\text{Subsidy } S_i} - F_i$$
$$- (1+g)\underbrace{[F_i - (1 - q_i - c_i)\, q_i]}_{\text{Subsidy } S_i}$$
$$= \frac{1}{2}q_i^2 + (1+g)(1 - q_i - c_i)\, q_i - (1+g)\, F_i.$$

An additional advantage of this rearranged welfare maximization problem is that we have now reduced the set of choice variables from two (q_i and S_i) to one, q_i.

- Differentiating with respect to q_i, and assuming interior solutions, we find

$$q_i + (1+g)(1 - 2q_i - c_i) = 0,$$

which, after rearranging and solving for q_i, yields the equilibrium number of students being served

$$q_i^{**} = \frac{(1+g)(1-c_i)}{1+2g},$$

which is decreasing in firm i's marginal production cost, c_i, and in the administrative cost, g, because

$$\frac{\partial q_i^{**}}{\partial g} = -\frac{1-c_i}{(1+2g)^2} < 0.$$

Intuitively, the more inefficient is the arrangement of the subsidy (higher administrative costs, thus entailing a higher g), the fewer students can be served. Note that when $g = 0$, the number of students simplifies to $q_i^{**} = (1 - c_i)$, which coincides with socially optimal output of a regulated monopoly.

- Substituting equilibrium output q_i^{**} into the participation constraint of the monopolist, we find

$$S_i^{**} = F_i - \left(1 - q_i^{**} - c_i\right) q_i^{**}$$
$$= F_i - \left(1 - \frac{(1+g)(1-c_i)}{(1+2g)} - c_i\right) \frac{(1+g)(1-c_i)}{1+2g}$$
$$= F_i - \frac{g\,(1+g)(1-c_i)^2}{(1+2g)^2},$$

which is decreasing in the administrative cost g because

$$\frac{\partial S_i^{**}}{\partial g} = -\frac{(1-c_i)^2}{(1+2g)^3} < 0.$$

Intuitively, the more inefficient is the arrangement of subsidy (higher g), the lower the subsidy that the university offers to the firm to provide catering services.

- *Summary.* Thus, the university offers a subsidy of $S_L^{**} = F_L - \frac{g(1+g)(1-c_L)^2}{(1+2g)^2}$ to the low-cost firm to serve $q_L^{**} = \frac{(1+g)(1-c_L)}{(1+2g)}$ students. When the firm's cost is high, however, the subsidy increases to $S_H^{**} = F_H - \frac{g(1+g)(1-c_H)^2}{(1+2g)^2}$, but fewer students, $q_H^{**} = \frac{(1+g)(1-c_H)}{(1+2g)}$, are served.

(c) *Asymmetric information.* Assume that the university president can only observe the probability distribution of the firm's costs but does not know whether the firm's cost is low or high. Set up the university's welfare maximization problem and characterize the firm's participation and incentive constraints. Then, argue which of these constraints hold with equality, and use your results to simplify the university's welfare maximization problem:

- The university solves the following expected welfare maximization problem:

$$\max_{q_L, q_H, S_L, S_H} p \left[\overbrace{\frac{1}{2}q_L^2}^{CS_L} + \overbrace{(1 - q_L - c_L)\, q_L + S_L - F_L}^{PS_L} - (1+g)\, S_L \right]$$

$$+ (1-p) \left[\overbrace{\frac{1}{2}q_H^2}^{CS_H} + \overbrace{(1 - q_H - c_H)\, q_H + S_H - F_H}^{PS_H} - (1+g)\, S_H \right]$$

subject to the following participation constraints for each firm type:

$$(1 - q_L - c_L)\, q_L + S_L - F_L \geq 0 \qquad (PC_L)$$

$$(1 - q_H - c_H)\, q_H + S_H - F_H \geq 0, \qquad (PC_H)$$

and the following incentive compatibility constraints for each firm type:

$$(1 - q_L - c_L)\, q_L + S_L - F_L \geq (1 - q_H - c_L)\, q_H + S_H - F_L \qquad (IC_L)$$

$$(1 - q_H - c_H)\, q_H + S_H - F_H \geq (1 - q_L - c_H)\, q_L + S_L - F_H, \qquad (IC_H)$$

where constraint IC_L (IC_H) mandates that the low-(high-) cost firm has no incentives to take on the contract designated for the high- (low-) cost firm.

- We now investigate whether PC_L is slack or not. From IC_L, we have

$$\begin{aligned}(1 - q_L - c_L)\, q_L + S_L - F_L &\geq (1 - q_H - c_L)\, q_H + S_H - F_L \\ &> (1 - q_H - c_H)\, q_H + S_H - F_H \geq 0,\end{aligned}$$

where the second inequality is due to the fact that $c_H > c_L$ and $F_H > F_L$ by definition. Using the first and last terms in the above inequality, we obtain

$$(1 - q_L - c_L)\, q_L + S_L - F_L > 0,$$

which implies that PC_L is slack (i.e., holds with strict inequality). Hence, PC_H must bind; otherwise, the university could further reduce the subsidy S_H to the high-cost firm and still induce its participation.

Substituting the binding PC_H into IC_H, the IC_H constraint becomes

$$0 \geq (1 - q_L - c_H)\, q_L + S_L - F_H$$

meaning that IC_H is slack.

Accordingly, IC_L must bind, so the low-cost firm does not have incentives to mimic the high-cost firm (otherwise, the university can further reduce the subsidy S_L to the low-cost firm and still induce this firm to choose the contract meant for its type).

In this context, let us rewrite the binding PC_H and IC_L constraints as follows:

$$\begin{aligned}
S_H &= F_H - (1 - q_H - c_H)\, q_H \\
S_L &= (1 - q_H - c_L)\, q_H + S_H - (1 - q_L - c_L)\, q_L \\
&= (1 - q_H - c_L)\, q_H + \overbrace{F_H - (1 - q_H - c_H)\, q_H}^{=S_H} - (1 - q_L - c_L)\, q_L \\
&= F_H + (c_H - c_L)\, q_H - (1 - q_L - c_L)\, q_L.
\end{aligned}$$

Substituting S_L and S_H into the university's objective function, we have

$$\max_{q_L, q_H \geq 0}\; p\left[\frac{1}{2}q_L^2 + (1 - q_L - c_L)q_L - F_L - g\overbrace{(F_H + (c_H - c_L)q_H - (1 - q_L - c_L)q_L)}^{=S_L}\right]$$
$$+ (1-p)\left[\frac{1}{2}q_H^2 + (1 - q_H - c_H)\, q_H - F_H - g\overbrace{(F_H - (1 - q_H - c_H)\, q_H)}^{=S_H}\right],$$

which reduces the number of choice variables, from four to two (q_L and q_H).

(d) Find the number of students served under incomplete information and the sponsorship when the firm's costs are (i) low and (ii) high.

- *Finding output levels.* Differentiating with respect to q_L and q_H, and assuming interior solutions, we find

$$q_L + (1+g)\,(1 - 2q_L - c_L) = 0$$
$$(1-p)\,[q_H + (1+g)\,(1 - 2q_H - c_H)] - pg\,(c_H - c_L) = 0.$$

Solving the system of equations above, we obtain equilibrium output levels under incomplete information:

$$q_L^* = \frac{(1+g)\,(1-c_L)}{1+2g}$$
$$q_H^* = \frac{(1-p)\,(1+g)\,(1-c_H) - pg\,(c_H - c_L)}{(1-p)\,(1+2g)}.$$

We compare these output levels against those under the complete information context in part (e) of the exercise.

- *Finding subsidies.* Substituting output levels q_L^* and q_H^* into the subsidy, we find that the equilibrium subsidies to each type of firm become

$$S_H^* = F_H - \left(1 - q_H^* - c_H\right) q_H^*$$
$$= F_H$$
$$- \frac{g\left[(1-p)(1-c_H) + p(c_H - c_L)\right]\left[(1-p)(1+g)(1-c_H) - pg(c_H - c_L)\right]}{(1-p)^2(1+2g)^2}$$

$$S_L^* = F_H + (c_H - c_L) q_H^* - \left(1 - q_L^* - c_L\right) q_L^*$$
$$= F_H + \frac{(c_H - c_L)\left[(1-p)(1+g)(1-c_H) - pg(c_H - c_L)\right]}{(1-p)(1+2g)}$$
$$- \frac{g(1+g)(1-c_L)^2}{(1+2g)^2}.$$

(e) *Comparison.* Compare your results in parts (b) and (d). Interpret:
- *Comparing output levels across information settings:*
  - The low-cost firm serves the same number of students across both information contexts, that is, $q_L^* = q_L^{**}$. In other words, asymmetric information does not cause this firm to change its output level relative to the complete information case, which is known as the "no distortion at the top" condition.
  - However, the high-cost firm serves fewer students when the university does not observe the firm's cost, since the output difference

$$q_H^* - q_H^{**} = \frac{(1-p)(1+g)(1-c_H) - pg(c_H - c_L)}{(1-p)(1+2g)} - \frac{(1+g)(1-c_H)}{(1+2g)}$$
$$= -\frac{pg(c_H - c_L)}{1+2g} < 0,$$

which is unambiguously negative. Therefore, $q_H^* < q_H^{**}$, and fewer students are served by the high-cost firm under incomplete than complete information.
- *Comparing subsidies across information settings:*
  - The low-cost firm receives a larger subsidy under asymmetric information, that is, $S_L^* > S_L^{**}$, since

$$F_H + (c_H - c_L) q_H^* > F_L,$$

which holds because $q_H^* \geq 0$, $F_H > F_L$, and $c_H > c_L$ by assumption. Intuitively, the university needs to offer an information premium of $S_L^* - S_L^{**}$ in order to induce the low-cost firm not to mimic the high-cost firm.
  - In contrast, the high-cost firm receives a smaller subsidy under asymmetric information, that is, $S_H^* < S_H^{**}$, when

$$\left(1 - q_H^* - c_H\right) q_H^* > \frac{g(1+g)(1-c_H)^2}{(1+2g)^2}$$

holds, which simplifies to $g < \overline{g} \equiv \frac{(1-p)(1-c_H)}{p(c_H-c_L)}$. As the high-cost firm serves fewer students under asymmetric information, its subsidy is reduced commensurately when the administrative cost is not too significant, that is, $g < \overline{g}$. Otherwise, this type of firm would receive a more generous subsidy.

(f) *Numerical example.* Consider parameter values $p = \frac{1}{2}$, $g = 1$, $c_L = F_L = 0$, $c_H = \frac{1}{4}$, and $F_H = \frac{1}{8}$. Find the firm's output and subsidies under the complete information setting in part (b) and the incomplete information context in part (d):

- *Complete information.* Substituting the values found in part (b) above, we find output levels

$$\begin{aligned} q_L^{**} &= \frac{(1+g)(1-c_L)}{1+2g} \\ &= \frac{1+1}{1+2} = \frac{2}{3} \text{ and} \\ q_H^{**} &= \frac{(1+g)(1-c_H)}{1+2g} \\ &= \frac{(1+1)\left(1-\frac{1}{4}\right)}{1+2} = \frac{1}{2}. \end{aligned}$$

Therefore, the low-cost firm serves more students than the high-cost firm. In addition, equilibrium subsidies in this setting become

$$\begin{aligned} S_L^{**} &= F_L - \frac{g(1+g)(1-c_L)^2}{(1+2g)^2} \\ &= 0 - \frac{(1+1)}{(1+2)^2} = -\frac{2}{9} \text{ and} \\ S_H^{**} &= F_H - \frac{g(1+g)(1-c_H)^2}{(1+2g)^2} \\ &= \frac{1}{8} - \frac{(1+1)\left(1-\frac{1}{4}\right)^2}{(1+2)^2} = 0. \end{aligned}$$

Interestingly, the university sets an entry fee to the low-cost firm but does not offer a subsidy to the high-cost firm and still induces the participation of both types of firms.

- *Asymmetric information.* Similarly, substituting the values found in part (d) above, we find output levels

$$q_L^* = \frac{(1+g)(1-c_L)}{1+2g} = \frac{1+1}{1+2} = \frac{2}{3} \text{ and}$$

$$q_H^* = \frac{(1-p)(1+g)(1-c_H) - pg(c_H - c_L)}{(1-p)(1+2g)} = \frac{(1+1)\left(1-\frac{1}{4}\right) - \frac{1}{4}}{1+2} = \frac{5}{12}.$$

While the low-cost firm still serves the same number of students, the high-cost firm serves fewer students than that of complete information. In addition, equilibrium subsidies in this setting become

$$S_H^* = F_H - \frac{g\left[(1-p)(1-c_H)+p(c_H-c_L)\right]\left[(1-p)(1+g)(1-c_H)-pg(c_H-c_L)\right]}{(1-p)^2(1+2g)^2}$$
$$= \frac{1}{8} - \frac{\left[\left(1-\frac{1}{4}\right)+\frac{1}{4}\right]\left[(1+1)\left(1-\frac{1}{4}\right)-\frac{1}{4}\right]}{(1+2)^2}$$
$$= -\frac{1}{72}, \text{ and}$$

$$S_L^* = F_H + \frac{(c_H-c_L)\left[(1-p)(1+g)(1-c_H)-pg(c_H-c_L)\right]}{(1-p)(1+2g)} - \frac{g(1+g)(1-c_L)^2}{(1+2g)^2}$$
$$= \frac{1}{8} + \frac{\frac{1}{4}\left[(1+1)\left(1-\frac{1}{4}\right)-\frac{1}{4}\right]}{1+2} - \frac{(1+1)}{(1+2)^2}$$
$$= \frac{1}{144}.$$

Since the high-cost firm is making positive profits, the university can still charge a small entry fee to this firm and make it indifferent to participation or not (that is, its participation constraint binds). Whereas, the low-cost firm is offered a subsidy so it does not have incentives to pay an entry fee and take up the high-cost contract (that is, its incentive constraint binds).

Exercise #5.7: Regulating a Cournot Oligopoly^A

5.7. Consider an industry with $N \geq 2$ symmetric firms competing à la Cournot, facing inverse demand function $p(Q) = 1 - Q$, where Q represents aggregate output, and having a common marginal cost of production c, where $1 > c \geq 0$. In the first stage, a social planner sets a per-unit subsidy s. In the second period, every firm i observes this per-unit subsidy s and responds independently and simultaneously choosing its output level q_i. For simplicity, assume that $c > s \geq 0$.

(a) *Second stage.* Find the best response function in the second period, $q_i(q_j)$.

- In the second stage, every firm i solves

$$\max_{q_1 \geq 0} (1 - q_i - Q_{-i})q_i - (c - s)\, q_i,$$

where $Q_{-i} = \sum_{j \neq i} q_j$ denotes the aggregate output from firm i's rivals. In addition, the above profit is evaluated at firm i's second-period cost, $c - s$, after receiving a per-unit subsidy s for every unit of output produced. Intuitively, this subsidy reduces the marginal cost of production from c to $c - s$.

- Differentiating with respect to q_i, we obtain

$$1 - 2q_i - Q_{-i} = c - s.$$

Solving for q_i, we find firm i's best response function

$$q_1(q_2) = \frac{1-(c-s)}{2} - \frac{1}{2}Q_{-i}.$$

Therefore, firm i's best response function originates at $\frac{1-(c-s)}{2}$ and decreases at a rate of 1/2 for every unit produced by any of its rivals. Graphically, this best response function shifts upward when the social planner sets a more generous subsidy s per unit of output in the first period of the game.

(b) Find the second-period output function, $q_i(s)$:

- In a symmetric equilibrium, all firms produce the same output, $q_i^* = q_j^* = q^*$ for every two firms $j \neq i$. Therefore, we must have that $Q_{-i}^* = (N-1)q^*$, which simplifies our above best response function to

$$q^* = \frac{1-(c-s)}{2} - \frac{(N-1)q^*}{2},$$

which, rearranging and solving for q_i, yields the equilibrium output of

$$q_i(s) = \frac{1-(c-s)}{N+1}.$$

This output level is positive since $1 > c > s$ by assumption. Furthermore, equilibrium output is decreasing in the effective marginal cost that firm i faces in the second period, $c - s$.

(c) *First stage.* Find the welfare maximizing subsidy s^* that the social planner sets in the first period, assuming that welfare is defined as $W = \frac{Q^2}{2} + N \times \pi(s) - sQ$, where the first term denotes the consumer surplus in this setting with linear inverse demand function, the second term represents aggregate profits as a function of subsidy s, and the third term indicates the total cost of providing the subsidy. For simplicity, we assume that there is no deadweight loss of raising the public funds dedicated to the subsidy:

- The social planner can first determine the socially optimal output at the aggregate level, Q^{SO}, which solves

$$\begin{aligned}\max_{Q \geq 0} W &= \underbrace{\frac{Q^2}{2}}_{CS} + \underbrace{[(1-Q)Q - (c-s)Q]}_{N\times\pi(s)} - sQ \\ &= \frac{Q^2}{2} + [(1-Q)Q - cQ].\end{aligned}$$

The last term in the welfare function (profits) does not include the subsidy s per unit of output.

Differentiating with respect to Q yields

$$1 - Q - c = 0,$$

and solving for Q, we obtain

$$Q^{SO} = 1 - c,$$

which coincides with the aggregate output under perfectly competitive markets. As a consequence, in this industry with N symmetric firms, each firm should produce

$$q^{SO} = \frac{Q^{SO}}{N} = \frac{1-c}{N}.$$

To determine the socially optimal subsidy, s^*, we only need to set the equilibrium output function we found in part (b), $q_i(s)$, equal to this socially optimal output level, q^{SO}, as follows:

$$\frac{1-(c-s)}{N+1} = \frac{1-c}{N},$$

and solving for s, we find an optimal subsidy

$$s^* = \frac{1-c}{N}.$$

(d) Is the optimal subsidy s^* increasing or decreasing in the number of firms? Interpret:

- The optimal subsidy s^* is decreasing in the number of firms competing in the industry, N, since

$$\frac{ds^*}{dN} = -\frac{1-c}{N^2} < 0.$$

Intuitively, as more firms compete, their aggregate output in equilibrium becomes closer to the socially optimal $Q^{SO} = 1 - c$, implying that the social planner needs to offer a lower subsidy. The optimal subsidy s^* is, however, positive (or approaches zero) for all values of N.

(e) Evaluate your above results in a Cournot duopoly with symmetric firms. What is the socially optimal subsidy s^*? What about the case of a monopolist?

- Evaluating the above subsidy s^* at $N = 2$, we obtain a duopoly subsidy of $s^* = \frac{1-c}{2}$.
- However, when dealing with a monopoly, the optimal subsidy becomes $s^* = 1-c$. In both cases, the aggregate output in equilibrium coincides with the socially optimal aggregate output $Q^{SO} = 1 - c$.

Exercise #5.8: Regulating a Polluting Cournot Oligopoly^B

5.8. Consider the industry in Exercise 5.7, but assuming that every unit of output generates one unit of pollution. The social planner considers the environmental damage of this pollution in the social welfare function, which now becomes

$$W = \left(\frac{Q^2}{2} + N \times \pi(s) - sQ\right) - dQ^2.$$

The first three terms (in brackets for clarity) coincide with those in the previous exercise. The last term, however, represents the environmental damage from aggregate pollution, which is increasing and convex in production (and thus in pollution) and parameter d satisfies $d > 0$:

(a) *Second stage.* Find the best response function in the second period, $q_i(q_j)$. Find the second-period output function, $q_i(s)$:
   - Every firm i still solves the same profit maximization problem as in the previous exercise, obtaining the same best response function, and yielding the same equilibrium output

$$q_i(s) = \frac{1-(c-s)}{N+1}.$$

(b) *First stage.* Find the welfare maximizing subsidy s^* that the social planner sets in the first period:
   - The social planner can first determine the socially optimal output at the aggregate level, Q^{SO}, which solves

$$\max_{Q \geq 0} W = \underbrace{\frac{Q^2}{2}}_{CS} + \underbrace{[(1-Q)Q - (c-s)Q]}_{N \times \pi(s)} - sQ - \underbrace{dQ^2}_{\text{Env. Damage}}.$$

Differentiating with respect to Q yields

$$1 - Q - c - 2dQ = 0,$$

and solving for Q, we obtain

$$Q^{SO} = \frac{1-c}{1+2d}.$$

Therefore, in this industry with N symmetric firms, each firm should produce

$$q^{SO} = \frac{Q^{SO}}{N} = \frac{1-c}{N(1+2d)}.$$

To determine the socially optimal subsidy, s^*, we only need to set the equilibrium output function we found in part (a), $q_i(s)$, equal to this socially optimal output level, q^{SO}, as follows:

$$\frac{1-(c-s)}{N+1} = \frac{1-c}{N(1+2d)},$$

and solving for s, we find an optimal subsidy

$$s^* = \frac{(1-c)(1-2dN)}{N(1+2d)}.$$

(c) Is the optimal subsidy s^* increasing or decreasing in d? What about in N?
   - Differentiating s^* with respect to d, we find

$$\frac{\partial s^*}{\partial d} = -\frac{2(1-c)(1+N)}{N(1+2d)^2} < 0,$$

and differentiating it with respect to N, we obtain

$$\frac{\partial s^*}{\partial N} = -\frac{1-c}{N^2(1+2d)} < 0.$$

Therefore, as the environmental damage from pollution becomes more severe (higher d), the social planner provides a less generous subsidy to the firms, thus inducing the production of fewer units. (We talk about this point in part d again, where we identify conditions under which the subsidy can become a tax, known in this context as an "emission fee.") The optimal subsidy s^* is also decreasing in the number of firms competing in the industry, thus exhibiting a similar comparative statics as in the previous exercise.

(d) Under which conditions on N and d does the optimal subsidy s^* become negative? Interpret:
- Subsidy s^* is positive for all $\frac{(1-c)(1-2dN)}{N(1+2d)} > 0$, which holds for all $d < \frac{1}{2N}$. Intuitively, the environmental damage from pollution cannot be extremely severe (relatively low values of d) as otherwise the social planner would prefer to set a negative subsidy (an emission fee per unit of pollution), to curb emissions. Another interpretation of this result highlights the presence of two market failures in this market: the underproduction in the Cournot oligopoly, relative to perfect competition, which would lead regulators to set subsidies that induce an increase in output levels, and the overproduction due to pollution, which would call for low (or negative) subsidies that induce output reductions:
  - When condition $d < \frac{1}{2N}$ holds, the first market failure dominates, leading to a positive subsidy, which induces firms to increase their production.
  - In contrast, when this condition is violated, the second market failure dominates, entailing a negative subsidy (a tax), ultimately inducing firms to decrease their production.
- The condition on d, $d < \frac{1}{2N}$, becomes more demanding as more firms compete in the industry. As depicted in Fig. 5.2, when $N = 2$ (at the vertical axis), the condition on d collapses to $d < \frac{1}{4}$, which still allows for positive subsidies when d lies in $d \in [0, \frac{1}{4})$. However, when $N = 10$, the condition becomes $d < \frac{1}{20}$, leaving a more restricted range $d \in [0, \frac{1}{20})$ where positive subsidies can be sustained in equilibrium.
- Needless to say, when only one firm operates in this market, $N = 1$, this condition on d simplifies to $d < \frac{1}{2}$, thus coinciding with the condition we found in Exercise 5.2, part (c), to sustain a positive subsidy on a monopolist.

(e) Evaluate social welfare at the socially optimal output level. How does it change with c, d, and N? Interpret:
- Social welfare in the presence of pollution, for a given Q, is

$$SW = (1-c)\,Q - \frac{Q^2}{2} - dQ^2$$
$$= \frac{[2\,(1-c) - (1+2d)\,Q]\,Q}{2}.$$

Substituting the socially optimal output, Q^{SO}, into the above expression yields the socially optimal welfare level

$$SW^{SO} = \left[2\,(1-c) - (1+2d)\,\frac{1-c}{1+2d}\right] \times \frac{1-c}{2\,(1+2d)}$$

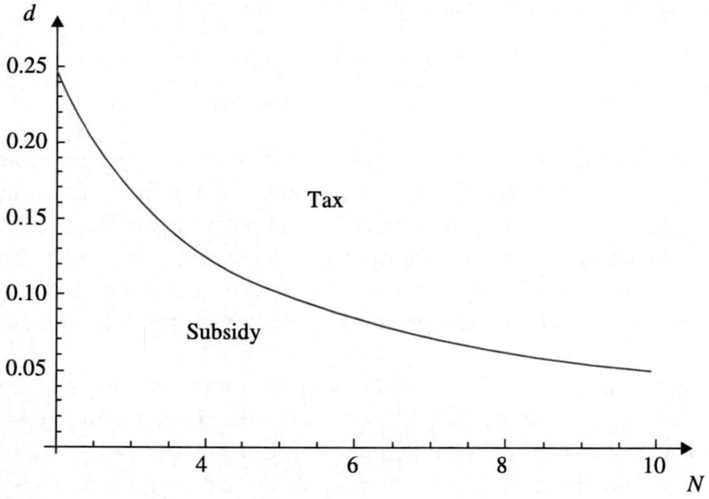

Fig. 5.2 Condition $d < \frac{1}{2N}$ for a positive subsidy

$$= \frac{(1-c)^2}{2(1+2d)},$$

which is decreasing in c and d, since a higher marginal cost or a more severe pollution unambiguously reduces welfare. However, since the socially optimal output Q^{SO} is independent of the number of firms, N, the socially optimal welfare is unaffected by N.

(f) Evaluate social welfare at the socially optimal output level. How does it change with c, d, and N? Interpret.

- Social welfare in the presence of pollution, for a given Q, is

$$SW = (1-c)\,Q - \frac{Q^2}{2} - dQ^2$$
$$= \frac{[2(1-c) - (1+2d)\,Q]\,Q}{2}.$$

Substituting the socially optimal output, Q^{SO}, into the above expression yields the socially optimal welfare level

$$SW^{SO} = \left[2(1-c) - (1+2d)\frac{1-c}{1+2d}\right] \times \frac{1-c}{2(1+2d)}$$
$$= \frac{(1-c)^2}{2(1+2d)},$$

which is decreasing in c and d, since a higher marginal cost or a more severe pollution unambiguously reduces welfare. However, since the socially optimal output Q^{SO} is independent of the number of firms, N, the socially optimal welfare is unaffected by N.

Exercise #5.9: Regulating a Polluting Cournot Oligopoly Under Complete Information – Pigouvian EPA^B

5.9. Consider the industry in Exercise 5.7, but assuming that every unit of output generates one unit of pollution. The social planner considers environmental damages of this pollution in the social welfare function, which now becomes

$$W = (1-\beta)\left[\frac{Q^2}{2} + \pi(s) - sQ\right] - \beta d Q^2.$$

The first three terms (in brackets for clarity) denote consumer and producer surplus ($CS + PS$) plus the cost of funding the subsidy program. The last term represents the environmental damages (ED) from aggregate pollution, which are increasing and convex in aggregate production (and, thus, in pollution) and parameter d satisfies $d > 0$.

In addition, parameter β satisfies $0 \leq \beta \leq 1$ and measures the weight that the regulator (e.g., EPA) assigns to environmental damages from output. For instance, when $\beta = 1$, the welfare function collapses to ED, indicating that the regulator only considers the external effects of pollution, setting Pigouvian emission fees to curb pollution; whereas when $\beta = 0$, the welfare function simplifies to $CS + PS$, representing that the regulator ignores the environmental damages from output. More compactly, parameter β denotes how "Pigouvian" the regulator is when setting emission fees (completely Pigouvian when $\beta = 1$, but not so Pigouvian otherwise). The timing coincides with that in Exercise 5.1, that is, in the first period, the regulator sets per-unit subsidy s; and, in the second period, every firm i responds choosing its output q_i:

(a) *Second stage.* Find the best response function in the second period, $q_i\left(q_j\right)$. Find the second-period output function, $q_i\,(s)$:
- Every firm i still solves the same profit maximization problem as in Exercise 5.7, obtaining the same best response function, and yielding the same equilibrium output

$$q_i\,(s) = \frac{a-(c-s)}{N+1}.$$

(b) *First stage.* Find the welfare maximizing subsidy s^* that the social planner sets in the first period:
- *First step, finding the socially optimal output.* We first find the socially optimal output Q^{SO}. This output solves

$$\max_{Q\geq 0}\ W = (1-\beta)\left\{\underbrace{\frac{Q^2}{2}}_{CS} + \underbrace{[(a-Q)\,Q-(c-s)\,Q]}_{N\times\pi(s)} - sQ\right\} - \underbrace{\beta d Q^2}_{ED},$$

where the first (second) term in the welfare function represents consumer (producer) surplus, the third term denotes the total cost of paying subsidy s to Q units of output, and the last term indicates the environmental damages from pollution. This welfare maximization problem simplifies to

$$\max_{Q\geq 0}\ W = (1-\beta)\,\frac{(2a-2c-Q)\,Q}{2} - \beta d Q^2.$$

Differentiating with respect to Q yields

$$(1-\beta)(a-c-Q)-2\beta dQ=0.$$

After rearranging, we find

$$(1-\beta)Q+2\beta dQ=(1-\beta)(a-c),$$

and solving for Q, we obtain the socially optimal output

$$Q^{SO}=\frac{(1-\beta)(a-c)}{1-\beta(1-2d)}.$$

Therefore, in this industry with N symmetric firms, each firm should produce

$$q^{SO}=\frac{Q^{SO}}{N}=\frac{(1-\beta)(a-c)}{N[1-\beta(1-2d)]}.$$

To gain some intuition of this result, consider the following special cases:

- Note that when environmental damages are absent, $d=0$, this output level simplifies to $q^{SO}=\frac{a-c}{N}$, thus coinciding with the socially optimal output in Exercise 5.6.
- The same result applies when the EPA assigns no weight on the environmental damage, $\beta=0$, which simplifies the above socially optimal output to $q^{SO}=\frac{a-c}{N}$.
- When the EPA is Pigouvian, $\beta=1$, this socially optimal aggregate output simplifies to $q^{SO}=0$, so the regulator seeks to reduce pollution as much as possible.
- Finally, when the EPA assigns the same weight to $CS+PS$ than to ED, $\beta=\frac{1}{2}$, this socially optimal output simplifies to $q^{SO}=\frac{a-c}{N(1+2d)}$, which is decreasing in the environmental damage parameter d.

• *Second step, finding the socially optimal subsidy.* We can now equate every firm i's second-stage output to the socially optimal output, $q_i(s)=q^{SO}$, which entails

$$\frac{a-(c-s)}{N+1}=\frac{(1-\beta)(a-c)}{N[1-\beta(1-2d)]}.$$

Solving for the per-unit subsidy s yields

$$s^*=\frac{(a-c)[1-\beta(1+2Nd)]}{N[1-\beta(1-2d)]},$$

which can also be expressed as $s^*=\frac{1-\beta(1+2Nd)}{1-\beta}q^{SO}$.

(c) Under which conditions is s^* positive? Is the socially optimal subsidy s^* increasing or decreasing in d, β, and N? Interpret:

• Subsidy s^* is positive if and only if $1-\beta(1+2Nd)>0$, or $d<\frac{1-\beta}{2\beta N}$. Intuitively, the regulator faces two market failures in this market: a typical underproduction due to imperfect competition, which would lead the regulator to set a subsidy to induce an increase in output, and an overproduction due to pollution, which would call for negative subsidies (emission fees) to induce an output reduction.

- When condition $d < \frac{1-\beta}{2\beta N}$ holds, indicating that pollution is not extremely damaging, the first market failure dominates, leading the regulator to set a positive subsidy, which induces every firm i to increase its production.
- In contrast, when $d > \frac{1-\beta}{2\beta N}$, pollution becomes more damaging, the second market failure dominates, and the regulator sets a negative subsidy (a tax), which induces every firm i to decrease its output.

- Subsidy s^* is unambiguously decreasing in the environmental damage parameter d since

$$\frac{\partial s^*}{\partial d} = -\frac{2\beta\,(1-\beta)\,(N+1)\,(a-c)}{N\,[1-\beta\,(1-2d)]^2} < 0.$$

In other words, the subsidy can be positive or negative depending on whether environmental damages from pollution are minor or severe, respectively, but a more damaging pollution produces a clear reduction in the per-unit subsidy that the regulator offers to every firm i.

- Subsidy s^* is unambiguously decreasing in the policy weight β since

$$\frac{\partial s^*}{\partial \beta} = -\frac{2\,(N+1)\,(a-c)\,d}{N\,[1-\beta\,(1-2d)]^2} < 0,$$

entailing that, as the regulator puts more emphasis on environmental damages, he offers a lower subsidy to every firm i.

- Subsidy s^* is unambiguously decreasing in the number of firms N since

$$\frac{\partial s^*}{\partial N} = -\frac{(1-\beta)\,(a-c)}{N^2\,[1-\beta\,(1-2d)]} < 0,$$

which suggests that as the market becomes more competitive, firms overproduce polluting goods, so a more stringent emission fee is needed to curb pollution.

(d) Find the social welfare level at the socially optimal output. How does this welfare level vary with β? Then, evaluate your results at $\beta = 0$, $\beta = 1/2$, and at $\beta = 1$. Interpret:

- Substituting Q^{SO} into the social welfare function, we find that

$$\begin{aligned} W^{SO} &= (1-\beta)\frac{\left(2a-2c-Q^{SO}\right)Q^{SO}}{2} - \beta d\left(Q^{SO}\right)^2 \\ &= \left[(1-\beta)\,(a-c) - \frac{1-\beta\,(1-2d)}{2}Q^{SO}\right]Q^{SO}. \end{aligned}$$

Substituting $Q^{SO} = \frac{(1-\beta)(a-c)}{1-\beta(1-2d)}$ into the above expression, we have that

$$\begin{aligned} W^{SO} &= \left[(1-\beta)\,(a-c) - \frac{1-\beta\,(1-2d)}{2}\frac{(1-\beta)\,(a-c)}{1-\beta\,(1-2d)}\right]\frac{(1-\beta)\,(a-c)}{1-\beta\,(1-2d)} \\ &= \frac{(1-\beta)^2\,(a-c)^2}{2\,[1-\beta\,(1-2d)]}. \end{aligned}$$

- Differentiating W^{SO} with respect to β, we obtain that

$$\frac{\partial W^{SO}}{\partial \beta} = \frac{(a-c)^2(1-\beta)\left[(1-\beta)(1-2d) - 2(1-\beta(1-2d))\right]}{2\left[1-\beta(1-2d)\right]^2}$$
$$= -\frac{(a-c)^2(1-\beta)\left[1-\beta+2(1+\beta)d\right]}{2\left[1-\beta(1-2d)\right]^2}$$

that is unambiguously negative, so that the more emphasis the regulator places on environmental damages, the lower is the resulting welfare level.

- When $\beta = 0$, this welfare level simplifies to

$$W^{SO} = \frac{(1-0)^2(a-c)^2}{2\left[1-0(1-2d)\right]} = \frac{(a-c)^2}{2},$$

which coincides to the setting without environmental damages.

- Whereas, when $\beta = 1/2$, this welfare level becomes

$$W^{SO} = \frac{\left(1-\frac{1}{2}\right)^2(a-c)^2}{2\left(1-\frac{1-2d}{2}\right)} = \frac{(a-c)^2}{4(1+2d)}$$

that decreases in environmental damages when the regulator is equally concerned about environmental damages as consumer and producer surplus.

- Finally, when $\beta = 1$, this welfare level is

$$W^{SO} = \frac{(1-1)^2(a-c)^2}{2\left[1-1(1-2d)\right]} = 0$$

since the regulator shuts down every firm i so that no output is produced.

- As a remark, the regulator should refrain from setting $\beta = 0$ and ignore environmental damages in polluting industries, since the resulting social welfare level does not take into account the pollution that ultimately reduces social welfare. Setting $\beta = 1$, however, where the regulator only considers environmental damages alone, would result in no output being produced, coming at the expense of consumers' gain and firms' profits. In this context, it is plausible for the regulator to set intermediate values of β, such as $\beta = 1/2$, to balance consumer and producer surplus with environmental damages.

(e) What happen to the socially optimal subsidy in part (b) when $N = 1$, $N = 2$, and when $N \to +\infty$? Interpret:

- When $N = 1$, the socially optimal subsidy in part (b) becomes

$$s^* = \frac{(a-c)\left[1-\beta(1+2d)\right]}{1-\beta(1-2d)},$$

thus coinciding with our results in part (c) of Exercise 5.3.

- When $N = 1$, the socially optimal subsidy in part (b) becomes

$$s^* = \frac{(a-c)\left[1-\beta(1+4d)\right]}{2\left[1-\beta(1-2d)\right]},$$

which does not coincide with our results in part (c) of Exercise 5.8 except for the parameter β, which measures how Pigouvian the EPA is.

- When $N \to +\infty$, the socially optimal subsidy in part (b) becomes

$$s^* = \frac{a-c}{1-\beta(1-2d)} \lim_{N\to+\infty} \frac{1-\beta(1+2Nd)}{N}.$$

Applying L'Hôpital's rule, we obtain that

$$s^* = \frac{a-c}{1-\beta(1-2d)} \times \frac{\frac{\partial[1-\beta(1+2Nd)]}{\partial N}}{\frac{\partial N}{\partial N}}$$
$$= -\frac{2(a-c)\beta d}{1-\beta(1-2d)},$$

which is unambiguously negative, entailing that in a perfectly competitive market, the regulator charges an emission fee to every firm i to curb pollution.

(f) Find the derivative $\frac{\partial s^*}{\partial \beta}$, and then find the cross partial $\frac{\partial^2 s^*}{\partial \beta \partial N}$. Interpret your results:

- Differentiating the socially optimal subsidy, s^*, with respect to β, we find that

$$\frac{\partial s^*}{\partial \beta} = -\frac{2(a-c)(N+1)d}{N[1-\beta(1-2d)]^2},$$

which is unambiguously negative, entailing that the more emphasis the regulator places on environmental damages (higher β), the less generous the subsidy becomes. Recall that this effect includes the possibility that $s^* < 0$ (an emission fee), which becomes more stringent as β increases.

- Further differentiating with respect to N, we obtain that

$$\frac{\partial^2 s^*}{\partial \beta \partial N} = \frac{2(a-c)d}{N^2[1-\beta(1-2d)]^2},$$

which is unambiguously positive, suggesting that the fall of subsidy in β is ameliorated as the market becomes more competitive.

(g) *Numerical example.* Assume an inverse demand function $p(q) = 1 - q$, marginal cost $c = 1/2$, $d = 1/4$, and $N = 10$. Evaluate the socially optimal subsidy in part (b) at $\beta = 0$, $\beta = 1/2$, and at $\beta = 1$. Interpret:

- The socially optimal subsidy in this context becomes

$$s^* = \frac{\left(1-\frac{1}{2}\right)\left[1-\beta\left(1+2\times 10\times\frac{1}{4}\right)\right]}{10\left[1-\beta\left(1-2\times\frac{1}{4}\right)\right]}$$
$$= \frac{1-6\beta}{10(2-\beta)}.$$

- When $\beta = 0$, this subsidy simplifies to

$$s^* = \frac{1-6\times 0}{10(2-0)} = \frac{1}{20}$$

entailing that every firm i is subsidized to produce more units of output as the regulator completely ignores environmental damages.

- Whereas, when $\beta = 1/2$, this subsidy becomes

$$s^* = \frac{1 - 6 \times \frac{1}{2}}{10\left(2 - \frac{1}{2}\right)} = -\frac{2}{15},$$

where the regulator charges an emission fee (i.e., negative subsidy) to every firm i when he considers environmental damages as equally important as consumer and producer surplus.
- Finally, when $\beta = 1$, this subsidy is

$$s^* = \frac{1 - 6 \times 1}{10\,(2 - 1)} = -\frac{1}{2},$$

which is negative, since now the regulator only considers environmental damages and taxes every firm i so it stops producing polluting output.
- Overall, the subsidy becomes less generous as the regulator becomes more Pigouvian (higher β), since he seeks to reduce pollution more significantly.

Exercise #5.10: Regulating Polluting Firms That Compete à la Stackelberg^B

5.10. Consider the setting in Exercise #5.8, but assume that two firms compete à la Stackelberg. That is, in the first stage, the regulator sets the emission fee t; in the second stage, firm 1 (leader) observes t and responds choosing its output level q_1; and, in the third stage, firm 2 (the follower) observes t and q_1 and responds by choosing its output level q_2:

(a) *Third stage.* Find the follower's best response function, $q_2\,(q_1, t)$:

- The follower chooses q_2 to solve

$$\max_{q_2 \geq 0} \pi_2\,(q_2) = (1 - q_1 - q_2)\,q_2 - (c + t)\,q_2.$$

Differentiating with respect to q_2, and assuming an interior solution, we find

$$1 - q_1 - 2q_2 - c - t = 0,$$

which we rearrange to yield

$$q_2\,(q_1, t) = \frac{1 - c - t}{2} - \frac{1}{2}q_1,$$

which has a vertical intercept of $\frac{1-c-t}{2}$ and a slope of $\frac{1}{2}$.

(b) *Second stage.* Find the leader's output level $q_1\,(t)$:

- The leader chooses q_1 to solve

$$\max_{q_1 \geq 0} \pi_1\,(q_1) = [1 - q_1 - q_2\,(q_1, t)]\,q_1 - (c + t)\,q_1.$$

Substituting $q_2\,(q_1, t) = \frac{1-c-t}{2} - \frac{1}{2}q_1$ into the above expression yields

$$\max_{q_1 \geq 0} \frac{(1 - q_1 - c - t)\,q_1}{2}.$$

Differentiating with respect to q_1, and assuming an interior solution, we find

$$1 - 2q_1 - c - t = 0,$$

which we rearrange to yield

$$q_1(t) = \frac{1 - c - t}{2},$$

which is decreasing in cost c and emission fee t.

(c) From part (b), find the follower's output level, $q_2(t)$, and the aggregate output in equilibrium.

- Substituting $q_1(t)$ into $q_2(q_1, t)$yields

$$\begin{aligned} q_2(t) &= \frac{1 - c - t}{2} - \frac{1}{2}\frac{1 - c - t}{2} \\ &= \frac{1 - c - t}{4}. \end{aligned}$$

Therefore, the aggregate output in equilibrium is

$$\begin{aligned} Q(t) &= q_1(t) + q_2(t) \\ &= \frac{1 - c - t}{2} + \frac{1 - c - t}{4} \\ &= \frac{3(1 - c - t)}{4}. \end{aligned}$$

(d) *First stage.* Find the optimal emission fee, t^{Seq}. [*Hint*: First, find the aggregate socially optimal output level, Q^{SO}, as found in Exercise 5.8, and then set Q^{SO} equal to the aggregate output that firms choose in equilibrium.]

- From Exercise 5.8, we know that $Q^{SO} = \frac{1-c}{1+2d}$. Setting $Q(t) = Q^{SO}$ yields

$$\frac{3(1 - c - t)}{4} = \frac{1 - c}{1 + 2d},$$

which simplifies to

$$(1 - c)[3(1 + 2d) - 4] = 3t(1 + 2d).$$

Solving for t, the optimal emission fee is

$$t^{Seq} = \frac{(1 - c)(6d - 1)}{3(1 + 2d)},$$

where *Seq* in the superscript denotes that firms compete sequentially.

(e) Under which conditions on d does the optimal emission fee, t^{Seq}, become positive?

- The emission fee is positive when

$$\frac{(1 - c)(6d - 1)}{3(1 + 2d)} > 0,$$

which happens if and only if

$$d > \frac{1}{6}$$

since $1 > c$ by assumption. Therefore, firms are taxed when environmental damages are relatively severe.

- If, instead, $d \leq \frac{1}{6}$ firms would receive a subsidy per unit of output, inducing them to increase their output level.
- In the special case where firms' production does not generate pollution, $d = 0$, they receive a subsidy of $\frac{1-c}{3}$ per unit of output, inducing them to produce the socially optimal output in this context, $Q^{SO} = 1 - c$.

(f) *Comparison.* Compare the optimal emission fee when firms compete sequentially (à la Stackelberg), t^{Seq} from part (d), and that when they compete simultaneously (à la Cournot), t^{Sim} from Exercise 5.8.

- The emission fee under simultaneous competition is

$$t^{Sim} = \frac{(1-c)(4d-1)}{2(1+2d)},$$

which falls below that under sequential competition since

$$\frac{(1-c)(4d-1)}{2(1+2d)} < \frac{(1-c)(6d-1)}{3(1+2d)}$$

becomes

$$2(6d-1) > 3(4d-1),$$

which further simplifies to $3 > 2$ that always holds. Therefore, emission fees are more stringent under sequential than under simultaneous competition.

- This is because, in the absence of emission fees, $t = 0$, aggregate output in equilibrium is larger under sequential than under simultaneous quantity competition, that is,

$$Q^{Seq}(0) = \frac{3(1-c)}{4} > \frac{2(1-c)}{3} = Q^{Sim}(0)$$

implying that the regulator needs to curb more pollution in the former than the latter.

Exercise #5.11: Endogenous Entry Decisions^B

5.11. Consider an industry with firms competing à la Cournot, and interacting in the following sequential-move game:

(1) In the first stage, every firm i simultaneously and independently chooses whether to enter the industry, incurring a fixed cost $F \in [0, 1]$, or not enter. If a firm does not enter, it makes zero profits.

(2) In the second stage, every firm i observes the number of firms that entered the industry, n, and responds simultaneously and independently chooses its output $q_i(n)$.

We solve this game by backward induction, assuming an inverse demand function $p(Q) = 1 - Q$, where Q denotes aggregate output, and a common marginal cost of production c, where $1 > c \geq 0$.

(a) *Second stage.* Find every firm i's best response function, $q_i(Q_{-i})$, and its output function $q_i(n)$:

- Every firm i solves

$$\max_{q_i \geq 0} (1 - q_i - Q_{-i})q_i - cq_i,$$

where $Q_{-i} = \sum_{j \neq i} q_j$ denotes the aggregate output from firm i's rivals. Differentiating with respect to q_i, we obtain

$$1 - 2q_i - Q_{-i} = c.$$

Solving for q_i, we find firm i's best response function

$$q_1(q_2) = \frac{1-c}{2} - \frac{1}{2}Q_{-i}.$$

Therefore, firm i's best response function originates at $\frac{1-c}{2}$ and decreases at a rate of 1/2 for every unit produced by any of its rivals.

- In a symmetric equilibrium, all firms produce the same output, $q_i^* = q_j^* = q^*$ for every two firms $j \neq i$. Therefore, we must have that $Q_{-i}^* = (n-1)q^*$, which simplifies our above best response function to

$$q^* = \frac{1-c}{2} - \frac{(n-1)q^*}{2},$$

which, rearranging and solving for q^*, yields the equilibrium output of

$$q^*(n) = \frac{1-c}{n+1}.$$

This output level is positive since $1 > c \geq 0$ by assumption, and it is decreasing in the number of firms competing in the industry.

(b) *First stage.* Find the equilibrium number of firms that enter the industry, n^*:

- Every firm i can anticipate its second-period profits as follows:

$$\begin{aligned}\pi(n) &= (1 - q^*(n) - (n-1)q^*(n))q^*(n) - cq^*(n) - F \\ &= \frac{(1-c)^2}{(n+1)^2} - F,\end{aligned}$$

where we included the cost of entering the industry, F. These profits are clearly decreasing in the number of firms entering the industry, n. Firms then keep entering until $\pi(n) = 0$, or

$$\frac{(1-c)^2}{(n+1)^2} = F,$$

which, solving for n, yields the equilibrium number of firms entering the industry

$$n^* = \frac{1-c}{\sqrt{F}} - 1.$$

(c) Is the equilibrium number of firms entering the industry, n^*, increasing in the fixed entry cost F?

- Differentiating n^* with respect to F, we find

$$\frac{\partial n^*}{\partial F} = -\frac{1-c}{2F^{3/2}} < 0.$$

Intuitively, as the entry becomes more expensive, the number of firms that decide to join the industry declines.

(d) *Social optimum.* Consider a social planner with welfare function $W = \frac{Q^2}{2} + N \times \pi$, where the first term denotes the consumer surplus in this setting with linear inverse demand function and the second term represents the aggregate profits. Find the socially optimal number of firms that maximize this welfare function, n^{SO}. For simplicity, you may assume $c = 0$:

- The social planner anticipates the output function that every firm i will choose in the second period, $q^*(n) = \frac{1-c}{n+1}$, the aggregate output that ensues, $Q(n) = \frac{n(1-c)}{n+1}$, and the profits of every firm i in that stage, $\pi(n) = \frac{(1-c)^2}{(n+1)^2} - F$. The social planner can then find the number of firms that maximizes social welfare as follows:

$$\max_{n>0} W = \underbrace{\frac{Q(n)^2}{2}}_{CS} + \underbrace{[(1 - Q(n))Q(n) - cQ(n) - nF]}_{N\times\pi(s)}$$

$$= \frac{n^2(1-c)^2}{2(n+1)^2} + \frac{n(1-c)^2}{(n+1)^2} - nF.$$

Differentiating with respect to n yields

$$\frac{(1-c)^2}{(n+1)^3} - F = 0.$$

Evaluating this equation at $c = 0$, and solving for n, we find the socially optimal number of firms entering the industry as follows:

$$n^{SO} = \frac{1}{F^{1/3}} - 1,$$

which is, as expected, decreasing in the fixed entry cost F.

(e) *Comparison.* Evaluate the equilibrium number of firms that you found in part (b), n^*, at a marginal cost $c = 0$. Compare the number of firms that enter in equilibrium found in part (b), n^*, against the socially optimal number of firms found in part (d), n^{SO}:

- We first evaluate n^* at a marginal cost $c = 0$ to obtain

$$n^* = \frac{1}{\sqrt{F}} - 1.$$

Comparing n^* and n^{SO}, we can see that $n^* > n^{SO}$ since

$$\frac{1}{\sqrt{F}} - 1 > \frac{1}{F^{1/3}} - 1$$

simplifies to $F^{1/3} > F^{1/2}$, which holds given that the fixed entry cost F satisfies $F \in [0, 1]$ by assumption. Therefore, the equilibrium number of firms is socially excessive (relative to n^{SO}).

- **Intuition:** The entry of an additional firm produces two welfare effects:
  - A positive effect, since the profits of the new firm increase social welfare (often known as the "appropriability effect"); and
  - A negative effect, because this entry reduces the profits of all incumbents in the industry (referred to as the "business- stealing effect")

  When a firm wonders whether to join the industry, it considers the profit it will earn from entering (thus considering the appropriability effect) but ignores the business-stealing effect that its entry entails on the incumbents. The social planner, however, considers both of these effects, implying that socially optimal entry n^{SO} is lower than entry in equilibrium, n^*.

Exercise #5.12: Tax Incidence in Monopoly^C

5.12. Consider the demand for liquor $Q_D(p)$, which is a twice differentiable downward-sloping function in price p. The supply of liquor is $Q_S(p-t)$, where t represents the per-unit tax that the city government charges for every quart of liquor that firms sell. For simplicity, assume that every firm faces a constant marginal cost c to produce every quart of liquor. Define supply elasticity to be $\varepsilon_S \equiv Q'_S \frac{p}{q}$, demand elasticity to be $\varepsilon_D \equiv -Q'_D \frac{p}{q}$, and $\rho \equiv \frac{dp}{dt}$, which measures how much percentage of the liquor tax is passed through to the consumers (often known as the "pass-through" rate):

(a) Show that in a perfectly competitive market the pass-through rate is $\rho = \frac{1}{1+\frac{\varepsilon_D}{\varepsilon_S}}$:

- In a competitive equilibrium, supply must equal demand,

$$Q_D(p) = Q_S(p-t).$$

Differentiating both sides of the above equality with respect to t, we obtain

$$\frac{dQ_D(p)}{dt} = \frac{dQ_S(p-t)}{dt}.$$

Applying the chain rule, we have

$$Q'_D \frac{dp}{dt} = Q'_S \left(\frac{dp}{dt} - 1\right).$$

Rearranging, we find

$$\frac{dp}{dt}\left(Q'_S - Q'_D\right) = Q'_S,$$

which is equivalent to

$$\frac{dp}{dt} = \frac{Q'_S}{Q'_S - Q'_D}.$$

Multiplying the numerator and denominator by $\frac{p}{q}$, and assuming $q \neq 0$, we obtain

$$\rho = \frac{\varepsilon_S}{\varepsilon_S + \varepsilon_D}$$

so that the pass-through rate becomes

$$\rho = \frac{1}{1 + \frac{\varepsilon_D}{\varepsilon_S}}$$

as required.

(b) What is the supply elasticity in a perfectly competitive market? Find the associated pass-through rate, and interpret your results.

- In a perfectly competitive market, supply is perfectly elastic, $\varepsilon_S = +\infty$, yielding

$$\rho^* = \frac{1}{1+0} = 1$$

implying that consumers bear 100% of the tax burden.

(c) Define marginal consumer surplus to be $MS = -\frac{dp}{dq}q$, and its elasticity to be

$$\varepsilon_{MS} = \frac{dq}{dMS}\frac{MS}{q}.$$

Write down the monopolist's profit maximization condition, and show that $\rho^m = \frac{1}{1+\frac{1}{\varepsilon_{MS}}}$:

- The profit maximization condition of the monopolist is

$$MR(q) = MC,$$

where the marginal revenue curve intersects the marginal cost curve. Differentiating the above condition with respect to t, and using the Chain rule, we obtain

$$\frac{dMR(q)}{dq}\frac{dq}{dt} = \frac{dMC}{dt}.$$

Since $MR'(q) = p + \frac{dp}{dq}q = p - MS$, we obtain

$$\frac{d[p - MS]}{dq}\frac{dq}{dt} = \frac{d[c+t]}{dt},$$

which is equivalent to

$$\frac{dq}{dt} = \frac{1}{\frac{dp}{dq} - \frac{dMS}{dq}}.$$

Rearranging, we have

$$\rho = \frac{dp}{dt} = \frac{dp}{dq}\frac{dq}{dt}$$
$$= \frac{dp}{dq}\frac{1}{\frac{dp}{dq} - \frac{dMS}{dq}}$$
$$= \frac{1}{1 - \frac{dMS}{dp}}.$$

Let us consider the term $\frac{dMS}{dp}$, which is expressed as

$$\frac{dMS}{dp} = \frac{1}{\frac{dp}{dMS}}$$
$$= -\frac{1}{\underbrace{-q\frac{dp}{dq}}_{=MS}\frac{1}{q}\frac{dq}{dMS}}$$
$$= -\frac{1}{\frac{MS}{q}\frac{dq}{dMS}}$$
$$= -\frac{1}{\varepsilon_{MS}}.$$

Substituting, we find the monopolist's pass-through rate as follows:

$$\rho^m = \frac{1}{1 - \frac{dMS}{dp}} = \frac{1}{1 - \left(-\frac{1}{\varepsilon_{MS}}\right)}$$
$$= \frac{1}{1 + \frac{1}{\varepsilon_{MS}}}.$$

(d) *Linear demand.* Consider a linear inverse demand function $p(q) = 1 - q$. What is the monopolist's pass-through rate?

- When $p(q) = 1 - q$, we have that $\frac{dp}{dq} = -1$, yielding a marginal surplus

$$MS = -\frac{dp}{dq}q = -(-1)q = q,$$

and an elasticity of the marginal surplus of

$$\varepsilon_{MS} = \frac{dq}{dMS}\frac{MS}{q} = \frac{dq}{dq}\frac{q}{q} = 1.$$

Substituting the above results into the pass-through rate in part (d), we obtain

$$\rho^m = \frac{1}{1+1} = \frac{1}{2},$$

which means that for every dollar of tax that the government charges, consumers pay an additional 50 cents for every quart of liquor.

(e) *Nonlinear demand.* Consider, in general, a nonlinear demand function $D \equiv Q_D(p)$. Show that

$$(\log D)'' = -\frac{1}{\varepsilon_{MS}}\frac{1}{MS^2}$$

and argue that if the demand is log-concave, the pass-through rate satisfies $\rho^m \in (0, 1)$, but if demand is log-convex, the pass -through rate satisfies $\rho^m \in (1, \infty)$.

- Differentiating $\log D$ with respect to p, we obtain

$$\begin{aligned}\frac{d\log Q_D(p)}{dp} &= \frac{Q'_D(p)}{Q_D(p)} \\ &= \frac{1}{\frac{dp}{dq}}\frac{1}{q} \\ &= -\frac{1}{MS}.\end{aligned}$$

Further differentiating the log-demand function with respect to p, we find

$$\frac{d^2\log Q_D(p)}{dp^2} = \frac{1}{MS^2}\frac{dMS}{dp}.$$

Applying $\frac{dMS}{dp} = -\frac{1}{\varepsilon_{MS}}$ from part (c), we have

$$(\log D)'' = -\frac{1}{\varepsilon_{MS}}\frac{1}{MS^2}.$$

- Therefore, we can identify three different cases:
  - If the demand is log-concave, $(\log D)''$ must be negative, so that $\frac{1}{\varepsilon_{MS}}$ must be positive. This means that

$$\rho^m = \frac{1}{1+\underbrace{\frac{1}{\varepsilon_{MS}}}_{>0}} \in (0, 1).$$

  - If the demand is log-convex, $(\log D)''$ must be positive, so that $\frac{1}{\varepsilon_{MS}}$ must be negative. This entails that

$$\rho^m = \frac{1}{1+\underbrace{\frac{1}{\varepsilon_{MS}}}_{<0}} \in (1, \infty),$$

since $\frac{1}{\varepsilon_{MS}} = \frac{dMS}{dq}\frac{q}{MS} < 0$ for a downward-sloping demand curve.

– In addition, if the demand is log-linear, then $(\log D)''$ must be zero, so that $\frac{1}{\varepsilon_{MS}}$ must be zero as well. This means that

$$\rho^m = \frac{1}{1 + \underbrace{\frac{1}{\varepsilon_{MS}}}_{=0}} = 1.$$

(f) *Parametric examples.* Consider the following demand functions, (i) $Q_D(p) = \frac{1}{p}$, and (ii) $Q_D(p) = \sqrt{1-p}$. Find the curvature of the log-demand function and the associated pass-through rates:

- The log-demand function is $\log Q_D(p) = -\log p$. Differentiating, we obtain

$$\frac{d \log Q_D(p)}{dp} = -\frac{1}{p} < 0$$
$$\frac{d^2 \log Q_D(p)}{dp^2} = \frac{1}{p^2} > 0$$

so that the log-demand function is decreasing at a decreasing rate and is log-convex.
Rearranging, we obtain the inverse demand function of $p = \frac{1}{q}$, so that

$$\begin{aligned} MS &= -\frac{dp}{dq}q \\ &= -\left(-\frac{1}{q^2}\right)q \\ &= \frac{1}{q}, \text{ and} \end{aligned}$$

$$\begin{aligned} \varepsilon_{MS} &= \left(\frac{dMS}{dq}\right)^{-1}\frac{MS}{q} \\ &= \left(-\frac{1}{q^2}\right)^{-1}\frac{\frac{1}{q}}{q} \\ &= -1, \end{aligned}$$

which gives the pass-through rate as follows:

$$p^m = \frac{1}{1 + \frac{1}{\varepsilon_{MS}}} = \frac{1}{1 + \frac{1}{-1}} \to +\infty.$$

- The log-demand function is $\log Q_D(p) = \frac{1}{2}\log(1-p)$. Differentiating, we have

$$\frac{d \log Q_D(p)}{dp} = -\frac{1}{2(1-p)} < 0$$
$$\frac{d^2 \log Q_D(p)}{dp^2} = -\frac{1}{2(1-p)^2} < 0$$

so that the log-demand function is decreasing at an increasing rate and is log-concave.

- Rearranging, we obtain the inverse demand function of $p = 1 - q^2$, so that

$$MS = -\frac{dp}{dq}q = -(-2q)\,q = 2q^2, \text{ and}$$

$$\varepsilon_{MS} = \left(\frac{dMS}{dq}\right)^{-1}\frac{MS}{q} = (4q)^{-1}\frac{2q^2}{q} = \frac{1}{2},$$

which gives the pass-through rate as follows:

$$p^m = \frac{1}{1+\frac{1}{\varepsilon_{MS}}} = \frac{1}{1+\frac{1}{\frac{1}{2}}} = \frac{1}{3}.$$

R&D Incentives 6

Introduction

In this chapter, we study firms' incentives to invest in research and development (R&D) with the goal of lowering their production costs in subsequent periods. Exercise 6.1 starts analyzing these incentives in a stylized setting, a monopoly where a single firm operates in all periods. In this context, the monopolist anticipates that any R&D investment today will lower its production costs in the next period, when it still maintains a monopolistic position. In other words, the monopolist does not have incentives to invest in R&D to become more competitive in the future, relative to its rivals. In oligopoly models, however, every firm considers this incentive to improve its relative cost advantage. Exercises 6.2 and 6.3 examine this incentive, first in an oligopoly where firms compete in quantities (Exercise 6.2) and then in one where firms compete in prices (Exercise 6.3). In both exercises, we evaluate whether firms may have more or less incentives to invest in R&D than what they would under monopoly, which help us identify whether the presence of more than one firm provides firm with an additional incentive to invest, relative to monopoly, or with less incentives. Exercise 6.4 then evaluates the welfare that arises in equilibrium, comparing whether that under monopoly is larger than under oligopoly, and Exercise 6.5 examines if R&D investments increase in oligopolistic markets that became more competitive (with more firms).

Exercises 6.6 and 6.7 consider two extensions of the standard R&D model studied in previous exercises. Specifically, we examine the role of "research joint ventures," which allow firms to coordinate their R&D investment decisions to maximize joint profits in a first stage, and then competing in quantities à la Cournot by independently selecting their output levels in the second stage in Exercise 6.6, or in prices à la Bertrand by independently choosing their prices in the second stage in Exercise 6.7. We analyze whether firms have more or less incentives to invest in R&D when operating in a research joint venture than when they independently choose their R&D investment. Exercise 6.8 considers a similar environment, where the R&D investment of every firm produces a positive "spillover effect," meaning that part of this R&D investment is unpatented (or can be copied without infringing patent law), leading to free-riding behavior between firms. In this setting, firms would have incentives to coordinate their R&D investment decisions to internalize the positive externality that spillover effects generate, inducing an overall increase in R&D investments.

The remainder of exercises in this chapter study different extensions of the above models. Exercise 6.9 analyzes a market with two firms, but, rather than investing in R&D to reduce their future costs, we assume that they sequentially choose whether to develop a new product or not and

P.-S. Choi et al., *Industrial Organization*, Springer Texts in Business and Economics,
https://doi.org/10.1007/978-3-031-38635-0_6

subsequently competing in quantities in the second stage. Exercise 6.10 considers R&D investment decisions again, but in a polluting market, where every firm invests in abatement technologies (which reduce the severity of the firm's pollution) rather than in cost-reducing technologies. This exercise is based on Lambertini et al. (2017) and allows for R&D spillovers, understood in this setting as that a share of every firm's investment in abatement can benefit other firms, reducing their emissions too. We show that aggregate investment in abatement is unambiguously decreasing in the number of firms competing in the industry when R&D spillovers are absent, but otherwise aggregate abatement exhibits an inverted-U shape, increasing when relatively few firms compete but eventually decreasing when the market becomes more competitive.

In previous exercises, every dollar invested in R&D produces a certain decrease in the firm's marginal production cost in the next stage of the game. In contrast, Exercise 6.11 – approach based on Delbono and Denicolo (1991) – assumes that every dollar invested in R&D increases the probability that a firm may be the winner of an innovation contest, but not winning with certainty, while all other firms are the losers in this contest. Exercises 6.12-6.14 examine entry-deterring investments by an incumbent, separately analyzing its R&D investment that leads to accommodating entry, which deters entry, and then comparing its profits to identify under which conditions the incumbent practices entry deterrence. Finally, Exercise 6.15 studies a sequential-move game where, in the first stage, a government patent office sets the duration of a patent (e.g., 20 years for a new drug). In the second stage, a firm responds choosing its investment in R&D, which can be interpreted as the probability that this firm innovates. In this setting based on Takalo (2001), we first identify the equilibrium R&D investment, as a function of the patent length provided by the patent office, and then characterize the socially optimal patent length that maximizes social welfare.

Exercise #6.1: Incentives to Innovate Under Monopoly^A

6.1 Consider a monopolist facing an inverse demand function $p(Q) = 1 - Q$, where Q denotes output, and with marginal production cost c, where $1 > c > 0$. In the first stage, the monopolist chooses whether to adopt a technology that lowers its marginal cost form c to $c - x$, where $x \in [0, c]$, so marginal costs are always weakly positive, at a fixed cost $F > 0$. This fixed cost is, then, only incurred if the firm invests in the cost-reducing technology. In the second stage, the monopolist chooses its output level Q.

(a) Solving the game by backward induction, find the output function of the monopolist in the second period, after investing in the innovation and when it did not.

- *After no innovation.* In the second stage, the monopolist solves

$$\max_{Q \geq 0} \; (1 - Q)\, Q - cQ.$$

Differentiating with respect to Q, we find:

$$1 - 2Q - c = 0$$

and solving for Q yields a monopoly output:

$$Q_m(c) = \frac{1 - c}{2},$$

where the subscript m denotes monopoly, and earns monopoly profits

$$\pi_m^{NI} = \frac{(1-c)^2}{4},$$

where superscript NI denotes "no innovation."

- *After innovation.* In this case, the monopolist solves a symmetric problem:

$$\max_{Q \geq 0} \; (1-Q)\,Q - (c-x)\,Q$$

since the monopolist's marginal cost is now $c - x$. Differentiating with respect to Q, we find:

$$1 - 2Q - (c-x) = 0$$

and solving for Q yields a monopoly quantity:

$$Q_m = \frac{1-c+x}{2}$$

earning monopoly profits

$$\pi_m^I = \left(1 - \frac{1-c+x}{2}\right)\frac{1-c+x}{2} - (c-x)\frac{1-c+x}{2}$$
$$= \frac{(1-c+x)^2}{4},$$

where superscript I denotes "innovation."

(b) Find under which values of fixed cost F the monopolist chooses to innovate in the first period.

- In the first period, the monopolist anticipates its own pricing strategies in the second stage, and then chooses to adopt the technology if and only if $\pi_m^I - F \geq \pi_m^{NI}$, or

$$\frac{(1-c+x)^2}{4} - F \geq \frac{(1-c)^2}{4}$$

which, solving for F, yields

$$F \leq \frac{x\,[x + 2\,(1-c)]}{4} \equiv F_m.$$

Therefore, the fixed cost from adopting the innovation, F, must be sufficiently low for the monopolist to adopt it.

- In addition, cutoff F_m increases in the cost-reducing effect of the innovation, x, indicating that the monopolist is willing to incur a larger fixed cost to adopt more efficient innovations, whereas if the firm becomes less efficient as marginal cost c increases, it obtains lower profit so that it can only afford a lower fixed cost F to support cost-reducing innovations.

(c) *Numerical example.* Assume that $c = 1/2$ and $x = 1/8$. Find under which values of fixed cost F the monopolist chooses to innovate in the first period.

- In this setting, the monopolist innovates in the first period if F satisfies

$$F \leq \frac{x\left[x+2\left(1-c\right)\right]}{4} = \frac{\frac{1}{8}\left[\frac{1}{8}+2\left(1-\frac{1}{2}\right)\right]}{4}$$

$$= \frac{9}{256} \simeq 0.035 \equiv F_m.$$

Otherwise, the monopolist does not innovate.

Exercise #6.2: Quantity Competition: More Incentives to Innovate Than Under Monopoly?^{B}

6.2 Consider an industry with two firms selling a homogeneous good. They face an inverse demand function $p(Q) = 1 - Q$, where Q denotes aggregate output, and both have the same marginal production cost $c > 0$. In the first stage, every firm simultaneously and independently chooses whether to adopt a technology that lowers its marginal cost to $c - x$, where $x \in [0, c]$, so marginal costs are always weakly positive, at a fixed cost $F > 0$. Then, a firm only incurs this fixed cost if it invests in the cost-reducing technology. In the second stage, every firm simultaneously and independently selects its output.

(a) *Monopoly*. Let us first consider, as a benchmark, a setting with only one firm (monopoly). Find the equilibrium price in this game, and under which conditions the monopolist innovates.

- From Exercise 6.1, we know that, in the first period, the monopolist invests in the innovation if and only if

$$F \leq \frac{x\left[x+2\left(1-c\right)\right]}{4} \equiv F_m$$

and, in the second period, it responds with an output level $Q_m(c) = \frac{1-c+x}{2}$ after investing in the innovation and with $Q_m(c) = \frac{1-c}{2}$ after not investing in the innovation.

(b) *Cournot duopoly*. Let us assume that two firms compete in quantities à la Cournot. Find equilibrium prices, and conditions under which each firm chooses to innovate.

- In the second period:
  - If no firm innovates, they play a standard Cournot game of quantity competition with symmetric marginal cost c, setting equilibrium quantity of $q_1 = q_2 = \frac{1-c}{3}$ and earning a profit of $\frac{1}{9}(1-c)^2$.
  - If both firms innovate, they also play a standard Cournot game with symmetric marginal cost $c - x$, so both firms set equilibrium quantity of $q_1 = q_2 = \frac{1-c+x}{3}$, earning a profit of $\frac{1}{9}(1-c+x)^2$. However, every firm pays a fixed cost F from investing in the innovation, yielding a net profit of $\frac{1}{9}(1-c+x)^2 - F$.
  - If only one firm i innovates, and the exercise assumes a non-drastic innovation, firm i has a best response function $q_i = \frac{1-c+x}{2} - \frac{1}{2}q_j$, while firm j has a best response function $q_j = \frac{1-c}{2} - \frac{1}{2}q_i$. In this context, we can simultaneously solve for these best response functions to obtain equilibrium output levels of

$$q_i^* = \frac{1-c+2x}{3} \text{ and } q_j^* = \frac{1-c-x}{3}$$

which entail equilibrium profits for each firm of

$$\pi_i^I = \frac{1}{9}(1 - c + 2x)^2, \text{ and}$$

$$\pi_j^{NI} = \frac{1}{9}(1 - c - x)^2.$$

- In the first stage, both firms anticipate their profits in the second stage. In this context, firms face the following payoff matrix, where I and NI denote innovation and no innovation, respectively.

		Firm 2	
		I	NI
Firm 1	I	$\frac{(1-c+x)^2}{9} - F, \frac{(1-c+x)^2}{9} - F$	$\frac{(1-c+2x)^2}{9} - F, \frac{(1-c-x)^2}{9}$
	NI	$\frac{(1-c-x)^2}{9}, \frac{(1-c+2x)^2}{9} - F$	$\frac{(1-c)^2}{9}, \frac{(1-c)^2}{9}$

- We can now use this payoff matrix to characterize the best responses. If firm j innovates (in the left column), then firm i responds innovating if and only if

$$\pi_i^I \geq \pi_i^{NI} \iff \frac{(1-c+x)^2}{9} - F \geq \frac{(1-c-x)^2}{9}$$

which, solving for F, simplifies to

$$F < \frac{4x(1-c)}{9} \equiv F_d,$$

where the subscript d denotes Cournot duopoly.
- If firm j does not innovate (in the right column), firm i responds innovating if and only if

$$\pi_i^I \geq \pi_i^{NI} \iff \frac{(1-c+2x)^2}{9} - F \geq \frac{(1-c)^2}{9}$$

which, solving for F, simplifies to

$$F < \frac{4x\,(1-c+x)}{9} \equiv F_d',$$

where cutoff F_d' satisfies $F_d' > F_d$. Given this clear ranking between cutoffs F_d and F_d', three regions of F arise, each leading to a different equilibrium result in the above payoff matrix:
  - *Low* F. When $F < F_d$, firm i responds innovating both when firm j innovates and when its rival does not. In other words, firm i finds innovation to be a strictly dominant strategy. Since this applies for every firm i, the unique Nash equilibrium of this game has both firms innovating (I, I).
  - *Intermediate* F. When $F_d \leq F < F_d'$, every firm i innovates when its rival does not, since $F < F_d'$, but does not innovate otherwise, since $F_d \leq F$. Best responses, therefore, resemble those in an anti-coordination game (e.g., chicken game), and the two Nash equilibria that arise are (I, NI) and (NI, I).

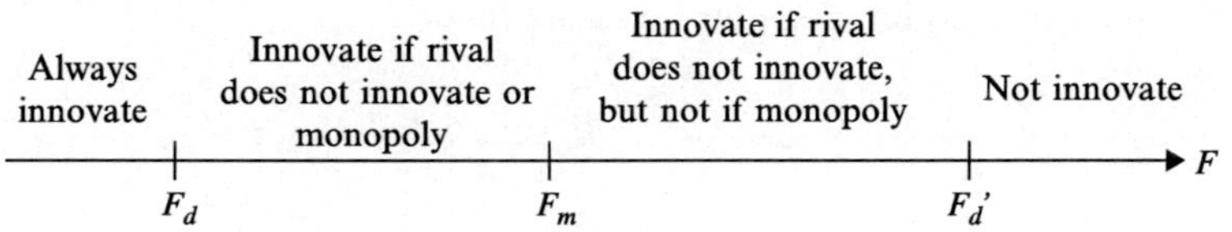

Fig. 6.1 Innovation decisions

– *High F*. When $F_d' \leq F$, every firm i responds not innovating regardless of its rival's choice, thus finding innovation to be a strictly dominated strategy for both firms. In this context, the only Nash equilibrium of the game is (NI, NI).

(c) Compare the conditions under which firms choose to innovate under monopoly (from part a) and under Cournot duopoly (from part b). In which market structure firms have stronger incentives to innovate? Interpret.

- Comparing cutoffs F_d and F_m yields:

$$F_m - F_d = \frac{x\left[x + 2(1-c)\right]}{4} - \frac{4x(1-c)}{9}$$
$$= \frac{x\left[9x + 18(1-c) - 16(1-c)\right]}{36}$$
$$= \frac{x\left[9x + 2(1-c)\right]}{36} > 0.$$

Comparing cutoffs F_d' and F_m yields:

$$F_d' - F_m = \frac{4x(1-c+x)}{9} - \frac{x\left[x + 2(1-c)\right]}{4}$$
$$= \frac{x\left[16(1-c+x) - 9x - 18(1-c)\right]}{36}$$
$$= \frac{x\left[7x - 2(1-c)\right]}{36}$$

so that when $x > \frac{2(1-c)}{7} \equiv \overline{x}$ for relatively efficient investments,[1] $F_d' > F_m$.

Adjoining the cutoffs, three cases emerge according to the value of fixed cost F, as depicted in Fig. 6.1. (For simplicity, the figure assumes $x > \frac{2(1-c)}{7} \equiv \overline{x}$ so that $F_d' > F_m$.)

– *High cost of innovation.* If $F > F_m > F_d' > F_d$ or $F > F_d' > F_m > F_d$, every firm i chooses not to innovate under any market structure.

– *Intermediate cost of innovation.* We have three subcases depending on the relative magnitude of the fixed cost as follows:

When $F_m > F > F_d' > F_d$, which happens if $x \leq \overline{x}$, every firm i innovates only if it monopolizes the market, but not whether its rival innovates or not, because the relatively inefficient cost-reducing investment can only support this firm's innovation only when it faces no competition in the market.

When $F_d' > F > F_m > F_d$, which happens if $x > \overline{x}$, every firm i innovates if it competes with a non-innovating rival, but not when its rival innovates, because firm i can take advantage of its rival's reduced output if the other firm does not innovate, but the

[1] A sufficient condition, under $c > x$, for this to hold is $c > \frac{2(1-c)}{7}$, which is equivalent to $c > \frac{2}{9}$.

fixed cost is too high to support both firms' innovation, whereas if firm i monopolizes the market, it does not innovate because it is already enjoying the monopolist's profit, and the relatively high fixed cost does not induce this firm to make the cost-reducing investments.

When $F_m > F_d' > F > F_d$ or $F_d' > F_m > F > F_d$, every firm i innovates only if it monopolizes the market or competes with a non-innovating rival, but not when its rival innovates.

– *Low cost of innovation.* If $F_m > F_d' > F_d > F$ or $F_d' > F_m > F_d > F$, every firm i innovates whether competing à la Cournot or not, that is, an innovation occurs under both market structures when the fixed cost of innovation is sufficiently low.

(d) *Numerical example.* Evaluate the conditions for fixed cost F that you found in parts (a) and (b) at parameter values $c = 1/2$ and $x = 1/8$. Interpret your results.

- Evaluating cutoff F_m at these parameter values yields:

$$F \leq \frac{x\left[x + 2(1-c)\right]}{4} = \frac{\frac{1}{8}\left[\frac{1}{8} + 2\left(1 - \frac{1}{2}\right)\right]}{4}$$
$$= \frac{9}{256} \simeq 0.035 \equiv F_m.$$

Similarly, evaluating cutoff F_d at these parameter values, we obtain:

$$F < \frac{4x(1-c)}{9} = \frac{4\frac{1}{8}(1-\frac{1}{2})}{9}$$
$$= \frac{1}{36} \simeq 0.028 \equiv F_d.$$

Finally, evaluating cutoff F_d' at these parameter values, we find:

$$F < \frac{4x\,(1-c+x)}{9} = \frac{4\frac{1}{8}(1-\frac{1}{2}+\frac{1}{8})}{9}$$
$$= \frac{5}{144} \simeq 0.034 \equiv F_d'.$$

Comparing the three cutoffs, we observe that they, in this case, are ranked as follows:

$$F_m = 0.035 > F_d' = 0.034 > F_d = 0.028.$$

Exercise #6.3: Price Competition: Less Incentives to Innovate Under Monopoly?^{B}

6.3 Consider an industry with two firms selling a homogeneous good. They face an inverse demand function $p(Q) = 1 - Q$, where Q denotes aggregate output, and both have the same marginal production cost $c > 0$. In the first stage, every firm simultaneously and independently chooses whether to adopt a technology that lowers its marginal cost to $c - x$, where $x \in [0, c]$, so marginal costs are always weakly positive, at a fixed cost $F > 0$. Then, a firm only incurs this fixed cost if it invests in the cost-reducing technology. In the second stage, every firm simultaneously and independently sets the price for its product.

For simplicity, assume that the size of the innovation, as captured by parameter x, is non-drastic, in the sense that, if only one firm innovates, its monopoly price evaluated at $c-x$, $p_m(c-x)$, lies above its rival's marginal cost c. As a consequence, the innovator sets a price slightly below the marginal cost of the non-innovating firm, $p = c - \varepsilon$, where $\varepsilon \to 0$.

(a) *Monopoly.* Let us first consider, as a benchmark, a setting with only one firm (monopoly). Find equilibrium prices in this game, and under which conditions the monopolist innovates.

- In the second stage, after no innovation, the monopolist solves

$$\max_{p_i>0} p\,(1-p) - c\,(1-p)$$

since $Q = 1 - p$. Differentiating with respect to p, we find:

$$1 - 2p + c = 0$$

and solving for p yields a monopoly price:

$$p_m(c) = \frac{1+c}{2},$$

where the subscript m denotes monopoly, and earns monopoly profits

$$\pi_m^{NI} = \frac{(1-c)^2}{4},$$

where the superscript NI denotes "no innovation."

- In the second stage, after innovation, the monopolist solves a symmetric problem:

$$\max_{p_i>0} p\,(1-p) - (c-x)\,(1-p)$$

since $Q = 1 - p$. Differentiating with respect to p, we find:

$$1 - 2p + (c-x) = 0$$

and solving for p yields a monopoly price:

$$p_m = \frac{1+(c-x)}{2}$$

earning monopoly profits

$$\pi_m^{I} = \frac{(1-c+x)^2}{4},$$

where the superscript I denotes "innovation."

- In the first stage, the monopolist anticipates its own pricing strategies in the second stage, and then chooses to adopt the technology if and only if $\pi_m^I - F \geq \pi_m^{NI}$, or

$$\frac{(1-c+x)^2}{4} - F \geq \frac{(1-c)^2}{4}$$

which, solving for F, yields

$$F \le \frac{x\,[x+2\,(1-c)]}{4} \equiv F_m.$$

Therefore, the fixed cost from adopting the innovation, F, must be sufficiently low for the monopolist to adopt it. In addition, note that cutoff F_m increases in the cost-reducing effect of the innovation, x, indicating that the monopolist is willing to pay a larger fixed cost to adopt more efficient innovations. Whereas, if the firm becomes less efficient as marginal cost c increases, it obtains lower profit so that it can only afford a lower fixed cost F to support cost-reducing innovations.

(b) *Bertrand duopoly*. Let us assume that two firms compete in prices à la Bertrand. Find equilibrium prices, and under which conditions each firm chooses to innovate.

- In the second period:
  - If no firm innovates, they play a standard Bertrand game of price competition with symmetric marginal cost c, setting equilibrium price of $p_1 = p_2 = c$ and earning zero profits.
  - If both firms innovate, they also play a standard Bertrand game with symmetric marginal cost $c - x$, so that both firms charge an equilibrium price of $p_1 = p_2 = c - x$, earning zero (gross) profit. However, every firm pays a fixed cost F from investing in the innovation, yielding a net profit $-F$.
  - If only one firm i innovates, and the exercise assumes a non-drastic innovation, firm i sets a price slightly below the marginal cost of the non-innovating rival, $p_i = c - \varepsilon$, where $\varepsilon \to 0$, entailing $p_i = c$. In this context, the innovator sells

$$Q = 1 - p_i = 1 - c \text{ units}$$

  and makes a profit

$$\begin{aligned}\pi_B^I &= c\,(1-c) - (c-x)\,(1-c)\\ &= x(1-c),\end{aligned}$$

  where the subscript B denotes the Bertrand duopoly. In contrast, the non-innovator captures no sales, and thus makes a zero profit, $\pi_B^{NI} = 0$.
- In the first stage, both firms anticipate their profits in the second stage. In this context, firms face the following payoff matrix, where I and NI denote innovation and no innovation, respectively.

		Firm 2	
		I	NI
Firm 1	I	$-F, -F$	$x\,(1-c) - F, 0$
	NI	$0, x\,(1-c) - F$	$0, 0$

  - If firm j innovates, firm i's best response is not to innovate since its profit from innovation, $-F$, is lower than from not innovating, $\pi_B^{NI} = 0$.
  - If firm j does not innovate, firm i's best response is to innovate if its profit from doing so, $\pi_B^I - F = x(1-c) - F$, exceeds that from not innovating, zero (in the Bertrand

game where both firms keep their initial marginal cost c). That is, firm i's best response to non-innovation is to innovate if $x\,(1-c) - F \geq 0$, or

$$F \leq x\,(1-c) \equiv F_B,$$

where the subscript B denotes a Bertrand duopoly.

- Summarizing, our above analysis identifies two cases:
  - When $F > F_B$, innovation is strictly dominated by no innovation, as depicted in the underlined best response payoffs in the following matrix. In this context, a unique equilibrium arises in the first stage of the game, (NI, NI), where no firm innovates, entailing zero profits in the subsequent Bertrand competition.

		Firm 2	
		I	NI
Firm 1	I	$-F, -F$	$x(1-c) - F, \underline{0}$
	NI	$\underline{0}, x(1-c) - F$	$\underline{0}, \underline{0}$

  - When $F \leq F_B$, every firm seeks to mis-coordinate with its rival's action in the first stage: innovating when its rival does not innovate, but not innovating otherwise. The matrix below depicts underlined best responses in this setting, showing similar incentives as in standard anti-coordination games. Therefore, two Nash equilibria emerge in the first stage: (I, NI) and (NI, I).

		Firm 2	
		I	NI
Firm 1	I	$-F, -F$	$\underline{x(1-c) - F}, \underline{0}$
	NI	$\underline{0}, \underline{x(1-c) - F}$	$0, 0$

(c) Compare the conditions under which firms choose to innovate under monopoly (from part a) and under Bertrand duopoly (from part b). In which market structure firms have stronger incentives to innovate? Interpret.

- Comparing the cutoffs for F, we found in parts (a) and (b) yields:

$$F_B - F_m = x(1-c) - \frac{x[x + 2(1-c)]}{4} = \frac{x[2(1-c) - x]}{4}$$

which is positive if $x \leq 2(1-c)$. Assuming that this condition holds (which is sufficient for $c < 0.5$), three cases emerge according to the value of fixed cost F, as depicted in Fig. 6.2:

  - *High cost of innovation.* If $F \geq F_B$, no firm innovates under any market structure.
  - *Intermediate cost of innovation.* If $F_B > F \geq F_m$, the monopolist does not innovate, but one firm innovates when competing à la Bertrand (as described in part b where we found asymmetric innovation profiles when $F_B > F$).
  - *Low cost of innovation.* If $F_m > F$, the monopolist and one of the firms competing à la Bertrand innovates. That is, an innovation occurs under both market structures.

The above cases highlight that the monopolist may have fewer incentives to innovate than when firms compete in prices. Intuitively, the monopolist already makes a positive

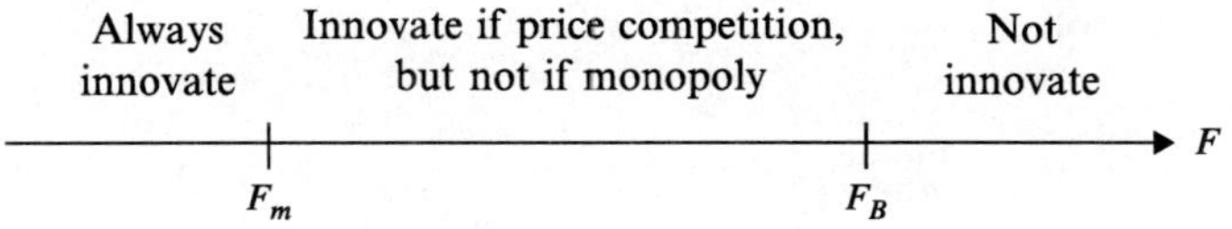

Fig. 6.2 Innovation profiles

profit, and will only increase its profit further if it innovates when the cost of innovation F is sufficiently low. When firms compete à la Bertrand, however, they make zero profits, and obtaining a cost advantage can lead to substantial profits, thus providing them with more incentives to innovate.

(d) *Numerical example.* Evaluate the conditions for fixed cost F that you found in parts (a) and (b) at parameter values $c = 1/2$ and $x = 1/8$. Interpret your results.

- Evaluating cutoff F_m at these parameter values yields:

$$F \leq \frac{x\,[x + 2\,(1 - c)]}{4} = \frac{\frac{1}{8}\left[\frac{1}{8} + 2\left(1 - \frac{1}{2}\right)\right]}{4}$$
$$= \frac{9}{256} \simeq 0.035 \equiv F_m.$$

Similarly, evaluating cutoff F_B at these parameter values, we obtain:

$$F \leq \frac{1}{8}\left(1 - \frac{1}{2}\right) = \frac{1}{16} = 0.0625 \equiv F_B$$

which, as expected, satisfies $F_B > F_m$.

Exercise #6.4: Larger R&D Under Monopoly or Duopoly? Welfare Evaluation^C

6.4 Consider an industry with two firms facing inverse demand function $p(Q) = 1 - Q$, where Q denotes aggregate output. Firm i's total cost from investing x_i dollars in R&D and producing q_i units of output is

$$C(q_i, x_i) = (c - x_i)q_i + \gamma \frac{1}{2} x_i^2,$$

where the first term denotes production costs, $1 > c > 0$, and the second captures R&D costs. Therefore, a lower parameter $\gamma > 1$ represents a higher R&D efficiency. Marginal production cost, $c - x_i$, is then decreasing in firm i's R&D investment, x_i.

Firms interact in the following two-stage game. In the first stage, every firm i simultaneously and independently chooses its investment in R&D, x_i. In the second stage, observing the profile of R&D investments (x_i, x_j), every firm i simultaneously and independently selects its output level q_i. Firms compete à la Cournot.

(a) *Second stage, Cournot.* Find equilibrium output levels, as a function of R&D investments x_i and x_j, $q_i(x_i, x_j)$ for every firm i.

- Every firm i solves

$$\max_{q_i \geq 0} (1 - q_i - q_j)q_i - \underbrace{(c - x_i)q_i}_{\text{Production cost}} - \underbrace{\frac{\gamma}{2}x_i^2}_{\text{R\&D cost}}.$$

Differentiating with respect to q_i yields:

$$1 - 2q_i - q_j - (c - x_i) = 0$$

and solving for q_i, we find firm i's best response function in the second stage of the game:

$$q_i(q_j) = \frac{1 - c + x_i}{2} - \frac{1}{2}q_j$$

which graphically shifts upward as firm i's R&D investment, x_i, increases. A symmetric expression applies to firm j:

$$q_j(q_i) = \frac{1 - c + x_j}{2} - \frac{1}{2}q_i.$$

Inserting $q_j(q_i)$ into $q_i(q_j)$ yields:

$$q_i = \frac{1 - c + x_i}{2} - \frac{1}{2}\left(\frac{1 - c + x_j}{2} - \frac{1}{2}q_i\right)$$

and, solving for q_i, we find the equilibrium output for firm i

$$q_i(x_i, x_j) = \frac{1 - c + 2x_i - x_j}{3}$$

being increasing in x_i (as firm i increases its cost advantage) but decreasing in x_j (as firm j increases its cost advantage).

(b) *First stage.* Anticipating output levels $q_i(x_i, x_j)$ and $q_j(x_i, x_j)$ in the second stage, find the equilibrium R&D investment that every firm i chooses in the first stage, x_i^*.

- In the first stage, every firm i chooses x_i to solve

$$\max_{x_i \geq 0} \pi_i = (1 - q_i(x_i, x_j) - q_j(x_i, x_j))q_i(x_i, x_j) - \underbrace{(c - x_i)q_i(x_i, x_j)}_{\text{Production cost}} - \underbrace{\frac{\gamma}{2}x_i^2}_{\text{R\&D cost}}$$

$$= \left(1 - \frac{1 - c + 2x_i - x_j}{3} - \frac{1 - c + 2x_j - x_i}{3}\right) \times \frac{1 - c + 2x_i - x_j}{3}$$

$$-(c - x_i)\frac{1 - c + 2x_i - x_j}{3} - \frac{\gamma}{2}x_i^2$$

$$= \left(\frac{1 + 2c - x_i - x_j}{3}\right) \times \frac{1 - c + 2x_i - x_j}{3} - (c - x_i)\frac{1 - c + 2x_i - x_j}{3} - \frac{\gamma}{2}x_i^2$$

which is only a function of R&D investments x_i and x_j. Rearranging the terms above, we have the profit function as follows:

$$\max_{x_i \geq 0} \pi_i = \frac{1 - c + 2x_i - x_j}{3} \times \left(\frac{1 + 2c - x_i - x_j}{3} - (c - x_i) \right) - \frac{\gamma}{2} x_i^2$$

$$= \left(\frac{1 - c + 2x_i - x_j}{3} \right)^2 - \frac{\gamma}{2} x_i^2.$$

Differentiating with respect to x_i, we find

$$\frac{4(1 - c + 2x_i - x_j)}{9} - \gamma x_i = 0$$

and, solving for x_i, we obtain firm i's best response function in the first stage of the game

$$x_i(x_j) = \frac{4(1 - c - x_j)}{9\gamma - 8}$$

which is decreasing in x_j since $\gamma \geq 1$ holds by assumption. Therefore, R&D investments are strategic substitutes, meaning that a higher investment by firm j induces firm i to reduce its investment. A symmetric expression applies to firm j,

$$x_j(x_i) = \frac{4(1 - c - x_i)}{9\gamma - 8}.$$

Invoking symmetry, where $x^* = x_i = x_j$, we have:

$$x^* = \frac{4(1 - c - x^*)}{9\gamma - 8}$$

and the equilibrium R&D investments become:

$$x_i^* = \frac{4}{9\gamma - 4}(1 - c).$$

- Therefore, equilibrium output in the second stage, evaluated at $x_i^* = x_j^* = x^*$, is

$$q^* = q_i(x_i^*, x_j^*) = q_j(x_i^*, x_j^*) = \frac{1 - c + 2x^* - x^*}{3} = \frac{1 - c + x^*}{3}$$

$$= \frac{1 - c + \frac{4}{9\gamma - 4}(1 - c)}{3} = \frac{3\gamma}{9\gamma - 4}(1 - c).$$

(c) Evaluate equilibrium profits, consumer surplus, and welfare. For simplicity, you may assume $\gamma = 1$.

- Equilibrium profits are

$$\pi_i^* = (1 - 2q^*) \times q^* - (c - x^*) \times q^* - \frac{1}{2}(x^*)^2$$

$$= q^*(1 - 2q^* - c + x^*) - \frac{1}{2}(x^*)^2$$

$$= \frac{3(1-c)}{5} \times \left(1 - c - 2 \times \frac{3(1-c)}{5} + \frac{4(1-c)}{5}\right) - \frac{1}{2}\left(\frac{4(1-c)}{5}\right)^2$$

$$= \frac{3(1-c)}{5} \times \left(\frac{5(1-c) - 6(1-c) + 4(1-c)}{5}\right) - \frac{16(1-c)^2}{50}$$

$$= \frac{9(1-c)^2}{25} - \frac{8(1-c)^2}{25}$$

$$= \frac{(1-c)^2}{25}.$$

- Consumer surplus is

$$CS^* = \int_0^{Q^*} \left[D(Q) - P^*(Q^*)\right] dQ = \int_0^{2q^*} \left[1 - Q - (1 - 2q^*)\right] dQ,$$

where $D(Q) = 1 - Q = 1 - q_i - q_j$ denotes the inverse demand function, Q is aggregate output, and $P^*(Q^*) = 1 - 2q^*$ is the equilibrium market price. Thus,

$$CS^* = \int_0^{2q^*} 1 - Q - (1 - 2q^*) dQ$$

$$= Q - \frac{1}{2}Q^2 - (1 - 2q^*)Q \Big|_0^{2q^*}$$

$$= 2(q^*)^2$$

$$= 2 \times \frac{9(1-c)^2}{25} = \frac{18(1-c)^2}{25}.$$

- Finally, $\pi_i^* = \pi_j^*$ due to symmetry, generating welfare of

$$W^* = CS^* + \left(\pi_i^* + \pi_j^*\right)$$

$$= \frac{18(1-c)^2}{25} + 2 \times \frac{(1-c)^2}{25}$$

$$= \frac{4(1-c)^2}{5}.$$

(d) *Monopoly*. Consider now that only one firm operates in the industry, choosing its R&D investment, x, in the first period, and its output level q in the second period. Find equilibrium output $q(x)$ in the second period, and equilibrium R&D investment, x^* in the first period. Then compare x^* against $x_i^* + x_j^*$ you found in part (b). Interpret.

- *Second stage*. Monopoly solves

$$\max_{q \geq 0} \pi = (1 - q) \times q - (c - x) \times q - \frac{1}{2}x^2.$$

Differentiating with respect to q yields

$$1 - 2q - (c - x) = 0$$

and solving for q, we find equilibrium output in monopoly:

$$q(x) = \frac{1-c+x}{2}.$$

- *First stage*. In the first stage, the monopoly chooses x to solve

$$\begin{aligned}\max_{x\geq 0}\ \pi &= (1-q(x))q(x) - (c-x)q(x) - \frac{1}{2}x^2 \\ &= \left(1-\frac{1-c+x}{2}\right)\times\frac{1-c+x}{2} - (c-x)\times\frac{1-c+x}{2} - \frac{1}{2}x^2 \\ &= \frac{(1-c+x)^2}{4} - \frac{1}{2}x^2.\end{aligned}$$

Differentiating with respect to x, we find:

$$\frac{1-c+x}{2} - x = 0.$$

Thus, we obtain the equilibrium R&D investment under monopoly:

$$x^* = 1-c.$$

Recall from the previous part that the aggregate R&D investment under Cournot competition is

$$x_i^* + x_j^* = 2x^* = 2\times\frac{4(1-c)}{5} = \frac{8(1-c)}{5},$$

and we compare these investment levels to obtain

$$\begin{aligned}x^* - (x_i^* + x_j^*) &= (1-c) - \frac{8(1-c)}{5} \\ &= -\frac{3(1-c)}{5} < 0\end{aligned}$$

meaning that R&D investment is lower under monopoly than under duopoly. Intuitively, the monopolist has less incentives to invest in R&D than the sum of investments made by Cournot duopolists since the latter can gain a cost advantage relative to its rival from investing in the R&D, while the monopolist cannot.

(e) Compare aggregate output, aggregate profits, consumer surplus, and welfare under monopoly and duopoly. Interpret your results.

- Based on the previous part, we can find that equilibrium output of the monopolist is

$$q^*(x^*) = \frac{1-c+x^*}{2} = \frac{1-c+\overbrace{(1-c)}^{x^*}}{2} = 1-c.$$

- *Profit comparison.* Equilibrium profits of the monopolist become:

$$\begin{aligned}\pi^* &= (1 - q^*(x^*)) \times q^*(x^*) - (c - x^*) \times q^*(x^*) - \frac{1}{2}(x^*)^2 \\ &= [1 - (1 - c)] \times (1 - c) - [c - (1 - c)] \times (1 - c) - \frac{1}{2}(1 - c)^2 \\ &= \frac{(1 - c)^2}{2}.\end{aligned}$$

Note that aggregate profits under duopoly (found in part c of the exercise) were

$$\pi_i^* + \pi_j^* = 2\pi^* = \frac{2(1 - c)^2}{25}.$$

Clearly, $\frac{(1-c)^2}{2} > \frac{2(1-c)^2}{25}$, so we conclude that aggregate profits are larger under monopoly than duopoly, $\pi^* > \pi_i^* + \pi_j^*$.
- *CS comparison.* Consumer surplus under monopoly is

$$\begin{aligned}CS_M^* &= \frac{1}{2}(1 - P(q^*)) \times q^* \\ &= \frac{1}{2}(1 - 1 + q^*) \times q^* \\ &= \frac{(q^*)^2}{2} = \frac{(1 - c)^2}{2}.\end{aligned}$$

Recall that consumer surplus under duopoly (found in part c of the exercise) was $CS_D^* = \frac{18(1-c)^2}{25}$, then

$$\begin{aligned}CS_M^* - CS_D^* &= \frac{(1 - c)^2}{2} - \frac{18(1 - c)^2}{25} \\ &= \frac{25(1 - c)^2}{50} - \frac{36(1 - c^2)}{50} \\ &= -\frac{11(1 - c)^2}{25} < 0\end{aligned}$$

implying that, as expected, consumer surplus is lower in monopoly than in duopoly.
- *Welfare comparison.* Welfare under monopoly is $W^* = CS_M^* + \pi^*$. We can directly compute the difference between the welfare under monopoly and duopoly as follows. That is,

$$\begin{aligned}\Delta W = W_M - W_D &= \overbrace{\left(CS_M^* - CS_D^*\right)}^{\text{Change in } CS} + \overbrace{\left(\pi^* - \left(\pi_i^* + \pi_j^*\right)\right)}^{\text{Change in profits}} \\ &= -\frac{11(1 - c)^2}{50} + \left(\frac{(1 - c)^2}{2} - \frac{2(1 - c)^2}{25}\right)\end{aligned}$$

$$= -\frac{11(1-c)^2}{50} + \frac{21(1-c)^2}{50}$$
$$= \frac{(1-c)^2}{5} > 0.$$

Therefore, social welfare is larger under monopoly than under duopoly. In particular, while consumer surplus is lower under monopoly, profits are significantly larger, yielding an overall increase in welfare. Intuitively, under duopoly firms invest in R&D more than what they would if they could coordinate their R&D investment, leading to relatively low profits and welfare.

Exercise #6.5: More Competitive Industries and R&D Investment^C

6.5 Consider an industry with $n \geq 1$ firms selling a homogeneous good and competing à la Cournot. They face linear inverse demand curve $p(Q) = 1 - Q$, with cost function $C(q) = cq$, where c represents the (constant) marginal cost of production, and $1 > c > 0$. Firms interact in the following two-stage game: in the first stage, every firm i independently and simultaneously chooses its investment in research and development (R&D), x_i; in the second period, observing the profile of R&D investments $x = (x_1, \ldots, x_n)$, every firm i independently and simultaneously responds with its output level q_i.

Every firm faces an R&D cost $\frac{\gamma}{2}x_i^2$ in the first period, and marginal production costs $c - x_i$ in the second period where $1 > c > 0$. Intuitively, parameter $\gamma > 0$ denotes the efficiency of R&D investment, c represents the marginal production cost if the firm does not invest in R&D (i.e., if $x_i = 0$). If the firm invests in R&D, however, second-period marginal costs decrease.

(a) *Second period.* Find every firm's best response function in the second period.

- In the second period, every firm i takes the profile of R&D investments $x = (x_1, \ldots, x_n)$ as given, and chooses its output level q_i to solve

$$\max_{q_i \geq 0} (1 - Q)q_i - (c - x_i)q_i,$$

where Q denotes aggregate output. Taking first-order conditions with respect to q_i, we obtain

$$1 - 2q_i - \sum_{j \neq i}^{n-1} q_j + x_i - c = 0$$

and solving for q_i yields the best response function

$$q_i(q_j) = \frac{1 - c + x_i - \sum_{j \neq i}^{n-1} q_j}{2}$$

which results coincide with Exercise 6.4 if we have only two firms competing.

(b) Find second-period equilibrium output and profits.

- Using the above best response function, we obtain an equilibrium output of

$$q(x_i, x_j) = \frac{1 - (c - x_i) + \sum_{j \neq i} (x_i - x_j)}{n + 1}$$

(For more details about how to manipulate best response functions in a Cournot model with n asymmetric firms to find the equilibrium output level, see Exercise 2.7 in Chap. 2). Equilibrium profits are then

$$\pi_i(x_i, x_j) = \left[\frac{1-(c-x_i)+\sum_{j\neq i}\left(x_i-x_j\right)}{n+1}\right]^2.$$

(c) *First period.* Set up every firm's problem in the first period, and differentiate with respect to x_i. Identify the three effects that a marginal increase in x_i produces in firm i's profits. [*Hint*: You can focus on symmetric equilibria where $x_j = x_k = x$ for every firm $j \neq k \neq i$ after differentiating with respect to x_i.]

- In the first period, every firm i anticipates equilibrium output levels in the second stage, and chooses its R&D investment, x_i, to solve

$$\max_{x_i \geq 0} \left[\frac{1-(c-x_i)+\sum_{j\neq i}\left(x_i-x_j\right)}{n+1}\right]^2 - \frac{\gamma}{2}x_i^2$$

which includes the cost of investing in R&D in the last term. Differentiating with respect to x_i yields

$$2n\frac{1-(c-x_i)+\sum_{j\neq i}\left(x_i-x_j\right)}{(n+1)^2} - \gamma x_i = 0.$$

In a symmetric equilibrium, firms choose the same R&D investments in the first stage, $x_j = x_k = x$ for all $j \neq k \neq i$, which entails that $\sum_{j\neq i}\left(x_i - x_j\right) = (n-1)(x_i - x)$. Therefore, the above first-order condition simplifies to

$$\underbrace{2n\frac{1-(c-x_i)}{(n+1)^2}}_{\text{Appropriability effect}} + \underbrace{2n\frac{(n-1)(x_i-x)}{(n+1)^2}}_{\text{Competition effect}} = \underbrace{\gamma x_i}_{\text{Marginal cost of R\&D}}.$$

The two terms in the left side of the equation represent the marginal benefits from investing one more dollar in R&D, while the right side measures the marginal cost from this additional R&D. We examine each marginal benefit separately:

- The first benefit, the "appropriability effect," denotes the increase in profits that firm i experiences when decreasing its second-period costs with R&D investments. Intuitively, this benefit is increasing in the per-unit margin $1-(c-x_i)$, and decreasing in the number of firms in the industry. Therefore, the appropriability effect from R&D is the highest when the firm operates as a monopoly, but decreases when more firms compete in the industry.
- The second benefit, the "competition effect," represents firm i's incentives to invest in R&D in order to enjoy a cost advantage relative to its rivals, which occurs when $x_i - x > 0$, as this implies that second-period marginal costs satisfy $c - x_i < c - x$. As opposed to the appropriability effect, which decreases in the number of firms in the

market, n, the competition effect increases in n since

$$\frac{\partial\left(2n\frac{(n-1)(x_i-x)}{(n+1)^2}\right)}{\partial n}=\frac{2(x-x_i)(2n(n-1)+(1-2n)(n+1))}{(n+1)^3}$$
$$=\frac{2(x-x_i)(1-3n)}{(n+1)^3}>0,$$

where $x - x_i < 0$ holds when firm i invests more in R&D than its rivals, $x < x_i$, and $1 - 3n < 0$ holds for all admissible values of n since $n \geq 1$ by assumption; but at a decreasing rate since

$$\frac{\partial^2\left(2n\frac{(n-1)(x_i-x)}{(n+1)^2}\right)}{\partial n^2}=\frac{12(n(x-x_i)-(x-x_i))}{(n+1)^4}$$
$$=\frac{12(n-1)(x-x_i)}{(n+1)^4}<0,$$

given that, as described above, $n \geq 1$ and $x < x_i$.

In addition, the competition effect is zero when the firm operates as a monopolist, $n = 1$, but positive otherwise. In other words, when firm i does not compete with other firms, it does not have incentives to lower its second-period marginal cost to gain a strategic advantage in the Cournot game. Otherwise, firm i has incentives to gain this strategic advantage, and these incentives are increasing in the number of rivals.

- For completeness, we here check second-order conditions. Differentiating the above first-order condition with respect to x_i again, we obtain:

$$\frac{2n}{(n+1)^2}+\frac{2n(n-1)}{(n+1)^2}-\gamma=\frac{2n^2}{(n+1)^2}-\gamma$$

which is negative if

$$\gamma>\overline{\gamma}\equiv\frac{2n^2}{(n+1)^2}.$$

Intuitively, firm i chooses x_i in that its marginal investment cost in the first period exactly offsets its marginal savings in production cost in the second period, which happens when R&D investments are relatively costly (otherwise, if $\gamma \leq \overline{\gamma}$, R&D investments are extremely efficient so that this firm can keep on making cost-reducing investments and further improve its profits.)

(d) Focusing on a symmetric equilibrium, find the equilibrium investment in R&D in the first period, x^*. Does x^* increase in n?

- In a symmetric equilibrium, $x_i = x_j = x$ for every firm $j \neq i$. Inserting this property in the first-order condition we identified in part (c), we find:

$$2n\frac{1-(c-x)}{(n+1)^2}+2n\frac{(n-1)(x-x)}{(n+1)^2}=\gamma x.$$

Simplifying, we obtain:

$$2n\,(1 - c + x) = \gamma x(n + 1)^2.$$

Solving for x, we find the equilibrium investment in R&D:

$$x^* = \frac{2n(1 - c)}{\gamma(n + 1)^2 - 2n}$$

which is decreasing in the R&D efficiency, γ, and in the number of firms, n, since

$$\frac{\partial x^*}{\partial n} = -\frac{2\gamma(1 - c)(n^2 - 1)}{\left[\gamma(n + 1)^2 - 2n\right]^2} \leq 0$$

given that $n^2 \geq 1$ while the competition effect cancels out in equilibrium since $x_i = x_j$ for every two firms i and j. Since in equilibrium all firms invest the same amount in R&D, the competition effect is nil, implying that the only benefit from investing in R&D originates, in equilibrium, from the appropriability effect. Because this effect is decreasing in n, while the cost of investing in R&D is unaffected by n, firms reduce their R&D investments.

(e) *Numerical example.* Evaluate the equilibrium investment in R&D of part (d) at parameter values $c = \gamma = 1/2$. Then, evaluate this equilibrium investment at $n = 2$, $n = 3$, and $n = 10$ firms.

- Equilibrium investment in R&D becomes:

$$x^* = \frac{2n(1 - \frac{1}{2})}{\frac{1}{2}(n + 1)^2 - 2n} = \frac{2n}{(n - 1)^2}.$$

Evaluating this investment at $n = 2$, we obtain $x^* = \frac{2\times2}{(2-1)^2} = 4$; when $n = 3$, it decreases to $x^* = \frac{2\times3}{(3-1)^2} = \frac{3}{2}$; and when $n = 10$, this investment further decreases to $x^* = \frac{2\times10}{(10-1)^2} = \frac{20}{81} \simeq 0.24$, confirming our comparative statics results in part (d).

Exercise #6.6: Research Joint Ventures in R&D^B

6.6 Firms are often allowed to form "research joint ventures" (RJVs) to coordinate their R&D investment. In this exercise, we examine how RJVs affect firms' investment decisions in R&D.

Consider an industry with two firms facing inverse demand function $p(Q) = 1 - Q$, where Q denotes aggregate output. Firm i's total cost from investing x_i dollars in R&D and producing q_i units of output is

$$C(q_i, x_i) = (c - x_i)q_i + \frac{1}{2}x_i^2,$$

where the first term denotes production costs, $1 > c > 0$, and the second captures R&D costs. Marginal production cost, $c - x_i$, is clearly decreasing in firm i's R&D investment, x_i.

Firms interact in the following two-stage game. In the first stage, every firm i simultaneously and independently chooses its investment in R&D, x_i. In the second stage, observing the profile of R&D investments (x_i, x_j), every firm i simultaneously and independently selects its output level q_i. Firms compete à la Cournot.

(a) *Second stage*. Find the equilibrium output levels, as a function of R&D investments x_i and x_j, $q_i(x_i, x_j)$ for every firm i.

- Every firm i solves

$$\max_{q_i \geq 0} (1 - q_i - q_j)q_i - \underbrace{(c - x_i)q_i}_{\text{Production cost}} - \underbrace{\frac{1}{2}x_i^2}_{\text{R\&D cost}}.$$

Differentiating with respect to q_i yields:

$$1 - 2q_i - q_j - (c - x_i) = 0$$

and, solving for q_i, we find firm i's best response function in the second stage of the game:

$$q_i(q_j) = \frac{1 - c + x_i}{2} - \frac{1}{2}q_j$$

which, graphically, shifts upward as firm i's R&D investment, x_i, increases. A symmetric expression applies to firm j:

$$q_j(q_i) = \frac{1 - c + x_j}{2} - \frac{1}{2}q_i.$$

Inserting $q_j(q_i)$ into $q_i(q_j)$ yields:

$$q_i(x_i, x_j) = \frac{1 - c + x_i}{2} - \frac{1}{2}\underbrace{\left(\frac{1 - c + x_j}{2} - \frac{1}{2}q_i\right)}_{q_j(q_i)}$$

and, solving for q_i, we find the equilibrium output for firm i:

$$q_i(x_i, x_j) = \frac{1 - c + 2x_i - x_j}{3}$$

which is increasing in x_i but decreasing in x_j.

(b) *First stage*. Anticipating output levels $q_i(x_i, x_j)$ and $q_j(x_i, x_j)$ in the second stage, find the equilibrium R&D investment that every firm i chooses in the first stage, x_i^*.

- In the first stage, every firm i chooses x_i to solve

$$\max_{x_i \geq 0} \pi_i = (1 - q_i(x_i, x_j) - q_j(x_i, x_j))q_i(x_i, x_j) - \underbrace{(c - x_i)q_i(x_i, x_j)}_{\text{Production cost}} - \underbrace{\frac{1}{2}x_i^2}_{\text{R\&D cost}}$$

$$= \left(1 - \frac{1 - c + 2x_i - x_j}{3} - \frac{1 - c + 2x_j - x_i}{3}\right) \times \frac{1 - c + 2x_i - x_j}{3}$$

$$-(c - x_i)\frac{1 - c + 2x_i - x_j}{3} - \frac{1}{2}x_i^2$$

$$= \left(\frac{1 + 2c - x_i - x_j}{3}\right) \times \frac{1 - c + 2x_i - x_j}{3} - (c - x_i)\frac{1 - c + 2x_i - x_j}{3} - \frac{1}{2}x_i^2$$

which is only a function of R&D investments x_i and x_j. Rearranging the terms above, we have the profit function as follows:

$$\max_{x_i \geq 0} \pi_i = \frac{1-c+2x_i-x_j}{3} \times \left(\frac{1+2c-x_i-x_j}{3} - (c-x_i)\right) - \frac{1}{2}x_i^2$$
$$= \left(\frac{1-c+2x_i-x_j}{3}\right)^2 - \frac{1}{2}x_i^2.$$

Differentiating with respect to x_i, we find:

$$\frac{4(1-c+2x_i-x_j)}{9} - x_i = 0$$

and solving for x_i, we obtain firm i's best response function in the first stage of the game:

$$x_i(x_j) = 4(1-c-x_j)$$

which is decreasing in x_j. Therefore, firm i benefits from its rival's innovation and reduces its own research in innovation. A symmetric expression applies to firm j:

$$x_j(x_i) = 4(1-c-x_i).$$

Inserting $x_j(x_i)$ into $x_i(x_j)$ yields:

$$x_i = 4(1-c-\underbrace{4(1-c-x_i)}_{x_j(x_i)})$$

and, solving for x_i, we find the equilibrium investment in R&D for firm i:

$$x_i^* = x_j^* = \frac{4}{5}(1-c).$$

- Therefore, equilibrium output in the second stage, evaluated at $x_i^* = x_j^* = x^*$, is

$$q^* = q_i(x_i^*, x_j^*) = q_j(x_i^*, x_j^*) = \frac{1-c+x^*}{3} = \frac{1-c+\overbrace{\frac{4}{5}(1-c)}^{x^*}}{3}$$
$$= \frac{9(1-c)}{15} = \frac{3(1-c)}{5}.$$

(c) Evaluate equilibrium profits.
- Equilibrium profits for every firm i are

$$\pi_i^* = (1-2q^*) \times q^* - (c-x^*) \times q^* - \frac{1}{2}(x^*)^2$$
$$= q^*(1-2q^*-c+x^*) - \frac{1}{2}(x^*)^2$$

and evaluating them at equilibrium output $q^* = \frac{3(1-c)}{5}$ yields

$$\pi_i^* = \frac{3(1-c)}{5} \times \left(1 - c - 2 \times \frac{3(1-c)}{5} + \frac{4(1-c)}{5}\right) - \frac{1}{2}\left(\frac{4(1-c)}{5}\right)^2$$
$$= \frac{9(1-c)^2}{25} - \frac{8(1-c)^2}{25}$$
$$= \frac{(1-c)^2}{25}.$$

(d) *Research joint venture*. Assume now that firms create a research joint venture (RJV) to coordinate their R&D investment decision in the first stage of the game maximizing their joint profits. Firms, however, do not coordinate their output levels in the second stage, so they still produce $q_i(x_i, x_j)$ and $q_i(x_i, x_j)$ as found in part (a). Find their R&D investments in the RJV, and label them x_i^{RJV} and x_j^{RJV}.

- In the first stage, the RJV maximizes the sum of both firms' profits as follows:

$$\max_{x_i, x_j \geq 0} \pi_i + \pi_j = \left(\frac{1 - c + 2x_i - x_j}{3}\right)^2 + \left(\frac{1 - c + 2x_j - x_i}{3}\right)^2 - \frac{1}{2}\left(x_i^2 + x_j^2\right).$$

Differentiating with respect to x_i and x_j, we obtain:

$$\frac{4\left(1 - c + 2x_i - x_j\right)}{9} - \frac{2\left(1 - c + 2x_i - x_j\right)}{9} - x_i = 0$$
$$\frac{4\left(1 - c + 2x_j - x_i\right)}{9} - \frac{2\left(1 - c + 2x_j - x_i\right)}{9} - x_j = 0.$$

Simultaneously solving for x_i and x_j, we obtain the equilibrium R&D investments in the RJV:

$$x_i^{RJV} = \frac{2(1-c)}{7}.$$

(e) Compare x_i^{RJV} against x_i^*. Interpret.

- Comparing x_i^{RJV} and x_i^*, we find:

$$x_i^{RJV} - x_i^* = \frac{2(1-c)}{7} - \frac{4(1-c)}{5} = -\frac{18(1-c)}{35}$$

which is negative since $1 - c > 0$. Intuitively, when firms coordinate their R&D investment in the RJV, they internalize the cost advantages that this investment produces in the firm investing and the cost disadvantage that such investment produces in its rival. As a result, when firms coordinate in the RJV they invest less than when they independently choose their R&D investment, $x_i^{RJV} < x_i^*$.

(f) Evaluate equilibrium output and profits under the RJV and compare them against those every firm earns when independently choosing its R&D investment.

- Equilibrium output levels in the RJV are

$$q_i^{RJV} = \frac{1-c+2x_i^{RJV}-x_j^{RJV}}{3} = \frac{1-c+\frac{2(1-c)}{7}}{3}$$
$$= \frac{3(1-c)}{7}$$

- Comparing output levels across these two settings, we find that firms produce less under the RJV since

$$q_i^{RJV} - q_i^* = \frac{3(1-c)}{7} - \frac{3(1-c)}{5}$$
$$= -\frac{6(1-c)}{35} < 0$$

Intuitively, as each firm internalizes the effect of their R&D on its rival, they invest less in R&D which leads to less overall savings. As a result, each firm produces less output.

- Equilibrium profits in the RJV are

$$\pi_i^{RJV} = \frac{(1-c+x_i^{RJV})^2}{9} - \frac{1}{2}(x_i^{RJV})^2$$

and evaluating these profits at $x_i^{RJV} = \frac{2(1-c)}{7}$, we obtain:

$$\pi_i^{RJV} = \frac{\left(1-c+\frac{2(1-c)}{7}\right)^2}{9} - \frac{1}{2}\left(\frac{2(1-c)}{7}\right)^2$$
$$= \frac{1}{9} \times \frac{81(1-c)^2}{49} - \frac{2(1-c)^2}{49}$$
$$= \frac{(1-c)^2}{7}$$

- Comparing profits across these two settings, we find that profits are higher under the RJV since

$$\pi_i^{RJV} - \pi_i^* = \frac{(1-c)^2}{7} - \frac{(1-c)^2}{25}$$
$$= \frac{18(1-c)^2}{175} > 0$$

Intuitively, under the RJV every firm internalizes the cost disadvantage that its individual R&D investment produces on its rival, helping them increase their profits relative to the setting where every firm independently chooses its R&D investment.

(g) Find the social welfare with and without the RJV, and compare to that when the social planner decides the optimal investment for every firm i. Interpret.

- Social welfare is the sum of consumer surplus and firms' profits, as follows:

$$SW\left(x_i, x_j\right) = \underbrace{\frac{1}{2}\left(\frac{1-c+2x_i-x_j}{3}+\frac{1-c+2x_j-x_i}{3}\right)^2}_{CS}$$
$$+\underbrace{\left(\frac{1-c+2x_i-x_j}{3}\right)^2+\left(\frac{1-c+2x_j-x_i}{3}\right)^2-\frac{1}{2}\left(x_i^2+x_j^2\right)}_{\pi_i+\pi_j}$$

Invoking symmetry where $x_i = x_j$, the above expression becomes:

$$SW(x) = \frac{4(1-c+x)^2}{9} - x^2$$

Differentiating with respect to x, and assuming interior solutions, we have:

$$\frac{8(1-c+x)}{9} - 2x = 0$$

which is rearranged to yield

$$x^{SO} = \frac{4(1-c)}{5} = x^*$$

where every firm competes to make socially optimal investments.

Substituting $x^* = \frac{4(1-c)}{5}$, social welfare under independent R&D is

$$SW^* = \frac{4\left(1-c+\frac{4(1-c)}{5}\right)^2}{9} - \left(\frac{4(1-c)}{5}\right)^2$$
$$= \frac{4(1-c)^2}{5}$$

which coincides with that under the social planner's directive, SW^{SO}.

Substituting $x_i^{RJV} = \frac{2(1-c)}{7}$, social welfare under RJV becomes:

$$SW^{RJV} = \frac{4\left(1-c+\frac{2(1-c)}{7}\right)^2}{9} - \left(\frac{2(1-c)}{7}\right)^2$$
$$= \frac{32(1-c)^2}{49}$$

- We find that $SW^{RJV} < SW^* = SW^{SO}$ since when firms coordinate to reduce R&D investments in the RJV, firms generate higher profits at the expense of consumer surplus that ultimately decreases social welfare.

(h) *Numerical example.* Evaluate the equilibrium R&D investment of part (b), the equilibrium investment under the RJV found in part (d), and the associated output, profits, and social welfare in each case, at marginal cost $c = 1/2$.

- Without the RJV, equilibrium investment in R&D is

$$x_i^* = \frac{4(1-c)}{5} = \frac{4 \times \left(1 - \frac{1}{2}\right)}{5} = \frac{4}{10} = 0.4$$

with equilibrium output of

$$q_i^* = \frac{3 \times \left(1 - \frac{1}{2}\right)}{5} = \frac{3}{10} = 0.3$$

and associated profits of

$$\pi_i^* = \frac{(1-c)^2}{25} = \frac{\left(1 - \frac{1}{2}\right)^2}{25} = \frac{1}{100} = 0.01$$

yielding social welfare of

$$SW^* = \frac{4 \times \left(1 - \frac{1}{2}\right)^2}{5} = \frac{1}{5} = 0.2$$

- Whereas, equilibrium investment under the RJV is

$$x_i^{RJV} = \frac{2(1 - \frac{1}{2})}{7} = \frac{1}{7} \simeq 0.14$$

with equilibrium output of

$$q_i^{RJV} = \frac{1 - \frac{1}{2} + \frac{1}{7}}{3} = \frac{3}{14} \simeq 0.21$$

and associated profits of

$$\pi_i^{RJV} = \frac{(1-c)^2}{7} = \frac{(1 - \frac{1}{2})^2}{7} = \frac{1}{28} \simeq 0.035$$

yielding social welfare of

$$SW^{RJV} = \frac{32 \times \left(1 - \frac{1}{2}\right)^2}{49} = \frac{8}{49} \simeq 0.16$$

- When firms form RJV to coordinate investments in the cost-reduction technology, every firm i makes fewer investments in the first stage and reduces equilibrium output in the second stage, even though firms continue to compete à la Cournot before and after the formation of the RJV. Accordingly, every firm obtains higher equilibrium profits, but that do not compensate for the loss of consumer surplus, so that social welfare declines.

Exercise #6.7: Research Joint Ventures in R&D When Firms Compete in Prices^B

6.7 Consider a similar industry as that in Exercise 6.6, where firms form a research joint venture (RJV) to coordinate their R&D investments, but assuming now that they compete in prices instead of quantities, and allowing for goods to be horizontally differentiated.

Consider a duopoly, where every firm i faces the demand function $q_i(p_i) = 1 - p_i + dp_j$, where d measures the responsiveness of firm i's output to a one-unit increase in its rival's price, p_j that satisfies $d < 1$, thus indicating that own-price effect dominates cross-price effect. Firm i's total costs from investing x_i dollars in R&D and producing q_i units of output coincide with those in exercise 6.6, that is,

$$C(q_i, x_i) = (c - x_i) q_i + \frac{1}{2}x_i^2$$

where the first term denotes production costs, $0 \leq c \leq 1$, and the second term captures R&D costs. Marginal production cost, $c - x_i$, is clearly decreasing in firm i's R&D investment, x_i.

Firms interact in the following two-stage game. In the first stage, every firm i simultaneously and independently chooses its investment in R&D, x_i. In the second stage, observing the profile of R&D investments (x_i, x_j), every firm i simultaneously and independently selects its price p_i.

(a) *Second stage*. Find firm i's price, output, and profits as a function of investments x_i and x_j. (*Hint*: Recall that firms competing à la Bertrand and selling differentiated products set prices above marginal cost.)

- Every firm i chooses price p_i to solve the following profit maximization problem:

$$\max_{p_i \geq 0} \left(1 - p_i + dp_j\right) p_i - \underbrace{(c - x_i)\left(1 - p_i + dp_j\right)}_{\text{Production cost}} - \underbrace{\frac{1}{2}x_i^2}_{\text{R\&D cost}}$$

Differentiating with respect to p_i, and assuming interior solutions, yields:

$$1 - 2p_i + dp_j + c - x_i = 0$$

Solving for p_i, we find firm i's best response function in the second stage of the game:

$$p_i(p_j) = \frac{1 + c - x_i}{2} + \frac{d}{2}p_j$$

which, graphically, originates at $\frac{1+c-x_i}{2}$ and increases in p_j at a rate of $\frac{d}{2}$. In addition, an increase in firm i's R&D investment decreases its marginal cost, $c - x_i$, graphically shifting its best response function downward. A symmetric expression applies to firm j, as follows:

$$p_j(p_i) = \frac{1 + c - x_j}{2} + \frac{d}{2}p_i$$

Inserting $p_j(p_i)$ into $p_i(p_j)$ yields:

$$p_i(x_i, x_j) = \frac{1+c-x_i}{2} + \frac{d}{2}\overbrace{\left(\frac{1+c-x_j}{2} + \frac{dp_i}{2}\right)}^{p_j(p_i)}$$

Solving for p_i, we find firm i's equilibrium price:

$$p_i(x_i, x_j) = \frac{(2+d)(1+c) - 2x_i - dx_j}{4-d^2}$$

which is decreasing in x_i and x_j at a rate of $\frac{2}{4-d^2}$ and $\frac{d}{4-d^2}$, respectively.

- *Comparison with output competition.* Recall that in the setting where firms compete in quantities (Exercise 6.6), an increase in x_j (x_j) made firm i relatively more (less) competitive, producing an increase (decrease) in its output. In contrast, now an increase in x_i (x_j) shifts firm i's (firm j's) best response function downward, reducing as a consequence the price that both firms set.
- *Finding equilibrium output and profits.* Substituting the equilibrium price $p_i(x_i, x_j)$ into firm i's demand function yields:

$$q_i(x_i, x_j) = 1 - \frac{(2+d)(1+c) - 2x_i - dx_j}{4-d^2} + \frac{d\left[(2+d)(1+c) - 2x_j - dx_i\right]}{4-d^2}$$
$$= \frac{(2+d)[1-(1-d)c] + (2-d^2)x_i - dx_j}{4-d^2}$$

Equilibrium output, $q_i(x_i, x_j)$, is increasing (decreasing) in firm i's R&D investment, x_i (x_j, respectively), thus following a similar result as equilibrium output in Exercise 6.6 (part a). Substituting $q_i(x_i, x_j)$ into firm i's profit function, we find:

$$\pi(x_i, x_j)$$
$$= \left[\frac{(2+d)(1+c) - 2x_i - dx_j}{4-d^2} - c + x_i\right]$$
$$\times \frac{(2+d)[1-(1-d)c] + (2-d^2)x_i - dx_j}{4-d^2} - \frac{1}{2}x_i^2$$
$$= \left[\frac{(2+d)(1-(1-d)c) + (2-d^2)x_i - dx_j}{4-d^2}\right]^2 - \frac{1}{2}x_i^2$$

(b) *First stage.* Anticipating price levels $p_i(x_i, x_j)$ and $p_j(x_i, x_j)$ in the second stage, find the R&D investment that every firm i makes in the first stage, x_i^*.

- In the first stage, every firm i chooses x_i to solve

$$\max_{x_i \geq 0} \pi_i(x_i) = \left[\frac{(2+d)(1-(1-d)c) + (2-d^2)x_i - dx_j}{4-d^2}\right]^2 - \frac{1}{2}x_i^2$$

Differentiating with respect to x_i, and assuming interior solutions, we have

$$\frac{2\left(2-d^2\right)\left[(2+d)\left(1-(1-d)\,c\right)+\left(2-d^2\right)x_i-dx_j\right]}{\left(4-d^2\right)^2}-x_i=0$$

Solving for x_i, we find that firm i s best response function in the first stage of the game is

$$x_i\left(x_j\right)=\frac{2\left(2-d^2\right)\left[(2+d)\left(1-(1-d)\,c\right)\right]}{8-d^4}-\frac{d}{8-d^4}x_j$$

which is decreasing in x_j (negative slope). Therefore, firm i benefits from its rival's innovation and reduces its own research efforts, so that investments in R&D can be deemed as *strategic substitutes*.

- At a symmetric equilibrium, $x_i^*=x_j^*$, yielding

$$x_i=\frac{2\left(2-d^2\right)\left[(2+d)\left(1-(1-d)\,c\right)-dx_i\right]}{8-d^4}$$

which simplifies to

$$x_i^*=\frac{2\left(2-d^2\right)\left[1-(1-d)\,c\right]}{4-d^3}.$$

(c) *Comparative statics.* How does the equilibrium investment x_i^* you identified in part (b) vary with parameters c and d? Explain.

- Differentiating x_i^* with respect to c, we have:

$$\frac{\partial x_i^*}{\partial c}=-\frac{2\left(2-d^2\right)(1-d)}{4-d^3}<0$$

so that when the per-unit production cost increases (higher c), every firm reduces its investments in R&D in anticipation of fewer units of output.

- Differentiating x_i^* with respect to d, we obtain

$$\frac{\partial x_i^*}{\partial d}=-\frac{2[d(8-6d+d^3)-(8+8d-18d^2+4d^3+d^4)c]}{(4-d^3)^2}$$

which is positive when $c>\frac{d(8-6d+d^3)}{8+8d-18d^2+4d^3+d^4}$. Intuitively, when the initial cost is high, a stronger cross-price effect (higher d) induces every firm to make more investments. In other words, when the cross-price effect increase, products become more homogeneous, thus expanding firms' incentives to invest in R&D.

(d) Evaluate the price, output, and profits in part (a) at the equilibrium investment level x_i^*.

- Equilibrium price p_i^* in the second stage, evaluated at x_i^*, is

$$p_i^*=\frac{(2+d)\,(1+c)-(2+d)\,\frac{2(2-d^2)[1-(1-d)c]}{4-d^3}}{4-d^2}$$

$$=\frac{d^2+\left(4-d^2\right)c}{4-d^3}$$

- Equilibrium output q_i^* in the second stage, evaluated at x_i^*, is

$$q_i^* = \frac{(2+d)\left[1-(1-d)\,c\right] + (1-d)\,(2+d)\,\frac{2(2-d^2)[1-(1-d)c]}{4-d^3}}{4-d^2}$$
$$= \frac{\left(4-d^2\right)\left[1-(1-d)\,c\right]}{4-d^3}$$

- Equilibrium profits π_i^* in the second stage, evaluated at x_i^*, are

$$\pi_i^* = \left[\frac{(2+d)(1-(1-d)c) + (1-d)(2+d)\frac{2(2-d^2)[1-(1-d)c]}{4-d^3}}{4-d^2}\right]^2$$
$$-\frac{1}{2}\left(\frac{2(2-d^2)[1-(1-d)c]}{4-d^3}\right)^2$$
$$= \frac{\left(8-d^4\right)\left[1-(1-d)\,c\right]^2}{(4-d^3)^2}$$

(e) Evaluate equilibrium output level and profits under the RJV and compare them against those every firm earns when independently choosing its R&D investment.

- Equilibrium output levels in the RJV are

$$q_i^{RJV} = \frac{1-c+2x_i^{RJV} - x_j^{RJV}}{3} = \frac{1-c+\frac{2(1-c)}{7}}{3}$$
$$= \frac{3(1-c)}{7}$$

- Comparing output levels across these two settings, we find that firms produce less under the RJV since

$$q_i^{RJV} - q_i^* = \frac{3(1-c)}{7} - \frac{3(1-c)}{5}$$
$$= -\frac{6(1-c)}{35} < 0$$

Intuitively, as each firm internalizes the effect of their R&D on its rival, they invest less in R&D which leads to less overall savings. As a result, each firm produces less output.

- Equilibrium profits in the RJV are

$$\pi_i^{RJV} = \frac{(1-c+x_i^{RJV})^2}{9} - \frac{1}{2}(x_i^{RJV})^2$$

and evaluating these profits at $x_i^{RJV} = \frac{2(1-c)}{7}$, we obtain

$$\pi_i^{RJV} = \frac{\left(1-c+\frac{2(1-c)}{7}\right)^2}{9} - \frac{1}{2}\left(\frac{2(1-c)}{7}\right)^2$$

$$= \frac{1}{9} \times \frac{81(1-c)^2}{49} - \frac{2(1-c)^2}{49}$$

$$= \frac{(1-c)^2}{7}$$

- Comparing profits across these two settings, we find that profits are higher under the RJV since

$$\pi_i^{RJV} - \pi_i^* = \frac{(1-c)^2}{7} - \frac{(1-c)^2}{25}$$

$$= \frac{18(1-c)^2}{175} > 0$$

Intuitively, under the RJV every firm internalizes the cost disadvantage that its individual R&D investment produces on its rival, helping them increase their profits relative to the setting where every firm independently chooses its R&D investment.

(f) Find the social welfare with and without the RJV, and compare to that when the social planner decides the optimal investment for every firm i. Interpret.

- Social welfare is the sum of consumer surplus and firms' profits, as follows.

$$SW\left(x_i, x_j\right) = \underbrace{\frac{1}{2}\left(\frac{1-c+2x_i-x_j}{3} + \frac{1-c+2x_j-x_i}{3}\right)^2}_{CS}$$

$$+ \underbrace{\left(\frac{1-c+2x_i-x_j}{3}\right)^2 + \left(\frac{1-c+2x_j-x_i}{3}\right)^2 - \frac{1}{2}\left(x_i^2+x_j^2\right)}_{\pi_i+\pi_j}$$

Invoking symmetry where $x_i = x_j$, the above expression becomes

$$SW(x) = \frac{4(1-c+x)^2}{9} - x^2$$

Differentiating with respect to x, and assuming interior solutions, we have

$$\frac{8(1-c+x)}{9} - 2x = 0$$

which is rearranged to yield

$$x^{SO} = \frac{4(1-c)}{5} = x^*$$

where every firm competes to make socially optimal investments.

Substituting $x^* = \frac{4(1-c)}{5}$, social welfare under independent R&D is

$$SW^* = \frac{4\left(1-c+\frac{4(1-c)}{5}\right)^2}{9} - \left(\frac{4(1-c)}{5}\right)^2$$

$$= \frac{4(1-c)^2}{5}$$

which coincides with that under the social planner's directive, SW^{SO}.

Substituting $x_i^{RJV} = \frac{2(1-c)}{7}$, social welfare under RJV becomes

$$SW^{RJV} = \frac{4\left(1-c+\frac{2(1-c)}{7}\right)^2}{9} - \left(\frac{2(1-c)}{7}\right)^2$$
$$= \frac{32\,(1-c)^2}{49}$$

- We find that $SW^{RJV} < SW^* = SW^{SO}$ since when firms coordinate to reduce R&D investments in the RJV, firms generate higher profits at the expense of consumer surplus that ultimately decreases social welfare.

(g) Compare x_i^{RJV}, as found in part (e), against the equilibrium investment, x_i^*, found in part (b). Interpret.

- Comparing x_i^{RJV} from part (e) against x_i^* from part (b), we find that

$$x_i^* - x_i^{RJV} = \frac{2\left(2-d^2\right)[1-(1-d)\,c]}{4-d^3} - \frac{2\,(1-d)\,[1-(1-d)\,c]}{(2+d)\left(2-d^2\right)}$$
$$= \frac{2\left(4+8d-8d^2-3d^3+d^4+d^5\right)[1-(1-d)\,c]}{(2+d)\left(2-d^2\right)\left(4-d^3\right)}$$

which is positive for all values of $d \in [0, 1)$. Therefore, when firms coordinate their R&D investments in the RJV, every firm i internalizes the cost advantage that this investment produces in its rival. As a consequence, every firm invests less than when they independently choose their investment levels, that is, $x_i^{RJV} < x_i^*$. This result goes in line with what we found in Exercise 6.6, part (e).

- Further considering the following ratio

$$\frac{x_i^*}{x_i^{RJV}} = \frac{\left(2-d^2\right)^2(2+d)}{(1-d)\left(4-d^3\right)}$$

which is positive and increasing in d for all values of $d \in [0, 1)$, entailing that when firms experience stronger cross-price effects, firms further reduce their investments in R&D to soften subsequent price competition.

(h) Evaluate the equilibrium price, output, and profits under the RJV and compare them to those when firms independently choose their investment levels in part (d). Interpret.

- Equilibrium price in the second stage, evaluated at x_i^{RJV}, is

$$p_i^{RJV} = \frac{(2+d)\,(1+c) - (2+d)\,\frac{2(1-d)[1-(1-d)c]}{(2+d)(2-d^2)}}{4-d^2}$$
$$= \frac{\left(2+4d-2d^2-d^3\right)+\left(6-2d-d^3\right)c}{\left(2-d^2\right)\left(4-d^2\right)}$$

which exceeds that when firms independently choose investment levels since

$$p_i^{RJV} = \frac{(2+4d-2d^2-d^3)+(6-2d-d^3)c}{\left(2-d^2\right)\left(4-d^2\right)} > \frac{d^2+\left(4-d^2\right)c}{4-d^3} = p_i^*$$

simplifies to

$$2\left[1-(1-d)c\right]\left(4+8d-8d^2-3d^3+d^4+d^5\right) > 0$$

which holds under all admissible parameters. Therefore, when firms coordinate to reduce investments, equilibrium price rises.

- Equilibrium output in the second stage, evaluated at x_i^{RJV}, is

$$q_i^{RJV} = \frac{(2+d)\left[1-(1-d)c\right]+(1-d)(2+d)\frac{2(1-d)[1-(1-d)c]}{(2+d)(2-d^2)}}{4-d^2}$$
$$= \frac{(6-2d-d^3)\left[1-(1-d)c\right]}{\left(2-d^2\right)\left(4-d^2\right)}$$

which falls below that when firms independently choose investment levels since

$$q_i^{RJV} = \frac{(6-2d-d^3)\left[1-(1-d)c\right]}{\left(2-d^2\right)\left(4-d^2\right)} < \frac{\left(4-d^2\right)\left[1-(1-d)c\right]}{4-d^3} = q_i^*$$

simplifies to

$$2(1-d)\left(4+8d-8d^2-3d^3+d^4+d^5\right) > 0$$

which holds. Then, when firms coordinate to reduce investments, equilibrium output falls.

- Finally, equilibrium profits in the second stage, evaluated at x_i^{RJV}, are

$$\pi_i^{RJV} = \left[\frac{(2+d)(1-(1-d)c)+(1-d)(2+d)\frac{2(1-d)[1-(1-d)c]}{(2+d)(2-d^2)}}{4-d^2}\right]^2$$
$$-\frac{1}{2}\left(\frac{2(1-d)[1-(1-d)c]}{(2+d)(2-d^2)}\right)^2$$
$$= \frac{\left(14-4d^2-d^4\right)\left[1-(1-d)c\right]^2}{\left(2-d^2\right)(2-d)^2(2+d)^2}$$

which lie above those when firms independently choose investment levels when

$$\pi_i^{RJV} = \frac{\left(14-4d^2-d^4\right)\left[1-(1-d)c\right]^2}{\left(2-d^2\right)(2-d)^2(2+d)^2} > \frac{\left(8-d^4\right)\left[1-(1-d)c\right]^2}{(4-d^3)^2} = \pi_i^*$$

which simplifies to $d > 0.513$. Therefore, firms benefit from coordinating their R&D investments only when the cross-price effect is relatively significant; otherwise, they are better off independently choosing their investments.

(i) *Numerical Example.* Evaluate your results in parts (b), (f), and (g) at $c = 1/4$ and $d = 1/3$. How are your results affected if d increases to $d = 2/3$? Explain.

- Substituting $c = 1/4$ and $d = 1/3$ into the equilibrium results in part (b), we obtain:

$$x_i^* = \frac{2\left(2 - \frac{1}{3^2}\right)\left[1 - \frac{1}{4}\left(1 - \frac{1}{3}\right)\right]}{4 - \frac{1}{3^3}} = \frac{85}{107} \simeq 0.794$$

$$p_i^* = \frac{\frac{1}{3^2} + \frac{1}{4}\left(4 - \frac{1}{3^2}\right)}{4 - \frac{1}{3^3}} = \frac{117}{428} \simeq 0.273$$

$$q_i^* = \frac{\left(4 - \frac{1}{3^2}\right)\left[1 - \frac{1}{4}\left(1 - \frac{1}{3}\right)\right]}{4 - \frac{1}{3^3}} = \frac{175}{214} \simeq 0.818$$

$$\pi_i^* = \frac{\left(8 - \frac{1}{3^4}\right)\left[1 - \frac{1}{4}\left(1 - \frac{1}{3}\right)\right]^2}{\left(4 - \frac{1}{3^3}\right)^2} = \frac{16175}{45796} \simeq 0.353$$

- Inserting $c = 1/4$ and $d = 1/3$ into the equilibrium results in part (g), we obtain:

$$x_i^{RJV} = \frac{2\left(1 - \frac{1}{3}\right)\left[1 - \frac{1}{4}\left(1 - \frac{1}{3}\right)\right]}{\left(2 + \frac{1}{3}\right)\left(2 - \frac{1}{3^2}\right)} = \frac{30}{119} \simeq 0.252$$

$$p_i^{RJV} = \frac{(2 + \frac{4}{3} - \frac{2}{3^2} - \frac{1}{3^3}) + \frac{1}{4}\left(6 - \frac{2}{3} - \frac{1}{3^3}\right)}{\left(2 - \frac{1}{3^2}\right)\left(4 - \frac{1}{3^2}\right)} = \frac{285}{476} \simeq 0.599$$

$$q_i^{RJV} = \frac{\left(6 - \frac{2}{3} - \frac{1}{3^3}\right)\left[1 - \frac{1}{4}\left(1 - \frac{1}{3}\right)\right]}{\left(2 - \frac{1}{3^2}\right)\left(4 - \frac{1}{3^2}\right)} = \frac{143}{238} \simeq 0.601$$

$$\pi_i^{RJV} = \frac{\left(14 - \frac{4}{3^2} - \frac{1}{3^4}\right)\left[1 - \frac{1}{4}\left(1 - \frac{1}{3}\right)\right]^2}{\left(2 - \frac{1}{3^2}\right)\left(2 - \frac{1}{3}\right)^2\left(2 + \frac{1}{3}\right)^2} = \frac{1097}{3332} \simeq 0.329$$

- Substituting $c = 1/4$ and $d = 2/3$ into the equilibrium results in part (b), we obtain:

$$x_i^* = \frac{2\left[2 - \left(\frac{2}{3}\right)^2\right]\left[1 - \frac{1}{4}\left(1 - \frac{2}{3}\right)\right]}{4 - \left(\frac{2}{3}\right)^3} = \frac{77}{100} \simeq 0.77$$

$$p_i^* = \frac{\left(\frac{2}{3}\right)^2 + \frac{1}{4}\left[4 - \left(\frac{2}{3}\right)^2\right]}{4 - \left(\frac{2}{3}\right)^3} = \frac{9}{25} = 0.36$$

$$q_i^* = \frac{\left[4-\left(\frac{2}{3}\right)^2\right]\left[1-\frac{1}{4}\left(1-\frac{2}{3}\right)\right]}{4-\left(\frac{2}{3}\right)^3} = \frac{22}{25} = 0.88$$

$$\pi_i^* = \frac{\left[8-\left(\frac{2}{3}\right)^4\right]\left[1-\frac{1}{4}\left(1-\frac{2}{3}\right)\right]^2}{\left[4-\left(\frac{2}{3}\right)^3\right]^2} = \frac{9559}{20000} \simeq 0.478$$

- Inserting $c = 1/4$ and $d = 2/3$ into the equilibrium results in part (g), we obtain:

$$x_i^{RJV} = \frac{2\left(1-\frac{2}{3}\right)\left[1-\frac{1}{4}\left(1-\frac{2}{3}\right)\right]}{\left(2+\frac{2}{3}\right)\left[2-\left(\frac{2}{3}\right)^2\right]} = \frac{33}{224} \simeq 0.147$$

$$p_i^{RJV} = \frac{\left[2+\frac{8}{3}-2\left(\frac{2}{3}\right)^2-\left(\frac{2}{3}\right)^3\right]+\frac{1}{4}\left[6-\frac{4}{3}-\left(\frac{2}{3}\right)^3\right]}{\left[2-\left(\frac{2}{3}\right)^2\right]\left[4-\left(\frac{2}{3}\right)^2\right]} = \frac{741}{896} \simeq 0.827$$

$$q_i^{RJV} = \frac{\left[6-\frac{4}{3}-\left(\frac{2}{3}\right)^3\right]\left[1-\frac{1}{4}\left(1-\frac{2}{3}\right)\right]}{\left[2-\left(\frac{2}{3}\right)^2\right]\left[4-\left(\frac{2}{3}\right)^2\right]} = \frac{649}{896} \simeq 0.724$$

$$\pi_i^{RJV} = \frac{\left[14-4\left(\frac{2}{3}\right)^2-\left(\frac{2}{3}\right)^4\right]\left[1-\frac{1}{4}\left(1-\frac{2}{3}\right)\right]^2}{\left[2-\left(\frac{2}{3}\right)^2\right]\left(2-\frac{2}{3}\right)^2\left(2+\frac{2}{3}\right)^2} = \frac{58927}{114688} \simeq 0.514$$

- *Comparison.*
  - Both when $d = 1/3$ and $d = 2/3$, we find that R&D investments satisfy $x_i^{RJV} < x_i^*$, prices satisfy $p_i^{RJV} > p_i^*$, and output satisfies $q_i^{RJV} < q_i^*$, so that when firms coordinate to reduce their R&D investments, they produce fewer units of output and raise prices.
  - Regarding profits, we find that when the cross-price effects are relatively weak (e.g., $d = 1/3$), profits satisfy $\pi_i^{RJV} < \pi_i^*$, so firms should make R&D investments independently. Intuitively, the cost-reduction effect of lowering its own production costs more than offsets the coordination effect of raising output prices. In contrast, when the cross-price effect is relatively strong (e.g., $d = 2/3$), profits satisfy $\pi_i^{RJV} > \pi_i^*$, indicating that firms should coordinate their R&D investments. In this context, the coordination effect of raising output prices more than offsets the cost-reduction effect of lowering its own production costs.

Exercise #6.8: Spillover Effects in R&D Investment^B

6.8 Firms investing in R&D often experience "spillover effects" since part of their innovation can be easily learned by other firms in the industry, perhaps because a worker moved across companies, parts of the innovation were not be patented, or simply because firms are concentrated in one region or city, where workers from different firms interact with each other, leading to the exchange of ideas. In this exercise, we seek to evaluate how the presence of R&D spillover reduces firms' incentives to innovate.

Consider two firms in an industry facing inverse demand function $p(Q) = 1 - Q$, where Q denotes aggregate output. Firm i's cost from investing x_i dollars in R&D is $\frac{1}{2}x_i^2$, and its total production cost is

$$(c - x_i - \beta x_j)q_i,$$

where $1 > c > 0$ and $\beta \in [0, 1]$. This production cost decreases in its own investment in R&D, x_i, and in its rival's investment, x_j. Intuitively, when $\beta = 0$, firm i's marginal cost of production is unaffected by its rival's R&D decision, x_j, but when $\beta > 0$, every unit of R&D by its rival generates a positive spillover effect, decreasing firm i's marginal cost of production.

Firms interact in the following two-stage game. In the first stage, every firm i simultaneously and independently chooses its investment in R&D, x_i. In the second stage, observing the profile of R&D investments (x_i, x_j), every firm i simultaneously and independently selects its output level q_i. Firms compete à la Cournot.

(a) *Second stage*. Operating by backward induction, let us start analyzing the second stage of the game. Find equilibrium output levels, as a function of R&D investments x_i and x_j, $q_i(x_i, x_j)$ for every firm i.

- Every firm i solves

$$\max_{q_i \geq 0} (1 - q_i - q_j)q_i - \underbrace{(c - x_i - \beta x_j)q_i}_{\text{Production cost}} - \underbrace{\frac{1}{2}x_i^2}_{\text{R\&D cost}}.$$

Differentiating with respect to q_i yields:

$$1 - 2q_i - q_j - (c - x_i - \beta x_j) = 0$$

and solving for q_i, we find firm i's best response function in the second stage of the game

$$q_i(q_j) = \frac{1 - (c - x_i - \beta x_j)}{2} - \frac{1}{2}q_j$$

which, graphically, shifts upward as firm i's R&D investment, x_i, increases, and as firm j's investment, x_j, increases (but in a smaller proportion since $\beta < 1$ by definition). A symmetric expression applies to firm j,

$$q_j(q_i) = \frac{1 - (c - x_j - \beta x_i)}{2} - \frac{1}{2}q_i.$$

Inserting $q_j(q_i)$ into $q_i(q_j)$ yields

$$q_i = \frac{1-(c-x_i-\beta x_j)}{2} - \frac{1}{2}\underbrace{\left(\frac{1-(c-x_j-\beta x_i)}{2} - \frac{1}{2}q_i\right)}_{q_j(q_i)}$$

and, solving for q_i, we find the equilibrium output for firm i:

$$q_i(x_i, x_j) = \frac{1-c+(2-\beta)x_i+(2\beta-1)x_j}{3}$$

which is unambiguously increasing in firm i's R&D investment, x_i. However, $q_i(x_i, x_j)$ is also increasing in its rival's R&D investment, x_j, if and only if $\beta \geq \frac{1}{2}$. Intuitively, when $\beta < \frac{1}{2}$, firm j's R&D investment produces a negative external effect on firm i's profits. (This includes the case where spillovers are absent, $\beta = 0$, as in previous exercises.) In contrast, when $\beta \geq \frac{1}{2}$, spillover effects are so intense that firm j's R&D investment generates a positive external effect on firm i's profits.

(b) *First stage*. Anticipating output levels $q_i(x_i, x_j)$ and $q_j(x_i, x_j)$ in the second stage, find the equilibrium R&D investment that every firm i chooses in the first stage, x_i^*.

- In the first stage, every firm i chooses x_i to solve

$$\max_{x_i \geq 0} \pi_i = (1-q_i(x_i,x_j)-q_j(x_i,x_j))q_i(x_i,x_j) - \underbrace{(c-x_i-\beta x_j)q_i(x_i,x_j)}_{\text{Production cost}} - \underbrace{\frac{1}{2}x_i^2}_{\text{R\&D cost}}$$

$$= \left(1 - \frac{1-c+(2-\beta)x_i+(2\beta-1)x_j}{3} - \frac{1-c+(2-\beta)x_j+(2\beta-1)x_i}{3}\right) \times$$
$$\frac{1-c+(2-\beta)x_i+(2\beta-1)x_j}{3}$$
$$-(c-x_i-\beta x_j) \times \frac{1-c+(2-\beta)x_i+(2\beta-1)x_j}{3} - \frac{1}{2}x_i^2$$
$$= \left(1 - \frac{2(1-c)+(1+\beta)(x_i+x_j)}{3} - (c-x_i-\beta x_j)\right)$$
$$\times \frac{1-c+(2-\beta)x_i+(2\beta-1)x_j}{3} - \frac{1}{2}x_i^2$$
$$= \left(\frac{3-2(1-c)-(1+\beta)(x_i+x_j)-3(c-x_i-\beta x_j)}{3}\right)$$
$$\times \frac{1-c+(2-\beta)x_i+(2\beta-1)x_j}{3} - \frac{1}{2}x_i^2$$
$$= \left(\frac{1-c+(2-\beta)x_i+(2\beta-1)x_j}{3}\right)^2 - \frac{1}{2}x_i^2$$

which is only a function of R&D investments x_i and x_j. Differentiating with respect to x_i, we find:

$$2 \times \frac{1-c+(2-\beta)x_i+(2\beta-1)x_j}{3} \times \frac{2-\beta}{3} - x_i = 0$$

which is simplified to yield

$$\left[9 - 2(2-\beta)^2\right] x_i = 2(2-\beta)\left[1 - c + (2\beta - 1)x_j\right]$$

and, solving for x_i, we obtain firm i's best response function in the first stage of the game:

$$x_i(x_j) = \frac{2(2-\beta)(1-c)}{1+8\beta-2\beta^2} + \frac{2(2-\beta)(2\beta-1)}{1+8\beta-2\beta^2}x_j.$$

We can check that this best response function is increasing (decreasing) in x_j when $\frac{1}{2} < \beta \leq 1$ ($0 < \beta \leq \frac{1}{2}$), so R&D investments are strategic complements (substitutes) when the spillover effect is strong (weak). In particular,

$$\frac{\partial x_i}{\partial x_j} = \frac{2(2-\beta)(2\beta-1)}{1+8\beta-2\beta^2} > 0,$$

where the terms in the numerator are clearly positive when $\beta \in (0.5, 1]$ for sufficiently strong spillover effects so that $2\beta - 1 > 0$ and $2 - \beta > 0$. The term in the denominator, $1 + 8\beta - 2\beta^2$, is also positive for all

$$\beta \in \left[2 - \frac{3}{\sqrt{2}}, 2 + \frac{3}{\sqrt{2}}\right] \simeq [-0.12, 4.12],$$

implying that the denominator is positive for all admissible values of β. Therefore, firm i responds to an increase in its rival's innovation by increasing its own investment in R&D.

- Invoking symmetry, where $x^* = x_i = x_j$, we obtain:

$$x^* = \frac{2(2-\beta)[1 - c + (2\beta - 1)x^*]}{1+8\beta-2\beta^2}$$

and simplifying, we find the equilibrium R&D for firm i:

$$x_i^* = \frac{2(2-\beta)(1-c)}{2\beta^2 - 2\beta + 5}.$$

which is unambiguously positive since the numerator is positive and the denominator is also positive for all admissible values of β. We can easily check that equilibrium R&D, x_i^*, is decreasing in the spillover effect from R&D investment, β, since

$$\frac{\partial x_i^*}{\partial \beta} = -\frac{2(1-c)\left[2\beta(4-\beta)+1\right]}{[5 - 2\beta(1-\beta)]^2} < 0$$

given that the denominator is positive, $1 - c > 0$, and the last term in the numerator, $2\beta(4-\beta) + 1$, is positive for all

$$\beta \in \left[2 - \frac{3}{\sqrt{2}}, 2 + \frac{3}{\sqrt{2}}\right] \simeq [-0.12, 4.12].$$

Therefore, $\frac{\partial x_i^*}{\partial \beta} < 0$, intuitively indicating that, as firm j can benefit from a larger share of firm i's investment in R&D, firm i chooses to invest less.

(c) *Research joint venture.* Assume that firms create a research joint venture (RJV) that could coordinate their R&D investment decision in the first stage of the game maximizing their joint profits. Firms, however, do not coordinate their output levels in the second stage, so they still produce $q_i(x_i, x_j)$ and $q_i(x_i, x_j)$ as found in part (a). Find their R&D investments in the RJV, and label them x_i^{RJV} and x_j^{RJV}. How does x_i^{RJV} change with β? Interpret.

- In the first stage, the RJV maximizes the sum of both firms' profits as follows:

$$\max_{x_i, x_j \geq 0} \pi_i + \pi_j = \left(\frac{1 - c + (2-\beta)x_i + (2\beta - 1)x_j}{3}\right)^2 - \frac{1}{2}x_i^2 + \left(\frac{1 - c + (2-\beta)x_j + (2\beta - 1)x_i}{3}\right)^2 - \frac{1}{2}x_j^2$$

which is only a function of R&D investments x_i and x_j. Differentiating with respect to x_i, we obtain

$$\frac{2(2-\beta)[1 - c + (2-\beta)x_i + (2\beta - 1)x_j]}{9} + \frac{2(2\beta - 1)[1 - c + (2-\beta)x_j + (2\beta - 1)x_i]}{9} - x_i = 0$$

and differentiating with respect to x_j, we find

$$\frac{2(2-\beta)[1 - c + (2-\beta)x_j + (2\beta - 1)x_i]}{9} + \frac{2(2\beta - 1)[1 - c + (2-\beta)x_i + (2\beta - 1)x_j]}{9} - x_j = 0.$$

In a symmetric equilibrium, both firms invest the same amount in R&D, that is, $x_i^{RJV} = x_j^{RJV}$. Inserting this property in any of the above first-order conditions yields:

$$\frac{2(2-\beta)[1 - c + (2-\beta)x_i + (2\beta - 1)x_i]}{9} + \frac{2(2\beta - 1)[1 - c + (2-\beta)x_i + (2\beta - 1)x_i]}{9} - x_i = 0.$$

Rearranging, and solving for x_i, we obtain the equilibrium R&D investments in the RJV as follows:

$$x_i^{RJV} = \frac{2(1+\beta)(1-c)}{7 - 4\beta - 2\beta^2}$$

which is positive given that the denominator, $7 - 4\beta - 2\beta^2$, is positive for all $\beta \in [0, 1]$.

- Differentiating x_i^{RJV} with respect to β, we have:

$$\frac{\partial x_i^{RJV}}{\partial \beta} = \frac{2\left[7 - 4\beta - 2\beta^2 + 4(1+\beta)^2\right](1-c)}{\left(7 - 4\beta - 2\beta^2\right)^2}$$

$$= \frac{2\left(11 + 4\beta + 2\beta^2\right)(1-c)}{\left(7 - 4\beta - 2\beta^2\right)^2} > 0$$

so that the RJV invests more when the spillover effect strengthens.

(d) *Comparison.* Compare investment level x_i^{RJV} against x_i^*. Interpret.

- Comparing x_i^{RJV} and x_i^*, we find:

$$x_i^{RJV} - x_i^* = \frac{2(1+\beta)(1-c)}{7 - 4\beta - 2\beta^2} - \frac{2(2-\beta)(1-c)}{2\beta^2 - 2\beta + 5}$$

$$= \frac{18(2\beta - 1)(1-c)}{\left(7 - 4\beta - 2\beta^2\right)\left(2\beta^2 - 2\beta + 5\right)}$$

which is positive (negative) when $\beta \in [0.5, 1]$ ($\beta \in [0, 0.5)$) since all terms in the denominator are positive, that is, $7 - 4\beta - 2\beta^2 > 0$ and $2\beta^2 - 2\beta + 5 > 0$.

- Intuitively, when $\beta < \frac{1}{2}$ (which includes the special case of $\beta = 0$, as in Exercise 6.6), firm i's R&D investment only generates a negative externality on its rival's costs. As a consequence, when firms form an RJV, they internalize this negative externality, choosing a lower investment in R&D. In contrast, when $\beta \geq \frac{1}{2}$, the positive externality from spillovers dominates the negative externality from R&D investment. In this context, the RJV internalizes the positive externality by investing more in R&D than when firms independently choose their own investments.
- Indeed, the difference, $x_i^{RJV} - x_i^*$, increases in β since

$$\frac{\partial\left[x_i^{RJV} - x_i^*\right]}{\partial \beta} = \frac{216\left(2\beta^4 - 3\beta^2 + 2\beta + 3\right)(1-c)}{\left(7 - 4\beta - 2\beta^2\right)^2\left(2\beta^2 - 2\beta + 5\right)^2} > 0$$

Therefore, the RJV makes more, while the independent firms make less investments, so that the difference in equilibrium investments, $x_i^{RJV} - x_i^*$, enlarges as the spillover effect, β, becomes stronger. In summary, the RJV internalizes the free-riding behavior stemming from the spillover that occurs when every firm independently chooses its investment. In addition, this internalization increases as the spillover effect (and its associated free-riding behavior) becomes more substantial.

(e) Find the equilibrium output and profits with and without the RJV, and compare.

- Equilibrium output before the formation of RJV is

$$q_i^* = \frac{1 - c + (2 - \beta + 2\beta - 1)\overbrace{\frac{2(2-\beta)(1-c)}{2\beta^2 - 2\beta + 5}}^{x_i^*}}{3}$$

$$= \frac{3(1-c)}{2\beta^2 - 2\beta + 5}$$

yielding equilibrium profits of

$$\pi_i^* = \left(\frac{1 - c + (2 - \beta + 2\beta - 1)\, \frac{2(2-\beta)(1-c)}{2\beta^2 - 2\beta + 5}}{3} \right)^2 - \frac{1}{2} \left(\frac{2\,(2 - \beta)\,(1 - c)}{2\beta^2 - 2\beta + 5} \right)^2$$

$$= \frac{\left[9 - 2\,(2 - \beta)^2\right](1 - c)^2}{\left(2\beta^2 - 2\beta + 5\right)^2}$$

- Equilibrium output after the formation of RJV is

$$q_i^{RJV} = \frac{1 - c + (2 - \beta + 2\beta - 1)\, \overbrace{\frac{2\,(1 + \beta)\,(1 - c)}{7 - 4\beta - 2\beta^2}}^{x_i^{RJV}}}{3}$$

$$= \frac{3\,(1 - c)}{7 - 4\beta - 2\beta^2}$$

yielding equilibrium profits of

$$\pi_i^{RJV} = \left(\frac{1 - c + (2 - \beta + 2\beta - 1)\, \frac{2(1+\beta)(1-c)}{7 - 4\beta - 2\beta^2}}{3} \right)^2 - \frac{1}{2} \left(\frac{2\,(1 + \beta)\,(1 - c)}{7 - 4\beta - 2\beta^2} \right)^2$$

$$= \frac{\left[9 - 2\,(1 + \beta)^2\right](1 - c)^2}{\left(7 - 4\beta - 2\beta^2\right)^2}$$

- Equilibrium output increases with the RJV, that is, $q_i^{RJV} > q_i^*$, if and only if

$$\frac{3\,(1 - c)}{7 - 4\beta - 2\beta^2} > \frac{3\,(1 - c)}{2\beta^2 - 2\beta + 5}$$

which reduces to

$$(2\beta - 1)\,(\beta + 1) > 0$$

that holds when $\beta > \frac{1}{2}$.

- Equilibrium profits increase with the RJV, that is, $\pi_i^{RJV} > \pi_i^*$, if and only if

$$\frac{\left[9 - 2\,(1 + \beta)^2\right](1 - c)^2}{\left(7 - 4\beta - 2\beta^2\right)^2} > \frac{\left[9 - 2\,(2 - \beta)^2\right](1 - c)^2}{\left(2\beta^2 - 2\beta + 5\right)^2}$$

which reduces to

$$18\,(1 - 2\beta)^2 \left(7 - 4\beta - 2\beta^2\right) > 0$$

that holds for all values of $\beta \in [0, 1]$.

(f) Evaluate the social welfare with and without the RJV, and compare. Interpret.

- Social welfare before the formation of RJV is

$$SW^* = \underbrace{\frac{1}{2}\left(2 \times \frac{3(1-c)}{2\beta^2 - 2\beta + 5}\right)^2}_{CS^*} + \underbrace{2 \times \frac{\left[9 - 2(2-\beta)^2\right](1-c)^2}{\left(2\beta^2 - 2\beta + 5\right)^2}}_{\pi_i^* + \pi_j^*}$$

$$= \frac{4\left[9 - (2-\beta)^2\right](1-c)^2}{\left(2\beta^2 - 2\beta + 5\right)^2}$$

- Social welfare after the formation of RJV is

$$SW^{RJV} = \underbrace{\frac{1}{2}\left(2 \times \frac{3(1-c)}{7 - 4\beta - 2\beta^2}\right)^2}_{CS^{RJV}} + \underbrace{2 \times \frac{\left[9 - 2(1+\beta)^2\right](1-c)^2}{\left(7 - 4\beta - 2\beta^2\right)^2}}_{\pi_i^{RJV} + \pi_j^{RJV}}$$

$$= \frac{4\left[9 - (1+\beta)^2\right](1-c)^2}{\left(7 - 4\beta - 2\beta^2\right)^2}$$

- Hence, RJV increases social welfare if and only if $SW^{RJV} > SW^*$, solving

$$\frac{4\left[9 - (1+\beta)^2\right](1-c)^2}{\left(7 - 4\beta - 2\beta^2\right)^2} > \frac{4\left[9 - (2-\beta)^2\right](1-c)^2}{\left(2\beta^2 - 2\beta + 5\right)^2}$$

which reduces to

$$9(2\beta - 1)\left(5 + 24\beta - 12\beta^2 - 4\beta^3\right) > 0$$

that holds if and only if $\beta > \frac{1}{2}$, since $5 + 24\beta - 12\beta^2 - 4\beta^3 > 0$ for all $\beta \in [0, 1]$. Intuitively, the RJV increases social welfare only when the spillover effect is relatively significant, since consumers can ultimately benefit from a larger output when $\beta > \frac{1}{2}$.

(g) *Numerical example.* Evaluate the equilibrium investment in R&D of part (b) and the equilibrium investment, output, profits, and social welfare under the RJV found in part (c), at parameter values $c = 1/2$ and $\beta = 2/3$.

- Equilibrium investment in R&D becomes:

$$x_i^* = \frac{2(2-\beta)(1-c)}{2\beta^2 - 2\beta + 5} = \frac{2(2-\frac{2}{3})(1-\frac{1}{2})}{2\left(\frac{2}{3}\right)^2 - 2\frac{2}{3} + 5} = \frac{12}{41} \simeq 0.29.$$

- However, equilibrium investment under the RJV is

$$x_i^{RJV} = \frac{2(1+\beta)(1-c)}{7 - 4\beta - 2\beta^2} = \frac{2(1+\frac{2}{3})(1-\frac{1}{2})}{7 - 4\frac{2}{3} - 2\left(\frac{2}{3}\right)^2} = \frac{15}{31} \simeq 0.48$$

which, as expected, satisfies $x_i^{RJV} > x_i^*$.

- Equilibrium output becomes

$$q_i^* = \frac{1 - \frac{1}{2} + \left(2 - \frac{2}{3} + 2 \times \frac{2}{3} - 1\right) \times \frac{12}{41}}{3} = \frac{27}{82} \simeq 0.33$$

and that under the RJV is

$$q_i^{RJV} = \frac{1 - \frac{1}{2} + \left(2 - \frac{2}{3} + 2 \times \frac{2}{3} - 1\right) \times \frac{15}{31}}{3} = \frac{27}{62} \simeq 0.44$$

so that firms produce more output under the RJV, that is, $q_i^{RJV} > q_i^*$.
- Substituting equilibrium R&D into the profit function, we obtain:

$$\begin{aligned}
\pi_i^* &= \left(\frac{1 - c + (2-\beta)\, x^* + (2\beta - 1)\, x^*}{3}\right)^2 - \frac{1}{2}\left(x^*\right)^2 \\
&= \frac{1}{9}\left[1 - \frac{1}{2} + \frac{12}{41}\left(1 + \frac{2}{3}\right)\right]^2 - \frac{1}{2}\left(\frac{12}{41}\right)^2 \\
&= \frac{441}{6724} \simeq 0.066 \\
\pi_i^{RJV} &= \left(\frac{1 - c + (2-\beta)\, x^{RJV} + (2\beta - 1)\, x^{RJV}}{3}\right)^2 - \frac{1}{2}\left(x^{RJV}\right)^2 \\
&= \frac{1}{9}\left[1 - \frac{1}{2} + \frac{15}{31}\left(1 + \frac{2}{3}\right)\right]^2 - \frac{1}{2}\left(\frac{15}{31}\right)^2 \\
&= \frac{1179}{3844} \simeq 0.307.
\end{aligned}$$

- Social welfare without the RJV is

$$SW^* = \frac{4\left[9 - \left(2 - \frac{2}{3}\right)^2\right]\left(1 - \frac{1}{2}\right)^2}{\left(2 \times \left(\frac{2}{3}\right)^2 - 2 \times \frac{2}{3} + 5\right)^2} = \frac{585}{1681} \simeq 0.35$$

and that under the RJV becomes:

$$SW^{RJV} = \frac{4\left[9 - \left(1 + \frac{2}{3}\right)^2\right]\left(1 - \frac{1}{2}\right)^2}{\left(7 - 4 \times \frac{2}{3} - 2 \times \left(\frac{2}{3}\right)^2\right)^2} = \frac{504}{961} \simeq 0.52$$

so that the RJV increases social welfare, that is, $SW^{RJV} > SW^*$. Comparing our results with Exercise 6.6, the existence of spillover effects reinforces the firms' incentives to form RJVs by making more investment in R&D and enjoying higher profit, that is, $\pi_i^{RJV} > \pi_i^*$, in equilibrium.

- Social welfare without the RJV is

$$SW^* = \frac{4\left[9 - \left(2 - \frac{2}{3}\right)^2\right]\left(1 - \frac{1}{2}\right)^2}{\left(2 \times \left(\frac{2}{3}\right)^2 - 2 \times \frac{2}{3} + 5\right)^2} = \frac{585}{1681} \simeq 0.35$$

and that under the RJV becomes

$$SW^{RJV} = \frac{4\left[9 - \left(1 + \frac{2}{3}\right)^2\right]\left(1 - \frac{1}{2}\right)^2}{\left(7 - 4 \times \frac{2}{3} - 2 \times \left(\frac{2}{3}\right)^2\right)^2} = \frac{504}{961} \simeq 0.52$$

so that the RJV increases social welfare, that is, $SW^{RJV} > SW^*$.

Exercise #6.9: Two Firms Sequentially Developing New Products^B

6.9 Consider an industry with two firms, 1 and 2, competing à la Cournot with inverse demand function $p(q_1, q_2) = a - q_1 - q_2$, where $a > 0$, and q_i stands for firm i's output, where $i \in \{1, 2\}$, and $j \neq i$. In the first stage, firm 1 chooses whether to develop a new product, at a fixed R&D cost of $F_1 > 0$, or not. If firm 1 does not develop the new product, the profits of all firms in this market are zero. If firm 1 develops the product, then in the second stage, firm 2 responds choosing whether to mimic the product innovation, at a lower R&D cost of F_2 where $F_2 < F_1$, or not mimicking it. For simplicity, assume that firm 2 cannot independently develop the new product if firm 1 did not develop it first. In the third stage, every firm i that developed the product simultaneously and independently chooses its output level q_i at a constant marginal cost of c, where $a > c \geq 0$.

(a) *Third stage.* Find the equilibrium output of every firm i in the third stage of the game.

- *Subgame A: Both firms develop the product.*

  The profit maximization problem of firm i is

$$\max_{q_i \geq 0} \pi_i(q_i) = \left(a - q_i - q_j\right) q_i - cq_i - F_i$$
$$= \left(a - q_i - q_j - c\right) q_i - F_i.$$

- Differentiating with respect to q_i, we obtain:

$$a - 2q_i - q_j - c = 0$$

rearranging, we find:

$$2q_i = a - q_j - c$$

and solving for q_i, we find firm i's best response function:

$$q_i\left(q_j\right) = \frac{a - c}{2} - \frac{1}{2} - q_j$$

which originates at $q_i = \frac{a-c}{2}$ and decreases in firm j's output, q_j, at a rate of $\frac{1}{2}$. Invoking symmetry, we have $q = q_i = q_j$, yielding

$$q = \frac{a-c-q}{2}$$

rearranging, we find the equilibrium output of every firm i:

$$q^* = \frac{a-c}{3}.$$

Substituting $q^* = \frac{a-c}{3}$ into the profit function, equilibrium profit becomes:

$$\begin{aligned}\pi_i^* &\equiv \pi_i\left(q_i^*\right) \\ &= \left(a - 2q^* - c\right) q^* - F_i \\ &= \left(a - \frac{2\,(a-c)}{3} - c\right) \times \frac{a-c}{3} - F_i \\ &= \frac{(a-c)^2}{9} - F_i.\end{aligned}$$

- *Subgame B: Only firm 1 develops the product.*
  In this context, the profit maximization problem of firm 1 is

$$\begin{aligned}\max_{q_1 \geq 0} \pi_1\left(q_1\right) &= \left(a - q_1\right) q_1 - c q_1 - F_1 \\ &= \left(a - q_1 - c\right) q_1 - F_1.\end{aligned}$$

- Differentiating with respect to q_1, we obtain:

$$a - 2q_1 - c = 0$$

and solving for q_1, we find firm 1's output of

$$q_1^m = \frac{a-c}{2}.$$

Substituting $q_1^m = \frac{a-c}{2}$ into firm 1's profit function, its equilibrium profit becomes:

$$\begin{aligned}\pi_1^m &\equiv \pi_1\left(q_1^m\right) \\ &= \left(a - q_1^m - c\right) q_1 - F_1 \\ &= \left(a - \frac{a-c}{2} - c\right) \frac{a-c}{2} - F_1 \\ &= \frac{(a-c)^2}{4} - F_1.\end{aligned}$$

- *Subgame C: Neither firm develops the product.* In this context, each firm produces zero output and obtains zero profits, that is, $\pi_i = q_i = 0$ where $i \in \{1, 2\}$.

(b) *Second stage.* Does firm 2 develop the new product in the second stage?
- In the second stage, firm 2 develops the new product if, upon observing that firm 1 developed in the first stage, firm 2's profits are larger developing the product than otherwise, that is, $\pi_2^* \geq 0$, or

$$\frac{(a-c)^2}{9} \geq F_2 \iff F_2 \leq \underline{F} \equiv \frac{(a-c)^2}{9}.$$

Otherwise, firm 2 does not develop the product, leaving firm 1 as the only developer in equilibrium.

(c) *First stage.* Does firm 1 develop the new product in the first stage?
- In the first stage, firm 1 anticipates whether firm 2 develops the new product in the second stage, which occurs if $F_2 \leq \underline{F}$, or not, which occurs if $F_2 > \underline{F}$.
  - In the first case, where $F_2 \leq \underline{F}$, firm 1 develops the product, anticipating that firm 2 will develop it too, if and only if $\pi_1^* \geq 0$, which simplifies to

$$\frac{(a-c)^2}{9} - F_1 \geq 0 \iff F_1 \leq \underline{F} \equiv \frac{(a-c)^2}{9}.$$

Since in this case, F_2 satisfies $F_2 \leq \underline{F}$, and we know that $F_2 < F_1$ by definition, we obtain a subgame perfect equilibrium where both firms develop if $F_1, F_2 \leq \underline{F}$. Intuitively, if firm 1 anticipates that its product will be mimicked by its rival, it is more profitable for firm 1 to develop the product (earning a positive profit) than not developing it at all. In contrast, when $F_2 < \underline{F} < F_1$, firm 2 would like to mimic its predecessor's product innovation that entails a profit loss for firm 1, so that firm 1 chooses to not develop the product in the first stage.
  - In the second case, where $F_2 > \underline{F}$, firm 1 anticipates that firm 2 will not develop the new product in the second stage, and thus develops if and only if

$$\frac{(a-c)^2}{4} - F_1 \geq 0$$

since firm 1 would be the only firm developing the product. Solving for F_1, we obtain

$$F_1 \leq \frac{(a-c)^2}{4} \equiv \overline{F}.$$

Therefore, we obtain two more subgame perfect equilibria: one in which firm 1 is the only firm developing the new product, which holds when $\underline{F} < F_2 < F_1 \leq \overline{F}$, and another in which no firm develops the product, which arises when $F_1 > \overline{F}$.
- Figure 6.3 summarizes the above three subgame perfect equilibria, as a function of firm 1's development cost, F_1, in the horizontal axis, and firm 2's development cost F_2 in the vertical axis. We focus on cost pairs below the 45-degree line since $F_2 \leq F_1$ by assumption. Intuitively, when both firms' costs are relatively low (close to the origin in the figure), both firms develop the new product. When firm 1's cost is intermediate but firm 2's is relatively high, only firm 1 develops the product. Finally, when both firms' costs are relatively high, no firm develops the new product. Note that, in (F_1, F_2)-pairs in the south area of this region, firm 2 is attracted to develop the product if firm 1 develops it in the first stage, but firm 1 suffers from a large cost disadvantage given that F_1 is significantly larger than F_2, leading firm 1 to not develop the product, which prevents firm 2 from developing too.

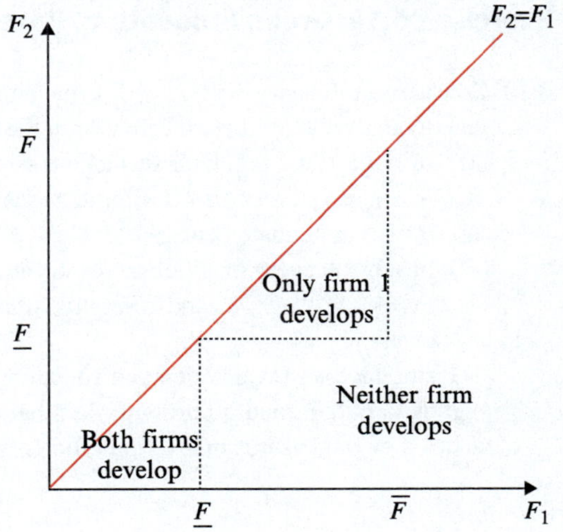

Fig. 6.3 Both or only one firm develops the product

(d) *Extreme cases.* Which equilibrium results can be sustained when firm 2's cost of mimicking firm 1's product is nil, i.e., $F_2 = 0$? Which can be sustained when both firms experience the same cost of developing the product, $F_1 = F_2$?

- Graphically, the horizontal axis in the above figure depicts the case in which $F_2 = 0$. In this setting, only two subgame perfect equilibria can be sustained: (1) both firms develop the new product, which can be sustained if $F_1 \leq \underline{F}$; and (2) no firm develops the product, which holds if $F_1 > \underline{F}$.
  - The equilibrium in which only firm 1 develops the product cannot be sustained in this context since, for that equilibrium to emerge, firm 2 must experience a relatively high (and similar) cost of development as firm 1 does, which cannot happen in this setting since $F_2 = 0$.
- Both firms experiencing the same cost of developing the product, $F_1 = F_2 = F$, is graphically depicted in the 45-degree line of the figure. In this case, three subgame perfect equilibrium can be supported: (1) both firms develop the new product if $F \leq \underline{F}$; (2) only firm 1 develops if $\underline{F} < F \leq \overline{F}$; and (3) no firm develops if $\overline{F} > F$.

(e) *Numerical example.* Consider an inverse demand function $p(Q) = 1 - Q$, where $Q = q_1 + q_2$ denotes aggregate output, and assume that $c = 1/2$. Under which conditions on fixed costs F_1 and F_2 only firm 1 develops the product, both firms develop it, or neither firm develop it?

- Evaluating the above two cutoffs of F at parameter values $a = 1$ and $c = 1/2$, we obtain:

$$\underline{F} \equiv \frac{(a-c)^2}{9} = \frac{\left(1-\frac{1}{2}\right)^2}{9} = \frac{1}{36} \simeq 0.028$$

and

$$\overline{F} \equiv \frac{(a-c)^2}{4} = \frac{\left(1-\frac{1}{2}\right)^2}{4} = \frac{1}{16} \simeq 0.062$$

which, as expected, satisfy $\overline{F} > \underline{F}$. These two cutoffs can be placed on the horizontal and vertical axis of Fig. 6.3 to obtain the regions of (F_1, F_2)-pairs described in part (c).

Exercise #6.10: Green Innovation, Based on Lambertini et al. (2017)^C

6.10 Consider an industry with $N \geq 2$ firms competing à la Cournot, each generating pollution. We seek to analyze equilibrium behavior in the following sequential-move game:

(i) In stage 1, the regulator sets an emission fee t on all firms in the industry.

(ii) In stage 2, every firm i simultaneously and independently responds to the emission fee investing in abatement, z_i.

(iii) In stage 3, every firm i observes the emission fee and the profile of investment in abatement $z = (z_1, \ldots, z_N)$, and responds simultaneously and independently choosing its output level q_i.

Firms face an inverse demand function of $p(Q) = 1 - Q$, where $Q = \sum_{i=1}^{N} q_i$ denotes aggregate output, and a constant marginal cost of c, where $c > 0$. Abatement costs are $\frac{\gamma}{2} z_i^2$, where $\gamma > 0$ measures innovation efficiency in the "green R&D," and firm i's net emissions are

$$e_i = q_i - z_i - \beta Z_{-i},$$

where $Z_{-i} \equiv \sum_{j \neq i} z_j$ denotes the aggregate abatement efforts from firm i's rivals, and parameter $\beta \in [0, 1]$ represents the degree of R&D spillovers. When $\beta = 0$, firm i does not benefit from the investment in the abatement efforts of its rivals, whereas, when $\beta > 1$, firm i's emissions also decrease when any of its rivals invests more resources in abatement efforts.

In the first stage, the regulator sets emission fee t to maximize social welfare:

$$W = CS + PS - Env,$$

where $CS = \frac{1}{2} Q^2$ denotes consumer surplus, $PS = \sum_{i=1}^{N} \pi_i$ represents producer surplus, and $Env = (e_1 + \ldots + e_N)^2$ indicates the environmental damage from emissions, which is increasing and convex in aggregate emissions.

(a) *Third stage.* Find the equilibrium output and profits in the third stage of the game.

- Every firm i solves the following profit maximization problem:

$$\max_{q_i \geq 0} \pi_i(q_i) = (1 - q_i - Q_{-i})q_i - cq_i - \frac{\gamma}{2} z_i^2 - t\overbrace{(q_i - z_i - \beta Z_{-i})}^{e_i},$$

where $Q_{-i} = \{q_1, \ldots, q_{i-1}, q_{i+1}, \ldots, q_N\}$ denotes the aggregate output of firm i's rivals.

Differentiating with respect to q_i, we obtain:

$$1 - 2q_i - Q_{-i} - c - t = 0.$$

Solving for q_i yields firm i's best response function:

$$q_i(Q_{-i}) = \frac{1 - c - t}{2} - \frac{1}{2} Q_{-i}.$$

In a symmetric equilibrium, all firms produce the same output level, $q_i = q_j$, which entails $Q_{-i} = (N - 1)q_i$. Inserting this property in the above best response function, we obtain:

$$q_i = \frac{1-c-t}{2} - \frac{1}{2}(N-1)q_i,$$

and solving for q_i, we find the output function:

$$q_i(t) = \frac{1-c-t}{N+1}$$

which is decreasing in the emission fee t that the regulator sets in the first period, the number of firms N competing in the industry, and the marginal production cost c.

- Aggregate output is

$$Q(t) = Nq_i(t) = \frac{N(1-c-t)}{N+1},$$

which increases in the number of firms in the industry, N, decreases in the emission fee t and the marginal production cost c. In addition, $Q(t)$ asymptotically approaches $1-c-t$ when the market becomes perfectly competitive.

- Substituting output into the firm i's profit function, we find that profits in this stage are

$$\pi_i(z_i, t) = (1 - q_i(t) - Q_{-i}(t))q_i(t) - cq_i(t) - \frac{\gamma}{2}z_i^2 - t(q_i(t) - z_i - \beta Z_{-i})$$

$$= \left(\frac{1-c-t}{N+1}\right)^2 - \frac{\gamma}{2}z_i^2 + t(z_i + \beta Z_{-i}).$$

(b) *Second stage.* Find the equilibrium abatement effort, and the associated emission and profits of every firm i, in the second stage of the game.

- In the second stage, every firm i anticipates the profit that it earns in the third stage, $\pi_i(z_i, t)$, and then chooses its abatement effort z_i to solve

$$\max_{z_i \geq 0} \pi_i(z_i, t).$$

Differentiating with respect to z_i, we obtain:

$$-\gamma z_i + t = 0$$

which, solving for z_i, yields equilibrium abatement efforts

$$z_i(t) = \frac{t}{\gamma}$$

which is increasing in the emission fee that regulator sets in the first stage, t. Intuitively, as the emission fee becomes more stringent, firms have stronger incentives to invest in abatement efforts in the second stage to save taxes in the next stage. Interestingly, firm i makes fewer investments in green R&D if innovation investments become less efficient (i.e., γ increases). In other words, if environmental regulation was absent, $t = 0$, all firms would choose to invest zero in green R&D.

- Aggregate abatement is

$$Z(t) = Nz_i(t) = \frac{Nt}{\gamma}$$

which increases in the number of firms, N. As suggested by Lambertini et al. (2017) "if taxation were exogenous, we obtain a clear-cut Arrowian result: increased competition (higher N) leads to an increase in aggregate R&D (innovation)." As we show in the last part of the exercise, when taxation (emission fees) are endogenous, the above result can be reversed.

- Invoking symmetry, the emission of every firm i becomes:

$$\begin{aligned} e_i(t) &= q_i(t) - z_i(t) - \beta(N-1)z_i(t) \\ &= \frac{1-c-t}{N+1} - \left(\frac{t}{\gamma} + \beta(N-1)\frac{t}{\gamma}\right) \\ &= \frac{1-c-t}{N+1} - \frac{t(1+\beta(N-1))}{\gamma}. \end{aligned}$$

- Therefore, equilibrium profits in the second stage are

$$\begin{aligned} \pi_i(t) &= \left(\frac{1-c-t}{N+1}\right)^2 - \frac{\gamma}{2}\left(\frac{t}{\gamma}\right)^2 + t\left(\frac{t}{\gamma} + \beta(N-1)\frac{t}{\gamma}\right) \\ &= \left(\frac{1-c-t}{N+1}\right)^2 + \frac{t^2(1+2\beta(N-1))}{2\gamma}. \end{aligned}$$

(c) *First stage.* Write the welfare expression that the regulator considers in the first stage of the game, and find the emission fee that maximizes social welfare. For simplicity, you can normalize marginal production cost to zero, $c = 0$, and assume innovation efficiency to be $\gamma = 5$.

- Social welfare is

$$\begin{aligned} W(t) &= \frac{1}{2}[Q(t)]^2 + N\pi_i(t) - N^2[e_i(t)]^2 \\ &= \frac{N^2}{2}\left(\frac{1-c-t}{N+1}\right)^2 + N\left(\frac{1-c-t}{N+1}\right)^2 + \frac{Nt^2[1+2\beta(N-1)]}{2\gamma} \\ &\quad - N^2\left[\left(\frac{1-c-t}{N+1}\right)^2 - \frac{2t(1-c-t)[1+\beta(N-1)]}{\gamma(N+1)} + \frac{t^2[1+\beta(N-1)]^2}{\gamma^2}\right] \\ &= \frac{2N^2[1+\beta(N-1)](1-c-t)t}{\gamma(N+1)} - \frac{N(N-2)(1-c-t)^2}{2(N+1)^2} \\ &\quad + \frac{N\left[\gamma(1+2\beta(N-1)) - 2N(1+\beta(N-1))^2\right]t^2}{2\gamma^2} \end{aligned}$$

Differentiating with respect to t, we obtain:

$$\frac{2N^2\left[1+\beta\left(N-1\right)\right]\left(1-c-2t\right)}{\gamma\left(N+1\right)}+\frac{N\left(N-2\right)\left(1-c-t\right)}{\left(N+1\right)^2}$$
$$+\frac{N\left[\gamma\left(1+2\beta\left(N-1\right)\right)-2N\left(1+\beta\left(N-1\right)\right)^2\right]t}{\gamma^2}=0$$

Solving for t, the optimal emission fee is

$$t\left(N\right)$$
$$=\frac{\gamma[2N(N+1)(1+\beta(N-1))+\gamma(N-2)](1-c)}{4\gamma N(N+1)[1+\beta(N-1)]+\gamma^2(N-2)-(N+1)^2\left[\gamma(1+2\beta(N-1))-2N(1+\beta(N-1))^2\right]}$$

Substituting $c = 0$ and $\gamma = 5$ into the above equation, we find that the optimal emission fee becomes:

$$t\left(N\right)$$
$$=\frac{5[2N(N+1)(1+\beta(N-1))+5(N-2)]}{20N(N+1)[1+\beta(N-1)]+25(N-2)-(N+1)^2\left[5(1+2\beta(N-1))-2N(1+\beta(N-1))^2\right]}$$

(d) Find the equilibrium output, $q_i\left(N\right)$, and abatement effort, $z_i\left(N\right)$, for every firm i.

- The equilibrium output of every firm i is

$$q_i\left(N\right)=\frac{1-t\left(N\right)}{N+1}$$

or

$$\frac{10N\left[1+\beta\left(N-1\right)\right]-\left(N+1\right)\left[5\left(1+2\beta\left(N-1\right)\right)-2N\left(1+\beta\left(N-1\right)\right)^2\right]}{\left[\begin{array}{c}20N\left(N+1\right)\left[1+\beta\left(N-1\right)\right]+25\left(N-2\right)-\left(N+1\right)^2\\ \times\left[5\left(1+2\beta\left(N-1\right)\right)-2N\left(1+\beta\left(N-1\right)\right)^2\right]\end{array}\right]}$$

with its associated abatement effort of

$$z_i\left(N\right)=\frac{t(N)}{\gamma}$$
$$=\frac{2N\left(N+1\right)\left(1+\beta\left(N-1\right)\right)+5\left(N-2\right)}{\left[\begin{array}{c}20N\left(N+1\right)\left[1+\beta\left(N-1\right)\right]+25\left(N-2\right)-\left(N+1\right)^2\\ \times\left[5\left(1+2\beta\left(N-1\right)\right)-2N\left(1+\beta\left(N-1\right)\right)^2\right]\end{array}\right]}$$

(e) Evaluate aggregate abatement $Z\left(N\right)$ at $\beta=0$, and show that $Z(N)$ is increasing in N.

- Evaluating $Z(N)=N\times z_i\left(N\right)$ at $\beta=0$, we find:

$$Z\left(N\right)=\frac{2N^2\left(N+1\right)+5N\left(N-2\right)}{20N\left(N+1\right)+25\left(N-2\right)-\left(N+1\right)^2\left(5-2N\right)}$$
$$=\frac{N(2N^2+7N-10)}{2N^3+19N^2+37N-55}.$$

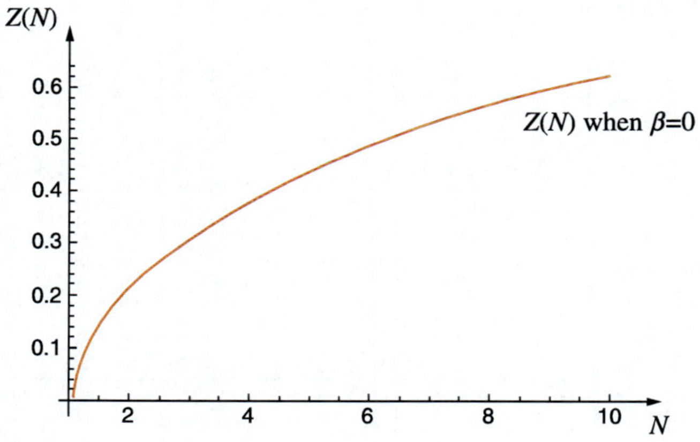

Fig. 6.4 Aggregate abatement when $\beta = 0$

Differentiating the above expression with respect to N yields:

$$\frac{\partial Z(N)}{\partial N} = \frac{24N^4 + 188N^3 + 119N^2 - 770N + 550}{\left(2N^3 + 19N^2 + 37N - 55\right)^2}$$

which is positive since $N \geq 2$, so that aggregate abatement increases with the number of firms N.

- Intuitively, when the market becomes more competitive, total output and abatement both increase since spillover effects of R&D are absent. This occurs because every firm does not benefit from the investment of its rivals, and thus responds increasing its own investment when more firms enter the industry. Figure 6.4 depicts $Z(N)$ as a function of N when $\beta = 0$.

(f) Evaluate aggregate abatement $Z(N)$ at $\beta = \frac{1}{2}$, and show that $Z(N)$ is initially increasing but eventually decreasing in N. For which value of N, the derivative of $Z(N)$ with respect to N is zero? Interpret your results.

- Evaluating $Z(N)$ at $\beta = \frac{1}{2}$, we find:

$$\begin{aligned} Z(N) &= \frac{2N^2(N+1)^2 + 10N(N-2)}{20N(N+1)^2 + 50(N-2) - (N+1)^2\left[10N - N(N+1)^2\right]} \\ &= \frac{2N^4 + 4N^3 + 12N^2 - 20N}{N^5 + 4N^4 + 16N^3 + 24N^2 + 61N - 100}. \end{aligned}$$

Differentiating with respect to N yields:

$$\frac{\partial Z(N)}{\partial N} = \frac{2\left(-N^8 - 4N^7 - 10N^6 + 40N^5 + 255N^4 + 164N^3 + 6N^2 - 1200N + 1000\right)}{\left(N^5 + 4N^4 + 16N^3 + 24N^2 + 61N - 100\right)^2}.$$

Setting this expression equal to zero, we find that $\frac{\partial Z(N)}{\partial N} > 0$ for all $N \in \left[2, \overline{N}\right]$, but $\frac{\partial Z(N)}{\partial N} < 0$ when $N > \overline{N} = 3.432$. This explains the inverted U-shape relationship between market competition and green innovation, as depicted in Fig. 6.5. Intuitively, when the market is represented by four or fewer firms, increasing the number of firms

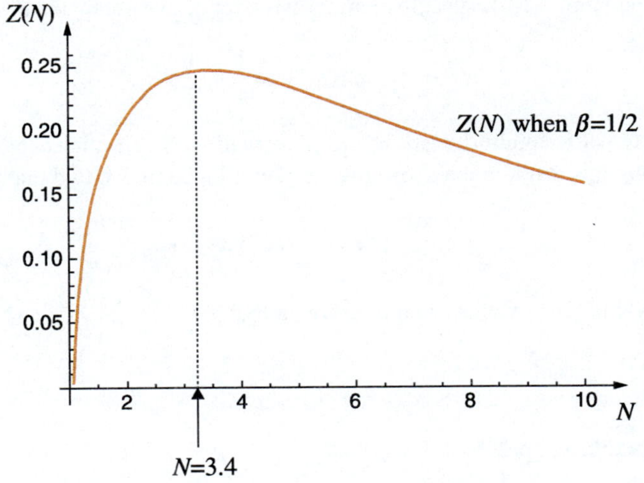

Fig. 6.5 Aggregate abatement when $\beta = 1/2$

in the industry (and their output) stimulates green innovation. However, when the market becomes increasingly competitive, every firm i exhibits stronger incentives to "free ride" on the investments of all other firms, and this, in turn, induces every firm to make fewer investments in green innovation.

Exercise #6.11: Incentives to Innovate in Cournot, Based on Delbono and Denicolo (1991)^C

6.11 Consider an industry with $N \geq 2$ firms competing à la Cournot. Firms face inverse demand function $p(Q) = 1 - Q$, where Q denotes aggregate output, and a marginal production cost $c = 1/2$.

We seek to analyze equilibrium behavior in the following sequential-move game:

(i) In stage 1, every firm i invests x_i in R&D, where $x_i \in [0, 1]$. With probability $\theta = \frac{1}{N}\sqrt{x_i}$, firm i is the first to innovate, which provides this firm the exclusive right to produce at cost $c^* = 1/4$ forever.

(ii) In stage 2, every firm i observes the outcome of the R&D race, and responds simultaneously and independently choosing its output level q_i.

We next evaluate the profits of the firm having the exclusive right to use the innovation ("winner"), the $N - 1$ firms that do not innovate ("losers"), which will help us compute firms' incentives to invest in R&D, and how these incentives are affected by the number of firms, N, initially competing in the industry.

(a) Find the pre-innovation profit, and label it π_i.

- Every firm i solves

$$\max_{q_i \geq 0} (1 - q_i - Q_{-i})q_i - \frac{1}{2}q_i,$$

where q_i denotes firm i's output and Q_{-i} represents the sum of its rivals' output. Note that this profit is evaluated at the pre-innovation cost $c = 1/2$ for all firms.

- Differentiating with respect to q_i and solving for q_i, we obtain the best response function:

$$q_i(Q_{-i}) = \frac{1}{4} - \frac{1}{2}Q_{-i}.$$

In a symmetric equilibrium, $q_i = q_j$ for every firm $i \neq j$, which entails $Q_{-i} = (N-1)q_i$. Inserting this result in the above best response function, we find that

$$q_i = \frac{1}{4} - \frac{1}{2}(N-1)q_i$$

and solving for q_i yields an equilibrium output

$$q_i = \frac{1}{2(N+1)}$$

with equilibrium profits

$$\begin{aligned}\pi_i &= \left(1 - \frac{1}{2(N+1)} - (N-1)\frac{1}{2(N+1)}\right)\frac{1}{2(N+1)} - \frac{1}{2}\frac{1}{2(N+1)} \\ &= \frac{1}{4(N+1)^2}.\end{aligned}$$

(b) Find the post-innovation profit for the winner, π_W, and for the losers, π_L.
- *Winning firm.* The winner of the innovation sees its marginal cost decrease from $c = 1/2$ to $c^* = 1/4$, and thus solves

$$\max_{q_i \geq 0} (1 - q_i - Q_{-i})q_i - \frac{1}{4}q_i.$$

Differentiating with respect to q_i and solving for q_i yield a best response function:

$$q_i(Q_{-i}) = \frac{3}{8} - \frac{1}{2}Q_{-i}.$$

Graphically, its best response function shifts upward as a result of the innovation, implying that this firm produces more units than before the innovation (see part a). In a symmetric equilibrium where all losing firms produce the same output level, $Q_{-i} = (N-1)q_L$, so we can rewrite the above best response function as

$$q_W(q_L) = \frac{3}{8} - \frac{1}{2}(N-1)q_L.$$

- *Losing firms.* The $N-1$ losers keep their marginal production cost unaffected at $c = 1/2$, and then solve

$$\max_{q_i \geq 0} (1 - q_i - Q_{-i})q_i - \frac{1}{2}q_i$$

which yields a best response function

$$q_i(Q_{-i}) = \frac{1}{4} - \frac{1}{2}Q_{-i}$$

since Q_{-i} includes the output of the winning firm, q_W, and those of all other losing firms, we can write it as $Q_{-i} = q_W + (N-2)q_L$. Inserting this result in the above best response function, we obtain:

$$q_L = \frac{1}{4} - \frac{q_W + (N-2)q_L}{2}$$

and, solving for q_j, this best response function can be rewritten as

$$q_L\left(q_W\right) = \frac{1}{2N} - \frac{1}{N}q_W.$$

- *Equilibrium output levels.* We can now combine the above best response functions for the winning firm and each of the losing firms to obtain equilibrium output levels. Inserting $q_L\left(q_W\right)$ into $q_W\left(q_L\right)$ yields:

$$q_W = \frac{3}{8} - \frac{N-1}{2}\underbrace{\left(\frac{1}{2N} - \frac{1}{N}q_W\right)}_{q_L(q_W)}$$

and solving for q_W, we find the equilibrium output for the winner of the innovation:

$$q_W^* = \frac{N+2}{4(N+1)}.$$

Inserting this output level in the losing firm's best response function yields:

$$q_L^* = q_L\left(q_W^*\right) = \frac{1}{2N} - \frac{1}{N}\underbrace{\frac{N+2}{4(N+1)}}_{q_W^*}$$

$$= \frac{1}{4(N+1)}.$$

- Therefore, the equilibrium profit for the winning firm is

$$\pi_W = \left[1 - q_W^* - (N-1)q_L^*\right]q_W^* - \frac{1}{4}q_W^*$$

$$= \frac{(N+2)^2}{16(N+1)^2}$$

and that of each losing firm is

$$\pi_L = \left[1 - q_L^* - q_W^* - (N-2)q_L^*\right]q_L^* - \frac{1}{2}q_L^*$$

$$= \frac{1}{16(N+1)^2}.$$

(c) Write firm i's expected discounted profits, V_i, as a function of the profits you found in parts (a) and (b). For simplicity, assume that the probability with which firm i innovates is $h(x_i) = \frac{1}{N}\sqrt{x_i}$, and a constant discount rate r that is the same across winner and losers.

- Expected discounted profits, V_i, are

$$V_i = \underbrace{\left(\frac{1}{N}\sqrt{x_i}\right)\frac{\pi_W}{r}}_{\text{Firm } i \text{ innovates}} + \underbrace{\left(\sum_{j\neq i}\frac{1}{N}\sqrt{x_j}\right)\frac{\pi_L}{r}}_{\text{Firm } j\neq i \text{ innovates}}$$

$$+ \underbrace{\left(1-\frac{1}{N}\sqrt{x_i}-\sum_{j\neq i}\frac{1}{N}\sqrt{x_j}\right)\left(\underbrace{\pi_i}_{\text{Profit}} + \underbrace{(1-r)V_i}_{\text{Continuation payoff}}\right)}_{\text{No firm innovates}} - \underbrace{x_i}_{\text{R\&D cost}},$$

where the first term represents the probability that firm i successfully innovates, times its discounted profits in that case, $\frac{\pi_W}{r}$; the second term denotes the probability that any firm $j \neq i$ innovates, leaving firm i with discounted losing profits $\frac{\pi_L}{r}$; and the third term represents the probability that no firm innovates, thus keeping firm i's profits unchanged during one period. When no firm innovates, firm i still has a chance to innovate in the following period, obtaining continuation payoff $(1-r)V_i$.

Rearranging, and solving for V_i, we obtain:

$$V_i = \frac{\frac{\sqrt{x_i}}{r}\pi_W + \frac{\sum_{j\neq i}\sqrt{x_j}}{r}\pi_L + \left(N-\sqrt{x_i}-\sum_{j\neq i}\sqrt{x_j}\right)\pi_i - Nx_i}{rN+(1-r)\left(\sqrt{x_i}+\sum_{j\neq i}\sqrt{x_j}\right)}.$$

(d) *Numerical example.* Evaluate your above results assuming that firms' discount factor is $r = 0.9$ and that only two firms compete in the industry, $N = 2$. Then, identify the equilibrium investment in R&D by every firm i.

- Current profit (when no firm innovates) is $\pi_i = \frac{1}{4(N+1)^2} = \frac{1}{36}$, winner profits are $\pi_W = \frac{(N+2)^2}{16(N+1)^2} = \frac{1}{9}$, and loser profits are $\pi_L = \frac{1}{16(N+1)^2} = \frac{1}{144}$. Then, assuming without loss of generality that firm 1 wins, its expected discounted profits become:

$$V_1 = \frac{\frac{10\sqrt{x_1}}{9}\frac{1}{9} + \frac{10\sqrt{x_2}}{9}\frac{1}{144} + \left(2-\sqrt{x_1}-\sqrt{x_2}\right)\frac{1}{36} - 2x_1}{\frac{9}{5}+\frac{1}{10}\left(\sqrt{x_1}+\sqrt{x_2}\right)}$$

$$= \frac{5\left(36+62\sqrt{x_1}-13\sqrt{x_2}-1296x_1\right)}{324\left(18+\sqrt{x_1}+\sqrt{x_2}\right)}.$$

- Differentiating with respect to x_1, and assuming interior solutions

$$\frac{\partial V_1}{\partial x_1} = \frac{5\left(360-432x_1-15552\sqrt{x_1}+25\sqrt{x_2}-864\sqrt{x_1x_2}\right)}{216\sqrt{x_1}\left(18+\sqrt{x_1}+\sqrt{x_2}\right)^2} = 0.$$

In a symmetric equilibrium, both firms invest the same amount in R&D, so we have $x^* = x_1^* = x_2^*$. Substituting in the above first-order condition, we obtain:

$$1296x + 15527\sqrt{x} - 360 = 0.$$

Solving for x, we find that the optimal investment for both firms is

$$x^* = \left(\frac{\sqrt{15527^2 + 4 \times 1296 \times 360} - 15527}{2 \times 1296}\right)^2 \simeq 0.0005.$$

Exercise #6.12: Cost-Reducing Investments That Accommodate Entry^B

6.12 Consider an industry with an incumbent (firm 1) and a potential entrant (firm 2). In the first stage, firm 1 can invest in cost-reducing technologies k at a cost of $\frac{1}{2}k^2$ to reduce the per-unit production cost from $1 > c > 0$ to $c - \alpha k$, where $\alpha \in [0, 1]$ denotes the effectiveness of investment in cost-reducing technologies. In the second stage, firm 2, having observed k, decides whether to enter into the market at an entry cost E or stay out of the market. The inverse demand function when firms compete à la Cournot on homogeneous goods is

$$p(Q) = 1 - Q$$

where $Q = q_1 + q_2$ represents aggregate output. We do not allow firm 2 to invest in cost-reducing technologies, so that its marginal cost of production is c. In this exercise, suppose firm 1 accommodates firm 2's entry. In the next exercise, we will assume that firm 1 deters firm 2 from entering the market, and then analyze under which conditions firm 1 profitably deters entry.

(a) *Second stage.* Set up the firms' profit maximization problem and derive the best response functions in the second period, $q_i(q_j)$. (Hint: Firms are asymmetric.)

- Firm 1 chooses q_1 to solve the following profit maximization problem:

$$\max_{q_1 \geq 0} \pi_1^2(k) = (1 - q_1 - q_2) q_1 - (c - \alpha k) q_1$$

which is evaluated at its second-period marginal cost, $c - \alpha k$, after investing k in cost-reducing technologies. Differentiating with respect to q_1, we obtain:

$$1 - 2q_1 - q_2 - (c - \alpha k) = 0$$

Solving for q_1, we find firm 1's best response function:

$$q_1(q_2) = \frac{1 - (c - \alpha k)}{2} - \frac{1}{2} q_2$$

- Firm 2 chooses q_2 to solve the following profit maximization problem:

$$\max_{q_2 \geq 0} \pi_2^2(k) = (1 - q_1 - q_2) q_2 - c q_2 - E$$

which is evaluated at its marginal production cost c, since this firm cannot invest in cost-reducing technologies. Differentiating with respect to q_2, we obtain:

$$1 - q_1 - 2q_2 - c = 0$$

Solving for q_2, we find firm 2's best response function:

$$q_2(q_1) = \frac{1-c}{2} - \frac{1}{2}q_1$$

Best response functions are, clearly, asymmetric, since firm 1's originates at a higher vertical intercept than firm 2's when $\alpha > 0$ and $k > 0$. Intuitively, when the investment in cost-reducing technologies reduces firm 1's marginal cost of production, it can produce more units than its rival.

(b) Find the second-stage output function $q_i(k)$, and associated profits $\pi_i(k)$, of every firm i. (Hint: Find the cutoff $\overline{\alpha}$ above which firm 2 does not produce positive units.)

- Plugging firm 2's best response function, $q_2(q_1)$, into firm 1's, $q_1(q_2)$, yields:

$$q_1 = \frac{1-c+\alpha k}{2} - \frac{1}{2}\overbrace{\left(\frac{1-c}{2} - \frac{q_1}{2}\right)}^{q_2(q_1)}$$
$$= \frac{1-c+2\alpha k}{4} + \frac{q_1}{4}$$

Solving for q_1, the equilibrium output function of firm 1 becomes:

$$q_1(k) = \frac{1-c+2\alpha k}{3}$$

which is positive since $1 > c > 0$ by definition.

- Inserting $q_1(k)$ into firm 2's best response function, $q_2(q_1)$, we obtain:

$$q_2 = \frac{1-c}{2} - \frac{1}{2}\overbrace{\left(\frac{1-c+2\alpha k}{3}\right)}^{q_1(k)}$$

Rearranging, the equilibrium output function of firm 2 becomes:

$$q_2(k) = \frac{1-c-\alpha k}{3}.$$

This output level is positive as long as $\alpha < \overline{\alpha} \equiv \frac{1-c}{k}$. Intuitively, the effectiveness of firm 1's investment cannot exceed $\overline{\alpha}$ or firm 2 would be so uncompetitive that it would stay out of the market, even if firm 1 accommodates its entry. This allows us to express firm 2's output function as follows:

$$q_2(k) = \begin{cases} \frac{1-c-\alpha k}{3} & \text{if } \alpha < \overline{\alpha}, \\ 0 & \text{otherwise.} \end{cases}$$

- Finally, substituting $q_1(k)$ and $q_2(k)$ into the profit function of firm 1 yields:

$$\pi_1^2(k) = \left[1 - \overbrace{\frac{1-c+2\alpha k}{3}}^{q_1(k)} - \overbrace{\frac{1-c-\alpha k}{3}}^{q_2(k)} - c + \alpha k\right] \overbrace{\frac{1-c+2\alpha k}{3}}^{q_1(k)}$$

$$= \left(\frac{1-c+2\alpha k}{3}\right)^2$$

- Similarly, substituting $q_1(k)$ and $q_2(k)$ into the profit function of firm 2 yields:

$$\pi_2^2(k) = \left[1 - \overbrace{\frac{1-c+2\alpha k}{3}}^{q_1(k)} - \overbrace{\frac{1-c-\alpha k}{3}}^{q_2(k)} - c\right] \overbrace{\frac{1-c-\alpha k}{3}}^{q_2(k)} - E$$

$$= \left(\frac{1-c-\alpha k}{3}\right)^2 - E$$

(c) *First stage.* Find firm 1's investment in cost-reducing technologies k^A, and use this information to evaluate the equilibrium output q_i^A and profits π_i^A of every firm i.

- Firm 1 chooses k to solve the following profit maximization problem:

$$\max_{k \geq 0} \pi_1^1(k) = \pi_1^2(k) - \frac{1}{2}k^2$$

$$= \left(\frac{1-c+2\alpha k}{3}\right)^2 - \frac{1}{2}k^2$$

Differentiating with respect to k, and assuming interior solutions, we find

$$\frac{4\alpha(1-c+2\alpha k)}{9} - k = 0$$

Rearranging and solving for k, the equilibrium investment of firm 1 becomes

$$k^A = \frac{4\alpha(1-c)}{9-8\alpha^2}$$

where the superscript A stands for entry accommodation, and is positive if and only if $9 - 8\alpha^2 > 0$, or $\alpha < \frac{3}{2\sqrt{2}} \simeq 1.06$, which holds given that $\alpha \in [0, 1]$ by assumption.

- Substituting k^A into $q_1(k)$, the equilibrium output of firm 1 becomes

$$q_1^A = \frac{1}{3}\left[1 - c + 2\alpha \overbrace{\frac{4\alpha(1-c)}{9-8\alpha^2}}^{k^A}\right]$$

$$= \frac{3(1-c)}{9-8\alpha^2}$$

- In this context, we are interested in evaluating cutoff $\overline{\alpha}$, where

$$\overline{\alpha} = \frac{1-c}{\frac{4\alpha(1-c)}{9-8\alpha^2}} = \frac{9-8\alpha^2}{4\alpha}$$

- Next, substituting k^A into $q_2(k)$, the equilibrium output of firm 2 becomes:

$$q_2^A = \begin{cases} \frac{1}{3}\left[1-c-\alpha\overbrace{\frac{4\alpha(1-c)}{9-8\alpha^2}}^{k^A}\right] & \text{if } \alpha < \overline{\alpha}, \\ 0 & \text{otherwise.} \end{cases}$$

Consider the range of α satisfying condition $\alpha < \frac{9-8\alpha^2}{4\alpha}$. Solving for α, we find:

$$\alpha < \frac{\sqrt{3}}{2} \simeq 0.87.$$

Simplifying, we obtain the equilibrium output of firm 2, as follows:

$$q_2^A = \begin{cases} \frac{(3-4\alpha^2)(1-c)}{9-8\alpha^2} & \text{if } \alpha < \frac{\sqrt{3}}{2}, \\ 0 & \text{otherwise.} \end{cases}$$

Intuitively, this result says that, in the case that firm 1 invests a relatively low amount in cost-reducing technologies, k^A, firm 2 responds producing a positive amount only when such investment is relatively ineffective (low values of α). In contrast, when such investment is very effective (α close to 1), firm 2 cannot profitably compete against firm 1.
- Finally, substituting k^A into $\pi_1^1(k)$, equilibrium profits of firm 1 are

$$\pi_1^A = \left(\frac{1-c+2\alpha\frac{4\alpha(1-c)}{9-8\alpha^2}}{3}\right)^2 - \frac{1}{2}\left(\frac{4\alpha(1-c)}{9-8\alpha^2}\right)^2$$
$$= \frac{(1-c)^2}{9-8\alpha^2}$$

- Similarly, substituting k^A into $\pi_2^2(k)$, equilibrium profits of firm 2 are

$$\pi_2^A = \begin{cases} \left(\frac{1-c-\alpha\frac{4\alpha(1-c)}{9-8\alpha^2}}{3}\right)^2 - E & \text{if } \alpha < \frac{\sqrt{3}}{2}, \\ 0 & \text{otherwise.} \end{cases}$$

which, after simplifying, yields

$$\pi_2^A = \begin{cases} \left(\frac{(3-4\alpha^2)(1-c)}{9-8\alpha^2}\right)^2 - E & \text{if } \alpha < \frac{\sqrt{3}}{2}, \\ 0 & \text{otherwise.} \end{cases}$$

(d) *Comparative statics.* How do variables k^A and q_i^A change in parameters α and c? Interpret.

- Differentiating equilibrium investment, k^A, with respect to α, we find:

$$\frac{\partial k^A}{\partial \alpha} = \frac{4\left(9+8\alpha^2\right)(1-c)}{\left(9-8\alpha^2\right)^2} > 0$$

so when the investment becomes more effective, firm 1 invests more in it.

- Differentiating equilibrium investment, k^A, with respect to c, we obtain:

$$\frac{\partial k^A}{\partial c} = -\frac{4\alpha}{9-8\alpha^2} < 0$$

so when marginal cost increases, firm 1 invests less in the technologies.

- Differentiating firm 1's equilibrium output, q_1^A, with respect to α, we have:

$$\frac{\partial q_1^A}{\partial \alpha} = \frac{48\alpha\,(1-c)}{\left(9-8\alpha^2\right)^2} > 0$$

so when the investment becomes more effective, firm 1 produces more output.

- Differentiating firm 1's equilibrium output, q_1^A with respect to c, we have:

$$\frac{\partial q_1^A}{\partial c} = -\frac{3}{9-8\alpha^2} < 0$$

so when marginal cost increases, firm 1 produces less output.

- Differentiating firm 2's equilibrium output, q_2^A, with respect to α, we have:

$$\frac{\partial q_2^A}{\partial \alpha} = -\frac{24\alpha\,(1-c)}{\left(9-8\alpha^2\right)^2} < 0$$

so when the investment becomes more effective, firm 2, in response to firm 1's increasing output, reduces its own output (since output are strategic substitutes).

- Differentiating firm 2's equilibrium output, q_2^A, with respect to c, we have

$$\frac{\partial q_2^A}{\partial c} = -\frac{3-4\alpha^2}{9-8\alpha^2} < 0$$

so that when marginal cost increases, firm 2 produces less output.

(e) Find the condition on α and E that firm 2 will enter if firm 1 accommodates.

- When $0 \leq \alpha < \frac{\sqrt{3}}{2}$, firm 2 earns positive profits if

$$\left(\frac{\left(3-4\alpha^2\right)(1-c)}{9-8\alpha^2}\right)^2 - E \geq 0$$

Solving for E, firm 2 will enter if firm 1 accommodates and entry cost E satisfies

$$E \leq \overline{E} \equiv \left(\frac{\left(3 - 4\alpha^2\right)(1 - c)}{9 - 8\alpha^2} \right)^2$$

- When $\frac{\sqrt{3}}{2} \leq \alpha \leq 1$, however, the investment is so effective that firm 2 does not enter into the market, and this result holds for all values of the entry cost $E \geq 0$.

(f) How does the cutoff you identified in part (e) change with α? Explain.

- Differentiating $\overline{E}$ with respect to α, we find:

$$\frac{\partial \overline{E}}{\partial \alpha} = 2(1 - c)^2 \left(\frac{3 - 4\alpha^2}{9 - 8\alpha^2} \right) \frac{-8\alpha\left(9 - 8\alpha^2\right) + 16\alpha\left(3 - 4\alpha^2\right)}{\left(9 - 8\alpha^2\right)^2}$$

$$= -\frac{48\alpha(1 - c)^2\left(3 - 4\alpha^2\right)}{\left(9 - 8\alpha^2\right)^3} < 0$$

so that when the investment becomes more effective (α increases), cutoff $\overline{E}$ decreases, indicating that it becomes more likely that entry by firm 2 is blockaded.

Exercise #6.13: Cost-Reducing Investments That Deter Entry^B

6.13 Consider the setting in the previous exercise, but suppose now that firm 1 deters firm 2's entry.

(a) *Second stage.* Find firm 1's output q_1^D that deters firm 2's entry. Does firm 1 produce above or below the level when it accommodates entry? Explain. (Hint: Solve firm 2's best response function and plug it back into the profit function to find q_1^D that causes $\pi_2(q_1) = 0$.)

- Firm 2 chooses q_2 to solve the following profit maximization problem:

$$\pi_2^2(q_2) = (1 - q_1 - q_2 - c)\, q_2 - E$$

Differentiating with respect to q_2, and assuming interior solutions, we obtain:

$$q_2(q_1) = \frac{1 - c}{2} - \frac{1}{2} q_1$$

Substituting firm 2's best response function back into its profit function, we find:

$$\pi_2^2(q_1) = \left(1 - q_1 - \overbrace{\left(\frac{1 - c}{2} - \frac{1}{2} q_1 \right)}^{q_2(q_1)} - c \right) q_2 - E$$

$$= \frac{(1 - q_1 - c)^2}{4} - E$$

Thus, if firm 1 makes firm 2 unprofitable to enter, we need $\pi_2^2(q_1) < 0$, where

$$\frac{(1-q_1-c)^2}{4} < E$$

Rearranging and solving for q_1, the minimum entry-deterring output from firm 1 that deters entry is

$$q_1^D = 1 - c - 2\sqrt{E}$$

- To deter entry, firm 1 will produce more than the accommodating level when

$$q_1^D = 1 - c - 2\sqrt{E} > \frac{3(1-c)}{9-8\alpha^2} = q_1^A$$

Solving for E, we obtain that

$$E < \left(\frac{(3-4\alpha^2)(1-c)}{9-8\alpha^2}\right)^2 \equiv \overline{E}$$

so firm 1 will produce above the accommodating level to deter entry. In contrast, when entry costs satisfy $E \geq \overline{E}$, firm 2 does not enter regardless of firm 1's actions. In this case, we say that entry is *blockaded*, and firm 1 produces at the profit-maximizing level, behaving as a monopolist in the market.

(b) *First stage.* Find firm 1's optimal investment k^D and associated profits π_1^D.

- Firm 1 chooses k to solve the following profit maximization problem:

$$\max_{k\geq 0} \pi_1^1(k) = \left[1 - \overbrace{\left(1-c-2\sqrt{E}\right)}^{q_1^D} - \overbrace{0}^{q_2} - c + \alpha k\right]\overbrace{\left(1-c-2\sqrt{E}\right)}^{q_1^D} - \frac{1}{2}k^2$$

$$= \left(\alpha k + 2\sqrt{E}\right)\left(1-c-2\sqrt{E}\right) - \frac{1}{2}k^2$$

where the monopolist balances the marginal benefit of lowering the cost of producing all the inframarginal units, $\alpha(1-c-2\sqrt{E})$, with the marginal cost of investing in additional units of the cost-reducing technologies, k, by taking the first-order condition with respect to k, as follows:

$$k^D = \alpha\left(1-c-2\sqrt{E}\right)$$

- Substituting $k^D = \alpha\left(1-c-2\sqrt{E}\right)$ into firm 1's profit function, we obtain:

$$\pi_1^D = \left(\alpha^2\left(1-c-2\sqrt{E}\right) + 2\sqrt{E}\right)\left(1-c-2\sqrt{E}\right) - \frac{\alpha^2}{2}\left(1-c-2\sqrt{E}\right)^2$$

$$= \frac{\left(1-c-2\sqrt{E}\right)\left[\alpha^2(1-c) + 2(2-\alpha^2)\sqrt{E}\right]}{2}$$

Exercise #6.14: Is Entry Deterrence Profitable?^{B}

6.14 Consider the results from the previous two exercises. Let us now characterize the optimal strategies of firm 1 and firm 2.

(a) Find the range of entry costs, E, for which firm 1 chooses to accommodate entry.

- Firm 1 will accommodate entry if and only if $\pi_1^A \geq \pi_1^D$, that is,

$$\frac{(1-c)^2}{9-8\alpha^2} \geq \frac{\left(1-c-2\sqrt{E}\right)\left[\alpha^2(1-c)+2\left(2-\alpha^2\right)\sqrt{E}\right]}{2}$$

which we can rearrange to yield

$$4\left(9-8\alpha^2\right)\left(2-\alpha^2\right)E-4\left(1-\alpha^2\right)\left(9-8\alpha^2\right)(1-c)\sqrt{E}$$
$$+\left[2-\alpha^2\left(9-8\alpha^2\right)\right](1-c)^2 \geq 0$$

Solving for E, we obtain two roots:

$$E \leq \underline{E} \equiv \left(\frac{(1-c)\left[\left(1-\alpha^2\right)\left(9-8\alpha^2\right)-\sqrt{\left(5-6\alpha^2\right)\left(9-8\alpha^2\right)}\right]}{2\left(2-\alpha^2\right)\left(9-8\alpha^2\right)}\right)^2$$

and

$$E \geq \overline{\overline{E}} \equiv \left(\frac{(1-c)\left[\left(1-\alpha^2\right)\left(9-8\alpha^2\right)+\sqrt{\left(5-6\alpha^2\right)\left(9-8\alpha^2\right)}\right]}{2\left(2-\alpha^2\right)\left(9-8\alpha^2\right)}\right)^2$$

which is defined on the real line when $\alpha < \sqrt{5/6} \simeq 0.91$.

- We next show that the latter inequality, $E \geq \overline{\overline{E}}$, is slack since $\overline{\overline{E}} > \overline{E}$ entails

$$\left(\frac{(1-c)\left[\left(1-\alpha^2\right)\left(9-8\alpha^2\right)+\sqrt{\left(5-6\alpha^2\right)\left(9-8\alpha^2\right)}\right]}{2\left(2-\alpha^2\right)\left(9-8\alpha^2\right)}\right)^2 > \left(\frac{\left(3-4\alpha^2\right)(1-c)}{9-8\alpha^2}\right)^2$$

that simplifies to

$$\sqrt{\left(5-6\alpha^2\right)\left(9-8\alpha^2\right)} > 3-5\alpha^2$$

yielding

$$36-64\alpha^2+23\alpha^4 > 0$$

- Similarly, we show that $\overline{E} > \underline{E}$, where

$$\left(\frac{(3-4\alpha^2)(1-c)}{9-8\alpha^2}\right)^2 > \left(\frac{(1-c)\left[(1-\alpha^2)(9-8\alpha^2)-\sqrt{(5-6\alpha^2)(9-8\alpha^2)}\right]}{2(2-\alpha^2)(9-8\alpha^2)}\right)^2$$

simplifies to

$$\sqrt{(5-6\alpha^2)(9-8\alpha^2)} > \left(1-\alpha^2\right)\left(9-8\alpha^2\right)-2\left(2-\alpha^2\right)\left(3-4\alpha^2\right)$$

that reduces to

$$36-64\alpha^2+23\alpha^4>0$$

Solving for α in the above inequalities, we obtain that

$$0 \le \alpha < \frac{3\sqrt{2}}{\sqrt{23}} \simeq 0.88$$

Hence, when $\alpha \in \left[0, \frac{3\sqrt{2}}{\sqrt{23}}\right)$ (which contains the range of $\alpha \in \left[0, \frac{\sqrt{3}}{2}\right]$ for firm 2 to earn positive profits), we have that $\overline{\overline{E}} > \overline{E} > \underline{E}$. Specifically, when $E \ge \overline{\overline{E}}$, even if firm 1 accommodates entry, firm 2 finds entry cost E too high to enter into the market, so entry is blockaded. In contrast, when $E \le \underline{E}$, the entry cost is so low that firm 1 finds it more profitable to accommodate than to deter entry.

(b) How does the binding cutoff you identified in part (a) change with α? Explain.

- Differentiating $\underline{E}$ with respect to α, we find:

$$\begin{aligned}\frac{\partial \underline{E}}{\partial \alpha} &= \frac{(1-c)^2\left[(1-\alpha^2)(9-8\alpha^2)-\sqrt{(5-6\alpha^2)(9-8\alpha^2)}\right]}{(2-\alpha^2)(9-8\alpha^2)}\times \\ &\left(\frac{\alpha(25-16\alpha^2)\left[(1-\alpha^2)(9-8\alpha^2)-\sqrt{(5-6\alpha^2)(9-8\alpha^2)}\right]}{(2-\alpha^2)^2(9-8\alpha^2)^2}\right. \\ &\left.-\frac{\alpha(2-\alpha^2)(9-8\alpha^2)\left[17-16\alpha^2-\frac{47-48\alpha^2}{\sqrt{(5-6\alpha^2)(9-8\alpha^2)}}\right]}{(2-\alpha^2)^2(9-8\alpha^2)^2}\right) \\ &= \underbrace{\frac{(1-c)^2\left[(1-\alpha^2)(9-8\alpha^2)-\sqrt{(5-6\alpha^2)(9-8\alpha^2)}\right]}{(2-\alpha^2)(9-8\alpha^2)}}_{>0}\times \\ &\underbrace{\frac{\alpha\left[\frac{2(1-\alpha^2)(1+24\alpha^2)}{\sqrt{(5-6\alpha^2)(9-8\alpha^2)}}-(9-8\alpha^2)^2\right]}{(2-\alpha^2)^2(9-8\alpha^2)^2}}_{<0}<0\end{aligned}$$

which holds for all $0 \leq \alpha < \frac{\sqrt{3}}{2}$, so that when the investment becomes more effective (α increases), cutoff $\underline{E}$ decreases, indicating that it becomes less likely that entry by firm 2 will be accommodated.

(c) Find the investment k^M, output q_1^M, and profits π_1^M of firm 1 if firm 2 does not enter.

- *Second stage.* Expecting that firm 2 does not enter, firm 1 chooses q_1 to solve

$$\max_{q_1 \geq 0} \pi_1^2(k) = (1 - q_1) q_1 - (c - \alpha k) q_1$$

Differentiating with respect to q_1, and assuming interior solutions, we obtain

$$1 - 2q_1 - (c - \alpha k) = 0$$

Solving for q_1, we find firm 1's output function, as follows:

$$q_1(k) = \frac{1 - (c - \alpha k)}{2}$$

- *First stage.* Firm 1 chooses k to solve the profit maximization problem:

$$\max_{k \geq 0} \pi_1^1(k) = \pi_1^2(k) - \frac{1}{2}k^2$$

$$= \left(\frac{1 - c + \alpha k}{2}\right)^2 - \frac{1}{2}k^2$$

Differentiating with respect to k, and assuming interior solutions, we find:

$$\frac{\alpha(1 - c + \alpha k)}{2} - k = 0$$

Rearranging and solving for k, the equilibrium investment of firm 1 becomes:

$$k^M = \frac{\alpha(1 - c)}{2 - \alpha^2}$$

where the superscript M stands for monopoly, and is positive since the denominator, $2 - \alpha^2$, is positive given that $\alpha \in [0, 1]$ by definition. Substituting k^M into $q_1(k)$, equilibrium output of firm 1 is

$$q_1^M = \frac{1}{2}\left[1 - c + \alpha \overbrace{\frac{\alpha(1 - c)}{2 - \alpha^2}}^{k^M}\right]$$

$$= \frac{1 - c}{2 - \alpha^2}$$

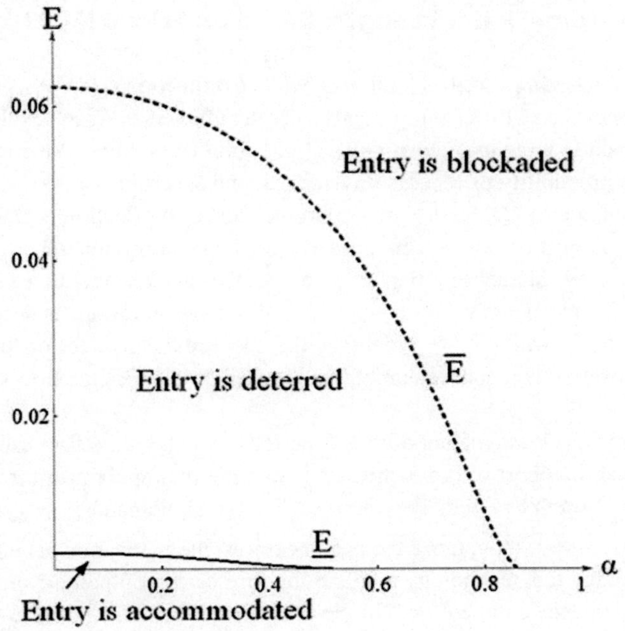

Fig. 6.6 Entry decisions as a function of E

Finally, substituting $k^M = \frac{\alpha(1-c)}{2-\alpha^2}$ into firm 1's profit function, we find:

$$\pi_1^M = \left(\frac{1 - c + \frac{\alpha^2(1-c)}{2-\alpha^2}}{2} \right)^2 - \frac{1}{2}\left(\frac{\alpha(1-c)}{2-\alpha^2}\right)^2$$

$$= \frac{(1-c)^2}{2(2-\alpha^2)}$$

(d) On a plane with α on the horizontal and E on the vertical axis, identify the incumbent firm's optimal strategies. Explain.

- Figure 6.6 depicts cutoffs $\underline{E}$ and $\overline{E}$, identifying three regions of entry cost E, as follows:
  - When entry cost E falls below cutoff $\underline{E}$, firm 2 finds entry less costly so firm 1 accommodates it, yielding equilibrium output and profits of q_1^A and π_1^A (q_2^A and π_2^A) for firm 1 (firm 2), respectively.
  - When entry cost E is intermediate, that is, within cutoffs $\underline{E}$ and $\overline{E}$, firm 1 finds it more profitable deterring than accommodating entry, so this firm produces q_1^D and earns π_1^D.
  - When entry cost E lies above cutoff $\overline{E}$, firm 2 finds it too costly to enter (i.e., entry is blockaded), leaving firm 1 as the monopoly producing q_1^M and earning π_1^M.
- In addition, the region of entry costs where entry is deterred shrinks as the cost-reducing investment becomes more effective (α increases), since it becomes more difficult for the entrant (firm 2) to compete with the more efficient incumbent (firm 1).

Exercise #6.15: Optimal Patent Length, Based on Takalo (2001)^C

6.15 Consider a firm investing $x \in [0, 1]$ in R&D, with cost function $C(x) = \frac{1}{2}\gamma x^2$, where γ denotes its R&D efficiency, i.e., a lower γ represents a greater efficiency. As in Takalo (2001), we assume that γ is sufficiently large to yield $x \in [0, 1]$ in equilibrium, thus allowing us to interpret x as the innovator's probability of success. Inverse demand function is $p(Q)$, where $Q \geq 0$ denotes aggregate output, and $p'(Q) < 0$. Firms' marginal cost of production is c, where $p(0) > c > 0$.

Without the innovation, every firm faces marginal cost of production c, where $p(0) > c > 0$, and we assume they interact in a perfectly competitive market, yielding zero profit, $\pi^{NI} = 0$, where superscript NI denotes "no innovation." When the innovator is successful, its marginal cost decreases to $\frac{c}{\alpha}$, where $\alpha \geq 1$ denotes the cost-reducing effect of the innovation. When $\alpha = 1$, the innovation is inconsequential, while when $\alpha \to +\infty$, the firm's marginal costs tend to zero.

The innovator receives a patent during T periods, which let the firm use the technology that decreases its marginal cost of production to $\frac{c}{\alpha}$, earning monopoly profit π^P, where superscript P represents a "patent" period. For instance, if inverse demand is $p(Q) = 1 - Q$, patent profits become $\pi^P = \frac{(1-\frac{c}{\alpha})^2}{4}$. Once the patent expires, the innovation becomes public, allowing other firms to enjoy this technology too, and the firm earns competitive profits $\pi^C = 0$, where superscript C denotes "competition." In the absence of patents, aggregate output occurs at the point where demand $p(Q)$ and marginal cost c cross, that is, Q_0 solves $p(Q_0) = c$. Aggregate output $Q_0 > 0$ since $c > 0$ and inverse demand function $p(Q)$ is downward sloping by definition. We consider the following time structure:

(1) In stage 1, the patent office (PO) chooses a patent length $T \geq 0$.
(2) In the second stage, the innovator observes the patent length T, and responds with its R&D investment, $x(T)$.
(3) In the third stage, every firm responds producing output q_i.

We solve the game operating by backward induction as follows:

(a) *Third stage.* Find equilibrium profits before and after the patent expires.
- After the patent expires (in periods $t \geq T$), every firm i enjoys production cost $\frac{c}{\alpha}$, where $\alpha \geq 1$ denotes the effectiveness of R&D at reducing marginal production costs. In this context, firms interact in a perfectly competitive market, with aggregate output Q^C, which solves $p(Q^C) = \frac{c}{\alpha}$, yielding zero profits for every firm, i.e., $\pi^C = 0$.
- Before the patent expires ($t < T$), the innovator enjoys production cost $\frac{c}{\alpha}$. We assume that the cost-reducing effect of the innovation is not "radical" in the following sense. Under the monopoly that provides the patent, the innovator could choose an output level q^m that solves

$$p(q^m) + p'(q^m)q^m = \frac{c}{\alpha},$$

so making marginal revenue and cost coincide. Inserting this monopoly output into the inverse demand function yields monopoly price $p^m \equiv p(q^m)$.

(b) *Second stage.* We define the cost-reducing effect of the innovation, as captured by α, to be "non-radical," which means that the monopoly price of the innovator during the patent period exceeds its rivals' common marginal cost, $p^m > c$. In this setting, for the innovator to be the only seller, it needs to set a price marginally below its rivals' common marginal cost, c, that is, $p^P = c - \varepsilon$, where $\varepsilon \to 0$. As a result, patent output q^P solves $p(q^P) = c$, which entails that $q^P = Q_0$. The innovator profits then become:

$$\pi^P = \left(p^P - \frac{c}{\alpha}\right) Q_0 = \left(c - \frac{c}{\alpha}\right) Q_0 = \left(\frac{\alpha - 1}{\alpha}\right) c Q_0$$

which are weakly positive since $\alpha \geq 1$ and $Q_0 > 0$. Therefore, profits are positive during the patent but zero afterward, $\pi^P > \pi^C = 0$. Intuitively, as the cost-reducing effect of the innovation increases (higher α), patent profits increase. For instance, in the context of inverse demand function $p(Q) = 1 - Q$, patent profits become $\pi^P = \frac{c(\alpha-1)(1-c)}{\alpha}$. Given this information, find the firm's equilibrium R&D investment as a function of the patent length, $x(T)$. Does it increase in T?

- Anticipating output decisions in the third stage, the innovator chooses its R&D investment, x, to solve

$$\max_{x \geq 0} \overbrace{x\Pi(T)}^{\text{Innov.}} + \overbrace{(1-x)\,\pi^{NI}}^{\text{No innov.}} - \frac{1}{2}\gamma x^2$$

where π^{NI} is the firm's profit without the innovation, and $\pi^{NI} = 0$ in a perfectly competitive market. In addition, $\Pi(T)$ denotes the innovator's return of a successful innovation (discounted in continuous time) as follows:

$$\begin{aligned}\Pi(T) &= \int_0^T e^{-rt}\pi^P dt + \int_T^{+\infty} e^{-rt}\pi^C dt \\ &= \int_0^T e^{-rt}\pi^P dt\end{aligned}$$

since $\pi^C = 0$. Note that $\Pi(T)$ is increasing in the patent length T, i.e., $\Pi'(T) \geq 0$. Intuitively, the innovator earns monopoly profit π^P during more periods. Differentiating with respect to x, and assuming interior solutions, yields:

$$x(T) = \frac{\Pi(T)}{\gamma}, \tag{6.1}$$

where $x'(T) = \frac{\pi^P e^{-rT}}{\gamma} \geq 0$ but $x''(T) = \frac{-r\pi^P e^{-rT}}{\gamma} < 0$. Therefore, when the firm invests more in R&D, $x(T)$ increases in the patent protection, T, but at a decreasing rate.

(c) *First stage.* Find the socially optimal patent length that the patent office (PO) chooses to maximize social welfare $W = CS + PS$, where CS denotes consumer surplus and PS represents producer surplus.

- The PO anticipates firms' decisions in subsequent stages and chooses the patent length, T, to solve

$$\max_{T \geq 0} \overbrace{x(T)\,S(T)}^{\text{Innov.}} + \overbrace{[1 - x(T)] \int_0^{+\infty} e^{-rt} W_0^N dt}^{\text{No innov.}} - \frac{1}{2}\gamma\,[x(T)]^2$$

where W_0^N is the welfare without the innovation, which is constant in T, thus not affecting the first-order conditions of the problem in expression (6.2). In addition, the term $x(T)$ comes from part (b) of the exercise, and $S(T)$ represents the social return of the innovation, defined as follows:

$$S(T) = \int_0^T e^{-rt} W^P dt + \int_T^{+\infty} e^{-rt} W^C dt$$
$$= \frac{e^{-rT}\left[W^C - (1 - e^{rT}) W^P\right]}{r}.$$

Specifically, W^P denotes the social welfare in each patent period, while W^C measures the social welfare after the patent expires. While profits during and after the patent satisfy $\pi^P > \pi^C = 0$, welfare ranking is the opposite, $W^P < W^C$, since a monopoly operates during patent periods.

Social return $S(T)$ satisfies

$$S'(T) = e^{-rT}(W^P - W^C) \leq 0 \text{ and}$$
$$S''(T) = e^{-rT}(W^C - W^P)r > 0$$

since $W^C \geq W^P$. Therefore, a longer patent decreases welfare at an increasing rate.

- Differentiating with respect to T in the PO's problem yields:

$$\underbrace{\frac{\partial x(T)}{\partial T} S(T)}_{MDG(T)} = \underbrace{x(T)\left(\gamma \frac{\partial x(T)}{\partial T} - \frac{\partial S(T)}{\partial T}\right)}_{MSL(T)}.$$

The left side measures the marginal dynamic gain, $MDG(T)$, from extending the patent for one more period. Intuitively, the firm increases its R&D intensity $x(T)$, which increases expected social welfare. The right side, in contrast, captures the marginal static loss, $MSL(T)$, of a longer patent, in the form of a larger R&D cost (first term) and a lower consumer surplus due to monopoly (second term).

In addition, we can check that second-order conditions hold since $MDG(T)$ is decreasing in T because

$$\frac{\partial MDG(T)}{\partial T} = \underbrace{\frac{\partial^2 x^*(T)}{\partial^2 T}}_{(-)} \underbrace{S(T)}_{(+)} + \underbrace{\frac{\partial x^*(T)}{\partial T}}_{(+)} \underbrace{\frac{\partial S(T)}{\partial T}}_{(-)} < 0$$

and the $MSL(T)$ is increasing in T if

$$\frac{\partial MSL(T)}{\partial T} = \underbrace{\underbrace{\frac{\partial x^*(T)}{\partial T}}_{(+)} \left(\gamma \underbrace{\frac{\partial x^*(T)}{\partial T}}_{(+)} - \underbrace{\frac{\partial S(T)}{\partial T}}_{(-)}\right)}_{(+)} + \underbrace{\underbrace{x^*(T)}_{(+)} \left(\gamma \underbrace{\frac{\partial^2 x^*(T)}{\partial^2 T}}_{(-)} - \underbrace{\frac{\partial^2 S(T)}{\partial^2 T}}_{(+)}\right)}_{(-)} > 0.$$

For an interior solution, $T^* > 0$, we also need that $MDG(T)$ originates above $MSL(T)$, which holds for all parameter values given that

$$MDG(0) = \left(\frac{\pi^P}{\gamma}\right) \int_0^{+\infty} e^{-rt} W^C dt = \frac{\pi^P W^C}{\gamma r} > 0,$$

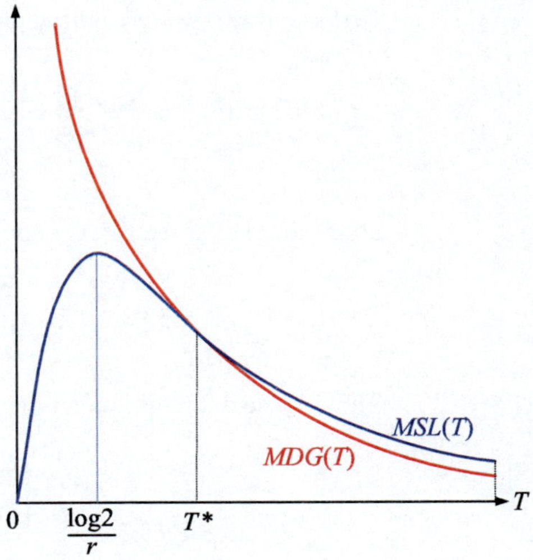

Fig. 6.7 Socially optimal patent

whereas $MSL(0) = 0$. Figure 6.7 depicts the $MDG(T)$ curve, which is decreasing in T, and the $MSL(T)$ curve, which is concave in T, and their crossing point at the socially optimal patent length, T^*. It is beyond the scope of this exercise, but you may check that, indeed, the $MSL(T)$ curve originates at $MSL(0) = 0$, increases in T, reaches a maximum at $T = \frac{\log 2}{r}$, and then decreases in T.

Exercise #6.16: Optimal Patent Length, An Application[B]

6.16 Consider the setting in Exercise 6.15 where a regulator sets the socially optimal patent length and the firm responds investing in R&D in the following period. For simplicity, assume an inverse demand function $p(Q) = 1 - Q$. In this exercise, we seek to find the socially optimal patent length in this parametric example. For simplicity, consider $\frac{1}{2} \leq c \leq 1$.

(a) For a given patent length T, find the firm's investment in R&D, $x(T)$.

- In this setting, patent profits become $\pi^P = \frac{c(\alpha-1)(1-c)}{\alpha}$, yielding a return from the innovation of

$$\Pi(T) = \frac{c(\alpha - 1)(1 - c)(1 - e^{-rT})}{\alpha r}.$$

Therefore, the innovator investment in R&D is

$$x(T) = \frac{c(\alpha - 1)(1 - c)(1 - e^{-rT})}{\gamma \alpha r}.$$

(b) Find social welfare during the periods when the patent is active, W^P, and when the patent expires, W^C. Use your results to find the social return from the innovation, $S(T)$.

- In this context, welfare during the patent periods becomes:

$$W^P = CS^P + PS^P = \underbrace{\frac{(1-c)^2}{2}}_{CS} + \underbrace{\frac{c(\alpha-1)(1-c)}{\alpha}}_{PS} = \frac{(1-c)(\alpha+\alpha c-2c)}{2\alpha}$$

and welfare after the patent expires is

$$W^C = CS^C = \frac{(1-\frac{c}{\alpha})^2}{2}.$$

Thus, the social return on innovation $S(T)$ is

$$S(T) = \frac{e^{-rT}\left[\frac{(1-\frac{c}{\alpha})^2}{2} - (1-e^{rT})\left(\frac{(1-c)[\alpha(1+c)-2c]}{2\alpha}\right)\right]}{r},$$

which, after simplifying, yields

$$S(T) = \frac{e^{-rT}}{2\alpha^2 r}\left[c^2(\alpha-1)^2 + e^{rT}\alpha(1-c)(\alpha+\alpha c-2c)\right].$$

(c) From your above results, find the expression of $MDG(T)$ and $MSL(T)$ in this setting. Then, identify the socially optimal patent length, T^*.

- From the following first-order conditions

$$x'(T) = \frac{e^{-rT}c(1-c)(\alpha-1)}{\gamma\alpha}$$

$$S'(T) = \frac{e^{-rT}}{2\alpha^2}\left[\alpha(1-c)(\alpha+\alpha c-2c) - (\alpha-c)^2\right]$$

A socially optimal patent occurs when $MDG(T) = MSL(T)$, where

$$MDG(T) = \frac{e^{-2rT}c(1-c)(\alpha-1)}{2\gamma\alpha^3 r}\left[c^2(\alpha-1)^2 + e^{rT}\alpha(1-c)(\alpha+\alpha c-2c)\right]$$

$$MSL(T) = \frac{e^{-rT}\left(1-e^{-rT}\right)c^2(1-c)(\alpha-1)^2}{2\gamma\alpha^3 r}\times$$
$$[\alpha(2-c)-c].$$

In this context, the optimal patent length T^* is

$$T^* = \left(\frac{1}{r}\right)\ln\left(\frac{2c(\alpha-1)}{2\alpha c-\alpha-c}\right)$$

(d) How does the optimal patent length, T^*, change with α and r? Explain.

- Differentiating T^* with respect to α, we obtain:

$$\frac{\partial T^*}{\partial \alpha} = -\frac{1-c}{r(\alpha-1)(2\alpha c-\alpha-c)} < 0$$

so the PO grants a shorter patent when the R&D becomes more effective.

- Differentiating T^* with respect to r, we have:

$$\frac{\partial T^*}{\partial r} = -\frac{1}{r^2}\log\left[\frac{2c(\alpha-1)}{2\alpha c-\alpha-c}\right] < 0$$

so the PO shortens the patent length when the discount factor increases.

(e) What is the innovator's equilibrium investment at T^*, and how does it change with α, γ, c, and r? Explain.

- The innovator's equilibrium investment is

$$x^* = \frac{c(\alpha-1)(1-c)\left[1-\frac{2\alpha c-\alpha-c}{2c(\alpha-1)}\right]}{\alpha\gamma r} = \frac{(1-c)(\alpha-c)}{2\alpha\gamma r}$$

which is decreasing in γ, c, and r since

$$\frac{\partial x^*}{\partial c} = -\frac{1+\alpha-2c}{2\alpha\gamma r}$$

that is negative since $c < 1 \leq \alpha$, and increasing in α because

$$\frac{\partial x^*}{\partial \alpha} = \frac{c(1-c)}{2\alpha^2\gamma r}$$

that is positive, entailing that the innovator makes more investment when the cost-reduction effect strengthens (higher α), but less investment when such investment becomes more costly (higher γ), the production process is less efficient (higher c), or the innovator is less patient (higher r).

(f) Evaluate the private and social return on innovation at T^*.

- The innovator's private return on innovation is

$$\Pi^* = \frac{c(\alpha-1)(1-c)\left[1-\frac{2\alpha c-\alpha-c}{2c(\alpha-1)}\right]}{\alpha r} = \frac{(1-c)(\alpha-c)}{2\alpha r}$$

- Whereas, the social return on innovation is

$$S^* = \frac{(2\alpha c - \alpha - c)\left[c^2(\alpha-1)^2 + \frac{\alpha(1-c)(\alpha+\alpha c-2c)(2\alpha c-\alpha-c)}{2c(\alpha-1)}\right]}{4\alpha^2 cr(\alpha-1)}$$

$$= \frac{(2\alpha c - \alpha - c)(\alpha - c)\left[(2\alpha c - \alpha - c)(\alpha - c) + c^2(\alpha-1)^2\right]}{8\alpha^2 c^2 r(\alpha-1)^2}$$

(g) *Numerical Example.* Evaluate the optimal patent length T^*, the innovator's equilibrium investment x^*, private and social return on innovation, Π^* and S^*, at $r = 0.025$, $c = 0.9$, $\alpha = 1.5$, and $\gamma = 0.1$. What if c increases to $c = 0.95$? Compare your results, and interpret.

- Substituting $r = 0.025$, $c = 0.9$, $\alpha = 1.5$, and $\gamma = 0.1$ into the optimal patent length T^* in part (c), equilibrium investment x^* in part (e), private and social return on innovation, Π^* and S^*, in part (f), we find:

$$T^* = \frac{1}{0.025}\log\left[\frac{2 \times 0.9 \times (1.5-1)}{2 \times 1.5 \times 0.9 - 1.5 - 0.9}\right]$$

$$= 40\log 3 \simeq 43.94$$

$$x^* = \frac{(1-0.9)(1.5-0.9)}{2 \times 1.5 \times 0.1 \times 0.025}$$

$$= 8$$

$$\Pi^* = \frac{(1-0.9)(1.5-0.9)}{2 \times 1.5 \times 0.025}$$

$$= 0.8$$

$$S^* = \frac{(2 \times 1.5 \times 0.9 - 1.5 - 0.9)(1.5-0.9) \times \left[(2 \times 1.5 \times 0.9 - 1.5 - 0.9)(1.5-0.9) + 0.9^2(1.5-1)^2\right]}{8 \times 1.5^2 \times 0.9^2 \times 0.025 \times (1.5-1)^2}$$

$$= \frac{34}{45} \simeq 0.76$$

- When c increases to $c = 0.95$, the above equilibrium results become:

$$T^* = \frac{1}{0.025}\log\left[\frac{2 \times 0.95 \times (1.5-1)}{2 \times 1.5 \times 0.95 - 1.5 - 0.95}\right]$$

$$= 40\log 2.375 \simeq 34.6$$

$$x^* = \frac{(1-0.95)(1.5-0.95)}{2 \times 1.5 \times 0.1 \times 0.025}$$

$$= \frac{11}{3} \simeq 3.67$$

$$\Pi^* = \frac{(1-0.95)(1.5-0.95)}{2 \times 1.5 \times 0.025}$$

$$= \frac{11}{30} \simeq 0.37$$

$$S^* = \frac{(2\times 1.5\times 0.95-1.5-0.95)(1.5-0.95)\times \left[(2\times 1.5\times 0.95-1.5-0.95)(1.5-0.95)+0.95^2(1.5-1)^2\right]}{8\times 1.5^2\times 0.95^2\times 0.025\times (1.5-1)^2}$$

$$= \frac{15686}{16425} \simeq 0.96$$

- Therefore, when c increases from 0.9 to 0.95, the innovator generates lower profits, makes fewer investments, and obtains a shorter patent. Interestingly, as the patent protection shortens, social welfare increases.

Mergers

7

Introduction

This chapter studies firms' incentives to merge and collude in imperfectly competitive markets. Exercise 7.1 begins with the basic setting of mergers between two firms into a monopoly. Exercise 7.2 extends into the setting of mergers between k firms in an industry with N firms. We show that when the number of firms that merge is sufficiently high, mergers become profitable as the gain in market power more than offsets the output increase of the firms that do not merge, and this is referred to as the "80% rule" after Salant et al. (1983). Exercise 7.3 suggests that when there are three or more firms merging together, the merger is unsustainable because every participating firm has incentives to leave and free ride on output reduction of the merger to increase its own output and profit levels. Exercise 7.4, in contrast, studies Bertrand mergers with three differentiated firms, showing that both partial and full mergers are sustainable when firms merge to soften price competition. Exercise 7.5 analyzes Cournot mergers with differentiated products, showing that as firms produce more homogeneous products, competition becomes more intense so more firms need to participate in order to sustain the merger. Exercise 7.6 shows that mergers between horizontally differentiated firms do not increase welfare when firms are relatively symmetric in costs, but are welfare-enhancing when firms are significantly cost-asymmetric in shifting output from the less to the more efficient firm.

Exercise 7.7 finds that companies have stronger incentives to merge when the merged firm benefits from cost-reduction effects. Exercise 7.8 suggests that when firms merge to gain economies of scope, they have more incentives to merge than when they suffer diseconomies of scope. Exercise 7.9 finds that if two firms in a homogeneous market of N firms merge to gain Stackelberg leadership, these two firms have incentives to merge under all parameter conditions. Exercise 7.10 analyzes mergers in a sequential-move game. We report that while a leader has incentives to acquire a follower, those incentives are weakened when the follower market becomes more competitive. To summarize, the existence of outsiders weakens firms' incentives to merge, unless the merged firm gains significant market power, enjoys cost advantage, or maintains output leadership.

Exercise 7.11 develops the case for the "tragedy of the anticommons," where firms, unlike a cartel which limits the extraction of common pool resources, increase the production of complementary goods relative to their competitive levels. Finally, Exercise 7.12 analyzes merger incentives in polluting markets, suggesting that emission fees can facilitate mergers of environmentally differentiated firms.

P.-S. Choi et al., *Industrial Organization*, Springer Texts in Business and Economics,
https://doi.org/10.1007/978-3-031-38635-0_7

Exercise #7.1: Mergers Between Two Firms^A

7.1 Consider an industry of two firms competing à la Cournot. In the first stage, every firm chooses whether or not to merge with its rival; and in the second stage, firms compete à la Cournot. If both firms choose to merge, they coordinate their production decision in the second stage to maximize their joint profits. If one or more firm does not choose to merge in the first stage, then the merger does not occur and firms compete à la Cournot in the second stage. Firms face an inverse demand curve $p(Q) = 1 - Q$, where $Q \geq 0$ denotes aggregate output. Firms are symmetric in their marginal cost of production c, where $1 > c \geq 0$.

(a) Find the profits of every firm i under no merger.

- If no merger occurs, every firm solves

$$\max_{q_i \geq 0} (1 - q_i - q_j)q_i - cq_i,$$

where q_j denotes the output of firm $j \neq i$. Differentiating with respect to q_i, and solving for q_i, we find firm i's best response function:

$$q_i(Q_{-i}) = \frac{1-c}{2} - \frac{1}{2}q_j,$$

which is negatively sloped in its rivals' output, q_j. In a symmetric equilibrium, $q_i = q_j = q$ for every two firms $i \neq j$. Therefore, the above best response function simplifies to

$$q = \frac{1-c}{2} - \frac{1}{2}q,$$

and solving for q, we obtain the equilibrium output of every firm i when no merger occurs:

$$q = \frac{1-c}{3}.$$

Then, equilibrium profits become

$$\pi^{NM} = \frac{(1-c)^2}{9},$$

where the superscript NM indicates "no merger."

(b) Find the profits that every merging firm earns.

- When firms merge, they coordinate their production decisions, so they maximize their joint profits

$$\max_{q_i, q_j \geq 0} \underbrace{\left[(1 - q_i - q_j)q_i - cq_i\right]}_{\pi_i} + \underbrace{\left[(1 - q_i - q_j)q_j - cq_j\right]}_{\pi_j}$$

or, after rearranging,

$$\max_{q_i, q_j \geq 0} (1 - q_i - q_j)\left(q_i + q_j\right) - c(q_i + q_j)$$

and using $Q = q_i + q_j$ to denote aggregate output, the above problem simplifies to

$$\max_{Q\geq 0}\,(1-Q)Q-cQ,$$

Intuitively, the merged firms now choose their aggregate output Q. After differentiating with respect to Q, we find:

$$1-2Q-c=0,$$

which yields an equilibrium aggregate output under the merger of

$$Q=\frac{1-c}{2}$$

Since firms are symmetric, every firm produces half of this output, that is, $q=\frac{Q}{2}=\frac{1-c}{4}$. As a result, the equilibrium profits under the merger are

$$\pi^M=\frac{(1-c)^2}{8},$$

where the superscript M indicates "merger."

(c) Do firms choose to merge in the first stage?
- Comparing profits, we find that $\pi^M>\pi^{NM}$, so every firm i chooses to merge in the first stage. This result holds for all values of marginal cost c.

(d) *Numerical example.* Evaluate firm profits when they do not merge, and when they do, at marginal cost $c=1/2$.
- Profits under no merger, as found in part (a), evaluated at $c=1/2$, become $\pi^{NM}=\frac{\left(1-\frac{1}{2}\right)^2}{9}=\frac{1}{36}$. Profits under the merger, as found in part (b), evaluated at $c=1/2$, are $\pi^M=\frac{\left(1-\frac{1}{2}\right)^2}{8}=\frac{1}{32}$, thus being higher than before the merger.

(e) Is this merger welfare improving? Explain.
- Generally, the welfare experienced in this exercise can be expressed as

$$W(Q)=\int_0^Q p(Q)dQ-cQ=(1-c)Q-\frac{1}{2}Q^2$$

- *No Merger.* Inserting the aggregate quantity, $Q^{NM}=\frac{2(1-c)}{3}$, into our welfare expression, yields:

$$\begin{aligned}W^{NM}&=(1-c)\left[\frac{2(1-c)}{3}\right]-\frac{1}{2}\left[\frac{2(1-c)}{3}\right]^2\\&=\frac{4(1-c)^2}{9}\end{aligned}$$

- *Merger*. Inserting the aggregate quantity, $Q^M = \frac{1-c}{2}$, into our welfare expression, yields:

$$W^M = (1-c)\left[\frac{1-c}{2}\right] - \frac{1}{2}\left[\frac{1-c}{2}\right]^2$$
$$= \frac{3(1-c)^2}{8}$$

- *Comparison*. We show that $W^{NM} > W^M$ holds for all values of $c \in [0, 1)$. Intuitively, it will never be welfare improving for firms to merge absent any efficiency gains.

Exercise #7.2: Mergers Between $n \geq 2$ Firms, Based on Salant et al. (1983)^B

7.2 Consider an industry of $n \geq 2$ firms competing à la Cournot. Let us analyze the following sequential-move game where, in the first stage, every firm chooses whether or not to merge with other $k-1$ firms (so k firms merge and the remaining $n-k$ firms do not); and, in the second stage, firms compete à la Cournot. The k firms that merge coordinate their production decision to maximize their joint profits, while the remaining $n-k$ firms do not coordinate. Firms face an inverse demand curve $p(Q) = a - Q$, where $a > 0$ and $Q \geq 0$ denotes aggregate output. Firms are symmetric in their marginal cost of production c, where $a > c \geq 0$.

(a) As a benchmark, find the profits of every firm i before the merger.

- If no merger occurs, every firm solves:

$$\max_{q_i \geq 0} \, (a - q_i - Q_{-i})q_i - cq_i,$$

where Q_{-i} denotes the aggregate output of firm i's rivals. Differentiating with respect to q_i, and solving for q_i, we find firm i's best response function:

$$q_i(Q_{-i}) = \frac{a-c}{2} - \frac{1}{2}Q_{-i}$$

which is negatively sloped in its rivals' aggregate output, Q_{-i}. In a symmetric equilibrium, $q_i = q_j = q$ for every two firms $i \neq j$, which entails $Q_{-i} = (n-1)q$. Therefore, the above best response function simplifies to

$$q = \frac{a-c}{2} - \frac{1}{2}(n-1)q$$

and, solving for q, we obtain the equilibrium output of every firm i when no merger occurs:

$$q = \frac{a-c}{n+1}.$$

Then, equilibrium profits become:

$$\pi^{NM} = \left(\frac{a-c}{n+1}\right)^2,$$

where the superscript NM indicates "no merger."

(b) Find the profits that every merging firm earns.

- When k firms merge, leaving $n - k$ unmerged firms, the total number of firms becomes $(n - k) + 1$. In this setting, equilibrium profits are

$$\pi^M = \left(\frac{a - c}{\underbrace{(n - k) + 1}_{\text{Number of firms}} + 1} \right)^2 = \left(\frac{a - c}{n - k + 2} \right)^2,$$

where the superscript M indicates "merger."

(c) For which values of n and k can a merger be sustained in equilibrium? Interpret.

- A merger between k out of n firms is profitable if the post-merger profits exceed the pre-merger profits (for all firms that merged, as a whole), that is,

$$\pi^M \geq k\pi^{NM}$$

or

$$\left(\frac{a - c}{n - k + 2} \right)^2 \geq k \left(\frac{a - c}{n + 1} \right)^2$$

which we can start simplifying as

$$(n + 1)^2 \geq k(n - k + 2)^2$$

or further rearrange as

$$(k - 1) \left[-k^2 + (2n + 3)k - (n + 1)^2 \right] \geq 0.$$

Since $k \geq 2$ by definition (a merger must include at least two firms), we can solve for k in the second term to obtain

$$k \geq \frac{2n + 3 - \sqrt{4n + 5}}{2} \equiv \widehat{k}.$$

Intuitively, for a merger to be profitable, it must be large enough, $k \geq \widehat{k}$. Otherwise, the merger is not profitable.

- **Interpretation.** Recall the positive and negative effect of a merger on profits when firms compete à la Cournot:
  - *Positive effect.* On the one hand, a merger generates a positive effect on profits since firms reduce their output when coordinating their actions, which raises prices, and thus margins.
  - *Negative effect.* On the other hand, however, this output reduction leads the unmerged $n - k$ firms to respond by increasing their output levels (since best response functions are negatively sloped in this setting), which decreases the profits of the k firms that merged.

  Our above result says that the positive effect of the merger dominates the negative effect when the number of firms merging is sufficiently high, $k \geq \widehat{k}$, since in that case the negative effect from the merger is relatively low.

- *Market share interpretation.* An alternative presentation of the above result divides both sides of the inequality by n, to obtain

$$\frac{k}{n} \geq \frac{\widehat{k}}{n}.$$

This rearranged inequality says that, for the merger to be profitable, the market share that the merged firms represent $\frac{k}{n}$ (left-hand side) must be larger than $\frac{\widehat{k}}{n}$ (right-hand side). In particular,

$$\frac{\widehat{k}}{n} = \frac{2n + 3 - \sqrt{4n + 5}}{2n}$$

which is a ratio extremely close to 0.8. Indeed, when $n = 2$, cutoff $\frac{\widehat{k}}{n}$ becomes 0.848, when $n = 3$, this cutoff is 0.812, when $n = 4$, the cutoff is 0.802, and when $n = 5$, the cutoff becomes 0.8. Intuitively, this means that, for the merger to be profitable, at least 80% of the firms must merge, which explains why this result is informally known as the "80% rule" after Salant et al. (1983).

(d) For what values of k are the mergers welfare improving? Explain.

- Generally, the welfare experienced in this exercise can be expressed as

$$W(Q) = \int_0^Q p(Q)dQ - cQ = (a - c)Q - \frac{1}{2}Q^2$$

- *No Merger.* Inserting the aggregate quantity, $Q^{NM} = \frac{n(a-c)}{n+1}$, into our welfare expression, yields:

$$\begin{aligned} W^{NM} &= (a - c)\left[\frac{n(a - c)}{n + 1}\right] - \frac{1}{2}\left[\frac{n(a - c)}{n + 1}\right]^2 \\ &= \frac{n(n + 2)(a - c)^2}{2(n + 1)^2} \end{aligned}$$

- *Merger.* Inserting the aggregate quantity, $Q^M = \frac{(n-k+1)(a-c)}{n-k+2}$, into our welfare expression, yields:

$$\begin{aligned} W^M &= (a - c)\left[\frac{(n - k + 1)(a - c)}{n - k + 2}\right] - \frac{1}{2}\left[\frac{(n - k + 1)(a - c)}{n - k + 2}\right]^2 \\ &= \frac{(n - k + 1)(n - k + 3)(a - c)^2}{2(n - k + 2)^2} \end{aligned}$$

- *Comparison.* We find that $W^{NM} > W^M$ holds if and only if

$$\frac{n(n + 2)(a - c)^2}{2(n + 1)^2} > \frac{(n - k + 1)(n - k + 3)(a - c)^2}{2(n - k + 2)^2}$$

which reduces to

$$(k - 1)(2n - k + 3) > 0$$

Since $k > 1$, the above inequality simplifies to

$$2n + 3 > k$$

which holds since $k \leq n$ by definition. Intuitively, it will never be welfare improving for any number of firms to merge absent any efficiency gains, since the loss in consumer surplus offsets the additional profits accrued to each firm. In other words, even if the merger is profitable (the merging entity represents a large market share), it is unambiguously welfare reducing.

(e) For what values of k are the mergers welfare improving? Explain.

- Generally, the welfare experienced in this exercise can be expressed as

$$W(Q) = \int_0^Q p(Q)dQ - cQ = (a - c)Q - \frac{1}{2}Q^2$$

- *No Merger*. Inserting the aggregate quantity, $Q^{NM} = \frac{n(a-c)}{n+1}$, into our welfare expression, yields:

$$W^{NM} = (a - c)\left[\frac{n(a - c)}{n + 1}\right] - \frac{1}{2}\left[\frac{n(a - c)}{n + 1}\right]^2$$
$$= \frac{n(n + 2)(a - c)^2}{2(n + 1)^2}$$

- *Merger*. Inserting the aggregate quantity, $Q^M = \frac{(n-k+1)(a-c)}{n-k+2}$, into our welfare expression, yields:

$$W^M = (a - c)\left[\frac{(n - k + 1)(a - c)}{n - k + 2}\right] - \frac{1}{2}\left[\frac{(n - k + 1)(a - c)}{n - k + 2}\right]^2$$
$$= \frac{(n - k + 1)(n - k + 3)(a - c)^2}{2(n - k + 2)^2}$$

- *Comparison*. We find that $W^{NM} > W^M$ holds if and only if

$$\frac{n(n + 2)(a - c)^2}{2(n + 1)^2} > \frac{(n - k + 1)(n - k + 3)(a - c)^2}{2(n - k + 2)^2}$$

which reduces to

$$(k - 1)(2n - k + 3) > 0$$

Since $k > 1$, the above inequality simplifies to

$$2n + 3 > k$$

which holds since $k \leq n$ by definition. Intuitively, it will never be welfare improving for any number of firms to merge absent any efficiency gains, since the loss in consumer surplus offsets the additional profits accrued to each firm. In other words, even if the merger is profitable (the merging entity represents a large market share), it is unambiguously welfare reducing.

Exercise #7.3: Unsustainable Mergers[A]

7.3 Consider an industry of $n \geq 2$ firms competing à la Cournot. Assume that $k \geq 2$ firms (out of n) merged and thus coordinate their production decisions, while the remaining $n - k$ firms did not so they operate à la Cournot. Firms face an inverse demand curve $p(Q) = a - Q$, where $a > 0$ and $Q \geq 0$ denotes aggregate output. Firms are symmetric in their marginal cost of production c, where $a > c \geq 0$.

(a) Show that when $n \geq 3$, the merger is not sustainable in equilibrium. Interpret your results.

- From Exercise 7.2, we know that the aggregate profits when firms merge are

$$\pi^M = \left(\frac{a-c}{n-k+2}\right)^2.$$

Therefore, every firm i that participates in the merger earns a share $\frac{1}{k}$ of them, that is,

$$\pi_i^M = \frac{1}{k}\left(\frac{a-c}{n-k+2}\right)^2.$$

When a firm competes à la Cournot, we know from Exercise 2.4 that its equilibrium profit is $\pi_i^{Cournot} = \left(\frac{a-c}{n+1}\right)^2$. Since the number of firms that remain out of the merger is $n-k+1$ (see Exercise 7.2 again for details), the profits of these firms become:

$$\pi_i^{Cournot} = \left(\frac{a-c}{(n-k+1)+2}\right)^2 = \left(\frac{a-c}{n-k+3}\right)^2.$$

- Comparing the above profits, every firm i has incentives to participate in the merger of k firms, rather than being one of the outside firms (so the merger includes now only $k - 1$ firms), if and only if

$$\frac{1}{k}\left(\frac{a-c}{n-k+2}\right)^2 \geq \left(\frac{a-c}{n-k+3}\right)^2$$

which simplifies to

$$(n-k)^2 + 6(n-k) + 9 \geq k(n-k)^2 + 4k(n-k) + 4k$$

and, further rearranging, we obtain:

$$\underbrace{\underbrace{(1-k)}_{-}\underbrace{(n-k)^2}_{+}}_{-} + \underbrace{2\underbrace{(3-2k)}_{-}\underbrace{(n-k)}_{+}}_{-} + (9-4k) \geq 0$$

where we marked the terms that are unambiguously positive or negative. To understand the signs, note that the number of merging firms, k, must satisfy $n \geq k \geq 2$, entailing that $1 - k < 0$, that $n - k > 0$, and that $3 - 2k < 0$ given that, solving for k, yields $k > 1.5$, which holds by definition. Finally, the last term, $9 - 4k$, is negative for all $k > 2.25$, i.e., when three or more firms merge.

- Therefore, in a market with $n \geq 3$ firms where three or more firms merge, $k \geq 3$, the above inequality cannot hold. In other words, regardless of the number of firms in the industry, n, every participant in the merger has unilateral incentives to leave, making the merger unsustainable in equilibrium.
- *Intuition*: The firms participating in the merger, by reducing their output levels, increase market prices, ultimately benefiting firms outside the cartel. Every firm then has incentives to free ride this benefit by leaving the cartel.

(b) Show that when $n = 2$, the merger can be supported in equilibrium.

- When the market only includes two firms, $n = 2$, the merger must contain these two firms, $k = 2$, entailing that $n = k$, and simplifying the above inequality to

$$9 - 4k \geq 0$$

which holds for all $k < 2.25$. Since $k = 2$ in this setting, the merger can be sustained in equilibrium.

Exercise #7.4: Bertrand Mergers with Differentiated Products^B

7.4 Consider an industry with $N = 3$ firms competing à la Bertrand. Every firm i faces a demand function:

$$q_i\left(p_i, \overline{p}_{-i}\right) = 1 - p_i + \gamma\overline{p}_{-i},$$

where $\overline{p}_{-i} \equiv \frac{\sum_{j \neq i} p_j}{2}$ represents the average price of all i's rivals. When $\gamma = 0$, products are completely differentiated, but when $\gamma \to 1$, products are homogeneous. For simplicity, assume that all firms have zero marginal costs.

(a) *No merger.* If firms do not merge, find their best response functions, equilibrium prices, and profits.

- Every firm i chooses p_i to solve the following profit maximization problem:

$$\max_{p_i > 0} \pi\left(p_i\right) = p_i\left(1 - p_i + \gamma\overline{p}_{-i}\right)$$

Differentiating with respect to p_i, we find:

$$1 - 2p_i + \gamma\overline{p}_{-i} = 0$$

Rearranging, the best response of firm i is

$$p_i\left(\overline{p}_{-i}\right) = \frac{1}{2} + \frac{\gamma}{2}\overline{p}_{-i}$$

which originates at a vertical intercept of $1/2$ and increases in the average price of all its rivals at a rate of $\gamma/2$. Intuitively, when goods are completely differentiated, $\gamma = 0$, this best response function collapses to $p_i\left(\overline{p}_{-i}\right)=\frac{1}{2}$, implying that firm i's pricing decision does not depend on its rivals' prices, and sets a price equal to $\frac{1}{2}$, as in standard monopolies with this linear demand curve. However, when goods are homogeneous, $\gamma \to 1$, the best response function becomes $p_i\left(\overline{p}_{-i}\right)=\frac{1}{2}+\frac{1}{2}\overline{p}_{-i}$.

Graphically, when $\gamma = 0$, the best response function is flat (unaffected by the pricing decisions of firm i's rivals, $\overline{p}_{-i}$). When γ increases, however, the best response function pivots upward, counterclockwise, thus exhibiting a positive slope, $\frac{\gamma}{2}$, which becomes the steepest when $\gamma \to 1$. This happens because prices are strategic complements, and price competition is softened when goods become more homogeneous (higher γ).

- In a symmetric equilibrium, prices satisfy $\overline{p}_{-i} = p_i$, entailing

$$p_i=\frac{1}{2}+\frac{\gamma}{2}p_i$$

which, solving for p_i, yields an equilibrium price

$$p^*=\frac{1}{2-\gamma}$$

which is increasing in γ. When goods are completely differentiated, $\gamma = 0$, this price becomes $p_i^*=\frac{1}{2}$, which coincides with the standard result in profit-maximizing monopolists. In contrast, when goods are homogeneous, $\gamma \to 1$, the price increases to $p_i^*=1$.

- Inserting this price, $p_i^*=\frac{1}{2-\gamma}$, into the profit function of firm i, we obtain an equilibrium price of

$$\begin{aligned}\pi^* &= p^*\left(1-p^{NM}+\gamma p^{NM}\right)\\ &=\frac{1}{2-\gamma}\left(1-\frac{1}{2-\gamma}+\gamma\frac{1}{2-\gamma}\right)\\ &=\frac{1}{(2-\gamma)^2}\end{aligned}$$

which is increasing in γ since firms charge a higher price on more homogeneous goods, softening price competition and, thus, increasing equilibrium profits.

(b) *Full merger.* Assume that all three firms merge. Find their equilibrium price and profits.

- Assume that all three firms merge, then it chooses p_i^M for every firm i to solve

$$\max_{p_i^M>0} \pi\left(p_i^M\right)=\sum_{i=1}^{3}p_i^M\left[1-p_i^M+\frac{\gamma}{2}\sum_{j\neq i}p_j^M\right]$$

where p_j^M is the price of every other merged firm $j \neq i$.

Differentiating with respect to p_i^M, and assuming interior solutions, we find:

$$1-2p_i^M+\gamma\sum_{j\neq i}p_j^M=0$$

Since every merged firm is symmetric, in equilibrium we must have $p^M = p_i^M = p_j^M$, yielding the best response function of every merged firm, as follows:

$$p^M = \frac{1}{2(1-\gamma)}$$

- When goods are completely differentiated, $\gamma = 0$, the best response function simplifies to $p^M = \frac{1}{2}$. In contrast, when γ increases, goods become more homogeneous, and we obtain that

$$\frac{\partial p^M}{\partial \gamma} = \frac{1}{2(1-\gamma)^2} > 0$$

so that p^M increases in γ.

(c) *Full merger, profit comparison.* Compare the equilibrium profits of the merged firms, as found in part (b), against their profits when unmerged, as found in part (a), to show the full merger can be sustained in equilibrium.

- Substituting p^M into π^M, every merged firm earns profits of

$$\pi^M = \frac{1}{2(1-\gamma)}\left[1 - \frac{1-\gamma}{2(1-\gamma)}\right] = \frac{1}{4(1-\gamma)}$$

- Every firm i has incentives to merge if and only if $\pi^M \geq \pi^*$, where

$$\frac{1}{4(1-\gamma)} \geq \frac{1}{(2-\gamma)^2}$$

simplifies to

$$\gamma^2 \geq 0$$

that holds, so that the full merger is supported for all values of $\gamma \in [0, 1)$.

(d) *Partial merger, merged firms.* Assume that two of the three firms merge. Find their best response function and compare it against that in the setting where every firm independently sets its price, as found in part (a).

- Assume that two of the three firms merge, then it chooses p_i^M for every firm i to solve

$$\max_{p_i^M > 0} \pi\left(p_i^M\right) = \sum_{i=1}^{2} p_i^M \left[1 - p_i^M + \frac{\gamma}{2}\left(p_j^M + p^{NM}\right)\right]$$

where p_j^M is the price of the other merged firm j and p^{NM} is the price of the unmerged firm.

Differentiating with respect to p_i^M, and assuming interior solutions, we find:

$$1 - 2p_i^M + \gamma p_j^M + \frac{\gamma}{2} p^{NM} = 0$$

Since every merged firm is symmetric, in equilibrium we must have $p^M = p_i^M = p_j^M$, yielding the best response function of every merged firm, as follows:

$$p^M\left(p^{NM}\right) = \frac{2+\gamma p^{NM}}{2(2-\gamma)}$$

which originates at a vertical intercept of $\frac{2}{2(2-\gamma)}$ and increases in the price of the unmerged firm at a rate of $\frac{\gamma}{2(2-\gamma)}$.

- When goods are completely differentiated, $\gamma = 0$, the best response function simplifies to $p^M = \frac{1}{2}$, implying that the equilibrium price of every merged firm i is unaffected by the pricing decisions of the unmerged firm, p^{NM}. In contrast, when γ increases, goods become more homogeneous, and we obtain that

$$\frac{\partial p^M(0)}{\partial \gamma} = \frac{1}{(2-\gamma)^2} > 0$$

$$\frac{\partial^2 p^M\left(p^{NM}\right)}{\partial p^{NM}\partial \gamma} = \frac{1}{(2-\gamma)^2} > 0$$

so that the best response function shifts upward and pivots counterclockwise, that is, both the vertical intercept and the slope increase in γ, respectively.

(e) *Partial merger, unmerged firms.* Still in the context of part (d), find the best response function of the unmerged firm. How do the vertical intercept and the slope of the best response function change in γ? Explain.

- The unmerged firm i chooses p_i^{NM} to solve

$$\max_{p_i^{NM}>0} \pi\left(p_i^{NM}\right) = p_i^{NM}\left(1 - p_i^{NM} + \gamma p^M\right)$$

Differentiating with respect to p_i^{NM}, and assuming interior solutions, we find:

$$1 - 2p_i^{NM} + \gamma p^M = 0$$

The best response function of the unmerged firm becomes:

$$p^{NM}\left(p^M\right) = \frac{1+\gamma p^M}{2}$$

which originates at a vertical intercept of $\frac{1}{2}$ and increases in the price of the merged firm at a rate of $\frac{\gamma}{2}$.

- When goods are completely differentiated, $\gamma = 0$, the best response function simplifies to $p^{NM} = \frac{1}{2}$, implying that the equilibrium price of the unmerged firm i is unaffected by the pricing decisions of the merged firms, p^M. In contrast, when γ increases, goods become more homogeneous, and we obtain that

$$\frac{\partial p^{NM}(0)}{\partial \gamma} = 0$$

$$\frac{\partial^2 p^{NM}\left(p^M\right)}{\partial p^M \partial \gamma} = \frac{1}{2} > 0$$

so that the best response function pivots counterclockwise, that is, the slope increases in γ.

(f) *Partial merger, equilibrium prices.* Use the best response functions you identified in parts (d) and (e) to find the equilibrium price and profits of the two merged firms and the unmerged firm.

- Substituting $p^M\left(p^{NM}\right)$ into $p^{NM}\left(p^M\right)$, we obtain:

$$p^{NM} = \frac{1}{2} + \frac{\gamma}{2}\overbrace{\frac{2+\gamma p^{NM}}{2(2-\gamma)}}^{p^M}$$

Solving for p^{NM}, equilibrium price of the unmerged firm becomes:

$$p^{NM} = \frac{4}{8-4\gamma-\gamma^2}$$

- Substituting $p^{NM}\left(p^M\right)$ into $p^M\left(p^{NM}\right)$, we obtain:

$$p^M = \frac{2}{2(2-\gamma)} + \frac{\gamma}{2(2-\gamma)}\overbrace{\frac{4}{8-4\gamma-\gamma^2}}^{p^{NM}}$$

Solving for p^M, equilibrium price of the merged firm becomes:

$$p^M = \frac{4+\gamma}{8-4\gamma-\gamma^2}$$

- Substituting p^M and p^{NM} into π^M, every merged firm earns profits of

$$\begin{aligned}\pi^M &= \frac{p^M}{2}\left[2-(2-\gamma)\,p^M+\gamma p^{NM}\right]\\ &= \frac{4+\gamma}{2\left(8-4\gamma-\gamma^2\right)}\left[2-\frac{(2-\gamma)\,(4+\gamma)}{8-4\gamma-\gamma^2}+\frac{4\gamma}{8-4\gamma-\gamma^2}\right]\\ &= \frac{(2-\gamma)(4+\gamma)^2}{2(8-4\gamma-\gamma^2)^2}\end{aligned}$$

- Substituting p^M and p^{NM} into π^{NM}, the unmerged firm earns profits of

$$\begin{aligned}\pi^{NM} &= p^{NM}\left(1-p^{NM}+\gamma p^M\right)\\ &= \frac{4}{8-4\gamma-\gamma^2}\left[1-\frac{4}{8-4\gamma-\gamma^2}+\frac{\gamma\,(4+\gamma)}{8-4\gamma-\gamma^2}\right]\\ &= \frac{16}{(8-4\gamma-\gamma^2)^2}\end{aligned}$$

(g) *Partial merger, profit comparison.* Compare the equilibrium profits of the merged firms, as found in part (f), against their profits when unmerged, as found in part (a), to show if the partial merger can be sustained in equilibrium.

- Every firm i has incentives to merge if and only if $\pi^M \geq \pi^*$, where

$$\frac{(2-\gamma)(4+\gamma)^2}{2(8-4\gamma-\gamma^2)^2} \geq \frac{1}{(2-\gamma)^2}$$

Solving for γ, we find that mergers can be sustained for any $0 \leq \gamma < 1$.

Exercise #7.5: Cournot Mergers with Differentiated Products and Symmetric Costs^B

7.5 Consider an industry with $N \geq 2$ firms competing à la Cournot. Every firm i, where $i \in \{1, \ldots, N\}$, faces an inverse demand function:

$$p_i(q_1, q_2, \ldots, q_N) = 1 - q_i - \gamma \sum_{j \neq i} q_j,$$

where $\gamma \in [0, 1]$ represents the degree of product differentiation. When $\gamma = 0$, products are completely differentiated, but when $\gamma = 1$, products are homogeneous. For simplicity, suppose that firms have constant marginal costs of production, $c = 0$.

(a) *Cournot competition.* Assume that every firm i independently and simultaneously chooses its output level, q_i. Find the equilibrium output, price, and profits for each firm.

- Every firm i simultaneously chooses q_i to solve

$$\max_{q_i \geq 0} \pi_i(q_i) = (1 - q_i - \gamma q_{-i}) q_i$$

where $q_{-i} \equiv \sum_{j \neq i} q_j$ represents the sum of output of all firm i's rivals.

Taking the first-order condition with respect to q_i, and assuming interior solutions

$$1 - 2q_i - \gamma q_{-i} = 0$$

Rearranging and solving for q_i, we obtain the best response function of firm i:

$$q_i(q_{-i}) = \frac{1}{2} - \frac{\gamma}{2} q_{-i}$$

which originates at $1/2$ and decreases in q_{-i} at a rate of $\frac{\gamma}{2}$. Intuitively, the more homogeneous are the products (higher γ), the more responsive is firm i's output to the output of the other $N-1$ firms.

Since firms are symmetric, in equilibrium we must have $q = q_i = q_j$, yielding

$$q = \frac{1}{2} - \frac{\gamma}{2}(N-1) q$$

Rearranging and solving for q, we find:

$$q^* = \frac{1}{2 + \gamma (N - 1)}$$

Substituting $q = \frac{1}{2+\gamma(N-1)}$ into the inverse demand function, equilibrium price becomes:

$$p^* = 1 - q^* - (N - 1)\gamma q^* = \frac{1}{2 + \gamma (N - 1)}$$

Finally, equilibrium profits of every firm i are

$$\pi^* = p^* q^* = \frac{1}{[2 + \gamma (N - 1)]^2}$$

(b) *Full merger.* Assume that all N firms merge to choose an output profile $(q_1, q_2, \ldots, q_N)$ that maximizes their joint profits. Find the equilibrium output, price, and profits in this setting.

- The merged firm chooses $(q_1, q_2, \ldots, q_N)$ to solve

$$\max_{q_1, q_2, \ldots, q_N \geq 0} \pi (q_1, q_2, \ldots, q_N) = \sum_{i=1}^{N} (1 - q_i - \gamma q_{-i})\, q_i$$

Taking first-order condition with respect to q_i, and assuming interior solutions

$$1 - 2q_i - 2\gamma q_{-i} = 0$$

Rearranging and solving for q_i, we obtain the best response function of firm i:

$$q_i (q_{-i}) = \frac{1}{2} - \gamma q_{-i}.$$

This best response function originates at $1/2$, as that in part (a) when firms independently choose their output levels, but decreases faster in q_{-i}, thus indicating that a given increase in the aggregate output of firm i's rivals yields twice as much reduction in i's output level. This happens because output is a strategic substitute, and by merging, every firm i internalizes the externality that increasing its own output induces all other firms to reduce their output.

Since firms are symmetric, in equilibrium we must have $q = q_i = q_j$, yielding

$$q = \frac{1}{2} - \gamma (N - 1)\, q$$

Rearranging and solving for q, we find

$$q^{FM} = \frac{1}{2\,[1 + \gamma (N - 1)]}$$

where the superscript *FM* stands for full merger.

Substituting $q^{FM} = \frac{1}{2[1+\gamma(N-1)]}$ into the inverse demand function, equilibrium price becomes:

$$p^{FM} = 1 - q - (N-1)\,\gamma q = \frac{1}{2}$$

which is same as the profit-maximizing price of a monopolist subject to an inverse demand function $p(Q) = 1 - Q$.

Finally, equilibrium profits of every firm i are

$$\pi^{FM} = p^{FM} q^{FM} = \frac{1}{4\,[1 + \gamma\,(N-1)]}$$

It is straightforward to verify that the every firm i enjoys higher profits when merged than not merged, since $\pi^{FM} \geq \pi^*$ simplifies to

$$\frac{1}{4\,[1 + \gamma\,(N-1)]} \geq \frac{1}{[2 + \gamma\,(N-1)]^2}$$

which is rearranged to yield

$$\gamma^2\,(N-1)^2 \geq 0$$

that holds since $\gamma \in [0, 1]$ by definition, and $N \geq 2$, so that every firm i has incentives to merge into one firm.

(c) *Partial merger.* Assume that a subset of firms, k out of N, merge, while the remaining $N - k$ firms stay out of the merger. Find the equilibrium output in this context.

- The merged firm chooses q_i to solve the profit maximization problem:

$$\max_{q_i \geq 0} \pi_i\,(q_i) = \sum_{i=1}^{k} \left(1 - q_i - \gamma \sum_{j \neq i} q_j - \gamma Q^{NM}\right) q_i$$

where q_j is the output of every other merged firm $j \neq i$ and Q^{NM} represents the total output of all unmerged firms.

Differentiating with respect to q_i, and assuming interior solutions, we obtain:

$$1 - 2q_i - 2\gamma \sum_{j \neq i} q_j - \gamma Q^{NM} = 0$$

Since every merged firm is symmetric, in equilibrium, we must have $q^M = q_i = q_j$, yielding the best response function of the merged firm as follows:

$$q^M\left(Q^{NM}\right) = \frac{1}{2\,[1 + \gamma\,(k-1)]} - \frac{\gamma}{2\,[1 + \gamma\,(k-1)]} Q^{NM}$$

When all firms merge, $k = N$, we have $Q^{NM} = 0$ for there are no outsiders, yielding $q^M(0) = \frac{1}{2[1+\gamma(N-1)]}$ as in part (b).

- Every unmerged firm chooses q_i to solve the profit maximization problem:

$$\max_{q_i \geq 0} \pi_i(q_i) = \left(1 - q_i - \gamma \sum_{l \neq i} q_l - \gamma Q^M\right) q_i$$

where q_l is the output of every other unmerged firm $l \neq i$ and Q^M represents the total output of all merged firms.

Differentiating with respect to q_i, and assuming interior solutions, we find:

$$1 - 2q_i - \gamma \sum_{l \neq i} q_l - \gamma Q^M = 0$$

Since every unmerged firm is symmetric, in equilibrium, we must have $q^{NM} = q_i = q_l$, yielding the best response function of every unmerged firm as follows:

$$q^{NM}\left(Q^M\right) = \frac{1}{2 + \gamma(N - k - 1)} - \frac{\gamma}{2 + \gamma(N - k - 1)} Q^M$$

- Since $Q^M = kq^M$ and $Q^{NM} = (N - k)\, q^{NM}$, the best response functions become:

$$q^M\left(q^{NM}\right) = \frac{1}{2[1 + \gamma(k - 1)]} - \frac{\gamma(N - k)}{2[1 + \gamma(k - 1)]} q^{NM}$$

$$q^{NM}\left(q^M\right) = \frac{1}{2 + \gamma(N - k - 1)} - \frac{\gamma k}{2 + \gamma(N - k - 1)} q^M$$

Substituting the first best response function into the second, we obtain

$$q^M = \frac{1}{2[1 + \gamma(k - 1)]} - \frac{\gamma(N - k)}{2[1 + \gamma(k - 1)]} \left[\frac{1}{2 + \gamma(N - k - 1)} - \frac{\gamma k}{2 + \gamma(N - k - 1)} q^M\right]$$

Simplifying, the equilibrium output of every merged firm becomes

$$q^M = \frac{2 - \gamma}{2[2 + \gamma(N + k - 3)] + \gamma^2[(k - 2)(N - k) - 2(k - 1)]}$$

Inserting q^M into $q^{NM}\left(q^M\right)$, the equilibrium output of every unmerged firm is

$$q^{NM} = \frac{2 + \gamma(k - 2)}{2[2 + \gamma(N + k - 3)] + \gamma^2[(k - 2)(N - k) - 2(k - 1)]}$$

(d) Evaluate the equilibrium output in part (c) when $\gamma = 0$ and when $\gamma = 1$. Interpret.

- When $\gamma = 0$, goods are completely differentiated, so $q^M = q^{NM} = 1/2$ as in standard monopolies where demands are independent of each other.
- When $\gamma = 1$, goods are homogeneous, so that every unmerged firm produces $q^{NM} = \frac{1}{N-k+2}$ which is k times the output of every merged firm, $q^M = \frac{1}{k(N-k+2)}$. This happens because the merged firm behaves as one firm producing one homogeneous good competing

with the other $N-k$ unmerged firms, in contrast with a multiproduct firm that has as many product lines as the number of merged firms when goods are nonhomogeneous, $\gamma \in (0, 1)$.

(e) Find the equilibrium price and profits of merged and unmerged firms.

- Substituting q^M and q^{NM} into p^{NM}, every unmerged firm charges a price of

$$p^{NM} = 1 - [1+\gamma(N-k-1)]\,q^{NM} - \gamma k q^M$$
$$= 1 - \frac{[1+\gamma(N-k-1)][2+\gamma(k-2)] - \gamma k(2-\gamma)}{2[2+\gamma(N+k-3)] + \gamma^2[(k-2)(N-k) - 2(k-1)]}$$
$$= \frac{2+\gamma(k-2)}{2[2+\gamma(N+k-3)] + \gamma^2[(k-2)(N-k) - 2(k-1)]}$$

- Substituting p^{NM} and q^{NM} into π^{NM}, every unmerged firm earns a profit of

$$\pi^{NM} = \frac{[2+\gamma(k-2)]^2}{\left[2(2+\gamma(N+k-3)) + \gamma^2((k-2)(N-k) - 2(k-1))\right]^2}$$

- Substituting q^M and q^{NM} into p^M, every merged firm charges a price of

$$p^M = 1 - [1+\gamma(k-1)]\,q^M - \gamma(N-k)\,q^{NM}$$
$$= 1 - \frac{(2-\gamma)[1+\gamma(k-1)] + \gamma(N-k)[2+\gamma(k-2)]}{2[2+\gamma(N+k-3)] + \gamma^2[(k-2)(N-k) - 2(k-1)]}$$
$$= \frac{(2-\gamma)[1+\gamma(k-1)]}{2[2+\gamma(N+k-3)] + \gamma^2[(k-2)(N-k) - 2(k-1)]}$$

- Substituting p^M and q^M into π^M, every merged firm earns a profit of

$$\pi^M = \frac{(2-\gamma)^2[1+\gamma(k-1)]}{\left[2(2+\gamma(N+k-3)) + \gamma^2((k-2)(N-k) - 2(k-1))\right]^2}$$

(f) Evaluate the equilibrium profits in part (e) when $\gamma = 0$ and when $\gamma = 1$:

- When $\gamma = 0$, goods are completely differentiated, so $\pi^M = \pi^{NM} = 1/4$ as in standard monopolies where demands are independent of each other.
- When $\gamma = 1$, goods are homogeneous, so that every unmerged firm earns $\pi^{NM} = \frac{1}{(N-k+2)^2}$ which is k times the profits of every merged firm, $\pi^M = \frac{1}{k(N-k+2)^2}$. Since the merged firm is in every aspect homogeneous to the unmerged firm, this firm evenly splits its profits among k member firms.

(g) Under what condition on k can the partial merger in part (c) be sustained?

- For the partial merger to be sustained, we need the profits of every firm that merges, π^M, to be greater than the profits that it earns if not merged, π^*, that is,

$$\frac{(2-\gamma)^2[1+\gamma(k-1)]}{\left[2(2+\gamma(N+k-3)) + \gamma^2((k-2)(N-k) - 2(k-1))\right]^2} \geq \frac{1}{[2+\gamma(N-1)]^2}$$

(h) Assume that $N = 10$. Find the minimum number of firms that merge, $\underline{k}$, which supports the merger when $\gamma = 0$, $\gamma = 0.25$, $\gamma = 0.5$, $\gamma = 0.75$, and $\gamma = 1$. Interpret.

- Substituting $N = 10$ and different values of γ into the results of part (g) yields:

$$\underline{k}\,(0) = 1$$

$$\underline{k}\,(0.25) = 4.39$$

$$\underline{k}\,(0.5) = 6.3$$

$$\underline{k}\,(0.75) = 7.38$$

$$\underline{k}\,(1) = 8.15$$

When goods are completely differentiated ($\gamma = 0$), firms are monopolies, so a merger between two or more firms does not create any spillover effects on rival firms. However, when goods become more homogeneous (γ increases), firms compete with each other more intensively, so a larger number of firms are required to merge in offsetting the competitive effect of the outsiders. In the limit where goods are homogeneous ($\gamma = 1$), the merger requires the largest number of participating firms when outputs are strongest strategic substitutes.

(i) Evaluate the condition you found in part (g) at $\gamma = 1$ and identify the cutoff on k, $\underline{k}(N)$, for which a merger with $k > \underline{k}$ firms can be sustained. How does cutoff $\underline{k}(N)$ vary with N? What happens to the ratio, $\frac{k}{N}$, when $N \to +\infty$? Interpret.

- When $\gamma = 1$, partial merger among k firms can be sustained when

$$k > \underline{k}(N) \equiv \left(\frac{\sqrt{4N+5}-1}{2}\right)^2 = \frac{2N+3-\sqrt{4N+5}}{2}.$$

This cutoff is monotonically increasing in N because

$$\begin{aligned}\frac{\partial \underline{k}}{\partial N} &= \frac{1}{2}\left[2 - \frac{4}{2\sqrt{4N+5}}\right] \\ &= 1 - \frac{1}{\sqrt{4N+5}}\end{aligned}$$

that is unambiguously positive for all values of $N \geq 2$, so that when the market has more firms (N increases), a larger number of firms are required to merge in order to make the merger profitable.

- When $N \to +\infty$, we find that the minimal market share that makes the merger profitable, $\frac{\underline{k}(N)}{N}$, satisfies:

$$\begin{aligned}\lim_{N\to+\infty} \frac{\underline{k}(N)}{N} &= \lim_{N\to+\infty}\left(1 + \frac{3}{2N} - \sqrt{\frac{4N+5}{N^2}}\right) \\ &= 1.\end{aligned}$$

Therefore, when the market becomes perfectly competitive, all firms must join the merger for it to be profitable. Alternatively, only a full merger can be sustained in equilibrium, as shown in part (b).

- The ratio $\frac{k(N)}{N}$ is referred to as the "80% rule" in Salant et al. (1983), where the merger among k firms can be sustained if a sufficiently large number of firms merge to gain market power and offset the output-increasing effect of the outsiders.

(j) *Numerical Example.* Evaluate the cutoff in part (h) when (i) $N = 11$ and (ii) $N = 29$.

- When $N = 11$, the minimum number of firms that merge becomes:

$$\underline{k} = \left(\frac{\sqrt{4 \times 11 + 5} - 1}{2}\right)^2 = 9$$

- When $N = 29$, the minimum number of firms that merge becomes:

$$\underline{k} = \left(\frac{\sqrt{4 \times 29 + 5} - 1}{2}\right)^2 = 25$$

In both cases, the number of firms that merge satisfies the " 80% rule" since when $N = 11$, the minimal market share becomes $\frac{k}{N} = 9/11 \simeq 81.8\%$ and when $N = 25$, this minimal market share is $\frac{k}{N} = 25/29 \simeq 86.2\%$.

Exercise #7.6: Cournot Mergers with Differentiated Products and Asymmetric Costs, Based on Gelves (2014)^C

7.6 Consider two firms producing differentiated products facing an inverse demand function:

$$p_i(q_i) = 1 - q_i - \gamma q_j,$$

where $\gamma \in [0, 1]$ represents the degree of product differentiation. When $\gamma = 0$, products are completely differentiated, but when $\gamma = 1$, products are completely homogeneous. Assume that firm 1 has a higher marginal production cost than firm 2, that is, $c_1 \geq c_2 \geq 0$.

(a) Find equilibrium output, profits, and social welfare if firms compete à la Cournot.

- Every firm i chooses q_i to solve the profit maximization problem:

$$\max_{q_i \geq 0} \pi_i(q_i) = \left(1 - q_i - \gamma q_j\right) q_i - c_i q_i$$

Differentiating with respect to q_i, and assuming interior solutions

$$1 - 2q_i - \gamma q_j - c_i = 0$$

Solving for q_i, we obtain the best response function of firm i,

$$q_i\left(q_j\right) = \frac{1 - c_i}{2} - \frac{\gamma}{2} q_j$$

which originates at a vertical intercept $\frac{1-c_i}{2}$ and decreases in the output of firm j at a rate of $\frac{\gamma}{2}$. When products become more homogeneous (γ increases), firm i's output decreases at a faster rate in firm j's output.

Substituting firm 2's best response function into firm 1's, we obtain:

$$q_1 = \frac{1-c_1}{2} - \frac{\gamma}{2}\overbrace{\left(\frac{1-c_2}{2} - \frac{\gamma}{2}q_1\right)}^{q_2}$$

which is simplified to yield

$$\left(1 - \frac{\gamma^2}{4}\right) q_1 = \frac{2\left(1-c_1\right) - \gamma\left(1-c_2\right)}{4}$$

Rearranging and solving for q_1, we find:

$$q_1^{NM} = \frac{2-\gamma-2c_1+\gamma c_2}{4-\gamma^2}$$

where the superscript *NM* denotes no merger.

Plugging q_1^{NM} into firm 2's best response function, we find

$$\begin{aligned} q_2^{NM} &= \frac{1-c_2}{2} - \frac{\gamma}{2}\overbrace{\left(\frac{2-\gamma-2c_1+\gamma c_2}{4-\gamma^2}\right)}^{q_1^{NM}} \\ &= \frac{2-\gamma-2c_2+\gamma c_1}{4-\gamma^2} \end{aligned}$$

- Substituting equilibrium output into the firm's profit function, we find:

$$\begin{aligned} \pi_1^{NM} &= \left(1 - q_1^{NM} - \gamma q_2^{NM} - c_1\right) q_1^{NM} \\ &= \left(\frac{2-\gamma-2c_1+\gamma c_2}{4-\gamma^2}\right)^2 \\ &= \left(q_1^{NM}\right)^2 \\ \pi_2^{NM} &= \left(1 - q_2^{NM} - \gamma q_1^{NM} - c_2\right) q_2^{NM} \\ &= \left(\frac{2-\gamma-2c_2+\gamma c_1}{4-\gamma^2}\right)^2 \\ &= \left(q_2^{NM}\right)^2 \end{aligned}$$

- From Belleflamme and Peitz (2015, p.23), consumer surplus is the area under the demand curve and above price, which is consumers' willingness-to-pay for how much they paid for the goods, i.e.,

$$\begin{aligned} CS &= q_1 + q_2 - \frac{q_1^2 + 2\gamma q_1 q_2 + q_2^2}{2} - \left(1 - q_1 - \gamma q_2\right) q_1 - \left(1 - q_2 - \gamma q_1\right) q_2 \\ &= \frac{q_1^2 + 2\gamma q_1 q_2 + q_2^2}{2} \end{aligned}$$

Finally, social welfare is the sum of consumer and producer surplus, given by

$$\begin{aligned} SW^{NM} &= CS + PS \\ &= \frac{\left(q_1^{NM}\right)^2 + 2\gamma q_1^{NM} q_2^{NM} + \left(q_2^{NM}\right)^2}{2} + \left(q_1^{NM}\right)^2 + \left(q_2^{NM}\right)^2 \\ &= \frac{3\left(q_1^{NM}\right)^2 + 2\gamma q_1^{NM} q_2^{NM} + 3\left(q_2^{NM}\right)^2}{2} \end{aligned}$$

(b) Find equilibrium output, profits, and social welfare if firms merge.

- If firms merge, they solve the following joint profit maximization problem:

$$\max_{q_1, q_2 \geq 0} \pi\left(q_1, q_2\right) = \left(1 - q_1 - \gamma q_2 - c_1\right) q_1 + \left(1 - q_2 - \gamma q_1 - c_2\right) q_2$$

Differentiating with respect to q_1, and assuming interior solutions, yields:

$$1 - 2q_1 - 2\gamma q_2 - c_1 = 0$$

which, after solving for q_1, we obtain:

$$q_1 = \frac{1 - c_1}{2} - \gamma q_2.$$

Similarly, differentiating with respect to q_2, we find that

$$1 - 2q_2 - 2\gamma q_1 - c_2 = 0$$

which, after solving for q_2, yields

$$q_2 = \frac{1 - c_2}{2} - \gamma q_1.$$

Inserting this result, $q_2 = \frac{1-c_2}{2} - \gamma q_1$, into $q_1 = \frac{1-c_1}{2} - \gamma q_2$, we obtain

$$q_1 = \frac{1 - c_1}{2} - \gamma \overbrace{\left(\frac{1 - c_2}{2} - \gamma q_1\right)}^{q_2}$$

which simplifies to

$$\left(1 - \gamma^2\right) q_1 = \frac{1 - c_1 - \gamma\left(1 - c_2\right)}{2}$$

Rearranging, and solving for q_1, yields the equilibrium output of firm 1 under the merger:

$$q_1^M = \frac{1 - \gamma - c_1 + \gamma c_2}{2\left(1 - \gamma^2\right)}$$

where the superscript M denotes merger.

Inserting this output level in $q_2 = \frac{1-c_2}{2} - \gamma q_1$, we obtain firm 2's equilibrium output under the merger:

$$q_2^M = \frac{1-c_2}{2} - \gamma \overbrace{\left(\frac{1-\gamma-c_1+\gamma c_2}{2\left(1-\gamma^2\right)}\right)}^{q_1^M}$$
$$= \frac{1-\gamma-c_2+\gamma c_1}{2\left(1-\gamma^2\right)}$$

- Substituting equilibrium output into the merged firm's profit function, we find:

$$\pi_1^M = \left(1 - q_1^M - \gamma q_2^M - c_1\right) q_1^M$$
$$= \frac{1-c_1}{2}\frac{1-\gamma-c_1+\gamma c_2}{2\left(1-\gamma^2\right)}$$
$$\pi_2^M = \left(1 - q_2^M - \gamma q_1^M - c_2\right) q_2^M$$
$$= \frac{1-c_2}{2}\frac{1-\gamma-c_2+\gamma c_1}{2\left(1-\gamma^2\right)}$$

- Finally, substituting q_1^M and q_2^M into the expression for social welfare, we have:

$$CS = \frac{\left(q_1^M\right)^2 + 2\gamma q_1^M q_2^M + \left(q_2^M\right)^2}{2}$$

Thus, social welfare is the sum of consumer and producer surplus, given by

$$SW^M = CS + PS$$
$$= \frac{\left(q_1^M\right)^2 + 2\gamma q_1^M q_2^M + \left(q_2^M\right)^2}{2} + \frac{1-c_1}{2} q_1^M + \frac{1-c_2}{2} q_2^M$$

(c) Assume that costs satisfy $c_1 = c_2 = 0$. How does welfare change with γ? Is the merger welfare improving? Explain.

- *No merger.* When $c_1 = c_2 = 0$, firms are symmetric, so output under no merger becomes

$$q_1^{NM} = q_2^{NM} = \frac{1}{2+\gamma}$$

yielding welfare under no merger of

$$SW^{NM} = \frac{1}{2}\left[3\left(\frac{1}{2+\gamma}\right)^2 + 2\gamma \frac{1}{2+\gamma}\frac{1}{2+\gamma} + 3\left(\frac{1}{2+\gamma}\right)^2\right]$$
$$= \frac{3+\gamma}{(2+\gamma)^2}$$

Differentiating SW^{NM} with respect to γ, we obtain:

$$\frac{\partial SW^{NM}}{\partial \gamma} = \frac{(2+\gamma)^2 - 2(2+\gamma)(3+\gamma)}{(2+\gamma)^4}$$
$$= -\frac{4+\gamma}{(2+\gamma)^3} < 0$$

so that welfare under no merger decreases as goods become more homogeneous.

- *Merger.* Similarly, when $c_1 = c_2 = 0$, output under merger becomes:

$$q_1^M = q_2^M = \frac{1}{2(1+\gamma)}$$

yielding welfare under merger of

$$SW^M = \frac{\left(\frac{1}{2(1+\gamma)}\right)^2 + 2\gamma \frac{1}{2(1+\gamma)} \frac{1}{2(1+\gamma)} + \left(\frac{1}{2(1+\gamma)}\right)^2}{2} + \frac{1}{2}\frac{1}{2(1+\gamma)} + \frac{1}{2}\frac{1}{2(1+\gamma)}$$
$$= \frac{3}{4(1+\gamma)}$$

Differentiating SW^M with respect to γ, we obtain:

$$\frac{\partial SW^M}{\partial \gamma} = -\frac{3}{4(1+\gamma)^2} < 0$$

so that welfare under merger decreases as goods become more homogeneous.

- *Comparison.* It is straightforward to verify that $SW^{NM} > SW^M$ since

$$\frac{3+\gamma}{(2+\gamma)^2} \geq \frac{3}{4(1+\gamma)}$$

simplifies to

$$\gamma(\gamma+4) > 0$$

which holds for all $\gamma \in [0, 1]$, so that mergers between cost-symmetric firms always reduce welfare. Importantly, this occurs regardless of how homogenous or differentiated their products are.

(d) Assume $\gamma = 1$. When will the merger be welfare improving? Explain.

- *No merger.* When $\gamma = 1$, goods are homogeneous, so output under no merger is

$$q_1^{NM} = \frac{1-2c_1+c_2}{3}$$
$$q_2^{NM} = \frac{1-2c_2+c_1}{3}$$

where $q_1^{NM} \geq 0$ if $c_1 \leq \frac{1+c_2}{2}$, and $q_2^{NM} \geq 0$ if $1 + c_1 - 2c_2 > 1 - c_2 > 0$ that is always satisfied. In this context, social welfare under no merger of

$$SW^{NM} = \frac{3\left(\frac{1-2c_1+c_2}{3}\right)^2 + 2\frac{1-2c_1+c_2}{3}\frac{1-2c_2+c_1}{3} + 3\left(\frac{1-2c_2+c_1}{3}\right)^2}{2}$$

$$= \frac{3(1-2c_1+c_2)^2 + 2(1-2c_1+c_2)(1-2c_2+c_1) + 3(1-2c_2+c_1)^2}{18}$$

$$= \frac{11(1-c_1)^2 - 14(1-c_1)(1-c_2) + 11(1-c_2)^2}{18}$$

- *Merger.* The merged firm produces all output at the low-cost firm 2, so that

$$q_1^M = 0$$

$$q_2^M = \frac{1-c_2}{2}$$

yielding social welfare under merger of

$$SW^M = \overbrace{\frac{1}{2}\left(q_2^M\right)^2}^{CS} + \overbrace{\left(1 - q_2^M - c_2\right)q_2^M}^{PS}$$

$$= \frac{1}{2}\left(\frac{1-c_2}{2}\right)^2 + \left(1 - \frac{1-c_2}{2} - c_2\right)\frac{1-c_2}{2}$$

$$= \frac{3(1-c_2)^2}{8}$$

- *Comparison.* Mergers reduce welfare if and only if $SW^{NM} \geq SW^M$, where

$$\frac{11(1-c_1)^2 - 14(1-c_1)(1-c_2) + 11(1-c_2)^2}{18} \geq \frac{3(1-c_2)^2}{8}$$

or, after rearranging,

$$44(1-c_1)^2 - 56(1-c_1)(1-c_2) + 17(1-c_2)^2 \geq 0$$

For example, when $c_2 = 0$, the above inequality simplifies to

$$(22c_1 - 5)(2c_1 - 1) \geq 0$$

Since $q_1^{NM} \geq 0$ requires $c_1 \leq \frac{1}{2}$, we find that mergers reduce (enhance) welfare if $0 \leq c_1 \leq \frac{5}{22}$ ($\frac{5}{22} < c_1 \leq \frac{1}{2}$). Intuitively, when firms are significantly cost-asymmetric, mergers that shut down the less efficient firm and shift all output to the more efficient firm can enhance welfare. In contrast, if firms are relatively symmetric in costs, mergers unambiguously reduce welfare.

Exercise #7.7: Cartels with Synergies^B

7.7 Consider an industry of $n \geq 2$ firms competing à la Cournot. Let us analyze the following sequential-move game where, in the first stage, every firm chooses whether or not to form a cartel; and, in the second stage, firms compete à la Cournot. The cartel coordinates their production decision to maximize their joint profits, while the remaining $n-k$ firms do not coordinate. Firms face an inverse demand curve $p(Q) = 1 - Q$, where $Q \geq 0$ denotes aggregate output. Firms are symmetric in their marginal cost of production c, where $1 > c \geq 0$.

In this setting, we allow for the k firms that form the cartel to benefit from a cost-reduction effect, so their marginal cost of production decreases to $c - x$. In contrast, the firms that did not form the cartel still face a marginal cost c.

(a) As a benchmark, find the profits of every firm i before the cartel.

- If no cartel occurs, every firm solves

$$\max_{q_i \geq 0} \ (1 - q_i - Q_{-i})q_i - cq_i,$$

where Q_{-i} denotes the aggregate output of firm i's rivals. Differentiating with respect to q_i, and solving for q_i, we find firm i's best response function:

$$q_i(Q_{-i}) = \frac{1-c}{2} - \frac{1}{2}Q_{-i}$$

which is negatively sloped in its rivals' aggregate output, Q_{-i}. In a symmetric equilibrium, $q_i = q_j = q$ for every two firms $i \neq j$, which entails $Q_{-i} = (N-1)q$. Therefore, the above best response function simplifies to

$$q = \frac{1-c}{2} - \frac{1}{2}(N-1)q$$

and, solving for q, we obtain the equilibrium output of every firm i when no cartel occurs:

$$q = \frac{1-c}{n+1}.$$

Then, equilibrium profits become:

$$\pi_{NC} = \left(\frac{1-c}{n+1}\right)^2,$$

where the subscript NC indicates "no cartel."

(b) Find the output that the cartel participants and nonparticipants produce.

- *Participants.* After the cartel, we need to solve a Cournot problem where firms face asymmetric costs. The cartel, as an entity, solves

$$\max_{q_P \geq 0} \ (1 - q_P - Q_{-i})q_P - (c - x)q_P,$$

where q_P denotes the production level of the cartel, Q_{-i} represents the aggregate output of all nonparticipants, and subscript P denotes cartel participants.

Differentiating with respect to q_P yields:

$$1 - 2q_P - Q_{-i} - c + x = 0$$

and solving for q_P, we obtain the best response function:

$$q_P(Q_{-i}) = \frac{1 - c + x}{2} - \frac{1}{2}Q_{-i}.$$

- *Nonparticipants.* For the nonparticipants, each firm chooses its individual production q_i to solve

$$\max_{q_i \geq 0} (1 - q_i - q_P - Q_{-i}^{NP})q_i - cq_i,$$

where Q_{-i}^{NP} denotes the aggregate production level of all other $(n - k) - 1$ firms that did not participate in the cartel, where superscript NP denotes nonparticipants.

Differentiating with respect to q_i, we find:

$$1 - 2q_i - q_P - Q_{-i}^{NP} - c = 0$$

and, solving for q_P, we obtain the best response function:

$$q_i(q_P, Q_{-i}^{NP}) = \frac{1 - c}{2} - \frac{1}{2}\left(q_P + Q_{-i}^{NP}\right)$$

At this point, we can invoke symmetry: in a symmetric equilibrium, all nonparticipants produce the same output, $q_i = q_j = q_{NP}$ for all $n - k$ firms. Therefore, Q_{-i}^{NP} can be expressed as $Q_{-i}^{NP} = [(n - k) - 1]\, q_{NP}$.

Inserting this result in the above best response function, we find that

$$q_{NP} = \frac{1 - c}{n - k + 1} - \frac{1}{n - k + 1}q_P$$

In addition, $Q_{-i} = (n - k)q_{NP}$, which inserted in the merged entity's best response function, yields:

$$q_P = \frac{1 - c + x}{2} - \frac{n - k}{2}q_{NP}.$$

- Combining the above two best response functions, we can simultaneously solve for the equilibrium values of q_P and q_{NP} as follows:

$$q_P = \frac{1 - c + (n - k + 1)x}{n - k + 2} \text{ and } q_{NP} = \frac{1 - c - x}{n - k + 2}.$$

(c) For which value of the cost-reduction effect, x, will the nonparticipants in the cartel produce a positive output?

- Setting $q_{NP} \geq 0$, we obtain:

$$\frac{1 - c - x}{n - k + 2} \geq 0$$

rearranging yields

$$\frac{x}{1-c} \leq 1.$$

Intuitively, the cost-reduction effect, x, as a percentage of per-unit margins, $1-c$, must be relatively small. Otherwise, the cost-reduction effect that the cartel experiences forces the exit of nonparticipants. For compactness, let $\theta \equiv \frac{x}{1-c}$, so we can write the above inequality as $\theta \leq 1$.

(d) Find the profits for the participants and nonparticipants in the cartel.

- Inserting the equilibrium values of q_P and q_{NP} into their corresponding profits, we obtain:

$$\begin{aligned}\pi_P &= (1 - q_P - (n-k)q_{NP})q_P - (c-x)q_P \\ &= \left(\frac{1-c+(n-k+1)x}{n-k+2}\right)^2,\end{aligned}$$

and

$$\begin{aligned}\pi_{NP} &= (1 - q_{NP} - q_P - [(n-k)-1]\,q_{NP})q_i - cq_i \\ &= \left(\frac{1-c-x}{n-k+2}\right)^2.\end{aligned}$$

(e) For which values of n and k is a cartel profitable? Interpret.

- A cartel between k out of n firms is profitable if the post-cartel profits exceed the pre-cartel profits (for all firms that form the cartel, as a whole), that is,

$$\pi_P \geq k\pi_{NC},$$

where π_{NC} was found in part (a). Therefore,

$$\left(\frac{1-c+(n-k+1)x}{n-k+2}\right)^2 \geq k\left(\frac{1-c}{n+1}\right)^2$$

which we can start simplifying as

$$\frac{1-c+(n-k+1)x}{n-k+2} \geq \sqrt{k}\frac{1-c}{n+1}$$

or further rearrange as

$$\theta \equiv \frac{x}{1-c} \geq \frac{(n-k+2)\sqrt{k}-(n+1)}{(n-k+1)(n+1)} \equiv \theta_C.$$

Intuitively, for the cartel to be profitable, the cost-reduction effect, as captured by θ in the left-hand side of the above inequality, must be sufficiently strong (higher than θ_C, where subscript C denotes that firms form a cartel).

(f) *No cost-reduction effects.* Evaluate your results in part (e) at $x = 0$. Interpret.

- When the cartel does not produce cost-reduction effects, $x = 0$, the left-hand side of our above inequality (θ) collapses to zero, and then the cartel is only profitable if $0 \geq \theta_C$ which, solving for the number of participants, k, yields

$$k \geq \frac{2n + 3 - \sqrt{4n + 5}}{2} \equiv \widehat{k}$$

as in our analysis of mergers between several firms in Exercise 7.2 (where the merger does not produce synergies).

(g) *Numerical example.* Assuming that $x > 0$, evaluate your equilibrium results in part (e) at $k = 2$. How does an increase in the number of firms n affect θ_C? Interpret.

- If the cost-reduction effect is positive, $x > 0$, and only two firms form a cartel, $k = 2$, condition $\theta \geq \theta_C$ becomes:

$$\theta \geq \frac{\left(\sqrt{2} - 1\right) n - 1}{n^2 - 1} \equiv \theta_C\left(2, n\right).$$

Differentiating cutoff $\theta_C(2, n)$ with respect to n, we find:

$$\frac{d\theta_C\left(2, n\right)}{dn} = \frac{\left(\sqrt{2} - 1\right)\left(n^2 - 1\right) - 2n^2\left(\sqrt{2} - 1\right) + 2n}{\left(n^2 - 1\right)^2}$$

$$= \frac{2n - \left(\sqrt{2} - 1\right)\left(1 + n^2\right)}{\left(n^2 - 1\right)^2}$$

which derivative is positive if

$$\left(\sqrt{2} - 1\right) n^2 - 2n + \left(\sqrt{2} - 1\right) > 0$$

that is equivalent to

$$n < \frac{1 + \sqrt{2\left(\sqrt{2} - 1\right)}}{\sqrt{2} - 1} \simeq 4.61$$

Cutoff $\theta_C(2, n)$ lies below 0.045 for most values of n.

- Intuitively, when the merger of two firms occurs in a less competitive market (small n), the merging firms need more substantial cost-reduction effect (higher θ_C) in order to offset the relatively significant output increase of outsiders. However, as the market becomes more competitive (large n), output increase of each outsider due to the merger becomes less significant, so that the two-firm merger is still profitable even under mild cost-reduction effect (lower θ_C).
- Figure 7.1 depicts cutoff $\theta_C(2, n)$ as a function of the number of firms, n, showing that it first increases and then decreases in n. For instance, when $n = 4$, cutoff $\theta_C(2, 4) = 0.043$, indicating that, for the cartel to be profitable, the cost-reduction effect (relative to the market size) must be larger than 4.3%.

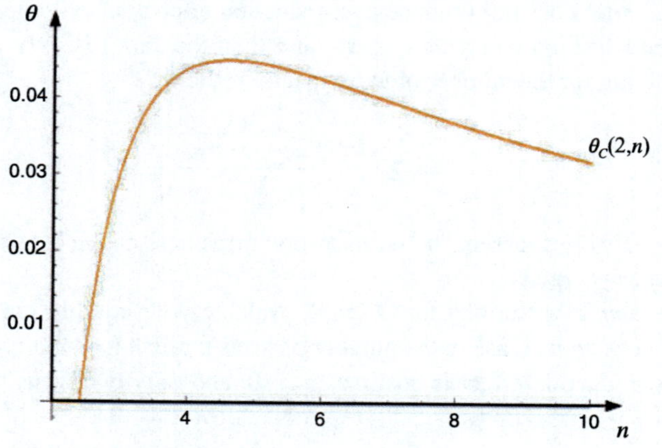

Fig. 7.1 Cutoff $\theta_C\,(2, n)$ as a function of n

Exercise #7.8: Merger to Gain Economies of Scope^B

7.8 Consider two firms selling differentiated goods and competing à la Cournot. The inverse demand function of every firm i is

$$p\,(q_i) = a - bq_i - dq_j,$$

where $i \in \{1, 2\}$, $j \neq i$, and $b > d \geq 0$ indicates that own-price effects dominate cross-price effects. When $d = 0$, products are completely differentiated but when $d \to b$, products are completely homogeneous. The total cost function of every unmerged firm i is

$$C\,(q_i) = \frac{c}{2}q_i^2$$

When firms merge and coordinate output, total cost of the merged firm becomes:

$$C\,(q_1, q_2) = \frac{c}{2}\left(q_1^2 + q_2^2\right) - \beta q_1 q_2$$

where $a > c \geq \beta$. When $\beta > 0$, the merged firm enjoys cost efficiencies in jointly producing the goods, which is commonly known as "economies of scope." However, when $\beta < 0$, the merged firm suffers cost inefficiencies, which is referred to as "diseconomies of scope." Finally, when $\beta = 0$, the merged firm does not enjoy any cost complementarities.

(a) *Before merger.* Set up firm i's profit maximization problem and solve for equilibrium output and profits. How do they change in parameters a, b, c, and d? Explain.

- Every firm i chooses q_i to solve the following profit maximization problem:

$$\max_{q_i \geq 0} \pi\,(q_i) = \left(a - bq_i - dq_j\right) q_i - C\,(q_i)$$

Differentiating with respect to q_i, and assuming interior solutions, we obtain

$$a - 2bq_i - dq_j = cq_i$$

Rearranging and solving for q_i, the best response function of firm i becomes:

$$q_i(q_j) = \frac{a}{2b+c} - \frac{d}{2b+c}q_j$$

which originates at an intercept of $\frac{a}{2b+c}$ and decreases in the output of firm j at a rate of $\frac{d}{2b+c}$.

In a symmetric equilibrium, both firms produce the same output level, $q_i = q_j$. Inserting this property in the above best response function yields:

$$q_i = \frac{a}{2b+c} - \frac{d}{2b+c}q_i$$

Rearranging and solving for q, equilibrium output becomes:

$$q_i^* = \frac{a}{2b+c+d}$$

which is increasing in market size a as every firm i produces more units when facing a larger demand. Equilibrium output, however, is decreasing in own-price effect b when consumers become more price-sensitive, and in cross-price effect d when goods are more homogeneous to each other. Finally, equilibrium output decreases in c as every unit becomes more costly to produce.

- Substituting $q_i^* = \frac{a}{2b+c+d}$ into the unmerged firm's profit function yields:

$$\pi_i^* = \left[a - (b+d)\frac{a}{2b+c+d}\right]\frac{a}{2b+c+d} - \frac{c}{2}\left(\frac{a}{2b+c+d}\right)^2$$
$$= \frac{a^2(2b+c)}{2(2b+c+d)^2}$$

which is increasing in a as every firm i generates more profits in a larger market, but decreasing in cross-price effect d as competition becomes more intense when goods are more homogeneous. In addition, we obtain that

$$\frac{\partial \pi_i^*}{\partial b} = -\frac{a^2(2b+c-d)}{(2b+c+d)^3}$$

which is negative since $b > d$ by assumption, indicating that when own-price effect becomes stronger (b increases), firm i earns less profits in equilibrium. Finally, differentiating the equilibrium profits with respect to c yields:

$$\frac{\partial \pi_i^*}{\partial c} = -\frac{a^2(2b+c-d)}{2(2b+c+d)^3}$$

which is negative since $b > d$ by definition, so that when it becomes more costly to produce every unit of the good, firm i earns less profits in equilibrium.

(b) *After merger.* Find equilibrium output and profits when firms merge into a monopoly. Compared to $\beta = 0$, how are your results affected when $\beta < 0$ and when $\beta > 0$?

- The merged firm chooses q_1 and q_2 to solve the joint profit maximization problem:

$$\max_{q_1, q_2 \geq 0} \pi(q_1, q_2) = (a - bq_1 - dq_2) q_1 + (a - bq_2 - dq_1) q_2 - C(q_1, q_2)$$

Differentiating with respect to q_1, and assuming interior solutions, we find that

$$a - 2bq_1 - 2dq_2 - cq_1 + \beta q_2 = 0$$

Differentiating with respect to q_2, and assuming interior solutions, we obtain

$$a - 2bq_2 - 2dq_1 - cq_2 + \beta q_1 = 0$$

Invoking symmetry, we obtain that $q_1 = q_2 = q^C$, yielding

$$a - 2bq - 2dq - cq + \beta q = 0$$

and, solving for q, we find that

$$q^C = \frac{a}{2(b+d) + c - \beta}$$

Inserting $q^C = \frac{a}{2(b+d)+c-\beta}$ into the merged firm's profit function yields:

$$\begin{aligned}
\pi^C &= \left(a - bq^C - dq^C\right) q^C + \left(a - bq^C - dq^C\right) q^C \\
&\quad - \left[\frac{c}{2}\left(\left(q^C\right)^2 + \left(q^C\right)^2\right) - \beta q^C q^C\right] \\
&= 2\left[a - (b+d) q^C\right] q^C - (c - \beta)\left(q^C\right)^2 \\
&= \frac{a^2}{2(b+d) + c - \beta}
\end{aligned}$$

- *Comparison.*
  - When $\beta = 0$, equilibrium output becomes:

$$q^C = \frac{a}{2(b+d) + c}$$

which coincides with that in mergers of differentiated duopolists with symmetric costs. In this context, equilibrium profits become $\pi^C = \frac{a^2}{2(b+d)+c}$.
  - When β increases, we find that equilibrium output q^C increases since

$$\frac{\partial q^C}{\partial \beta} = \frac{a}{[2(b+d) + c - \beta]^2}$$

is unambiguously positive. Therefore, when $\beta > 0$, the merged firm enjoys economies of scope, so that it produces more units of output and earns higher profits. In contrast, when $\beta < 0$, the merged firm suffers diseconomies of scope, produces fewer units of output, and earns lower profits.

(c) *Merger incentives.* Do firms have incentives to merge into a monopoly? Explain.

- Firms have incentives to merge if and only if their joint profits exceed the sum of unmerged firms' profits, that is,

$$\pi^C \geq \pi_i^* + \pi_j^*$$

where, given cost symmetry, $\pi_i^* + \pi_j^* = 2\pi_i^*$. Substituting the results from parts (a) and (b), we obtain:

$$\frac{a^2}{2(b+d)+c-\beta} \geq \frac{a^2(2b+c)}{(2b+c+d)^2}$$

which is rearranged to yield

$$(2b+c)^2 + 2d(2b+c) + d^2 \geq (2b+c)^2 + (2b+c)(2d-\beta)$$

Simplifying, we obtain:

$$\beta(2b+c) + d^2 \geq 0.$$

- To interpret this result, it is convenient to separately consider economies and diseconomies of score:
  - When firms enjoy economies of scope ($\beta > 0$), the above inequality unambiguously holds since $b, c, d > 0$ by assumption.
  - However, when firms suffer diseconomies of scope ($\beta < 0$), firms have incentives to merge only if these diseconomies are not too significant, namely,

$$\beta \geq -\frac{d^2}{2b+c}$$

  Informally, this condition means that parameter β cannot be too negative. Otherwise, remaining unmerged would a more profitable option, as firms would avoid the significant cost inefficiencies that arise from the merger. For instance, if $b = 1$, $d = c = \frac{1}{2}$, this condition becomes $\beta > -\frac{\left(\frac{1}{2}\right)^2}{2+\frac{1}{2}} = -\frac{1}{10}$.

- Finally, the ratio $-\frac{d^2}{2b+c}$ decreases (becoming more negative) when d increases and when b and c decrease. Graphically, the region of β that satisfies the condition $\beta > -\frac{d^2}{2b+c}$ expands, making firms more likely to merge: (i) when their products become more homogeneous (higher d); (ii) when sales become less price sensitive (lower b); and (iii) when their original costs decrease (lower c).

Exercise #7.9: Merger to Gain Stackelberg Leadership^B

7.9 Consider a market with N symmetric firms competing à la Cournot. Every firm i, where $i \in \{1, \ldots, N\}$, produces a homogeneous good at a constant marginal cost c. The inverse demand function is $p(Q) = a - bQ$, where $Q = \sum_{i=1}^{N} q_i$ represents aggregate output, $a > c > 0$, and $b > 0$.

(a) *Cournot competition.* Set up firm i's profit maximization problem and find equilibrium output and profits.

- Every firm i chooses q_i to solve the following profit maximization problem:

$$\max_{q_i \geq 0} \pi_i(q_i) = (a - bq_i - bq_{-i}) q_i - cq_i$$

where $q_{-i} = \sum_{j \neq i} q_j$ represents the total output of all other $N - 1$ firms.
Differentiating with respect to q_i, and assuming interior solutions, we obtain:

$$a - 2bq_i - bq_{-i} - c = 0$$

Solving for q_i, we obtain the best response function of firm i:,

$$q_i(q_{-i}) = \frac{a-c}{2b} - \frac{1}{2} q_{-i}$$

which originates at $\frac{a-c}{2b}$ and is decreasing in firm i's rivals output at a rate of $1/2$.
Since firms are symmetric, we have $q_i = q_j$ for all $j \neq i$, so that

$$q_i = \frac{a-c}{2b} - \frac{N-1}{2} q_i$$

Rearranging, the equilibrium output of firm i becomes:

$$q_i^* = \frac{a-c}{b(N+1)}$$

which is decreasing in the number of firms N.
Substituting $q_i^* = \frac{a-c}{b(N+1)}$ into firm i's profit function yields equilibrium profits:

$$\begin{aligned}
\pi_i^* &= \left(a - Nbq_i^* - c\right) q_i^* \\
&= \left(a - \frac{N(a-c)}{N+1} - c\right) \frac{a-c}{b(N+1)} \\
&= \frac{1}{b} \left(\frac{a-c}{N+1}\right)^2
\end{aligned}$$

which is increasing in the market size a but decreasing in the slope of the inverse demand curve b, marginal cost c, and number of firms N.

(b) *Stackelberg merger.* Assume that two of N firms merge and they obtain Stackelberg leadership. In the first stage, the merged firm chooses output q^l. In the second stage, every firm i that is not merged observes q^l and responds choosing its output q^f simultaneously. Use backward induction to find the equilibrium output and profits of the leader and the follower.

- *Second stage.* At this stage, we have $N - 2$ symmetric followers, each firm independently choosing q_i^f to solve the following profit maximization problem:

$$\max_{q_i^f \geq 0} \pi_i \left(q_i^f\right) = \left(a - bq^l - bq_{-i}^f - bq_i^f\right) q_i^f - cq_i^f$$

where $q_{-i}^f = \sum_{j \neq i} q_j^f$ represents the total output of all other $N - 3$ followers.
Differentiating with respect to q_i^f, and assuming interior solutions, we obtain:

$$a - bq^l - bq_{-i}^f - 2bq_i^f - c = 0$$

Solving for q_i^f, we obtain firm i's best response function,

$$q_i^f \left(q_i^f, q^l\right) = \frac{a-c}{2b} - \frac{1}{2}q^l - \frac{1}{2}q_{-i}^f$$

Since followers are symmetric, we have $q_i^f = q_j^f$ for all followers, so that

$$q_i^f = \frac{a-c}{2b} - \frac{1}{2}q^l - \frac{N-3}{2}q_i^f$$

Rearranging, every follower's best response function becomes:

$$q^f \left(q^l\right) = \frac{a-c}{b\,(N-1)} - \frac{1}{N-1}q^l.$$

Intuitively, when the leader increases output by one unit, every follower reduces output by $1/(N-1)$ units.

- *First stage.* The leader chooses q^l to solve the following problem to maximize merger profits:

$$\max_{q^l \geq 0} \pi^l \left(q^l\right) = \left[a - bq^l - (N-2)\,bq^f\left(q^l\right)\right] q^l - cq^l$$

Since the leader anticipates how the followers respond to its output, we can substitute the follower's best response function into the leader's profit function, yielding

$$\max_{q^l \geq 0} \pi \left(q^l\right) = \left[a - bq^l - (N-2)\,b\left(\frac{a-c}{b\,(N-1)} - \frac{1}{N-1}q^l\right) - c\right] q^l$$
$$= \frac{\left(a - c - bq^l\right) q^l}{N-1}.$$

Differentiating with respect to q^l, and assuming interior solutions, we obtain

$$\frac{a - c - 2bq^l}{N-1} = 0$$

Rearranging and solving for q^l, the leader's equilibrium output becomes

$$q^l = \frac{a-c}{2b}$$

which coincides with the output level that a monopolist would choose in this setting, and every firm that merges produces half of that output, i.e., $q_i^l = \frac{a-c}{4b}$.

Substituting the leader's equilibrium output, $q^l = \frac{a-c}{2b}$, into the follower's best response function, we find that each follower's equilibrium output is

$$\begin{aligned} q^f &= \frac{a-c}{b(N-1)} - \frac{1}{N-1}\frac{a-c}{2b} \\ &= \frac{a-c}{2b(N-1)} \end{aligned}$$

- In this context, the leader's equilibrium profits become

$$\begin{aligned} \pi^l &= \frac{\left(a-c-b\overbrace{\frac{a-c}{2b}}^{q^l}\right)\overbrace{\frac{a-c}{2b}}^{q^l}}{N-1} \\ &= \frac{(a-c)^2}{4b(N-1)} \end{aligned}$$

Similarly, every follower earns equilibrium profits, as follows:

$$\begin{aligned} \pi^f &= \left[a - b\overbrace{\frac{a-c}{2b}}^{q^l} - (N-2)\,b\overbrace{\frac{a-c}{2b(N-1)}}^{q^f} - c\right]\overbrace{\frac{a-c}{2b(N-1)}}^{q^f} \\ &= \frac{(a-c)^2}{4b(N-1)^2} \end{aligned}$$

(c) *Profit comparison.* Does the leader earn more profits than the follower? Explain.
- Comparing profits of the leader and the follower, we find that $\pi^l > \pi^f$, since

$$\frac{(a-c)^2}{4b(N-1)} > \frac{(a-c)^2}{4b(N-1)^2}$$

which simplifies to

$$1 > \frac{1}{N-1}$$

which holds for all $N > 2$, implying that the leader earns more profits than the follower. Intuitively, since the leader chooses output before the followers, the leader can set an output as if it was a monopolist, leaving the residual demand for the followers (first-mover advantage).

(d) *Merger incentives.* Do firms have incentives to merge and become a leader?

- Two firms have incentives to merge and become a leader if

$$\pi^l \geq 2\pi_i^*$$

which says that the merged firm earns more profits by becoming a leader than two unmerged firms competing simultaneously with other firms in quantities.

From parts (a) and (b), we have that

$$\pi_i^* = \frac{1}{b}\left(\frac{a-c}{N+1}\right)^2 \text{ and } \pi^l = \frac{(a-c)^2}{4b\,(N-1)}.$$

Substituting into the left-hand side of the above inequality yields:

$$\frac{(a-c)^2}{4b\,(N-1)} \geq \frac{2}{b}\left(\frac{a-c}{N+1}\right)^2$$

which simplifies to

$$\frac{(a-c)^2}{4b}\left[\frac{1}{N-1} - \frac{8}{(N+1)^2}\right] \geq 0$$

$$\frac{(a-c)^2\,(N-3)^2}{4b\,(N-1)\,(N+1)^2} \geq 0$$

so that any two firms have incentives to merge and become a leader.

Exercise #7.10: Merger Between a Leader and Follower, Based on Huck et al. (2001)^C

7.10 Consider the Stackelberg game between m leaders and $n-m$ followers in Exercise 4.7. Recall that the market has n symmetric firms with inverse demand function:

$$P(Q) = 1 - Q,$$

where $Q = \sum_{i=1}^{N} q_i$ represents the total output and q_i is firm i's individual output. For simplicity, assume zero production costs. We now seek to examine incentives to merge between only two firms.

(a) Do two firms have incentives to merge if both firms are leaders, and the merged firm remains as a leader?

- When two leaders merge, there are $n-1$ firms in the market; within which, $m-1$ firms are leaders and $n-m$ firms are followers, yielding

$$\pi^l\,(n-1, m-1) = \frac{1}{m^2\,(n-m+1)}$$

Therefore, two leaders have incentives to merge if

$$\pi^{l}(n-1, m-1) - 2\pi^{l}(n, m) > 0$$

$$\frac{1}{m^2(n-m+1)} - \frac{2}{(m+1)^2(n-m+1)} > 0$$

which simplifies to

$$\frac{m^2+2m+1-2m^2}{m^2(m+1)^2(n-m+1)} > 0$$

$$-\frac{m^2-2m-1}{m^2(m+1)^2(n-m+1)} > 0$$

which holds if

$$m^2 - 2m - 1 < 0$$

or, after solving for m in this quadratic equation,

$$\frac{2-\sqrt{4+4}}{2} < m < \frac{2+\sqrt{4+4}}{2}$$

$$\Longrightarrow 1-\sqrt{2} < m < 1+\sqrt{2}$$

which is satisfied only if $m = 2$, that is, a merger between two leaders is profitable if there are only two leaders in the market.

(b) Do two firms have incentives to merge if both firms are followers, and the merged firm remains as a follower?

- When two followers merge, there are $n-1$ firms in the market; within which, m firms are leaders and $n-m-1$ firms are followers, yielding

$$\pi^{f}(n-1, m) = \frac{1}{(m+1)^2(n-m)^2}$$

Therefore, two followers have incentives to merge if

$$\pi^{f}(n-1, m) - 2\pi^{f}(n, m) > 0$$

$$\frac{1}{(m+1)^2(n-m)^2} - \frac{2}{(m+1)^2(n-m+1)^2} > 0$$

which simplifies to

$$\frac{(n-m+1)^2 - 2(n-m)^2}{(m+1)^2(n-m)^2(n-m+1)^2} > 0$$

$$-\frac{(n-m)^2 - 2(n-m) - 1}{(m+1)^2(n-m)^2(n-m+1)^2} > 0$$

which holds if

$$(n-m)^2 - 2(n-m) - 1 < 0$$

or, after solving for m in this quadratic equation,

$$1-\sqrt{2} < n-m < 1+\sqrt{2}$$

which is satisfied only if $n - m = 2$, that is, a merger between two followers is profitable if there are only two followers in the market.

(c) Still in the context of a merger between two firms, do they have incentives to merge if one firm is the leader and the other firm is the follower, and the merged firm becomes a leader?

- When a leader merges with a follower to become a leader, there are $n - 1$ firms in the market; within which, m firms are leaders and $n - m - 1$ firms are followers, yielding

$$\pi^l(n-1, m) = \frac{1}{(m+1)^2(n-m)}$$

Therefore, a leader and a follower have incentives to merge if

$$\pi^l(n-1, m) - \pi^l(n, m) - \pi^f(n, m) > 0$$

$$\frac{1}{(m+1)^2(n-m)} - \frac{1}{(m+1)^2(n-m+1)} - \frac{1}{(m+1)^2(n-m+1)^2} > 0$$

which simplifies to

$$\frac{(n-m+1)^2 - (n-m)(n-m+1) - (n-m)}{(m+1)^2(n-m)(n-m+1)^2} > 0$$

$$\frac{1}{(m+1)^2(n-m)(n-m+1)^2} > 0$$

which always holds for any m and n, such that a leader and a follower always find it profitable to merge into a leader to take advantage of output leadership.

(d) When a leader merges with a follower to become a leader, does the profit gain increase or decrease in the number of firms in the industry n (and you can hold m constant)? Show that when n becomes infinite, the profit gain from the merger approaches zero.

- Differentiating the profit gain in part (c) with respect to n, we obtain

$$\frac{\partial}{\partial n}\left[\pi^l(n-1, m) - \pi^l(n, m) - \pi^f(n, m)\right]$$

$$= -\frac{n-m+1+2(n-m)}{(m+1)^2(n-m)^2(n-m+1)^3}$$

$$= -\frac{3(n-m)+1}{(m+1)^2(n-m)^2(n-m+1)^3} < 0$$

Since $n > m$, when the number of firms in the industry grows, and we control for the number of leaders m, the number of followers $n - m$ must increase. In this setting, the profit gain from the merger decreases, meaning that a more competitive market for the followers reduces the leader's incentives to merge with a follower.

- *Perfectly competitive market.* Furthermore, the profit gain from the merger converges to zero when the number of firms in the industry grows to infinity, since

$$\lim_{n\to\infty} \frac{1}{(m+1)^2 (n-m)(n-m+1)^2} = 0$$

Intuitively, when the market becomes perfectly competitive that leaves both leaders and followers with zero profit, the merger between a leader and a follower does not generate enough market power to yield positive profit.

Exercise #7.11: The "Tragedy of the Anticommons," Heller and Eisenberg (1998)^B

7.11 Consider a setting with two firms, A and B, selling complementary goods. Every firm i faces a linear inverse demand function:

$$p_i(q_i, q_j) = a - bq_i + dq_j,$$

where $b > d \geq 0$. Intuitively, parameter $b > 0$ indicates that firm i faces a downward sloping demand curve for its product. When $d = 0$, the products of firm i and j are regarded as independent by consumers, but when $d > 0$, customers see the goods as complementary, so more sales by firm j increase the demand for firm i's product. In addition, we assume that own-price effects dominate cross-price effects, that is, $b > d$. For simplicity, assume that firms face a symmetric marginal cost of production c, where $a > c \geq 0$.

(a) *Nash equilibrium.* Assuming that every firm i simultaneously and independently chooses its output q_i, find its best response function, the equilibrium output, price, and profits.

- Firm i maximizes its profit:

$$\max_{q_i \geq 0} \ (a - bq_i + dq_j)q_i - cq_i$$

Differentiating with respect to q_i, we find:

$$a - 2bq_i + dq_j - c = 0$$

Solving for q_i, we find the best response function of firm i:

$$q_i(q_j) = \frac{a-c}{2b} + \frac{d}{2b}q_j \qquad (\text{BRF}_i)$$

which increases in the output of firm j when products are complements in consumption ($d > 0$) but reduces to the monopoly output $q_i = \frac{a-c}{2b}$ when consumers see the goods as independent ($d = 0$). By symmetry, best response function for firm j is

$$q_j(q_i) = \frac{a-c}{2b} + \frac{d}{2b}q_i \qquad (\text{BRF}_j)$$

In a symmetric equilibrium, $q_i^* = q_j = q^*$, which implies

$$q^* = \frac{a-c}{2b} + \frac{d}{2b}q^*.$$

Rearranging and solving for output q^*, we obtain an equilibrium output:

$$q^* = \frac{a-c}{2b-d}$$

- In this context, equilibrium prices are

$$\begin{aligned} p^* &= a - b\frac{a-c}{2b-d} + d\frac{a-c}{2b-d} \\ &= \frac{ab + (b-d)c}{2b-d}. \end{aligned}$$

and equilibrium profits of every firm i are

$$\begin{aligned} \pi_i^* &= \frac{ab + (b-d)c}{2b-d}\frac{a-c}{2b-d} - c\frac{a-c}{2b-d} \\ &= \frac{b(a-c)^2}{(2b-d)^2}. \end{aligned}$$

(b) *Coordinating output decisions in a cartel.* Assume now that both firms coordinate their production decisions to maximize their joint profits. Find their equilibrium output, price, and profits.

- The joint profit maximization problem of the cartel is

$$\max_{q_1, q_2 \geq 0} \ [(a - bq_1 + dq_2)q_1 - cq_1] + [(a - bq_2 + dq_1)q_2 - cq_2]$$

Differentiating with respect to q_1, we obtain:

$$a - 2bq_1 + dq_2 - c + dq_2 = 0$$

Solving for q_1 yields:

$$q_1 = \frac{a-c}{2b} + \frac{d}{b}q_2$$

This function informs us that, for an increase in output level q_2, firm 1 responds increasing its output more significantly in the cartel than when every firm independently chooses its output level in part (a), where firm 1's best response function was $q_i(q_j) = \frac{a-c}{2b} + \frac{d}{2b}q_j$.

- Differentiating with respect to q_2 and solving for q_2, we find

$$q_2 = \frac{a-c}{2b} + \frac{d}{b}q_1$$

In symmetric equilibrium, $q_1^{Cartel} = q_2^{Cartel} = q^{Cartel}$, which implies

$$q^{Cartel} = \frac{a-c}{2b} + \frac{d}{b}q^{Cartel}.$$

Rearranging and solving for q^{Cartel}, we obtain

$$q^{Cartel} = \frac{a-c}{2(b-d)}$$

which is larger than the equilibrium output we found in part (a), $q^* = \frac{a-c}{2b-d}$, since the difference

$$\begin{aligned} q^{Cartel} - q^* &= \frac{a-c}{2(b-d)} - \frac{a-c}{2b-d} \\ &= \frac{(a-c)d}{2(b-d)(2b-d)} \end{aligned}$$

which is positive since $b > d$. Therefore,

$$q^{Cartel} > q^*,$$

implying that the cartel produces more units than the equilibrium level in part (a) since the cartel internalizes the complementarity in consumption, which firms ignored when independently choosing their individual output. That explains why this type of markets are often referred to as the "anticommons" because, unlike common pool resources where cartels help reduce individual appropriation of the resource, cartels in this context help firms to increase the production of complementary goods.

- The prices in the cartel are

$$\begin{aligned} p^{Cartel} &= a - b\frac{(a-c)}{2(b-d)} + d\frac{(a-c)}{2(b-d)} \\ &= \frac{a+c}{2}. \end{aligned}$$

Therefore, equilibrium profit of every firm i in the cartel is

$$\begin{aligned} \pi_i^{Cartel} &= \frac{a+c}{2}\frac{(a-c)}{2(b-d)} - c\frac{(a-c)}{2(b-d)} \\ &= \frac{(a-c)^2}{4(b-d)}. \end{aligned}$$

(c) Compare equilibrium profits in the cartel against those in the Nash equilibrium of part (a).
- Comparing profits, we find that

$$\begin{aligned} \pi_i^{Cartel} - \pi_i^* &= \frac{(a-c)^2}{4(b-d)} - \frac{b(a-c)^2}{(2b-d)^2} \\ &= \frac{(a-c)^2 d^2}{4(b-d)(2b-d)^2} \end{aligned}$$

which is unambiguously positive, entailing that $\pi_i^{Cartel} > \pi_i^*$, and every firm i finds it profitable to form a cartel in coordinating the sales of complementary goods.

(d) *Numerical example.* Evaluate equilibrium profits in the cartel and in the Nash equilibrium of the game at $a = b = 1$ and $c = 1/2$. Then, evaluate them at $d = 0$ and at $d = 1/2$. Interpret your results.

- Evaluating profits at $a = b = 1$ and $c = 1/2$, we obtain

$$\pi_i^{Cartel} = \frac{(1 - \frac{1}{2})^2}{4(1 - d)} = \frac{1}{16\,(1 - d)}$$

and

$$\pi_i^* = \frac{(1 - \frac{1}{2})^2}{(2 - d)^2} = \frac{1}{4\,(2 - d)^2}.$$

Therefore, when products are regarded as independent by consumers, $d = 0$, these profits become $\pi_i^{Cartel} = \pi_i^* = \frac{1}{16}$, whereas when consumers see products as complementary, $d = 1/2$, these profits are $\pi_i^{Cartel} = \frac{1}{8} > \frac{1}{9} = \pi_i^*$. More generally, we can show that the profit gain of forming a cartel in this context:

$$\pi_i^{Cartel} - \pi_i^* = \frac{\left(1 - \frac{1}{2}\right)^2 d^2}{4\,(1 - d)\,(2 - d)^2} = \frac{d^2}{16\,(1 - d)\,(2 - d)^2}$$

is increasing in the complementary between the two products since

$$\frac{d\left(\pi_i^{Cartel} - \pi_i^*\right)}{dd} = \frac{d\left(4 - 2d - d^2\right)}{16\,(1 - d)^2\,(2 - d)^3} > 0$$

given that $b = 1 > d$ by assumption. Intuitively, as goods become more complementary, independent sales generate a larger complementarity in consumption, which firms could internalize if they form a cartel.

Exercise #7.12: Mergers in Polluting Markets, Based on Fikru and Gautier (2016)^C

7.12 Consider an industry with two firms competing in quantities. Every firm i faces inverse demand curve:

$$p_i\left(q_i, q_j\right) = 1 - q_i - \gamma q_j,$$

where $\gamma \in [0, 1]$ represents the degree of product differentiation, i.e., when $\gamma = 0$ ($\gamma = 1$) products are completely differentiated (homogeneous). The marginal cost of production is c, where $1 > c \geq 0$. Output generates environmental damage dQ^2, which is increasing and convex in aggregate output $Q \equiv \sum_{i=1}^2 q_i$, and $d \geq \frac{1}{2}$ indicates the intensity of polluting emissions. Firms interact in the following sequential-move game:

(a) In the first stage, every firm i chooses whether to merge with firm j or not. If one firm opposes, the merger does not ensue. If both firms approve, then the merger occurs.
(b) In the second stage, the regulator sets an emission fee per unit of output, t^k, where superscript $k = \{M, NM\}$ denotes merger (no merger, respectively), to maximize welfare $W = CS + PS + tax - Env$, where *CS (PS)* denotes consumer (producer) surplus, *tax* represents tax revenue, and $Env = d\,(q_1 + q_2)^2$ denotes total environmental damages.
(c) In the third stage, if the merger does not ensue, every firm i independently and simultaneously chooses its output level q_i^{NM} and engages in Cournot competition. If the merger occurs, then firms coordinate their output levels q_i^M to maximize their joint profits.

We solve the game by backward induction, starting in the last stage.

(a) *Third stage.* Find equilibrium output and associated profit levels when firms do not merge, $\left(q_1^{NM}, q_2^{NM}\right)$, and equilibrium output levels when firms merge, $\left(q_1^{M}, q_2^{M}\right)$.

- *No merger.* For $i = \{1, 2\}$, every firm i chooses q_i to solve

$$\max_{q_i \geq 0} \pi\,(q_i) = \left(1 - q_i - \gamma q_j\right) q_i - \left(c + t\right) q_i$$

which is evaluated at any given emission fee t that the regulator sets in the second stage. Differentiating with respect to q_i, and assuming interior solutions, we obtain:

$$1 - 2q_i - \gamma q_j - c - t = 0$$

which yields firm i's best response function

$$q_i\left(q_j\right) = \frac{1 - c - t}{2} - \frac{\gamma}{2} q_j$$

In a symmetric equilibrium output, $q^{NM} = q_1^{NM} = q_2^{NM}$, entailing that the above best response function simplifies to

$$q^{NM} = \frac{1 - c - t}{2} - \frac{\gamma}{2} q^{NM}$$

Solving for q, we find the equilibrium output under no merger as follows:

$$q^{NM}\,(t) = \frac{1 - c - t}{2 + \gamma}$$

which is decreasing in the firm's marginal cost c and in the emission fee t that the regulator charges. When γ decreases, products become more differentiated so that the firm's output is less affected by its competitor and its equilibrium output increases.

Substituting equilibrium output, $q^{NM}\,(t)$, into firm i's profit function, we find:

$$\pi_i\left(t^{NM}\right) = \left(1 - (1 + \gamma)\, q^{NM}\right) q^{NM} - \left(c + t\right) q^{NM}$$
$$= \left(\frac{1 - c - t}{2 + \gamma}\right)^2$$

- *Merger.* The merged firm chooses q_1 and q_2 to maximize its joint profits:

$$\max_{q_1,q_2\geq 0} \pi(q_1,q_2) = \overbrace{(1-q_1-\gamma q_2)\,q_1-(c+t)q_1}^{\pi_1}$$
$$+\underbrace{(1-q_2-\gamma q_1)\,q_2-(c+t)q_2}_{\pi_2}$$

which is also evaluated at any given emission fee t that the regulator sets in the second stage. Differentiating with respect to q_1 and q_2, and assuming interior solutions, we obtain:

$$1-2q_1-2\gamma q_2-c-t=0$$
$$1-2q_2-2\gamma q_1-c-t=0$$

Invoking symmetry, $q^M = q_1^M = q_2^M$, we find that the equilibrium output under the merger is

$$q^M(t) = \frac{1-c-t}{2(1+\gamma)}$$

which exhibits similar comparative statics as the equilibrium output without the merger, namely, $q^M(t)$ is decreasing in the firm's marginal cost c and in the emission fee t that the regulator charges. When γ decreases, firm's products are more differentiated so that increasing output of one product causes less impact on the profitability of the other product, and this allows the merged firm to increase output of both products.

Comparing equilibrium output with and without the merger, we see that, for a given fee t, $q^{NM}(t) \geq q^M(t)$ since

$$\frac{1-c-t}{2+\gamma} \geq \frac{1-c-t}{2(1+\gamma)}.$$

Intuitively, the merged entity reduces individual output levels to maximize joint profits.

Substituting equilibrium output, $q^M(t)$, into the merged firm's profit function, we find

$$\pi\left(t^M\right) = 2\left[\left(1-(1+\gamma)\,q^M\right)q^M-\left(c+t\right)q^M\right]$$
$$= \frac{(1-c-t)^2}{2(1+\gamma)}.$$

(b) *Second stage.* Find the emission fee that the regulator sets when the firms merge, t^M, and when they do not, t^{NM}.

- *Roadmap.* To identify the emission fee in this context, we first find the socially optimal output, q^{SO}, and then set it equal to the equilibrium output function that each firm chooses under no merger, $q^{NM}(t) = q^{SO}$, to solve for fee t.

- *Finding socially optimal output levels.* Social welfare is the sum of consumer surplus and producer surplus minus environmental damages:

$$\max_{q_1,q_2\geq 0} W = \underbrace{\frac{q_1^2+2\gamma q_1 q_2+q_2^2}{2}}_{=CS}$$

$$+\underbrace{\overbrace{[(1-q_1-\gamma q_2)q_1-(c+t)\,q_1]}^{\pi_1}+\overbrace{[(1-q_2-\gamma q_1)q_2-(c+t)\,q_2]}^{\pi_2}}_{=PS}$$

$$+t(q_1+q_2)-\underbrace{d\,(q_1+q_2)^2}_{=Env}$$

where the consumer surplus in a duopoly with product differentiation is given by $CS = \frac{q_1^2+2\gamma q_1 q_2+q_2^2}{2}$. Simplifying this welfare function, we obtain:

$$\max_{q_1,q_2\geq 0} W = \left(1-\frac{1+2d}{2}q_1-c\right)q_1+\left(1-\frac{1+2d}{2}q_2-c\right)q_2-(\gamma+2d)\,q_1q_2$$

Differentiating with respect to q_1 and q_2, and assuming interior solutions, we find:

$$1-(1+2d)\,q_1-c-(\gamma+2d)\,q_2=0$$
$$1-(1+2d)\,q_2-c-(\gamma+2d)\,q_1=0$$

Invoking symmetry, $q^{SO}=q_1^{SO}=q_2^{SO}$, we find that socially optimal output for every firm is

$$q^{SO}=\frac{1-c}{1+\gamma+4d}$$

- *Finding equilibrium fees, No merger.* When firms do not merge, the regulator seeks to induce firms to produce the socially optimal output q^{SO} in equilibrium. To find the emission fee that induces firms to choose q^{SO} in this context, t^{NM}, we set $q^{NM}\left(t^{NM}\right)=q^{SO}$ as follows:

$$\frac{1-c-t^{NM}}{2+\gamma}=\frac{1-c}{1+\gamma+4d}$$

After rearranging and solving for t yields

$$t^{NM}=\frac{(1-c)\,(4d-1)}{1+\gamma+4d}$$

which is positive since $d>1/2$ holds by assumption. In addition, this fee is increasing in d but decreasing in γ since

$$\frac{\partial t^{NM}}{\partial d}=\frac{4(1-c)(2+\gamma)}{(1+\gamma+4d)^2}>0 \text{ and } \frac{\partial t^{NM}}{\partial \gamma}=-\frac{(1-c)(4d-1)}{(1+\gamma+4d)^2}<0$$

Hence, the emission fee becomes more stringent when output generates a more intense environmental damage (higher d) but less stringent when products are more homogeneous (higher γ) since firms respond reducing their individual output when γ increases.

- *Finding equilibrium fees, Merger.* In a similar fashion, when firms do not merge, the regulator seeks to induce the socially optimal output q^*. We then set $q^M\left(t^M\right) = q^*$, which entails

$$\frac{1-c-t^M}{2\left(1+\gamma\right)} = \frac{1-c}{1+\gamma+4d}$$

which, after rearranging, yields

$$t^M = \frac{\left(1-c\right)\left(4d-1-\gamma\right)}{1+\gamma+4d}.$$

which is positive if d satisfies $d > \frac{1+\gamma}{4}$, which holds given our initial assumption $d > 1/2$. Indeed, when $\gamma = 0$, condition $d > \frac{1+\gamma}{4}$ simplifies to $d > \frac{1}{4}$; and if $\gamma = 1$, condition $d > \frac{1+\gamma}{4}$ becomes: $d > \frac{1}{2}$, which also holds in our setting. Emission fee t^M exhibits similar comparative statics as t^{NM} does since

$$\frac{\partial t^M}{\partial d} = \frac{8(1-c)(1+\gamma)}{(1+\gamma+4d)^2} > 0 \text{ and } \frac{\partial t^M}{\partial \gamma} = -\frac{8d(1-c)}{(1+\gamma+4d)^2} < 0$$

implying that fee t^M becomes more stringent in the environmental damage, d, but less stringent when products are more homogeneous.

(c) Compare the equilibrium fees you found in part (b). Which one is more stringent?

- Comparing fees t^{NM} and t^M, we can see that $t^{NM} > t^M$, since

$$\frac{\left(1-c\right)\left(4d-1\right)}{1+\gamma+4d} > \frac{\left(1-c\right)\left(4d-1-\gamma\right)}{1+\gamma+4d}$$

simplifies to $\gamma \geq 0$, which holds by definition. Therefore, the regulator sets more stringent emission fees when firms do not merge than when they merge. Intuitively, when firms merge, they reduce their production, which moves their equilibrium output in this context, $q^M\left(t^M\right)$, closer to the socially optimal output q^*. As a result, firms need less severe regulation when they merge than otherwise.

(d) *First stage.* Identify under which parameter conditions firms have incentives to merge.

- Anticipating the regulator's emission fees in the second stage, the sum of firm i and firm j's profits if the two firms do not merge becomes

$$\pi^{NM} = \frac{2}{(2+\gamma)^2}\left(1-c-\overbrace{\frac{\left(1-c\right)\left(4d-1\right)}{1+\gamma+4d}}^{=t^{NM}}\right)^2 = 2\left(\frac{1-c}{1+\gamma+4d}\right)^2$$

Whereas, if firms merge, total profits become

$$\pi^M = \frac{1}{2\left(1+\gamma\right)}\left(1-c-\overbrace{\frac{\left(1-c\right)\left(4d-1-\gamma\right)}{1+\gamma+4d}}^{=t^M}\right)^2 = 2\left(1+\gamma\right)\left(\frac{1-c}{1+\gamma+4d}\right)^2.$$

Therefore, firms' profit gain from merging becomes:

$$\pi^M - \pi^{NM} = 2\left[(1+\gamma) - 1\right]\left(\frac{1-c}{1+\gamma+4d}\right)^2$$

which is positive because $(1+\gamma) \geq 1$ holds given that $\gamma \geq 0$ by definition.

- **Intuition:** The merger brings two benefits:
  - First, aggregate output is reduced, thus increasing equilibrium price and profits (as in standard merger models we considered in previous exercises in this chapter).
  - Second, the reduction in aggregate output leads to a less stringent emission fee in the second stage (as shown in part c).

  In other words, firms have more incentive to merge when they are subject to environmental regulation (as they capture both benefits listed above) than when they are not subject to this regulation (since firms only capture the first benefit).

(e) *Comparison.* How would your results be affected if regulation did not exist in the second stage of the game? Consider (i) profit levels, (ii) social welfare, and (iii) profit gains with and without regulation.

- *Profit gain from the merger.* Setting $t^M = t^{NM} = 0$ in part (a), we find that a merger is always profitable because it forms a monopoly, that is,

$$\pi^M = \frac{(1-c)^2}{2(1+\gamma)} > 2\left(\frac{1-c}{2+\gamma}\right)^2 = \pi^{NM}$$

$$(2+\gamma)^2 > 4(1+\gamma)$$

$$4 + 4\gamma + \gamma^2 > 4 + 4\gamma$$

$$\gamma^2 > 0$$

  which holds for all $\gamma \in [0, 1]$.

- *Welfare gain from the merger.* In terms of social welfare, we find that

$$W^k = \left(1 - \frac{1+2d}{2} q_1^k - c\right) q_1^k + \left(1 - \frac{1+2d}{2} q_2^k - c\right) q_2^k - (\gamma + 2d)\, q_1^k q_2^k$$

$$= \left[2(1-c) - (1+\gamma+4d)\, q^k\right] q^k$$

  In this context, welfare under no merger becomes:

$$W^{NM} = \left[2(1-c) - (1+\gamma+4d)\overbrace{\left(\frac{1-c}{2+\gamma}\right)}^{=q^{NM}}\right]\overbrace{\frac{1-c}{2+\gamma}}^{=q^{NM}}$$

$$= \frac{(1-c)^2(3+\gamma-4d)}{(2+\gamma)^2}$$

while merger welfare becomes

$$W^M = \left[2(1-c) - (1+\gamma+4d)\overbrace{\left(\frac{1-c}{2(1+\gamma)}\right)}^{=q^M} \right] \overbrace{\frac{1-c}{2(1+\gamma)}}^{=q^M}$$

$$= \frac{(1-c)^2(3+3\gamma-4d)}{4(1+\gamma)^2}$$

Therefore, merger under no regulation enhances welfare if

$$\frac{(1-c)^2(3+3\gamma-4d)}{4(1+\gamma)^2} > \frac{(1-c)^2(3+\gamma-4d)}{(2+\gamma)^2}$$

which simplifies to

$$\frac{3+3\gamma-4d}{4(1+\gamma)^2} > \frac{3+\gamma-4d}{(2+\gamma)^2}$$

Solving for d, we find that the merger is welfare improving if

$$d > \overline{d} \equiv \frac{(1+\gamma)(4+\gamma)}{4(4+3\gamma)}$$

Note that cutoff $\overline{d}$ increases in γ because

$$\frac{\partial \overline{d}}{\partial \gamma} = \frac{8+8\gamma+3\gamma^2}{4(4+3\gamma)^2} > 0$$

In other words, when regulation is not in place, mergers that reduce the production of more homogeneous goods (higher γ) can enhance welfare only when the reduction of environmental damages from producing fewer units of the more polluting goods (higher d), together with the profit gain from increased market power, more than offset the reduction in consumer surplus.

- *Profit gain from the merger, with and without regulation.* Let us also analyze the difference in profits under different regulatory regimes. When firms merge in the absence of regulation, profit gains are

$$\Delta\pi^{NR} = \pi^{M,NR} - \pi^{NM,NR} = \frac{\gamma^2(1-c)^2}{2(1+\gamma)(2+\gamma)^2},$$

where the superscript *NR* denotes no regulation. In contrast, when firms are subject to regulation, profit gains from the merger become:

$$\Delta\pi^{R} = \pi^{M,R} - \pi^{NM,R} = \frac{2\gamma(1-c)^2}{(1+\gamma+4d)^2},$$

where the superscript *R* denotes regulation.

Therefore, the presence of regulation makes the merger less attractive for firms if $\Delta\pi^{NR} > \Delta\pi^{R}$, which entails

$$\frac{\gamma^2}{2(1+\gamma)(2+\gamma)^2} > \frac{2\gamma}{(1+\gamma+4d)^2}$$

and, after rearranging and solving for d, we obtain:

$$d > \overline{d}^R \equiv \frac{2+\gamma}{2}\sqrt{\frac{1+\gamma}{\gamma}} - \frac{1+\gamma}{4}$$

Cutoff $\overline{d}^R$ is positive because $3\gamma^2 + 15\gamma + 16 > 0$ holds for all $\gamma \in [0, 1]$.

Intuitively, when goods are relatively homogeneous (higher γ), profit gain under regulation falls below that under no regulation if, in addition, goods are relatively polluting (high d). This happens because firms have stronger incentives to merge when producing more homogeneous goods (relative to the case of completely differentiated goods where pre- and post-merger profits are the same), and such incentives are weakened only when the regulator charges more stringent emission fees on the more polluting goods.

- Further differentiating cutoff $\overline{d}^R$ with respect to γ, we obtain:

$$\frac{d\overline{d}^R}{d\gamma} = \frac{1}{4\gamma\sqrt{\gamma(1+\gamma)}}\left[2\gamma\sqrt{1+\gamma}\left(\sqrt{1+\gamma} + \frac{2+\gamma}{2\sqrt{1+\gamma}}\right) - (1+\gamma)(2+\gamma)\right] - \frac{1}{4}$$
$$= \frac{2\gamma^2 + \gamma\left(1 - \sqrt{\gamma(1+\gamma)}\right) - 2}{4\gamma\sqrt{\gamma(1+\gamma)}}$$

which is negative for all $\gamma \in [0, 1]$, so that cutoff $\overline{d}^R$ decreases in γ.

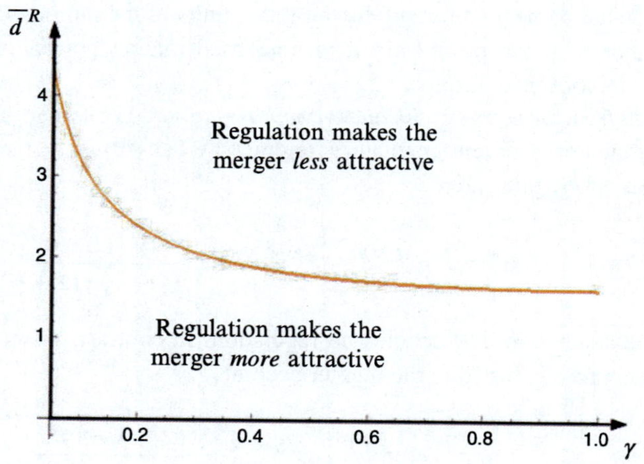

Fig. 7.2 Profit gain from merging, with and without regulation

(f) *Comparative statics.* In part (e), we analyzed under which conditions firms experience a larger profit gain from the merger when they face regulation or when they do not. How are these conditions affected by parameter γ? Interpret.

- Figure 7.2 plots cutoff $\overline{d}^R$ (found in part e) as a function of γ, showing that $\overline{d}^R$ decreases in γ. When goods become more homogeneous (higher γ), regulation makes the merger less attractive under a larger range of d. For instance, at $\gamma = 1/4$, cutoff $\overline{d}^R$ becomes $\overline{d}^R = 2.20$, and when $\gamma = 1/2$, this cutoff decreases to $\overline{d}^R = 1.79$.

Collusion 8

Introduction

This chapter studies firms' incentives to collude in imperfectly competitive markets. Exercise 8.1 identifies the range of discount factors that support the collusion of two firms. Extending into a multiple firm context, Exercise 8.2 indicates that it becomes more difficult for firms to collude as the relative profit gain from cheating increases with the number of firms, so that every firm needs to assign a sufficiently high value on future profits to sustain collusion.

We are also interested in analyzing firms' repeated interaction when competing in prices. Exercise 8.3 investigates firms' incentives to collude in prices when two firms sell homogeneous products. Exercise 8.4 considers Nash reversion strategies in more general settings, and Exercise 8.5 allows for product differentiation, and we show that firms can collude under a wider range of discount rates when products become more differentiated. Exercise 8.6 examines markets with time-varying demand, showing that in markets with growing (shrinking) demand, collusion can be sustained under less (more) restrictive conditions on discount rates as the deviating firm faces larger (smaller) punishments in future periods. In sum, collusion is easier when goods are more differentiated and markets are more concentrated and expanding.

Exercises 8.7 and 8.8 examine how collusive incentives are affected when firms' production costs are convex, instead of linear. In particular, Exercise 8.7 considers that firms' costs are symmetric, while Exercise 8.8 allows for cost asymmetries. Intuitively, all firms produce a positive output in the cartel, which becomes more attractive when firms are more asymmetric, since they can shift a larger output share to the most efficient firm, ultimately increasing cartel profits.

We further consider the effects of prosecution on the stability of a cartel. Exercise 8.9 shows that collusion becomes more difficult when firms are subject to prosecution that dissolves the cartel with a positive probability in every period. Interestingly, a heavier penalty makes collusion easier to sustain since firms now have lower expected gains from output coordination that makes deviation less profitable. Exercise 8.10 illustrates that the same results hold when firms compete on prices.

Exercise 8.11 investigates collusive behavior when firms exhibit economies of scope, while Exercise 8.12 examines collusive behavior under temporary punishment. We show that the tit-for-tat strategy cannot be sustained in equilibrium because one-period reversal to the Nash equilibrium output does not offset the firm's instantaneous gain from deviation. However, the longer the punishment phrase, the more future profits that the firm loses so that collusion can be sustained under a wider range of discount rates. Exercise 8.13 studies the condition for which firms find collusion sustainable

P.-S. Choi et al., *Industrial Organization*, Springer Texts in Business and Economics,
https://doi.org/10.1007/978-3-031-38635-0_8

and punishment credible, where full (partial) collusion can be supported when discount factors are relatively high (low). Exercise 8.14 considers intermittent collusion, which requires higher discount factors when firms collude less frequently.

Exercise 8.15 demonstrates that collusion is more difficult to sustain when prices are inflexible. Exercise 8.16, based on Exercise 7.9, suggests that when a firm deviates from a Stackelberg cartel and loses output leadership, this firm is less likely to deviate than when it retains output leadership after the deviation, as in Exercise 8.17. Exercise 8.18 reveals that cost-reduction effects may facilitate (hinder) collusion when firms have stronger incentives to collude (defect).

Exercise 8.19 evaluates leniency programs, where antitrust authorities seek to induce one of the colluding firms to provide evidence against the cartel in exchange for a pecuniary reward, such as a reduced fine or shorter prison sentences. Exercise 8.20 further shows that when the antitrust authority is able to prove firms guilty of collusion when an investigation is opened, firms still have incentives to collude when they consider investigation to be a rare event. However, once an investigation is opened, firms may pledge guilty and pay a reduced fine to provide evidence against the cartel.

Exercise #8.1: Collusion with Two Firms Competing à la Cournot^B

8.1 Consider an industry with two identical firms competing à la Cournot and facing an inverse demand function $p(Q) = 1 - Q$, where $Q \geq 0$ denotes aggregate output. Every firm $i = \{1, 2\}$ faces a constant marginal cost of production c, where $0 \leq c < 1$.

(a) Find the equilibrium output and profit level if both firms collude.

- If the firms collude, they maximize their joint profits, which is equivalent to pricing as if they were a monopoly. Setting up a monopolist's profit maximization function in this regard,

$$\max_{Q \geq 0} \ (1 - Q)Q - cQ$$

with accompanying first-order condition

$$\frac{\partial \pi}{\partial Q} = 1 - 2Q - c = 0.$$

Solving this expression for Q provides our equilibrium output level for a monopolist in this setting (i.e., our aggregate output)

$$Q^C = \frac{1-c}{2}.$$

Since there are 2 identical firms, they each produce $q_i^C = \frac{Q^C}{2}$ of the aggregate quantity, or

$$q_i^C = \frac{Q^C}{2} = \frac{1-c}{4}.$$

To find our equilibrium profit level, we must first derive our equilibrium price, which we obtain by inserting the aggregate output into the inverse demand function

$$p^C = 1 - Q^C = \frac{1+c}{2}.$$

Lastly, we insert our equilibrium price and output into the profit function to find the equilibrium profit level of every firm i

$$\pi_i^C = (p^C - c)q_i^C = \left(\frac{1+c}{2} - c\right)\left(\frac{1-c}{4}\right) = \frac{(1-c)^2}{8}.$$

(b) Find the minimum value of the discount factor, $\underline{\delta}$, that firms are able to sustain collusion if each firm implements a Grim-Trigger Strategy.

- *Profits from collusion.* If each firm colludes, they receive profits of $\pi_i^* = \frac{(1-c)^2}{8}$ in each period, leading to a present discounted value of lifetime profits of

$$\begin{aligned} &\pi_i^C + \delta\pi_i^C + \delta^2\pi_i^C + \ldots \\ &= \pi_i^C(1 + \delta + \delta^2 + \ldots) \\ &= \pi_i^C\left(\frac{1}{1-\delta}\right) \\ &= \frac{(1-c)^2}{8(1-\delta)}. \end{aligned}$$

- *Profits from deviating.* To find our minimum discount factor, we must first determine the profit level of firm i's most profitable deviation if the other firm colludes, π_i^D, and then the profit level of firm i for every period after, π_i^N. Setting up firm i's profit maximization problem,

$$\max_{q_i \geq 0} \; (1 - q_i - q_j)q_i - cq_i$$

with accompanying first-order condition

$$\frac{\partial \pi_i}{\partial q_i} = 1 - 2q_i - q_j - c = 0,$$

which we can arrange into firm i's best response function to the quantities produced by the other firm

$$q_i = \frac{1-c}{2} - \frac{1}{2}q_j.$$

If the other firm colludes, then $q_j^D = \frac{1-c}{4}$. Substituting this expression into firm i's best response function, we obtain

$$q_i = \frac{1-c}{2} - \frac{1}{2}\left(\frac{1-c}{4}\right),$$

and simplifying, we find firm i's optimal deviation

$$q_i^D = \frac{3(1-c)}{8}.$$

This leads to a price of

$$\begin{aligned} p^D &= 1 - Q = 1 - q_i^D - q_j^D \\ &= 1 - \frac{3(1-c)}{8} - \frac{1-c}{4} \\ &= \frac{3+5c}{8} \end{aligned}$$

and profits

$$\begin{aligned} \pi_i^D = (p^D - c)q_i^D &= \left(\frac{3+5c}{8} - c\right)\left(\frac{3(1-c)}{8}\right) \\ &= \frac{9(1-c)^2}{64}. \end{aligned}$$

- *Profits from reversion to the Nash equilibrium.* Next, we calculate the profits that firm i receives after each firm returns to the Nash equilibrium of the stage game. In a symmetric equilibrium, every firm produces the same output level, $q_i = q_j = q$. Inserting this property into firm i's best response function,

$$q_i = \frac{1-c}{2} - \frac{1}{2}q_j,$$

we find

$$q = \frac{1-c}{2} - \frac{1}{2}q.$$

Solving this expression for q yields each firm's equilibrium output level

$$q^N = \frac{1-c}{3}$$

with aggregate output

$$Q^N = 2q^N = \frac{2(1-c)}{3},$$

equilibrium price

$$p^N = 1 - Q^N = 1 - \frac{2(1-c)}{3} = \frac{1+2c}{3},$$

and lastly, equilibrium profits

$$\pi_i^N = (p^N - c)q^N = \left(\frac{1+2c}{3} - c\right)\left(\frac{1-c}{3}\right) = \frac{(1-c)^2}{9}.$$

Now, if firm i deviates in the first period, it earns a profit of π_i^D in that period, but a profit of π_i^N in every period afterward as each firm reverts back to the Nash equilibrium of the stage game. Its present discounted value of lifetime profits is

$$\begin{aligned}&\pi_i^D + \delta\pi_i^N + \delta^2\pi_i^N + \ldots \\ &= \pi_i^D + \delta\pi_i^N(1 + \delta + \delta^2 + \ldots) \\ &= \pi_i^D + \pi_i^N\left(\frac{\delta}{1-\delta}\right) \\ &= \frac{9(1-c)^2}{64} + \frac{\delta(1-c)^2}{9(1-\delta)}.\end{aligned}$$

- *Calculating the minimal discount factor, $\underline{\delta}$.* Firm i has incentives to collude if its present discounted value of lifetime profits is at least as high as deviating, i.e.,

$$\begin{aligned}\pi_i^C\left(\frac{1}{1-\delta}\right) &\geq \pi_i^D + \pi_i^N\left(\frac{\delta}{1-\delta}\right) \\ \frac{(1-c)^2}{8(1-\delta)} &\geq \frac{9(1-c)^2}{64} + \frac{\delta(1-c)^2}{9(1-\delta)}.\end{aligned}$$

Rearranging, we find

$$\begin{aligned}\frac{1}{8} &\geq \frac{9(1-\delta)}{64} + \frac{\delta}{9} \\ \left(\frac{9}{64} - \frac{1}{9}\right)\delta &\geq \frac{9}{64} - \frac{1}{8} \\ \frac{17}{576}\delta &\geq \frac{1}{64}.\end{aligned}$$

Finally, solving for δ, we obtain the minimal discount factor sustaining collusion as follows:

$$\delta \geq \frac{9}{17} \equiv \underline{\delta}.$$

Therefore, collusion can be sustained if firms assign a sufficiently high value to future profits. Otherwise, every firm wants to deviate and collusion cannot be sustained.

Exercise #8.2: Collusion with n Firms Competing à la Cournot^C

8.2 Consider an industry with $n \geq 2$ identical firms that compete à la Cournot, facing an inverse demand function $p(Q) = 1 - Q$, where $Q \geq 0$ denotes aggregate output. Every firm i faces a constant marginal cost of production c, where $0 \leq c < 1$.

(a) Find the equilibrium output and profit level if all n firms collude.

- If the firms collude, they maximize their joint profits, which is equivalent to pricing as if they were a monopoly. Setting up a monopolist's profit maximization function in this regard,

$$\max_{Q \geq 0} \; (1-Q)Q - cQ$$

with accompanying first-order condition

$$\frac{\partial \pi}{\partial Q} = 1 - 2Q - c = 0.$$

Solving this expression for Q provides our equilibrium output level for a monopolist in this setting (i.e., our aggregate output)

$$Q^C = \frac{1-c}{2}.$$

Since there are n identical firms, they each produce $q_i^C = \frac{Q^C}{n}$ of the aggregate output, or

$$q_i^C = \frac{Q^C}{n} = \frac{1-c}{2n}.$$

To find our equilibrium profit level, we must first derive our equilibrium price, which we obtain by inserting the aggregate output into the inverse demand function

$$p^C = 1 - Q^C = \frac{1+c}{2}.$$

Lastly, we insert our equilibrium price and output into our profit function to find the equilibrium profit level of firm i

$$\pi_i^C = (p^C - c)q_i^C = \left(\frac{1+c}{2} - c\right)\left(\frac{1-c}{2n}\right) = \frac{(1-c)^2}{4n}.$$

(b) Find the minimum value of the discount factor, $\underline{\delta}$, that firms are able to sustain collusion if each firm implements a Grim-Trigger Strategy.

- *Profits from collusion.* If firms collude, they each receive profits of $\pi_i^* = \frac{(1-c)^2}{4n}$ in each period, leading to a present discounted value of lifetime profits of

$$\begin{aligned} &\pi_i^C + \delta\pi_i^C + \delta^2\pi_i^C + \ldots \\ &= \pi_i^C(1 + \delta + \delta^2 + \ldots) \\ &= \pi_i^C\left(\frac{1}{1-\delta}\right) \\ &= \frac{(1-c)^2}{4n(1-\delta)}. \end{aligned}$$

- *Profits from deviating.* To find our minimum discount factor, we must first determine the profit level of firm i's most profitable deviation if all other firms collude, π_i^D, and then the profit level of firm i for every period after, π_i^N. Setting up firm i's profit maximization problem,

$$\max_{q_i \geq 0} \; (1 - q_i - \sum_{j \neq i} q_j)q_i - cq_i$$

with accompanying first-order condition

$$\frac{\partial \pi_i}{\partial q_i} = 1 - 2q_i - \sum_{j \neq i} q_j - c = 0,$$

which we can arrange into firm i's best response function to the quantities produced by all other firms

$$q_i = \frac{1-c}{2} - \frac{1}{2}\left(\sum_{j \neq i} q_j\right).$$

If every other firm colludes, then $q_j = \frac{1-c}{2n}$ for all $j \neq i$, and

$$Q_{-i}^D = \sum_{j \neq i} q_j = \left(\frac{n-1}{n}\right)\left(\frac{1-c}{2}\right).$$

Substituting this expression into firm i's best response function,

$$q_i = \frac{1-c}{2} - \frac{1}{2}\left(\frac{n-1}{n}\right)\left(\frac{1-c}{2}\right),$$

and simplifying, we find firm i's optimal deviation

$$\begin{aligned} q_i^D &= \left(\frac{1-c}{2}\right)\left(1 - \frac{n-1}{2n}\right) \\ &= \left(\frac{n+1}{2n}\right)\left(\frac{1-c}{2}\right). \end{aligned}$$

This leads to a price of

$$\begin{aligned} p^D &= 1 - Q = 1 - q_i^D - Q_{-i}^D \\ &= 1 - \left(\frac{n+1}{2n}\right)\left(\frac{1-c}{2}\right) - \left(\frac{n-1}{n}\right)\left(\frac{1-c}{2}\right) \\ &= 1 - \left(\frac{3n-1}{2n}\right)\left(\frac{1-c}{2}\right) \\ &= \frac{n+1+c(3n-1)}{4n} \end{aligned}$$

and profits

$$\begin{aligned} \pi_i^D = (p^D - c)q_i^D &= \left(\frac{n+1+c(3n-1)}{4n} - c\right)\left(\frac{n+1}{2n}\right)\left(\frac{1-c}{2}\right) \\ &= \frac{[(n+1)(1-c)]^2}{16n^2}. \end{aligned}$$

- *Profits from reversion to the Nash equilibrium.* Next, we calculate the profits that firm i receives after each firm returns to the Nash equilibrium of the stage game. In a symmetric equilibrium, every firm produces the same output level, $q_i = q_j = q$. Inserting this property into firm i's best response function

$$q_i = \frac{1-c}{2} - \frac{1}{2}\left(\sum_{j \neq i} q_j\right),$$

we find

$$q = \frac{1-c}{2} - \frac{1}{2}(n-1)q.$$

Solving this expression for q yields each firm's equilibrium output level

$$q^N = \frac{1-c}{n+1}$$

with aggregate output

$$Q^N = nq^N = \frac{n(1-c)}{n+1},$$

and equilibrium price

$$p^N = 1 - Q^N = 1 - \frac{n(1-c)}{n+1} = \frac{1+nc}{n+1},$$

and equilibrium profits

$$\pi_i^N = (p^N - c)q^N = \left(\frac{1+nc}{n+1} - c\right)\left(\frac{1-c}{n+1}\right) = \frac{(1-c)^2}{(n+1)^2}.$$

Now, if firm i deviates in the first period, it earns a profit of π_i^D in that period, but a profit of π_i^N in every period afterward as each firm reverts back to the Nash equilibrium of the stage game. Its present discounted value of lifetime profits is

$$\begin{aligned} &\pi_i^D + \delta\pi_i^N + \delta^2\pi_i^N + \ldots \\ &= \pi_i^D + \delta\pi_i^N(1 + \delta + \delta^2 + \ldots) \\ &= \pi_i^D + \pi_i^N\left(\frac{\delta}{1-\delta}\right) \\ &= \frac{[(n+1)(1-c)]^2}{16n^2} + \frac{\delta(1-c)^2}{(n+1)^2(1-\delta)}. \end{aligned}$$

- *Calculating the minimal discount factor,* $\underline{\delta}$. Firm i has incentives to collude if its present discounted value of lifetime profits is at least as high as deviating, i.e.,

$$\pi_i^C\left(\frac{1}{1-\delta}\right) \geq \pi_i^D + \pi_i^N\left(\frac{\delta}{1-\delta}\right)$$

$$\frac{(1-c)^2}{4n(1-\delta)} \geq \frac{[(n+1)(1-c)]^2}{16n^2} + \frac{\delta(1-c)^2}{(n+1)^2(1-\delta)}.$$

Rearranging, we find

$$\frac{1}{4n(1-\delta)} \geq \frac{(n+1)^4(1-\delta)+16n^2\delta}{16n^2(n+1)^2(1-\delta)}$$

$$\delta\left[(n+1)^2+4n\right]\left[(n+1)^2-4n\right] \geq (n+1)^4 - 4n(n+1)^2.$$

Finally, solving for δ, we obtain the minimal discount factor sustaining collusion as follows:

$$\delta \geq \frac{(n+1)^2}{n^2+6n+1} \equiv \underline{\delta},$$

which has a value of $\underline{\delta} = \frac{9}{17}$ when there are only 2 firms, as shown in Exercise 8.1.

(c) What happens to the minimal discount factor, $\underline{\delta}$, as the number of firms increases?

- Differentiating cutoff $\underline{\delta}$ with respect to n, we find

$$\frac{\partial \underline{\delta}}{\partial n} = \frac{2\left(n^2-1\right)}{\left(n^2+6n+1\right)^2} > 0$$

Therefore, $\underline{\delta}$ is increasing as the number of firms increases. Intuitively, the profits from colluding fall relative to the profits from deviating as more firms are required to collude. In other words, cheating becomes more attractive when the deviating firm can free ride on the collusion of more firms. This, in turn, requires each individual firm to be more patient in order to sustain collusion. This result is presented in Fig. 8.1.

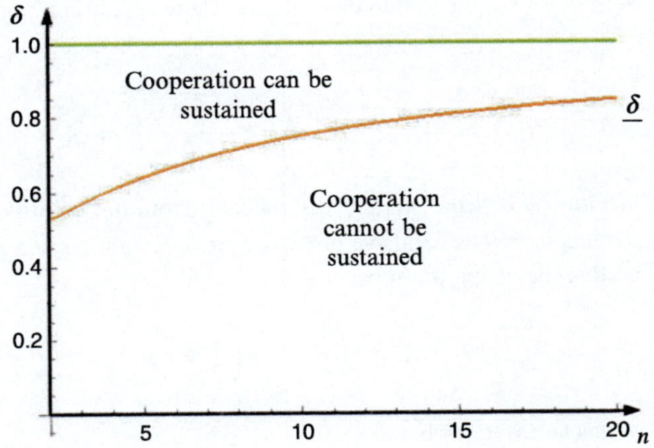

Fig. 8.1 Minimal discount factor $\underline{\delta}$

Exercise #8.3: Collusion in the Price of a Homogeneous Product^A

8.3 Consider two firms competing à la Bertrand on homogeneous goods. The demand function of firm i, where $i, j \in \{1, 2\}$, is

$$q_i\left(p_i, p_j\right) = \begin{cases} 1 - p_i & \text{if } p_i < p_j \\ \frac{1-p_i}{2} & \text{if } p_i = p_j \\ 0 & \text{if } p_i > p_j. \end{cases}$$

For simplicity, assume zero production costs.

(a) Find the price and associated profit when firms (i) compete against each other, (ii) form a price cartel, and (iii) unilaterally deviate.

- *Price competition.* As shown in Exercise 3.1, when firms compete in prices and sell a homogeneous good, the Nash equilibrium price pair is $\left(p_i^*, p_j^*\right) = (c, c)$; that is, both firms set a price equal to their common marginal cost. In this setting, where $c = 0$, the Nash equilibrium price and profit become

$$p^n = 0$$
$$\pi^n = 0.$$

- *Collusion.* If firms collude with each other, they set a collusive price equal to that under monopoly. This price is found by solving the monopolist's profit maximization problem as follows:

$$\max_{p_i \geq 0} \, (p_i - c)(1 - p_i).$$

Differentiating with respect to p_i yields $1 - 2p_i + c = 0$, and solving for p_i, we obtain the monopoly price $p^m = \frac{1+c}{2}$. Therefore, the collusive price is

$$p^c = \frac{1}{2} = p^m,$$

so collusive profit becomes half of monopoly profit, $\pi^c = \frac{(p^c-c)(1-p^c)}{2}$, which simplifies to $\pi^c = \frac{(p^c-0)(1-p^c)}{2}$ since $c = 0$ in this context. Therefore, firm i's profit under collusion is

$$\pi^c = \frac{p^c\left(1 - p^c\right)}{2} = \frac{\frac{1}{2}\left(1 - \frac{1}{2}\right)}{2} = \frac{1}{8}.$$

- *Unilateral deviation.* If firm i unilaterally deviates from the collusive price $p^c = \frac{1}{2}$, it slightly undercuts its rival's collusive price, setting a deviating price $p^d = \frac{1}{2} - \varepsilon$, where $\varepsilon \to 0$. Therefore, deviating profit becomes

$$\pi^d = \left(\frac{1}{2} - \varepsilon\right)\left(\frac{1}{2} + \varepsilon\right),$$

which converges to $\pi^d = \frac{1}{4}$ when $\varepsilon \to 0$.

(b) What is the minimum discount rate $\underline{\delta}$ to sustain collusion, assuming that firms interact infinitely and implement a Grim-Trigger Strategy once defection is detected?

- The minimum discount factor to sustain collusion solves

$$\pi^c + \delta\pi^c + \delta^2\pi^c + \ldots \geq \pi^d + \delta 0 + \delta^2 0 + \ldots,$$

which indicates that every firm i prefers to earn collusive profits in every period, π^c, rather than deviating in the current period (earning the higher profit π^d) but then earning no profits in all subsequent periods (once its rival detects its deviation). The above inequality simplifies to

$$\frac{\pi^c}{1-\delta} \geq \pi^d,$$

and after solving for δ, we find that

$$\delta \geq \frac{\pi^d - \pi^c}{\pi^d} \equiv \underline{\delta}.$$

Inserting the value of each of these profits in our setting, we find that the minimum discount factor sustaining collusion is

$$\underline{\delta} = \frac{\frac{1}{4} - \frac{1}{8}}{\frac{1}{4}} = \frac{1}{2}.$$

Intuitively, firms must assign a sufficiently high value to future payoffs to collude in prices rather than unilaterally deviating in the current period and earning zero profit in all future periods.

(c) *Asymmetric profit shares.* Consider now a more general setting where, instead of evenly sharing cartel profits, firm i earns a share $\alpha \in [0, 1]$ and firm j earns a share $(1-\alpha)$. Parameter α can be interpreted as firm i's bargaining power, relative to firm j's, as in Nash bargaining. Therefore, $\alpha = 1$ indicates that firm i keeps all cartel profits; $\alpha = \frac{1}{2}$ means that both firms have equal bargaining power, evenly sharing cartel profits, and $\alpha = 0$ represents that firm j captures all cartel profits. Find the minimum discount factor supporting collusion, $\underline{\delta}(\alpha)$. How is it affected by a marginal increase in α? Interpret.

- Firm i's cartel profits in this context become

$$\pi_i^c = \alpha\left(p^c - c\right)\left(1 - p^c\right) = \frac{\alpha}{4}$$

leaving firm j with profits

$$\pi_j^c = (1-\alpha)\left(p^c - c\right)\left(1 - p^c\right) = \frac{1-\alpha}{4}.$$

- In this context, firm i's minimum discount factor sustaining collusion (following our analysis in part b) becomes

$$\underline{\delta}_i(\alpha) = \frac{\pi^d - \pi_i^c}{\pi^d} = \frac{\frac{1}{4} - \frac{\alpha}{4}}{\frac{1}{4}} = 1 - \alpha.$$

Intuitively, $\underline{\delta}_i(\alpha)$ is decreasing in α, meaning that, as firm i keeps a larger share of cartel profits, it is willing to collude under larger parameter values. Similarly, firm j's minimum discount factor is

$$\underline{\delta}_j(\alpha) = \frac{\pi^d - \pi_j^c}{\pi^d} = \frac{\frac{1}{4} - \frac{1-\alpha}{4}}{\frac{1}{4}} = \alpha,$$

which is increasing in α. Intuitively, as firm j earns a smaller share of cartel profits, it becomes less willing to collude.

- Therefore, collusion can be sustained in equilibrium when both firms' discount factors exceed the maximum of the two cutoffs found above, $\underline{\delta}_i(\alpha)$ and $\underline{\delta}_j(\alpha)$, that is, we need that δ satisfies

$$\delta \geq \max\left\{\underline{\delta}_i(\alpha), \underline{\delta}_j(\alpha)\right\} \text{ for every firm } i.$$

We can alternatively write this condition as $\delta \geq \max\{1-\alpha, \alpha\}$ or as follows:

$$\delta \geq \underline{\delta} \equiv \begin{cases} 1-\alpha & \text{if } \alpha \leq \frac{1}{2} \\ \alpha & \text{otherwise.} \end{cases}$$

Graphically, this condition says that δ must exceed the upper bound of two diagonal lines: one that coincides with the 45-degree line, $\underline{\delta} = \alpha$, and another one that originates at $\delta = 1$ and decreases at a rate of one, $\underline{\delta} = 1 - \alpha$; as depicted in Fig. 8.2.

- Therefore, when $\alpha \leq 1/2$, an increase in firm i's share in cartel profits, α, decreases the height of this upper bound, expanding the region of δ for which collusion can be sustained (i.e., facilitating collusion). In contrast, when $\alpha > 1/2$, a further increase in α increases the upper bound, ultimately hindering collusion. Overall, collusion is more difficult to emerge

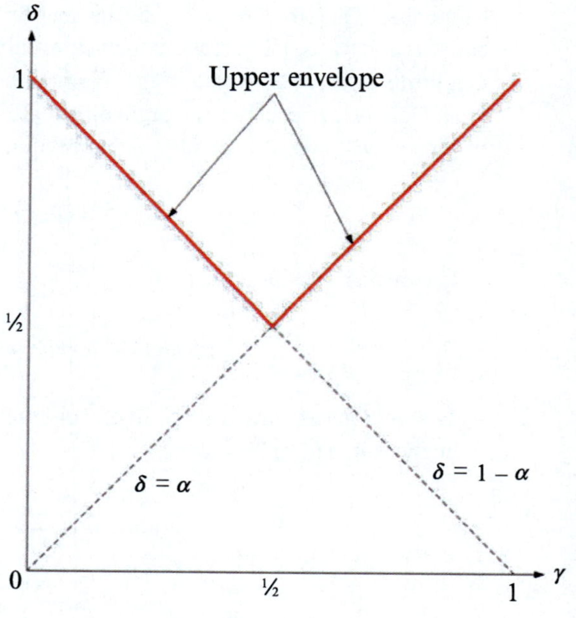

Fig. 8.2 Upper envelope of $1-\alpha$ and α

when profits are unevenly shared across the two firms (α close to 0 or 1) but is facilitated when profits are evenly distributed (α close to $1/2$).

(d) Assume that firm i earns a share $\alpha = 3/4$ of cartel profits. What is the minimum discount factor supporting collusion in this setting? Interpret.

- When $\alpha = 3/4$, the minimum discount factor supporting collusion for firm i is $1/4$ while that for firm j is $3/4$. Since firm i (j) takes a larger (smaller) share of cartel profits, it has incentives to collude under lower (higher) discount factors. Therefore, the minimum discount factor for both firms to collude is the maximum of these two cutoffs, $3/4$, which means that both firms must be sufficiently patient in order to find collusion profitable under asymmetric profit shares.

Exercise #8.4: Collusion, a General Approach^B

8.4 Consider an industry where, for generality, we do not assume whether firms compete in quantities or prices, nor the inverse demand function or costs they face. Consider that firms are symmetric and in the Nash equilibrium of the unrepeated game, every firm earns profits π^N, so we label the present value of the non-collusive stream as

$$V^N \equiv \pi^N + \delta\pi^N + \ldots = \frac{1}{1-\delta}\pi^N.$$

When firms collude, each of them earns profit π^C, where $\pi^C > \pi^N$. When a firm unilaterally deviates from the collusive outcome, it earns a deviating profit of π^D, where $\pi^D > \pi^C$ in that period. Consider a standard Grim-Trigger Strategy (GTS) where every firm chooses to collude in period $t = 1$ and continues to do so in subsequent periods $t > 1$ if all firms colluded in previous periods. If one firm did not cooperate in any previous period, however, all firms revert to the Nash equilibrium of the unrepeated game, earning π^N thereafter (permanent punishment scheme). For simplicity, assume that all firms exhibit the same discount factor $\delta \in (0, 1)$.

(a) Find the minimal discount factor that sustains this GTS as a subgame perfect equilibrium of the game, $\widehat{\delta}$.

- After a history of cooperation, every firm i keeps cooperating as long as

$$\underbrace{\frac{1}{1-\delta}\pi^C}_{\text{Cooperation}} \geq \underbrace{\pi^D}_{\text{Deviation}} + \underbrace{\frac{\delta}{1-\delta}\pi^N}_{\text{Permanent punishment}},$$

which, after solving for discount factor δ, yields

$$\delta \geq \widehat{\delta} \equiv \frac{\pi^D - \pi^C}{\pi^D - \pi^N}.$$

The above number is a positive number less than one since the profits from deviating, π^D, satisfy $\pi^D > \pi^C$, $\pi^D > \pi^N$, and $\pi^C > \pi^N$ by definition.

(b) Does the minimal discount factor $\widehat{\delta}$ increase in the profits from cooperation, π^C, those from deviation, π^D, or those from the Nash equilibrium of the unrepeated game, π^N? Interpret.

- Cutoff $\widehat{\delta}$ is increasing in π^N. Intuitively, as the profits from reverting to the Nash equilibrium of the unrepeated game π^N increase, the punishment from deviation becomes less severe, ultimately making the deviation more attractive.
- In contrast, cutoff $\widehat{\delta}$ is decreasing in π^C. In other words, this indicates that, when the profits from cooperation π^C increase, deviation becomes less attractive and can be sustained under larger values of discount factor δ.
- Finally, cutoff $\widehat{\delta}$ is increasing in the deviating profit π^D since

$$\frac{\partial \widehat{\delta}}{\partial \pi^D} = \frac{\pi^C - \pi^N}{(\pi^D - \pi^N)^2} > 0$$

Because the profits from cooperation π^C satisfy $\pi^C > \pi^N$. Intuitively, when deviation becomes more attractive, the GTS can only be sustained under more restrictive conditions on discount factor δ.

Exercise #8.5: Collusion in the Price of a Differentiated Product^B

8.5 Consider two firms competing à la Bertrand on differentiated goods. The demand function of firm i, where $i \in \{1, 2\}$, is

$$q_i\left(p_i, p_j\right) = 1 - p_i + \gamma p_j,$$

where $\gamma \in (-1, 1)$ represents the degree of product differentiation, approaching perfect substitutes (homogeneous goods) as $\gamma \to 1$, completely differentiated products when $\gamma = 0$, and perfect complements as $\gamma \to -1$. For simplicity, we assume zero production costs.

(a) *Price competition (no collusion).* Find and plot the best response functions. What are the Nash equilibrium price and associated profit of each firm, and how do they vary with parameter γ?

- Every firm i, where $i \in \{1, 2\}$, chooses p_i to solve the following profit maximization problem:

$$\max_{p_i \geq 0} \pi_i\left(p_i\right) = p_i\left(1 - p_i + \gamma p_j\right).$$

Differentiating with respect to p_i, we obtain

$$\frac{\partial \pi_i\left(p_i\right)}{\partial p_i} = 1 - 2p_i + \gamma p_j = 0.$$

Solving for p_i, we find that the best response function of firm i is

$$p_i\left(p_j\right) = \frac{1}{2} + \frac{\gamma}{2} p_j,$$

which originates at $\frac{1}{2}$ and increases in p_j at a rate of $\frac{\gamma}{2}$, as depicted in Fig. 8.3 which considers $\gamma > 0$. If, instead, $\gamma < 0$, the best response function would be negatively sloped, indicating that an increase in firm j's price induces a reduction in firm i's price, which occurs because goods are complements when $\gamma < 0$.

Fig. 8.3 Firm i's best response function with $\gamma > 0$

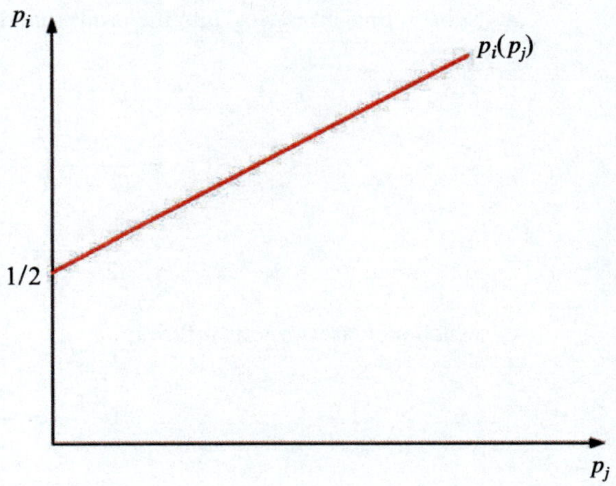

Fig. 8.4 Equilibrium price pairs

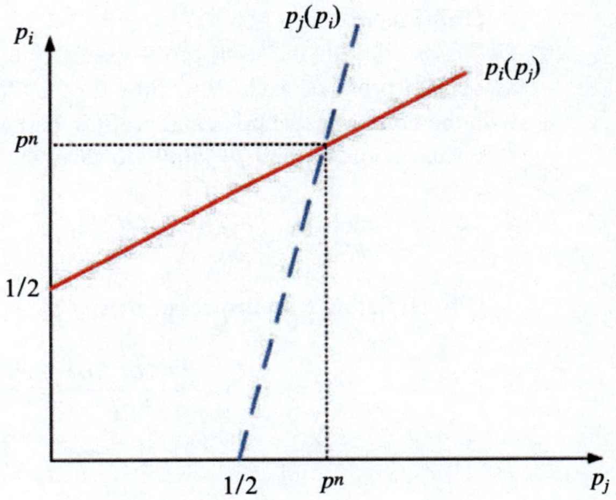

Invoking symmetry, where $p = p_1 = p_2$, we have that

$$p = \frac{1}{2} + \frac{\gamma}{2}p.$$

Simplifying, we obtain Nash equilibrium price of every firm i

$$p^n = \frac{1}{2-\gamma},$$

which is depicted in Fig. 8.4 where both best response functions cross each other (assuming, as in Fig. 8.3, that $\gamma > 0$). The equilibrium price p^n increases in γ because

$$\frac{dp^n}{d\gamma} = \frac{1}{(2-\gamma)^2} > 0.$$

Intuitively, as products become closer substitutes, the firm charges a higher price on its consumers who do not switch to its rival firm.

- Substituting $p^n = \frac{1}{2-\gamma}$ into the profit function, the Nash equilibrium profit becomes

$$\pi^n = p^n\left(1 - p^n + \gamma p^n\right)$$
$$= \frac{1}{2-\gamma}\left(1 - \frac{1}{2-\gamma} + \gamma\frac{1}{2-\gamma}\right)$$
$$= \frac{1}{(2-\gamma)^2},$$

which increases in γ as follows:

$$\frac{d\pi^n}{d\gamma} = \frac{2}{(2-\gamma)^3} > 0$$

so that as the goods become more substitutable (γ increases), the equilibrium prices and profits increase.

(b) *Collusion.* Set up the joint profit maximization problem, and find the collusive price and associated profit of each firm. How do they vary with parameter γ? Compare your results with the non-collusive price and profit in part (a).

- Firms choose p_1 and p_2 jointly to solve the profit maximization problem

$$\max_{p_1, p_2 \geq 0} \left[\pi_1(p_1) + \pi_2(p_2)\right] = p_1(1 - p_1 + \gamma p_2) + p_2(1 - p_2 + \gamma p_1).$$

Differentiating with respect to p_1 and p_2, respectively, we obtain

$$\frac{\partial\pi(p_1, p_2)}{\partial p_1} = 1 - 2p_1 + 2\gamma p_2 = 0$$
$$\frac{\partial\pi(p_1, p_2)}{\partial p_2} = 1 - 2p_2 + 2\gamma p_1 = 0.$$

In a symmetric price profile, $p_1 = p_2 = p^c$. Inserting this property in any of the above first-order conditions, we obtain

$$1 - 2p^c + 2\gamma p^c = 0.$$

Solving for p^c yields

$$p^c = \frac{1}{2(1-\gamma)},$$

which is higher than the Nash equilibrium price in part (a) when the goods are substitutes, because

$$\frac{1}{2(1-\gamma)} > \frac{1}{2-\gamma}$$
$$2 - \gamma > 2 - 2\gamma$$
$$\Rightarrow \gamma > 0.$$

Intuitively, firms collude to raise prices and internalize the negative impact of price competition on the profit of the other firm when the goods are substitutes. On the contrary, when $\gamma < 0$, the goods are complements and each firm lowers their prices under collusion relative to competition in order to increase the sales of their complementary product.

- Substituting $p^c = \frac{1}{2(1-\gamma)}$ into the profit function, collusive profit becomes

$$\begin{aligned}\pi^c &= p_1^c\left(1 - p_1^c + \gamma p_2^c\right)\\ &= \frac{1}{2(1-\gamma)}\left(1 - \frac{1}{2(1-\gamma)} + \frac{\gamma}{2(1-\gamma)}\right)\\ &= \frac{1}{4(1-\gamma)},\end{aligned}$$

which is higher than the Nash equilibrium profit in part (a), because

$$\begin{aligned}\frac{1}{4(1-\gamma)} &> \frac{1}{(2-\gamma)^2}\\ 4 - 4\gamma + \gamma^2 &> 4 - 4\gamma\\ &\Rightarrow \gamma^2 > 0,\end{aligned}$$

which holds regardless of the sign of γ. Intuitively, every firm obtains larger profit under collusion.

- Lastly, the collusive price and profit level also increase with the degree of substitutability, as shown below:

$$\begin{aligned}\frac{dp^c}{d\gamma} &= \frac{1}{2(1-\gamma)^2} > 0\\ \frac{d\pi^c}{d\gamma} &= \frac{1}{4(1-\gamma)^2} > 0.\end{aligned}$$

Intuitively, as the goods become more substitutable (γ increases), firms respond by increasing both their prices and corresponding profits.

(c) *Deviation.* Does firm i have incentive to deviate from the collusive price? If so, find its deviating price, and compare the associated profit level with profit levels found in parts (a) and (b).

- Without loss of generality, assume that firm 1 deviates and firm 2 keeps its price at the collusive price p^c. Firm 1 now chooses p_1 to solve the following profit maximization problem:

$$\begin{aligned}\max_{p_1 \geq 0} \pi\left(p_1, p_2^c\right) &= p_1\left(1 - p_1 + \gamma p_2^c\right)\\ &= p_1\left(1 - p_1 + \frac{\gamma}{2(1-\gamma)}\right).\end{aligned}$$

Differentiating with respect to p_1, we obtain

$$\frac{\partial \pi\left(p_1, p_2^c\right)}{\partial p_1} = \frac{2-\gamma}{2(1-\gamma)} - 2p_1 = 0.$$

Rearranging the above equation, we obtain the deviating price of firm 1

$$p^d = \frac{2-\gamma}{4(1-\gamma)},$$

which is higher than the Nash equilibrium price in part (a), because

$$\frac{2-\gamma}{4(1-\gamma)} > \frac{1}{2-\gamma}$$
$$4-4\gamma+\gamma^2 > 4-4\gamma$$
$$\Rightarrow \gamma^2 > 0.$$

This price is lower (higher) than the collusive price found in part (b) if goods are substitutes (complements). In the case of substitutable goods, we obtain

$$\frac{2-\gamma}{4(1-\gamma)} < \frac{1}{2(1-\gamma)}$$
$$4-6\gamma+2\gamma^2 < 4-4\gamma$$
$$\Rightarrow -\gamma(1-\gamma) < 0,$$

which holds for all parameter values. Therefore, prices satisfy $p^n < p^d < p^c$, so the deviating firm sets a price lower than the collusive price. However, when prices are strategic complements, the other firm keeps setting a price $p^c > p^n$, and this firm can set an above-equilibrium price $p^d > p^n$.

- Substituting $p^d = \frac{2-\gamma}{4(1-\gamma)}$ into the profit function, deviating profit becomes

$$\pi^d = p_1^d\left(1-p_1^d+\gamma p_2^c\right)$$
$$= \frac{2-\gamma}{4(1-\gamma)}\left(1-\frac{2-\gamma}{4(1-\gamma)}+\frac{\gamma}{2(1-\gamma)}\right)$$
$$= \left(\frac{2-\gamma}{4(1-\gamma)}\right)^2.$$

- We further check that firm 1 has incentives to deviate by

$$\left(\frac{2-\gamma}{4(1-\gamma)}\right)^2 > \frac{1}{4(1-\gamma)}$$
$$\gamma^2-4\gamma+4 > 4-4\gamma$$
$$\gamma^2 > 0,$$

which holds, so that every firm i has incentives to deviate from the collusive price. Finally, note that every firm i does not have incentives to deviate to the Nash equilibrium price since its profit from doing so, π_i^n, satisfies $\pi_i^d > \pi_i^c > \pi_i^n$.

(d) What is the minimum discount factor $\underline{\delta}$ to sustain collusion, assuming that firms interact infinitely and implement a Grim-Trigger Strategy once defection is detected?

- From Exercise 8.4, the minimum discount factor to sustain collusion is

$$\underline{\delta} = \frac{\pi^d - \pi^c}{\pi^d - \pi^n}$$

$$= \frac{\left(\frac{2-\gamma}{4(1-\gamma)}\right)^2 - \frac{1}{4(1-\gamma)}}{\left(\frac{2-\gamma}{4(1-\gamma)}\right)^2 - \frac{1}{(2-\gamma)^2}}$$

$$= \frac{\frac{1}{16(1-\gamma)^2}}{\frac{1}{16(1-\gamma)^2(2-\gamma)^2}} \times \frac{(2-\gamma)^2 - 4(1-\gamma)}{(2-\gamma)^4 - 16(1-\gamma)^2}$$

$$= \frac{(2-\gamma)^4 - 4(2-\gamma)^2(1-\gamma)}{(2-\gamma)^4 - 16(1-\gamma)^2}$$

$$= \frac{(2-\gamma)^2}{8 - 8\gamma + \gamma^2}.$$

(e) How does the minimal discount factor $\underline{\delta}$ you found in part (d) vary with γ? Explain and graph your results.

- The minimum discount factor $\underline{\delta}$ is increasing in γ at an increasing rate because

$$\frac{d\underline{\delta}}{d\gamma} = \frac{-2(2-\gamma)(8 - 8\gamma + \gamma^2) + 2(4-\gamma)(2-\gamma)^2}{(8 - 8\gamma + \gamma^2)^2}$$

$$= \frac{-2(2-\gamma)(8 - 8\gamma + \gamma^2 + 6\gamma - \gamma^2 - 8)}{(8 - 8\gamma + \gamma^2)^2}$$

$$= \frac{4\gamma(2-\gamma)}{(8 - 8\gamma + \gamma^2)^2},$$

which is positive (negative) if goods are substitutes (complements).

Further differentiating with respect to γ, we obtain

$$\frac{d^2\underline{\delta}}{d\gamma^2} = \frac{8(1-\gamma)(8 - 8\gamma + \gamma^2)^2 + 16\gamma(2-\gamma)(4-\gamma)(8 - 8\gamma + \gamma^2)}{(8 - 8\gamma + \gamma^2)^4}$$

$$= \frac{8\left[(1-\gamma)(8 - 8\gamma + \gamma^2) + 2\gamma(2-\gamma)(4-\gamma)\right]}{(8 - 8\gamma + \gamma^2)^3}$$

$$= \frac{8(8 - 3\gamma^2 + \gamma^3)}{(8 - 8\gamma + \gamma^2)^3} > 0.$$

- Figure 8.5 illustrates cutoff $\underline{\delta}$ as a function of parameter γ. Intuitively, as goods become more homogeneous (γ increases), firms compete more intensely, so that a higher discount rate is required to sustain collusion.

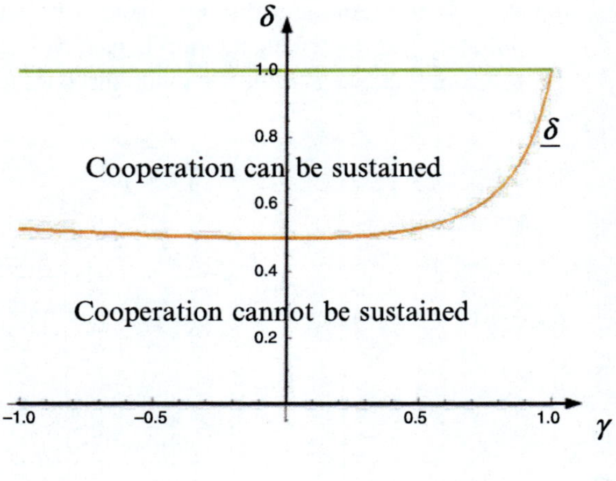

Fig. 8.5 Minimal discount factor as a function of γ

Exercise #8.6: Collusion with Time-Varying Demand^B

8.6 Consider an industry with two identical firms that compete à la Cournot, facing an inverse demand function $p(Q) = 1 - Q$ in period $t = 0$ and

$$p(Q) = g^t(1 - Q)$$

in all subsequent periods $t = 1, 2, \ldots$, where $Q \geq 0$ denotes aggregate output and g^t represents that the demand for this market grows at rate g in each period, i.e., $g > 1$ when the demand grows, $g \in (0, 1)$ when demand declines, and $g = 1$ when it stays constant across periods. For simplicity, assume that every firm faces a zero marginal cost of production, $c = 0$.

(a) Find the equilibrium output and profit level if both firms collude.

- If the firms collude, they maximize their joint profits, which is equivalent to producing as if they were a monopoly. Hence, they solve

$$\max_{Q \geq 0} \underbrace{g^t(1 - Q)}_{p(Q)} Q.$$

Differentiating with respect to Q yields

$$g^t(1 - 2Q) = 0.$$

Solving for Q provides our equilibrium output level for a monopolist in this setting (i.e., our aggregate output)

$$Q^C = \frac{1}{2}.$$

Since there are 2 identical firms, they each produce $q_i^C = \frac{Q^C}{2}$ of the aggregate output, or

$$q_i^C = \frac{Q^C}{2} = \frac{1}{4}.$$

To find our equilibrium profit level, we first derive our equilibrium price as follows:

$$p^C = 1 - Q^C = 1 - \frac{1}{2} = \frac{1}{2}$$

in period 0, and

$$p^C = g^t(1 - Q^C) = g^t\left(1 - \frac{1}{2}\right) = \frac{g^t}{2}$$

in period $t = 1, 2, \ldots$. Lastly, we insert our equilibrium price and aggregate output into our profit function to find our equilibrium profit level for firm i

$$\pi_i^C = p^C q_i^C = \left(\frac{1}{2}\right)\left(\frac{1}{4}\right) = \frac{1}{8}$$

in period 0, and

$$\pi_i^C = p^C q_i^C = \left(\frac{g^t}{2}\right)\left(\frac{1}{4}\right) = \frac{g^t}{8}$$

in period $t = 1, 2, \ldots$.

(b) Find the minimum discount factor, $\underline{\delta}$, that sustains collusion if each firm implements a Grim-Trigger Strategy. For simplicity, assume that $g\delta < 1$.

- After a history of $T - 1$ periods of cooperation, where all firms chose $q_i^C = \frac{1}{4}$ in all previous rounds, every firm i must decide whether to keep colluding or to deviate, keeping firm j's Grim-Trigger Strategy. We next separately evaluate the profits from collusion and from deviation and then compare these profits.
- *Profits from collusion.* If after a history of $T - 1$ periods of collusion, firm i colludes, it earns profits of $\pi_i^C = \frac{g^{T+t}}{8}$, where $t \geq 0$ (that is, in period T and in all subsequent periods), leading to a present discounted profits of

$$\begin{aligned}\pi_i^C + \delta\pi_i^C + \delta^2\pi_i^C + \ldots &= \frac{g^T}{8} + \delta\left(\frac{g^{T+1}}{8}\right) + \delta^2\left(\frac{g^{T+2}}{8}\right) + \ldots \\ &= \frac{g^T}{8}(1 + g\delta + (g\delta)^2 + \ldots) \\ &= \frac{g^T}{8}\left(\frac{1}{1 - g\delta}\right) \\ &= \frac{g^T}{8(1 - g\delta)}.\end{aligned}$$

Note that this discounted payoff is increasing in the number of previous periods when firms colluded, $T - 1$.

- *Profits from deviating.* If, after a history of $T - 1$ periods of collusion, firm i instead unilaterally deviates, we must first find its most profitable deviation. Setting up firm i's profit maximization problem at period T,

$$\max_{q_i \geq 0} \; g^T(1 - q_i - q_j)q_i,$$

which is evaluated at period T. Differentiating with respect to q_i yields

$$g^T\left(1 - 2q_i - q_j\right) = 0,$$

and solving for q_i, we find firm i's best response function

$$q_i(q_j) = \frac{1}{2} - \frac{1}{2}q_j.$$

If the other firm colludes, producing $q_j^C = \frac{1}{4}$ units, firm i's most profitable deviation becomes

$$q_i^D \equiv q_i(q_j^C) = \frac{1}{2} - \frac{1}{2}\left(\frac{1}{4}\right) = \frac{3}{8} \text{ units,}$$

which leads to a deviating profit of

$$\pi_i^D = \underbrace{g^T\left(1 - \frac{3}{8} - \frac{1}{4}\right)}_{p(Q)}\frac{3}{8} = \frac{9g^T}{64}.$$

This discounted payoff is also increasing in the number of previous periods when firms colluded, $T-1$.

- *Profits from reversion to the Nash equilibrium.* Next, we calculate the profits that firm i receives after each firm returns to the Nash equilibrium of the stage game (punishment stage). To find the Nash equilibrium of the stage game, we can use firm i's best response function found above, $q_i(q_j) = \frac{1}{2} - \frac{1}{2}q_j$. In a symmetric Nash equilibrium, every firm produces the same output level, $q_i = q_j = q$. Inserting this property into firm i's best response function, we find

$$q = \frac{1}{2} - \frac{1}{2}q.$$

Solving for q yields each firm's equilibrium output level

$$q^N = \frac{1}{3}$$

with aggregate output

$$Q^N = 2q^N = \frac{2}{3}.$$

Therefore, the equilibrium price in the first punishment period T is

$$p^N = g^T\left(1 - Q^N\right) = g^T\left(1 - \frac{2}{3}\right) = \frac{g^T}{3},$$

and lastly, equilibrium profits in this period T are

$$\pi_i^N = p^N q^N = \left(\frac{g^T}{3}\right)\left(\frac{1}{3}\right) = \frac{g^T}{9}.$$

Therefore, if firm i deviates in period T, it earns deviating profits π_i^D in that period, but π_i^N profits in all subsequent periods as each firm reverts back to the Nash equilibrium of the stage game after detecting firm i's deviation. Then, firm i's present discounted profit from deviating is

$$\begin{aligned}\pi_i^D + \underbrace{\delta\pi_i^N + \delta^2\pi_i^N + \ldots}_{\text{Punishment}} &= \frac{9g^T}{64} + \delta^1\left(\frac{g^{T+1}}{9}\right) + \delta^2\left(\frac{g^{T+2}}{9}\right) + \ldots \\ &= \frac{9g^T}{64} + \frac{g^{T+1}\delta}{9}\left(1 + g\delta + (g\delta)^2 + \ldots\right) \\ &= \frac{9g^T}{64} + \frac{g^{T+1}\delta}{9(1-g\delta)}.\end{aligned}$$

- *Calculating the minimal discount factor, $\underline{\delta}$.* Combining our previous results, we can claim that, after $T-1$ periods of collusion, every firm i colludes in period T if its present discounted profit is at least as high as that from deviating, that is,

$$\frac{g^T}{8(1-g\delta)} \geq \frac{9g^T}{64} + \frac{g^{T+1}\delta}{9(1-g\delta)}.$$

Rearranging, we find

$$\begin{aligned}\frac{1}{8} &\geq \frac{9(1-g\delta)}{64} + \frac{g\delta}{9} \\ g\delta\left(\frac{9}{64} - \frac{1}{9}\right) &\geq \frac{9}{64} - \frac{1}{8} \\ \frac{17}{576}g\delta &\geq \frac{1}{64}.\end{aligned}$$

Finally, solving for δ, we obtain the minimal discount factor sustaining collusion as follows:

$$\delta \geq \frac{9}{17g} \equiv \underline{\delta}.$$

(c) *Comparative statics.* How does the minimal discount factor $\underline{\delta}$ change when demand grows ($g > 1$), when demand declines ($g \in (0, 1)$), and when it stays constant, $g = 1$?

- *Growing demand.* When $g > 1$, this market's demand (and its profits) increases in each period, which makes collusion easier to sustain. Intuitively, deviating today entails a larger punishment in future periods, as firms would give up large future profits with a growing demand. In this context, deviation from collusion becomes less attractive, ultimately helping to sustain collusion under a larger range of discount factors in $\delta \in [0, 1]$.

- *Declining demand.* In contrast, when $g < 1$, demand shrinks in each period (as well as profits), making collusion more difficult to sustain in equilibrium. In this setting, deviation receives a smaller future punishment because future profits decline in time. As a consequence, deviation at any period T becomes more attractive, hindering the emergence of collusion in equilibrium. Note that when $g < \frac{9}{17}$, $\underline{\delta} > 1$ so that collusion cannot be sustained under any discount rates δ.
- *Constant demand.* Finally, if $g = 1$, demand does not change across time periods, simplifying the above condition on δ to that in standard oligopoly games, $\delta \geq \frac{9}{17}$, as we already found in Exercise 8.6.

Exercise #8.7: Collusion with Convex Costs and Symmetric Firms^A

8.7 Consider a duopoly selling homogeneous goods competing à la Cournot with an inverse demand function $p(Q) = 1 - Q$, where Q denotes the aggregate output, $Q = q_i + q_j$. Every firm i's total cost function is $C(q_i) = \frac{cq_i^2}{2}$, where q_i denotes its output and parameter c satisfies $0 \leq c \leq 1$.

(a) *Cournot competition.* Find the equilibrium output and profits when firms compete.

- *Finding best responses.* Every firm i chooses q_i to solve

$$\max_{q_i \geq 0} \; \pi_i(q_i) = \left(1 - q_i - q_j\right) q_i - \frac{cq_i^2}{2}.$$

Differentiating with respect to q_i and assuming interior solutions, we have

$$1 - 2q_i - q_j - cq_i = 0$$

yielding a best response function of

$$q_i(q_j) = \frac{1}{2+c} - \frac{1}{2+c} q_j,$$

which originates at $\frac{1}{2+c}$ and decreases in q_j at a rate of $\frac{1}{2+c}$. Graphically, as the common marginal cost increases (higher c), the best response function originates at a lower vertical intercept and becomes flatter. Therefore, an increase in c induces every firm to produce fewer units, but to reduce its output less significantly when its rival increases its own.

- *Finding equilibrium output.* Invoking symmetry where $q^* = q_i^* = q_j^*$ yields

$$q^* = \frac{1}{3+c}$$

yielding equilibrium profits of

$$\pi_i^* = \frac{\left[2(1 - 2q^*) - cq^*\right] q^*}{2}$$
$$= \frac{2+c}{2(3+c)^2}.$$

(b) *Cartel.* Find the equilibrium output and profits when firms coordinate.

- The cartel chooses q_i and q_j to solve

$$\max_{q_i, q_j \geq 0} \pi\left(q_i, q_j\right) = \overbrace{\left(1 - q_i - q_j\right) q_i - \frac{cq_i^2}{2}}^{\pi_i(q_i, q_j)} + \overbrace{\left(1 - q_i - q_j\right) q_j - \frac{cq_j^2}{2}}^{\pi_j(q_i, q_j)}.$$

Differentiating with respect to q_i and assuming interior solutions, we have

$$1 - 2q_i - 2q_j = cq_i,$$

and a similar expression applies to firm j, as follows:

$$1 - 2q_i - 2q_j = cq_j.$$

Since firms are symmetric, in equilibrium we must have $q^c = q_i^c = q_j^c$, yielding

$$q^c = \frac{1}{4 + c}$$

with associated profits of

$$\pi_i^c = \frac{\left[2\left(1 - 2q^c\right) - cq^c\right] q_i^c}{2} = \frac{1}{2\left(4 + c\right)}.$$

(c) Does firm i earn more profits under the cartel, as found in part (b), than under Cournot competition, as found in part (a)? Interpret your results.

- We find that firm i earns more profits from the cartel when

$$\pi_i^c = \frac{1}{2\left(4 + c\right)} \geq \frac{2 + c}{2\left(3 + c\right)^2} = \pi_i^*,$$

which we rearrange to

$$c^2 + 6c + 9 \geq c^2 + 6c + 8,$$

which holds for all $c \in [0, 1]$, so every firm i has incentives to join the cartel which internalizes the cost externality that firms impose on each other.

(d) *Deviation.* Find the optimal deviation and profits for every firm i.

- Substituting q^c into firm i's best response function found in part (a) yields

$$q_i(q_j^c) = \frac{1}{2 + c} - \frac{1}{2 + c} \underbrace{\frac{1}{4 + c}}_{q_j^c}.$$

Rearranging, we find a deviation output, q_i^d, of

$$q_i^d = \frac{3 + c}{\left(2 + c\right)\left(4 + c\right)}$$

with associated profits of

$$\pi_i^d = \frac{\left[2\left(1 - q_i^d - q_j^c\right) - cq_i^d\right] q_i^d}{2}$$
$$= \frac{(3+c)^2}{2\,(2+c)\,(4+c)^2}.$$

(e) Does firm i have incentives to deviate if firms interact only once?
- We find that firm i has incentives to deviate since $\pi_i^d > \pi_i^c$ yields

$$\frac{(3+c)^2}{2\,(2+c)\,(4+c)^2} > \frac{1}{2\,(4+c)},$$

which simplifies to

$$c^2 + 6c + 9 \geq c^2 + 6c + 8,$$

which holds for all $c \in [0, 1]$, so every firm i finds it profitable to deviate from q_i^c to q_i^d.

(f) *Minimal discount factor*. Find the minimum discount factor that sustains cooperation and denote it by $\underline{\delta}\,(c)$.For simplicity, you may assume that firms deploy the Grim-Trigger Strategy (GTS), where firms continue to cooperate when they have a history of cooperation, but revert to the Nash equilibrium found in part (a) forever once defection is detected.
- The profits from cooperation for firm i are

$$PV_i^c = \pi_i^c + \delta\pi_i^c + \delta^2\pi_i^c + \ldots = \frac{1}{1-\delta}\frac{1}{2\,(4+c)}$$

since $\pi_i^c = \frac{1}{2(4+c)}$, as found in part (b). This is also known as the present value of cooperation.
- If, instead, firm i deviates, its discounted stream of profits becomes

$$PV_i^d = \pi_i^d + \delta\pi_i^* + \delta^2\pi_i^* + \ldots = \frac{(3+c)^2}{2\,(2+c)\,(4+c)^2} + \frac{\delta}{1-\delta}\frac{2+c}{2\,(3+c)^2}$$

from our results in part (d), where we found $\pi_i^d = \frac{(3+c)^2}{2(2+c)(4+c)^2}$, and in part (a), where we found $\pi_i^* = \frac{2+c}{2(3+c)^2}$. This is also known as the present value of defection.
- Therefore, firm i has no incentives to deviate when $PV_i^c \geq PV_i^d$, solving

$$\frac{1}{1-\delta}\frac{1}{2(4+c)} \geq \frac{(3+c)^2}{(2+c)(4+c)^2} + \frac{\delta}{1-\delta}\frac{2+c}{2(3+c)^2}.$$

Solving for δ, we find that the minimum discount factor $\underline{\delta}\,(c)$ sustaining cooperation is

$$\delta \geq \underline{\delta}\,(c) \equiv \frac{(3+c)^2}{2c^2 + 12c + 17}.$$

(g) *Comparative statics.* How is the minimum discount factor you found in part (f) affected by c? What happens if $c = 0$. Explain and compare your results to Exercise 8.1.

- Differentiating $\underline{\delta}(c)$ with respect to c, we find

$$\frac{\partial \underline{\delta}(c)}{\partial c} = -\frac{2(c+3)}{\left(2c^2+12c+17\right)^2} < 0$$

so that the higher the cost parameter c, the lower is the minimum discount factor $\underline{\delta}(c)$. Intuitively, every firm i finds the good more costly to produce, so that both firms have incentives to collude under less restrictive conditions.

- Specifically, when $c = 0$, the above discount factor becomes

$$\underline{\delta}(0) = \frac{9}{17},$$

which coincides with the minimum discount factor we found in Exercise 8.1, where firms face linear production costs (and where we found that $\underline{\delta}$ was unaffected by c).

Exercise #8.8: Collusion with Convex Costs and Firm Asymmetry^B

8.8 Consider the industry in the previous exercise but assume, for simplicity, that every firm i's total cost function is $C(q_i) = \frac{c_i q_i^2}{2}$, where q_i denotes the output and parameter c_i satisfies $0 \le c_i \le 1$.

(a) *Cournot competition.* Find the equilibrium output and profits when firms compete.

- *Finding best responses.* Every firm i chooses q_i to solve the following profit maximization problem:

$$\max_{q_i \ge 0} \ \pi_i(q_i) = \left(1 - q_i - q_j\right) q_i - \frac{c_i q_i^2}{2}.$$

Differentiating with respect to q_i and assuming interior solutions, we have

$$1 - 2q_i - q_j - c_i q_i = 0,$$

which, solving for q_i, yields firm i's best response function

$$q_i\left(q_j\right) = \frac{1}{2+c_i} - \frac{1}{2+c_i} q_j,$$

which has a vertical intercept of $\frac{1}{2+c_i}$ and decreases in the output of firm j at a rate of $\frac{1}{2+c_i}$. Since $q_j \in [0,1]$, firm i always produces positive units of output.

- *Finding equilibrium output.* Substituting best response function $q_j(q_i) = \frac{1}{2+c_j} - \frac{q_i}{2+c_j}$ into $q_i\left(q_j\right) = \frac{1}{2+c_i} - \frac{q_j}{2+c_i}$, we obtain

$$q_i = \frac{1}{2+c_i} - \frac{\frac{1}{2+c_j} - \frac{q_i}{2+c_j}}{2+c_i},$$

which we rearrange as

$$q_i^* = \frac{1+c_j}{(2+c_i)\left(2+c_j\right) - 1}$$

so that firm i earns equilibrium profits of

$$\pi_i^* = \frac{\left[2\left(1 - q_i^* - q_j^*\right) - c_i q_i^*\right] q_i^*}{2}$$
$$= \frac{(2 + c_i)\left(1 + c_j\right)^2}{2\left[(2 + c_i)\left(2 + c_j\right) - 1\right]^2}.$$

(b) *Cartel.* Find the equilibrium output and profits when firms coordinate output. Which firm will produce more output in the cartel? Explain.

- The cartel chooses q_i and q_j to solve the joint profit maximization problem

$$\max_{q_i, q_j \geq 0} \pi\left(q_i, q_j\right) = \overbrace{\left(1 - q_i - q_j\right) q_i - \frac{c_i q_i^2}{2}}^{\pi_i(q_i, q_j)} + \overbrace{\left(1 - q_i - q_j\right) q_j - \frac{c_j q_j^2}{2}}^{\pi_j(q_i, q_j)}.$$

Differentiating with respect to q_i and assuming interior solutions, we have

$$1 - 2q_i - 2q_j = c_i q_i,$$

and a similar expression applies to firm j, as follows:

$$1 - 2q_i - 2q_j = c_j q_j$$

so that the cartel balances the marginal cost of production between firm i and j, that is, $c_i q_i = c_j q_j$, yielding the following system of linear equations:

$$\begin{bmatrix} 2 + c_i & 2 \\ 2 & 2 + c_j \end{bmatrix} \begin{bmatrix} q_i \\ q_j \end{bmatrix} = \begin{bmatrix} 1 \\ 1 \end{bmatrix}.$$

Solving for q_i and q_j, we find that

$$q_i^c = \frac{c_j}{2\left(c_i + c_j\right) + c_i c_j}$$
$$q_j^c = \frac{c_i}{2\left(c_i + c_j\right) + c_i c_j}$$

so that firm i produces more output in the cartel when

$$\frac{c_j}{2\left(c_i + c_j\right) + c_i c_j} > \frac{c_i}{2\left(c_i + c_j\right) + c_i c_j},$$

which simplifies to $c_j > c_i$, where it has a lower cost parameter than the rival firm j.

- Substituting q_i^c and q_j^c into the profit function, firm i earns profits of

$$\pi_i^c = \frac{\left[2\left(1 - q_i^c - q_j^c\right) - c_i q_i^c\right] q_i^c}{2}$$
$$= \frac{c_j}{2\left[2\left(c_i + c_j\right) + c_i c_j\right]}.$$

(c) Find the condition for firm i to earn more profits under the cartel, as found in part (b), than under Cournot competition, as found in part (a). Interpret your results.

- We find that firm i earns more profits from the cartel when

$$\frac{c_j}{2\left[(2+c_i)\left(2+c_j\right)-4\right]} \geq \frac{(2+c_i)\left(1+c_j\right)^2}{2\left[(2+c_i)\left(2+c_j\right)-1\right]^2},$$

which holds when

$$(2+c_i)^2\left(2+c_j\right)-2\left(2+c_i\right)\left(c_j^2+2c_j+2\right)-c_j \leq 0,$$

which we rearrange to

$$c_i \leq \overline{c}_i \equiv \frac{c_j^2+2c_j+2+\sqrt{\left[c_j\left(2+c_j\right)+1\right]\left[c_j\left(2+c_j\right)+4\right]}}{2+c_j}-2$$

so that firm i has incentives to join the cartel when its cost is relatively equal to firm j, and a symmetric argument applies to firm j. In summary, forming a cartel is individually rational for both firms when $c_i \leq \overline{c}_i$ holds for firm i and $c_j \leq \overline{c}_j$ holds for firm j. Figure 8.6 plots cutoff $\overline{c}_i$ as a function of c_j, which lies above the 45-degree line, where $c_i = c_j$. Similarly, the figure also includes cutoff $\overline{c}_j$, as a function of c_i, which is the mirror image of cutoff $\overline{c}_i$, about the 45-degree line. Graphically, (c_i, c_j)-pairs below cutoff $\overline{c}_i\left(c_j\right)$ and above cutoff

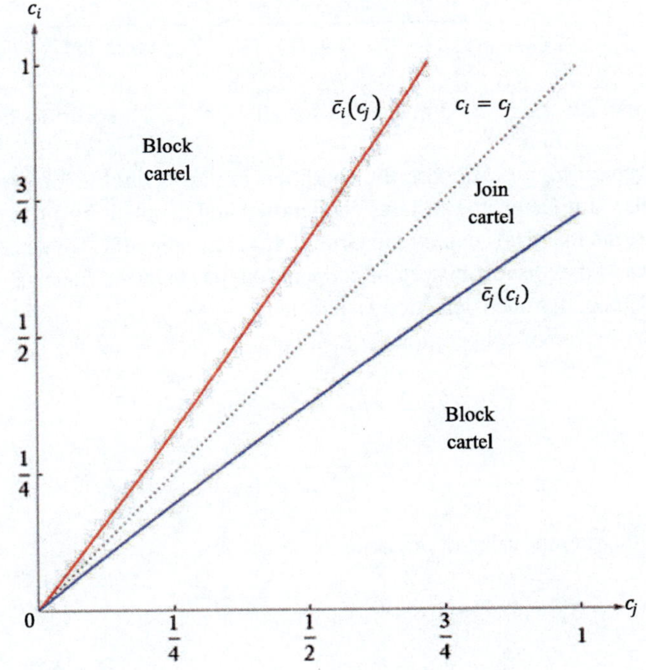

Fig. 8.6 Cartel profits in different cost regions

$\overline{c}_j\,(c_i)$ indicate that firms are relatively symmetric in their production costs and both of them find the cartel more profitable than Cournot competition.

(d) *Deviation.* Find the optimal deviation and profits for every firm i.

- Suppose $c_i \leq \overline{c}_i$ holds for every firm i. Substituting q_j^c into $q_i\left(q_j\right)$ yields

$$q_i\left(q_j^c\right) = \frac{1}{2+c_i} - \frac{\frac{c_i}{2(c_i+c_j)+c_i c_j}}{2+c_i},$$

which, simplifying, gives the deviation output of firm i

$$q_i^d = \frac{c_i + 2c_j + c_i c_j}{(2+c_i)\left[2\left(c_i+c_j\right)+c_i c_j\right]}$$

with associated profits of

$$\pi_i^d = \frac{\left[2\left(1-q_i^d-q_j^c\right)-c_i q_i^d\right] q_i^d}{2}$$

$$= \frac{\left(c_i+2c_j+c_i c_j\right)^2}{2\,(2+c_i)\left[2\left(c_i+c_j\right)+c_i c_j\right]^2}.$$

(e) Does firm i have incentives to deviate if firms interact only once?

- We find that firm i has incentives to deviate since $\pi_i^d > \pi_i^c$ yields

$$\frac{\left(c_i+2c_j+c_i c_j\right)^2}{2\,(2+c_i)\left[2\left(c_i+c_j\right)+c_i c_j\right]^2} > \frac{c_j}{2\left[2\left(c_i+c_j\right)+c_i c_j\right]},$$

which simplifies to $c_i^2 > 0$ that holds for all $c_i \in [0, 1]$, so firm i finds it profitable to deviate from q_i^c to q_i^d.

(f) *Minimal discount factor.* Suppose the condition in part (c) holds for every firm i. Find the minimum discount factor that sustains cooperation and denote it by $\underline{\delta}\left(c_i, c_j\right)$. For simplicity, you may assume that firms deploy the Grim-Trigger Strategy (GTS), where firms continue to cooperate when they have a history of cooperation, but revert to the Nash equilibrium found in part (a) perpetually once defection is detected.

- The present value of cooperation for firm i is

$$PV_i^c = \pi_i^c + \delta\pi_i^c + \delta^2\pi_i^c + \ldots$$

$$= \frac{c_j}{2\,(1-\delta)\left[2\left(c_i+c_j\right)+c_i c_j\right]}.$$

- Whereas the present value of deviation for firm i is

$$PV_i^d = \pi_i^d + \delta\pi_i^* + \delta^2\pi_i^* + \ldots$$

$$= \frac{\left(c_i+2c_j+c_i c_j\right)^2}{2\,(2+c_i)\left[2\left(c_i+c_j\right)+c_i c_j\right]^2} + \frac{\delta}{1-\delta}\frac{(2+c_i)\left(1+c_j\right)^2}{2\left[(2+c_i)\left(2+c_j\right)-1\right]^2}.$$

- Therefore, every firm i has no incentives to deviate when $PV_i^c \geq PV_i^d$, solving

$$\frac{c_j}{2(1-\delta)\left[2(c_i+c_j)+c_ic_j\right]} \geq \frac{\left(c_i+2c_j+c_ic_j\right)^2}{2(2+c_i)\left[2(c_i+c_j)+c_ic_j\right]^2} + \frac{\delta}{1-\delta}\frac{(2+c_i)\left(1+c_j\right)^2}{2\left[(2+c_i)(2+c_j)-1\right]^2}.$$

Solving for δ, the minimum discount factor $\underline{\delta}_i(c_i, c_j)$ sustaining cooperation is

$$\delta \geq \underline{\delta}_i(c_i, c_j) \equiv \frac{z^2\left[y_i^2 - x(2+c_i)c_j\right]}{\left[y_iz + x(2+c_i)(1+c_j)\right]\left[y_iz - x(2+c_i)(1+c_j)\right]},$$

where, for compactness, we define $x \equiv 2(c_i + c_j) + c_ic_j$, $y_i \equiv c_i + 2c_j + c_ic_j$, and $z \equiv (2+c_i)(2+c_j) - 1$.

(g) *Numerical simulation.* Evaluate the minimal discount factor found in part (f), $\underline{\delta}_i(c_i, c_j)$, at $c_j = 1/4$, at $c_j = 1/2$, and at $c_j = 3/4$. Show that, for a given c_j, $\underline{\delta}_i(c_i, c_j)$ increases in c_i. Interpret your results.

- Figure 8.7 depicts $\underline{\delta}_i(c_i, c_j)$ as a function of firm i's costs (c_i on the horizontal axis) and evaluated at three different values of its rival's cost: $c_j = 1/4$, $c_j = 1/2$, and $c_j = 3/4$. The figure shows that, for a given value of c_j, as firm i's cost increases (firms become more symmetric in costs), collusion is hindered. In contrast, cost asymmetry facilitates collusion because it helps firms shift a larger share of cartel output toward the most cost-

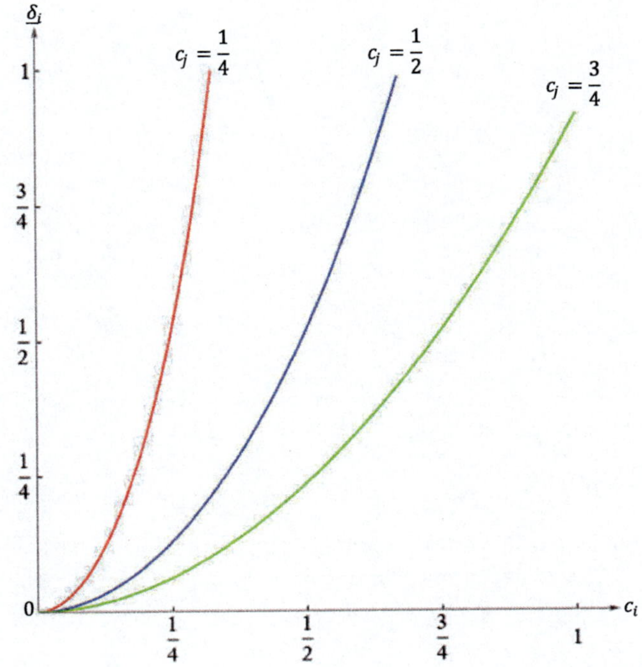

Fig. 8.7 Minimal discount factor sustaining collusion, $\underline{\delta}_i(c_i, c_j)$

efficient firm, increasing the overall profits of the cartel and, ultimately, making the cartel more attractive. In the extreme case where firm i's costs are zero ($c_i = 0$, in the origin of the figure), the minimal discount factor collapses to $\underline{\delta}_i(0, c_j) = 0$ for all c_j, implying that collusion can be sustained for all values of δ.

The figure also indicates that, as c_j increases, the curve representing $\underline{\delta}_i(c_i, c_j)$ shifts downward, indicating that collusion is facilitated. Intuitively, the same argument as above applies, but with the firms' identities switched.

- Finally, we can conclude that both firms cooperate under GTS if and only if

$$\delta \geq \max\left\{\underline{\delta}_i(c_i, c_j), \underline{\delta}_j(c_j, c_i)\right\},$$

which, graphically, indicates that δ must exceed the upper-envelope of the minimal discount factor sustaining collusion for firm i, $\underline{\delta}_i(c_i, c_j)$ and for firm j's, $\underline{\delta}_j(c_j, c_i)$. As an illustration, Fig. 8.8 depicts minimal discount factor $\underline{\delta}_i(c_i, \frac{1}{2})$, evaluated $c_j = \frac{1}{2}$, and $\underline{\delta}_j(\frac{1}{2}, c_j)$ evaluated at $c_i = \frac{1}{2}$, which is the mirror image of $\underline{\delta}_i(c_i, \frac{1}{2})$ around the 45-degree line. The upper-envelope, in this example, would be the region between $\underline{\delta}_i(c_i, \frac{1}{2})$ and $\underline{\delta}_j(\frac{1}{2}, c_j)$ for all $c_i, c_j \leq \frac{1}{2}$.

(h) *Symmetric costs.* Evaluate your results in parts (c) and (f) assuming $c = c_i = c_j$.

- Substituting $c = c_i = c_j$ into the cost cutoff in part (c) yields

$$c \leq \frac{c^2 - 2 + (1+c)\sqrt{c^2 + 2c + 4}}{2+c},$$

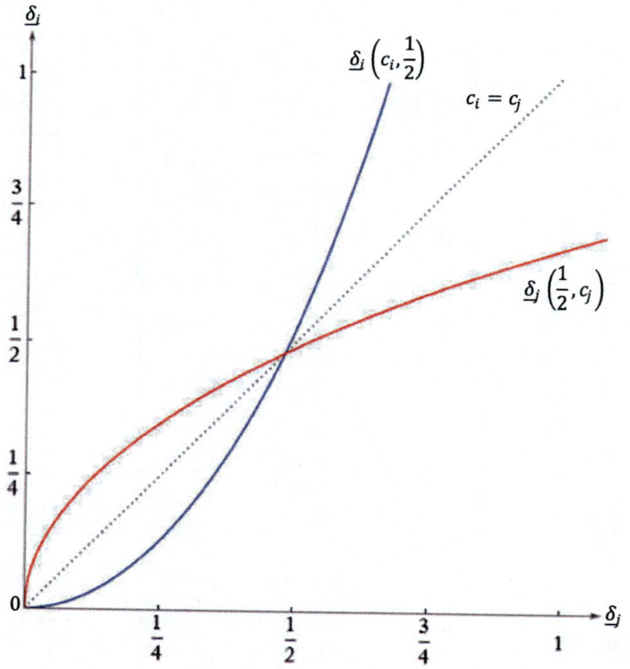

Fig. 8.8 $\underline{\delta}_i\left(c_i, \frac{1}{2}\right)$ and $\underline{\delta}_j\left(\frac{1}{2}, c_j\right)$

which simplifies to

$$, c\,(2 + c) \geq 0$$

which holds, so every firm has incentives to join the cartel when costs are equal.

- Substituting $c = c_i = c_j$ into the minimum discount factor in part (f) yields

$$\underline{\delta}\,(c, c) = \frac{(3 + c)^2}{2c^2 + 12c + 17},$$

which coincides with the minimum discount factor $\underline{\delta}\,(c)$ we found in Exercise 8.7, part (f).

Exercise #8.9: Collusion with Probability of Being Caught, Based on Harrington (2014)^C

8.9 Consider the setting in Exercise 8.6 with N firms, but assume that the cartel faces an exogenous probability p of being discovered, prosecuted, and convicted, by a regulatory agency, such as the US Federal Trade Commission. If caught and convicted in period t, a firm must pay a fine F_t, where $F_t = \beta F_{t-1} + f$. Parameter $1 - \beta$ can be understood as the depreciation rate, which we assume to satisfy $\beta \in (0, 1)$ to guarantee that the penalty is bounded, and f can be considered as the fixed penalty for every period that firms collude. In addition, assume that $F_0 = 0$, so that $F_1 = f$ and $F_2 = \beta F_1 + f$ and similarly for subsequent periods.

(a) For the first approach, let us consider a situation where the fine if convicted is constant, that is, $F_t = F$ for all t. Find the collusive value $V^C(F)$ as a function of penalty F. Interpret your results. [*Hint*: Solve for $V^C(F)$ recursively.]

- The collusive value $V^C(F)$ is defined recursively, at any given period t, as follows:

$$V^C(F) = \pi^C + \underbrace{p\left[\frac{\delta}{1-\delta}\pi^N - F\right]}_{\text{Detected}} + \underbrace{(1-p)\delta V^C(F)}_{\text{Not detected}}.$$

Therefore, the collusive value $V^C(F)$ includes the collusive profit, π^C, and in the next period payoffs are the following:

  – If the cartel is detected, which happens with probability p, every firm must pay a penalty F after being detected and generate a stream of Nash equilibrium profits π^N thereafter since we assumed that after being detected firms can never form a cartel in the future.
  – If the cartel is undetected, which occurs with probability $1 - p$, the stream of collusive payoffs starts in the next period, giving rise to continuation payoff $\delta V^C(F)$. Informally, we can understand this case as if the game starts again in period $t + 1$.

- This is a recursive function for $V^C(F)$, which shows up on both sides of the equality, so we can rearrange

$$(1-\delta)V^C(F) = (1-\delta)\pi^C + p\left[\delta\pi^N - (1-\delta)F\right] + (1-p)\delta(1-\delta)V^C(F)$$

and then solve for $V^C(F)$ obtaining that the collusive value is

$$V^C(F) = \frac{\pi^C(1-\delta) - p\left[F - \delta(F + \pi^N)\right]}{(1-\delta)\left[1 - \delta(1-p)\right]},$$

which can be rearranged as follows:

$$V^C(F) = \underbrace{\frac{\pi^C + p\pi^N \frac{\delta}{1-\delta}}{1 - \delta(1-p)}}_{\substack{\text{Expected present value} \\ \text{of profits from the product market}}} - \underbrace{p\frac{F}{1 - \delta(1-p)}}_{\text{Expected discounted penalty}}.$$

Intuitively, the first term represents the expected present value of profits from the product market, which includes the possibility of earning collusive profits π^C for the periods that the cartel is undetected, or Nash equilibrium profits π^N for all periods after the cartel is detected. The second term indicates the expected discounted penalty once the cartel is discovered, prosecuted, and convicted.

(b) Let us now extend our analysis to nonconstant fines. Find the collusive value $V^C(F)$ given an accumulated penalty F. [*Hint*: Solve for $V^C(F)$ recursively.]

- The collusive value $V^C(F)$ is defined recursively, for period $t = 1$ and $t = 2$, as follows:

$$V^C(F_1) = \pi^C + \underbrace{p\left[\frac{\delta}{1-\delta}\pi^N - F_2\right]}_{\text{Detected}} + \underbrace{(1-p)\delta V^C(F_2)}_{\text{Not detected}},$$

where penalties are $F_1 = \beta F_0 + f = f$ and $F_2 = \beta F_1 + f = \beta F + f$. More generally for any two periods t and $t+1$, the collusive value is defined as

$$V^C(F) = \pi^C + \underbrace{p\left[\frac{\delta}{1-\delta}\pi^N - (\beta F + f)\right]}_{\text{Detected}} + \underbrace{(1-p)\delta V^C(\beta F + f)}_{\text{Not detected}}.$$

In other words, the collusive value includes the collusive profit, π^C, and in the next period payoffs are the following:

  - If the cartel is detected, which happens with probability p, every firm must pay a penalty $\beta F + f$ after being detected and generate a stream of Nash equilibrium profits π^N thereafter since we assumed that after being detected firms can never form a cartel in the future.
  - If the cartel is undetected, which occurs with probability $1 - p$, the stream of collusive payoffs starts in the next period, giving rise to continuation payoff $\delta V^C(\beta F + f)$.

- Since the collusive value $V^C(F)$ shows up in both the left- and right-hand sides of the above expression, we can solve for $V^C(F)$ to obtain

$$V^C(F) = \pi^C + p\left[\frac{\delta}{1-\delta}\pi^N - (\beta F + f)\right] + (1-p)\delta V^C(\beta F + f),$$

which can be rearranged as follows:

$$V^C(F) = \pi^C + p\left[\frac{\delta}{1-\delta}\pi^N - (\beta F + f)\right] + (1-p)\delta\pi^C + p(1-p)\delta\frac{\delta}{1-\delta}\pi^N$$
$$- p(1-p)\delta\left(\beta\left(\beta F + f\right) + f\right) + (1-p)^2\delta^2 V^C\left(\beta\left(\beta F + f\right) + f\right)$$
$$= \left(\pi^C + p\pi^N\frac{\delta}{1-\delta}\right)\left[1 + \delta\left(1-p\right)\right] - pF\left[\beta + (1-p)\delta\beta^2\right]$$
$$- pf\left[1 + \delta\left(1-p\right) + \delta\left(1-p\right)\beta\right] + (1-p)^2\delta^2 V^C\left(\beta\left(\beta F + f\right) + f\right)$$
$$= \left(\pi^C + p\pi^N\frac{\delta}{1-\delta}\right)\sum_{t=0}^{T}\left[\delta\left(1-p\right)\right]^t - p\beta F\sum_{t=0}^{T}\left[(1-p)\,\beta\delta\right]^t$$
$$- pf\sum_{t=0}^{T}\delta^t\left(1-p\right)^t\left(\sum_{s=0}^{t}\beta^s\right) + (1-p)^{T+1}\delta^{T+1}V^C\left(\beta^{T+1}F + f\sum_{t=0}^{T}\beta^t\right).$$

The transversality condition is

$$\lim_{T\to\infty}(1-p)^{T+1}\delta^{T+1}V^C\left(\beta^{T+1}F + f\sum_{t=0}^{T}\beta^t\right) = 0,$$

which means the present discounted value of collusion at infinity is zero.

Inserting this result in our above expression for $V^C(F)$, we obtain

$$V^C(F) = \left(\pi^C + p\pi^N\frac{\delta}{1-\delta}\right)\frac{1 - \left[\delta\left(1-p\right)\right]^{T+1}}{1 - \delta\left(1-p\right)} - p\beta F\frac{1 - \left[\beta\delta\left(1-p\right)\right]^{T+1}}{1 - \beta\delta\left(1-p\right)}$$
$$- pf\sum_{t=0}^{T}\delta^t\left(1-p\right)^t\frac{1-\beta^{t+1}}{1-\beta}$$
$$= \left(\pi^C + p\pi^N\frac{\delta}{1-\delta}\right)\frac{1 - \left[\delta\left(1-p\right)\right]^{T+1}}{1 - \delta\left(1-p\right)} - p\beta F\frac{1 - \left[\beta\delta\left(1-p\right)\right]^{T+1}}{1 - \beta\delta\left(1-p\right)}$$
$$- \frac{pf}{1-\beta}\frac{1 - \left[\delta\left(1-p\right)\right]^{T+1}}{1 - \delta\left(1-p\right)} + \frac{p\beta f}{1-\beta}\frac{1 - \left[\beta\delta\left(1-p\right)\right]^{T+1}}{1 - \beta\delta\left(1-p\right)}.$$

For an infinite horizon game, we take $T \to \infty$, such that

$$V^C(F) = \left(\pi^C + p\pi^N\frac{\delta}{1-\delta}\right)\frac{1}{1 - \delta\left(1-p\right)} - \frac{p\beta F}{1 - \beta\delta\left(1-p\right)}$$
$$- \frac{pf}{1-\beta}\frac{1}{1 - \delta\left(1-p\right)} + \frac{p\beta f}{1-\beta}\frac{1}{1 - \beta\delta\left(1-p\right)}$$
$$= \frac{\pi^C + p\pi^N\frac{\delta}{1-\delta}}{1 - \delta\left(1-p\right)} -$$
$$\frac{p\beta F\left(1-\beta\right)\left[1 - \delta\left(1-p\right)\right] + pf\left[1 - \beta\delta\left(1-p\right)\right] - p\beta f\left[1 - \delta\left(1-p\right)\right]}{\left(1-\beta\right)\left[1 - \delta\left(1-p\right)\right]\left[1 - \beta\delta\left(1-p\right)\right]}$$

$$= \underbrace{\frac{\pi^C + p\pi^N \frac{\delta}{1-\delta}}{1-\delta(1-p)}}_{\substack{\text{Expected present value} \\ \text{of profits from the product market}}} - \underbrace{\frac{p(\beta[1-\delta(1-p)]F + f)}{[1-\delta(1-p)][1-\beta\delta(1-p)]}}_{\text{Expected discounted penalty}}.$$

Intuitively, the first term represents the expected present value of profits from the product market, which includes the possibility of earning collusive profits π^C for the periods that the cartel is undetected, or Nash equilibrium profits π^N for all periods after the cartel is detected. The second term indicates the expected discounted penalty once the cartel is discovered, prosecuted, and convicted.

(c) Write down the condition (inequality) expressing that every firm has incentives to collude, obtaining $V^C(F)$ rather than deviating. For simplicity, you can assume that if the cartel is convicted during the deviation period, it has no chances of being caught during the permanent punishment phase.

- Every firm cooperates if and only if

$$\underbrace{V^C(F)}_{\text{Collusion}} \geq \underbrace{\pi^D - p(\beta F + f)}_{\text{Expected profit from deviation}} + \underbrace{\frac{\delta}{1-\delta}\pi^N}_{\text{Permanent punishment}}.$$

Intuitively, when the firm deviates from the collusive price, the cartel can still be detected by regulatory authorities, which occurs with probability p, and thus charged with a penalty $\beta F + f$.

(d) The steady-state penalty is $F = \frac{f}{1-\beta}$, which is found by solving $F = \beta F + f$. Evaluate the collusive value $V^C(F)$ at this penalty, and compare your result to the minimal discount factor $\widehat{\delta}$ found in Exercise 8.8. Rearrange and interpret.

- Evaluating the collusive value $V^C(F)$ at the steady-state penalty $F = \frac{f}{1-\beta}$, we obtain

$$V^C\left(\frac{f}{1-\beta}\right) = \frac{\pi^C + p\pi^N\left(\frac{\delta}{1-\delta}\right) - \frac{pf}{1-\beta}}{1-\delta(1-p)}.$$

- Substituting the above results into part (b) yields

$$\frac{\pi^C + \frac{p\delta\pi^N}{1-\delta} - \frac{pf}{1-\beta}}{1-\delta(1-p)} \geq \pi^D - \frac{pf}{1-\beta} + \frac{\delta\pi^N}{1-\delta}.$$

Rearranging, we find

$$\pi^C - \delta(1-p)\frac{pf}{1-\beta} + \delta(1-p)\pi^D \geq \pi^D + \frac{\delta\pi^N}{1-\delta}[1-\delta(1-p)-p],$$

and solving for the discount factor δ, we obtain

$$\delta \geq \widehat{\delta}(p) \equiv \frac{\pi^D - \pi^C}{(1-p)\left(\pi^D - \pi^N - \frac{pf}{1-\beta}\right)}.$$

Note that cutoff $\widehat{\delta}(p)$ collapses to $\frac{\pi^D-\pi^C}{\pi^D-\pi^N}$ when the probability of cartel detection is zero ($p = 0$) as in part (a) of the exercise. In contrast, when detection is perfect, $p = 1$, cutoff $\widehat{\delta}(p)$ approaches infinity, thus indicating that condition $\delta \geq \widehat{\delta}(p)$ cannot hold for any admissible discount factor $\delta \in [0, 1]$.

(e) Does the minimal discount factor $\widehat{\delta}(p)$ increase in probability p? What about in penalty f?

- First, we differentiate cutoff $\widehat{\delta}(p)$ with respect to p

$$\frac{\partial \widehat{\delta}(p)}{\partial p} = -\frac{\left(\pi^D - \pi^C\right)\left[\pi^D - \pi^N - (1-2p)\frac{f}{1-\beta}\right]}{\left[(1-p)\left(\pi^D - \pi^N - \frac{pf}{1-\beta}\right)\right]^2},$$

which is positive for relative high levels of penalty, that is, $f > \frac{(1-\beta)(\pi^D - \pi^N)}{1-2p}$. Intuitively, as detection becomes more frequent, the cartel is more likely to be dissolved that makes collusion less attractive, such that the minimal discount factor sustaining collusion $\widehat{\delta}(p)$ increases.

- Second, we differentiate cutoff $\widehat{\delta}(p)$ with respect to f

$$\frac{\partial \widehat{\delta}(p)}{\partial f} = \frac{p\left(\pi^D - \pi^C\right)}{(1-\beta)(1-p)\left(\pi^D - \pi^N - \frac{pf}{1-\beta}\right)^2} \geq 0.$$

Intuitively, as the penalty becomes more severe, the minimal discount factor sustaining collusion $\widehat{\delta}(p)$ increases, since firms have to face a larger penalty in expectation and have less to gain from collusion.

Exercise #8.10: Collusion with Probability of Being Caught–Bertrand Competition^B

8.10 Consider the setting in Exercise 8.9, but assume that firms compete à la Bertrand, selling homogeneous products with inverse demand function $P(Q) = 1 - Q$, where Q denotes aggregate output. All firms face a symmetric marginal cost $c > 0$. In this setting, every firm obtains zero profits in the Nash equilibrium of the unrepeated game, entailing $\pi^N = 0$. If a firm unilaterally deviates from the collusive price (charging a price infinitely close, but below, the collusive price), it captures all industry sales, earning a profit $\pi^D = N\pi^C$ during the deviating period.

(a) Evaluate your results from part (d) of Exercise 8.3 in this context.

- We first need to find the profits that firms obtain from cooperating by setting a collusive price. Since the inverse demand function is $P(Q) = 1 - Q$, the direct demand function becomes $Q = 1 - P$, implying that the joint profit maximization problem in this setting is

$$\max_{P>0} PQ - cQ = (P - c)(1 - P).$$

Differentiating with respect to price P yields $1 - 2P + c = 0$, and solving for P we obtain a collusive price of $P^C = \frac{1+c}{2}$. Therefore, collusive profits for the industry are

$$N\pi^C = (P^C - c)(1 - P^C) = \left(\frac{1+c}{2} - c\right)\left(1 - \frac{1+c}{2}\right) = \frac{(1-c)^2}{4},$$

which implies that every firm's collusive profit, π^C, is $\frac{(1-c)^2}{4N}$.

- Evaluating our above results from part (c) of the previous exercise at profits $\pi^N = 0$, $\pi^C = \frac{(1-c)^2}{4N}$, and $\pi^D = N\pi^C = \frac{(1-c)^2}{4}$, we find that every firm cooperates after a history of cooperation if and only if

$$\delta \geq \overline{\delta} = \frac{\frac{(1-c)^2}{4} - \frac{(1-c)^2}{4N}}{(1-p)\left[\frac{pf}{1-\beta} + \frac{(1-c)^2}{4}\right]}$$

$$= \frac{1}{1-p} \times \frac{\frac{(1-c)^2}{4}\left(1 - \frac{1}{N}\right)}{\frac{4pf+(1-\beta)(1-c)^2}{4(1-\beta)}}$$

$$= \frac{N-1}{N(1-p)} \times \frac{(1-\beta)(1-c)^2}{4pf + (1-\beta)(1-c)^2}.$$

(b) Does collusion becomes easier to sustain when the penalty f increases? Or when the number of firms N increases?

- Differentiating the cutoff $\overline{\delta}$ with respect to N, we obtain

$$\frac{\partial \overline{\delta}}{\partial N} = \frac{1}{N^2(1-p)} \times \frac{(1-\beta)(1-c)^2}{4pf + (1-\beta)(1-c)^2} \geq 0$$

so that as the number of firm increases, a higher discount rate is needed to sustain collusion because the firm can capture a larger profit from deviation, π^D, relative to their collusion profit, π^C.

- Differentiating the cutoff $\overline{\delta}$ with respect to f, we obtain

$$\frac{\partial \overline{\delta}}{\partial f} = -\frac{N-1}{N(1-p)} \frac{4p(1-\beta)(1-c)^2}{\left[4pf + (1-\beta)(1-c)^2\right]^2} \leq 0.$$

Intuitively, as the penalty becomes more severe, the minimal discount factor sustaining collusion decreases, since firms have to face a larger penalty in expectation and have less to gain from collusion.

(c) *Numerical example.* Assume parameter values $\beta = 4/5$, $c = 1/2$, and $f = 1/80$. Find the minimal discount factor $\underline{\delta}$ that supports collusion. How does N interact with probability p to affect $\underline{\delta}$? Interpret.

- Substituting $\beta = 4/5$, $c = 1/2$, and $f = 1/80$ into the minimal discount factor found in part (a), $\underline{\delta}$, we obtain

$$\underline{\delta} = \frac{N-1}{N(1-p)} \times \frac{\left(1-\frac{4}{5}\right)\left(1-\frac{1}{2}\right)^2}{4p \times \frac{1}{80} + \left(1-\frac{4}{5}\right)\left(1-\frac{1}{2}\right)^2}$$

$$= \frac{N-1}{N(1-p)^2}.$$

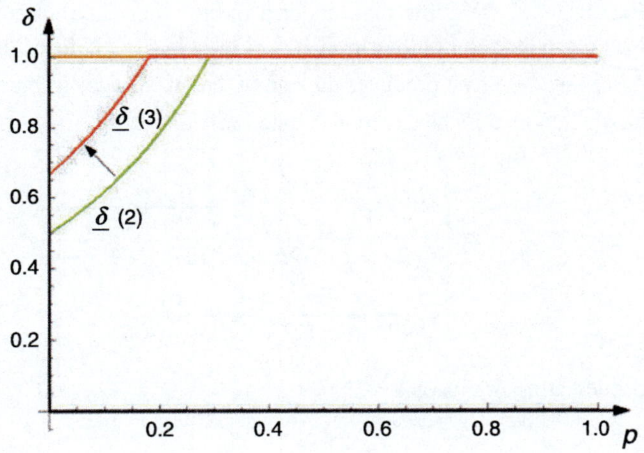

Fig. 8.9 Minimal discount factor $\underline{\delta}$ as a function of p

- Applying Implicit Function Theorem, we find

$$\frac{\partial N}{\partial p} = -\frac{\frac{\partial \delta}{\partial p}}{\frac{\partial \delta}{\partial N}} = -\frac{\frac{2(N-1)}{N(1-p)^3}}{\frac{1}{N^2(1-p)^2}} = -\frac{2N\,(N-1)}{1-p} < 0$$

so that holding $\underline{\delta}$ constant, when the probability of detection p increases, collusion can only be sustained when there are fewer firms colluding (N decreases).

- Figure 8.9 depicts cutoff $\underline{\delta}$ as a function of p, first evaluated at $n = 2$ firms and then at $n = 3$ firms. The figure shows that, when the number of colluding firms increases, cutoff $\underline{\delta}$ shifts to the northwest direction, thereby shrinking the range of (δ, p)-pair that supports collusion, making collusion more difficult to sustain.

Exercise #8.11: Collusion When Firms Exhibit Economies of Scope^C

8.11 Consider the same setting as in Exercise 7.8, but consider an infinitely repeated game. Firms use a standard GTS: every firm starts cooperating in period $t = 1$ and continues doing so if both firms cooperated in all previous periods. If one or both firms deviate in a previous period, however, every firm i responds producing the Cournot output thereafter. For simplicity, assume a common discount factor $\delta \in [0, 1]$.

(a) *Deviation*. When firm i deviates, what are its output, q_i^D, and associated profits, π_i^D? [*Hint*: You may consider that when firm i deviates, it does not enjoy economies of scope.]

- From Exercise 7.8, we know that, under a merger that maximizes firms' joint profits, every firm i produces an output level $q^C = \frac{a}{2(b+d)+c-\beta}$. If firm i unilaterally deviates from $q^C = \frac{a}{2(b+d)+c-\beta}$, it produces the output that maximizes its current profit (found by inserting q^C into firm i's best response function)

$$q_i^D = \frac{a}{2b+c} - \frac{d}{2b+c}\overbrace{\frac{a}{2(b+d)+c-\beta}}^{q^C}$$
$$= \frac{a(2b+d+c-\beta)}{(2b+c)(2(b+d)+c-\beta)}$$

and earns deviating profits of

$$\pi_i^D = \left(a - b\overbrace{\frac{a(2b+d+c-\beta)}{(2b+c)(2(b+d)+c-\beta)}}^{q_i^D} - d\overbrace{\frac{a}{2(b+d)+c-\beta}}^{q^C}\right) \times$$
$$\overbrace{\frac{a(2b+d+c-\beta)}{(2b+c)(2(b+d)+c-\beta)}}^{q_i^D} - \frac{c}{2}\left(\overbrace{\frac{a(2b+d+c-\beta)}{(2b+c)(2(b+d)+c-\beta)}}^{q_i^D}\right)^2$$
$$= \frac{a^2(2b+d+c-\beta)^2}{2(2b+c)(2(b+d)+c-\beta)^2}.$$

(b) *Present value*. Find the present value of firm i's profits if it (i) continues to cooperate with firm j and (ii) deviates from the cooperative agreement.

- *Cooperation*. When firm i continues to cooperate with firm j, every firm produces output $q^C = \frac{a}{2(b+d)+c-\beta}$, yielding

$$PV^C = \frac{\pi^C}{2} + \delta\frac{\pi^C}{2} + \delta^2\frac{\pi^C}{2} + \ldots$$
$$= \frac{1}{1-\delta}\frac{a^2}{2(2(b+d)+c-\beta)},$$

where $\pi^C = \frac{a^2}{2(b+d)+c-\beta}$ denotes the profits from cooperating (merger).

- *Deviation*. When firm i unilaterally deviates from cooperating with firm j, it earns π_i^D in the current period (as found in part a), but Cournot profit $\pi_i^* = \frac{a^2(2b+c)}{2(2b+c+d)^2}$ in all future periods, yielding

$$PV^D = \pi_i^D + \delta\pi_i^* + \delta^2\pi_i^* + \ldots$$
$$= \frac{a^2(2b+d+c-\beta)^2}{2(2b+c)(2(b+d)+c-\beta)^2} + \frac{\delta}{1-\delta}\frac{a^2(2b+c)}{2(2b+c+d)^2}$$

(c) Find the minimum discount factor $\underline{\delta}$ that sustains collusion.

- In order to sustain collusion, we need that $PV^C \geq PV^D$, such that

$$\frac{1}{1-\delta}\frac{a^2}{2(2(b+d)+c-\beta)} \geq \frac{a^2(2b+d+c-\beta)^2}{2(2b+c)(2(b+d)+c-\beta)^2} + \frac{\delta}{1-\delta}\frac{a^2(2b+c)}{2(2b+c+d)^2}.$$

Multiplying both sides by $\frac{2(1-\delta)(2b+c)(2(b+d)+c-\beta)^2(2b+c+d)^2}{a^2}$ yields

$$\begin{aligned}&(2b+c)\left(2\left(b+d\right)+c-\beta\right)\left(2b+c+d\right)^2\\ &\geq\left(1-\delta\right)\left(2b+c+d-\beta\right)^2\left(2b+c+d\right)^2+\delta\left(2b+c\right)^2\left(2\left(b+d\right)+c-\beta\right)^2.\end{aligned}$$

Simplifying the above expression, we obtain

$$\begin{aligned}&\delta\left[\left(2b+c+d-\beta\right)^2\left(2b+c+d\right)^2-\left(2b+c\right)^2\left(2\left(b+d\right)+c-\beta\right)^2\right]\\ &\geq\left(2b+c+d\right)^2\left[\left(2b+c+d-\beta\right)^2-\left(2b+c\right)\left(2\left(b+d\right)+c-\beta\right)\right].\end{aligned}$$

Rearranging and solving for δ, we find the minimum discount factor $\underline{\delta}$, where

$$\delta\geq\underline{\delta}\equiv\frac{\left(2b+c+d\right)^2\left(d^2-\beta\left(2\left(b+d\right)+c\right)+\beta^2\right)}{d\left(d-\beta\right)\left[2\left(2b+c\right)^2+2\left(2d-\beta\right)\left(2b+c\right)+d\left(d-\beta\right)\right]}.$$

(d) How does the minimum discount factor $\underline{\delta}$ change with β? Interpret your results.

- Differentiating cutoff $\underline{\delta}$ with respect to β yields

$$\begin{aligned}&\frac{\partial\underline{\delta}}{\partial\beta}\\ &=-\frac{(2b+c)t(2b+c+d)^2\left[2(2b+c)^2+2(3d-2\beta)(2b+c)+(d-\beta)(3d-\beta)\right]}{(d-\beta)^2\left[2(2b+c)^2+2(2d-\beta)(2b+c)+d(d-\beta)\right]^2},\end{aligned}$$

which is negative if and only if

$$2\left(2b+c\right)^2+2\left(3d-2\beta\right)\left(2b+c\right)+\left(d-\beta\right)\left(3d-\beta\right)\geq 0.$$

Rearranging and solving for β, we obtain that

$$\beta\leq\overline{\beta}\equiv 2\left(2b+c+d\right)-\sqrt{d^2+2d\left(2b+c\right)+2\left(2b+c\right)^2}.$$

(When solving for β, we actually find two roots, but we reject the case where $\beta \geq 2\left(2b+c+d\right)+\sqrt{d^2+2d\left(2b+c\right)+2\left(2b+c\right)^2}$ since $\beta \leq c$ by definition.)

- Next, we find the condition for $\overline{\beta} > c$ where

$$2\left(d+2b+c\right)-\sqrt{d^2+2d\left(2b+c\right)+2\left(2b+c\right)^2}>c$$

reduces to

$$3d^2 + 12bd + 2cd + 8b^2 - c^2 > 0,$$

which holds when the cost parameter c is not too high. In this context, it suffices for $\beta \leq c$ to satisfy the condition $\beta < \overline{\beta}$ yielding $\frac{\partial \delta}{\partial \beta} < 0$. In other words, when firms benefit from more significant cost complementarities (higher β), cutoff $\underline{\delta}$ decreases, ultimately sustaining collusion over a wider range of discount factors.

(e) Evaluate your results in part (c) assuming $\beta = 0$. What if $b = d = 1$ and $c = 0$? Interpret.

- When $\beta = 0$, the minimum discount factor becomes

$$\underline{\delta} = \frac{(2b + c + d)^2}{2\,(2b + c)^2 + 4d\,(2b + c) + d^2}.$$

- Substituting $b = d = 1$ and $c = 0$ into the above expression, we obtain

$$\underline{\delta} = \frac{(2 + 0 + 1)^2}{2\,(2 + 0)^2 + 4\,(2 + 0) + 1^2} = \frac{9}{17},$$

which coincides with the minimum discount factor for Cournot firms producing homogeneous goods and not benefiting from economies of scope.

(f) *Numerical example.* Evaluate the minimum discount factor $\underline{\delta}$ at $b = 1$, $c = d = \frac{1}{2}$, and $\beta = \frac{1}{20}$. What if, instead, $\beta = -\frac{1}{20}$? Interpret your results.

- When $b = 1$, $c = d = \frac{1}{2}$, and $\beta = \frac{1}{20}$, we find that

$$\begin{aligned}\underline{\delta} &= \frac{\left(2 \times 1 + \frac{1}{2} + \frac{1}{2}\right)^2 \left(\frac{1}{2^2} - \frac{1}{20}\left(2\left(1 + \frac{1}{2}\right) + \frac{1}{2}\right) + \frac{1}{20^2}\right)}{\frac{1}{2}\left(\frac{1}{2} - \frac{1}{20}\right)\left[2\left(2 \times 1 + \frac{1}{2}\right)^2 + 2\left(2 \times \frac{1}{2} - \frac{1}{20}\right)\left(2 \times 1 + \frac{1}{2}\right) + \frac{1}{2}\left(\frac{1}{2} - \frac{1}{20}\right)\right]} \\ &= \frac{124}{699} \simeq 0.177,\end{aligned}$$

whereas when $b = 1$, $c = d = \frac{1}{2}$, and $\beta = -\frac{1}{20}$, we obtain that

$$\begin{aligned}\underline{\delta} &= \frac{\left(2 \times 1 + \frac{1}{2} + \frac{1}{2}\right)^2 \left(\frac{1}{2^2} + \frac{1}{20}\left(2\left(1 + \frac{1}{2}\right) + \frac{1}{2}\right) + \frac{1}{20^2}\right)}{\frac{1}{2}\left(\frac{1}{2} + \frac{1}{20}\right)\left[2\left(2 \times 1 + \frac{1}{2}\right)^2 + 2\left(2 \times \frac{1}{2} + \frac{1}{20}\right)\left(2 \times 1 + \frac{1}{2}\right) + \frac{1}{2}\left(\frac{1}{2} + \frac{1}{20}\right)\right]} \\ &= \frac{6156}{7931} \simeq 0.776.\end{aligned}$$

- Figure 8.10 depicts $\underline{\delta}$ as a function of β, showing that when firms suffer diseconomies of scope, they need a higher discount factor to sustain collusion than when they experience economies of scope. Specifically, when $\beta > 0.073$ ($\beta < -0.1$), the economies (diseconomies) of scope are so significant that firms (do not) have incentives to collude for all discount factors, whereas when $\beta = 0$, firms do not have economies or diseconomies of scope, and the minimum discount factor is $\underline{\delta} = 0.507$. This falls below $\frac{9}{17} \simeq 0.53$ that

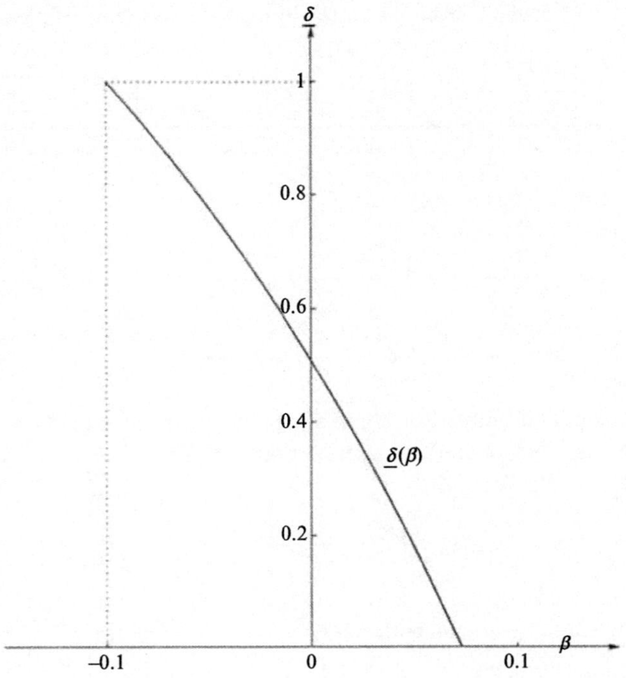

Fig. 8.10 Minimal discount factor $\underline{\delta}$ as a function of β

we found in part (e), since firms produce differentiated goods ($d = \frac{1}{2}$) thus facing less intensive competition, implying that collusion can be supported over a wider range of δ relative to the case of homogeneous goods ($d = 1$).

Exercise #8.12: Temporary Punishments in Cournot Collusion^C

8.12 Consider a Cournot duopoly with firms facing inverse demand function $p(Q) = 1 - Q$, where Q denotes aggregate output. Firms have a common marginal production cost c, where $1 > c \geq 0$, and a common discount factor $\delta \in (0, 1)$.

Assume a collusive agreement in which firms use a punishment scheme where they revert to the Nash equilibrium of the stage game during T consecutive periods, and after that punishment phase they return to cooperation. Under which conditions on the discount factor δ can collusion be sustained as an SPNE of the infinitely repeated game when firms rely on this temporary punishment? Interpret. [*Hint*: Rather than solving for the minimal discount factor sustaining cooperation, δ, solve for the length of the temporary punishment T.]

- In a collusive agreement, each firm produces half of the monopoly output, $\frac{1}{2}\frac{1-c}{2}$, and earns half of the monopoly profit, i.e., $\frac{(1-c)^2}{8}$.
- After a history of cooperation, every firm i's discounted stream of payoffs from continuing its cooperation is

$$\frac{(1-c)^2}{8} + \delta \frac{(1-c)^2}{8} + \delta^2 \frac{(1-c)^2}{8} + \ldots + \delta^T \frac{(1-c)^2}{8} + \ldots.$$

If, instead, firm i deviates from the cartel, its discounted stream of payoffs becomes

$$\underbrace{\frac{9(1-c)^2}{64}}_{\text{Deviation}} + \underbrace{\delta\frac{(1-c)^2}{9} + \ldots + \delta^T\frac{(1-c)^2}{9}}_{\text{Punishment for } T \text{ periods}} + \underbrace{\delta^{T+1}\frac{(1-c)^2}{8} + \ldots}_{\text{Back to cooperation}}.$$

Therefore, firm i prefers to cooperate if

$$\frac{(1-c)^2}{8} + \delta\frac{(1-c)^2}{8} + \delta^2\frac{(1-c)^2}{8} + \ldots + \delta^T\frac{(1-c)^2}{8}$$
$$\geq \frac{9(1-c)^2}{64} + \delta\frac{(1-c)^2}{9} + \ldots + \delta^T\frac{(1-c)^2}{9},$$

where we simplified both sides eliminating the stream of payoffs that coincide after the punishment phase is over (i.e., after period T). The above expression can be further simplified to

$$(1+\delta+\delta^2+\ldots+\delta^T)\frac{(1-c)^2}{8} \geq \frac{9(1-c)^2}{64} + \delta(1+\delta+\delta^2+\ldots+\delta^{T-1})\frac{(1-c)^2}{9}$$

or, after canceling $(1-c)^2$ on both sides,

$$(1+\delta+\delta^2+\ldots+\delta^T)\frac{1}{8} \geq \frac{9}{64} + \delta(1+\delta+\delta^2+\ldots+\delta^{T-1})\frac{1}{9}$$

At this point, note that $(1+\delta+\delta^2+\ldots+\delta^T)$ is a finite geometric progression which can be expressed as $\sum_{t=0}^{T}\delta^t = \frac{1-\delta^{T+1}}{1-\delta}$. This helps us simplify the above inequality as

$$\frac{1-\delta^{T+1}}{8(1-\delta)} \geq \frac{9}{64} + \frac{\delta\left(1-\delta^T\right)}{9(1-\delta)}.$$

Further rearranging, we obtain

$$\frac{9-9\delta^{T+1}-8\delta+8\delta^{T+1}}{72(1-\delta)} \geq \frac{9}{64}$$
$$\Rightarrow 8\left(9-8\delta-\delta^{T+1}\right) \geq 81(1-\delta)$$
$$\Rightarrow 8\delta^{T+1}-17\delta+9 \leq 0.$$

- In similar exercises, we solve for discount factor δ to find the minimal discount factor that supports cooperation in the infinitely repeated game. However, in this case, our expression is highly nonlinear in δ and does not allow for such an approach. We can nonetheless gain some intuition of our results by solving for T, which represents the length of the punishment phase as follows:

$$T \geq \widehat{T} \equiv \frac{\ln\left(\frac{17\delta-9}{8}\right)}{\ln\delta} - 1.$$

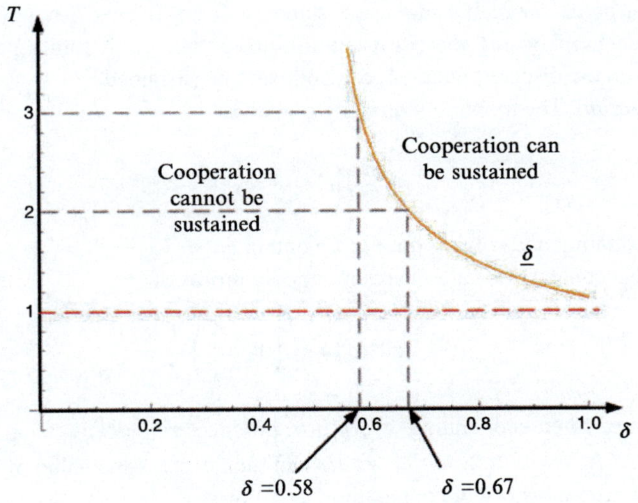

Fig. 8.11 Cutoff $\widehat{T}$ as a function of δ

Plotting cutoff $\widehat{T}$ in the vertical axis and the discount factor $\delta \in (0, 1)$ in the horizontal axis, we obtain Fig. 8.11. Intuitively, the punishment phase must be long enough and firms must care enough about their future profits (as indicated by T and δ pairs in the northwest of the figure) for the cartel with temporary punishment to be sustained as an SPNE of the infinitely repeated game. For illustration purposes, the figure also includes a dotted line at a height of $T = 1$, which does not cross with the curve representing cutoff $\widehat{T}$. This indicates that, under the tit-for-tat strategy when punishment only lasts for one period, cooperation cannot be sustained in the infinitely repeated game even if players assign full weight to their future payoffs (i.e., even if $\delta = 1$). When the punishment phase lasts for two periods, as indicated by the dotted line at a height of $T = 2$, cooperation can be sustained for discount factors satisfying $\delta \geq 0.67$, graphically represented by the range of δ to the right-hand side of $\delta = 0.67$ in the figure. A similar argument applies when the punishment phase lasts for $T = 3$ periods, where we obtain that cooperation can be supported as long as $\delta \geq 0.58$.

- After a history in which at least one firm deviated from cooperation, the cartel prescribes that every firm i implements the punishment during T rounds. This is firm i's best response to firm j implementing the punishment, so there are no further conditions on the discount factor, δ, or the length of the punishment phase, T, that we need to impose.
- When $T \to \infty$, we obtain our standard results under the Grim-Trigger Strategy that are found in Exercise 8.6. Specifically, collusion can be sustained under the widest range of discount rates because the firm that deviates suffers the most severe punishment from the permanent reversal to the Nash equilibrium output.

Exercise #8.13: Sustaining Collusion with One-Period Punishment^B

8.13 Consider a duopoly with two symmetric firms competing à la Cournot and facing inverse demand function $p(Q) = 1 - Q$, where Q denotes aggregate output. For simplicity, assume that firms have a common discount factor, δ, where $\delta \in (0, 1]$, and no production costs, $c = 0$.

(a) *Permanent punishment.* If firms use a standard Grim-Trigger Strategy, where a deviation is punished by choosing the Nash equilibrium output in all future periods, under which condition on the discount factor, δ, can collusion be sustained?

- *Cooperation.* The monopoly quantity q_m solves

$$\max_{q_m \geq 0} \pi^m = (1 - q_m)q_m.$$

Differentiating and solving for q_m, we obtain $q_m = \frac{1}{2}$.

In this context, the cartel earns monopoly profits of

$$\pi^m (q_m) = \left(1 - \frac{1}{2}\right)\frac{1}{2} = \frac{1}{4}.$$

Therefore, when cooperating, every firm produces $q^* = \frac{1}{2}q_m = \frac{1}{4}$ and earns collusive profits $\pi^* (q^*) = \frac{1}{2}\pi^m (q_m) = \frac{1}{8}$. We can then express the value of cooperation, at any period t after a history of cooperation, as follows:

$$\begin{aligned} V^* &= \frac{1}{8} + \frac{1}{8}\delta + \frac{1}{8}\delta^2 + \ldots \\ &= \frac{1}{8\,(1 - \delta)}. \end{aligned}$$

- *Punishment (Cournot equilibrium).* Every firm i chooses quantity q_i to solve

$$\max_{q_i \geq 0} \pi^c = (1 - q_i - q_j)q_i.$$

Differentiating with respect to q_i, we obtain $1 - 2q_i - q_j = 0$.

The best response function of firm i then becomes

$$q_i\left(q_j\right) = \frac{1}{2} - \frac{1}{2}q_j,$$

so that the profit of firm i in response to its rival's output is

$$\begin{aligned} \pi_i\left(q_j\right) &= \left(1 - q_i\left(q_j\right) - q_j\right) q_i\left(q_j\right) \\ &= \frac{\left(1 - q_j\right)^2}{4}, \end{aligned}$$

which is monotonically decreasing in the output of firm j.

In addition, we obtain a symmetric best response function that applies to firm j. Simultaneously solving for q_i and q_j, we find the Cournot equilibrium output $q_c = \frac{1}{3}$ for every firm i, entailing equilibrium profits of

$$\begin{aligned} \pi^c\left(q_c\right) &= \left(1 - \frac{1}{3} - \frac{1}{3}\right)\frac{1}{3} \\ &= \frac{1}{9}. \end{aligned}$$

- *Deviation.* If firm i deviates in period t, after a history of cooperation, its optimal deviation solves

$$\max_{q_d \geq 0} \pi^d = \left(1 - q^* - q_d\right) q_d$$
$$= \left(1 - \frac{1}{4} - q_d\right) q_d$$
$$= \frac{3}{4} q_d - q_d^2.$$

Differentiating with respect to q_d, we obtain

$$\frac{3}{4} - 2q_d = 0.$$

Solving for q_d yields

$$q_d = \frac{3}{8}.$$

Plugging q_d into π^d, deviation profits become

$$\pi^d = \frac{3}{4} \times \frac{3}{8} - \left(\frac{3}{8}\right)^2$$
$$= \frac{9}{64}.$$

Therefore, the value that firm i obtains from deviating at any period t is

$$V^d = \pi^d + \delta \pi^c + \delta^2 \pi^c + \ldots$$
$$= \frac{9}{64} + \frac{1}{9}\delta + \frac{1}{9}\delta^2 + \ldots$$
$$= \frac{9}{64} + \frac{\delta}{9(1-\delta)}.$$

- *Incentives to cooperate.* For cooperation to be sustained, we need $V^* \geq V^d$, that is,

$$\frac{1}{8(1-\delta)} \geq \frac{9}{64} + \frac{\delta}{9(1-\delta)}.$$

Rearranging yields

$$\frac{1}{8} \geq \frac{9}{64}(1-\delta) + \frac{1}{9}\delta,$$

and solving for δ, we obtain

$$\delta \geq \frac{9}{17}.$$

Hence, firms need to assign sufficiently high importance to future payoffs, in particular, $\delta \in \left[\frac{9}{17}, 1\right]$, to sustain the cartel.

(b) *Temporary punishment.* Assume now that firms only punish deviations by either firm choosing a punishing output during one period. If both firms choose the punishing output, $\hat{q}$, they go back to the collusive output in the next period. In addition, firms seek to produce a collusive output q that lies between the Nash equilibrium output, q_c, and that under monopoly, q_m. For each discount factor, δ, find the punishing output that firms must use to sustain collusion.

- In this setting, we need that every firm has incentives to apply the punishment during one period after observing that another firm deviated from the agreement (often known as the "credibility condition"). After a history of cooperation, every firm has incentives to continue the cooperation rather than unilaterally deviate (the "sustainability condition"). We analyze each condition below.
- *Credibility condition.* Denote the optimal collusive output as q^*. The payoff of complying with the punishment quantity $\hat{q}$ is

$$\begin{aligned} V^p &= \pi\left(\hat{q}\right) + \delta\pi\left(q^*\right) + \delta^2\pi\left(q^*\right) + \ldots \\ &= \pi\left(\hat{q}\right) + \frac{\delta}{1-\delta}\pi\left(q^*\right). \end{aligned}$$

For cooperation to be sustained, we need $V^p \geq \pi^d(\hat{q}) + \delta V^p$, or $(1-\delta)\,V^p \geq \pi^d(\hat{q})$, which is equivalent to

$$\delta[\pi(q^*) - \pi(\hat{q})] \geq \pi^d(\hat{q}) - \pi(\hat{q}).$$

Intuitively, the future payoff increase of returning to cooperation (moving from the punishing output $\hat{q}$ to the collusive output q^*) exceeds the current payoff gain of deviating from the punishing output $\hat{q}$ when its rival produces $\hat{q}$ units.

From part (a), we know that $\pi\left(q^*\right) = \frac{1}{8}$, $\pi(\hat{q}) = (1-2\hat{q})\hat{q}$, and $\pi^d(\hat{q}) = \frac{(1-\hat{q})^2}{4}$, which allows us to rewrite the credibility condition as follows:

$$\delta\left[\frac{1}{8} - (1-2\hat{q})\hat{q}\right] \geq \frac{(1-\hat{q})^2}{4} - (1-2\hat{q})\hat{q}.$$

Rewriting the above equation yields

$$2\,(8\delta - 9)\,\hat{q}^2 - 4\,(2\delta - 3)\,\hat{q} + (\delta - 2) \geq 0.$$

Solving for $\hat{q}(\delta)$, we obtain that

$$\hat{q} \leq \frac{2\,(2\delta - 3) - \sqrt{4\,(2\delta - 3)^2 - 2\,(8\delta - 9)\,(\delta - 2)}}{2\,(8\delta - 9)}$$

or

$$\hat{q} \geq \frac{2\,(2\delta - 3) + \sqrt{4\,(2\delta - 3)^2 - 2\,(8\delta - 9)\,(\delta - 2)}}{2\,(8\delta - 9)},$$

which is simplified to

$$\hat{q} \leq \frac{2(3-2\delta)+\sqrt{2\delta}}{2(9-8\delta)} \quad \text{or} \quad \hat{q} \geq \frac{2(3-2\delta)-\sqrt{2\delta}}{2(9-8\delta)}.$$

Since price must be non-negative, that is, $p = 1 - 2\hat{q} \geq 0$, we find that $0 \leq \hat{q} \leq \frac{1}{2}$. Now we have to determine which value of $\hat{q}$ satisfies both conditions.

i. First, let us consider $\hat{q} \leq \frac{2(3-2\delta)+\sqrt{2\delta}}{2(9-8\delta)}$, where

$$0 \leq \frac{2(3-2\delta)+\sqrt{2\delta}}{2(9-8\delta)} \leq \frac{1}{2}.$$

The left inequality, $0 \leq \frac{2(3-2\delta)+\sqrt{2\delta}}{2(9-8\delta)}$, holds for all $\delta \in (0, 1]$. The right inequality, $\frac{2(3-2\delta)+\sqrt{2\delta}}{2(9-8\delta)} \leq \frac{1}{2}$, simplifies to $(8\delta - 9)(2\delta - 1) \geq 0$, or $\delta \leq \frac{1}{2}$.

ii. Next, let us consider $\hat{q} \geq \frac{2(3-2\delta)-\sqrt{2\delta}}{2(9-8\delta)}$, where

$$0 \leq \frac{2(3-2\delta)-\sqrt{2\delta}}{2(9-8\delta)} \leq \frac{1}{2}.$$

The left inequality, $0 \leq \frac{2(3-2\delta)-\sqrt{2\delta}}{2(9-8\delta)}$, yields $(2-\delta)(9-8\delta) \geq 0$, or $\delta \leq \frac{9}{8}$, which is satisfied for all $\delta \in (0, 1]$. The right inequality, $\frac{2(3-2\delta)-\sqrt{2\delta}}{2(9-8\delta)} \leq \frac{1}{2}$, simplifies to $(9-8\delta)(2\delta-1) \geq 0$, or $\delta \geq \frac{1}{2}$.

In sum, the punishing output that satisfies the credibility condition becomes

$$\hat{q}(\delta) = \begin{cases} \frac{2(3-2\delta)+\sqrt{2\delta}}{2(9-8\delta)} & \text{if } 0 < \delta \leq \frac{1}{2} \\ \frac{2(3-2\delta)-\sqrt{2\delta}}{2(9-8\delta)} & \text{if } \frac{1}{2} < \delta \leq 1. \end{cases}$$

- *Sustainability condition.* After a history of cooperation, firms continue cooperating if

$$(1+\delta)\pi(q^*) \geq \pi^d(q^*) + \delta\pi(\hat{q}),$$

which is equivalent to

$$\delta[\pi(q^*) - \pi(\hat{q})] \geq \pi^d(q^*) - \pi(q^*).$$

Intuitively, given that firm j chooses q^*, firm i has more incentives to cooperate on q^* than deviating to $\hat{q}$. Specifically, choosing q^* in this period increases its future payoff by $\pi(q^*) - \pi(\hat{q})$ relative to deviating, which exceeds its current payoff gain from deviating, $\pi^d(q^*) - \pi(q^*)$, when its rival produces q^* units.

From part (a), we know that $\pi(q^*) = \frac{1}{8}$, $\pi(\hat{q}) = (1-2\hat{q})\hat{q}$, and $\pi^d(q^*) = \frac{(1-q^*)^2}{4} = \frac{9}{64}$. We can then rewrite the sustainability condition as follows:

$$\delta\left[\frac{1}{8} - \left(1 - 2\hat{q}(\delta)\right)\hat{q}(\delta)\right] \geq \frac{1}{64}.$$

Plugging in $\hat{q}(\delta)$ allows us to rewrite the sustainability condition as follows:

i. When $0 < \delta \leq \frac{1}{2}$, $\hat{q}(\delta) = \frac{2(3-2\delta)+\sqrt{2\delta}}{2(9-8\delta)}$, yielding the sustainability condition

$$\delta\left[\frac{1}{8}-\left(1-\frac{2(3-2\delta)+\sqrt{2\delta}}{9-8\delta}\right)\frac{2(3-2\delta)+\sqrt{2\delta}}{2(9-8\delta)}\right]\geq\frac{1}{64}.$$

Rewriting this equation, we obtain

$$\frac{\delta\left(2\sqrt{2\delta}+3\right)^2}{(9-8\delta)^2}\geq\frac{1}{8}.$$

Rearranging, we find

$$96\delta\sqrt{2\delta}+216\delta-81\geq 0,$$

which holds for $\delta \geq \frac{9}{32}$, so that $\hat{q}(\delta) = \frac{2(3-2\delta)+\sqrt{2\delta}}{2(9-8\delta)}$ is sustainable for $\delta \geq \frac{9}{32}$.

ii. When $\frac{1}{2} < \delta \leq 1$, $\hat{q}(\delta) = \frac{2(3-2\delta)-\sqrt{2\delta}}{2(9-8\delta)}$, yielding the sustainability condition

$$\delta\left[\frac{1}{8}-\left(1-\frac{2(3-2\delta)-\sqrt{2\delta}}{9-8\delta}\right)\frac{2(3-2\delta)-\sqrt{2\delta}}{2(9-8\delta)}\right]\geq\frac{1}{64}.$$

Rewriting this equation yields

$$\frac{\delta\left(2\sqrt{2\delta}-3\right)^2}{(9-8\delta)^2}\geq\frac{1}{8}.$$

Rearranging, we obtain

$$96\delta\sqrt{2\delta}-216\delta+81\leq 0,$$

which does not hold for any $\delta \in (0, 1]$. Thus $\hat{q}(\delta) = \frac{2(3-2\delta)-\sqrt{2\delta}}{2(9-8\delta)}$, despite being credible for $\frac{1}{2} < \delta \leq 1$, is not sustainable for all $0 < \delta \leq 1$.

In this case, let us consider $\hat{q}(\delta) = \frac{2\left(3-2\times\frac{1}{2}\right)+\sqrt{2\times\frac{1}{2}}}{2\left(9-8\times\frac{1}{2}\right)} = \frac{1}{2}$ for all $\frac{1}{2} < \delta \leq 1$, which is credible since substituting $\hat{q}(\delta) = \frac{1}{2}$ into the credibility condition yields $2(8\delta-9)\frac{1}{2^2}-4(2\delta-3)\frac{1}{2}+(\delta-2)>0$, which simplifies to $\delta-\frac{1}{2}>0$ that holds. Plugging $\hat{q}(\delta) = \frac{1}{2}$ into the sustainability condition, we find

$$\delta\left[\frac{1}{8}-\left(1-2\times\frac{1}{2}\right)\frac{1}{2}\right]\geq\frac{1}{64}.$$

Rewriting this equation, we obtain

$$\frac{1}{8}\delta\geq\frac{1}{64}$$

Since the premise $\delta > \frac{1}{2}$ satisfies the weaker condition $\delta \geq \frac{1}{8}$, $\hat{q}(\delta) = \frac{1}{2}$ can be sustained as the punishing output for all $\frac{1}{2} < \delta \leq 1$.

- In summary, punishing output that is both credible and sustainable becomes

$$\hat{q}(\delta) = \begin{cases} \frac{2(3-2\delta)+\sqrt{2\delta}}{2(9-8\delta)} & \text{if } \frac{9}{32} \leq \delta \leq \frac{1}{2} \\ \frac{1}{2} & \text{if } \frac{1}{2} < \delta \leq 1. \end{cases}$$

Note that when $\delta < \frac{9}{32}$, we do not identify any punishing output that sustains the one-period punishment in our model. Intuitively, the discount factor is so low that every firm i has incentives to deviate from the collusive output q^*.

- *Partial collusion.* If $0 < \delta < \frac{9}{32}$, full collusion can no longer be sustained, but partial collusion can be sustained, that is, each firm's output should be larger than $\frac{1}{4}$ (the cooperative level) but less than $\frac{1}{3}$ (the competitive level). In order to find the optimal output q^* and corresponding punishing output $\hat{q}$, we must solve the binding credibility and sustainability conditions together, where

$$\delta\left[(1-2q^*)\,q^* - (1-2\hat{q})\,\hat{q}\right] = \frac{(1-\hat{q})^2}{4} - (1-2\hat{q})\,\hat{q}$$

$$\delta\left[(1-2q^*)\,q^* - (1-2\hat{q})\,\hat{q}\right] = \frac{(1-q^*)^2}{4} - (1-2q^*)\,q^*.$$

Equating the first and second equalities, and factorizing, yields

$$\frac{1}{4}\left[(1-\hat{q}) - 2\hat{q}\right]^2 = \frac{1}{4}\left[(1-q^*) - 2q^*\right]^2.$$

Since $\frac{1}{4} < q^* < \frac{1}{3} < \hat{q}$, we obtain that $1 - 3q^* > 0 > 1 - 3\hat{q}$, yielding

$$3\hat{q} - 1 = 1 - 3q^*,$$

which is rearranged to give

$$\hat{q} = \frac{2}{3} - q^*.$$

In words, it means that when firms choose to cooperate on a higher q^*, a lower punishing output $\hat{q}$ suffices to sustain collusion.

Substituting $\hat{q} = \frac{2}{3} - q^*$ into either of the above equalities yields

$$\delta\left[(1-2q^*)\,q^* - \left(1 - 2\left(\frac{2}{3} - q^*\right)\right)\left(\frac{2}{3} - q^*\right)\right] = \frac{(1-3q^*)^2}{4},$$

which is simplified to give

$$81\,(q^*)^2 - 6\,(9-4\delta)\,q^* + (9-8\delta) = 0.$$

Applying the quadratic formula, we obtain $q^* = \frac{9-8\delta}{27}$ or $q^* = \frac{1}{3}$. If $q^* = \frac{1}{3}$, then $\hat{q} = \frac{1}{3}$, and firms are not punished for coordinating on the Cournot output. Therefore, we have

$q^*(\delta) = \frac{9-8\delta}{27}$, which, substituting into $\hat{q} = \frac{2}{3} - q^*$, yields

$$\hat{q}(\delta) = \frac{9+8\delta}{27}.$$

- *Summary.* We can now report under which conditions the discount factor δ, the collusive output q^*, and the punishing output $\hat{q}$ can sustain collusion.
  i. If $\delta \in \left(0, \frac{9}{32}\right)$, partial collusion can be supported if

$$q^*(\delta) = \frac{9-8\delta}{27}$$
$$\hat{q}(\delta) = \frac{9+8\delta}{27}.$$

  ii. If $\delta \in \left[\frac{9}{32}, \frac{1}{2}\right]$, full collusion arises if

$$q^* = \frac{1}{4}$$
$$\hat{q}(\delta) = \frac{2(3-2\delta)+\sqrt{2\delta}}{2(9-8\delta)}.$$

  iii. If $\delta \in \left(\frac{1}{2}, 1\right]$, full collusion arises if

$$q^* = \frac{1}{4}$$
$$\hat{q} = \frac{1}{2}.$$

(c) How do the collusive output q^* and punishing output $\hat{q}$ vary with δ? Explain.

- When $\delta \in \left(0, \frac{9}{32}\right)$, it is straightforward to verify that $q^*(\delta)$ decreases with δ, but $\hat{q}(\delta)$ increases with δ (see the left side of Fig. 8.12).
- When $\delta \in \left[\frac{9}{32}, \frac{1}{2}\right]$, the output level q^* does not change with δ, while $\hat{q}(\delta)$ increases with δ since

$$\frac{\partial \hat{q}(\delta)}{\partial \delta} = \frac{9+8\delta+12\sqrt{2\delta}}{2\sqrt{2\delta}(9-8\delta)^2} > 0,$$

  as depicted in the figure, for intermediate values of δ.
- When $\delta \in \left(\frac{1}{2}, 1\right]$, both q^* and $\hat{q}(\delta)$ level off and do not vary with δ, as illustrated in the two flat lines of the figure for all $\delta > \frac{1}{2}$.
- *Intuition.* Starting with a low value of δ, when firms assign larger weights to future payoffs (δ increases), firms need more severe punishments (i.e., further output expansion to punish the deviating firm, implying a larger $\hat{q}(\delta)$) in order to make punishment:
  – Credible, so that the deviating firm will not deviate from $\hat{q}$ in the punishment phrase and

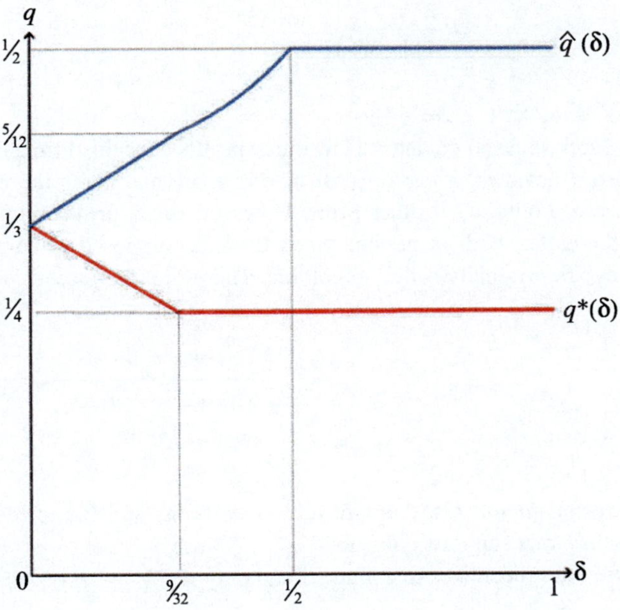

Fig. 8.12 Output as a function of discount factor

– Sustainable, so the deviating firm will not deviate from q^* in the cooperation phrase. When δ becomes sufficiently large, in particular, $\delta > \frac{1}{2}$, producing at the perfectly competitive level $\hat{q} = \frac{1}{2}$, with aggregate output of $Q = 2\hat{q} = 1$ and zero profits for every firm, firms do not to deviate from the collusive output $q^* = \frac{1}{4}$.

Exercise #8.14: Intermittent Collusion^B

8.14 Consider an industry with n firms selling homogeneous goods. All firms form a cartel every k period, equally spliting the monopoly profit π^m among themselves. In all other periods, firms compete on prices and obtain zero profits. When firm i deviates, it captures all market shares in that period, but all firms respond reverting to marginal cost pricing thereafter. For simplicity, assume that all firms share a common discount factor δ, where $\delta \in [0, 1]$.

(a) *Collusion.* Show that the lifetime profits of firm i if all firms collude are $\frac{\pi^m}{n(1-\delta^k)}$.

- As all firms collude every k period, the lifetime profits of firm i are

$$PV^c = \overbrace{\frac{\pi^m}{n}}^{\text{Collusion}} + \overbrace{\left(\delta \times 0 + \ldots + \delta^{k-1} \times 0\right)}^{\text{Price competition}} + \overbrace{\delta^k \frac{\pi^m}{n}}^{\text{Collusion}}$$

$$+ \overbrace{\left(\delta^{k+1} \times 0 + \ldots + \delta^{2k-1} \times 0\right)}^{\text{Price competition}} + \overbrace{\delta^{2k} \frac{\pi^m}{n}}^{\text{Collusion}} + \ldots$$

$$= \frac{\pi^m}{n}\left(1 + \delta^k + \delta^{2k} + \ldots\right)$$

$$= \frac{\pi^m}{n\left(1-\delta^k\right)}$$

since $1 + \delta^k + \delta^{2k} + \ldots = \frac{1}{1-\delta^k}$.

(b) *Deviation.* Suppose firm i deviates. How much profits does this firm earn?

- When firm i deviates, it captures all market share and earns the monopoly profit π^m, leaving zero profits to all other firms. However, every firm responds by marginal cost pricing thereafter, both in periods when they are supposed to compete and in periods when they are originally called to collude. Therefore, the present profits from deviating become

$$PV^d = \overbrace{\pi^m}^{\text{Deviation}} + \overbrace{\delta \times 0 + \delta^2 \times 0 + \ldots}^{\text{Price competition}}$$
$$= \pi^m.$$

(c) *Minimal discount factor.* Use your results in parts (a) and (b) to identify the minimum discount factor $\underline{\delta}$ that supports collusion.

- Every firm i has incentives to collude if and only if $PV^c \geq PV^d$, that is,

$$\frac{\pi^m}{n\left(1-\delta^k\right)} \geq \pi^m.$$

Rearranging, we obtain

$$1 - \delta^k \leq \frac{1}{n}.$$

Solving for δ, the minimum discount factor that supports collusion is

$$\delta \geq \underline{\delta} \equiv \left(1 - \frac{1}{n}\right)^{\frac{1}{k}}.$$

(d) *Comparative statics.* How does the minimal discount factor sustaining collusion, $\underline{\delta}$, change with parameters n and k? Explain.

- Differentiating $\underline{\delta}$ with respect to n, we find

$$\frac{\partial \underline{\delta}}{\partial n} = \frac{1}{kn^2}\left(1 - \frac{1}{n}\right)^{\frac{1}{k}-1} > 0$$

so that a higher discount factor is required to support more firms to collude. In other words, collusion is more difficult to hold if several firms seek to collude.

- Differentiating $\underline{\delta}$ with respect to k, we have

$$\frac{\partial \underline{\delta}}{\partial k} = -\frac{1}{k^2}\left(1 - \frac{1}{n}\right)^{\frac{1}{k}} \log\left(1 - \frac{1}{n}\right) > 0$$

since $\log\left(1 - \frac{1}{n}\right) < 0$ for all $n > 1$, so that a higher discount factor is required when firms collude less frequently. Therefore, collusion is more difficult to hold if firms plan to collude in few periods.

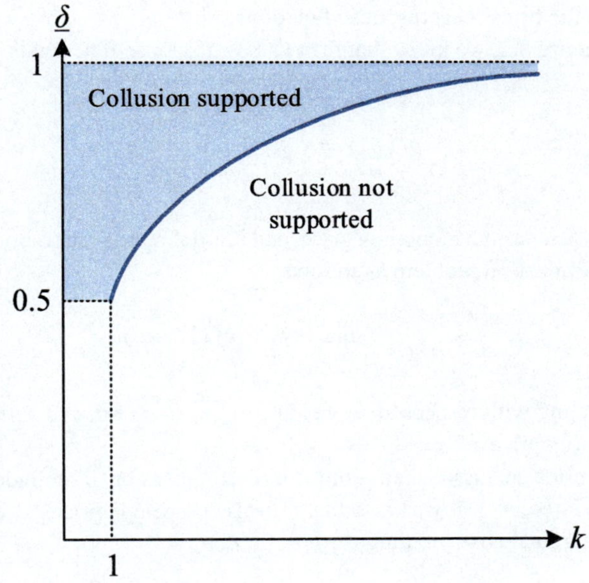

Fig. 8.13 Range of δ and k supporting collusion

(e) *Numerical example.* Assume $n = 2$. Find the minimum discount factor $\underline{\delta}$, and plot how it varies with k with k on the horizontal and $\underline{\delta}$ on the vertical axis. Interpret.

- Inserting $n = 2$ into $\underline{\delta}$, the minimum discount factor becomes

$$\underline{\delta} = \frac{1}{2^{\frac{1}{k}}}.$$

- When $k = 1$, firms collude every period, so $\underline{\delta} = \frac{1}{2}$ as in the standard collusion model. When $k \to \infty$, however, firms never collude again, since $\underline{\delta}$ becomes

$$\lim_{k\to\infty} \frac{1}{2^{\frac{1}{k}}} = \frac{1}{2^{\lim_{k\to\infty} \frac{1}{k}}} = \frac{1}{2^0} = 1$$

where we apply the Continuous Mapping Theorem to the continuous function $\frac{1}{2^{\frac{1}{k}}}$ in the limit, so that firms will not collude if they never collude again. Figure 8.13 depicts the minimum discount factor $\underline{\delta}$ as a function of k, illustrating our above results.

Exercise #8.15: Multi-period Collusion and Inflexible Prices^C

8.15 Consider two firms competing à la Bertrand on homogeneous goods produced at a constant marginal cost of c. Firm i's demand, where $i, j \in \{1, 2\}$, is

$$Q_i(p_i) = \begin{cases} 1 - p_i & \text{if } p_i < p_j \\ \frac{1}{2}(1 - p_i) & \text{if } p_i = p_j \\ 0 & \text{if } p_i > p_j. \end{cases}$$

(a) Characterize the firms' best response functions.
- From Exercise 3.1, we know that firm i's best response function is

$$p_i(p_j) = \begin{cases} p^m & \text{if } p_j > p^m \\ p_j - \varepsilon & \text{if } c < p_j \leq p^m \\ c & \text{if } p_j \leq c, \end{cases}$$

where p^m denotes the monopoly price, and it is defined as the solution to the monopolist's profit maximization problem as follows:

$$\max_{p_i \geq 0} \; (p_i - c)(1 - p_i).$$

Differentiating with respect to p_i yields $1 - 2p_i + c = 0$, and solving for p_i, we obtain the monopoly price $p^m = \frac{1+c}{2}$.

(b) Find firm i's price and associated profit if it (i) competes or (ii) colludes with firm j.
- Also from Exercise 3.1, we know that if firms compete in prices, every firm i sets its price equal to marginal cost, yielding zero profit; that is,

$$p^n = c$$
$$\pi^n = 0,$$

where subscript n stands for Nash equilibrium outcomes.
- If firms collude with each other, every firm i sets the monopolist's price

$$p^c = p^m = \frac{1+c}{2}$$

earning half of the monopoly profit

$$\pi^c = \frac{(p^c - c)(1 - p^c)}{2} = \frac{\left(\frac{1+c}{2} - c\right)\left(1 - \frac{1+c}{2}\right)}{2} = \frac{(1-c)^2}{8}.$$

(c) What are firm i's price and associated profit if it deviates from collusion with firm j?
- If firm i unilaterally deviates from $p^c = \frac{1+c}{2}$, its deviating price is to undercut its rival's price by ε, so that

$$p^d = \frac{1+c}{2} - \varepsilon,$$

which converges to $p^d = \frac{1+c}{2}$ when $\varepsilon \to 0$. Therefore, its deviating profits become

$$\pi^d = \left(p^d - c\right)\left(1 - p^d\right)$$

$$= \left(\frac{1+c}{2} - c\right)\left(1 - \frac{1+c}{2}\right)$$

$$= \frac{(1-c)^2}{4}.$$

(d) Suppose firm j, having detected firm i's defection, has its price fixed at the collusive level for k periods before reverting to the Nash equilibrium level at the $k+1$th period and thereafter. What is the minimum discount rate $\underline{\delta}$ that sustains collusion, and how does it change with k? Explain and graph your results.

- If firm i commits to p^c, its present value of colluding profit becomes

$$V^c = \underbrace{\pi^c}_{\text{this period}} + \underbrace{\delta\pi^c + \delta^2\pi^c + \ldots}_{\text{all future periods}}$$

$$= \left(1 + \delta + \delta^2 + \ldots\right)\frac{(1-c)^2}{8}$$

$$= \lim_{k\to\infty} \frac{1-\delta^{k+1}}{1-\delta}\frac{(1-c)^2}{8}$$

$$= \frac{1}{1-\delta} \times \frac{(1-c)^2}{8}.$$

- If firm i deviates from p^c to p^d, its present value of deviating profit becomes

$$V^d = \underbrace{\pi^d}_{\text{this period}} + \underbrace{\delta\pi^d + \delta^2\pi^d + \ldots + \delta^k\pi^d}_{k \text{ periods}} + \underbrace{\delta^{k+1}\pi^n + \delta^{k+2}\pi^n + \ldots}_{k+1 \text{ period onwards}}$$

$$= \left(1 + \delta + \ldots + \delta^k\right)\frac{(1-c)^2}{4} + 0$$

$$= \frac{1-\delta^{k+1}}{1-\delta} \times \frac{(1-c)^2}{4}.$$

- Therefore, for firm i to honor its commitment, we need $V^c > V^d$, or

$$\frac{1}{1-\delta} \times \frac{(1-c)^2}{8} > \frac{1-\delta^{k+1}}{1-\delta} \times \frac{(1-c)^2}{4}$$

$$\frac{1}{2} > 1 - \delta^{k+1}$$

$$\delta^{k+1} > \frac{1}{2}$$

$$\Rightarrow \delta > \underline{\delta}(k) \equiv \frac{1}{2^{\frac{1}{k+1}}}.$$

- Taking the first and second derivatives of $\underline{\delta}(k) = 2^{-\frac{1}{k+1}} = \exp\left(-\frac{1}{k+1} \times \log 2\right)$ with respect to k,

$$\begin{aligned}\frac{\partial \underline{\delta}}{\partial k} &= \frac{\log 2}{(k+1)^2} \exp\left(-\frac{1}{k+1} \times \log 2\right) \\ &= \log 2 \times 2^{-\frac{1}{k+1}} (k+1)^{-2} > 0 \\ \frac{\partial^2 \underline{\delta}}{\partial k^2} &= (\log 2)^2 \times 2^{-\frac{1}{k+1}} (k+1)^{-4} - 2 \log 2 \times (k+1)^{-3} 2^{-\frac{1}{k+1}} \\ &= \frac{\log 2}{(k+1)^4} 2^{-\frac{1}{k+1}} (\log 2 - 2k - 2) < 0\end{aligned}$$

so that $\underline{\delta}(k)$ is strictly increasing and concave (increases at a decreasing rate) in k.

We further evaluate the limiting values of $\underline{\delta}(k)$, finding that

$$\underline{\delta}(0) = \frac{1}{2^{\frac{1}{0+1}}} = \frac{1}{2}, \text{ and}$$

$$\begin{aligned}\lim_{k\to\infty} \underline{\delta}(k) &= \lim_{k\to\infty} \frac{1}{2^{\frac{1}{k+1}}} \\ &= \frac{1}{2^0} = 1.\end{aligned}$$

Figure 8.14 depicts that collusion is sustained when $\delta \geq \underline{\delta}$, but it is not sustained otherwise. The minimum discount rate $\underline{\delta}$ increases in k because if firms are relatively inflexible to price adjustments, a higher discount rate is required to sustain collusion. Specifically, when $k = 0$, firms can revert to the Nash equilibrium price instantaneously, so collusion can be sustained under the least restrictive parameter conditions, for which $\delta \geq \frac{1}{2}$. However, as k approaches infinity, the firm's price is fixed infinitely and collusion cannot be sustained for any value of $\delta \in [0, 1]$.

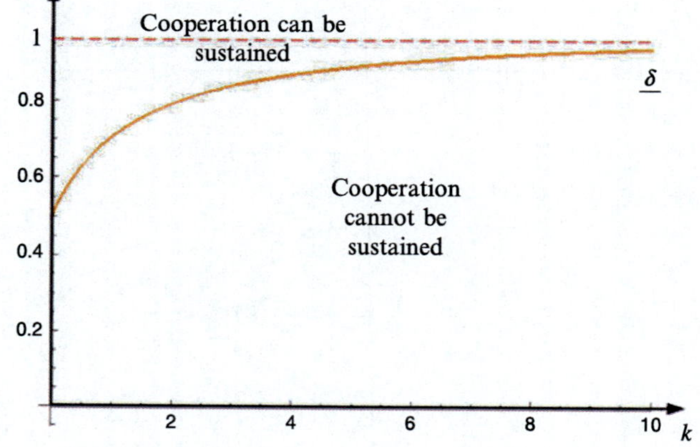

Fig. 8.14 Minimal discount factor increases in k

(e) Suppose firm j, having detected firm i's defection, retaliates by pricing at the Nash equilibrium level for k periods and has incentive to cooperate with firm i again at the $k+1$th period and thereafter. What is the minimum discount rate $\underline{\delta}$ that sustains collusion, and how does it change with k? Explain and graph your results.

- If firm i deviates from p^c to p^d, its present value of deviating profit becomes

$$PV^d = \underbrace{\pi^d}_{\text{this period}} + \underbrace{0 + \ldots + 0}_{k \text{ periods}} + \underbrace{\delta^{k+1}\pi^c + \delta^{k+2}\pi^c + \ldots}_{k+1 \text{ period onwards}}$$

$$= \frac{(1-c)^2}{4} + 0 + \delta^{k+1}\left(1 + \delta + \delta^2 + \ldots\right)\frac{(1-c)^2}{8}$$

$$= \frac{(1-c)^2}{4} + \frac{\delta^{k+1}}{1-\delta} \times \frac{(1-c)^2}{8}.$$

- Therefore, for firm i to honor its commitment, we need

$$PV^c > PV^d$$

$$\frac{1}{1-\delta} \times \frac{(1-c)^2}{8} > \frac{(1-c)^2}{4} + \frac{\delta^{k+1}}{1-\delta} \times \frac{(1-c)^2}{8}$$

$$\frac{1-\delta^{k+1}}{1-\delta} \times \frac{1}{2} > 1$$

$$1 - \delta^{k+1} > 2 - 2\delta$$

$$\delta^{k+1} - 2\delta + 1 < 0.$$

- Defining $\delta(k) \equiv \delta^{k+1} - 2\delta + 1 = \exp\left[(k+1)\log\delta\right] - 2\delta + 1$, we take the first and second derivatives of $\delta(k)$ with respect to k

$$\frac{\partial \delta(k)}{\partial k} = \log\delta \times \exp\left((k+1)\log\delta\right)$$

$$= \log\delta \times \delta(k) < 0$$

$$\frac{\partial^2 \delta(k)}{\partial k^2} = \log\delta \times \delta'(k)$$

$$= (\log\delta)^2 \times \delta(k) > 0$$

so that $\delta(k)$ is strictly decreasing and convex (decreases at a decreasing rate) in k.

We further evaluate the limiting values of $\delta(k)$, finding that

$$\delta(1) = \delta^{1+1} - 2\delta + 1 = (1-\delta)^2 > 0, \text{ and}$$

$$\lim_{k\to\infty} \delta(k) = \lim_{k\to\infty} \delta^{k+1} - 2\delta + 1$$

$$= 1 - 2\delta > 0 \text{ if } \delta > \frac{1}{2}.$$

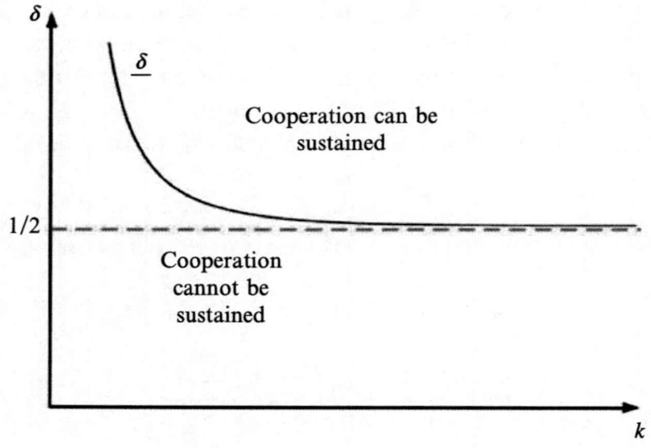

Fig. 8.15 Minimal discount factor decreases in k

- Let $\underline{\delta}$ be the minimal discount factor that solves $\delta(k) = 0$. Figure 8.15 depicts that collusion is (not) sustained when $\delta \geq \underline{\delta}$ ($\delta < \underline{\delta}$). The minimum discount rate $\underline{\delta}$ decreases in k because if firm i punishes the deviating firm j for more periods, collusion can be sustained under less restrictive parameter conditions. Specifically, when $k = 1$, collusion cannot be sustained for any values of δ under the tit-for-tat strategy (i.e., deviations are only punished during one period since $k = 1$). However, when deviations are punished thereafter $k \to \infty$, collusion can be sustained for $\delta \geq \frac{1}{2}$ as in the standard the Grim-Trigger Strategy.

Exercise #8.16: Deviation Punished by Losing the Output Leadership^B

8.16 Continue with the setting of Exercise 7.9, but consider an infinitely repeated game. Two firms forming a cartel will continue coordinating their output decisions as long as they have a history of cooperation. However, if firm i deviates in period t, then beginning period $t + 1$, firm j will remain as the Stackelberg leader while firm i will compete in quantities with the other $N - 2$ followers simultaneously. For simplicity, assume that all firms discount future payoff at a common discount factor of $\delta \in [0, 1]$.

(a) *Deviation.* Assume that firm i deviates. What are its deviating output, q_i^d, and associated profits, π_i^d?

- From part (b) of Exercise 7.9, firm i takes $q^f = \frac{a-c}{2b(N-1)}$ and $q_j^l = \frac{a-c}{4b}$ and solves the following profit maximization problem:

$$\max_{q_i^d \geq 0} \pi_i\left(q_i^d\right) = \left[a - bq_j^l - (N-2)\,bq^f - bq_i^d\right] q_i^d - cq_i^d$$

$$= \left[a - b\frac{a-c}{4b} - (N-2)\,b\frac{a-c}{2b\,(N-1)} - bq_i^d - c\right] q_i^d$$

$$= \left[\frac{(a-c)\,(N+1)}{4\,(N-1)} - bq_i^d\right] q_i^d.$$

Differentiating with respect to q_i^d and assuming interior solutions,

$$\frac{(a-c)(N+1)}{4(N-1)} - 2bq_i^d = 0.$$

Rearranging and solving for q_i^d, firm i's deviating output becomes

$$q_i^d = \frac{(a-c)(N+1)}{8b(N-1)}.$$

Substituting $q_i^d = \frac{(a-c)(N+1)}{8b(N-1)}$ into firm i's profit function, deviating profits are

$$\begin{aligned}\pi_i^d &= \left[\frac{(a-c)(N+1)}{4(N-1)} - b\overbrace{\frac{(a-c)(N+1)}{8b(N-1)}}^{q_i^d}\right]q_i^d \\ &= \frac{(a-c)^2(N+1)^2}{64b(N-1)^2}.\end{aligned}$$

(b) *Punishment.* Consider that firm i loses its output leadership. What are the output and associated profits of the deviating firm i, non-deviating firm j, and the followers?

- Since firm j now becomes the only Stackelberg leader, we learn from part (b) of Exercise 7.9 that its output is $q^l = \frac{a-c}{2b}$ with associated profits of $\pi^l = \frac{(a-c)^2}{4bN}$. In comparison, firm i, which becomes a follower, competes with the other $N-2$ followers (for there are now altogether $N-1$ followers), where each follower in this setting produces $q^f = \frac{a-c}{2bN}$ with associated profits of $\pi^f = \frac{(a-c)^2}{4bN^2}$.

(c) *Collusion.* Find the minimum discount factor $\underline{\delta}^f$ that sustains collusion.

- Having observed a history of cooperation, if firm i continues to cooperate with firm j, it splits the leader's profits with firm j, entailing a present value of

$$\begin{aligned}PV^c &= \frac{\pi^l}{2} + \delta\frac{\pi^l}{2} + \delta^2\frac{\pi^l}{2} + \ldots \\ &= \frac{1}{1-\delta}\frac{(a-c)^2}{8b(N-1)}.\end{aligned}$$

If firm i deviates, it earns π_i^d in the current period and π^f in all future periods, entailing a present value of

$$\begin{aligned}PV^d &= \pi_i^d + \delta\pi^f + \delta^2\pi^f + \ldots \\ &= \frac{(a-c)^2(N+1)^2}{64b(N-1)^2} + \frac{\delta}{1-\delta}\frac{(a-c)^2}{4bN^2}.\end{aligned}$$

Therefore, firm i has incentives to cooperate if and only if $PV^c \geq PV^d$, where

$$\frac{1}{1-\delta}\frac{(a-c)^2}{8b(N-1)} \geq \frac{(a-c)^2(N+1)^2}{64b(N-1)^2} + \frac{\delta}{1-\delta}\frac{(a-c)^2}{4bN^2}.$$

Solving for δ, we obtain the minimum discount factor $\underline{\delta}^f$, as follows:

$$\delta \geq \underline{\delta}^f \equiv \frac{N^2 (N-3)^2}{\left(N^2 - 3N + 4\right)\left(N^2 + 5N - 4\right)}.$$

(d) *Comparative statics.* How does $\underline{\delta}^f$ change with N? Explain.

- Differentiating $\underline{\delta}^f$ with respect to N, we obtain

$$\frac{\partial \underline{\delta}^f}{\partial N} = \frac{8N(N-3)\left(N^4 - 3N^3 + 12N^2 - 20N + 12\right)}{\left(N^2 - 3N + 4\right)^2 \left(N^2 + 5N - 4\right)^2},$$

which is positive for $N \geq 3$, so that when the number of firms increases, every firm i needs a higher discount factor δ to sustain collusion.

Exercise #8.17: Deviation and Retain Output Leadership^B

8.17 Consider the same setting as in Exercise 8.16, but allow the deviating firm to retain its output leadership. Two firms forming a cartel will continue coordinating output as long as they have a history of cooperation. However, if firm i deviates in period t, then beginning period $t+1$, the game is evolved as follows: In the first stage, the two firms that dissolved the cartel will compete in quantities. In the second stage, the other $N-2$ firms, having observed the leaders' output q^l, respond choosing its output q^f simultaneously. For simplicity, assume that all firms discount future payoff at a common discount factor of $\delta \in [0, 1]$.

(a) *Deviation.* Assume that firm i deviates. What are its deviating output, q_i^d, and associated profits, π_i^d?

- From part (a) of Exercise 8.16, we have that $q_i^d = \frac{(a-c)(N+1)}{8b(N-1)}$ and $\pi_i^d = \frac{(a-c)^2(N+1)^2}{64b(N-1)^2}$.

(b) *Punishment.* Consider that firm i retains its output leadership. What are the output and associated profits of the deviating firm i, non-deviating firm j, and the followers.

- Every leader i chooses q_i^l to solve the following profit maximization problem:

$$\max_{q_i^l \geq 0} \pi_i^l\left(q_i^l\right) = \left[a - bq_i^l - bq_j^l - (N-2)\, bq^f\left(q^l\right)\right] q_i^l - cq_i^l,$$

where $q^l = q_i^l + q_j^l$, and q_j^l is the output of the other Stackelberg leader, firm j.
From part (b) of Exercise 7.9, every follower's best response function is

$$q^f\left(q^l\right) = \frac{a-c}{b(N-1)} - \frac{1}{N-1} q^l.$$

Since the leader anticipates how the followers respond to its output, we substitute the followers' best response function into the leader's profit function, yielding

$$\max_{q_i^l \geq 0} \pi\left(q_i^l\right) = \left[a - bq_i^l - bq_j^l - (N-2)\, b\left(\frac{a-c}{b(N-1)} - \frac{q_i^l + q_j^l}{N-1}\right) - c\right] q_i^l$$

$$= \frac{\left(a - c - bq_i^l - bq_j^l\right) q_i^l}{N-1}.$$

Differentiating with respect to q_i^l and assuming interior solutions, we obtain

$$\frac{a-c-2bq_i^l-bq_j^l}{N-1}=0.$$

Rearranging and solving for q_i^l, the leader's best response function is

$$q_i^l\left(q_j^l\right)=\frac{a-c}{2b}-\frac{1}{2}q_j^l.$$

Since the two leaders are symmetric, in equilibrium we have $q_i^l=q_j^l$, so

$$q_i^l=\frac{a-c}{3b},$$

which coincides with the output level that a Cournot duopolist would choose in this setting, yielding the sum of leaders' output of $q^l=q_i^l+q_j^l=\frac{2(a-c)}{3b}$.

Substituting the leaders' total output, $q^l=\frac{2(a-c)}{3b}$, into the follower's best response function, we find that every follower's equilibrium output is

$$\begin{aligned}q^f&=\frac{a-c}{b\,(N-1)}-\frac{1}{N-1}\frac{2\,(a-c)}{3b}\\&=\frac{a-c}{3b\,(N-1)}.\end{aligned}$$

- In this context, the leader's equilibrium profits become

$$\begin{aligned}\pi_i^l&=\frac{\left(a-c-2b\overbrace{\frac{a-c}{3b}}^{q_i^l}\right)\overbrace{\frac{a-c}{3b}}^{q_i^l}}{N-1}\\&=\frac{(a-c)^2}{9b\,(N-1)}.\end{aligned}$$

Similarly, every follower earns equilibrium profits, as follows:

$$\begin{aligned}\pi^f&=\left[a-b\overbrace{\frac{2\,(a-c)}{3b}}^{q^l}-(N-2)\,b\overbrace{\frac{a-c}{3b\,(N-1)}}^{q^f}-c\right]\overbrace{\frac{a-c}{3b\,(N-1)}}^{q^f}\\&=\frac{(a-c)^2}{9b\,(N-1)^2}.\end{aligned}$$

(c) *Collusion.* Find the minimum discount factor $\underline{\delta}^l$ that sustains collusion.

- Having observed a history of cooperation, if firm i continues to cooperate with firm j, it splits the leader's profits with firm j, entailing a present value of

$$PV^c = \frac{\pi^l}{2} + \delta\frac{\pi^l}{2} + \delta^2\frac{\pi^l}{2} + \ldots$$
$$= \frac{1}{1-\delta}\frac{(a-c)^2}{8b(N-1)}.$$

If firm i deviates, it earns π_i^d in the current period and π_i^l in all future periods, entailing a present value of

$$PV^d = \pi_i^d + \delta\pi_i^l + \delta^2\pi_i^l + \ldots$$
$$= \frac{(a-c)^2(N+1)^2}{64b(N-1)^2} + \frac{\delta}{1-\delta}\frac{(a-c)^2}{9b(N-1)}.$$

Therefore, firm i has incentives to cooperate if and only if $PV^c \geq PV^d$, where

$$\frac{1}{1-\delta}\frac{(a-c)^2}{8b(N-1)} \geq \frac{(a-c)^2(N+1)^2}{64b(N-1)^2} + \frac{\delta}{1-\delta}\frac{(a-c)^2}{9b(N-1)}.$$

Solving for δ, we obtain the minimum discount factor $\underline{\delta}^l$, as follows:

$$\delta \geq \underline{\delta}^l \equiv \frac{9(N-3)^2}{9N^2-46N+73}.$$

(d) *Comparative statics.* How does $\underline{\delta}^l$ change with N? Explain.

- Differentiating $\underline{\delta}^l$ with respect to N, we obtain

$$\frac{\partial\underline{\delta}^l}{\partial N} = \frac{72(N+1)(N-3)}{\left(9N^2-46N+73\right)^2},$$

which is positive for $N \geq 3$, so that when the number of firms increases, every firm i needs a higher discount factor δ to sustain collusion.

(e) Compare the discount factors in this exercise and Ex 8.16. Interpret.

- We check that $\underline{\delta}^l > \underline{\delta}^f$ where

$$\frac{9(N-3)^2}{9N^2-46N+73} > \frac{N^2(N-3)^2}{\left(N^2-3N+4\right)\left(N^2+5N-4\right)}$$

simplifies to

$$(N-1)\left(4N^2-9N+9\right) > 0,$$

which is satisfied for $N > 1$, so that collusion can only be sustained under more restrictive conditions when the deviating firm retains its output leadership position in the punishment phase. In other words, when the deviating firm becomes a follower, it earns lower profits, so that collusion can be sustained even if the firms assign lower importance to future payoffs, that is, $\underline{\delta}^l > \underline{\delta}^f$.

Exercise #8.18: Can Mergers Facilitate Collusion?[C]

8.18 Consider an industry with $n \geq 2$ firms producing a homogeneous good at the same constant marginal cost c and competing in prices. For simplicity, assume that the demand is perfectly price-inelastic; all consumers have a unit demand and the same reservation price r, where $r > c$. So, for any price below r, firms can sell an aggregate output of q.

(a) *No merger.* If firms do not merge, find the minimal discount factor sustaining collusion and label it δ_{pre}, where the subscript pre indicates "pre-merger."

- *Collusion.* When firms collude, every firm charges the monopoly price $p_m = r$, earning collusive profits

$$\pi^{Col} = \left((r-c)\frac{q}{n}\right) + \delta\left((r-c)\frac{q}{n}\right) + \ldots = \frac{1}{1-\delta}(r-c)\frac{q}{n},$$

where in each period $r - c$ denotes the firm's per-unit margin, and $\frac{q}{n}$ represents its individual sales.

- *Deviation.* If a firm, instead, unilaterally deviates from this collusive agreement, it charges a deviating price p^{Dev} that slightly undercuts the collusive price, that is, $p^{Dev} = r - \varepsilon$, where $\varepsilon > 0$. Since ε can be made arbitrarily close to zero, $\varepsilon \to 0$, the deviating price can be considered, for practical reasons, $p^{Dev} = r$, entailing a deviating profit of $\pi^{Dev} = (r-c)\,q$ since the deviating firm captures all the market, selling q units.
- *Punishment.* After the deviation is detected, all firms punish such defection setting equilibrium prices under the unrepeated version of the game (Bertrand competition), that is, $p = c$, yielding zero profits thereafter.
- Putting together our above results, we can say that collusion is sustained if

$$\frac{1}{1-\delta}(r-c)\frac{q}{n} \geq (r-c)q + \frac{\delta}{1-\delta}0, \tag{8.1}$$

and after solving for the discount factor δ, we find

$$\delta \geq 1 - \frac{1}{n} \equiv \delta_{pre}.$$

Cutoff δ_{pre} satisfies $\delta_{pre} > 0$ since $n \geq 2$ by assumption and $\delta_{pre} < 1$ since $1 - \frac{1}{n} = \frac{n-1}{n} < 1$ simplifies to $n - 1 < n$, which always holds.

(b) *Merger—Outsiders.* Assume now that k firms merge (where k satisfies $2 \leq k < n$). Because of synergies, merging firms benefit from a lower marginal cost of production, namely $c - x$, where $x \geq 0$ can be interpreted as the cost-reducing effect of the merger. Find the minimal discount factor sustaining collusion for outsiders, label it $\delta_{outsider}$, and compare it against the minimal discount factor before the merger, δ_{pre}.

- Recall that k firms merge, and $n - k$ firms are outsiders, implying that the total number of firms after the merger is $n - k + 1$. Outsiders keep their marginal production cost at c (that is, they do not benefit from the cost-reducing effects of the merger).

- *Collusion.* Colluding, every outsider earns discounted profits

$$\frac{1}{1-\delta}(r-c)\frac{q}{n-k+1},$$

since the number of firms after the merger is $n-k+1$. Intuitively, outsiders make the same per-unit margin as before the merger, but sales are distributed among fewer firms ($n-k+1$ rather than n).
- *Deviation.* If an outsider deviates, it charges a deviating price $p^{Dev}=r$ (as in part (a)) earning the same deviating profits as in part (a), that is, $\pi^{Dev}=(r-c)\,q$.
- *Punishment.* Finally, during the punishment stage, all firms revert to the Nash equilibrium of the unrepeated Bertrand competition game, $p=c$, yielding zero profits thereafter.
- Combining our above results, outsiders are willing to collude if

$$\frac{1}{1-\delta}(r-c)\frac{q}{n-k+1}\geq(r-c)q+\frac{\delta}{1-\delta}0. \tag{8.2}$$

Comparing expression (8.2) against (8.1), we can see that outsiders have stronger incentives to collude, since the left-hand side of (8.2) is larger than that of (8.1), while the right-hand sides coincide in both expressions. After solving for the discount factor δ, we find

$$\delta\geq 1-\frac{1}{n-k+1}\equiv\delta_{outsider},$$

where, as predicted, $\delta_{outsider}<\delta_{pre}$. Intuitively, if a firm has incentives to collude before the merger, it must still have incentives to collude after the merger when it is an outsider.

(c) *Merger—Insiders.* Find now the minimal discount factor sustaining collusion for insiders, labeling it $\delta_{insider}$.

- *Collusion.* Colluding, the merged entity (the combination of k firms) earns discounted profits

$$\frac{1}{1-\delta}\underbrace{[r-(c-x)]}_{\text{per-unit margin}}\frac{q}{n-k+1},$$

since merged firms see their marginal production cost decrease to $c-x$, which increases their per-unit margin from $r-c$ before the merger to $r-(c-x)$ after the merger. In addition, the number of firms after the merger is $n-k+1$, increasing sales relative to the pre-merger levels.
- *Deviation.* If the merged entity deviates, it charges a deviating price $p^{Dev}=r$ (as in part (a)) earning deviating profits $\pi^{Dev}=[r-(c-x)]\,q$.
- *Punishment.* Finally, during the punishment stage, all firms revert to the Nash equilibrium of the unrepeated Bertrand competition game, $p=c$. Such price profile, however, does not yield zero profits for the merged entity, since its marginal costs are lower, $c-x$. In particular, its profits in each period of the punishment phase are $[c-(c-x)]q=xq$.
- Combining our above results, the merged entity colludes if

$$\frac{1}{1-\delta}[r-(c-x)]\frac{q}{n-k+1}\geq[r-(c-x)]\,q+\frac{\delta}{1-\delta}xq. \tag{8.3}$$

The merged entity (of k firms) operates like one firm competing à la Bertrand against the $n-k$ unmerged firms (outsiders). Therefore, on the left-hand side of expression (8.3), the merged firm (one firm) shares profits with the outsiders ($n-k$ unmerged firms), implying that collusive profits are divided by $n-k+1$ in the denominator. However, in the punishment phrase (the last term on the right-hand side of the expression), since the merged firm has lower per-unit cost than the unmerged firms, the former captures all market profits, leaving the latter with zero profits.

Comparing expression (8.3) against (8.1), we can see that the profits that the merged entity earns from colluding (left-hand side) are larger after the merger, but the deviation profit and the profit during the punishment phase are also larger after the merger. As a consequence, we cannot unambiguously rank whether the merged entity colludes under larger parameter conditions before or after the merger. Simplifying expression (8.3), we obtain

$$\frac{r-c+x}{n-k+1} \geq (1-\delta)(r-c+x)+\delta x$$
$$= r-c+x-\delta\,(r-c)\,,$$

and solving for the discount factor δ, we find

$$\delta \geq \left(\frac{n-k}{n-k+1}\right)\left(\frac{r-c+x}{r-c}\right) \equiv \delta_{insider}.$$

- Cutoff $\delta_{insider}$ is positive, $\delta_{insider}>0$, since $n \geq k$ and $r>c$ by assumption. In addition, cutoff $\delta_{insider}$ satisfies $\delta_{insider} \leq 1$ when

$$r-c+x \leq \frac{n-k+1}{n-k}(r-c).$$

Solving for the ratio $\frac{x}{r-c}$, as in our previous analysis of mergers with synergies in Exercise 7.7, yields

$$\theta \equiv \frac{x}{r-c} \leq \frac{1}{n-k} \equiv \overline{\theta}.$$

(d) *Comparing minimal discount factors.* Compare $\delta_{insider}$ from part (c) against the minimal discount factor before the merger, δ_{pre}, found in part (a).

- We can compare the merged entity's minimal discount factor supporting collusion before the merger, δ_{pre}, and after the merger, $\delta_{insider}$. Since collusion is sustained when $\delta \geq \delta_{pre}$ before the merger and when $\delta \geq \delta_{insider}$ after the merger, we can claim that the merger facilitates collusion if

$$\delta_{insider} \leq \delta_{pre}.$$

Comparing these cutoffs, we obtain

$$\left(\frac{n-k}{n-k+1}\right)\left(\frac{r-c+x}{r-c}\right) \leq \frac{n-1}{n}.$$

which simplifies to

$$n(n-k)x \leq (r-c)\left[(n-1)(n-k+1) - n(n-k)\right].$$

Solving for the ratio $\frac{x}{r-c}$, we obtain

$$\theta \equiv \frac{x}{r-c} \leq \frac{k-1}{n(n-k)} \equiv \widehat{\theta}.$$

In a line representing the cost-reduction effect from the merger, θ, cutoff $\widehat{\theta}$ divides the line into two regions: one to the left of cutoff $\widehat{\theta}$ where mergers facilitate collusion and one to the right of cutoff $\widehat{\theta}$ where mergers hinder collusion.

- *Remark:* Note that cutoff $\widehat{\theta}$ satisfies $\widehat{\theta} < \overline{\theta}$ since $\frac{k-1}{n(n-k)} < \frac{1}{n-k}$ simplifies to $k < n+1$, which holds by definition given that $n > k$. This implies that the line of θ values is divided into the following segments:
  - When $\theta \leq \widehat{\theta}$, we have that $\delta_{insider} \leq \delta_{pre}$ and $\delta_{insider} \leq 1$.
  - When $\widehat{\theta} < \theta \leq \overline{\theta}$, we have that $\delta_{insider} > \delta_{pre}$, but we still have $\delta_{insider} \leq 1$.
  - When $\overline{\theta} < \theta$, we have that $\delta_{insider} > \delta_{pre}$ and $\delta_{insider} > 1$.
- **Interpretation of cutoff $\widehat{\theta}$:**
  - When the cost-reduction effect, as captured by the ratio $\theta \equiv \frac{x}{r-c}$, is relatively small, $\theta \leq \widehat{\theta}$, the merged entity and outsiders remain relatively similar after the merger, facilitating collusion. In terms of expression (8.3), the merged entity is not too attracted to deviate since its cost advantage relative to outsiders is small. In this case, we say that the merger facilitates collusion.
  - In contrast, when the cost-reduction effect is sufficiently large, $\theta > \widehat{\theta}$, then the merged entity is more attracted to defect after than before the merger, sustaining collusion under more restrictive conditions after the merger. In this case, we say that the merger hinders collusion.
- **Comparative statics:** We can finally do comparative statics of cutoff $\widehat{\theta}$ as follows:

$$\frac{\partial \widehat{\theta}}{\partial k} = \frac{n-1}{n(n-k)^2} > 0,$$

since $n > k$ by assumption. Therefore, cutoff $\widehat{\theta}$ shifts rightward as more firms merge (higher k for a given n), expanding the range of cost efficiencies for which the merger facilitates collusion. This holds because more firms merge to gain cost efficiencies that makes the merger more attractive. Graphically, the range $\theta \leq \widehat{\theta}$ expands as $\widehat{\theta}$ moves rightward.

In contrast, cutoff $\widehat{\theta}$ shifts leftward as the industry grows (higher n for a given number of firms merging k) since

$$\frac{\partial \widehat{\theta}}{\partial n} = -\frac{\overbrace{(k-1)}^{+}\overbrace{(2n-k)}^{+}}{n^2(n-k)^2} < 0.$$

In this context, the range of cost efficiencies for which the merger facilitates collusion, $\theta \leq \widehat{\theta}$, shrinks. This holds because the merged firm obtains a smaller share of the market that makes the merger less attractive.

Exercise #8.19: Collusion and Leniency Programs^B

8.19 Consider two firms, which sell a homogenous product competing à la Bertrand, face a linear demand function $Q(p) = 1 - p$, where Q denotes aggregate output and p represents price. Firms are symmetric in their marginal production cost, c, which we normalize to $c = 0$.

(a) *No fines*. In a setting where the competition authority does not monitor collusion activities, find the minimal discount factor supporting collusion.

- Let π^c denote the stage profits that every firm i earns from colluding, $\pi^c = \frac{\pi^m}{2}$, where π^m represents the monopoly profits, that is,

$$\pi^m \equiv \max_{p \geq 0} p(1-p)$$

so that $p^m = \frac{1}{2}$ and monopoly profits are $\pi^m = \frac{1}{4}$. Similarly, let π^d denote the profits that firm i earns from unilaterally deviating from the collusive agreement, where $\pi^d = \pi^m$, since firm i can capture all sales by setting a price marginally below the monopoly price, $p = p^m - \varepsilon$. Since ε can be arbitrarily small, deviation profits are approximated with monopoly profits π^m. Finally, let π^n represent the profits that every firm earns when they behave as in the Nash equilibrium of the stage game (Bertrand model), where every firm sets a price equal to their common marginal cost, so that $\pi^n = 0$.

- In this setting, if firms use a standard Grim-Trigger Strategy (GTS), collusion can be sustained in the subgame perfect equilibrium of the infinitely repeated game if

$$\frac{1}{1-\delta}\underbrace{\pi^c}_{1/8} \geq \underbrace{\pi^d}_{\pi^m=\frac{1}{4}} + \frac{\delta}{1-\delta}\underbrace{\pi^n}_{0}$$

or, after rearranging, if and only if $\delta \geq \frac{1}{2}$.

(b) *Fines*. Assume now that the competition authority monitors collusion in every period with probability $\alpha \in [0, 1]$ and, if the industry is colluding, every firm receives a fine $F > 0$. Find the minimal discount factor that sustains collusion in this context, $\widehat{\delta}(\alpha, F)$. How is it affected by an increase in α or in F? Interpret.

- In this setting, collusion is sustained if

$$\frac{1}{1-\delta}\left(\pi^c - \alpha F\right) \geq \left(\pi^d - \alpha F\right) + \frac{\delta}{1-\delta}\underbrace{\pi^n}_{0}.$$

Since $\pi^c = \frac{1}{8}$ and $\pi^d = \frac{1}{4}$, the above expression simplifies to

$$\frac{1}{8} - \alpha F \geq (1-\delta)\left(\frac{1}{4} - \alpha F\right)$$

or, after solving for δ, we obtain

$$\delta \geq \widehat{\delta}(\alpha, F) \equiv \frac{1}{2(1-4\alpha F)}.$$

The minimal discount factor sustaining collusion, $\widehat{\delta}(\alpha, F)$, is increasing in both the severity of the fine, F, and the likelihood that firms are caught colluding, α. Both effects hinder collusion, as the range of discount factor sustaining collusion, $\delta \geq \widehat{\delta}(\alpha, F)$, shrinks.

(c) *Comparison.* Compare your results in parts (a)-(b). Is collusion facilitated or hindered by fines?

- In the absence of fines, collusion holds when $\delta \geq \frac{1}{2}$, while in the presence of fines, collusion can be sustained if $\delta \geq \widehat{\delta}$. Comparing these minimal discount factors, we obtain that collusion is deterred by fines if and only if

$$\frac{1}{2} \leq \frac{1}{2(1-4\alpha F)},$$

which simplifies to

$$4\alpha F \geq 0,$$

which holds since $\alpha \geq 0$ and $F > 0$ by definition, such that collusion occurs under more restrictive conditions when firms face fines than when they do not.

(d) *Leniency program.* Consider that now the competition authority offers a leniency program in which a firm that provides evidence about the industry collusion receives a reward, R, where $R > -\alpha F$, but it can still be negative (if $0 > R > -\alpha F$, when the firm managers enjoy, for instance, a reduced sentence than in the absence of the leniency program) or positive (if $R > 0 > -\alpha F$, if the firm manager receive a government subsidy and no jail time for their help with the investigation). Find the minimal discount factor that sustains collusion in this context, $\widehat{\delta}(\alpha, F, R)$. How is it affected by an increase in R? Interpret. (For simplicity, you may assume that the fine F satisfies $F \geq \frac{1}{8\alpha}$ for the rest of the exercise.)

- In this setting, collusion is sustained if and only if

$$\frac{1}{1-\delta}\left(\pi^c - \alpha F\right) \geq \left(\pi^d + R\right) + \frac{\delta}{1-\delta}\underbrace{\pi^n}_{0}.$$

Since $\pi^c = \frac{1}{8}$ and $\pi^d = \frac{1}{4}$, the above expression simplifies to

$$\frac{1}{8} - \alpha F \geq (1-\delta)\left(\frac{1}{4} + R\right)$$

or, after solving for δ,

$$\delta \geq \widehat{\delta}(\alpha, F, R) \equiv \frac{1 + 8(R + \alpha F)}{2(1+4R)},$$

which exhibits similar comparative statics as in part (b), that is, a more severe fine (higher F) or a more likely fine (higher α) increases the minimal discount factor $\widehat{\delta}$, thus discouraging collusion. In addition, we have that

$$\frac{\partial\widehat{\delta}}{\partial R} = \frac{2(1-8\alpha F)}{(1+4R)^2},$$

which is positive when $F < \frac{1}{8\alpha}$, which holds by assumption. Intuitively, a more generous reward R increases the minimal discount factor $\widehat{\delta}$, ultimately deterring collusion.

(e) *Comparison.* Compare your results in parts (b) and (d). Is collusion facilitated or hindered by leniency programs?

- Comparing the minimal discount factor sustaining collusion in part (b), $\widehat{\delta}(\alpha, F)$, and that in part (d), $\widehat{\delta}(\alpha, F, R)$, we find that

$$\frac{1}{2(1-4\alpha F)} < \frac{1+8(R+\alpha F)}{2(1+4R)}$$

simplifies to

$$1+4R < [1+8(R+\alpha F)](1-4\alpha F),$$

which we can further rearrange as follows:

$$8\alpha F(R+\alpha F) < R+\alpha F.$$

Since $R+\alpha F > 0$, the above inequality becomes

$$F < \frac{1}{8\alpha},$$

which holds by assumption. Therefore, the leniency program hinders collusion, that is, $\widehat{\delta}(\alpha, F, R) > \widehat{\delta}(\alpha, F)$.

Exercise #8.20: Leniency Programs, Based on Motta (2003)^C

8.20 Consider n identical firms. In every period t, every firm simultaneously and independently chooses whether to collude, earning monopoly profit π_M, or unilaterally deviate, earning profit π_D, where $\pi_D > \pi_M$. Deviations trigger an infinite punishment (standard Grim-Trigger Strategy) where all firms choose the Nash equilibrium of the unrepeated game, earning profit π_N, where $\pi_D > \pi_M > \pi_N$. For simplicity, assume that firms' common discount factor, δ, satisfies $\delta \geq \frac{\pi_D - \pi_M}{\pi_D - \pi_N}$, which guarantees that collusion emerges in equilibrium when firms are not subject to fines from the antitrust authority (AA).

In this exercise, we seek to understand the effects of leniency programs, where the AA lowers the fine that collusive firms have to pay when found guilty of collusion, from F to R, where $R < F$. Assume that the AA can perfectly observe collusion, but it needs hard evidence when bringing the case to court. Consider the following time structure: In period $t = 1$, every firm i independently and simultaneously chooses whether to collude, facing the above Grim-Trigger Strategies. In all subsequent periods, $t \geq 2$, the AA then opens an investigation with probability $\alpha \in [0, 1]$ against the collusive firms.

- If the inquiry is not open, every collusive firm earns profit π_M.
- If the inquiry is opened, every firm simultaneously and independently chooses whether or not to reveal information to the AA.
- If at least one firm reveals, the AA is able to prove them all guilty of collusion.

- A firm that cooperates with the AA pays a reduced fine of R, whereas every firm that does not cooperate with the AA pays the full fine F.
- If no firm reveals information, the AA is able to prove the firms guilty of collusion with probability $p \in [0, 1]$.

If the AA has not been able to prove the firms guilty of collusion at the end of this inquiry period t, the firms will never be investigated again in the future. However, if the firms are found guilty in period t, they will never collude again in the future, earning π_N in all subsequent periods.

(a) *Reveal game.* Show that, in the subgame that ensues after the AA opens an investigation, we can sustain two Nash equilibria: one where all firms reveal and one where all do not reveal information to the AA. Discuss under which parameter conditions each Nash equilibrium can be sustained. Interpret your results.

- i. If a firm reveals, it gets a payoff of $\frac{\Pi_N}{1-\delta} - R$.
  ii. If a firm does not reveal any information but at least one other firm does, the former gets a payoff of $\frac{\Pi_N}{1-\delta} - F$.
  iii. If no firm reveals any information, each firm gets an expected payoff of $p\left(\frac{\Pi_N}{1-\delta} - F\right) + (1-p)\frac{\Pi_M}{1-\delta}$.
- Suppose there are 2 firms in the market, the payoff table is

		Firm2	
		Reveal	Not reveal
Firm1	Reveal	$\frac{\Pi_N}{1-\delta} - R, \frac{\Pi_N}{1-\delta} - R$	$\frac{\Pi_N}{1-\delta} - R, \frac{\Pi_N}{1-\delta} - F$
	Not reveal	$\frac{\Pi_N}{1-\delta} - F, \frac{\Pi_N}{1-\delta} - R$	$p\left(\frac{\Pi_N}{1-\delta} - F\right) + (1-p)\frac{\Pi_M}{1-\delta}$, $p\left(\frac{\Pi_N}{1-\delta} - F\right) + (1-p)\frac{\Pi_M}{1-\delta}$.

i. If firm 2 reveals, firm 1's best response is reveal since $\frac{\Pi_N}{1-\delta} - R > \frac{\Pi_N}{1-\delta} - F$; similarly, if firm 1 reveals, firm 2's best response is reveal.

ii. If firm 2 does not reveal, firm 1's best response is not reveal if

$$\frac{\Pi_N}{1-\delta} - R \leq p\left(\frac{\Pi_N}{1-\delta} - F\right) + (1-p)\frac{\Pi_M}{1-\delta},$$

which we rearrange to yield

$$pF - R \leq \frac{1-p}{1-\delta}(\Pi_M - \Pi_N).$$

If $pF \leq R$, firm 1's best response is always not reveal; if $pF \geq R$, firm 1's best response is not reveal if

$$p\left[\frac{\Pi_M - \Pi_N}{1-\delta} + F\right] \leq R + \frac{\Pi_M - \Pi_N}{1-\delta}.$$

Solving for p, we obtain

$$p \leq \frac{\Pi_M - \Pi_N + R(1-\delta)}{\Pi_M - \Pi_N + F(1-\delta)} \equiv \widetilde{p}(\delta, R, F).$$

- Similarly, in the n-firm case, (reveal,..., reveal) is always an NE. In contrast, (not reveal,..., not reveal) is an NE if
  i. $pF \leq R$ always holds or
  ii. If $pF \geq R$ and $p \leq \widetilde{p}(\delta, R, F)$
- Therefore, firms reveal information only if $p > \widetilde{p}$. And when no leniency program exists, $R = F$ and $\widetilde{p} = 1$, firms will never collaborate with the AA even after an investigation has started. In this context, to induce revelation, the best the AA can do is to set $R = 0$.

(b) *Collude and reveal.* In the first stage, identify under which conditions on probability α every firm chooses to collude in $t = 1$ and reveal in $t = 2$ (rather than deviating in $t = 1$ which triggers π_N profits thereafter).

- $p > \widetilde{p}$. In this case firms reveal if an investigation is opened by the AA. Define Π_R as the expected profit immediately before an investigation is opened, then

$$\Pi_R = \alpha\left(\frac{\Pi_N}{1-\delta} - R\right) + (1-\alpha)(\Pi_M + \delta\Pi_R),$$

which, solving for Π_R, yields

$$\Pi_R = \frac{(1-\alpha)(1-\delta)\Pi_M + \alpha\left[\Pi_N - (1-\delta)R\right]}{(1-\delta)\left[1-\delta(1-\alpha)\right]}.$$

- If a firm decides to set the collusive price, its expected discounted payoff becomes

$$\begin{aligned} V_{CR} &= \Pi_M + \delta\Pi_R \\ &= \Pi_M + \delta\frac{(1-\alpha)(1-\delta)\Pi_M + \alpha\left[\Pi_N - (1-\delta)R\right]}{(1-\delta)\left[1-\delta(1-\alpha)\right]} \\ &= \frac{(1-\delta)\Pi_M + \alpha\delta\left[\Pi_N - (1-\delta)R\right]}{(1-\delta)\left[1-\delta(1-\alpha)\right]}. \end{aligned}$$

- If a firm decides to deviate from the collusive strategy, then its payoff is

$$V_D = \Pi_D + \frac{\delta\Pi_N}{1-\delta}.$$

- Thus collusion can arise if and only if $V_{CR} \geq V_D$, that is,

$$\frac{(1-\delta)\Pi_M + \alpha\delta\left[\Pi_N - (1-\delta)R\right]}{(1-\delta)\left[1-\delta(1-\alpha)\right]} \geq \Pi_D + \frac{\delta\Pi_N}{1-\delta}.$$

Solving for α yields

$$\alpha \leq \frac{\Pi_M - \delta\Pi_N - (1-\delta)\Pi_D}{\delta\left(\Pi_D - \Pi_N + R\right)} \equiv \alpha_{CR}(\delta, R).$$

(c) *Collude and not reveal.* In the first stage, identify under which conditions on probability α every firm chooses to collude in $t = 1$ and not reveal in $t = 2$ (rather than deviating in $t = 1$ which triggers π_N profits thereafter).

- $p \leq \widetilde{p}$. In this case, firms anticipate that even if an investigation is started, no firm will reveal any information. Only by beginning an investigation and then proving the firms are guilty can the AA break the cartel.
- The expected profit immediately before knowing if an investigation is opened is

$$\Pi_{NR} = \alpha \left[p \left(\frac{\Pi_N}{1-\delta} - F \right) + (1-p) \left(\frac{\Pi_M}{1-\delta} \right) \right] + (1-\alpha)(\Pi_M + \delta \Pi_{NR}),$$

which is rearranged to yield

$$\Pi_{NR} = \frac{\alpha p \left[\Pi_N - (1-\delta) F \right] + \left[1 - \delta - \alpha (p - \delta) \right] \Pi_M}{(1-\delta) \left[1 - \delta (1-\alpha) \right]}.$$

- If a firm follows the collusive strategy, its expected discounted payoff is

$$\begin{aligned} V_{CNR} &= \Pi_M + \delta \Pi_{NR} \\ &= \Pi_M + \delta \frac{\alpha p \left[\Pi_N - (1-\delta) F \right] + \left[1 - \delta - \alpha (p - \delta) \right] \Pi_M}{(1-\delta) \left[1 - \delta (1-\alpha) \right]} \\ &= \frac{\left[1 - \delta + \alpha \delta (1-p) \right] \Pi_M + \alpha \delta p \left[\Pi_N - (1-\delta) F \right]}{(1-\delta) \left[1 - \delta (1-\alpha) \right]}. \end{aligned}$$

- A firm that deviates obtains a payoff

$$V_D = \Pi_D + \frac{\delta \Pi_N}{1-\delta}.$$

- Therefore, a firm will collude and not reveal if and only if $V_{CNR} \geq V_D$, that is,

$$\frac{\left[1 - \delta + \alpha \delta (1-p) \right] \Pi_M + \alpha \delta p \left[\Pi_N - (1-\delta) F \right]}{(1-\delta) \left[1 - \delta (1-\alpha) \right]} \geq \frac{(1-\delta) \Pi_D + \delta \Pi_N}{1-\delta}.$$

  Solving for α, we obtain

$$\alpha \leq \frac{(1-\delta) \left[\Pi_M - (1-\delta) \Pi_D - \delta \Pi_N \right]}{\delta \left[p (1-\delta) F + (1-\delta) \Pi_D - (1-p) \Pi_M - (p-\delta) \Pi_N \right]} \equiv \alpha_{CNR} (\delta, p, F).$$

(d) Draw a figure with probability α in the vertical axis and probability p in the horizontal axis, and identify the regions of parameter values that sustain the subgame perfect equilibria found in parts (b) and (c). Interpret your results.

- From the cutoff

$$\alpha_{CR} (\delta, R) = \frac{\Pi_M - \delta \Pi_N - (1-\delta) \Pi_D}{\delta (\Pi_D - \Pi_N + R)},$$

  we see that $\alpha_{CR}(\delta, R)$ does not depend on p since in the region considered here firms cooperate with the AA once an investigation is opened. Below this cutoff, firms prefer to collude even though they anticipate that, if an investigation is opened, collusion would collapse because firms would have incentives to reveal information to the AA. Above

Fig. 8.16 Ranges of α and p supporting collusion

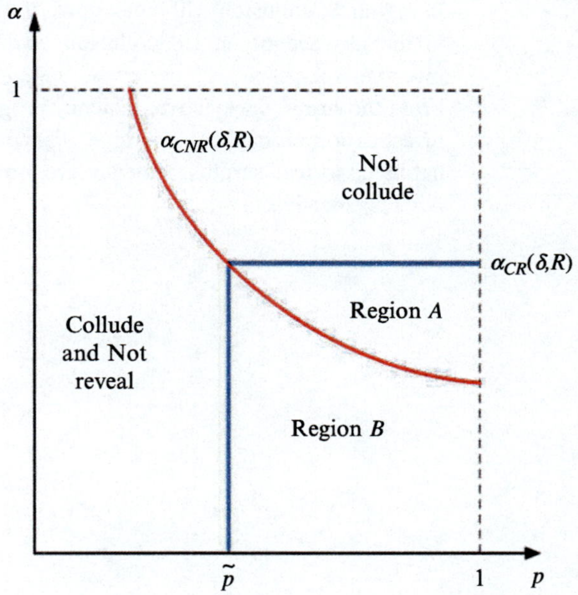

this cutoff, however, firms, anticipating revelation, prefer to deviate, and the collusive outcome never occurs (Fig. 8.16).

- Differentiating cutoff $\alpha_{CNR}(\delta, p, F)$ with respect to p, we obtain

$$\frac{\partial\alpha_{CNR}(\delta, p, F)}{\partial p} = -\frac{(1-\delta)\left[(1-\delta)F + \Pi_M - \Pi_N\right]\left[\Pi_M - (1-\delta)\Pi_D - \delta\Pi_N\right]}{\delta\left[p(1-\delta)F + (1-\delta)\Pi_D - (1-p)\Pi_M - (p-\delta)\Pi_N\right]^2},$$

which is unambiguously negative since $\delta \geq \frac{\pi_D - \pi_M}{\pi_D - \pi_N}$ entails that $\Pi_M - (1-\delta)\Pi_D - \delta\Pi_N \geq 0$, so that cutoff $\alpha_{CNR}(\delta, p, F)$ is decreasing in p. Intuitively, as detection probability p increases, collusion can only occur when investigation probability α decreases. In the limit where $p \to 0$, $V_{CNR} = \frac{\Pi_M}{1-\delta} > \frac{(1-\delta)\Pi_D + \delta\Pi_N}{1-\delta}$, so that collusion is preferred when the likelihood of found guilty of collusion is low.

(e) Using the figure in part (d), what happens when no leniency programs are in place (i.e., $R = F$)? How does a reduction in R, for a given F, affect the regions supporting each equilibrium outcome? Interpret. Is a leniency program always optimal?

- If $R = F$, firms have no reason to reveal information to AA once an investigation is opened. The equilibrium outcome would be defined uniquely by $\alpha_{CNR}(\delta, p, F)$. Below the line, firms would collude; above it, firms would not since any proposed agreement would break down immediately.
- If there is a reduction in R, i.e., $R < F$:
  i. If $p < \tilde{p}$, firms do not reveal. $\alpha_{CNR}(\delta, p, F)$ distinguishes the equilibrium where firms collude and not reveal from that where no collusion occurs.
  ii. If $p \geq \tilde{p}$, firms reveal information, and they prefer not to collude above $\alpha_{CR}(\delta, R)$, and below $\alpha_{CR}(\delta, R)$ will initially collude and then reveal if investigated.
- In region A, no collusion can be sustained in the industry if full fines are given, but under an LP firms would engage in collusion, and if monitored, they would reveal and pay the reduced fine $R < F$.

- In region B, collusion will break down if the AA starts monitoring the industry, whereas in the absence of an LP collusion could stop only after a successfully completed investigation.
- From the firms' perspective, leniency program induces collusion in contexts where the investigation probability p is high (i.e., $p > \widetilde{p}$). However, since collusion, in general, harms consumer surplus, leniency programs may not promote social welfare from the AA's perspective.

Bundling Incentives

9

Introduction

This chapter explores settings where a monopolist offers two goods to customers who exhibit correlated valuations for each good. For instance, in a market with two consumers, i and j, and two goods, A and B, customer i is the individual with the highest valuation for good A but he is the one with the lowest value for good B. In this context, the monopolist can offer to sell each good, A and B, at a different price or, alternatively, sell the bundle of both goods at a single price. This is the setting that we consider in Exercise 9.1, showing that, when consumer valuations are negatively correlated, a firm has incentives to only offer the bundle and make a higher profit than selling each good separately. Exercise 9.2 considers bundling decisions when the valuations for the two products, A and B, are positively correlated, Exercise 9.3 provides a numerical example, and Exercise 9.4 extends our analysis to a setting where valuations can be negatively or positively correlated. Interestingly, we show that bundling is only profitable when valuations are negatively correlated.

Exercise 9.5 considers a more subtle environment, where the firm offers both the bundle and each good sold separately, letting each consumer choose the option he prefers, known as "mixed bundling." We evaluate whether the firm can increase its profits as a result of this richer list of options to consumers. Exercise 9.6 then presents a numerical example to illustrate the above results. Exercise 9.7 considers bundling under exponentially distributed values, where the monopolist offers the pure bundle at lower prices and earns more profits than separate selling. Finally, Exercise 9.8 analyzes a pay-as-you-go contract, where firms offer consumers the option of buying or leasing the product. Surprisingly, firms have incentives to offer the product for free (or at a very low price) and charge consumers for using the product. Intuitively, firms obtain rental income to subsidize the capital cost of the main product, which is common in markets such as printer and cartridges or razors and razor blades.

Exercise #9.1: Bundling with Negatively Correlated Values^A

9.1 Consider a restaurant selling hamburger and French fries. For simplicity, assume that its consumers are uniformly distributed on the unit line, whose preferences are stated as follows. For instance, consumers in group 1 assign a value of 6 to the hamburger and only 2 to the French fries, while consumers in group 2 assign the same value, x, to both food items, where $0 \leq \lambda \leq 1$

P.-S. Choi et al., *Industrial Organization*, Springer Texts in Business and Economics,
https://doi.org/10.1007/978-3-031-38635-0_9

stands for the proportion of consumers in group 1 and $2 \leq x \leq 6$ represents the willingness to pay of group 2 consumers.

Group	Proportion	Hamburger	French fries
1	λ	6	2
2	$1-\lambda$	x	x

Determine the optimal prices and the associated profits if the restaurant sells hamburger and French fries (a) separately, (b) as a set menu only, and (c) separately and as a set menu. What pricing strategy will the restaurant use? Interpret.

(a) *Separate Selling.* Let p_h^s and p_f^s be the prices of hamburger and French fries, respectively. The restaurant will charge 6 for hamburger if it earns more by selling only to group 1 than to both groups at a lower price of x, which happens if $6\lambda > x$ that is rearranged to yield $\lambda > \frac{x}{6}$. Similarly, it will charge x for French fries if it earns more by selling only to group 2 than to both groups at a lower price of 2, which happens if $(1-\lambda)\,x > 2$ that is rearranged to yield $\lambda < 1-\frac{2}{x}$. It is straightforward to verify that $\frac{x}{6} > 1-\frac{2}{x}$ for $2 \leq x \leq 6$, so that the restaurant's pricing strategy under separate selling is

$$\left(p_h^s, p_f^s\right) = \begin{cases} (x,x) & \text{if } \lambda \in \left[0, 1-\frac{2}{x}\right) \\ (x,2) & \text{if } \lambda \in \left[1-\frac{2}{x}, \frac{x}{6}\right) \\ (6,2) & \text{if } \lambda \in \left[\frac{x}{6}, 1\right] \end{cases}$$

generating profits of

$$\pi^s \equiv \pi_h^s + \pi_f^s = \begin{cases} x\,(2-\lambda) & \text{if } \lambda \in \left[0, 1-\frac{2}{x}\right) \\ x+2 & \text{if } \lambda \in \left[1-\frac{2}{x}, \frac{x}{6}\right) \\ 6\lambda+2 & \text{if } \lambda \in \left[\frac{x}{6}, 1\right]. \end{cases}$$

(b) *Pure Bundling.* Let p^b be the price of the set menu, and we consider two cases: (i) $2 \leq x \leq 4$ and (ii) $4 < x \leq 6$. In case (i), group 1 has a higher willingness to pay for the bundle than group 2, so that the restaurant will set a price of 8 for group 1 instead of $2x$ for both groups if $8\lambda > 2x$ that is rearranged to yield $\lambda > \frac{x}{4}$. In case (ii), however, group 2 has a higher willingness to pay for the bundle than group 1, so that it will set a price of $2x$ for group 2 instead of 8 for both groups if $2\,(1-\lambda)\,x > 8$ that is rearranged to yield $\lambda < 1-\frac{4}{x}$. Therefore, the restaurant's pricing strategy under pure bundling is

$$p^b = \begin{cases} 2x & \text{if } \lambda \in \left[0, \frac{x}{4}\right) \\ 8 & \text{if } \lambda \in \left[\frac{x}{4}, 1\right] \end{cases} \text{ if } 2 \leq x \leq 4, \text{ and}$$

$$p^b = \begin{cases} 2x & \text{if } \lambda \in \left[0, 1-\frac{4}{x}\right) \\ 8 & \text{if } \lambda \in \left[1-\frac{4}{x}, 1\right] \end{cases}, \text{ if } 4 < x \leq 6.$$

generating profits of

$$\pi^b = \begin{cases} 2x & \text{if } \lambda \in \left[0, \frac{x}{4}\right) \\ 8\lambda & \text{if } \lambda \in \left[\frac{x}{4}, 1\right] \end{cases} \text{ if } 2 \leq x \leq 4, \text{ and}$$

$$\pi^b = \begin{cases} 2(1-\lambda)x & \text{if } \lambda \in \left[0, 1-\frac{4}{x}\right) \\ 8 & \text{if } \lambda \in \left[1-\frac{4}{x}, 1\right] \end{cases}, \text{ if } 4 < x \leq 6.$$

(c) *Mixed Bundling*. Let p_h^m, p_f^m, and p^m be the prices of hamburger, French fries, and the set menu, respectively. Again, we consider two cases: (i) $2 \leq x \leq 4$ and (ii) $4 < x \leq 6$.

In case (i), the restaurant will set $p_h^m > 8 - x$, $p_f^m = x$, and $p^m = 8$ to induce group 1 consumers to buy the set menu instead of hamburger and French fries separately, while group 2 consumers will only buy French fries. This happens if the profit generated from this menu pricing exceeds that from the bundling price of $2x$ for both consumer groups, which satisfy the following inequality:

$$8\lambda + (1-\lambda)x > 2x$$

$$\lambda(8-x) > x$$

$$\lambda > \frac{x}{8-x}$$

In case (ii), however, the restaurant will charge $p_h^m = 6$, $p_f^m > 2x - 6$, and $p^m = 2x$ to induce group 2 consumers to buy the set menu instead of hamburger and French fries separately, while group 1 consumers will only buy hamburgers. This happens if the profit generated from this menu pricing exceeds that from the bundling price of 8 for both consumer groups, which satisfy the following inequality:

$$6\lambda + 2(1-\lambda)x > 8$$

$$2(x-4) > 2\lambda(x-3)$$

$$\lambda < \frac{x-4}{x-3}$$

Therefore, the restaurant's pricing strategy under mixed bundling is

$$p_h^m = 8 - x,\ p_f^m = x \text{ , and } p^m = \begin{cases} 2x & \text{if } \lambda \in \left[0, \frac{x}{8-x}\right) \\ 8 & \text{if } \lambda \in \left[\frac{x}{8-x}, 1\right] \end{cases} \text{ and } 2 \leq x \leq 4, \text{ and}$$

$$p_h^m = 6,\ p_f^m = 2x - 6 \text{ , and } p^m = \begin{cases} 2x & \text{if } \lambda \in \left[0, \frac{x-4}{x-3}\right) \\ 8 & \text{if } \lambda \in \left[\frac{x-4}{x-3}, 1\right] \end{cases} \text{ and } 4 < x \leq 6.$$

generating profits of

$$\pi^m = \begin{cases} 2x & \text{if } \lambda \in \left[0, \frac{x}{8-x}\right) \\ 8\lambda + (1-\lambda)x & \text{if } \lambda \in \left[\frac{x}{8-x}, 1\right] \end{cases} \text{ and } 2 \leq x \leq 4, \text{ and}$$

$$\pi^m = \begin{cases} 6\lambda + 2(1-\lambda)x & \text{if } \lambda \in \left[0, \frac{x-4}{x-3}\right) \\ 8 & \text{if } \lambda \in \left[\frac{x-4}{x-3}, 1\right] \end{cases} \text{ and } 4 < x \leq 6.$$

(d) *Comparison.* Compare the profits from the pure and mixed bundling.
- It is obvious that mixed bundling weakly dominates pure bundling because the restaurant can always serve more customers by separately selling rather than selling the set menu only. Let us compare separate selling to mixed bundling.
- In case (i), we consider four regions:
  - First, when $\lambda \in \left[0, 1 - \frac{2}{x}\right)$, profits satisfy

$$\pi^m = 2x > x(2 - \lambda) = \pi^s$$

reducing to $\lambda > 0$ that is true.
  - Second, when $\lambda \in \left[1 - \frac{2}{x}, \frac{x}{6}\right)$, we find

$$\pi^m = 2x > x + 2 = \pi^s$$

reducing to $x > 2$ that holds.
  - Third, when $\lambda \in \left[\frac{x}{6}, \frac{x}{8-x}\right)$, we have

$$\pi^m = 2x > 6\lambda + 2 = \pi^s$$

which becomes

$$x > 3\lambda + 1 > 3\left(\frac{x}{6}\right) + 1$$
$$x > \frac{x}{2} + 1$$
$$x > 2$$

that is satisfied.
  - Fourth, when $\lambda \in \left[\frac{x}{8-x}, 1\right]$, profits satisfy

$$\pi^m = 8\lambda + (1 - \lambda)x > 6\lambda + 2 = \pi^s$$

which becomes

$$(1 - \lambda)x > 2(1 - \lambda)$$
$$x > 2$$

that is satisfied.
- In case (ii), we also consider four regions:
  - First, when $\lambda \in \left[0, \frac{x-4}{x-3}\right)$, we obtain

$$\pi^m = 6\lambda + 2(1 - \lambda)x > x(2 - \lambda) = \pi^s$$

which becomes

$$6\lambda > x(2 - \lambda - 2 + 2\lambda)$$
$$x < 6$$

that holds.

- Second, when $\lambda \in \left[\frac{x-4}{x-3}, 1 - \frac{2}{x}\right)$, we find that

$$\pi^m = 8 > x(2 - \lambda) = \pi^s$$

which becomes

$$8 > x\left(2 - 1 + \frac{2}{x}\right)$$
$$x < 6$$

that is satisfied.

- Third, when $\lambda \in \left[1 - \frac{2}{x}, \frac{x}{6}\right)$, profits satisfy

$$\pi^m = 8 > x + 2 = \pi^s$$

which reduces to $x < 6$ that is satisfied.

- Fourth, when $\lambda \in \left[\frac{x}{6}, 1\right]$, $8 > 6\lambda + 2$ reducing to $\lambda < 1$ that is true.

• In all cases, we see that mixed bundling generates larger profits than separate selling, such that the restaurant will sell both the set menu and separate items. This allows the restaurant to serve those consumers who only buy one item but not the set menu.

Exercise #9.2: Bundling to a Single Consumer Type^A

9.2 Consider a monopolist selling two goods, A and B. Consumers' valuation of good A is uniformly distributed on the unit line, that is, $\theta_A \sim U[0, 1]$, and valuation for good B is positively correlated with good A as follows:

$$\theta_B = \rho\theta_A + (1 - \rho)(1 - \theta_A)$$

where $2/3 \leq \rho \leq 1$ measures the degree of correlation between goods A and B. When $\rho = 1$, valuation of good B is perfectly correlated with good A, that is, $\theta_B = \theta_A$. For simplicity, assume that production costs are zero.

(a) Find out the monopolist's profits if goods A and B are sold separately.

• *Separate Selling*. Consumers buy good A if their valuation of the good exceeds its price, where

$$\theta_A > p_A$$

Given uniform distribution of consumers' valuation of good A, demand for good A becomes

$$Q_A(p_A) = 1 - p_A$$

• Therefore, for good A, the monopolist's profit maximization problem is

$$\max_{p_A \geq 0} \pi(p_A) = p_A Q_A(p_A)$$
$$= p_A(1 - p_A)$$

Differentiating the profit function with respect to price p_A, we obtain

$$\frac{d\pi(p_A)}{dp_A} = 1 - 2p_A = 0$$

yielding the optimal price of good A of

$$p_A^* = \frac{1}{2}$$

and equilibrium profit from good A of

$$\begin{aligned}\pi_A^* &= p_A^*(1 - p_A^*)\\ &= \frac{1}{2}\left(1 - \frac{1}{2}\right)\\ &= \frac{1}{4}\end{aligned}$$

- Consumers buy good B if their valuation of the good exceeds its price, where

$$\rho\theta_A + (1-\rho)(1-\theta_A) > p_B$$

which is rearranged to yield

$$\theta_A > \frac{p_B + \rho - 1}{2\rho - 1}$$

Given uniform distribution of valuation of good A, demand for good B becomes

$$Q_B(p_B) = 1 - \frac{p_B + \rho - 1}{2\rho - 1}$$

- Therefore, for good B, the monopolist's profit maximization problem is

$$\begin{aligned}\max_{p_B \geq 0} \pi(p_B) &= p_B Q_B(p_B)\\ &= p_B\left(1 - \frac{p_B + \rho - 1}{2\rho - 1}\right)\\ &= \frac{p_B(\rho - p_B)}{2\rho - 1}\end{aligned}$$

Differentiating the profit function with respect to price p_B, we obtain

$$\frac{d\pi(p_B)}{dp_B} = \rho - 2p_B = 0$$

yielding the optimal price of good B of

$$p_B^* = \frac{\rho}{2}$$

and equilibrium profit from good B of

$$\pi_B^* = \frac{p_B^*\left(\rho - p_B^*\right)}{2\rho - 1}$$
$$= \frac{\frac{\rho}{2}\left(\rho - \frac{\rho}{2}\right)}{2\rho - 1}$$
$$= \frac{\rho^2}{4\left(2\rho - 1\right)}$$

- As a result, the monopolist's profits of selling the goods separately are

$$\pi_A^* + \pi_B^* = \frac{1}{4} + \frac{\rho^2}{4\left(2\rho - 1\right)}$$
$$= \frac{\rho^2 + 2\rho - 1}{4\left(2\rho - 1\right)}.$$

(b) Find out the monopolist's profits if only the pure bundle *AB* is sold.
- *Pure Bundling*. Consumers buy bundle *AB* if its valuation exceeds its price, where

$$\theta_A + \theta_B > p_{AB}$$
$$2\rho\theta_A + 1 - \rho > p_{AB}$$
$$\Rightarrow \theta_A > \frac{p_{AB} + \rho - 1}{2\rho}$$

Given uniform distribution of valuation of good A, demand for the bundle is

$$Q_{AB}\left(p_{AB}\right) = 1 - \frac{p_{AB} + \rho - 1}{2\rho}$$

- Therefore, for the bundle *AB*, the monopolist's profit maximization problem is

$$\max_{p_{AB} \geq 0} \pi\left(p_{AB}\right) = p_{AB} Q_{AB}\left(p_{AB}\right)$$
$$= p_{AB}\left(1 - \frac{p_{AB} + \rho - 1}{2\rho}\right)$$
$$= \frac{p_{AB}\left(1 + \rho - p_{AB}\right)}{2\rho}$$

Differentiating the profit function with respect to price p_{AB}, we obtain

$$\frac{d\pi\left(p_{AB}\right)}{dp_{AB}} = 1 + \rho - 2p_{AB} = 0$$

yielding the optimal price of bundle *AB* of

$$p_{AB}^* = \frac{1 + \rho}{2}$$

and equilibrium profit of selling bundle AB of

$$\pi^*_{AB} = \frac{p^*_{AB}\left(1+\rho-p^*_{AB}\right)}{2\rho} = \frac{\frac{1+\rho}{2}\left(1+\rho-\frac{1+\rho}{2}\right)}{2\rho} = \frac{(1+\rho)^2}{8\rho}.$$

(c) For what values of ρ does the monopolist obtain larger profits from separate selling than pure bundling? Explain.

- The monopolist obtains more profits from separate selling than pure bundling if

$$\pi^*_A + \pi^*_B \geq \pi^*_{AB}$$

which entails

$$\frac{\rho^2+2\rho-1}{4(2\rho-1)} \geq \frac{(1+\rho)^2}{8\rho}$$
$$2\rho\left(\rho^2+2\rho-1\right) \geq (2\rho-1)(1+\rho)^2$$
$$2\rho^3+4\rho^2-2\rho \geq 2\rho^3+3\rho^2-1$$
$$1-2\rho+\rho^2 \geq 0$$
$$\Rightarrow (1-\rho)^2 \geq 0$$

which holds for all $2/3 \leq \rho \leq 1$. In other words, the monopolist always obtains more profits from separate selling than pure bundling. Intuitively, the monopolist can serve those consumers, who buy either good A or good B but not the pure bundle AB, if the goods are sold separately. Since the price of the bundle is equal to the sum prices of separate goods, where $p^*_{AB} = p^*_A + p^*_B$, separate selling leads to larger demand and, thus, larger profits than pure bundling.

Exercise #9.3: Bundling to a Single Consumer Type, a Numerical Example^A

9.3 Consider a monopolist selling two goods, A and B. Consumers' valuation of good A is uniformly distributed on the unit line, that is, $\theta_A \sim U[0,1]$, and valuation for good B is positively correlated with good A as follows:

$$\theta_B = \frac{1}{4} + \frac{1}{2}\theta_A$$

so that $\theta_B \sim U\left[\frac{1}{4}, \frac{3}{4}\right]$. For simplicity, assume that production costs are zero.

(a) Find out the monopolist's profits if goods A and B are sold separately.

- *Separate Selling*. Consumers buy good A if their valuation of the good exceeds its price, where

$$\theta_A > p_A$$

Given uniform distribution of consumers' valuation of good A, demand for good A becomes

$$Q_A(p_A) = 1 - p_A$$

- Therefore, for good A, the monopolist's profit maximization problem is

$$\begin{aligned}\max_{p_A \geq 0} \pi(p_A) &= p_A Q_A(p_A) \\ &= p_A(1 - p_A)\end{aligned}$$

Differentiating the profit function with respect to price p_A, we obtain

$$\frac{d\pi(p_A)}{dp_A} = 1 - 2p_A = 0$$

yielding the optimal price of good A of

$$p_A^* = \frac{1}{2}$$

and equilibrium profit from good A of

$$\begin{aligned}\pi_A^* &= p_A^*(1 - p_A^*) \\ &= \frac{1}{2}\left(1 - \frac{1}{2}\right) \\ &= \frac{1}{4}\end{aligned}$$

- Consumers buy good B if their valuation of the good exceeds its price, where

$$\theta_B = \frac{1}{4} + \frac{1}{2}\theta_A > p_B$$

which is rearranged to yield

$$\begin{aligned}\frac{1}{2}\theta_A &> \frac{4p_B - 1}{4} \\ \Rightarrow \theta_A &> \frac{4p_B - 1}{2}\end{aligned}$$

Given uniform distribution of valuation of good A, the demand for good B becomes

$$\begin{aligned} Q_B(p_B) &= 1 - \frac{4p_B - 1}{2} \\ &= \frac{3}{2} - 2p_B \end{aligned}$$

- Therefore, for good B, the monopolist's profit maximization problem is

$$\begin{aligned} \max_{p_B \geq 0} \pi(p_B) &= p_B Q_B(p_B) \\ &= p_B\left(\frac{3}{2} - 2p_B\right) \end{aligned}$$

Differentiating the profit function with respect to price p_B, we obtain

$$\frac{d\pi(p_B)}{dp_B} = \frac{3}{2} - 4p_B = 0$$

yielding the optimal price of good B of

$$p_B^* = \frac{3}{8}$$

and equilibrium profit from good B of

$$\begin{aligned} \pi_B^* &= p_B^*\left(\frac{3}{2} - 2p_B^*\right) \\ &= \frac{3}{8} \times \left(\frac{3}{2} - 2 \times \frac{3}{8}\right) \\ &= \frac{9}{32} \end{aligned}$$

- As a result, the monopolist's profits of selling the goods separately are

$$\pi_A^* + \pi_B^* = \frac{1}{4} + \frac{9}{32} = \frac{17}{32}.$$

(b) Find out the monopolist's profits if only the pure bundle AB is sold.

- *Pure Bundling*. Consumers buy bundle AB if its valuation exceeds its price, where

$$\begin{aligned} \theta_A + \theta_B &> p_{AB} \\ \theta_A + \frac{1}{4} + \frac{1}{2}\theta_A &> p_{AB} \\ \frac{3}{2}\theta_A &> p_{AB} - \frac{1}{4} \end{aligned}$$

which simplifies to

$$\theta_A > \frac{4p_{AB} - 1}{6}$$

Since the valuation of good A is uniformly distributed, the demand for the bundle is

$$Q_{AB}(p_{AB}) = 1 - \frac{4p_{AB} - 1}{6} = \frac{7 - 4p_{AB}}{6}$$

- Therefore, for the bundle AB, the monopolist's profit maximization problem is

$$\begin{aligned}\max_{p_{AB} \geq 0} \ \pi(p_{AB}) &= p_{AB} Q_{AB}(p_{AB}) \\ &= p_{AB}\left(\frac{7 - 4p_{AB}}{6}\right)\end{aligned}$$

Differentiating the profit function with respect to price p_{AB}, we obtain

$$\frac{d\pi(p_{AB})}{dp_{AB}} = 7 - 8p_{AB} = 0$$

yielding the optimal price of bundle AB of

$$p^*_{AB} = \frac{7}{8}$$

and equilibrium profit of selling bundle AB of

$$\begin{aligned}\pi^*_{AB} &= p^*_{AB}\left(\frac{7 - 4p^*_{AB}}{6}\right) \\ &= \frac{7}{8} \times \left(\frac{7 - 4 \times \frac{7}{8}}{6}\right) \\ &= \frac{49}{96}.\end{aligned}$$

(c) Does the monopolist obtain larger profits from separate selling than pure bundling? Explain.

- The monopolist obtains more profits from separate selling than pure bundling as

$$\pi^*_A + \pi^*_B = \frac{51}{96} > \frac{49}{96} = \pi^*_{AB}$$

Intuitively, the monopolist can serve those consumers, who buy either good A or good B but not the pure bundle AB, if the goods are sold separately. Since the price of the bundle is equal to the sum prices of separate goods, where $p^*_{AB} = p^*_A + p^*_B$, separate selling leads to larger demand and, thus, larger profits than pure bundling.

Exercise #9.4: Bundling to a Single Consumer Type—Negatively and Positively Correlated Valuations[A]

9.4 Consider a monopolist selling two goods, A and B. Consumers' valuation of good A is uniformly distributed on the unit line, that is, $\theta_A \sim U[0, 1]$, and valuation for good B is positively correlated with good A as follows:

$$\theta_B = \rho\theta_A + (1-\rho)(1-\theta_A)$$

where $0 \leq \rho \leq 1$ measures the degree of correlation between goods A and B. When $\rho = 0$, valuation of good B degenerates to $\theta_B = 1-\theta_A$, meaning that valuations are perfectly negatively correlated. In contrast, when $\rho = 1$, valuation of good B simplifies to $\theta_B = \theta_A$, implying that valuations are perfectly positively correlated. For simplicity, assume that production costs are zero.

(a) Find out the monopolist's profits if goods A and B are sold separately.

- *Good A*. Consumers buy good A if their valuation of the good exceeds its price, that is,

$$\theta_A > p_A$$

Given uniform distribution of consumers' valuation of good A, demand for good A becomes

$$Q_A(p_A) = 1 - p_A$$

Therefore, for good A, the monopolist's profit maximization problem is

$$\max_{p_A \geq 0} \pi(p_A) = p_A Q_A(p_A)$$
$$= p_A(1-p_A)$$

Differentiating the profit function with respect to price p_A, we obtain

$$\frac{d\pi(p_A)}{dp_A} = 1 - 2p_A = 0$$

yielding an equilibrium price of good A of

$$p_A^* = \frac{1}{2}$$

and equilibrium profit from good A of

$$\pi_A^* = p_A^*(1-p_A^*)$$
$$= \frac{1}{2}\left(1-\frac{1}{2}\right)$$
$$= \frac{1}{4}$$

- *Good B*. Consumers buy good B if their valuation of the good exceeds its price, that is,

$$\rho\theta_A + (1-\rho)(1-\theta_A) > p_B$$

which is rearranged to yield

$$\begin{cases} \theta_A < \frac{1-p_B-\rho}{1-2\rho} & \text{if } \rho < \frac{1}{2}, \\ p_B < \frac{1}{2} & \text{if } \rho = \frac{1}{2}, \text{ and} \\ \theta_A > \frac{p_B+\rho-1}{2\rho-1} & \text{if } \rho > \frac{1}{2}. \end{cases}$$

Therefore, we separate our analysis into three cases: (1) $0 \leq \rho < \frac{1}{2}$, (2) $\rho = \frac{1}{2}$, and (3) $\frac{1}{2} < \rho \leq 1$.

– *Case 1.* When $0 \leq \rho < \frac{1}{2}$, the demand for good B becomes

$$Q_B(p_B) = \frac{1 - p_B - \rho}{1 - 2\rho}.$$

Since valuations are negatively correlated, consumers with low valuation for good A have a high valuation for good B and vice versa.

In this context, the monopolist's profit maximization problem is

$$\begin{aligned} \max_{p_B \geq 0} \pi(p_B) &= p_B Q_B(p_B) \\ &= p_B \left(\frac{1 - p_B - \rho}{1 - 2\rho} \right) \end{aligned}$$

Differentiating the profit function with respect to price p_B, we obtain

$$\frac{d\pi(p_B)}{dp_B} = \frac{1 - 2p_B - \rho}{1 - 2\rho} = 0$$

yielding the optimal price of good B of

$$p_B^* = \frac{1 - \rho}{2}$$

and equilibrium profit from good B of

$$\begin{aligned} \pi_B^* &= \frac{p_B^*\left(1 - p_B^* - \rho\right)}{1 - 2\rho} \\ &= \frac{\frac{1-\rho}{2}\left(1 - \frac{1-\rho}{2} - \rho\right)}{1 - 2\rho} \\ &= \frac{(1 - \rho)^2}{4(1 - 2\rho)}. \end{aligned}$$

– *Case 2.* When $\rho = \frac{1}{2}$, the demand for good B becomes

$$Q_B(p_B) = \begin{cases} 1 & \text{if } p_B \leq \frac{1}{2}, \text{ and} \\ 0 & \text{otherwise.} \end{cases}$$

In this context, the monopolist will set $p_B^* = \frac{1}{2}$ that yields equilibrium profit of

$$\pi_B^* = p_B^* Q_B\left(p_B^*\right) = \frac{1}{2} \times 1 = \frac{1}{2},$$

since the monopolist serves all consumers whose valuation is degenerated to $\theta_B = \frac{1}{2}$.

– *Case 3*. When $\frac{1}{2} < \rho \leq 1$, the demand for good *B* becomes

$$Q_B\left(p_B\right) = 1 - \frac{p_B + \rho - 1}{2\rho - 1}$$

In this case, valuations are positively correlated, and consumers with high valuation for good *A* also have a high valuation for good *B*.

In this context, the monopolist's profit maximization problem is

$$\begin{aligned} \max_{p_B \geq 0} \pi\left(p_B\right) &= p_B Q_B\left(p_B\right) \\ &= p_B\left(1 - \frac{p_B + \rho - 1}{2\rho - 1}\right) \\ &= \frac{p_B\left(\rho - p_B\right)}{2\rho - 1} \end{aligned}$$

Differentiating the profit function with respect to price p_B, we obtain

$$\frac{d\pi\left(p_B\right)}{dp_B} = \frac{\rho - 2p_B}{2\rho - 1} = 0$$

yielding an equilibrium price of good *B* of

$$p_B^* = \frac{\rho}{2}$$

and equilibrium profit from good *B* of

$$\begin{aligned} \pi_B^* &= \frac{p_B^*\left(\rho - p_B^*\right)}{2\rho - 1} \\ &= \frac{\frac{\rho}{2}\left(\rho - \frac{\rho}{2}\right)}{2\rho - 1} \\ &= \frac{\rho^2}{4\left(2\rho - 1\right)}. \end{aligned}$$

- Combining the above cases, the monopolist's profits of selling the goods separately are

$$\pi_A^* + \pi_B^* = \begin{cases} \frac{1}{4} + \frac{(1-\rho)^2}{4(1-2\rho)} = \frac{\rho^2 - 4\rho + 2}{4(1-2\rho)} & \text{if } \rho < \frac{1}{2}, \\ \frac{1}{4} + \frac{1}{2} = \frac{3}{4} & \text{if } \rho = \frac{1}{2}, \text{ and} \\ \frac{1}{4} + \frac{\rho^2}{4(2\rho-1)} = \frac{\rho^2 + 2\rho - 1}{4(2\rho-1)} & \text{if } \rho > \frac{1}{2}. \end{cases}$$

(b) Find out the monopolist's profits if only the pure bundle *AB* is sold.

- Consumers buy bundle *AB* if their valuation exceeds the price, p_{AB}, that is,

$$\theta_A + \theta_B > p_{AB}$$
$$2\rho\theta_A + 1 - \rho > p_{AB}$$
$$\Rightarrow \theta_A > \frac{p_{AB} + \rho - 1}{2\rho}$$

Given uniform distribution of valuation of good *A*, demand for the bundle is

$$Q_{AB}(p_{AB}) = 1 - \frac{p_{AB} + \rho - 1}{2\rho}.$$

- Therefore, for bundle *AB*, the monopolist's profit maximization problem is

$$\max_{p_{AB} \geq 0} \pi(p_{AB}) = p_{AB} Q_{AB}(p_{AB})$$
$$= p_{AB}\left(1 - \frac{p_{AB} + \rho - 1}{2\rho}\right)$$
$$= \frac{p_{AB}(1 + \rho - p_{AB})}{2\rho}$$

Differentiating the profit function with respect to price p_{AB}, we obtain

$$\frac{d\pi(p_{AB})}{dp_{AB}} = 1 + \rho - 2p_{AB} = 0,$$

yielding the equilibrium price of bundle *AB* of

$$p^*_{AB} = \frac{1+\rho}{2},$$

and equilibrium profit of selling bundle *AB* of

$$\pi^*_{AB} = \frac{p^*_{AB}(1 + \rho - p^*_{AB})}{2\rho}$$
$$= \frac{\frac{1+\rho}{2}\left(1 + \rho - \frac{1+\rho}{2}\right)}{2\rho}$$
$$= \frac{(1+\rho)^2}{8\rho}.$$

(c) For what values of ρ does the monopolist obtain larger profits from separate selling than pure bundling? Explain.

- The monopolist obtains more profits from separate selling than pure bundling if

$$\pi^*_A + \pi^*_B \geq \pi^*_{AB}$$

which again we separate into three cases.

- *Case 1*, $0 \leq \rho < \frac{1}{2}$. The monopolist prefers separate selling if

$$\frac{\rho^2 - 4\rho + 2}{4(1-2\rho)} > \frac{(1+\rho)^2}{8\rho}$$
$$2\rho\left(\rho^2 - 4\rho + 2\right) > (1-2\rho)(1+\rho)^2$$
$$16\rho^3 - 20\rho^2 + 16\rho - 4 > 0$$

or $\rho > 0.37$. Intuitively, goods are negatively correlated in this case, but if this correlation is weak (strong), the monopolist prefers separate selling to pure bundling (otherwise). When valuations are sufficiently negatively correlated, where $\rho < 0.37$, consumers who buy good *A* (*B*) have low incentives to buy good *B* (*A*) as well, implying that the monopolist can improve its profits if the goods are bundled together.
- *Case 2*, $\rho = \frac{1}{2}$. The monopolist prefers separate selling if

$$\frac{3}{4} > \frac{\left(1+\frac{1}{2}\right)^2}{8 \times \frac{1}{2}}$$

which is simplified to $\frac{3}{4} > \frac{9}{16}$ that holds. Therefore, the monopolist prefers separate selling to pure bundling when $\rho = \frac{1}{2}$, since the firm captures all consumers when it sells good *B* separately from good *A*, as valuations are sufficiently positively correlated, but only those consumers whose combined valuation for both goods exceeds $p^*_{AB} = \frac{1+\frac{1}{2}}{2} = \frac{3}{4}$ when offering the bundle.
- *Case 3*, $\frac{1}{2} < \rho \leq 1$. The monopolist prefers separate selling if

$$\frac{\rho^2 + 2\rho - 1}{4(2\rho - 1)} \geq \frac{(1+\rho)^2}{8\rho}$$
$$2\rho\left(\rho^2 + 2\rho - 1\right) \geq (2\rho - 1)(1+\rho)^2$$
$$1 - 2\rho + \rho^2 \geq 0$$

which simplifies to $(1-\rho)^2 \geq 0$, which holds for all values of ρ. In other words, the monopolist always earns more profits from separate selling than pure bundling when good valuation is positively correlated. Intuitively, the monopolist can serve those consumers, who buy either good *A* or good *B* but not the pure bundle *AB*, if the goods are sold separately. Since in this case the price of the bundle is equal to the aggregate prices of the separate goods, $p^*_{AB} = p^*_A + p^*_B$, selling the goods separately leads to larger demand and, thus, larger profits than pure bundling.

(d) *Numerical example.* Evaluate the monopolist's profits when correlation is (i) $\rho = 1/4$ and (ii) $\rho = 3/4$. What pricing strategy will the monopolist adopt? Explain your results.

- When $\rho = \frac{1}{4}$, valuations are negatively correlated. In this context, profits from pure bundling are

$$\pi^*_{AB} = \frac{\left(1+\frac{1}{4}\right)^2}{8 \times \frac{1}{4}} = \frac{25}{32} \simeq 0.78$$

which exceed the profits from separate selling:

$$\pi_A^* + \pi_B^* = \frac{\left(\frac{1}{4}\right)^2 - 4 \times \frac{1}{4} + 2}{4\left(1 - 2 \times \frac{1}{4}\right)} = \frac{17}{32} \simeq 0.53.$$

In this setting, the monopolist enjoys higher profits by selling bundled products than offering the products separately.

- When $\rho = \frac{3}{4}$, valuations are positively correlated. In this case, profits from pure bundling are

$$\pi_{AB}^* = \frac{\left(1 + \frac{3}{4}\right)^2}{8 \times \frac{3}{4}} = \frac{49}{96} \simeq 0.51$$

which fall below the profits from separate selling:

$$\pi_A^* + \pi_B^* = \frac{\left(\frac{3}{4}\right)^2 + 2 \times \frac{3}{4} - 1}{4\left(2 \times \frac{3}{4} - 1\right)} = \frac{17}{32} \simeq 0.53.$$

In this context, the monopolist enjoys higher profits by selling the goods separately than offering the bundle. Intuitively, the monopolist not only serves those with a high valuation for both products but also captures the demand of those who only buy one product.

Exercise #9.5: Pure vs. Mixed Bundling^B

9.5 Consider a restaurant selling bento boxes and bubble tea. For simplicity, assume a unit mass of consumers uniformly distributed on the unit line, whose preferences (i.e., willingness to pay) are heterogeneous and stated as follows:

Group	Proportion	Bento boxes	Bubble tea
Food lover	λ	120	30
Tea lover	$1 - \lambda$	60	60

where λ $(1 - \lambda)$ represents the proportion of consumers who are food (tea) lover and $0 \leq \lambda \leq 1$. Assume away production and transportation costs. Determine the optimal prices and the associated profits if the restaurant sells bento boxes and bubble tea (i) separately, (ii) as a set menu only, and (iii) both separately and as a set menu. What pricing strategies will this restaurant adopt at different intervals of λ? Interpret.

(a) *Separate selling*. Let p_1^s and p_2^s be the prices of bento boxes and bubble tea, respectively.
   - If the restaurant sets the price of bento boxes at $p_1^s = 120$, it can only sell to food lovers, but if it sets $p_1^s = 60$, it can sell to both groups of consumers. Therefore, it sets a price of 120 if doing so generates more profit than setting a price of 60, that is, $120\lambda > 60$, which is rearranged to $\lambda > \frac{1}{2}$.
   - If the restaurant sets the price of bubble tea at $p_2^s = 60$, it can only sell to tea lovers, but if it sets $p_2^s = 30$, it can sell to both groups of consumers. Therefore, it sets a price of 60 if doing so generates more profit than setting a price of 30, that is, $60(1 - \lambda) > 30$, which is rearranged to $\lambda < \frac{1}{2}$.

- To sum up, the restaurant's price strategy under separate selling is

$$\left(p_1^s, p_2^s\right) = \begin{cases} (60, 60) & \text{if } \lambda \in \left[0, \frac{1}{2}\right) \\ (120, 30) & \text{if } \lambda \in \left[\frac{1}{2}, 1\right] \end{cases}$$

generating profits of

$$\pi^s \equiv \pi_1^s + \pi_2^s = \begin{cases} 120 - 60\lambda & \text{if } \lambda \in \left[0, \frac{1}{2}\right) \\ 120\lambda + 30 & \text{if } \lambda \in \left[\frac{1}{2}, 1\right]. \end{cases}$$

(b) *Pure bundling*. Let p^b be the price of the set menu under pure bundling.

- The restaurant sets the price of the set menu at $p^b = 150$ if there is a sufficient demand from food lovers; otherwise, it sets $p^b = 120$ to satisfy the demand of both groups of consumers. This means for the restaurant to restrict sales to only food lovers, demand from this group of consumers must satisfy $150\lambda > 120$ that is simplified to $\lambda > \frac{4}{5}$.
- In other words, the restaurant's price strategy under pure bundling is

$$p^b = \begin{cases} 120 & \text{if } \lambda \in \left[0, \frac{4}{5}\right) \\ 150 & \text{if } \lambda \in \left[\frac{4}{5}, 1\right] \end{cases}$$

generating profits of

$$\pi^b = \begin{cases} 120 & \text{if } \lambda \in \left[0, \frac{4}{5}\right) \\ 150\lambda & \text{if } \lambda \in \left[\frac{4}{5}, 1\right]. \end{cases}$$

(c) *Mixed bundling*. Let p_1^m, p_2^m, and p^m be the prices of bento boxes, bubble tea, and the set menu, respectively, under mixed bundling.

- Suppose the set menu is priced at $p^m = 150$ to food lovers. Then the restaurant can set the price of bento boxes at $p_1^m = 120$ and the price of bubble tea at $p_2^m = 60$ to prevent food lovers from buying bento boxes and bubble tea separately, while tea lovers will still buy the bubble tea at that price. In this context, profit becomes

$$\pi^m = 150\lambda + 60\,(1 - \lambda) = 60 + 90\lambda.$$

- Otherwise, if the restaurant sets a lower price for the set menu at $p^m = 120$ to tea lovers, its profit is 120 because food lovers will also buy the set menu and enjoy a positive surplus of $150 - 120 = 30$ that can be understood as buying the set menu at the price of bento box and getting the bubble tea for free.
- Thus, the restaurant will price the set menu at $p^m = 150$ if $60 + 90\lambda \geq 120$, which holds for $\lambda \in \left[\frac{2}{3}, 1\right]$. Its price strategy under mixed bundling becomes

$$p_1^m = 120, \ p_2^m = 60, \ p^m = \begin{cases} 120 & \text{if } \lambda \in \left[0, \frac{2}{3}\right) \\ 150 & \text{if } \lambda \in \left[\frac{2}{3}, 1\right] \end{cases}$$

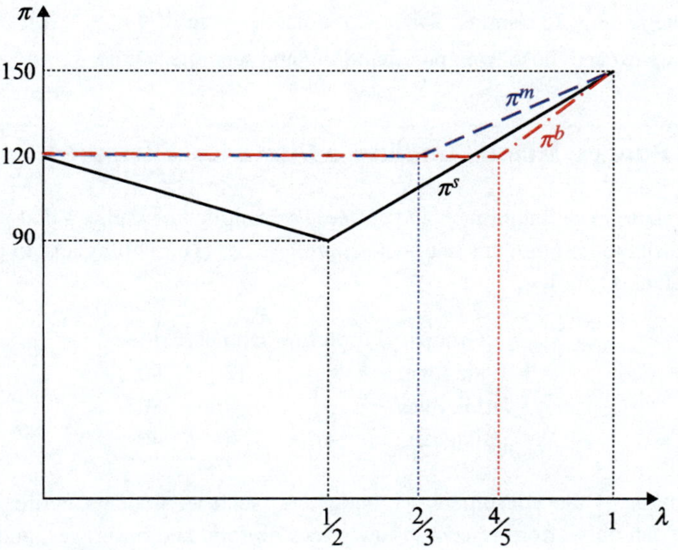

Fig. 9.1 Pricing decisions as a function of λ

generating profits of

$$\pi^m = \begin{cases} 120 & \text{if } \lambda \in \left[0, \frac{2}{3}\right) \\ 60 + 90\lambda & \text{if } \lambda \in \left[\frac{2}{3}, 1\right]. \end{cases}$$

(d) *Comparison.* Compare the profits that the firm makes with the pure and mixed form of bundling, and determine under which conditions each pricing strategy is preferred.

- Figure 9.1 plots the profits from separately selling to each type of customer, π^s, from bundling, π^b, and from mixed bundling, π^m, as a function of λ.
  - When $0 \leq \lambda < \frac{2}{3}$, pure and mixed bundling yields the same profit since both food lovers and tea lovers buy the set menu, which is better than separate selling for

    $$\pi^b = \pi^m = 120 > \max\{120 - 60\lambda, 120\lambda + 30\} = \pi^s$$

    because the restaurant does not satisfy the demand of food lovers for bubble tea when $0 \leq \lambda < \frac{1}{2}$ and the demand of tea lovers for bento boxes when $\frac{1}{2} \leq \lambda < \frac{2}{3}$ under separate selling.
  - In contrast, when $\frac{2}{3} \leq \lambda \leq 1$, mixed bundling yields the highest profit because $\pi^m = 60 + 90\lambda \geq 120 = \pi^b$ when $\frac{2}{3} \leq \lambda < \frac{4}{5}$, $\pi^m = 60 + 90\lambda \geq 150\lambda = \pi^b$ when $\frac{4}{5} \leq \lambda \leq 1$, and in addition,

    $$\pi^m = 60 + 90\lambda \geq 120\lambda + 30 = \pi^s$$

    that simplifies to $\lambda \leq 1$.
- In summary, mixed bundling yields the highest profit under all parameter values.

Figure 9.1 indicates that when the proportion of food lovers λ is relatively low $\left(0 \leq \lambda < \frac{2}{3}\right)$, profits from mixed and pure bundling are the same and are both higher than separate selling. In contrast, when λ is relatively high $\left(\frac{2}{3} \leq \lambda < 1\right)$, profits from mixed bundling exceed those from pure bundling and separate selling.

Exercise #9.6: Pure vs. Mixed Bundling, a Numerical Example^A

9.6 Consider a restaurant selling brunch and coffee. For simplicity, assume a unit mass of consumers uniformly distributed on the unit line, whose preferences (i.e., willingness to pay) are heterogeneous and stated as follows:

Group	Proportion	Brunch	Coffee
Food lover	$\frac{3}{8}$	120	40
Drink lover	$\frac{3}{8}$	40	80
Budgeter	$\frac{1}{4}$	40	40

where one-fourth of the consumers are budgeters (such as students, retirees, or unemployed persons) who can only afford relatively low prices on food and beverages, and the remainder of consumers are equally split between food lovers and drink lovers. Assume away production and transportation costs, and determine the optimal prices and the associated profits if the restaurant sells brunch and coffee (i) separately, (ii) as a set menu only, and (iii) both separately and as a set menu. What pricing strategies will this restaurant adopt? Interpret.

(a) *Separate selling*. Let p_1^s and p_2^s be the prices of brunch and coffee, respectively.
- If the restaurant sets the price of brunch at $p_1^s = 120$, it can only sell to food lovers and obtains an expected profit of $\pi_1^s(120) = \frac{3}{8} \times 120 = 45$. However, if it sets $p_1^s = 40$, it can sell to all consumers and obtains an expected profit of $\pi_1^s(40) = 1 \times 40 = 40$.
- If the restaurant sets the price of coffee at $p_2^s = 80$, it can only sell to drink lovers and obtains an expected profit of $\pi_2^s(80) = \frac{3}{8} \times 80 = 30$. However, if it sets $p_2^s = 40$, it can sell to all consumers and obtains an expected profit of $\pi_2^s(40) = 1 \times 40 = 40$.
- Therefore, the restaurant's price strategy under separate selling is

$$\left(p_1^s, p_2^s\right) = (120, 40)$$

and obtains a profit of

$$\begin{aligned}\pi^s &= \pi_1^s(120) + \pi_2^s(40)\\ &= 45 + 40 = 85.\end{aligned}$$

(b) *Pure bundling*. Let p^b be the price of the set menu under pure bundling.
- If the restaurant sets the price of the set menu at $p^b = 160$, it obtains an expected profit of $\pi^b(160) = \frac{3}{8} \times 160 = 60$. However, if it sets $p^b = 120$ to sell the set menu to food and drink lovers, it obtains an expected profit of $\pi^b(120) = \frac{3+3}{8} \times 120 = 90$. Lastly, if it sets $p^b = 80$ to sell the set menu to all consumers, it obtains an expected profit of $\pi^b(80) = 1 \times 80 = 80$.
- Since $\pi^b(120) = 90 > 80 = \pi^b(80)$, and in turn, $\pi^b(80) = 80 > 60 = \pi^b(160)$, the restaurant sets the price of the set menu at 120 and obtains a profit of 90.

(c) *Mixed bundling*. Let p_1^m, p_2^m, and p^m be the prices of brunch, coffee, and the set menu, respectively, under mixed bundling.

- Suppose the set menu is priced at $p^m = 160$ for food lovers, and then the restaurant sets the price of brunch and coffee at $p_1^m = 120$ and $p_2^m = 80$, respectively, to prevent food lovers from buying food and drink separately, while drink lovers will only buy coffee, yielding an expected profit of $\pi^m(120, 80, 160) = \frac{3}{8} \times 160 + \frac{3}{8} \times 80 = 90$.
- If the set menu is priced at $p^m = 120$ for drink lovers, then the restaurant sets the price of brunch at $p_1^m = 80$ and $p_2^m = 80$, respectively, to prevent drink lovers from buying food and drink separately, while food lovers who buy the set menu obtain a surplus of $120 + 40 - 120 = 40$ that can be understood as buying the set menu at the price of brunch and get coffee for free (check that food lovers obtain the same surplus if they only buy the brunch). In this context, budgeters are not served, and the restaurant obtains an expected profit of $\pi^m(80, 80, 120) = \frac{3}{4} \times 120 = 90$.
- Since both sets of prices yield the same profit of 90, the restaurant is indifferent between $\left(p_1^m, p_2^m, p^m\right) = (120, 80, 160)$ and $\left(p_1^m, p_2^m, p^m\right) = (80, 80, 120)$.

(d) *Comparison.* Compare the profits from pure and mixed bundling.

- Since $\pi^s = 85 < 90 = \pi^b = \pi^m$, the restaurant is indifferent between selling the set menu only or along with brunch and coffee. However, selling brunch and coffee separately without the set menu does not maximize the profit of the restaurant.

Exercise #9.7: Bundling Under Exponentially Distributed Values^C

9.7 Consider a monopolist producing two goods, A and B, at zero marginal costs. There is mass one of identical consumers, whose valuation of good i, θ_i, is independent and identically distributed according to the exponential distribution, $F(\theta_i) = 1 - \exp(-\lambda\theta_i)$, where $\theta_i \in [0, \infty)$, $\lambda > 0$, and $i \in \{A, B\}$, with an associated density function $f(\theta_i) = \lambda \exp(-\lambda\theta_i)$.

(a) *Separate markets.* Find the equilibrium price and profits when the monopolist practices separate selling.

- *Finding demand.* Every consumer will buy good i if and only if his valuation exceeds its price, where $\theta_i \geq p_i$, which occurs with probability:

$$\begin{aligned}
\Pr(\theta_i \geq p_i) &= 1 - \Pr(\theta_i < p_i) \\
&= 1 - F(p_i) \\
&= \exp(-\lambda p_i).
\end{aligned}$$

Therefore, the demand of consumers who purchase good i is

$$\begin{aligned}
D_i(p_i) &= \Pr(\theta_i \geq p_i) \\
&= \exp(-\lambda p_i)
\end{aligned}$$

- *Profit maximization problem.* The monopolist chooses price p_i to maximize its profits from good i, as follows:

$$\begin{aligned}
\max_{p_i > 0} \pi(p_i) &= p_i D_i(p_i) \\
&= p_i \underbrace{\exp(-\lambda p_i)}_{D_i(p_i)}
\end{aligned}$$

Differentiating with respect to p_i, we obtain

$$\exp(-\lambda p_i) - \lambda p_i \exp(-\lambda p_i) = 0$$

which is rearranged to yield

$$(1 - \lambda p_i) \exp(-\lambda p_i) = 0$$

Solving for p_i, the equilibrium price becomes

$$p_i^s = \frac{1}{\lambda}$$

where the superscript s denotes separate selling.

- *Equilibrium profits.* Substituting price $p_i^s = \frac{1}{\lambda}$ into the profit function, equilibrium profits become

$$\pi_i^s = \overbrace{\frac{1}{\lambda}}^{p_i} \exp\left(-\lambda \times \overbrace{\frac{1}{\lambda}}^{p_i}\right)$$
$$= \frac{\exp(-1)}{\lambda}$$

To conclude, the monopolist sets equilibrium prices of

$$p_A^s = p_B^s = \frac{1}{\lambda}$$

and obtains total profits of

$$\pi^s = \pi_A^s + \pi_B^s = \frac{2\exp(-1)}{\lambda} \simeq \frac{0.74}{\lambda}.$$

Note that profits are decreasing in parameter λ, since an increase in λ indicates that a larger mass of consumers have low valuations.

(b) *Pure bundling, finding demand.* Show that the mass of consumers who purchase the pure bundle is

$$D_{AB}(p_{AB}) = (1 + \lambda p_{AB}) \exp(-\lambda p_{AB})$$

- We first consider those consumers who *do not* buy the bundle, if and only if the sum of their valuations of good A and good B falls below the price of the bundle, i.e., $\theta_A + \theta_B < p_{AB}$, with an associated mass of

$$\int_0^{p_{AB}} \left(\int_0^{p_{AB}-\theta_A} dF(\theta_B)\right) dF(\theta_A)$$
$$= \int_0^{p_{AB}} |1 - \exp(-\lambda\theta_B)|_0^{p_{AB}-\theta_A} \, dF(\theta_A)$$

$$= \int_0^{p_{AB}} \left[1 - \exp\left(-\lambda\left(p_{AB} - \theta_A\right)\right)\right] dF\left(\theta_A\right)$$

$$= \int_0^{p_{AB}} dF\left(\theta_A\right) - \int_0^{p_{AB}} \exp\left(-\lambda\left(p_{AB} - \theta_A\right)\right) f\left(\theta_A\right) d\theta_A$$

$$= F\left(p_{AB}\right) - F(0) - \lambda \int_0^{p_{AB}} \exp\left(-\lambda\left(p_{AB} - \theta_A\right)\right) \exp\left(-\lambda\theta_A\right) d\theta_A$$

$$= F\left(p_{AB}\right) - \lambda \int_0^{p_{AB}} \exp\left(-\lambda p_{AB}\right) d\theta_A$$

$$= F\left(p_{AB}\right) - \lambda \exp\left(-\lambda p_{AB}\right) \left|\theta_A\right|_0^{p_{AB}}$$

$$= 1 - \exp\left(-\lambda p_{AB}\right) - \lambda p_{AB} \exp\left(-\lambda p_{AB}\right)$$

$$= 1 - \left(1 + \lambda p_{AB}\right) \exp\left(-\lambda p_{AB}\right)$$

Therefore, the mass of consumers who buy the bundle is simply 1 minus the mass of consumers who do not buy the bundle, given by

$$D_{AB}\left(p_{AB}\right) = 1 - \underbrace{\left[1 - \left(1 + \lambda p_{AB}\right) \exp\left(-\lambda p_{AB}\right)\right]}_{\text{Mass of consumers who do not buy}}$$

$$= \left(1 + \lambda p_{AB}\right) \exp\left(-\lambda p_{AB}\right).$$

(c) *Pure bundling, finding profits.* Find the equilibrium price and profits when the monopolist practices pure bundling.

- In this context, the monopolist chooses p_{AB} to maximize its profits, as follows:

$$\max_{p_{AB}>0} \pi\left(p_{AB}\right) = p_{AB} D_{AB}\left(p_{AB}\right)$$

$$= p_{AB} \underbrace{\left(1 + \lambda p_{AB}\right) \exp\left(-\lambda p_{AB}\right)}_{D_{AB}\left(p_{AB}\right)}$$

Differentiating with respect to p_{AB}, we obtain

$$\left(1 + 2\lambda p_{AB}\right) \exp\left(-\lambda p_{AB}\right) - \lambda p_{AB}\left(1 + \lambda p_{AB}\right) \exp\left(-\lambda p_{AB}\right) = 0$$

which is rearranged to yield

$$\left(1 + \lambda p_{AB} - \lambda^2 p_{AB}^2\right) \exp\left(-\lambda p_{AB}\right) = 0$$

and further simplifies to

$$\lambda^2 p_{AB}^2 - \lambda p_{AB} - 1 = 0$$

Using the quadratic formula, we obtain that

$$p_{AB}^b = \frac{\lambda + \sqrt{\lambda^2 + 4\lambda^2}}{2\lambda^2} = \frac{1 + \sqrt{5}}{2\lambda}$$

where the superscript b denotes pure bundling.

Finally, substituting $p^b_{AB} = \frac{1+\sqrt{5}}{2\lambda}$ into the profit function, equilibrium profits are

$$\pi^b = \overbrace{\frac{1+\sqrt{5}}{2\lambda}}^{p^b_{AB}}\left(1+\lambda\overbrace{\frac{1+\sqrt{5}}{2\lambda}}^{p^b_{AB}}\right)\exp\left(-\lambda\overbrace{\frac{1+\sqrt{5}}{2\lambda}}^{p^b_{AB}}\right)$$

$$= \frac{\left(1+\sqrt{5}\right)\left(3+\sqrt{5}\right)}{4\lambda}\exp\left(-\frac{1+\sqrt{5}}{2}\right)$$

$$\simeq \frac{0.84}{\lambda}.$$

As with separate selling in part (a), profits from pure bundling are also decreasing in λ, since a higher parameter λ indicates that a larger mass of consumers have low valuations.

(d) *Profit comparison.* Compare your results in parts (a) and (c). Which selling strategy does the monopolist prefer? Explain.

- The monopolist sets a lower price for the pure bundle than the total price of the goods being sold separately, i.e., $p^b_{AB} < p^s_A + p^s_B$, since

$$\frac{1+\sqrt{5}}{2\lambda} < \frac{2}{\lambda}$$

  simplifies to $4 > 1+\sqrt{5}$, or $4 > 3.23$.
- Despite setting a lower price under pure bundling, the monopolist earns higher profits than with separate selling, i.e., $\pi^b > \pi^s$, since

$$\frac{\left(1+\sqrt{5}\right)\left(3+\sqrt{5}\right)}{4\lambda}\exp\left(-\frac{1+\sqrt{5}}{2}\right) > \frac{2\exp(-1)}{\lambda}$$

  holds for any $\lambda > 0$. This profit comparison is more visual by comparing the approximated profits we found in parts (a) and (c), as follows:

$$\frac{0.84}{\lambda} > \frac{0.73}{\lambda}.$$

  Intuitively, the larger demand that the monopolist captures with pure bundling offsets its lower profit margin per unit, ultimately yielding larger overall profits than under separate selling.
- Finally, note that prices and profits are decreasing in λ at a decreasing rate since when λ increases, the rate of decay of consumers' valuation increases. That is, the mass of consumers concentrated at low valuations increases in λ at an increasing rate, implying that the monopolist charges a lower price and obtains lower profits. However, a marginal increase in λ produces a larger decrease in the profits from pure bundling, approximated by $-\frac{0.84}{\lambda^2}$, than in those from separate selling, $-\frac{0.73}{\lambda^2}$.

Exercise #9.8: Pay-as-You-Go Contract^B

9.8 Consider a cellular company that offers you a locked phone at the price p_p and charges you p_c for every gigabyte of data used. Currently, you use your own unlocked phone and pay p_v for every gigabyte of data used to a virtual mobile network operator (VMNO). Given that you use q gigabytes per month, where $q \geq 0$, you are considering whether to switch from the VMNO to the local cellular company, which is a monopoly in the provision of locked phones, or not. For simplicity, assume that the costs for the local cellular company to provide locked phones and data services are negligible.

(a) If you want to switch and save money, how many gigabytes of data would you use?

- I will switch to the local cellular company if the total cost of obtaining the locked phone and using its data services, $p_p + p_c q$, is lower than the current charges of $p_v q$ paid to the virtual mobile network operator (VMNO), that is,

$$p_p + p_c q \leq p_v q$$

which, after rearranging, becomes

$$q \geq \hat{q} \equiv \frac{p_p}{p_v - p_c}$$

Therefore, if I use a sufficiently large amount of data, where $q \geq \hat{q}$, it makes sense for me to switch from the VMNO to the local cellular company and save money.

(b) Assume that there is a mass x of identical consumers, whose data usage is uniformly distributed on the interval $q \in [0, x]$, where $x > 0$ represents the maximum amount of data that the most data-savvy consumer uses. Write down the monopolist's profit maximization problem, and solve for the optimal prices p_p and p_c, and profit.

- The demand for locked phones is

$$Q_p(p_p, p_c) = x - \hat{q} = x - \frac{p_p}{p_v - p_c}$$

The demand for cellular data is

$$Q_c(p_p, p_c) = \int_{\hat{q}}^{x} q\,dq = \left[\frac{q^2}{2}\right]_{\hat{q}}^{x} = \frac{1}{2}\left(x^2 - \left(\frac{p_p}{p_v - p_c}\right)^2\right)$$

- Therefore, the monopolist's profit maximization problem becomes

$$\max_{p_p, p_c \geq 0} \pi(p_p, p_c) = p_p Q_p(p_p, p_c) + p_c Q_c(p_p, p_c)$$

$$= p_p\left(x - \frac{p_p}{p_v - p_c}\right) + \frac{p_c}{2}\left(x^2 - \left(\frac{p_p}{p_v - p_c}\right)^2\right)$$

Differentiating the profit with respect to p_p, and focusing on interior solutions,

$$\frac{\partial \pi(p_p, p_c)}{\partial p_p} = x - \frac{2p_p}{p_v - p_c} - \frac{p_p p_c}{(p_v - p_c)^2} = 0$$

which, after rearranging, we obtain

$$x\,(p_v - p_c)^2 = 2p_P p_v - p_P p_c$$

$$p_P = \frac{x\,(p_v - p_c)^2}{2p_v - p_c}$$

Differentiating the profit with respect to p_c, and focusing on interior solutions,

$$\frac{\partial \pi\,(p_P, p_c)}{\partial p_c} = -\left(\frac{p_P}{p_v - p_c}\right)^2 + \frac{1}{2}\left(x^2 - \left(\frac{p_P}{p_v - p_c}\right)^2\right) - \frac{p_c p_P^2}{(p_v - p_c)^3}$$

$$= \frac{x^2}{2} - \frac{p_P^2}{2\,(p_v - p_c)^3}\,[2\,(p_v - p_c) + (p_v - p_c) + 2p_c]$$

$$= \frac{x^2}{2} - \frac{(3p_v - p_c)\,p_P^2}{2\,(p_v - p_c)^3}$$

Substituting $p_P = \frac{x(p_v - p_c)^2}{2p_v - p_c}$ into the above expression, we find

$$\left.\frac{\partial \pi\,(p_P, p_c)}{\partial p_c}\right|_{p_P = \frac{x(p_v - p_c)^2}{2p_v - p_c}} = \frac{x^2}{2} - \frac{(3p_v - p_c)}{2\,(p_v - p_c)^3} \times \frac{x^2\,(p_v - p_c)^4}{(2p_v - p_c)^2}$$

$$= \frac{x^2}{2}\left[1 - \frac{(3p_v - p_c)\,(p_v - p_c)}{(2p_v - p_c)^2}\right]$$

$$= \frac{x^2\,(4p_v^2 - 4p_v p_c + p_c^2 - 3p_v^2 + 4p_v p_c - p_c^2)}{2\,(2p_v - p_c)^2}$$

$$= \frac{x^2}{2}\left(\frac{p_v}{2p_v - p_c}\right)^2 > 0$$

so that profit is monotonically increasing in p_c. In this context, the monopolist charges $p_c = p_v - \varepsilon$, where $\varepsilon > 0$, yielding the price of locked phone of

$$p_P = \frac{x\,(p_v - p_v + \varepsilon)^2}{2p_v - p_v + \varepsilon} = \frac{x\varepsilon^2}{p_v + \varepsilon}$$

such that by taking $\varepsilon \to 0$, the monopolist's optimal prices in the limit become

$$p_c^* = \lim_{\varepsilon\to 0} p_c = \lim_{\varepsilon\to 0}\,(p_v - \varepsilon) = p_v$$

$$p_P^* = \lim_{\varepsilon\to 0} p_P = \lim_{\varepsilon\to 0} \frac{x\varepsilon^2}{p_v + \varepsilon} = 0$$

- Intuitively, the monopolist offers the locked phone for free and charges the same price per gigabyte of cellular data as the MNVO, earning equilibrium profit of

$$\pi\left(p_p^*, p_c^*\right) = p_p^*\left(x - \frac{p_p^*}{p_v^* - p_c^*}\right) + \frac{p_c^*}{2}\left(x^2 - \left(\frac{p_p^*}{p_v^* - p_c^*}\right)^2\right)$$
$$= 0 + \frac{p_v}{2}\left(x^2 - 0\right) = \frac{1}{2}p_v x^2.$$

(c) Suppose otherwise data is free ($p_c = 0$). How much will you pay for the phone p_p? Evaluate the monopolist's profit in this context, and compare your results to part (b).

- The demand for locked phones now becomes

$$Q_p\left(p_p\right) = x - \hat{q} = x - \frac{p_p}{p_v}$$

The monopolist chooses p_p to solve the following profit maximization problem:

$$\max_{p_p \geq 0} \pi\left(p_p\right) = p_p\left(x - \frac{p_p}{p_v}\right)$$

Differentiating the profit with respect to p_p, and focusing on interior solutions,

$$\frac{\partial \pi\left(p_p\right)}{\partial p_p} = x - \frac{2p_p}{p_v} = 0$$

which, after rearranging, we obtain

$$p_p' = \frac{xp_v}{2}$$

Substituting $p_p' = \frac{xp_v}{2}$ into the profit function, equilibrium profit becomes

$$\pi\left(p_p'\right) = p_p'\left(x - \frac{p_p'}{p_v}\right)$$
$$= \frac{xp_v}{2}\left(x - \frac{xp_v}{2p_v}\right) = \frac{1}{4}p_v x^2$$

- Accordingly, if data is free, the monopolist will charge the locked phone at a price of $\frac{xp_v}{2}$, earning equilibrium profit of $\frac{1}{4}p_v x^2$ that is half of that in part (b).

(d) Suppose there are two packages, one in part (b) and another one in part (c), available. Find the range of q in which you will have incentives to switch to the local operator, and explain how your decision will be affected by the market size x.

- The cost of using the package in part (b) is

$$p^* = p_p^* + p_c^* q$$
$$= p_v q$$

The cost of using the package in part (c) is

$$\begin{aligned} p' &= p'_p + p'_c q \\ &= \frac{x p_v}{2} \end{aligned}$$

Therefore, I prefer the package in part (c) to the package in part (b) when

$$\begin{aligned} p^* &> p' \\ p_v q &> \frac{x p_v}{2} \\ q &> \frac{x}{2} \end{aligned}$$

- Accordingly, my decision rule is to (i) obtain a free phone and pay for data if I use data less intensively, that is, the package in part (b) if $q \leq \frac{x}{2}$, and (ii) pay for the locked phone and use free data if I use data more intensively, that is, the package in part (c) if $q > \frac{x}{2}$.
  - In case (i), I pay the same data charges compared to the current operator, so that I have no incentives to switch.
  - In case (ii), however, I can save money once I buy the locked phone And, thus, have incentives to switch to the local operator.

  Note that incentives for the once-and-for-all plan in part (c) are strengthened (weakened) if the market size shrinks (grows), since I have to use data more intensively than half of the population in order for the free data plan in part (c) to be less costly than the free phone plan in part (b).

(e) Which plan yields the highest profit? Explain.

- The free phone plan in part (b) entails a profit of $\frac{1}{2} p_v x^2$, which is twice as much as the profit from the free data plan of $\frac{1}{4} p_v x^2$ in part (c). Intuitively, the monopolist has incentives to offer the phone for free and charge for data, and this is commonly observed with VMNOs providing free locked phones and data subscription plans.

Incomplete Information, Signaling, and Competition

10

Introduction

This chapter analyzes strategic interaction of firms under incomplete information. Exercise 10.1 studies entry decisions when the incumbent's cost is unobservable to the entrant. We show that the low-cost firm can strategically increase output relative to the complete information setting, to the level that the high-cost firm cannot profitably imitate, in order to deter entry. Exercise 10.2 examines the firm's incentives to offer damaged goods at an extra cost. We find that consumers are better off since high-value consumers can buy the undamaged version of the good at a lower price than when the damaged good is unavailable, while low-value consumers can buy the damaged good and would otherwise not be served. Exercise 10.3 considers firms' incentives to invest in corporate social responsibility (CSR) when consumers do not receive an accurate signal on product quality. We report that the high-quality firm can invest in CSR to signal its product differentiation from low-quality rivals, and CSR investments are usually observed in markets with noisy signals such as fashionable clothes, cosmetics, and electronics to distinguish from counterfeit or inferior products. Exercise 10.4 identifies firms' intertemporal pricing decisions when they can advertise and poach consumers from one another. We find that in a symmetric setting, every firm obtains an equal share of the market in the first period and sells to its rival's consumers at a lower price than that of its own consumers in the second period.

Exercise 10.5 inspects the firm's incentives to undercut its rival's price in the presence of consumer loyalty. Shaffer and Zhang (2000) find that firms offer discounts to its own (rival's) consumers if consumers are less loyal to this (the other) firm, which are referred to as "pays to stay" ("pays to switch"). Exercise 10.6 seeks to understand the effect of different pricing schemes on the monopolist's profits and social welfare when the monopolist is faced with two types of consumers (high- and low-valuation), with whom the monopolist only knows the distribution but not the exact realization of consumer types. We show that profits and welfare are maximized when the monopolist practices menu pricing, in which consumers self-select into different menus. This happens because low-valuation consumers are not served with uniform pricing entailing a welfare loss, and similarly, high-valuation consumers are not price-discriminated that yield lower profits for the monopolist who practices uniform pricing.

Exercise 10.7 investigates return policies of experience goods. We find that when consumers' valuations are uniformly distributed, and these valuations are realized after the good is purchased, allowing consumers to return the product improves the monopolist's profits. This occurs because under no return policy, consumers are subject to the risk of overpaying for the product, so that risk-

P.-S. Choi et al., *Industrial Organization*, Springer Texts in Business and Economics,
https://doi.org/10.1007/978-3-031-38635-0_10

averse consumers would pay a certainty equivalent that is below the profit-maximizing price of the product when return is allowed. Exercise 10.8 reviews a firm's price randomization strategy when it sets a competitive price for the informed consumers and randomizes its prices to the uninformed consumers. Exercise 10.9 considers vertically differentiated products, where firms can use pricing or advertising to convey quality information to their consumers. Interestingly, the high-quality firm relies on price alone and does not make any advertisements, since advertising is equally costly for both firms and any advertising made by the high-quality firm can be easily mimicked by the low-quality firm.

Exercise 10.10 studies the strategic interaction between the US Environmental Protection Agency (EPA) and a nongovernment organization (NGO), where low (high) social pressure exerted by the NGO induces more (less) monitoring effort of the EPA, and reciprocally, low (high) monitoring effort of the EPA produces more (less) social pressure from the NGO, thus being regarded as strategic complements (substitutes). Exercise 10.11 shows that firms willingly disclose their quality information when the underlying quality is sufficiently high. Finally, Exercise 10.12 analyzes behavior-based price discrimination (BBPD), where the small firm has incentives to acquire information on consumers to price-discriminate between those who previously purchased from the medium and the large firm, while the medium and the large firm will compete on selling the customer list to the small firm at a price of zero.

Exercise #10.1: Signaling and Limit Pricing, Based on Milgrom and Roberts (1982)^B

10.1 Consider a market with inverse demand function $p(Q) = 1 - Q$, where $Q = q_1 + q_2$ denotes aggregate output. Let us analyze an entry game with an incumbent monopolist (Firm 1) and an entrant (Firm 2) who analyzes whether or not to join the market. The incumbent's marginal costs are either high H or low L, i.e., $c_1^H = \frac{1}{2} > c_1^L = \frac{1}{3}$, while it is common knowledge that the entrant's marginal costs are high, i.e., $c_2 = c_1^H = \frac{1}{2}$. To make the entry decision interesting, assume that when the incumbent's costs are low, entry is unprofitable, whereas when the incumbent's costs are high, entry is profitable. (Otherwise, the entrant would enter regardless of the incumbent's cost or stay out regardless of the incumbent's cost.) For simplicity, assume no discounting of future payoffs throughout this exercise.

(a) *Complete information.* Let us first examine the case in which entrant and incumbent are informed about each other's marginal costs. Consider a two-stage game where, in the first stage, the incumbent has monopoly power and selects an output level. In the second stage, a potential entrant decides whether or not to enter. If entry occurs, firms compete as Cournot duopolists, simultaneously and independently selecting production levels. If entry does not occur, the incumbent maintains its monopoly power over both periods. Find the subgame perfect Nash equilibrium (SPNE) of this complete information game.

- We next apply backward induction, starting from the second-period game.
- *Second period.* When no entry occurs, the incumbent solves

$$\max_{q_1 \geq 0} (1 - q_1)q_1 - c_1^K q_1,$$

thus selecting monopoly output $q_1^{K,m} = \frac{1-c_1^K}{2}$ for every incumbent type $K = \{H, L\}$, yielding profits $\pi_1^{K,m} = \left(\frac{1-c_1^K}{2}\right)^2$. If entry occurs, every firm $i = \{1, 2\}$ solves

$$\max_{q_i \geq 0} (1 - q_i - q_j)q_i - c_i^K q_i$$

which yields the best response function of

$$q_i = \frac{1 - c_i^K}{2} - \frac{1}{2} q_j$$

that originates at $\frac{1-c_i^K}{2}$ and decreases in the rival firm's output q_j at a rate of $\frac{1}{2}$. When its own cost c_i^K increases, firm i's best response function shifts downward that intersects the best response function of firm j at lower (higher) output level for this (the rival) firm. Simultaneously solving for incumbent and entrant's output and profits yields equilibrium output $q_1^{K,d} = \frac{1+c_2-2c_1^K}{3}$ and profits $\pi_1^{K,d} = \left(\frac{1+c_2-2c_1^K}{3}\right)^2$ for the incumbent and $q_2^{K,d} = \frac{1-2c_2+c_1^K}{3}$ and $\pi_2^{K,d} = \left(\frac{1-2c_2+c_1^K}{3}\right)^2$ for the entrant.

- *First period.* Regardless of the entrant's entry decision during the second period, the incumbent selects the standard monopoly output $q^{K,Info} = \frac{1-c_1^K}{2}$ in the first period. This is because the incumbent's output choice in this complete information setting does not affect the entrant's entry decision.

(b) *Incomplete information.* In this section, we investigate the case where the incumbent is privately informed about its marginal costs, while the entrant only observes the incumbent's first-period output which the entrant uses as a signal to infer the incumbent's cost. The time structure of this signaling game is as follows:

1. Nature decides the realization of the incumbent's marginal costs, either high or low, with probabilities $p \in (0, 1)$ and $1 - p$, respectively. The incumbent privately observes this realization but the entrant does not.
2. The incumbent chooses its first-period output level, q.
3. Observing the incumbent's output decision, the entrant forms beliefs about the incumbent's initial marginal costs. Let $\mu(c_1^H|q)$ denote the entrant's posterior belief about the initial costs being high after observing a particular first-period output from the incumbent q.
4. Given the above beliefs, the entrant decides whether or not to enter the industry.
5. If entry does not occur, the incumbent maintains its monopoly power, whereas if entry occurs, both agents compete as Cournot duopolists and the entrant observes the incumbent's type.

Write down the incentive compatibility conditions that must hold for a separating perfect Bayesian equilibrium (PBE) to be sustained. Then find the set of separating PBEs.

- In a separating equilibrium in which the high-cost firm selects q^H while the low-cost firm chooses q^L, information about the incumbent's type is conveyed to the potential entrant, who responds entering after observing the incumbent producing q^H and does not enter after observing q^L. For simplicity, we assume that all other output levels $q \neq q^H \neq q^L$ (i.e., off-the-equilibrium outputs) also lead the entrant to enter the industry, as depicted in Fig. 10.1. Let us next separately analyze each type of incumbent.
- *High-cost incumbent.* Since, by selecting q^H, this type of incumbent attracts entry, this firm selects the output that maximizes its first-period (monopoly) profits, that is, q^H coincides with its output under complete information $q^{H,Info} = \frac{1-c_1^H}{2}$. If, instead, the incumbent deviates toward the low-cost incumbent's output q^L, it conceals its type from the entrant and deters entry. Hence, the high-cost incumbent selects its equilibrium output

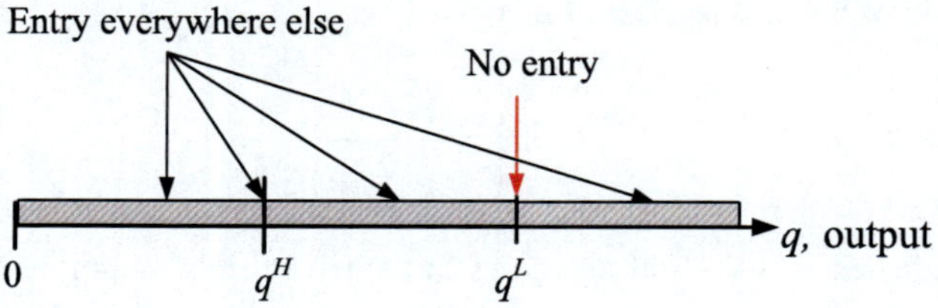

Fig. 10.1 Output choices and entry in the separating PBE

q^H rather than deviating if $M_1^H(q^{H,Info}) + \delta D_1^H \geq M_1^H(q^L) + \delta \overline{M}_1^H$, where

$$M_1^H(q) = (1-q)q - c^H q \quad \text{for every output } q$$

denotes the incumbent's first-period monopoly profits, D_1^H represents second-period duopoly profits when the incumbent's costs are high, and $\overline{M}_1^H$ indicates the second-period monopoly profits for the incumbent (in the case of no entry) when its costs are high. We can now rewrite the above incentive compatibility condition as follows:

$$M_1^H(q^{H,Info}) - M_1^H(q^L) \geq \delta \left[\overline{M}_1^H - D_1^H \right] \qquad (IC_H)$$

(where we grouped first-period profits on the left-hand side and discounted second-period profits on the right-hand side). In other words, the above inequality means that the first-period gain in producing q^H more than offsets the discounted second-period loss of inducing entry. For our parameter values, we obtain profits of $M_1^H(q^{H,Info}) = \overline{M}_1^H = \frac{1}{16}$ since $c_1^H = \frac{1}{2}$, and $D_1^H = \frac{1}{36}$ given that $c_1^H = c_2 = \frac{1}{2}$. Hence, condition IC_H reduces to

$$\frac{1}{16} - \left[(1-q^L)q^L - \frac{1}{2}q^L \right] \geq \delta \left[\frac{1}{16} - \frac{1}{36} \right]$$

Figure 10.2 depicts IC_H. Specifically, the curve depicting the difference in first-period profits, $M_1^H(q^{H,Info}) - M_1^H(q^L)$, becomes zero at $q^L = q^{H,Info}$ since at that point $M_1^H(q^{H,Info}) = M_1^H(q^L)$, but otherwise is positive since $M_1^H(q^{H,Info}) > M_1^H(q^L)$ for all $q^L \neq q^{H,Info}$. In contrast, the difference in discounted second-period profits, $\delta \left[\overline{M}_1^H - D_1^H \right]$, is constant in first-period output q^L. Hence, IC_H holds if output q^L lies in the range depicted in the horizontal axis of Fig. 10.2.

- *Low-cost incumbent.* If the low-cost incumbent chooses the equilibrium output q^L, it deters entry. If instead the incumbent deviates toward the high-cost incumbent's output, q^H, it attracts entry. Conditional on attracting entry, the low-cost incumbent would select output $q^{L,Info}$, since such output maximizes its first-period profits, yielding $M_1^L(q^{L,Info}) + \delta D_1^L$. Thus, the low-cost incumbent selects its equilibrium output of q^L if $M_1^L(q^{L,Info}) + \delta D_1^L \leq M_1^L(q^L) + \delta \overline{M}_1^L$, or equivalently,

$$M_1^L(q^{L,Info}) - M_1^L(q^L) \leq \delta \left[\overline{M}_1^L - D_1^L \right]. \qquad (IC_L)$$

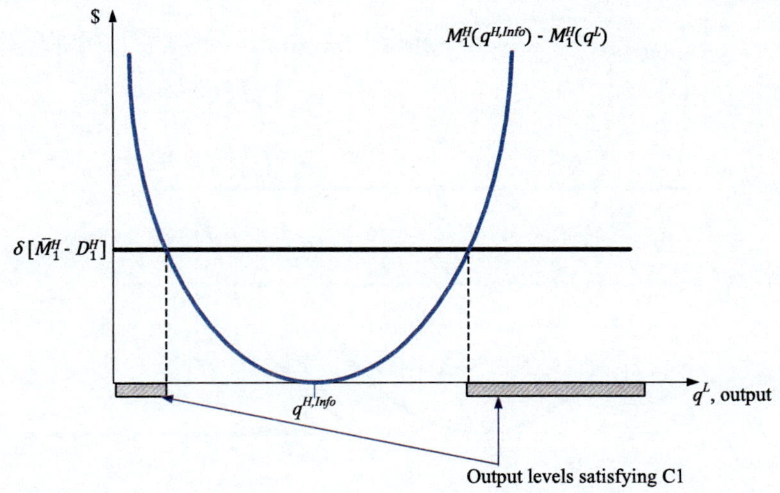

Fig. 10.2 Incentive compatibility condition IC_H

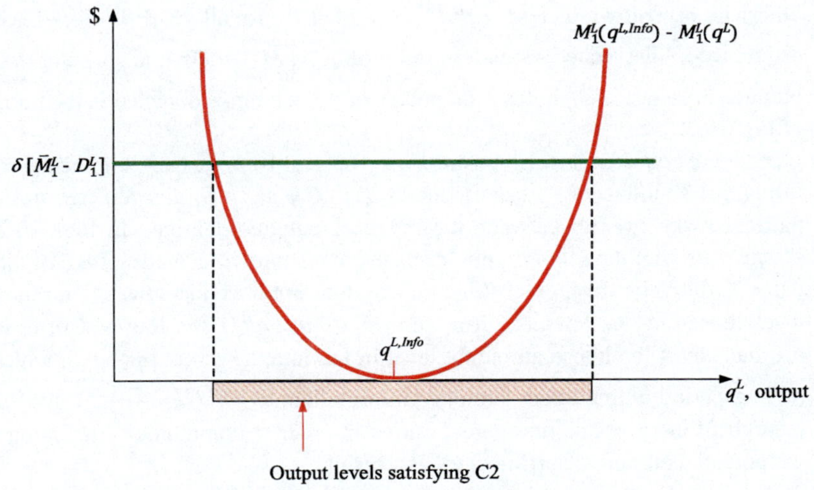

Fig. 10.3 Incentive compatibility condition IC_L

In other words, the above inequality means that the first-period gain in producing output other than q^L does not offset the discounted second-period loss of inducing entry. For our parameter values, yields $M_1^L(q^{L,Info}) = \overline{M}_1^L = \frac{1}{9}$ and $D_1^L = \frac{25}{324}$ given that $c_1^L = \frac{1}{3}$ and $c_2 = \frac{1}{2}$. Hence, condition IC_L reduces to

$$\frac{1}{9} - \left[(1 - q^L)q^L - \frac{1}{3}q^L\right] \leq \delta\left[\frac{1}{9} - \frac{25}{324}\right].$$

An argument similar to IC_H applies to the graphical representation of IC_L. As Fig. 10.3 illustrates, the curve depicting the difference in first-period profits, $M_1^L(q^{L,Info}) - M_1^L(q^L)$, becomes zero at $q^L = q^{L,Info}$ since at that point $M_1^L(q^{L,Info}) = M_1^L(q^L)$, but

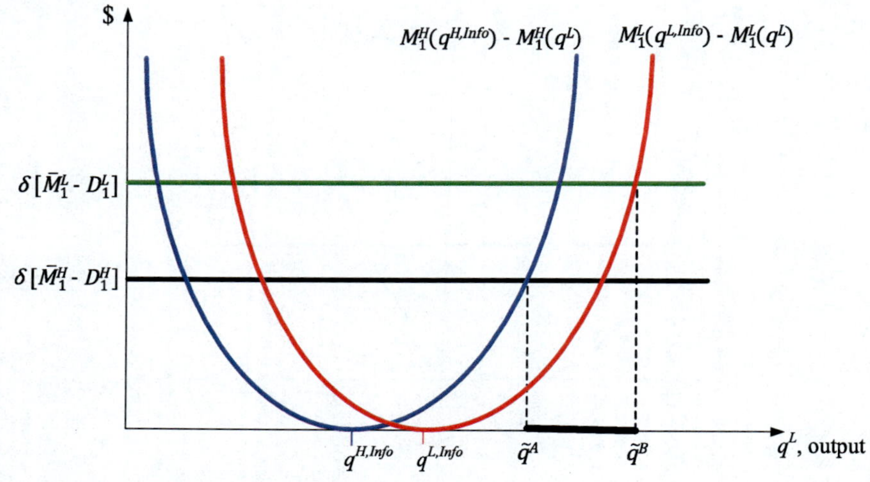

Fig. 10.4 Separating equilibria in the limit pricing model

otherwise is positive since $M_1^L(q^{L,Info}) > M_1^L(q^L)$ for all $q^L \neq q^{L,Info}$. In contrast, the difference in discounted second-period profits, $\delta\left[\overline{M}_1^L - D_1^L\right]$, is constant in first-period output q^L. Hence, IC_L holds if output q^L lies in the range depicted in the horizontal axis of Fig. 10.3.

- *Combining both ICs*. Superimposing Figs. 10.2 and 10.3, we can examine the set of output levels that simultaneously satisfy conditions IC_H and IC_L, as depicted in Fig. 10.4. In particular, the overlap between the range of outputs identified in Figs. 10.2 and 10.3 provides us with the set of output levels that constitute a separating PBE of the signaling game. Intuitively, the high-cost incumbent does not have incentives to mimic the output level chosen by the low-cost firm, i.e., $q^L \in [q^A, q^B]$. The low-cost firm, in contrast, has incentives to choose an output level in the interval $q^L \in [q^A, q^B]$, which is above its first-period output under complete information, $q^{L,Info} = \frac{1-c_1^L}{2}$. Thus, the low-cost incumbent increases its first-period output in order to communicate its efficient costs to the potential entrant, deterring entry as a result.

In particular, the lower-bound output q^A solves condition IC_H with equality, and the upper-bound output q^B solves condition IC_L with equality. Rearranging condition IC_H, and assuming that there is no discounting, $\delta = 1$, we obtain

$$\frac{1}{36} = (1 - q^L)q^L - \frac{1}{2}q^L$$

or

$$36\left(q^L\right)^2 - 18q^L + 1 = 0$$

and solving for output q^L, we obtain two roots for the lower bound q^A:

$$q^A = \frac{3 - \sqrt{5}}{12} = 0.06 \quad \text{and} \quad q^A = \frac{3 + \sqrt{5}}{12} = 0.44.$$

Similarly operating with condition IC_L in order to obtain the upper bound q^B, and since there is no discounting, $\delta = 1$, IC_L simplifies to

$$324\left(q^L\right)^2 - 216q^L + 25 = 0.$$

Solving for output q^L, we obtain two roots for the upper bound q^B,

$$q^B = \frac{6-\sqrt{11}}{18} = 0.15 \quad \text{and} \quad q^A = \frac{6+\sqrt{11}}{18} = 0.52.$$

Hence, the set of separating output levels for the low-cost firm must lie on the interval $q^L \in [0.44, 0.52]$.

(c) Which separating PBEs of those you found in part (b) survive Cho and Kreps' (1987) Intuitive Criterion?

- Starting from the separating PBE in which the low-cost incumbent chooses the highest output level $q^L = q^B$, a deviation toward any output level in $q^L \in [q^A, q^B)$ can only be profitable for the low-cost incumbent (but not for the high-cost firm). Formally, deviating toward $q^L \in \left[q^A, q^B\right)$ is "equilibrium dominated" for the high-cost incumbent alone. Hence, the potential entrant would update its beliefs accordingly, making such a deviation profitable for the low-cost firm. A similar argument applies to all other separating PBEs in the interval $q^L \in (q^A, q^B)$ but not for $q^L = q^A$, the least-costly separating PBE (also known as the "Riley outcome") that survives Cho and Kreps' (1987) Intuitive Criterion.
- Summarizing, the low-cost incumbent raises its first-period output from $q^{L,Info} = \frac{1-c_1^L}{2} = \frac{1-\frac{1}{3}}{2} = 0.33$, under complete information, to $q_1^A = 0.44$, under the separating equilibrium. Hence, the "separating output" that this firm must produce in order to reveal its type to the potential entrant (and thus deter entry) is measured by the distance $q^A - q^{L,Info} = 0.44 - 0.33 = 0.11$.
- *Remark:* Importantly, the low-cost incumbent chooses his output level q^A by considering the incentive compatibility condition of the high-cost incumbent (condition IC_H). This is not a mistake! The low-cost firm uses IC_H to identify its optimal output q^A because this is the lowest output level that would make the high-cost firm indifferent between mimicking its output decision q^A and produce $q^{H,Info}$. Output levels above q^A would be strictly unprofitable to imitate by the high-cost firm, while output levels strictly lower than q^A would be profitably mimicked by the high-cost incumbent. The low-cost firm then increases its output from $q^{L,Info}$ to q^A to successfully make mimicking unprofitable for the high-cost firm (often referred to as that the low-cost firm "separates" from the high-cost firm), thus conveying its type to the potential entrant, who is deterred from entering the market.

Exercise #10.2: Selling a Damaged Good at an Extra Cost, Based on Deneckere and McAfee (1996)^B

10.2 A firm sells a product in a market where there are two types of consumers, high- and low-valuation consumers. There are q high-value consumers and $1 - q$ low-value consumers, and the total number of consumers is normalized to 1, so q can be understood as the proportion of high-value consumers. The product has value v_H to the high-value consumers and value v_L to the low-value consumers. We assume that the proportion of high-value buyers satisfies $q > \frac{v_L}{v_H}$. All consumers have unit demand, i.e., they either buy one unit or do not participate. The product is produced at a zero marginal cost.

(a) *No damaged version.* Find the profit-maximizing price and calculate the firm's profit. Find welfare in this setting.
- The seller may sell either to a share of q consumers at a price v_H or to all consumers at v_L. Comparing the profits from each pricing strategy, we find that $qv_H > v_L$ because q satisfies $q > \frac{v_L}{v_H}$ by assumption, yielding a profit qv_H. Intuitively, high-value consumers are so likely that the firm prefers to sell to them only (at a relatively high price v_H) than selling to all consumers at a lower price v_L.
- Welfare in this setting is just

$$W = CS + \pi = 0 + qv_H = qv_H$$

since high-value consumers keep no surplus.

(b) *Damaged version.* The firm considers introducing a damaged version of the good. The damaged version is produced at the constant marginal cost equal to c. It results in a utility of w_L to the low-value consumers and of w_H to the high-value consumers where $1 > w_H > w_L > c$. Find the optimal price of the normal and damaged version of the good and total profits.
- *Low-value consumers.* The firm charges w_L to the low-value consumers for the damaged version of the good.
- *High-value consumers.* The firm makes the high-value consumers indifferent between buying the undamaged and damaged version of the good as follows:

$$v_H - p = w_H - w_L.$$

Solving for p in the above equation, we obtain the price for the undamaged good:

$$p^* = v_H - w_H + w_L.$$

Then, the profit is

$$\pi^D = \underbrace{qp^*}_{\text{High-value cons.}} + \underbrace{(1-q)(w_L - c)}_{\text{Low-value cons.}},$$

where the first term denotes the expected profit from selling the undamaged version to the high-value consumers at a price p^*, making a profit equal to qp^* since the firm incurs no cost to produce this version of the good. The second term represents the expected profit from selling the damaged version of the good to the low-value consumers at a price w_L, making a profit margin $w_L - c$.

Rearranging, the profit from offering the damaged good, π^D, simplifies to

$$\begin{aligned}\pi^D &= qp^* + (1-q)(w_L - c)\\ &= q(v_H - w_H + w_L) + (1-q)(w_L - c).\end{aligned}$$

(c) Does the firm increase its profits by introducing the damaged version?

- The firm has incentives to offer the damaged version of the good if its profits from doing so, π^D (from part b), exceed those from offering the undamaged version alone, qv_H (which we found in part a); that is,

$$\pi^D > qv_H,$$

which entails

$$q(v_H - w_H + w_L) + (1-q)(w_L - c) > qv_H.$$

Solving for q, we find that the firm offers the damaged version of the good if

$$\frac{w_L - c}{w_H - c} > q$$

which happens when high-value consumers are relatively infrequent. In other words, the firm offers the damaged version of the good (which is targeted to low-value consumers alone) if this type of consumers is sufficiently prevalent in the population.

(d) Does the introduction of damaged goods improve social welfare?

- The welfare when the seller offers the damaged good is

$$W^D = \pi^D + \underbrace{q(v_H - p^*)}_{CS_H} + \underbrace{(1-q)(w_L - w_L)}_{CS_L},$$

where the second and third terms indicate consumer surplus (zero for low-value consumers but positive for high-value consumers), which simplifies to

$$\begin{aligned} W^D &= \pi^D + q\left[v_H - \overbrace{(v_H - w_H + w_L)}^{p^*}\right] \\ &= \pi^D + q(w_H - w_L). \end{aligned}$$

Based on the setting in part (a), when the seller does not offer the damaged good, welfare increases from $W = qv_H$ to W^D since the difference $W^D - W$ is positive:

$$\begin{aligned} W^D - W &= \pi^D + q(w_H - w_L) - qv_H \\ &= \underbrace{[q(v_H - w_H + w_L) + (1-q)(w_L - c)]}_{\pi^D} + q(w_H - w_L) - qv_H \\ &= (1-q)(w_L - c) > 0 \quad \text{where} \quad w_L > c. \end{aligned}$$

However, welfare is not maximized since the low-value consumers purchase a damaged good that requires an additional cost for the firm. In other words, the firm offers the damaged good as a tool to attract the low-value consumers to buy some of its products, which the firm did not profitably achieve when the damaged good is absent. Intuitively, the damaged good operates as a discriminating tool that the firm uses to sort consumer types.

(e) *Numerical example.* Assume valuations $v_L = 1$ and $v_H = 3$, probability $q = 1/2$, and valuations $w_H = 1/2$ and $w_L = 1/4$ for the damaged version, and a cost $c = 1/10$ to develop this version of the product. Evaluate your equilibrium results in parts (a), (b), and (c).

- When there is no damaged version of the good, the firm changes a price $v_H = \$3$, making an expected profit $qv_H = \frac{1}{2}3 = \$1.5$, which exceeds the profit of selling to both customers at $v_L = \$1$, earning a profit of $q1 + (1-q)1 = \$1$. That is, the initial condition $q > \frac{v_L}{v_H}$ holds in this numerical example because $\frac{1}{2} > \frac{1}{3}$.
- When the firm introduced a damaged version of its product, it sets a price:

$$\begin{aligned} p^* &= v_H - w_H + w_L \\ &= 3 - \frac{1}{2} + \frac{1}{4} \\ &= \frac{11}{4} = 2.75. \end{aligned}$$

The firm does not have incentives to develop the damaged version because the condition that we found in part (c), $\frac{w_L - c}{w_H - c} > q$, does not hold in this setting, that is,

$$\frac{\frac{1}{4} - \frac{1}{10}}{\frac{1}{2} - \frac{1}{10}} = \frac{3}{8} \not> \frac{1}{2}.$$

Exercise #10.3: Investing in Product Quality, Based on Calveras and Ganuza (2018)^C

10.3 Consider a perfectly competitive industry with firms selling a homogeneous product. One of the firms, which we refer to as "the firm," faces the following sequential-move game:

(1) The firm privately observes its fixed cost $F \geq 0$ to produce a high-quality good, q_H. All other players do not observe this realization, but know that $F = +\infty$ with probability $1 - r$, and $0 \leq F < +\infty$ with probability r.

(2) Observing the realization of F, the firm chooses to produce a high-quality good, q_H, at a cost F, or keep producing its low-quality good, q_L.

(3) Consumers and the firm observe the realization of signal s, where $s \in \{s_L, s_H\}$, with the following conditional probabilities:

$$\Pr(s_H | q_H) = 1 \text{ and } \Pr(s_L | q_H) = 0$$

when the firm chooses q_H, and

$$\Pr(s_H | q_L) = 1 - \gamma \text{ and } \Pr(s_L | q_L) = \gamma$$

when the firm chooses q_L, where $\gamma \in [0, 1]$ represents the signal's information accuracy. Intuitively, when $\gamma = 1$, the signal is perfectly informative, whereas when $\gamma = 0$, the signal conveys no information about the firm's product quality.

(4) The firm sets its price p.

(5) Observing the signal, consumers update their beliefs about the firm's product quality and respond buying from the firm at price p or from any other firm at a normalized price of zero. Assume that consumer utility is $u = v - p$, where valuation $v = v_H$ for the high-quality good but $v = v_L = 0$ for the low-quality good.

We consider a separating perfect Bayesian equilibrium (PBE) where the firm with infinitely high fixed cost $F = +\infty$ produces a low-quality good but a high-quality good when its cost is finite and does not exceed the consumers' valuation for this type of good, that is, $0 \leq F \leq v_H$.

(a) Find consumers' updated belief in this equilibrium.

- Invoking the Bayes' rule, consumers, having received a high-quality signal, infer the probability that the product is of high quality with a posterior probability of

$$\Pr(q_H|s_H) = \frac{\Pr(s_H|q_H)\Pr(q_H)}{\Pr(s_H|q_L)\Pr(q_L) + \Pr(s_H|q_H)\Pr(q_H)}$$
$$= \frac{r}{(1-\gamma)(1-r)+r}$$
$$= \frac{r}{1-\gamma(1-r)}.$$

(b) Under which condition on the accuracy parameter γ can the above PBE be sustained?

- Upon receiving the high-quality signal, consumers expect with probability $\frac{r}{1-\gamma(1-r)}$ that the product is of high quality, so their willingness to pay becomes

$$p^*(s_H) = \frac{r}{1-\gamma(1-r)} v_H.$$

If the firm invests in quality, its profits become

$$\pi_H = p^*(s_H) - F$$

which is the price it receives from the consumers less the investment cost F.

Otherwise, if the firm does not invest in quality, its profits become

$$\pi_L = \Pr(s_H|q_L)\, p^*(s_H)$$

which is the probability that consumers pay high-quality price for low-quality good. Therefore, the firm has incentives to invest in quality if and only if $\pi_H \geq \pi_L$, that is,

$$p^*(s_H) - F \geq (1-\gamma)\, p^*(s_H).$$

Substituting $p^*(s_H) = \frac{r}{1-\gamma(1-r)} v_H$ into the above inequality, we find

$$\gamma \frac{r}{1-\gamma(1-r)} v_H \geq F.$$

Solving for γ, we obtain that the PBE can be sustained if

$$\gamma \geq \overline{\gamma} \equiv \frac{F}{r v_H + (1-r)F}.$$

(c) How are equilibrium conditions affected by an increase in F, r, and v_H? Interpret your results.

- It is straightforward to verify that cutoff $\overline{\gamma}$ decreases in v_H, meaning that the higher is the valuation that the consumers assign to the high-quality good, the less restrictive is the range of accuracy parameter γ that supports the above separating PBE.
- Next, we take the first-order condition of cutoff $\overline{\gamma}$ with respect to F

$$\frac{\partial \overline{\gamma}}{\partial F} = \frac{r v_H}{\left(r v_H + F\,(1-r)\right)^2} > 0$$

which indicates that cutoff $\overline{\gamma}$ increases with F. Intuitively, the more costly is the quality investment F, the more restrictive are the parameter conditions on accuracy parameter γ to sustain the above separating PBE.

Similarly, we take the first-order condition of cutoff $\overline{\gamma}$ with respect to r:

$$\frac{\partial \overline{\gamma}}{\partial r} = -\frac{v_H - F}{\left(r v_H + F\,(1-r)\right)^2} \leq 0$$

which holds because $v_H \geq F$ by assumption, suggesting that cutoff $\overline{\gamma}$ decreases with r. Intuitively, the more likely is the probability that the firm realizes a finite fixed cost, the larger is the set of the above separating PBE.

(d) *Introducing external CSR.* Assume now that the firm, besides producing a high- or low-quality product, can invest in corporate social responsibility (CSR) by, for instance, donating to a local hospital or to a charity, which is often referred to as external CSR since this investment does not directly affect firm inputs or productivity. External CSR may affect sales, however, as we study in this part of the exercise. Assume that if the firm practices CSR, it incurs a cost c, while if it does not, its cost of CSR is normalized to zero. Consumers observe both the firm's CSR decision (denoted as C or NC for investments and no investment in CSR, respectively) and signal s and update its valuation of the product, v.

Intuitively, these valuations are increasing in product quality (i.e., for a given CSR decision, valuations are higher for the high- than the low-quality product), and in CSR (i.e., for a given product quality, valuations are higher when the firm invests in CSR than when it does not). In addition, these valuations indicate that CSR is more valuable in the high-quality product since $v_{H,C} - v_{H,NC} > v_{L,C} - v_{L,NC}$. For simplicity, we assume that, upon observing NC, the consumer believes that the firm must be $F = +\infty$ so that it does not invest in high quality, that is, $v_{L,NC} = 0$.

Find the consumers' updated beliefs in this setting, and identify under which conditions on the accuracy parameter γ and on the CSR cost c can the above PBE be sustained.

- First, for the low-cost firm to invest in quality (conditional on this firm investing in CSR), we need the following incentive compatibility condition to hold:

$$v_{H,C} - F - c \geq (1-\gamma)\, v_{H,C} + \gamma v_{L,C} - c.$$

The left side of the above inequality says that if the firm invests in quality at a cost F, consumers receive the signal with probability one that its product is of high quality. In contrast, the right side suggests that if the firm does not invest in quality, consumers infer that the product is of low (high) quality with probability γ $(1-\gamma)$.

Rearranging, we obtain the threshold on accuracy parameter γ as follows:

$$\gamma \geq \overline{\gamma}^{CSR} \equiv \frac{F}{v_{H,C} - v_{L,C}}.$$

- Then, for the low-cost firm to invest in CSR (conditional on this firm investing in quality), we also need the following condition:

$$v_{H,C} - F - c \geq v_{L,NC} - F,$$

where consumers assume that the firm is of low-quality absent CSR, yielding

$$v_{H,C} - c \geq 0$$

which gives the upper bound on the cost of CSR as follows:

$$c \leq \overline{c} \equiv v_{H,C}.$$

- Third, the high-cost firm will never find it optimal to invest in quality, because $F = +\infty$, and for this firm not to invest in CSR, we need

$$v_{L,NC} \geq (1 - \gamma)\, v_{H,C} + \gamma v_{L,C} - c.$$

The left side of the above inequality says that consumers assign a reservation utility of zero to the firm not investing in CSR, while the right side means that consumers assign a probability of γ $(1 - \gamma)$ to the firm being low (high) quality observing CSR.

Rearranging, we identify the lower bound on the cost of CSR as follows:

$$c \geq \underline{c} \equiv (1 - \gamma)\, v_{H,C} + \gamma v_{L,C}.$$

- Combining the above conditions, a separating PBE can be sustained when

$$\underline{c} \leq c \leq \overline{c} \quad \text{and} \quad \gamma \geq \overline{\gamma}^{CSR}.$$

(e) Compare your results in part (b) and (d). Can the separating equilibrium be sustained under more general parameter conditions when the firm has access to external CSR?

- We now show that the separating PBE be sustained under more general parameter conditions if $\overline{\gamma} > \overline{\gamma}^{CSR}$ holds, which entails

$$\overline{\gamma} \equiv \frac{F}{rv_H + F\,(1 - r)} > \frac{F}{v_{H,C} - v_{L,C}} \equiv \overline{\gamma}^{CSR}.$$

To begin with, let us recall the single-crossing condition

$$v_{H,C} - v_{H,NC} > v_{L,C} - v_{L,NC}$$

which we can rearrange in the following way:

$$v_{H,C} - v_{L,C} > \underbrace{v_{H,NC} - v_{L,NC}}_{v_H}$$

and the right side is the valuation for high-quality good without CSR. Furthermore,

$$v_H > rv_H + F(1-r)$$

since $v_H > F$ by assumption. Combining the above inequalities, we find

$$v_{H,C} - v_{L,C} > rv_H + F(1-r)$$

yielding

$$\frac{F}{rv_H + F(1-r)} > \frac{F}{v_{H,C} - v_{L,C}}$$

so that the inequality $\overline{\gamma} > \overline{\gamma}^{CSR}$ holds.

Intuitively, CSR enables the firm to signal its high quality even if the market is less informative (i.e., under less restrictive parameter conditions on γ). In particular, we can identify three regions according to where γ lies relative to cutoffs $\overline{\gamma}$ and $\overline{\gamma}^{CSR}$:

- When $\overline{\gamma}^{CSR} \leq \gamma < \overline{\gamma}$, the firm that (does not) invests in CSR (fails to) can signal its high quality.
- If $\gamma \geq \overline{\gamma}$, the firm can signal its high quality even without CSR.
- Lastly, if $\gamma < \overline{\gamma}$, the market is so noisy that whether the firm invests in CSR or not, consumers do not believe that the firm's product is of high quality.

(f) *Numerical example.* Evaluate your results at $F = 1/3$, $v_H = 2$, $v_{H,C} = 2$, $v_{H,NC} = 1$, $v_{L,C} = 1/2$, and $v_{L,NC} = 1/4$, and probability $r = 2/3$.

- Evaluating cutoff $\overline{\gamma}$ at these parameter values yields

$$\overline{\gamma} = \frac{\frac{1}{3}}{\frac{2}{3} \times 2 + \left(1 - \frac{2}{3}\right) \times \frac{1}{3}} = \frac{3}{13}.$$

And evaluating cutoff $\overline{\gamma}^{CSR}$, we obtain

$$\overline{\gamma}^{CSR} = \frac{\frac{1}{3}}{2 - \frac{1}{2}} = \frac{2}{9},$$

where, as found in part (e), these cutoffs rank as follows: $\overline{\gamma} = \frac{3}{13} > \frac{2}{9} = \overline{\gamma}^{CSR}$. Furthermore, condition $v_{H,C} - v_{H,NC} > v_{L,C} - v_{L,NC}$ holds in this case since

$$2 - 1 = 1 > \frac{1}{4} = \frac{1}{2} - \frac{1}{4}.$$

- In addition, cutoffs $\overline{c}$ and $\underline{c}$ become $\overline{c} = 2$ and

$$\underline{c} = 2(1-\gamma) + \frac{\gamma}{2} = \frac{4 - 3\gamma}{2},$$

so that condition $\underline{c} \leq c \leq \overline{c}$ becomes $2 - \frac{3\gamma}{2} \leq c \leq 2$. Therefore, the range of c where the separating PBE can be sustained shrinks in γ.

Exercise #10.4: Horizontal Differentiation with Imperfectly Informed Consumers, Based on Esteves and Cerqueira (2017)^C

10.4 Consider an industry with two firms, A and B, each located at the endpoints of a unit line (firm A is located at 0 and firm B is at 1). Consumers are uniformly distributed over the line, $x \sim U[0, 1]$, and enjoy utility $v - p_A - tx$ when purchasing from firm A and $v - p_B - t(1-x)$ when purchasing from firm B, where $v > 0$ represents the reservation price that the consumer obtains and $t > 0$ denotes the unit transportation cost. Firms interact in the following sequential-move game.

(i) At period 1, every firm i simultaneously and independently sets price p_i, and its advertising intensity, $\phi_i \in [0, 1]$. At the end of period 1, a proportion ϕ_i receive the ad from firm i. That is, a proportion $\phi_i(1 - \phi_j)$ of consumers becomes "captive" of firm i, meaning that they purchase firm i's product as long as their utility is positive. Similarly, a proportion $\phi_i \phi_j$ of consumers receive the ad from both firms i and j and purchase from the firm that is more convenient, which we call "selective" consumers. Finally, a proportion $(1 - \phi_i)(1 - \phi_j)$ receive no ad from either firm and, for simplicity, do not purchase the good.

(ii) At period 2, every firm i simultaneously and independently sets price pair $\{p_i^O, p_i^R\}$, where superscript O represents "own customers" and R denotes "rival customers." Intuitively, after observing whether a customer purchased from firm i or j, firm i can price-discriminate, charging a price p_i^O to customers who bought from firm i in the previous periods but a price p_i^R to those customers who did not (stealing them from its rival).

The advertising cost for every firm i is $A_i(\phi_i) = \frac{a}{2}\phi_i^2$, where $a > 0$ denotes the effectiveness of advertising effort ϕ_i in reaching more customers. In the following parts of the exercise, we solve the game by backward induction.

(a) *Second period.* For a given pair of first-period prices p_A and p_B: (i) for firm i's captive consumers, find under which condition they buy from firm i again; and (ii) for selective consumers buying from firm i in the first period, find the indifferent consumer.

- If the consumer is captive to firm A, he keeps buying from this firm in the second period rather than not buying at all if $v - p_A - tx \geq 0$, which simplifies to

$$p_A + tx \leq v.$$

Similarly, if the consumer is captive to firm B, he keeps buying from firm B as long as $v - p_B - t(1 - x) \geq 0$ or, after rearranging, if

$$p_B + t(1 - x) \leq v.$$

- Among all selective consumers who bought from firm A in the first period, the indifferent consumer $\hat{x}_A$ is different between buying again from firm A in the second period at a price p_A^O and switching to firm B in the second period at a price p_B^R. Therefore, $\hat{x}_A$ solves

$$v - p_A^O - t\hat{x}_A = v - p_B^R - t(1 - \hat{x}_A)$$

which solving for $\hat{x}_A$ yields

$$\hat{x}_A = \frac{p_B^R - p_A^O + t}{2t}.$$

Similarly, among all selective consumers who bought from firm B in the first period, the indifferent consumer $\hat{x}_B$ is indifferent between buying again from firm B in the second period at a price p_B^O and switching to firm A at a price p_A^R, where $\hat{x}_B$ solves

$$v - p_A^R - t\hat{x}_B = v - p_B^O - t(1 - \hat{x}_B)$$

which solving for $\hat{x}_B$ yields

$$\hat{x}_B = \frac{p_B^O - p_A^R + t}{2t}.$$

(b) Evaluate the profits that every firm i earns from its own customers in the second period, π_i^O, and from stealing its rival's customers in that period, π_i^R.

- Firm A's second-period profits from selling to its own customers are

$$\begin{aligned}\pi_A^O &= \overbrace{\phi_A\phi_B p_A^O \hat{x}_A}^{\text{Selective cons.}} + \overbrace{\phi_A(1-\phi_B)p_A^O}^{\text{Captive cons.}} \\ &= \phi_A p_A^O\left(\phi_B\hat{x}_A + 1 - \phi_B\right) \\ &= \phi_A p_A^O\left(1 - \phi_B\frac{p_A^O - p_B^R + t}{2t}\right).\end{aligned}$$

The first part represents the profits from selective consumers, which occur with probability $\phi_A\phi_B$ because consumers receive ads from both firms. The second part denotes the profits from firm A's captive consumers (i.e., those who purchased from this firm in the first period and buy again from the same firm in the second period), which occur with probability $\phi_A\left(1 - \phi_B\right)$ as consumers only receive ads from this firm.

- Firm A's second-period profits from stealing its rival's customers are

$$\begin{aligned}\pi_A^R &= \phi_A\phi_B p_A^R(\hat{x}_B - \hat{x}) \\ &= \phi_A\phi_B p_A^R\left(\frac{p_B^O - p_A^R + t}{2t} - \hat{x}\right)\end{aligned}$$

which are a fraction of selective consumers (those receiving ads from both firms). In particular, those in the interval $\hat{x}_B - \hat{x}$ imply that they bought from firm B in the first period (they are to the right-hand side of the first-period indifferent consumer $\hat{x}$) but switch to firm A in the second period (since they are to the left-hand side of the second-period indifferent consumer $\hat{x}_B$).

- Similarly, firm B's second-period profits from selling to its own customers are

$$\begin{aligned}\pi_B^O &= \overbrace{\phi_B(1-\phi_A)p_B^O}^{\text{Captive cons.}} + \overbrace{\phi_A\phi_B p_B^O(1-\hat{x}_B)}^{\text{Selective cons.}} \\ &= \phi_B p_B^O\left(1 - \phi_A + \phi_A - \phi_A\hat{x}_B\right) \\ &= \phi_B p_B^O\left(1 - \phi_A\frac{p_B^O - p_A^R + t}{2t}\right).\end{aligned}$$

The first part denotes the profits from firm B's captive consumers, which occur with probability $\phi_B\left(1 - \phi_A\right)$ because consumers only receive ads from this firm. The second part are the profits from selective consumers who bought from firm B in the first period and keep buying from the same firm in the second period.

- Firm B's second-period profits from stealing its rival's customers are

$$\pi_B^R = \phi_A \phi_B p_B^R (\hat{x} - \hat{x}_A)$$
$$= \phi_A \phi_B p_B^R \left(\hat{x} - \frac{p_B^R - p_A^O + t}{2t} \right)$$

which are a fraction of selective customers in interval $\hat{x} - \hat{x}_A$. Recall that, in this interval, consumers bought from firm A in the first period (since they are to the left-hand side of the first-period indifferent consumer $\hat{x}$) but switch to firm B in the second period (since they are to the right-hand side of the second-period indifferent consumer $\hat{x}_B$).

(c) Find equilibrium prices in the second stage of the game for every firm i, $\{p_i^O, p_i^R\}$.

- Every firm i chooses a price pair $\{p_i^O, p_i^R\}$ to maximize its profits. Specifically, profit $\pi_i^k(p_i^k)$, where $k = \{O, R\}$, is only a function of the price that firm i charges to customers in segment k, but not those in the other segment $r \neq k$, where $r = \{O, R\}$. Therefore, we can simplify firm i's problem into choosing the price p_i^O that maximizes profit π_i^O and choosing the price p_i^R that maximizes profit π_i^R.
- Differentiating, we obtain the following first-order conditions for firm A:

$$\frac{\partial \pi_A^O}{\partial p_A^O} = \frac{\phi_A}{2t} \left[2t - \phi_B \left(2p_A^O - p_B^R + t \right) \right] = 0$$
$$\frac{\partial \pi_A^R}{\partial p_A^R} = \frac{\phi_A \phi_B}{2t} \left(p_B^O - 2p_A^R + t - 2t\hat{x} \right) = 0.$$

Operating similarly for firm B yields first-order conditions:

$$\frac{\partial \pi_B^O}{\partial p_B^O} = \frac{\phi_B}{2t} \left[2t - \phi_A \left(2p_B^O - p_A^R + t \right) \right] = 0$$
$$\frac{\partial \pi_B^R}{\partial p_B^R} = \frac{\phi_A \phi_B}{2t} \left(2t\hat{x} - 2p_B^R + p_A^O - t \right) = 0.$$

Hence, we obtain

$$p_A^O(p_B^R) = \frac{p_B^R - t}{2} + \frac{t}{\phi_B} \quad \text{and}$$
$$p_A^R(p_B^O) = \frac{p_B^O + t - 2t\hat{x}}{2}$$

and

$$p_B^O(p_A^R) = \frac{p_A^R - t}{2} + \frac{t}{\phi_A} \quad \text{and}$$
$$p_B^R(p_A^O) = \frac{p_A^O - t + 2t\hat{x}}{2}.$$

After solving for the above four prices (two from each firm) in the first-order conditions, we obtain the following second-period prices for firm A:

$$p_A^O = \frac{t\left[4+\phi_B\left(2\hat{x}-3\right)\right]}{3\phi_B} \text{ and } p_A^R = \frac{t\left[2+\phi_A\left(1-4\hat{x}\right)\right]}{3\phi_A}$$

and the following second-period prices for firm B:

$$p_B^O = \frac{t\left[4-\phi_A\left(1+2\hat{x}\right)\right]}{3\phi_A} \text{ and } p_B^R = \frac{t\left[2+\phi_B\left(4\hat{x}-3\right)\right]}{3\phi_B}.$$

(d) Write down the second-period equilibrium profits, evaluated at the equilibrium prices you found in part (c).

- Inserting the above second-period prices into the profit function of firm A, we find that its second-period profits are

$$\begin{aligned}\pi_A^2 &= \pi_A^O + \pi_A^R \\ &= \frac{t\phi_A}{18\phi_B}\left[4+\phi_B\left(2\hat{x}-3\right)\right]^2 + \frac{t\phi_B}{18\phi_A}\left[2+\phi_A\left(1-4\hat{x}\right)\right]^2 \\ &= t\frac{\phi_A^2\left[4+\phi_B\left(2\hat{x}-3\right)\right]^2 + \phi_B^2\left[2+\phi_A\left(1-4\hat{x}\right)\right]^2}{18\phi_A\phi_B}\end{aligned}$$

and those of firm B are

$$\begin{aligned}\pi_B^2 &= \pi_B^O + \pi_B^R \\ &= \frac{t\phi_B}{18\phi_A}[4-\phi_A(1+2\hat{x})]^2 + \frac{t\phi_A}{18\phi_B}[2+\phi_B(4\hat{x}-3)]^2 \\ &= t\frac{\phi_B^2[4-\phi_A(1+2\hat{x})]^2 + \phi_A^2[2+\phi_B(4\hat{x}-3)]^2}{18\phi_A\phi_B}.\end{aligned}$$

(e) *First period.* Find the indifferent consumer in the first period. [*Hint*: In this two-period setting, he must be indifferent between (i) buying from firm A in period 1 at price p_A and then buying from firm B in period 2 at the poaching price p_B^R, or (ii) buying from firm B in period 1 at price p_B and then buying from firm A at the poaching price p_A^R in the second period.]

- In the first period, the indifferent consumer $\hat{x}$ solves

$$\underbrace{v - p_A - t\hat{x}}_{\text{Buy from } A \text{ today}} + \underbrace{\delta[v - p_B^R - t(1-\hat{x})]}_{\text{Buy from } B \text{ tomorrow}} = \underbrace{v - p_B - t(1-\hat{x})}_{\text{Buy from } B \text{ today}} + \underbrace{\delta(v - p_A^R - t\hat{x})}_{\text{Buy from } A \text{ tomorrow}}.$$

Rearranging the above expression, we obtain

$$2t\left(1-\delta\right)\hat{x} = \left(t - p_A + p_B\right) - \delta\left(t - p_A^R + p_B^R\right).$$

Simplifying and solving for $\hat{x}$ yield

$$\hat{x} = \frac{t\delta(2\phi_B - 2\phi_A + \phi_A\phi_B) + 3\phi_A\phi_B(t - p_A + p_B)}{2t\phi_A\phi_B(\delta + 3)}.$$

(f) Write the sum of first- and second-period profits. For simplicity, you can assume no discounting of future payoffs.

- Since we assume no discounting, we need to evaluate the indifferent consumer $\hat{x}$ we found in part (e) at $\delta = 1$, to obtain

$$\hat{x} = \frac{t(2\phi_B - 2\phi_A + \phi_A\phi_B) + 3\phi_A\phi_B(t - p_A + p_B)}{8t\phi_A\phi_B}.$$

Hence, firm A's first-period profits, evaluated at $\hat{x}$, are

$$\begin{aligned}\pi_A^1 &= p_A\phi_A(1 - \phi_B) + p_A\phi_A\phi_B\hat{x} - \frac{a\phi_A^2}{2} \\ &= \frac{2p_At\left(3\phi_A + \phi_B - 2\phi_A\phi_B\right) - 3p_A\phi_A\phi_B\left(p_A - p_B\right) - 4at\phi_A^2}{8t}\end{aligned}$$

and firm B's first-period profits, evaluated at $\hat{x}$ too, are

$$\begin{aligned}\pi_B^1 &= p_B\phi_B(1 - \phi_A) + p_B\phi_A\phi_B(1 - \hat{x}) - \frac{a\phi_B^2}{2} \\ &= \frac{2p_Bt\left(\phi_A + 3\phi_B - 2\phi_A\phi_B\right) - 3p_B\phi_A\phi_B\left(p_B - p_A\right) - 4at\phi_B^2}{8t}.\end{aligned}$$

- The total profit of firm i across both periods is $\pi_i = \pi_i^1 + \pi_i^2$. Since we assume no discounting, second-period profits of each firm, π_A^2 and π_B^2, are defined in part (d).

(g) Find the equilibrium first-period prices, p_i^*, advertising intensities, ϕ_i^*, and second-period prices, p_i^{O*} and p_i^{R*}.

- Let us first consider firm A's profit maximization problem. To maximize total profits $\pi_A = \pi_A^1 + \pi_A^2$, we differentiate with respect to first-period price p_A and advertising intensity ϕ_A, finding the first-order conditions:

$$\frac{\partial \pi_A^1}{\partial p_A} + \frac{\partial \pi_A^2}{\partial p_A} = 0 \text{ and } \frac{\partial \pi_A^1}{\partial \phi_A} + \frac{\partial \pi_A^2}{\partial \phi_A} = 0$$

or, using the Chain rule:

$$\frac{\partial \pi_A^1}{\partial p_A} + \frac{\partial \pi_A^2}{\partial \hat{x}}\frac{\partial \hat{x}}{\partial p_A} = 0 \text{ and } \frac{\partial \pi_A^1}{\partial \phi_A} + \frac{\partial \pi_A^2}{\partial \hat{x}}\frac{\partial \hat{x}}{\partial \phi_A} = 0.$$

We now separately find each term:

$$\begin{aligned}\frac{\partial \pi_A^1}{\partial p_A} &= \frac{2t\left(3\phi_A + \phi_B - 2\phi_A\phi_B\right) - 3\phi_A\phi_B\left(2p_A - p_B\right)}{8t}, \\ \frac{\partial \pi_A^2}{\partial \hat{x}} &= \frac{2t\left[4\left(\phi_A - \phi_B\right) + 5\phi_A\phi_B\left(2\hat{x} - 1\right)\right]}{9},\end{aligned}$$

$$\frac{\partial \hat{x}}{\partial p_A} = -\frac{3}{8t},$$

$$\frac{\partial \pi_A^1}{\partial \phi_A} = \frac{2p_A t\,(3-2\phi_B) - 3p_A\phi_B\,(p_A - p_B) - 8at\phi_A}{8t}, \text{ and}$$

$$\frac{\partial \hat{x}}{\partial \phi_A} = -\frac{1}{4\phi_A^2}.$$

Since the game is symmetric, in a subgame perfect Nash equilibrium, we must have that $p_A = p_B = p^*$ and $\phi_A = \phi_B = \phi^*$, yielding

$$\hat{x} = \frac{t\,(2\phi^* - 2\phi^* + (\phi^*)^2) + 3\phi^2(t - p^* + p^*)}{8t\,(\phi^*)^2} = \frac{1}{2}$$

so that the indifferent consumer is located exactly in the middle of the unit line.

Evaluating the partial derivatives in the symmetric equilibrium, we obtain

$$\frac{\partial \pi_A^1}{\partial p_A} = \frac{2t\left(3\phi^* + \phi^* - 2\left(\phi^2\right)^*\right) - 3\left(\phi^2\right)^*(2p^* - p^*)}{8t} = \frac{4t\phi^*(2-\phi^*) - 3p^*\left(\phi^2\right)^*}{8t},$$

$$\frac{\partial \pi_A^2}{\partial \hat{x}} = \frac{2t\left[4\,(\phi^* - \phi^*) + 5\left(\phi^2\right)^*\left(2 \times \frac{1}{2} - 1\right)\right]}{9} = 0,$$

$$\frac{\partial \hat{x}}{\partial p_A} = -\frac{3}{8t},$$

$$\frac{\partial \pi_A^1}{\partial \phi_A} = \frac{2p^*t\,(3-2\phi^*) - 3p^*\phi^*\,(p^* - p^*) - 8at\phi^*}{8t} = \frac{p^*\,(3-2\phi^*) - 4a\phi^*}{4}, \text{ and}$$

$$\frac{\partial \hat{x}}{\partial \phi_A} = -\frac{1}{4\left(\phi^2\right)^*}.$$

Substituting the above expressions into the first-order conditions, we find

$$\frac{4t\phi^*\,(2-\phi^*) - 3p^*\left(\phi^2\right)^*}{8t} + 0\left(-\frac{3}{8t}\right) = 0$$

$$\Longrightarrow p^*\left(\phi^*\right) = \frac{4t\,(2-\phi^*)}{3\phi^*}$$

and in addition,

$$\frac{p^*\,(3-2\phi^*) - 4a\phi^*}{4} + 0\left(-\frac{1}{4\left(\phi^2\right)^*}\right) = 0$$

$$\Longrightarrow \phi^*\left(p^*\right) = \frac{3p^*}{2\,(2a+p^*)}.$$

Solving the simultaneous equations above, we have

$$(3a-2t)\left(\phi^*\right)^2+7t\phi^*-6t=0.$$

Since ϕ^* must be positive, equilibrium advertising intensity ϕ^* becomes

$$\phi^*=\frac{\sqrt{t^2+72at}-7t}{2\left(3a-2t\right)}.$$

Plugging ϕ^* into $p^*\left(\phi^*\right)$, the first-period equilibrium price p^* becomes

$$p^*=\frac{4t\left(12a-t-\sqrt{t^2+72at}\right)}{3\left(\sqrt{t^2+72at}-7t\right)}.$$

In this context, we can find equilibrium second-period prices, which are symmetric across firms, that is, $p_A^{O*}=p_B^{O*}=p^{O*}$ and $p_A^{R*}=p_B^{R*}=p^{R*}$, where

$$p^{O*}=\frac{2t\left(2-\phi^*\right)}{3\phi^*}=\frac{2t\left(12a-t-\sqrt{t^2+72at}\right)}{3\left(\sqrt{t^2+72at}-7t\right)}$$

$$p^{R*}=\frac{t\left(2-\phi^*\right)}{3\phi^*}=\frac{t\left(12a-t-\sqrt{t^2+72at}\right)}{3\left(\sqrt{t^2+72at}-7t\right)}.$$

It is straightforward to show that $p^*=2p^{O*}=4p^{R*}$, implying that firms set the highest price in the first period. In the second period, however, every firm has incentives to set a lower price to retain its consumers and to lower the price further to poach consumers from the other firm.

(h) *Numerical example.* Evaluate equilibrium prices in both periods, and the advertising intensity, at parameter values $a=1$ and $t=\frac{1}{4}$.

- When $a=1$ and $t=\frac{1}{4}$, we find an advertising intensity of

$$\phi^*=\frac{\sqrt{\left(\frac{1}{4}\right)^2+72\times1\times\frac{1}{4}}-7\times\frac{1}{4}}{2\left(3\times1-2\times\frac{1}{4}\right)}=\frac{1}{2}$$

with equilibrium price in the first period of

$$p^*=\frac{4\times\frac{1}{4}\left(12\times1-\frac{1}{4}-\sqrt{\left(\frac{1}{4}\right)^2+72\times1\times\frac{1}{4}}\right)}{3\left(\sqrt{\left(\frac{1}{4}\right)^2+72\times1\times\frac{1}{4}}-7\times\frac{1}{4}\right)}=\$1$$

and equilibrium prices in the second period of

$$p^{O*} = \frac{2 \times \frac{1}{4}\left(12 \times 1 - \frac{1}{4} - \sqrt{\left(\frac{1}{4}\right)^2 + 72 \times 1 \times \frac{1}{4}}\right)}{3\left(\sqrt{\left(\frac{1}{4}\right)^2 + 72 \times 1 \times \frac{1}{4}} - 7 \times \frac{1}{4}\right)} = \$\frac{1}{2}\text{, and}$$

$$p^{R*} = \frac{\frac{1}{4}\left(12 \times 1 - \frac{1}{4} - \sqrt{\left(\frac{1}{4}\right)^2 + 72 \times 1 \times \frac{1}{4}}\right)}{3\left(\sqrt{\left(\frac{1}{4}\right)^2 + 72 \times 1 \times \frac{1}{4}} - 7 \times \frac{1}{4}\right)} = \$\frac{1}{4}.$$

Exercise #10.5: Pay to Switch or Pay to Stay? Based on Shaffer and Zhang (2000)^C

10.5 Consider an industry with two firms, A and B, selling a homogeneous good, and both facing the same marginal cost of production, $c > 0$. Consumers buy at most one unit of either firm. Consumers are partitioned in the following way: Group a of consumers represents a fraction $\theta \in \left[\frac{1}{2}, 1\right]$ of all consumers, while group b represents the remaining fraction $1 - \theta$ of all consumers. If both firms charge the same price, all consumers in group a (group b) purchase from firm A (firm B, respectively). This entails that, if both firms charge the same price to both groups, firm A captures a larger market share.

In addition, let $F_k(x)$ denote the fraction of consumers in group $k = \{a, b\}$, with loyalty l_k less or equal to x, and

$$F_k(x) = \begin{cases} 0 \text{ if } x < 0 \\ \frac{x}{l_k} \text{ if } 0 \leq x \leq l_k \\ 1 \text{ if } x > l_k. \end{cases}$$

To understand the intuition behind this probability, consider, as an illustration, consumers in group a. If firm A charges a price premium $x = p_A - p_B$, then consumers in group a purchase from firm A if and only if $x < l_a$ (i.e., if $p_A - p_B < l_a$ or $p_A < l_a + p_B$), which occurs with probability $1 - F_a(x)$. In other words, the fraction of consumers in group a who purchase from firm A is $1 - F_a(x)$, while the remaining fraction of consumers purchase from firm B, that is, $F_a(x)$. In contrast, the fraction of consumers in group b who purchase from firm A is $F_b(x)$, while those buying from firm B is $1 - F_b(x)$.

Graphically, $F_k(x)$ originates at $x = 0$, and increases in price premium x with a constant slope of $\frac{1}{l_k}$, reaching a height of 1 at $x = l_k$. Therefore, an increase in loyalty parameter l_k makes function $F_k(x)$ flatter, expanding the range of price premia for which group k purchases units. Let p_i denote the price that firm $i = \{A, B\}$ sets on group a, and let $\widetilde{p}_i$ represent the price that firm i sets on group b. Firm A's demand is then

$$D_A = \theta\left[1 - F_a(p_A - p_B)\right] + (1 - \theta) F_b(\widetilde{p}_B - \widetilde{p}_A),$$

where the first (second) term indicates the sales from group a (group b). Similarly, firm B's demand is

$$D_B = \theta F_a(p_A - p_B) + (1-\theta)\left[1 - F_b(\widetilde{p}_B - \widetilde{p}_A)\right],$$

where, similarly, the first (second) term indicates the sales from group a (group b).

(a) *No price discrimination.* If firms cannot price-discriminate, $p_i = \widetilde{p}_i$ for every firm i show that firm A sets a strictly higher price than firm B if and only if $\theta > \frac{1}{2}$.

- Let us prove by contradiction. Suppose a pure strategy Nash equilibrium exists in which $(\overline{p}_A, \overline{p}_B)$ are the equilibrium prices and $\overline{p}_B > \overline{p}_A$. It follows that

$$\overline{p}_A = \underset{p_A \geq 0}{\arg\max}\left(\theta(p_A - c) + (1-\theta)(p_A - c)\frac{\overline{p}_B - p_A}{l_b}\right)$$

$$\overline{p}_B = \underset{p_B \geq 0}{\arg\max}\left((1-\theta)(p_B - c)\frac{l_b + \overline{p}_A - p_B}{l_b}\right)$$

because firm A (B) cannot lose (poach) consumers to (from) firm B (A).

The associated first-order conditions are

$$\frac{\partial \pi_A}{\partial p_A} = \frac{l_b\theta + (1-\theta)(\overline{p}_B - 2p_A + c)}{l_b} = 0$$

$$\frac{\partial \pi_B}{\partial p_B} = \frac{1-\theta}{l_b}(l_b + \overline{p}_A - 2p_B + c) = 0.$$

Solving for prices $\overline{p}_A$ and $\overline{p}_B$, we obtain

$$\overline{p}_A = \frac{1+\theta}{3(1-\theta)}l_b + c$$

$$\overline{p}_B = \frac{2-\theta}{3(1-\theta)}l_b + c.$$

Comparing these prices, it is straightforward to verify that $\overline{p}_B > \overline{p}_A$ if and only if $\theta < \frac{1}{2}$, which contradicts our assumption $\theta \in \left[\frac{1}{2}, 1\right]$. Therefore, we must have that $\overline{p}_A \geq \overline{p}_B$ and firm A cannot set a price lower than that of firm B, given that we assume group a to be weakly larger than group b.

- Next, we show that the above prices coincide, $\overline{p}_B = \overline{p}_A$, if and only if $\theta = \frac{1}{2}$.
  - *"If" part.* Substituting $\theta = \frac{1}{2}$ into equilibrium prices above, $\overline{p}_A = \overline{p}_B = l_b + c$.
  - *"Only if" part.* From $p_B \geq \overline{p}_B = \overline{p}_A \geq p_A$, we have $\frac{\partial \pi_A}{\partial p_A} \geq 0$ and $\frac{\partial \pi_B}{\partial p_B} \leq 0$ to the left and right of the firms' marginal profit functions, respectively, yielding

$$2p_A - \overline{p}_B \leq \frac{\theta}{1-\theta}l_b + c$$

$$2p_B - \overline{p}_A \geq l_b + c.$$

By the continuous mapping theorem and twice differentiability of the firms' profit functions in prices, partial derivatives of the profit function in the limit exist. Taking

$p_A \to \overline{p}_A$ and $p_B \to \overline{p}_B$ in the first and second expressions above, respectively, and using the condition $\overline{p}_B = \overline{p}_A$, we obtain

$$\overline{p}_A \leq \frac{\theta}{1-\theta} l_b + c$$
$$\overline{p}_B \geq l_b + c.$$

By symmetry, we obtain the following inequalities:

$$\overline{p}_B \leq \frac{1-\theta}{\theta} l_a + c$$
$$\overline{p}_A \geq l_a + c.$$

Combining the four inequalities regarding $\overline{p}_A$ and $\overline{p}_B$ above, we have

$$l_a + c \leq \overline{p}_A \leq \frac{\theta}{1-\theta} l_b + c$$
$$l_b + c \leq \overline{p}_B \leq \frac{1-\theta}{\theta} l_a + c.$$

Given the premise $\overline{p}_A = \overline{p}_B$, we rearrange the above inequalities as follows:

$$l_b + c \leq \overline{p}_A \leq \frac{\theta}{1-\theta} l_b + c$$
$$l_a + c \leq \overline{p}_B \leq \frac{1-\theta}{\theta} l_a + c.$$

When $\theta < \frac{1}{2}$, we obtain that $\frac{\theta}{1-\theta} < 1$, which contradicts $l_b + c \leq \overline{p}_A \leq \frac{\theta}{1-\theta} l_b + c$. Similarly, when $\theta > \frac{1}{2}$, we find that $\frac{1-\theta}{\theta} < 1$, which contradicts $l_a + c \leq \overline{p}_B \leq \frac{1-\theta}{\theta} l_a + c$. Therefore, it must be that $\theta = \frac{1}{2}$. (QED)

(b) Still in the setting of part (a), write down each firm's profit maximization problem and find equilibrium prices and profits.

- Firm A's profits are

$$\pi_A = \theta (p_A - c) \frac{l_a + p_B - p_A}{l_a}$$

which only originate from consumers in group a, since firm A does not poach consumers from group b given that $p_A > p_B$ (from part a). Differentiating π_A with respect to p_A yields

$$\frac{\partial \pi_A}{\partial p_A} = \frac{\theta}{l_a} (l_a + p_B - 2p_A + c) = 0$$

and solving for p_A, we obtain firm A's best response function:

$$p_A (p_B) = \frac{p_B + c + l_a}{2}$$

which is increasing in firm B's price, p_B, in the marginal production cost, c, and in group a's loyalty to firm A's product, l_a.

- Firm B's profits are

$$\pi_B = \underbrace{(1-\theta)(p_B - c)}_{\text{From group } b} + \underbrace{\theta(p_B - c)\frac{p_A - p_B}{l_a}}_{\text{From group } a}$$

which in this case depend on both group a (since firm B is undercutting A) and on group b. Differentiating π_B with respect to p_B yields

$$\frac{\partial \pi_B}{\partial p_B} = \frac{l_a(1-\theta) + \theta(p_A - 2p_B + c)}{l_a} = 0$$

and solving for p_B, we obtain firm B's best response function:

$$p_B(p_A) = \frac{1}{2}\left(p_A + c + l_a\frac{1-\theta}{\theta}\right).$$

Combining the above best response functions, we find

$$p_B = \frac{1}{2}\left(\underbrace{\frac{p_B + c + l_a}{2}}_{p_A(p_B)} + c + l_a\frac{1-\theta}{\theta}\right)$$

which, rearranging, yields

$$4\theta p_B = \theta p_B + \theta c + \theta l_a + 2\theta c + 2(1-\theta)l_a$$

that gives the equilibrium price of firm B as follows:

$$p_B^* = \frac{2-\theta}{3\theta}l_a + c.$$

Substituting $p_B^* = \frac{2-\theta}{3\theta}l_a + c$ into $p_A(p_B)$, we find the equilibrium price of firm A:

$$\begin{aligned} p_A^* &= \frac{p_B^* + c + l_a}{2} \\ &= \frac{1}{2}\left(\frac{2-\theta}{3\theta}l_a + c + c + l_a\right) \\ &= \frac{1+\theta}{3\theta}l_a + c. \end{aligned}$$

Substituting p_A^* and p_B^* into the respective profit functions, equilibrium profits are

$$\begin{aligned} \pi_A^* &= \theta(p_A^* - c)\frac{l_a + p_B^* - p_A^*}{l_a} \\ &= \frac{\theta}{l_a}\left(\frac{1+\theta}{3\theta}l_a + c - c\right)\left(l_a + \frac{2-\theta}{3\theta}l_a + c - \frac{1+\theta}{3\theta}l_a - c\right) \end{aligned}$$

$$= \frac{\theta}{l_a}\left(\frac{1+\theta}{3\theta}l_a\right)\left(\frac{3\theta+2-\theta-1-\theta}{3\theta}\right)l_a$$

$$= \frac{(1+\theta)^2}{9\theta}l_a$$

$$\pi_B^* = (1-\theta)\left(p_B^* - c\right) + \theta\left(p_B^* - c\right)\frac{p_A^* - p_B^*}{l_a}$$

$$= (1-\theta)\left(\frac{2-\theta}{3\theta}l_a + c - c\right) + \frac{\theta}{l_a}\left(\frac{2-\theta}{3\theta}l_a + c - c\right)$$

$$\times\left(\frac{1+\theta}{3\theta}l_a + c - \frac{2-\theta}{3\theta}l_a - c\right)$$

$$= \frac{(1-\theta)(2-\theta)}{3\theta}l_a - \frac{(2-\theta)(1-2\theta)}{9\theta}l_a$$

$$= \frac{(2-\theta)^2}{9\theta}l_a.$$

(c) Evaluate the equilibrium prices you found in part (b) at $\theta = \frac{1}{2}$. Interpret your results.
- At $\theta = \frac{1}{2}$, equilibrium price of firm A simplifies to

$$p_A = \frac{1+\frac{1}{2}}{3\times\frac{1}{2}}l_a + c = l_a + c$$

and, similarly, equilibrium price of firm B becomes

$$p_B = \frac{2-\frac{1}{2}}{3\times\frac{1}{2}}l_a + c = l_a + c$$

which coincide with equilibrium prices of the Hotelling model. Intuitively, $\theta = \frac{1}{2}$ means that firms equally share the market if they set the same price. In a market where price discrimination is prohibited by law, the above result means that both firms set the same price, $p_A = p_B = l_a + c$, where the loyalty parameter l_a plays the same role as the transportation cost in the Hotelling linear city model.

(d) Evaluate equilibrium prices you found in part (b) at $\theta = 1$. Interpret your results.
- At $\theta = 1$, equilibrium price of firm A simplifies to

$$p_A = \frac{1+1}{3}l_a + c = \frac{2}{3}l_a + c$$

and similarly, the equilibrium price of firm B becomes

$$p_B = \frac{2-1}{3}l_a + c = \frac{1}{3}l_a + c.$$

In this setting, $\theta = 1$ means that firm A can charge a higher price than firm B if all consumers belong to group A.

(e) *Price discrimination*. When firms can price-discriminate, where $p_i \neq \widetilde{p}_i$ for every firm i, show that $p_A > p_B$ but $\widetilde{p}_B > \widetilde{p}_A$.

- Suppose a pure strategy Nash equilibrium exists in which $p_B \geq p_A$. Then firm A's profit from group a consumers is $(p_A - c)\,\theta$, and firm B's profit from group a consumers is zero. For this to be true in equilibrium, it must be that firm A cannot price-discriminate; otherwise, it could increase its profit by charging a slightly higher price in group a.

  It must also be that firm B cannot price-discriminate, or it could charge a slightly lower price than p_A in group a and earn positive profit. Since p_A must strictly exceed marginal cost in any equilibrium given nonzero royalty, if neither firm can price-discriminate, then we establish a contradiction. Using a similar reasoning, it can be shown that no pure strategy Nash equilibrium exists in which $\widetilde{p}_A \geq \widetilde{p}_B$.

(f) Still in the setting of part (e), write down each firm's profit maximization problem and find equilibrium prices and profits.

- Firm A's profits are

$$\pi_A = \underbrace{\theta\,(p_A - c)\,\frac{l_a + p_B - p_A}{l_a}}_{\text{From group } a} + \underbrace{(1-\theta)\,(\widetilde{p}_A - c)\,\frac{\widetilde{p}_B - \widetilde{p}_A}{l_b}}_{\text{From group } b}$$

  which originate from both groups of consumers. Differentiating π_A with respect to p_A and $\widetilde{p}_A$ yields

$$\frac{\partial \pi_A}{\partial p_A} = \frac{\theta}{l_a}\,(l_a - 2p_A + p_B + c) = 0$$

$$\frac{\partial \pi_A}{\partial \widetilde{p}_A} = \frac{1-\theta}{l_b}\,(\widetilde{p}_B - 2\widetilde{p}_A + c) = 0$$

  and solving for p_A and $\widetilde{p}_A$, we obtain firm A's best response functions as follows:

$$p_A\,(p_B) = \frac{l_a + p_B + c}{2}$$

$$\widetilde{p}_A\,(\widetilde{p}_B) = \frac{\widetilde{p}_B + c}{2}.$$

- Firm B's profits are

$$\pi_B = \underbrace{(1-\theta)\,(\widetilde{p}_B - c)\,\frac{l_b + \widetilde{p}_A - \widetilde{p}_B}{l_b}}_{\text{From group } b} + \underbrace{\theta\,(p_B - c)\,\frac{p_A - p_B}{l_a}}_{\text{From group } a}$$

  which depend on both groups of consumers. Differentiating π_B with respect to p_B and $\widetilde{p}_B$ yields

$$\frac{\partial \pi_B}{\partial p_B} = \frac{\theta}{l_a}\,(p_A - 2p_B + c) = 0$$

$$\frac{\partial \pi_B}{\partial \widetilde{p}_B} = \frac{1-\theta}{l_b}\,(l_b - 2\widetilde{p}_B + \widetilde{p}_A + c) = 0$$

and solving for p_B and $\widetilde{p}_B$, we obtain firm B's best response functions as follows:

$$p_B(p_A) = \frac{p_A + c}{2}$$

$$\widetilde{p}_B(\widetilde{p}_A) = \frac{l_b + \widetilde{p}_A + c}{2}.$$

Combining the above best response functions, we find

$$p_A = \frac{l_a + c}{2} + \frac{1}{2}\left(\frac{p_A + c}{2}\right)$$

$$\widetilde{p}_A = \frac{c}{2} + \frac{1}{2}\left(\frac{l_b + \widetilde{p}_A + c}{2}\right)$$

which, rearranging and solving for p_A and $\widetilde{p}_A$, yields

$$p_A^{**} = \frac{2}{3}l_a + c$$

$$\widetilde{p}_A^{**} = \frac{1}{3}l_b + c.$$

Therefore, firm B's equilibrium prices become

$$p_B^{**} = \frac{1}{3}l_a + c$$

$$\widetilde{p}_B^{**} = \frac{2}{3}l_b + c.$$

And equilibrium profits are

$$\begin{aligned}
\pi_A^{**} &= \theta\left(p_A^{**} - c\right)\frac{l_a + p_B^{**} - p_A^{**}}{l_a} + (1-\theta)\left(\widetilde{p}_A^{**} - c\right)\frac{\widetilde{p}_B^{**} - \widetilde{p}_A^{**}}{l_b} \\
&= \frac{\theta}{l_a}\left(\frac{2}{3}l_a + c - c\right)\left(l_a + \frac{1}{3}l_a + c - \frac{2}{3}l_a - c\right) \\
&+ \frac{1-\theta}{l_b}\left(\frac{1}{3}l_b + c - c\right)\left(\frac{2}{3}l_b + c - \frac{1}{3}l_b - c\right) \\
&= \frac{4}{9}\theta l_a + \frac{1}{9}(1-\theta)\, l_b \\
\pi_B^{**} &= (1-\theta)\left(\widetilde{p}_B^{**} - c\right)\frac{l_b + \widetilde{p}_A^{**} - \widetilde{p}_B^{**}}{l_b} + \theta\left(p_B^{**} - c\right)\frac{p_A^{**} - p_B^{**}}{l_a} \\
&= \frac{1-\theta}{l_b}\left(\frac{2}{3}l_b + c - c\right)\left(l_b + \frac{1}{3}l_b + c - \frac{2}{3}l_b - c\right) \\
&+ \frac{\theta}{l_a}\left(\frac{1}{3}l_a + c - c\right)\left(\frac{2}{3}l_a + c - \frac{1}{3}l_a - c\right) \\
&= \frac{1}{9}\theta l_a + \frac{4}{9}(1-\theta)\, l_b.
\end{aligned}$$

(g) *Comparison*. Compare equilibrium prices from part (f) to compute the price discount that firm A offers, $d_A = p_A - \widetilde{p}_A$. When positive, firm A applies a discount to consumers in group b, which Shaffer and Zhang (2000) refer to as firm A "pays to switch." However, when $d_A < 0$, firm A applies a discount to consumers in its own group a, informally known as "pays to stay." Show that firm A practices "pay to stay" only when the loyalty ratio $\frac{l_b}{l_a}$ satisfies $\frac{l_b}{l_a} > 2$.

- The price discount that firm A offers is

$$d_A = p_A^{**} - \widetilde{p}_A^{**}$$
$$= \frac{2}{3}l_a + c - \frac{1}{3}l_b - c$$
$$= \frac{2l_a - l_b}{3}.$$

Therefore, firm A "pays to stay" if $d_A < 0$, that is,

$$\frac{2l_a - l_b}{3} < 0$$

which, after simplifying, becomes

$$\frac{l_b}{l_a} > 2.$$

(h) Compute the price discount that firm B offers, $d_B = \widetilde{p}_B - p_B$, showing that it "pays to stay" only when the loyalty ratio $\frac{l_b}{l_a}$ satisfies $\frac{l_b}{l_a} < \frac{1}{2}$. Combine your results with those in part (g) to argue that it is never optimal for both firms to practice "pay to stay." Interpret.

- The price discount that firm B offers is

$$d_B = \widetilde{p}_B^{**} - p_B^{**}$$
$$= \frac{2}{3}l_b + c - \frac{1}{3}l_a - c$$
$$= \frac{2l_b - l_a}{3}.$$

Therefore, firm B "pays to stay" if $d_B < 0$, that is,

$$\frac{2l_b - l_a}{3} < 0$$

which, after simplifying, becomes

$$\frac{l_b}{l_a} < \frac{1}{2}.$$

- We can therefore identify three cases, depending on the value of the loyalty ratio $\frac{l_b}{l_a}$, each of them depicted in Fig. 10.5 as follows:
  - When $\frac{l_b}{l_a} > 2$, firm A "pays to stay" and firm B "pays to switch" (see right side of Fig. 10.1). Since the loyalty of group a consumers is less than half of that of group b consumers, firm B can charge a higher price to group b consumers and still maintain

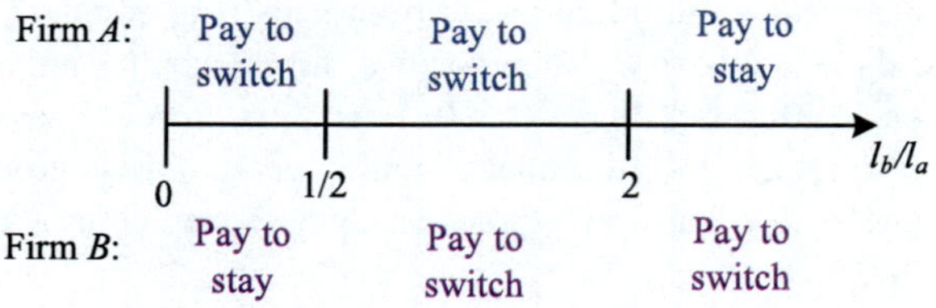

Fig. 10.5 Firms' pricing behavior as a function of consumer loyalty

its market share while charging a lower price to group a consumers to induce the less loyal group a consumers to switch to this firm. Accordingly, firm A has to lower its price to group a consumers in order to defend its market share.

- When $\frac{1}{2} \leq \frac{l_b}{l_a} \leq 2$, both firms practice "pay to switch" (at the center of Fig. 10.1). When consumers are similarly loyal to their respective firms, each firm will offer discounts to the consumers of the other group in order to poach consumers from the other firm.
- When $\frac{l_b}{l_a} < \frac{1}{2}$, firm A "pays to switch" and firm B "pays to stay" (at the left end of Fig. 10.1). Since the loyalty of group a consumers is more than double than that of group b consumers, firm A can offer discount to group b consumers to induce them to switch to this firm, so that firm B has to lower its price to the less loyal group b consumers in order to defend its market share.

Exercise #10.6: Nonlinear Pricing in Monopoly, Based on Maskin and Riley (1984)^C

10.6 Consider a monopolist seeking to sell a product to a consumer with quasilinear utility function:

$$u_i(x, y) = \theta_i\sqrt{x} + y,$$

where θ_i denotes his preference for quality (where $\theta_H > \theta_L$), x represents the quality of a particular good the monopolist sells, and y is money (representing the consumption of all other goods). The probability of a low-valuation consumer is α, while that of a high valuation is $1-\alpha$, where $\alpha \in [0, 1]$.

The monopolist offers the item at a price p and its cost of producing one unit of quality is c, where c satisfies $\theta_L > c > 0$. Assume that the reservation utility of both types of consumers is zero when they do not purchase the good. Since the monopolist cannot observe each type of consumer (i.e., the realization of parameter θ_i), it needs to screen customers by offering a menu that induces each type to self-select the offer meant for him. In the following sections of the exercise, we first show that pricing strategies such as linear pricing or single two-part tariff yield a lower profit than a menu of two-part tariffs.

(a) *Uniform pricing.* Suppose that the monopolist can only offer one price p to every type of customer. What would be the profit-maximizing price and profit?
   - We can use backward induction in this setting, by first finding consumer i's demand for a given price p (second stage), and then identifying the price that the seller sets in the first stage.

- *Second stage.* After observing a price and quality pair (p, x), consumer type i solves the utility maximization problem:

$$\max_{x\geq 0} \theta_i\sqrt{x} + y - px.$$

Differentiating with respect to x, we obtain

$$\frac{\theta_i}{2\sqrt{x}} - p = 0.$$

Solving for x, we find that consumer i's demand for quality is

$$x_i(p) = \frac{\theta_i^2}{4p^2}.$$

Such demand increases in consumer i's preference for quality, θ_i, but decreases in the price p set by the monopolist. Hence, aggregate demand becomes

$$(1-\alpha)\,x_H(p) + \alpha x_L(p) = \frac{(1-\alpha)\,\theta_H^2 + \alpha\theta_L^2}{4p^2}.$$

- *First stage.* Anticipating such an aggregate demand, the monopolist sets a uniform price (the same price to all types of customers) that maximizes its profits as follows:

$$\max_{p\geq 0} \underbrace{(p-c)}_{\text{Margin}}\underbrace{\frac{\left((1-\alpha)\,\theta_H^2 + \alpha\theta_L^2\right)}{4p^2}}_{\text{Aggregate demand}}.$$

Differentiating with respect to p, we obtain

$$-\frac{\left[(1-\alpha)\,\theta_H^2 + \alpha\theta_L^2\right](p-2c)}{4p^3} = 0.$$

Solving for p, we find a uniform price of

$$p^U = 2c$$

which increases in cost c.
- In this context, equilibrium profits become

$$\begin{aligned}\pi^U &= \left(p^U - c\right)\frac{\left[(1-\alpha)\,\theta_H^2 + \alpha\theta_L^2\right]}{4\left(p^U\right)^2}\\ &= (2c-c)\,\frac{\left[(1-\alpha)\,\theta_H^2 + \alpha\theta_L^2\right]}{4\,(2c)^2}\\ &= \frac{(1-\alpha)\,\theta_H^2 + \alpha\theta_L^2}{16c}\end{aligned}$$

which is decreasing in cost c and the expected value (see numerator).

(b) *Single two-part tariff.* Suppose the monopolist can offer a single two-part tariff consisting of an initial fee T and a unit price p (which does not depend on the quantity sold). What is the profit-maximizing (T, p)-pair for the monopolist?

- We first identify the profit-maximizing two-part tariff when the monopolist serves both types of customers, or only the high-value customers, and subsequently compare the profits from each option.
- *Serving both types of customers.* When the monopolist sells to all types of consumers, we need that the low-type consumer participates, i.e., $T \leq S_L(p)$ where $S_L(p)$ represents the surplus for the low-valuation customers, as we next define for any type i:

$$\begin{aligned} S_i(p) &\equiv u_i(x_i(p), y) - p x_i(p) \\ &= \theta_i \sqrt{\frac{\theta_i^2}{4p^2}} - p\frac{\theta_i^2}{4p^2} \\ &= \frac{\theta_i^2}{4p}. \end{aligned}$$

We also require a similar condition to hold for the high-type consumer: $T \leq S_H(p) = \frac{\theta_H^2}{4p}$. However, $S_L(p) \leq S_H(p)$ for every $p \geq 0$ given that $\theta_L < \theta_H$ by definition, implying that we only need to impose the condition on the low-type consumer. In other words, the participation constraint of the high-type consumer becomes slack.

- In this context, the monopolist's profit maximization problem becomes

$$\max_{T, p \geq 0} \; T + \underbrace{(p - c)}_{\text{Margin}} \underbrace{[\alpha x_L(p) + (1 - \alpha) x_H(p)]}_{\text{Expected sales}}$$

$$\text{subject to } T \leq S_L(p) \qquad (PC_L)$$

In addition, note that PC_L must bind. Otherwise, the monopolist would have incentives to increase the fee T and still achieve participation of the low type. Hence, $T = S_L(p) = \frac{\theta_L^2}{4p}$, implying that the above problem can be expressed as the following unconstrained maximization problem:

$$\max_{p \geq 0} \underbrace{\frac{\theta_L^2}{4p}}_{T} + (p - c)\left[\frac{\alpha\theta_L^2 + (1 - \alpha)\theta_H^2}{4p^2}\right].$$

Differentiating with respect to p, we obtain

$$-\frac{\theta_L^2}{4p^2} - \frac{(p - 2c)[\alpha\theta_L^2 + (1 - \alpha)\theta_H^2]}{4p^3} = 0.$$

Solving for p, we obtain

$$p^{ST} = \frac{4c\left(\alpha\theta_L^2 + (1 - \alpha)\theta_H^2\right)}{(1 + 2\alpha)\theta_L^2 + 2(1 - \alpha)\theta_H^2},$$

where the superscript ST denotes "single two-part tariff." Plugging p^{ST} into the fee $T^{ST} = \frac{\theta_L^2}{4p^{ST}}$ yields

$$T^{ST} = \frac{\theta_L^2}{4p^{ST}} = \frac{\theta_L^2\left[(1+2\alpha)\,\theta_L^2 + 2\,(1-\alpha)\,\theta_H^2\right]}{16c\left[\alpha\theta_L^2 + (1-\alpha)\,\theta_H^2\right]}$$

entailing profits of

$$\begin{aligned}\pi^{ST} &= \frac{\theta_L^2}{4p^{ST}} + \left(p^{ST} - c\right)\left[\frac{\alpha\theta_L^2 + (1-\alpha)\,\theta_H^2}{4\left(p^{ST}\right)^2}\right] \\ &= \frac{\theta_L^2 p^{ST} + \left(p^{ST} - c\right)\left(\alpha\theta_L^2 + (1-\alpha)\,\theta_H^2\right)}{4\left(p^{ST}\right)^2} \\ &= \frac{\left[(3+2\alpha)\,\theta_L^2 + 2\,(1-\alpha)\,\theta_H^2\right]\left[(1+2\alpha)\,\theta_L^2 + 2\,(1-\alpha)\,\theta_H^2\right]}{64c\left[\alpha\theta_L^2 + (1-\alpha)\,\theta_H^2\right]}.\end{aligned}$$

- *Serving only the high-type customer.* We now check if the monopolist's profits from serving both types of customers are larger than its profits from selling to the high-type customer alone. We next find the tariff and price if the monopolist only sells to the high-type customer. In this context, its profit maximization problem becomes

$$\max_{T,p\geq 0}\ (1-\alpha)\left[T + (p-c)\,x_H(p)\right]$$

$$\text{subject to } T \leq S_H(p) \qquad (PC_H)$$

Note that now the only PC constraint we consider is that of the high type, where $S_H(p) = \frac{\theta_H^2}{4p}$. In addition, PC_H must bind (otherwise the monopolist could increase the fee T and still achieve participation of this type of consumer), so we must have that $T = S_H(p) = \frac{\theta_H^2}{4p}$. The monopolist's problem then becomes the following unconstrained program:

$$\max_{p\geq 0}\ (1-\alpha)\left[\frac{\theta_H^2}{4p} + (p-c)\,\frac{\theta_H^2}{4p^2}\right].$$

Differentiating with respect to p, we obtain

$$(1-\alpha)\left[\frac{(c-p)\theta_H^2}{2p^3}\right] = 0.$$

Solving for p yields

$$p^H = c,$$

where superscript H denotes that the monopolist only serves the high-type consumer. Then, the fee in this context becomes

$$T^H = \frac{\theta_H^2}{4p^H} = \frac{\theta_H^2}{4c},$$

entailing profits of

$$\pi^H = (1-\alpha)\left[\overbrace{\frac{\theta_H^2}{4c}}^{T^H} + (c-c)\frac{\theta_H^2}{4c^2}\right]$$

$$= \frac{(1-\alpha)\,\theta_H^2}{4c}.$$

- *Comparison.* Finally, we can now compare the profits of using a single two-part tariff that serves both customers, π^{ST}, against those from selling to the high-type customer alone, π^H, finding that $\pi^{ST} < \pi^H$ if

$$\frac{\left[(3+2\alpha)\,\theta_L^2 + 2\,(1-\alpha)\,\theta_H^2\right]\left[(1+2\alpha)\,\theta_L^2 + 2\,(1-\alpha)\,\theta_H^2\right]}{64c\left[\alpha\theta_L^2 + (1-\alpha)\,\theta_H^2\right]} < \frac{(1-\alpha)\,\theta_H^2}{4c}.$$

This inequality simplifies to

$$12(1-\alpha)^2\theta_H^4 - 8(1-\alpha)^2\theta_H^2\theta_L^2 - (1+2\alpha)(3+2\alpha)\theta_L^4 > 0,$$

which we can rearrange as

$$12(1-\alpha)^2\left(\frac{\theta_H}{\theta_L}\right)^4 - 8(1-\alpha)^2\left(\frac{\theta_H}{\theta_L}\right)^2 - (1+2\alpha)(3+2\alpha) > 0.$$

Solving for ratio $\frac{\theta_H}{\theta_L}$, we obtain

$$\frac{\theta_H}{\theta_L} > \sqrt{\frac{2(1-\alpha)+\sqrt{13+16\alpha+16\alpha^2}}{6(1-\alpha)}}.$$

For example, when both types of consumers are equally likely, $\alpha = \frac{1}{2}$, the above inequality simplifies to $\frac{\theta_H}{\theta_L} > \sqrt{2} \approx 1.414$, meaning that when the high-type consumers place a sufficiently higher valuation than the low-type consumers, it is more profitable for the monopolist to only serve the high type than to serve both types. More generally, the right-hand side of the above inequality is increasing and convex in α. Intuitively, this means that, as the proportion of low-type consumers increases, focusing on high-type buyers alone becomes less attractive, and the "value differential" that high-type consumers assign to the good, as captured by $\frac{\theta_H}{\theta_L}$, must be higher.

(c) *Menu of two-part tariffs.* Let us now find the menu of offers (contracts), (x_L, T_L) and (x_H, T_H), meant for low-type and high-type customer, respectively.
- The problem for the monopolist is now the following (as usual in screening problems, we need to include the participation constraint for both types of customers, along with the incentive compatibility conditions for both of them):

$$\max_{x_H, T_H, x_L, T_L \geq 0} \quad (1-\alpha)\,(T_H - cx_H) + \alpha\,(T_L - cx_L)$$

subject to

$$\theta_L\sqrt{x_L} - T_L \geq 0 \qquad (PC_L)$$

$$\theta_H\sqrt{x_H} - T_H \geq 0 \qquad (PC_H)$$

$$\theta_L\sqrt{x_L} - T_L \geq \theta_L\sqrt{x_H} - T_H \qquad (IC_L)$$

$$\theta_H\sqrt{x_H} - T_H \geq \theta_H\sqrt{x_L} - T_L \qquad (IC_H)$$

Since

$$\theta_H\sqrt{x_H} - T_H \underbrace{\geq}_{\text{From } IC_H} \theta_H\sqrt{x_L} - T_L \underbrace{>}_{\text{From } \theta_H > \theta_L} \theta_L\sqrt{x_L} - T_L \geq 0,$$

combining the first and last inequality yields $\theta_H\sqrt{x_H} - T_H > 0$ so that the PC_H constraint must be slack. In other words, the PC_L constraint binds, as the monopolist can charge a higher tariff to this type of consumer and still achieve participation.

- Substituting the binding PC_L into IC_L, we obtain $0 \geq \theta_L\sqrt{x_H} - T_H$, meaning that it is unprofitable for the low type to take on a high-type contract, so that IC_L is slack. From the binding PC_L and IC_H, we obtain

$$\theta_L\sqrt{x_L} - T_L = 0$$
$$\theta_H\sqrt{x_H} - T_H = \theta_H\sqrt{x_L} - T_L.$$

Rearranging, we obtain

$$T_L = \theta_L\sqrt{x_L}$$

and using this expression in $\theta_H\sqrt{x_H} - T_H = \theta_H\sqrt{x_L} - T_L$ yields

$$\begin{aligned} T_H &= \theta_H\sqrt{x_H} - \theta_H\sqrt{x_L} + \overbrace{\theta_L\sqrt{x_L}}^{=T_L} \\ &= \theta_H\left(\sqrt{x_H} - \sqrt{x_L}\right) + \theta_L\sqrt{x_L}. \end{aligned}$$

Inserting T_L and T_H, we simplify the monopolist's expected profit maximization problem to the following unconstrained program which, in addition, has only two choice variables (x_H and x_L) rather than the original four choice variables:

$$\max_{x_H, x_L \geq 0} \; (1-\alpha)\left(\overbrace{\theta_H\left(\sqrt{x_H} - \sqrt{x_L}\right) + \theta_L\sqrt{x_L}}^{=T_H} - cx_H\right) + \alpha\left(\overbrace{\theta_L\sqrt{x_L}}^{=T_L} - cx_L\right).$$

Differentiating with respect to x_H and x_L yields

$$\frac{\partial E[\pi]}{\partial x_H} = \frac{\theta_H}{2\sqrt{x_H}} - c = 0$$
$$\frac{\partial E[\pi]}{\partial x_L} = \frac{\theta_L - (1-\alpha)\theta_H}{2\sqrt{x_L}} - \alpha c = 0.$$

Simplifying, we obtain

$$x_H = \frac{\theta_H^2}{4c^2}$$
$$x_L = \frac{(\theta_L - (1-\alpha)\theta_H)^2}{4\alpha^2 c^2}.$$

- Substituting these results, we find that optimal tariffs are

$$T_L = \theta_L\sqrt{x_L} = \theta_L\sqrt{\underbrace{\frac{(\theta_L - (1-\alpha)\theta_H)^2}{4\alpha^2 c^2}}_{x_L}} = \frac{\theta_L(\theta_L - (1-\alpha)\theta_H)}{2\alpha c}$$

and

$$\begin{aligned} T_H &= \theta_H\left(\sqrt{x_H} - \sqrt{x_L}\right) + \theta_L\sqrt{x_L} \\ &= \theta_H\left(\sqrt{\underbrace{\frac{\theta_H^2}{4c^2}}_{x_H}} - \sqrt{\underbrace{\frac{(\theta_L - (1-\alpha)\theta_H)^2}{4\alpha^2 c^2}}_{x_L}}\right) + \theta_L\sqrt{\underbrace{\frac{(\theta_L - (1-\alpha)\theta_H)^2}{4\alpha^2 c^2}}_{x_L}} \\ &= \theta_H\left(\frac{\theta_H}{2c} - \frac{\theta_L - (1-\alpha)\theta_H}{2\alpha c}\right) + \theta_L\frac{\theta_L - (1-\alpha)\theta_H}{2\alpha c} \\ &= \frac{\theta_H^2 - (2-\alpha)\theta_H\theta_L + \theta_L^2}{2\alpha c}. \end{aligned}$$

Therefore, the monopolist's profits from the menu of two-part tariffs are

$$\begin{aligned} \pi^{MT} &= (1-\alpha)(T_H - cx_H) + \alpha(T_L - cx_L) \\ &= (1-\alpha)\left(\frac{\theta_H^2 - (2-\alpha)\theta_H\theta_L + \theta_L^2}{2\alpha c} - \frac{c\theta_H^2}{4c^2}\right) \\ &+ \alpha\left(\frac{\theta_L(\theta_L - (1-\alpha)\theta_H)}{2\alpha c} - \frac{c(\theta_L - (1-\alpha)\theta_H)^2}{4\alpha^2 c^2}\right) \\ &= \frac{(1-\alpha)\theta_H^2 - 2(1-\alpha)\theta_L\theta_H + \theta_L^2}{4\alpha c}. \end{aligned}$$

(d) Rank the profits of the monopolist in parts (a)–(c). Interpret your results. For simplicity, assume that both types of consumers are equally likely, $\alpha = \frac{1}{2}$.

- We have relaxed the constraints in steps. As we move from part (a) to part (c), the monopolist derives more and more rent (surplus) as he develops more sophisticated pricing strategies. We can then have two cases, depending on whether $\frac{\theta_H}{\theta_L} > \sqrt{2}$ or $\frac{\theta_H}{\theta_L} \leq \sqrt{2}$.
- *Case 1, High-type consumers assign a relatively high value to quality.* When condition $\frac{\theta_H}{\theta_L} > \sqrt{2}$ holds, in part (b) we found that the monopolist earns a higher profit when serving high-type customers alone, so that $\pi^{ST} < \pi^H$. Therefore, we can rank profits in parts (a)–(c) as follows:

$$\pi^U < \pi^H < \pi^{MT}$$

which holds if

$$\frac{\theta_L^2 + \theta_H^2}{32c} < \frac{\theta_H^2}{8c} < \frac{\theta_H^2 - 2\theta_L\theta_H + 2\theta_L^2}{4c}.$$

For the first inequality to hold, we need $4\theta_H^2 > \theta_L^2 + \theta_H^2$, or $\frac{\theta_H}{\theta_L} > \frac{1}{\sqrt{3}}$, which is satisfied since $\frac{\theta_H}{\theta_L} > 1$ given that $\theta_L < \theta_H$. For the second inequality to hold, we need $\theta_H^2 - 4\theta_H\theta_L + 4\theta_L^2 > 0$, which holds because this is factorized into $(\theta_H - 2\theta_L)^2 > 0$.
- *Case 2, High-type consumers assign a relatively low value to quality.* When condition $\frac{\theta_H}{\theta_L} \leq \sqrt{2}$ holds, in part (b) we found that the monopolist earns a higher profit when serving both types of consumers, so that $\pi^{ST} > \pi^H$. Therefore, we can rank profits in parts (a)–(c) as follows:

$$\pi^U < \pi^{ST} < \pi^{MT}$$

which holds if

$$\frac{\theta_L^2 + \theta_H^2}{32c} < \frac{\left(4\theta_L^2 + \theta_H^2\right)\left(2\theta_L^2 + \theta_H^2\right)}{32c\left(\theta_L^2 + \theta_H^2\right)} < \frac{\theta_H^2 - 2\theta_L\theta_H + 2\theta_L^2}{4c}.$$

For the first inequality to hold, we need $\theta_L^2\left(7\theta_L^2 + 4\theta_H^2\right) > 0$, which is satisfied. For the second inequality to hold, we need

$$7\left(\frac{\theta_H}{\theta_L}\right)^4 - 16\left(\frac{\theta_H}{\theta_L}\right)^3 + 18\left(\frac{\theta_H}{\theta_L}\right)^2 - 16\left(\frac{\theta_H}{\theta_L}\right) + 8 > 0,$$

which holds for all values of $\frac{\theta_H}{\theta_L} \leq \sqrt{2}$.
- Combining the above cases, the monopolist obtains the highest (lowest) profit in practicing menu (uniform) pricing for all values of θ_L and θ_H.

(e) *Welfare comparison.* Assume again that both types of consumers are equally likely, $\alpha = \frac{1}{2}$, and that $c = \frac{1}{4}$, and $\frac{\theta_H}{\theta_L} > \sqrt{2}$. Evaluate the expected social welfare that emerges when the monopolist practices uniform pricing (as in part a), limited pricing by serving only high-type customers (as in part b), and offers a menu of two-part tariffs (as in part c). Which pricing strategy yields the highest expected social welfare? Compare your welfare ranking with the profit ranking obtained in part (d). Interpret.

- *Uniform pricing.* Expected social welfare is

$$W^U = \overbrace{(1-\alpha)\, x_H\left(p^U\right)(\theta_H - c)}^{\text{High-type cons.}} + \overbrace{\alpha x_L\left(p^U\right)(\theta_L - c)}^{\text{Low-type cons.}}$$

$$= (1-\alpha)\overbrace{\frac{\theta_H^2}{4p^2}}^{=x_H(p^U)}(\theta_H - c) + \alpha \overbrace{\frac{\theta_L^2}{4p^2}}^{=x_L(p^U)}(\theta_L - c)$$

$$= \frac{\theta_H^2(4\theta_H - 1) + \theta_L^2(4\theta_L - 1)}{32p^2}$$

$$= \frac{\theta_H^2\,(4\theta_H - 1) + \theta_L^2\,(4\theta_L - 1)}{8},$$

where the last step considers that $p^U = 2c = \$\frac{1}{2}$.

- *Limited pricing.* Expected social welfare in this case, as the monopolist only serves the high-type consumers, is given by

$$W^H = (1-\alpha)\, x_H\left(p^H\right)(\theta_H - c)$$

$$= (1-\alpha)\overbrace{\frac{\theta_H^2}{4p^2}}^{=x_H(p^H)}(\theta_H - c)$$

$$= \frac{\theta_H^2(4\theta_H - 1)}{32p^2}$$

$$= \frac{\theta_H^2\,(4\theta_H - 1)}{2},$$

where the last step inserts $p_H = c = \$\frac{1}{4}$.

- *Menu pricing.* Expected social welfare in this case, where the monopolist sells to both types of consumers, is

$$W^{MT} = \overbrace{(1-\alpha)\, x_H^{MT}(\theta_H - c)}^{\text{High-type cons.}} + \overbrace{\alpha x_L^{MT}(\theta_L - c)}^{\text{Low-type cons.}}$$

$$= (1-\alpha)\overbrace{\left(\frac{\theta_H^2}{4c^2}\right)}^{=x_H^{MT}}(\theta_H - c) + \alpha \overbrace{\left(\frac{(\theta_L - (1-\alpha)\,\theta_H)^2}{4\alpha^2 c^2}\right)}^{=x_L^{MT}}(\theta_L - c)$$

$$= \frac{\theta_H^2\,(4\theta_H - 1) + (2\theta_L - \theta_H)^2\,(4\theta_L - 1)}{2}.$$

- *Welfare comparison.* Comparing across different pricing strategies, we find that:
  - $W^H > W^U$, that is, social welfare under limited pricing is unambiguously higher than uniform pricing because

$$\frac{\theta_H^2 (4\theta_H - 1)}{2} > \frac{\theta_H^2 (4\theta_H - 1) + \theta_L^2 (4\theta_L - 1)}{8}$$

holds if $3\theta_H^2 (4\theta_H - 1) > \theta_L^2 (4\theta_L - 1)$, which is true for $\theta_H > \theta_L$ by definition.
  - $W^{MT} > W^H$, that is, social welfare under limited pricing falls below menu pricing because

$$\frac{\theta_H^2 (4\theta_H - 1) + (2\theta_L - \theta_H)^2 (4\theta_L - 1)}{2} > \frac{\theta_H^2 (4\theta_H - 1)}{2}$$

which simplifies to

$$(2\theta_L - \theta_H)^2 (4\theta_L - 1) > 0$$

that holds as long as $\theta_L > c = \frac{1}{4}$ that holds by definition.
  - In summary, we obtain that social welfare satisfies

$$W^{MT} > W^H > W^U,$$

which coincides with the profit ranking when $\frac{\theta_H}{\theta_L} > \sqrt{2}$, where

$$\pi^{MT} > \pi^H > \pi^U.$$

because the monopolist can extract the most consumer surplus from all consumers when they are served under menu pricing, which is more preferred from the social and the firm's perspective of only serving high-type consumers in limit pricing. This is, in turn, better off than uniform pricing in which the loss in the monopolist's profits does not offset the gains in consumer surplus.

Exercise #10.7: Return Policies, Based on Che (1998)^B

10.7 Consider a monopolist selling one unit of a good to a consumer. Assume that the good is an "experience" good, implying that the consumer only knows his valuation v after purchasing it, such as with goods he does not know from previous experiences, or gifts that the consumer makes to another individual who does not know well (e.g., correct size, style, etc.)

Valuation v is drawn from distribution $F(v)$ in the interval $[0, 1]$, with positive density $f(v) > 0$ in all its support. The consumer's von Neumann-Morgenstern utility function $U(\cdot)$ is strictly increasing and concave. For simplicity, assume that $U(\cdot)$ exhibits constant absolute risk-aversion (CARA) form.

(a) *No return policy.* Show that the optimal price that the monopolist charges when it does not offer return policy (i.e., the seller does not refund the price p to the buyer under any circumstances) is $p_{NR} = v_{CE}$, where subscript NR denotes "no return" policy and v_{CE} represents the buyer's certainty equivalent, i.e., v_{CE} solves $U(v_{CE}) = E[U(v)]$.

- The seller offers a price p to the buyer, who decides whether to accept it or not. Operating by backward induction, we start analyzing the buyer's decision rule.
- *Buyer.* For a given price p, the consumer purchases the good if and only if his expected utility is positive, that is,

$$E[U(v-p)] \geq 0.$$

- *Seller.* The seller sets a price p_{NR} that makes the consumer indifferent between buying and not buying the good, that is, p_{NR} solves $E[U(v-p_{NR})]=0$. We next show that such a price coincides with the certainty equivalent v_{CE}. In particular,

$$E[U(v-v_{CE})] = kE[U(v)-U(v_{CE})] = 0 \text{ for any constant } k \in \mathbb{R}.$$

The second equality holds by the definition of the certainty equivalent, i.e., v_{CE} solves $U(v_{CE}) = E[U(v)]$. The first inequality is shown as follows. First, define $z \equiv U(v) - U(v_{CE})$. Then, $U(v-v_{CE}) = \phi(z)$, where

$$\phi(z) \equiv U\left(U^{-1}(z+U(v_{CE})) - v_{CE}\right).$$

We then only need to show that function $\phi(\cdot)$ is linear, which follows since its second derivative, $\phi''(z)$, is

$$\frac{U''\left(U^{-1}(z+U(v_{CE})) - v_{CE}\right)}{U'\left(U^{-1}(z+U(v_{CE})) - v_{CE}\right)} - \frac{U''\left(U^{-1}(z+U(v_{CE}))\right)}{U'\left(U^{-1}(z+U(v_{CE}))\right)}$$

or, more compactly,

$$\frac{U''(a)}{U'(a)} - \frac{U''(b)}{U'(b)},$$

where $a \equiv U^{-1}(z+U(v_{CE})) - v_{CE}$, and $b \equiv U^{-1}(z+U(v_{CE}))$. Since, by definition, $U(\cdot)$ exhibits a CARA form, the Arrow-Pratt coefficient of absolute risk-aversion, $r_A(x) \equiv -\frac{U''(x)}{U'(x)}$, is constant in x, entailing that

$$\frac{U''(a)}{U'(a)} = \frac{U''(b)}{U'(b)},$$

implying that $\phi''(z) = 0$ and, therefore, $\phi(z)$ is linear, as required. As a result, the monopolist sets a price equal to the consumer's certainty equivalent, $p_N = v_{CE}$.

(b) How is price p_N affected by the consumer's degree of risk aversion?
- As the consumer becomes more risk averse, the certainty equivalent, v_{CE}, decreases (approaching the valuation's lower bound, 0, when the consumer is infinitely risk averse), decreasing as a result of the price that the monopolist can charge for the good. Intuitively, under no return policy, consumers bear the entire risk associated with their uncertain ex post valuation.

(c) *Return policy*. Assume now that the buyer can bring the good back to the seller and be reimbursed the price p that he paid for it. Find the optimal price that the monopolist charges when it offers return policy and label this price p_R, where subscript R denotes "return" policy.

- *Buyer*. After observing a price p, the buyer purchases the good if his valuation (i.e., his ex post valuation) v satisfies $v > p$, but returns it otherwise.
- *Seller*. The seller does not observe the buyer's valuation, but can calculate the probability that he keeps or returns the good. Specifically, the buyer returns the good when $v < p$, which occurs with probability $F(p)$ but keeps it when $v > p$, which occurs with probability $1 - F(p)$. Graphically, price p then divides the valuation line [0, 1] into two regions: one to the left of p which occurs with probability $F(p)$ and one to the right of p which occurs with probability $1 - F(p)$.

Therefore, the seller's expected profit maximization problem is

$$\max_{p \geq 0} \; (p - c)[1 - F(p)],$$

where $p - c$ denotes the per-unit margin and $1 - F(p)$ represents the probability that the buyer keeps the good. Differentiating with respect to p yields

$$1 - F(p) - (p - c)\, f(p) = 0$$

with second-order conditions

$$-2f(p) - (p - c)\, f'(p) \leq 0.$$

Assuming $p \geq c$ throughout all our analysis, a sufficient condition for concavity is that the density function is weakly increasing, $f'(p) \geq 0$.

Solving for price p, we obtain the seller's optimal price under the return policy:

$$p_R = \frac{1 - F(p)}{f(p)} + c.$$

- *Comparative statics of* p_R. First, price p_R is not a function of the consumer's risk aversion. Intuitively, the return policy eliminates his risk of paying more than his ex post valuation, that is, $v < p$ does not occur in equilibrium. Second, the seller screens consumers, so only high-value buyers purchase the good, which did not occur in the absence of the return policy.

(d) *Parametric example*. Evaluate equilibrium prices in parts (a) and (c) assuming that valuations are uniformly distributed in [0, 1] and that the consumer's CARA utility function is $u(x) = -e^{-ax}$, where $a > 0$. Assuming that $a = \frac{1}{2}$, does the seller offer a return policy?

- *Return policy.* Since valuations are uniformly distributed, $F(p) = p$ and $f(p) = 1$, implying that the equilibrium price with return policy, p_R, that we found in part (c) becomes

$$p_R = (1 - p_R) + c$$

which, solving for p_R, yields

$$p_R = \frac{1 + c}{2}.$$

The seller's expected profits in this context are

$$\mathbb{E}[\pi_R] = (p_R - c) \times [1 - F(p_R)] = \left(\frac{1+c}{2} - c\right) \times \left(1 - \frac{1+c}{2}\right) = \frac{(1-c)^2}{4}.$$

- *No return policy*. In this context, $p_{NR} = v_{CE}$, where v_{CE} solves $U(v_{CE}) = E[U(v)]$, implying

$$\begin{aligned}\overbrace{-e^{-a \times v_{CE}}}^{U(v_{CE})} &= \overbrace{\int_0^1 -e^{-ax} dx}^{E[U(v)]} \\ &= \left[\frac{e^{-ax}}{a}\right]_0^1 \\ &= \frac{e^{-a} - 1}{a}.\end{aligned}$$

Solving for v_{CE}, we obtain a highly nonlinear expression. At this point, we can evaluate the above expression at $a = \frac{1}{2}$, as required in this exercise, which yields

$$-e^{-\frac{v_{CE}}{2}} = \frac{e^{-\frac{1}{2}} - 1}{\frac{1}{2}}.$$

Rearranging and simplifying, we find

$$v_{CE} = -2\log\left[-2\left(e^{-\frac{1}{2}} - 1\right)\right],$$

so that

$$p_{NR} = v_{CE} = 0.479,$$

which falls below the price under refund policy, p_R, for all values of $c \in [0, 1]$, since risk-averse consumers discount the price of the product, p_{NR}, to its certainty equivalent value, v_{CE}, when refund cannot be made.

Defining $x \equiv \log\left[-2\left(e^{-\frac{1}{2}} - 1\right)\right]$, the seller's expected profits become

$$\mathbb{E}[\pi_{NR}] = (v_{CE} - c) \times [1 - F(v_{CE})] = -(2x + c)(1 + 2x),$$

implying that the seller earns a higher profit offering return policy if and only if

$$\begin{aligned}\frac{(1-c)^2}{4} &> -(2x + c)(1 + 2x) \\ c^2 + 2c + 1 &> -4[4x^2 + 2x(1 + c)] \\ (1 + c)^2 + 8(1 + c)x + 16x^2 &> 0 \\ (1 + c + 4x)^2 &> 0,\end{aligned}$$

which is always positive. Thus, the seller is more profitable offering a return policy for its product. Intuitively, by allowing consumers to return the product after they have experienced it, the seller is able to charge a higher price to those consumers whose valuation is above the price of the product.

- The seller may, however, not find it profitable to offer a return policy in other contexts (e.g., when valuations are not uniformly distributed or when buyers have different risk-aversion preferences). For a more general presentation, see Proposition 1 in Che (1998).

Exercise #10.8: A Model of Sales, Based on Varian (1980)^C

10.8 Firms offer sales at different times. In this exercise, we show that offering sales (or, more generally, randomizing over prices) is a strategy that helps firms maximize their expected profits. This exercise belongs to the literature on "price dispersion" where firms face a share of consumers who are uninformed about prices, and offer different prices, either at different locations (spatial price dispersion) or at different points in time (temporal price dispersion, as we analyze in this exercise). Price discrimination models, in contrast, assume that consumers can perfectly observe prices.

Consider an industry with N firms and free entry, so firms enter until the profits from doing so are zero. Consumers have a reservation price r for a homogeneous good and purchase at most one unit. A share α^I of consumers is informed about prices, buying from the cheapest firm, and a share $1 - \alpha^I$ are uninformed, who purchase from any firm. Therefore, there are $\alpha^U = \frac{1-\alpha^I}{N}$ uninformed consumers per firm. Firms face a symmetric cost function $C(q) = F + cq$, where $F > 0$ denotes fixed costs and c represents its marginal cost. Every firm can only charge one price for its product.

As a reference, note that $C(\alpha^I + \alpha^U) = F + c(\alpha^I + \alpha^U)$ denotes the cost from serving the maximum amount of customers (both informed and uninformed consumers). Therefore, the ratio

$$p_L \equiv \frac{F + c(\alpha^I + \alpha^U)}{\alpha^I + \alpha^U}$$

represents the average cost in this setting.

We next show that, in the above context, every firm has incentives to randomize its pricing over a certain interval. The following questions should help you find the specific cumulative distribution function $F(p)$ that every firm uses in the mixed-strategy Nash equilibrium of the game.

(a) Show that $F(p) = 0$ for all $p < p_L$ and that $F(p) = 1$ for all $p > r$.

- This question essentially asks us to "trim" the support of price randomization in $F(p)$ and characterize its lower and upper bounds.
- *Lower bound.* When charging prices below p_L, a firm must be making losses, since its price lies below its cost in the most favorable scenario (when all types of consumers purchase the good). Therefore, the firm does not assign a probability weight on prices below p_L.
- *Upper bound.* If a firm charges a price above the reservation price r, no customer buys from it, regardless of whether he is informed or uninformed. The firm then has no incentives to assign a probability weight on prices above r. Combining our above results, the price p in $F(p)$ must lie in the interval $[p_L, r]$.

(b) Show that the cumulative distribution function $F(p)$ is nondegenerated, that is, there is no pure strategy Nash equilibrium.
- If firm i uses a pure strategy, charging price $p_i = p_L$, it makes a loss, thus having incentives to exit the industry. (Recall that, in equilibrium, firms make zero profits.) If, instead, the firm sets a higher price p_i that satisfies $r \geq p_i > p_L$, other firms would have incentives to undercut firm i's price by a small ε. Therefore, firm i does not use a pure strategy.

(c) For simplicity, assume that $F(p)$ is continuous.[1] Find expected profits from the pricing strategy $F(p)$.
- If a firm sets the lowest price, it attracts all consumers, and its profit is

$$\pi_s(p) = p(\alpha^I + \alpha^U) - F - c(\alpha^I + \alpha^U),$$

where the subscript s denotes that the firm is successful at attracting all consumers.

If, instead, the firm is unsuccessful, it only sells its product to uninformed consumers, earning

$$\pi_f(p) = p\alpha^U - F - c\alpha^U,$$

where the subscript f denotes "failure."

The probability that firm i sets a price p higher than its rival $j \neq i$ is

$$F(p) = \text{Prob}\left\{p \geq p_j\right\},$$

so the probability that $p < p_j$ is the converse, $1 - F(p)$. As a result, the probability that p is lower than the prices of all its $N - 1$ rivals is

$$[1 - F(p)]^{N-1},$$

which represents the probability that firm i sells to informed consumers. Finally, the probability that firm i does not sell to informed consumers is

$$1 - [1 - F(p)]^{N-1}.$$

- We are now ready to write firm i's expected profit:

$$\int_{p_L}^{r} \left[\underbrace{\pi_s(p)\,[1 - F(p)]^{N-1}}_{\text{Success}} + \underbrace{\pi_f(p)\left[1 - [1 - F(p)]^{N-1}\right]}_{\text{Failure}} \right] f(p)dp.$$

(d) Using the no entry condition, find the cumulative distribution function $F(p)$ with which every firm randomizes.

[1] That is, there is no "mass point" in the pricing strategy $F(p)$ that every firm uses. Intuitively, the firm chooses all prices in the $[p_L, r]$ interval with positive probability. More compactly, this means that the density function $f(p) > 0$ for all $p \in [p_L, r]$.

- Since firms make no profits in equilibrium (otherwise entry or exit would still be profitable), the above expected profit must be equal to zero, which entails

$$\pi_s(p)\left[1-F(p)\right]^{N-1}+\pi_f(p)\left[1-\left[1-F(p)\right]^{N-1}\right]=0.$$

Rearranging,

$$F(p)=1-\left(\frac{\pi_f(p)}{\pi_f(p)-\pi_s(p)}\right)^{\frac{1}{N-1}}.$$

The denominator is negative since $\pi_f(p)<\pi_s(p)$ for any price $p\in[p_L,r]$. Therefore, the numerator must also be negative, $\pi_f(p)<0$.

(e) Show that the cumulative distribution function $F(p)$ has full support in $p\in[p_L,r]$. That is, $F(p_L+\varepsilon)>0$ and $F(r-\varepsilon)<1$ for any $\varepsilon>0$.

- *Prices slightly above p_L.* If, instead, $F(p_L+\varepsilon)=0$, firm i is assigning no probability weight to prices slightly higher than the lower bound p_L. Therefore, firm i assigns probability weight to prices strictly above $p_L+\varepsilon$. In that case, another firm j could undercut firm i's price and set, for instance, a price $p_L+\frac{\varepsilon}{2}$ to make positive profits. Hence, $F(p_L+\varepsilon)>0$ for any $\varepsilon>0$.
- *Prices slightly below r.* If, instead, $F(r-\varepsilon)=1$, firm i assigns no probability to prices slightly below r. At $\widetilde{p}<r$, only uninformed consumers purchase the good and the firm earns $\widetilde{p}\alpha^U-F-c\alpha^U$, yielding zero profits. However, a deviation to price $p=r$ yields $r\alpha^U-F-c\alpha^U$ which is positive, thus making such deviation profitable. Therefore, $F(r-\varepsilon)<1$ for any $\varepsilon>0$.

(f) Taking into account that $\pi_f(r)=0$, find the equilibrium number of firms in the industry, n^*.

- Condition $\pi_f(r)=0$ entails

$$r\alpha^U-F-c\alpha^U=0.$$

Substituting $\alpha^U=\frac{1-\alpha^I}{N}$ into the above expression yields

$$F=(r-c)\underbrace{\frac{1-\alpha^I}{N}}_{\alpha^U}.$$

Solving for N, we obtain

$$N^*=\frac{(r-c)\left(1-\alpha^I\right)}{F}.$$

Therefore, the higher the profit margin $r-c$, the larger the share of the uninformed consumers $1-\alpha^I$, and the lower the entry cost F, the more firms in equilibrium.

(g) Taking into account that $\pi_s(p_L)=0$, and the equilibrium number of firms N^*, find the lower bound of firms' randomization strategy, p_L.

- Condition $\pi_s(p_L)=0$ entails

$$p_L(\alpha^I+\alpha^U)-F-c(\alpha^I+\alpha^U)=0.$$

Substituting $\alpha^U = \frac{1-\alpha^I}{N}$ into the above expression yields

$$F = (p_L - c)\left(\alpha^I + \frac{1-\alpha^I}{N}\right).$$

Further inserting the equilibrium number of firms, N^*, found in part (f), we have

$$F = (p_L - c)\left(\alpha^I + \frac{1-\alpha^I}{\frac{(r-c)(1-\alpha^I)}{F}}\right).$$

Rearranging, we obtain

$$F = (p_L - c)\left(\frac{(r-c)\,\alpha^I + F}{r-c}\right).$$

Solving for p_L, we find the lower bound of firms' randomization strategy:

$$p_L = \frac{c\,(r-c)\,\alpha^I + rF}{(r-c)\,\alpha^I + F}.$$

(h) Evaluate your above results in the special case in which all consumers are uninformed.

- When all consumers are uninformed, $\alpha^I = 0$, the lower bound of firms' randomization strategy, p_L, becomes

$$p_L = \frac{rF}{F} = r,$$

which coincides with the upper bound of firms' randomization strategy. In other words, firms put full probability weight on one price, $p = r$, with every firm extracting all surplus from a share $\frac{1}{N}$ of consumers.

(i) *Numerical example*. Evaluate your results in parts (d), (f), and (g) at parameter values $r = 1$, $F = \frac{2}{9}$, $c = 0$, and $\alpha^I = \frac{1}{3}$.

- In this setting, the equilibrium number of firms becomes

$$N^* = \frac{(1-0)\left(1-\frac{1}{3}\right)}{\frac{2}{9}} = 3.$$

In addition, the lower bound is

$$p_L = \frac{0\,(1-0)\,\frac{1}{3} + \frac{2}{9}\times 1}{(1-0)\,\frac{1}{3} + \frac{2}{9}} = \frac{2}{5}.$$

In this context, the share of uninformed consumers for every firm becomes

$$\begin{aligned}\alpha^U &= \frac{1-\alpha^I}{N^*}\\ &= \frac{1-\frac{1}{3}}{3}\\ &= \frac{2}{9}.\end{aligned}$$

- Finally, the cumulative distribution function is

$$F(p) = 1 - \left(\frac{\pi_f(p)}{\pi_f(p) - \pi_s(p)}\right)^{\frac{1}{N-1}},$$

where profits from successfully attracting all customers are

$$\begin{aligned}\pi_s(p) &= (p-c)\left(\alpha^I + \alpha^U\right) - F\\ &= (p-0)\left(\frac{1}{3} + \frac{2}{9}\right) - \frac{2}{9}\\ &= \frac{5p-2}{9},\end{aligned}$$

and profits from only attracting uninformed consumers are

$$\begin{aligned}\pi_f(p) &= (p-c)\,\alpha^U - F\\ &= (p-0) \times \frac{2}{9} - \frac{2}{9}\\ &= -\frac{2\,(1-p)}{9}.\end{aligned}$$

Therefore, the above function $F\,(p)$ becomes

$$\begin{aligned}F\,(p) &= 1 - \left(\frac{2\,(1-p)}{5p-2-2\,(1-p)}\right)^{\frac{1}{2}}\\ &= 1 - \sqrt{\frac{2\,(1-p)}{7p-4}},\end{aligned}$$

which is distributed between the lower bound $p_L = \frac{2}{9}$ and the upper bound $r = 1$.
Differentiating $F(p)$ with respect to p, we find its probability density function:

$$f\,(p) = \frac{3\,(7p-4)^{-\frac{3}{2}}}{\sqrt{2\,(1-p)}},$$

which is positive so that firms randomize over the full support of the interval $\left[\frac{2}{9}, 1\right]$.

Exercise #10.9: Using Price and Advertising as Signals, Based on Milgrom (1986)^B

10.9 Consider a monopolist selling a good of either high or low quality. The production cost of the high-quality good is c_H, while that of the low-quality good is c_L, where $0 \leq c_L < c_H < 1$. For compactness, we normalize their costs, so that $c \equiv c_H - c_L$ denotes the cost differential. The monopolist can invest \$$a$ in advertising.

Every consumer buys at most one unit. When the good's price is p, the consumer's utility is $u_L = 1 - p$ when the good's quality is low and $u_H = (1 + \gamma x) - p$ when the good's quality is high, where x is uniformly distributed in $[0, 1]$. Parameter $\gamma > 0$ represents how much the consumer values the high-quality good, relative to the low-quality product, thus capturing vertical differentiation between qualities of the good.

(a) *Complete information.* If consumers can perfectly observe the good's quality (high or low), find the demand for each good's quality and equilibrium prices.

- *Low-quality good.* If the good is of low quality, the consumer purchases one unit if and only if $u_L = 1 - p \geq 0$, which holds only if price p is sufficiently low, $p \leq 1$. Otherwise, the consumer does not purchase from this firm. More compactly, demand is

$$D_L(p) = \begin{cases} 1 \text{ if } p \leq 1 \\ 0 \text{ otherwise.} \end{cases}$$

Therefore, the low-quality monopolist sets the highest possible price, $p_L^{CI} = 1$, to extract the most surplus from the consumers. As a consequence, sales in equilibrium are $D_L(p_L^{CI}) = D_L(1) = 1$ unit.

- *High-quality good.* If the good is of high quality, the consumer purchases one unit if and only if $u_H = (1 + \gamma x) - p \geq 0$ or, after solving for x, if $x \geq \frac{p-1}{\gamma}$. Intuitively, the consumer must care enough about this good's quality differential to buy one unit. Therefore, demand is given by all consumers to the right side of the indifferent consumer, $x = \frac{p-1}{\gamma}$, that is, $D_H(p) = 1 - \frac{p-1}{\gamma}$, as in similar models where consumers are uniformly distributed in the unit interval. This demand is, however, positive if and only if $\frac{1-p+\gamma}{\gamma} \geq 0$, or $p \leq 1 + \gamma$, and less than or equal to 1 (the entire unit interval, which is the upper bound on sales) if and only if $\frac{1-p+\gamma}{\gamma} \leq 1$, or $1 - p \leq 0$, which holds for all $p \geq 1$. We can then summarize the demand in this setting as follows:

$$D_H(p) = \begin{cases} \min\left\{\frac{1-p+\gamma}{\gamma}, 1\right\} \text{ if } p \leq 1 + \gamma \\ 0 \text{ otherwise.} \end{cases}$$

In this context, the high-quality monopolist sets a price that solves

$$\max_{p>0} D_H(p)(p - c) = \frac{(p - c)(1 - p + \gamma)}{\gamma}$$

which first-order condition in assuming an interior solution is

$$1 - p + \gamma - p + c = 0$$

that yields the equilibrium price under complete information of

$$p_H^{CI} = \frac{1+c+\gamma}{2}$$

if $\frac{1-p+\gamma}{\gamma} < 1$; but a price $p_H^{CI} = 1$ otherwise.

(b) *Incomplete information.* Assume now that consumers cannot observe the good's quality. However, they can observe its price, p, and the advertising investment, a, by the monopolist. Using the pair of signals (p, a), the consumer updates his beliefs about the firm's type and decides whether to purchase the good or not. Under which conditions can a separating equilibrium be supported where the high-quality firm chooses a price-advertising pair $\sigma_H = (p_H, a_H)$, while the low-quality firm selects $\sigma_L = (p_L^{CI}, 0)$, where p_L^{CI} denotes the complete information price found in part (a)? For simplicity, assume that the price that the high-quality firm chooses under complete information, p_H^{CI}, satisfies $p_H^{CI} > 1$.

- *Consumer beliefs.* In a separating strategy profile, upon observing σ_H, the consumer updates his beliefs, according to Bayes' rule, so that $\mu(H|\sigma_H) = 1$ and $\mu(L|\sigma_H) = 0$. Intuitively, upon observing σ_H, the consumer puts full probability weight on facing a high-quality firm. Similarly, upon observing σ_L, the consumer updates his beliefs, according to Bayes' rule, so that $\mu(H|\sigma_L) = 0$ and $\mu(L|\sigma_L) = 1$, implying that the consumer assigns full probability to facing a low-quality firm.
- *High-quality firm's decision.* This firm chooses $\sigma_H = (p, a)$, as prescribed by the strategy profile we are testing, instead of deviating to the low-type strategy, σ_L, if and only if

$$(p-c)D_H(p) - a \geq \left(p_L^{CI} - c\right) D_L(p_L^{CI})$$

where the left side indicates its profits from setting price p, which yields sales of $D_H(p)$, and investing a in advertising. The right side represents the profits from deviating to $\sigma_L = (p_L^{CI}, 0)$, where the firm still incurs the production cost c (as it still produces the high-quality good) but saves the advertising cost a. Since $p_L^{CI} = \$1$ from part (a), the equilibrium demand is $D_L(p_L^{CI}) = 1$ unit (also found in part a), recalling that $c < 1$ by definition. We can then rewrite the above incentive compatibility condition as follows:

$$(p-c)D_H(p) - a \geq 1 - c$$

or, after rearranging,

$$(p-c)D_H(p) - (1-c) \geq a.$$

Intuitively, when the profit gain from truthfully revealing its type is larger than the advertising cost, the high-quality firm chooses the price-advertising pair (p, a) that conveys its type to consumers.
- *Low-quality firm's decision.* This firm type chooses $\sigma_L = (p_L^{CI}, 0)$, as prescribed by this strategy profile, instead of deviating to σ_H if and only if

$$p_L^{CI} D_L(p_L^{CI}) \geq p D_H(p) - a$$

where the left side measures the profits from setting price p_L^{CI} and not investing in advertising, while the right side reflects its profits from deviating to $\sigma_H = (p, a)$, where the firm still does not incur any production costs (as it still produces a low-quality good) but invests a in advertising. Given that $p_L^{CI} = \$1$ and $D_L(p_L^{CI}) = 1$ unit (referring to part a), we can then rewrite this inequality as follows:

$$1 \geq pD_H(p) - a$$

or, after rearranging,

$$a \geq pD_H(p) - 1.$$

Intuitively, when advertising costs are sufficiently high, relative to the profit gain that the low-quality firm earns by pretending to be a high-quality firm, it has incentives to truthfully reveal its type, choosing $\sigma_L = (p_L^{CI}, 0)$, that is, the same price as under complete information, $p_L^{CI} = \$1$, and does not make any investments in advertising.

- Combining the above incentive compatibility conditions, we obtain that

$$(p - c)D_H(p) - (1 - c) \geq a \geq pD_H(p) - 1.$$

Therefore, (p, a)-pairs satisfying this double inequality can be supported as separating PBEs of this signaling game.

(c) *Equilibrium refinement.* Apply Cho and Kreps' (1987) Intuitive Criterion to rule out all pooling equilibria, and all but one separating equilibria. Interpret your results.

- Starting with (p, a) pairs that make the high-quality firm indifferent, that is, $(p - c)D_H(p) - (1 - c) = a$. Now, consider an off-the-equilibrium strategy profile $\sigma' = (p, a')$-pair where $a' = a - \varepsilon$ and $\varepsilon \to 0$ (marginal decrease in advertising but charging the same price).
- *First step of the Intuitive Criterion.* Upon observing the off-the-equilibrium strategy (p, a'), consumers update their (off-the-equilibrium beliefs) to $\mu\left(H|\sigma'\right) = 1$ and $\mu\left(L|\sigma'\right) = 0$, implying that they assign full probability weight to σ' originating from the high-quality firm. Indeed, condition $(p - c)D_H(p) - (1 - c) > a'$ now holds with strict inequality, and in addition, we can always find an $\varepsilon > 0$ such that for a' to be in the ε-neighborhood of a, where $a' = a - \varepsilon$, the condition $a' \geq pD_H(p) - 1$ still holds.
- *Second step of the Intuitive Criterion.* Given the above updated off-the-equilibrium beliefs, the high-quality firm has incentives to decrease its advertising, as that improves its equilibrium payoff at (p, a). In summary, the separating PBE that we initially considered, where the high-type firm chooses (p, a), violates the Intuitive Criterion. A similar argument applies to all separating PBEs where $a > 0$, so only separating PBEs where $(p, 0)$ survive this refinement criterion. In particular, p solves

$$(p - c)D_H(p) - (1 - c) \geq 0 \geq pD_H(p) - 1$$

since $a = 0$. This inequality yields two conditions on the price p:
– $(p - c)D_H(p) - (1 - c) \geq 0$ or, after solving for p, $p \geq \frac{1-c[1-D_H(p)]}{D_H(p)}$; and
– $0 \geq pD_H(p) - 1$, or $p \leq \frac{1}{D_H(p)}$.

- In summary, we identified an upper and a lower bound on the price p that the high-quality firm uses to convey its type to consumers. From part (a), we know that if $p \leq 1 + \gamma$,

$D_H(p) = 1 - \frac{p-1}{\gamma}$, which we can insert in the above conditions for p to obtain

$$p \geq \frac{1 - c\,[1 - D_H(p)]}{D_H(p)} = \frac{1 - c\left[1 - \left(1 - \frac{p-1}{\gamma}\right)\right]}{1 - \frac{p-1}{\gamma}} = \frac{c(1-p) + \gamma}{1 - p + \gamma}$$

which yields the equation $(p-1)(\gamma + c - p) \geq 0$, that holds if $p \geq 1$, which is satisfied by assumption, and $p \leq c + \gamma$.

Similarly, inserting $D_H(p) = 1 - \frac{p-1}{\gamma}$ into the upper bound for p found above, we find that

$$\frac{1}{D_H(p)} = \frac{1}{1 - \frac{p-1}{\gamma}} = \frac{\gamma}{1 - p + \gamma} \geq p$$

which yields the equation $(p-1)(p-\gamma) \geq 0$, that holds if $p \geq 1$, which is satisfied by assumption, and $p \geq \gamma$. Combining our results from both equations, we obtain that the high-quality firm's price, p, satisfies

$$c + \gamma \geq p \geq \gamma.$$

Finally, we confirm that the initial condition $p \leq 1+\gamma$ holds, because condition $p \leq c+\gamma$ is more stringent, that is, if $p \leq c + \gamma$ holds, then $p \leq 1 + \gamma$ must be satisfied too since $c < 1$ by definition.

- **Intuition:** The high-quality firm, then, signals its type by relying on prices alone, without the need to invest in advertising. In other words, despite having two possible tools to convey its type to consumers, the high-quality firm uses only one of them in equilibrium. Intuitively, this happens because both types of firms face the same unit cost of advertising (one dollar), implying that advertising can be mimicked by the low-quality firm more easily than prices. As a consequence, the high-quality firm uses the least costly signaling tool (prices) to convey its type to consumers.

Exercise #10.10: Social Pressure Jointly Produced by EPA and NGO, Based on Heyes and Oestreich (2018)^C

10.10 Consider an industry with a large number of heterogeneous firms, each deciding to comply with an environmental norm or not. Assume that the cost of compliance is $c \in [0, 1]$, which is uniformly distributed. Social pressure to comply comes from two sources: a government agency, such as EPA, and a nongovernment organization, NGO. The EPA puts equal weights on compliance, operating costs, and environmental damage. The EPA detects noncompliance with probability p when it inspects the firm, which is costly for the EPA with the following cost function $v(p) = \frac{1}{2}\alpha p^2$, where $\alpha \in (0, 1)$. The EPA's objective is to minimize the social loss function that includes the expected cost of compliance, social damage from noncompliance $D(\cdot)$, and the costs of EPA and NGO. Assume that $D(\cdot) = \frac{1}{2}\gamma(1 - p\theta)^2$ and $\gamma, \theta \in (0, 1)$.

The NGO assigns a weight λ on industry costs and $(1 - \lambda)$ on environmental quality, where $\lambda \in [0, 1/2)$, which means that the NGO assigns a weakly larger weight to environmental quality than the EPA does. The NGO relies on information from EPA to influence public opinion. Influencing public opinion is costly and described by $k(\theta) = \frac{1}{2}\beta\theta^2$, where $\beta \in (0, 1)$. The NGO's objective is to minimize its loss function that is comprised of NGO's cost, expected cost of compliance, and social damage.

Firms and regulators move sequentially, as follows: At *stage one*, firms decide whether to comply. At *stage two*, the EPA and NGO choose simultaneously the inspection probability, p, and the social atmosphere, θ.

(a) Write the EPA's social loss minimization problem and find its best response function, $p(\theta)$.

- *First stage.* Since the social loss function contains compliance cost, we have to find the proportion of firms that comply given the cost, c. For a firm of type c, the payoff from compliance, $a = 1$, is $\Pi(a = 1) = -c$, while the payoff from noncompliance and instead facing the expected social penalty in case of detection is $\Pi(a = 0) = -p\theta$. As a result, a firm complies if $-c \geq -p\theta$, or $c \leq p\theta$, which intuitively means that complying is less costly than the social cost of not complying.

  Given c is uniformly distributed on $[0, 1]$, the probability that a firm complies is

$$\Pr(c \leq p\theta) = F(p\theta) = p\theta.$$

  Hence, the fraction of firms that do not comply is $(1 - p\theta)$.

- *Second stage.* Now we can find the EPA's best response function by minimizing the social loss function W:

$$W = \min_p \; \underbrace{-\int_0^{p\theta} cf(c)dc}_{\text{Expected cost of compliance}} \; \underbrace{-\frac{1}{2}\gamma(1-p\theta)^2}_{\text{Social damage}} - \underbrace{\frac{1}{2}\beta\theta^2}_{\text{NGO's cost}} - \underbrace{\frac{1}{2}\alpha p^2}_{\text{EPA's cost}}$$

  Differentiating with respect to p yields

$$\frac{\partial W}{\partial p} = -p\theta^2 + \gamma\theta(1 - p\theta) - \alpha p = 0$$

  and solving for p, we obtain the EPA's best response function:

$$p(\theta) = \frac{\gamma\theta}{\alpha + \theta^2(1+\gamma)}$$

- *Numerical example.* Evaluating this best response function at $\gamma = 1$ and $\alpha = 1/2$, we find

$$p(\theta) = \frac{2\theta}{1 + 4\theta^2}.$$

(b) Find the value of θ that maximizes $p(\theta)$. Is the best response function $p(\theta)$ increasing or decreasing in θ? Interpret.

- Differentiating $p(\theta)$ with respect to θ, we obtain

$$\frac{\partial p(\theta)}{\partial \theta} = \frac{\gamma\left[\alpha - \theta^2(1+\gamma)\right]}{\left[\alpha + \theta^2(1+\gamma)\right]^2}.$$

- The value of θ that maximizes $p(\theta)$ is found by setting $\frac{\partial p(\theta)}{\partial \theta} = 0$, where

$$\frac{\gamma\left[\alpha - \theta^2(1+\gamma)\right]}{\left[\alpha + \theta^2(1+\gamma)\right]^2} = 0$$

yielding

$$\overline{\theta} = \sqrt{\frac{\alpha}{1+\gamma}}.$$

We further check that $\overline{\theta}$ maximizes $p(\theta)$ since

$$\frac{\partial^2 p(\theta)}{\partial \theta^2} = -\frac{2\gamma(1+\gamma)\theta\left[3\alpha - \theta^2(1+\gamma)\right]}{\left[\alpha + \theta^2(1+\gamma)\right]^3}$$

which, when evaluating at $\overline{\theta}$, yields

$$\left.\frac{\partial^2 p(\theta)}{\partial \theta^2}\right|_{\theta=\overline{\theta}} = -\frac{2\gamma(1+\gamma)\sqrt{\frac{\alpha}{1+\gamma}}\left[3\alpha - \frac{\alpha}{1+\gamma}(1+\gamma)\right]}{\left[\alpha + \frac{\alpha}{1+\gamma}(1+\gamma)\right]^3}$$
$$= \frac{-\gamma\sqrt{1+\gamma}}{2\alpha^{\frac{3}{2}}}$$

which is negative for all values of α and γ, so that $\overline{\theta}$ maximizes $p(\theta)$.

Depending on the values of α, γ, and θ, the best response function $p(\theta)$ can be either increasing or decreasing in θ:

– If $\theta < \overline{\theta}$, an increase in θ leads to an increase in p, implying that the EPA's best response function is positively sloped. In other words, the EPA regards the NGO's action, as captured by θ, as a strategic complement.

– In contrast, if $\theta \geq \overline{\theta}$, an increase in θ leads to a decrease in p. In this case, the EPA's best response function is negatively sloped, meaning that the EPA regards the probability of inspection p and the NGO's social atmosphere, θ, as a strategic substitute.

- *Numerical example.* Evaluating our above results at $\gamma = 1$ and $\alpha = 1/2$, the derivative $\frac{\partial p(\theta)}{\partial \theta}$ becomes

$$\frac{\partial p(\theta)}{\partial \theta} = \frac{2\left(1 - 4\theta^2\right)}{\left(1 + 4\theta^2\right)^2}$$

and cutoff $\overline{\theta}$ in this context is

$$\overline{\theta} = \sqrt{\frac{\frac{1}{2}}{1+1}} = \frac{1}{2}.$$

(c) Write the NGO's problem and find its best response function, $\theta(p)$.

- The NGO's objective is to minimize the loss function Z:

$$Z = \min_{\theta} \underbrace{-\lambda \int_0^{p\theta} cf(c)dc}_{\text{Expected cost of compliance}} \underbrace{-(1-\lambda)\frac{1}{2}\gamma(1-p\theta)^2}_{\text{Social damage}} - \underbrace{\frac{1}{2}\beta\theta^2}_{\text{NGO's cost}}$$

Differentiating with respect to θ, we obtain

$$\frac{\partial Z}{\partial \theta} = -p^2\lambda\theta + (1-\lambda)p\gamma(1-p\theta) - \beta\theta = 0$$

and solving for θ, we find the NGO's best response function:

$$\theta(p) = \frac{(1-\lambda)p\gamma}{\beta + p^2[\lambda + (1-\lambda)\gamma]}.$$

- *Numerical example.* Evaluating this best response function at parameter values $\gamma = 1$ and $\beta = 1/2$ yields

$$\theta(p) = \frac{2p(1-\lambda)}{1+2p^2}.$$

(d) Find the value of p that maximizes $\theta(p)$. Is the best response function $\theta(p)$ increasing or decreasing in p? Is it increasing or decreasing in λ? Interpret.

- Differentiating $\theta(p)$ with respect to p, we find

$$\frac{\partial\theta(p)}{\partial p} = \frac{(1-\lambda)\gamma\left[\beta - p^2(\lambda + (1-\lambda)\gamma)\right]}{\left[\beta + p^2(\lambda + (1-\lambda)\gamma)\right]^2}$$

- The value of p that maximizes $\theta(p)$ is found by setting $\frac{\partial\theta(p)}{\partial p} = 0$, that is,

$$\frac{(1-\lambda)\gamma\left[\beta - p^2(\lambda + (1-\lambda)\gamma)\right]}{\left[\beta + p^2(\lambda + (1-\lambda)\gamma)\right]^2} = 0$$

which, solving for p, yields

$$\overline{p} = \sqrt{\frac{\beta}{\lambda + (1-\lambda)\gamma}}.$$

We further check that $\overline{p}$ maximizes $\theta(p)$ since

$$\frac{\partial^2\theta(p)}{\partial p^2} = -\frac{2\gamma p(1-\lambda)(\lambda + (1-\lambda)\gamma)\left[3\beta - p^2(\lambda + (1-\lambda)\gamma)\right]}{\left[\beta + p^2(\lambda + (1-\lambda)\gamma)\right]^3}$$

which, when evaluating at $\overline{p}$, yields

$$\frac{\partial^2\theta(p)}{\partial p^2} = \Big|_{p=\overline{p}}$$

$$= -\frac{2\gamma\sqrt{\frac{\beta}{\lambda+(1-\lambda)\gamma}}\,(1-\lambda)\,(\lambda+(1-\lambda)\,\gamma)\left[3\beta-\frac{\beta}{\lambda+(1-\lambda)\gamma}\,(\lambda+(1-\lambda)\,\gamma)\right]}{\left[\beta+\frac{\beta}{\lambda+(1-\lambda)\gamma}\,(\lambda+(1-\lambda)\,\gamma)\right]^3}$$

$$= \frac{-\gamma\,(1-\lambda)\,\sqrt{\lambda+(1-\lambda)\,\gamma}}{2\beta^{\frac{3}{2}}}$$

which is negative for all values of β and γ, so that $\overline{p}$ maximizes $\theta\,(p)$.

In this context, we identify two segments in the NGO's best response function:

- If $p < \overline{p}$, an increase in p leads to an increase in $\theta\,(p)$, meaning that the NGO treats the EPA's inspection policy as a strategic complement, i.e., the NGO increases its social pressure θ when the EPA inspects more.
- In contrast, if $p \geq \overline{p}$, an increase in p leads to a decrease in $\theta\,(p)$, so the NGO treats the EPA's inspection policy as a strategic substitute, i.e., it reduces the social pressure θ when the EPA inspects more.

Finally, differentiating the best response function $\theta(p)$ with respect to λ, we obtain

$$\frac{\partial\theta\,(p)}{\partial\lambda} = \frac{-2p}{1+2p^2}$$

that is unambiguously negative. Intuitively, the more NGO cares about firm profits (higher λ), the lower the social penalty that it induces (lower θ).

(e) *Numerical example.* Evaluating our above results at parameter values $\gamma = 1$ and $\beta = 1/2$, the derivative $\frac{\partial\theta(p)}{\partial p}$ becomes

$$\frac{\partial\theta\,(p)}{\partial p} = \frac{2(1-\lambda)(1-2p^2)}{\left(1+2p^2\right)^2}$$

and, therefore, cutoff $\overline{p}$ is

$$\overline{p} = \sqrt{\frac{\frac{1}{2}}{\lambda+1-\lambda}} = \frac{\sqrt{2}}{2}.$$

(f) Find the equilibrium pair (p^*, θ^*) using parameter values $\gamma = 1$ and $\alpha = \beta = 1/2$.

- Inserting the EPA's best response function in this context, $p\,(\theta) = \frac{2\theta}{1+4\theta^2}$, into the NGO's best response function, $\theta\,(p) = \frac{2p(1-\lambda)}{1+2p^2}$, we find

$$\theta = \frac{2\overbrace{\frac{2\theta}{1+4\theta^2}}^{p(\theta)}(1-\lambda)}{1+2\underbrace{\left(\frac{2\theta}{1+4\theta^2}\right)^2}_{p(\theta)}}.$$

which is rearranged to yield

$$16\theta^4 + 16\lambda\theta^2 + (4\lambda - 3) = 0$$

Invoking the quadratic formula, the equilibrium social pressure solves

$$\theta^2 = \frac{-16\lambda + \sqrt{256\lambda^2 - 64\,(4\lambda - 3)}}{32}$$

which is rearranged to yield

$$\theta^* = \frac{\sqrt{\sqrt{4\lambda^2 - 4\lambda + 3} - 2\lambda}}{2}$$

Inserting θ^* into the EPA's best response function, we obtain the equilibrium inspection probability of the EPA, as follows:

$$p^* = \frac{\sqrt{\sqrt{4\lambda^2 - 4\lambda + 3} - 2\lambda}}{1 - 2\lambda + \sqrt{4\lambda^2 - 4\lambda + 3}}$$

(g) Assuming a weight $\lambda = 1/3$, find the equilibrium pair (p^*, θ^*). Do the players' best response functions intersect at a point where they regard their rival's strategies as strategic substitutes or complements?

- Inserting $\lambda = 1/3$ in the equilibrium social pressure θ^* in part (e), we find

$$\theta^* = \frac{\sqrt{\sqrt{4 \times \frac{1}{3^2} - 4 \times \frac{1}{3} + 3} - 2 \times \frac{1}{3}}}{2}$$

$$= \frac{\sqrt{\sqrt{19} - 2}}{2\sqrt{3}} \approx 0.443$$

and for the inspection probability, we have

$$p^* = \frac{\sqrt{\sqrt{4 \times \frac{1}{3^2} - 4 \times \frac{1}{3} + 3} - 2 \times \frac{1}{3}}}{1 - 2 \times \frac{1}{3} + \sqrt{4 \times \frac{1}{3^2} - 4 \times \frac{1}{3} + 3}}$$

$$= \frac{\sqrt{3\left(\sqrt{19} - 2\right)}}{\sqrt{19} + 1} \approx 0.496$$

- To understand the relationship between p and θ, let us compare them to cutoff points $\overline{p}$ and $\overline{\theta}$. Since $p^* < \overline{p} = \frac{\sqrt{2}}{2}$, and $\theta^* < \overline{\theta} = \frac{1}{2}$, players view their actions as strategic complements, as shown in parts (b) and (d).

(h) Assuming now that $\lambda = 0$, find the equilibrium pair (p^*, θ^*). Do players' best response functions intersect at a point where they regard their rival's strategies as strategic substitutes or complements?

- Inserting $\lambda = 0$ in the equilibrium social pressure θ^* in part (e), we find

$$\theta^* = \frac{\sqrt{\sqrt{4 \times 0^2 - 4 \times 0 + 3} - 2 \times 0}}{2}$$
$$= \frac{3^{\frac{1}{4}}}{2} \approx 0.658$$

and for the inspection probability, we find

$$p^* = \frac{\sqrt{\sqrt{4 \times 0^2 - 4 \times 0 + 3} - 2 \times 0}}{1 - 2 \times 0 + \sqrt{4 \times 0^2 - 4 \times 0 + 3}}$$
$$= \frac{3^{\frac{1}{4}}}{1 + \sqrt{3}} \approx 0.482$$

- Again, to understand the relationship between p and θ, let us compare them to cutoff points $\overline{p}$ and $\overline{\theta}$. Since $p^* < \overline{p} = \frac{\sqrt{2}}{2}$ and $\theta^* > \overline{\theta} = \frac{1}{2}$, EPA actions are strategic substitutes, while NGO actions are strategic complements.

Exercise #10.11: Quality Disclosure and Competition, Based on Levin, Peck and Ye (2009)^C

10.11 Consider a duopoly selling a differentiated product, where we allow for vertical and horizontal differentiation. Firm A is located at one end of the unit line (0) and firm B is located at the opposite end (1). Consumers are uniformly distributed in the unit interval, so if a consumer located at x purchases from firm A, its utility is $(2 + q_A) - x$, where q_A denotes firm A's quality, whereas if he buys from firm B, his utility becomes $(2 + q_B) - (1 - x)$, where q_B represents firm B's quality. For simplicity, marginal production costs are normalized to zero for both firms.

The time structure of the game is the following: in the first stage, every firm $i = \{A, B\}$ privately observes its quality, q_i, and decides whether or not to disclose it to consumers, at a cost $1/10$. (We assume that, if a firm discloses its quality, it cannot lie about it, and consumers perfectly observe q_i.) In the second stage, every firm i observes the disclosing strategy profile that happened in the first stage and responds setting a price for its product, p_i. In the third stage, consumers observe both the disclosing strategy profile from the first stage and the price pair from the second stage and respond buying one unit of the product from firm A, firm B, or none.

If a firm discloses its quality q_i, consumers are perfectly informed about this quality, but otherwise they must update their beliefs about the firm's quality. In other words, consumers form a "perceived quality" based on firm i's disclosure behavior, which we denote as $\widetilde{q}_i$. We are interested in identifying a symmetric PBE where firm i discloses its quality, q_i, if and only if this quality is sufficiently high. Otherwise, firm i does not disclose its quality.

(a) *Third stage.* Given the price pair firms select in the second stage, (p_A, p_B), and consumers' perceived qualities $(\widetilde{q}_A, \widetilde{q}_B)$, find the indifferent consumer x^* in the last stage of the game. Find every firm's equilibrium profits in this stage.

- The indifferent consumer x^* satisfies

$$(2+\widetilde{q}_A) - p_A - x^* = (2+\widetilde{q}_B) - p_B - (1-x^*)$$

which, solving for x^*, yields

$$x^* = \frac{1}{2} + \frac{(\widetilde{q}_A - \widetilde{q}_B) - (p_A - p_B)}{2}$$

which identifies firm A's demand that is increasing in its perceived quality differential, $\widetilde{q}_A - \widetilde{q}_B$, but decreasing in its price differential, $p_A - p_B$. Accordingly, firm B's demand is $1 - x^*$.

- Given prices (p_A, p_B), firm A's profits are (recall that production is costless)

$$\pi_A(p_A, p_B) = p_A x^* = p_A \left[\frac{1}{2} + \frac{(\widetilde{q}_A - \widetilde{q}_B) - (p_A - p_B)}{2}\right]$$

while firm B's are

$$\pi_B(p_A, p_B) = p_B(1-x^*) = p_B \left[\frac{1}{2} - \frac{(\widetilde{q}_A - \widetilde{q}_B) - (p_A - p_B)}{2}\right].$$

(b) *Second stage.* Anticipating the profits you identified in part (a), and for given perceived qualities $(\widetilde{q}_A, \widetilde{q}_B)$, find equilibrium prices p_A^* and p_B^*.

- Firm A sets its price p_A to solve

$$\max_{p_A \geq 0} \pi_A(p_A, p_B) = p_A \left[\frac{1}{2} + \frac{(\widetilde{q}_A - \widetilde{q}_B) - (p_A - p_B)}{2}\right]$$

Differentiating with respect to p_A yields

$$\frac{1}{2} + \frac{(\widetilde{q}_A - \widetilde{q}_B) - (p_A - p_B)}{2} - \frac{p_A}{2} = 0$$

and solving for p_A, we obtain its best response function:

$$p_A(p_B) = \frac{1 + \widetilde{q}_A - \widetilde{q}_B}{2} + \frac{p_B}{2}.$$

- Similarly, firm B sets its price p_B to solve

$$\max_{p_B \geq 0} \pi_B(p_A, p_B) = p_B \left[\frac{1}{2} - \frac{(\widetilde{q}_A - \widetilde{q}_B) - (p_A - p_B)}{2}\right]$$

Differentiating with respect to p_B yields

$$\frac{1}{2} - \frac{(\widetilde{q}_A - \widetilde{q}_B) - (p_A - p_B)}{2} - \frac{p_B}{2} = 0$$

and solving for p_B, we find its best response function:

$$p_B(p_A) = \frac{1 + \widetilde{q}_B - \widetilde{q}_A}{2} + \frac{p_A}{2}.$$

- Inserting best response function $p_B(p_A)$ into $p_A(p_B)$, we obtain

$$p_A = \frac{1 + \widetilde{q}_A - \widetilde{q}_B}{2} + \frac{1}{2}\left(\frac{1 + \widetilde{q}_B - \widetilde{q}_A}{2} + \frac{p_A}{2}\right)$$

and rearranging, we find firm A's equilibrium price:

$$p_A^* = 1 + \frac{\widetilde{q}_A - \widetilde{q}_B}{3}$$

Substituting this price into $p_B(p_A)$, we obtain firm B's equilibrium price:

$$\begin{aligned} p_B(p_A^*) &= \frac{1 + \widetilde{q}_B - \widetilde{q}_A}{2} + \frac{1}{2}\left(1 + \frac{\widetilde{q}_A - \widetilde{q}_B}{3}\right) \\ &= 1 + \frac{\widetilde{q}_B - \widetilde{q}_A}{3}. \end{aligned}$$

- Therefore, firm A's equilibrium profits become

$$\begin{aligned} \pi_A(\widetilde{q}_A, \widetilde{q}_B) &\equiv \pi_A(p_A^*, p_B^*) \\ &= \left(1 + \frac{\widetilde{q}_A - \widetilde{q}_B}{3}\right)\left[\frac{1}{2} + \frac{(\widetilde{q}_A - \widetilde{q}_B) - \left(\left(1 + \frac{\widetilde{q}_A - \widetilde{q}_B}{3}\right) - \left(1 + \frac{\widetilde{q}_B - \widetilde{q}_A}{3}\right)\right)}{2}\right] \\ &= \frac{(3 + \widetilde{q}_A - \widetilde{q}_B)^2}{18} \end{aligned}$$

which are only a function of perceived qualities $(\widetilde{q}_A, \widetilde{q}_B)$, and thus we denote as $\pi_A(\widetilde{q}_A, \widetilde{q}_B)$. Similarly, firm B's equilibrium profits become

$$\begin{aligned} \pi_B(\widetilde{q}_A, \widetilde{q}_B) &\equiv \pi_B(p_A^*, p_B^*) \\ &= \left(1 + \frac{\widetilde{q}_B - \widetilde{q}_A}{3}\right)\left[\frac{1}{2} - \frac{(\widetilde{q}_A - \widetilde{q}_B) - \left(\left(1 + \frac{\widetilde{q}_A - \widetilde{q}_B}{3}\right) - \left(1 + \frac{\widetilde{q}_B - \widetilde{q}_A}{3}\right)\right)}{2}\right] \\ &= \frac{(3 + \widetilde{q}_B - \widetilde{q}_A)^2}{18}. \end{aligned}$$

(c) *First stage.* Find a PBE where every firm i discloses its quality q_i if and only if $q_i > \overline{q}$.
- In this strategy profile, a firm does not disclose its quality q_i if it satisfies $q_i \leq \overline{q}$, which happens with probability $\overline{q}$ since qualities are uniformly distributed (probability of q_i being below cutoff $\overline{q}$). In contrast, a firm discloses its quality if $q_i > \overline{q}$, which occurs with probability $1 - \overline{q}$.

- *Disclosing*. Therefore, if firm A discloses its quality q_A, its expected profits are

$$E\pi_A(Disc) = \underbrace{\overline{q}\frac{(3+q_A-\frac{\overline{q}}{2})^2}{18}}_{\text{Firm } B \text{ does not disclose}} + \underbrace{(1-\overline{q})\int_{\overline{q}}^{1}\frac{(3+q_A-q)^2}{18}dq}_{\text{Firm } B \text{ discloses}} - \underbrace{\frac{1}{10}}_{\text{Disclosure cost}}$$

In the first term, firm B does not disclose, which entails that consumers assign a perceived quality $\widetilde{q}_B$ equal to the average non-disclosed quality (in the interval between 0 and $\overline{q}$), that is, $\widetilde{q}_B = E[q_B|q_B < \overline{q}] = \frac{\overline{q}-0}{2} = \frac{\overline{q}}{2}$. In the second term, firm B discloses its quality, but firm A does not know this quality before choosing whether to disclose its own quality or not, which explains why we need to take expectation over q_B in between $\overline{q}$ and 1. Finally, the last term represents firm A's cost of disclosing its quality, which is $1/10$ by definition.

- *Not disclosing*. If, instead, firm A does not disclose its quality, its expected profit becomes

$$E\pi_A(NoDisc) = \underbrace{\overline{q}\frac{1}{2}}_{\text{Firm } B \text{ does not disclose}} + \underbrace{(1-\overline{q})\int_{\overline{q}}^{1}\frac{\left(3+\frac{\overline{q}}{2}-q\right)^2}{18}dq}_{\text{Firm } B \text{ discloses}}.$$

In the first term, no firm discloses its quality, and thus both firms evenly share the market. In the second term, firm B discloses but A does not, which entails that consumers' perceived quality from firm A is $\widetilde{q}_A = E[q_A|q_A < \overline{q}] = \frac{\overline{q}-0}{2} = \frac{\overline{q}}{2}$ given that it chose to not disclose (so that its quality must lie in the interval between 0 and $\overline{q}$).

- For a given cutoff $\overline{q}$, note that $E\pi_A(Disc)$ is strictly increasing in quality q_A, while $E\pi_A(NoDisc)$ is not a function of q_A, which entails that $q_A = \overline{q}$ produces an interior solution, meaning that cutoff $\overline{q}$ satisfies $0 < \overline{q} < 1$. Inserting this property, $q_A = \overline{q}$, into $E\pi_A(Disc)$, yields

$$\begin{aligned} E\pi_A(Disc) &= \overline{q}\frac{(3+\overline{q}-\frac{\overline{q}}{2})^2}{18} + (1-\overline{q})\int_{\overline{q}}^{1}\frac{(3+\overline{q}-q)^2}{18}dq - \frac{1}{10} \\ &= \frac{\overline{q}(6+\overline{q})^2}{72} - \frac{(1-\overline{q})}{54}\left(19-4\overline{q}-2\overline{q}^2-\overline{q}^3\right) - \frac{1}{10} \\ &= \frac{272+80\overline{q}+220\overline{q}^2+35\overline{q}^3+20\overline{q}^4}{1080} \end{aligned}$$

while $E\pi_A(NoDisc)$ simplifies to

$$\begin{aligned} E\pi_A(NoDisc) &= \frac{\overline{q}}{2} + \frac{1-\overline{q}}{432}\left[(6-\overline{q})^3-(4+\overline{q})^3\right] \\ &= \frac{76-46\overline{q}+81\overline{q}^2-4\overline{q}^3+\overline{q}^4}{216} \end{aligned}$$

Setting them equal to each other, we obtain

$$272+80\overline{q}+220\overline{q}^2+35\overline{q}^3+20\overline{q}^4 = 380-230\overline{q}+405\overline{q}^2-20\overline{q}^3+5\overline{q}^4$$

Rearranging, we find that this expression simplifies to

$$15\overline{q}^4 + 55\overline{q}^3 - 185\overline{q}^2 + 310\overline{q} - 108 = 0$$

Finally, solving for cutoff $\overline{q}$ yields $\overline{q} \simeq 0.452$ (since $0 \leq \overline{q} \leq 1$), so that every firm i discloses its quality q_i if $q_i > 0.452$, but otherwise it does not.

Exercise #10.12: Behavior-Based Price Discrimination and Customer Information Sharing, Based on De Nijs (2017)^C

10.12 Consider a model with two periods of consumption and repeated purchase. There is mass one of identical customers willing to buy at most one product each period. A customer has a valuation $v = 4$ if the product matches her preferences, and zero otherwise. In each period, customers buy the lowest-price product that matches their preferences.

There are three firms, $i = 1, 2, 3$, in the market that compete on prices. Every firm has zero marginal cost and players do not discount future payoffs. The good that a customer purchases matches his preferences with probability $\theta = 1/2$, giving rise to three types of customers: (1) "captive" customers, who only consider buying from one firm, which occur with probability $\theta(1-\theta)^2 = 0.125$; (2) "local" shoppers, who consider buying only from two of the three firms, which happen with probability $\theta^2(1-\theta) = 0.125$; and (3) "global" shoppers, who consider buying from all three firms, with probability $\theta^3 = 0.125$.

The time structure of the game is the following. In period 1, firms set uniform prices; in period 2, when information sharing is not allowed, firms can engage in behavior-based price discrimination (BBPD), where they can charge a different price to their new and past customers. When information sharing is allowed, each firm decides simultaneously whether and at which prices to sell its past customers' list. Also, each firm decides which databases it purchases from its rivals. Depending on what type of information is shared, a firm can price-discriminate in the second period.

(a) *Uniform pricing.* Assume that BBPD is not allowed. Find a symmetric equilibrium in which firms randomize their prices in each period.

- Let $F_u(p)$ be the cumulative distribution function (CDF) that a firm's price p is the lowest among all competing firms, where $p \in [(1-\theta)^2 v, v]$ and the subscript u denotes uniform pricing. In this context, the probability that the firm's product matches its customers' preferences is θ, that a competing firm does not match those preferences is $(1-\theta)$, and that the competing firm matches those preferences but charges a price higher than p is $\theta(1 - F_u(p))$.
- Since firms are symmetric and every firm faces two rivals, first-period profits become

$$\Pi_u^{1st}(p) = p\theta\left[(1-\theta) + \theta(1 - F_u(p))\right]^2$$

- In the second period, every firm can charge its captive customers their full valuation v with a probability mass of $\theta(1-\theta)^2$. This happens whether the firm's first-period price fell below all its rivals with probability $F_u^2(p)$, beat only one of its rivals with probability $2F_u(p)[1 - F_u(p)]$, or charged the highest price among all firms with probability $[1 - F_u(p)]^2$. In this context, second-period profits become

$$\begin{aligned}\Pi_u^{2nd}(p) &= \theta(1-\theta)^2 v\left[F_u^2(p) + 2F_u(p)(1 - F_u(p)) + (1 - F_u(p))^2\right] \\ &= \theta(1-\theta)^2 v\end{aligned}$$

- Whereas, when customers are charged at their valuation, $p = v$, total profits across both periods become

$$\begin{aligned}\Pi_u^{tot}(v) &= \Pi_u^{1st}(v) + \Pi_u^{2nd}(v)\\ &= 2\theta(1-\theta)^2 v\end{aligned}$$

- Therefore, to support randomization strategy, every firm must be indifferent between all prices in the support $p \in [(1-\theta)^2 v, v]$, entailing that

$$\underbrace{p\theta[(1-\theta)+\theta(1-F_u(p))]^2}_{\Pi_u^{1st}(p)} + \underbrace{\theta(1-\theta)^2 v}_{\Pi_u^{2nd}(p)} = \underbrace{2\theta(1-\theta)^2 v}_{\Pi_u^{tot}(v)}$$

Simplifying the above expression, we obtain that

$$(1-\theta)+\theta\,(1-F_u(p)) = (1-\theta)\sqrt{\frac{v}{p}}.$$

Rearranging and solving for $F_u\,(p)$ yields

$$F_u(p) = 1 - \frac{1-\theta}{\theta}\left(\sqrt{\frac{v}{p}} - 1\right)$$

- Finally, substituting $\theta = \frac{1}{2}$ and $v = 4$, there exists a symmetric equilibrium where, in each period, firms randomize their prices according to the CDF found above:

$$\begin{aligned}F_u(p) &= 1 - \frac{1-\frac{1}{2}}{\frac{1}{2}}\left(\sqrt{\frac{4}{p}} - 1\right)\\ &= 1 - \left(\sqrt{\frac{4}{p}} - 1\right)\end{aligned}$$

with support $\left[4\left(1-\frac{1}{2}\right)^2, 4\right] = [1, 4]$ that satisfies the regularity conditions, that is, $F_u\,(1) = 1 - \left(\sqrt{\frac{4}{1}} - 1\right) = 0$ at the lower bound of the randomization interval and $F_u\,(4) = 1 - \left(\sqrt{\frac{4}{4}} - 1\right) = 1$ at its upper bound.

- In this setting, every firm earns total expected profits of

$$\begin{aligned}\Pi_u^{tot} &= 2 \times \frac{1}{2}\left(1 - \frac{1}{2}\right)^2 \times 4\\ &= 1.\end{aligned}$$

(b) *Second period, no data sharing*. Let us first consider the case in which data sharing is not allowed. In this context, each firm can charge a different price to its new and past customers (often referred to as behavior-based price discrimination, BBPD). Find equilibrium prices and profits in the second period for the small, medium, and large firm. Which firm earns the largest second-period profit? Explain.

- Let $F_i^j(p)$ be the cumulative distribution function characterizing the price randomization strategy of firm i in market j in the second period, where $i \in \{s, m, l\}$ represents the small, medium, and large firm, while $j \in \{o, n\}$ denotes the market of old and new customers, respectively, and $\Pi_i^j(p)$ be the associated expected profits.
- When BBPD is allowed without data sharing, the small firm earns a profit of

$$\Pi_s^o(p) = p\theta(1-\theta)^2$$

from its captive customers, with a density mass of $\theta(1-\theta)^2$ by charging a price p.

Also, the small firm wins local customers over from the medium (large) firm, each with a density mass of $\theta^2(1-\theta)$, when its price falls below the respective competitor with a probability of $1 - F_m^o(p)$ ($1 - F_l^{o,n}(p)$, respectively), where superscript o, n indicates both old and new customers. In addition, the small firm attracts global customers, with a density mass of θ^3, when its price falls below both the medium and the large firm with a joint probability of $\left(1 - F_m^n(p)\right)\left(1 - F_l^{o,n}(p)\right)$. This allows us to write the small firm's profits from new customers, as follows:

$$\Pi_s^n(p) = p\theta^2(1-\theta)\left[\left(1 - F_m^o(p)\right) + \left(1 - F_l^{o,n}(p)\right)\right] + p\theta^3\left(1 - F_m^n(p)\right)\left(1 - F_l^{o,n}(p)\right)$$

Summing up, the small firm's total profits in the second period becomes

$$\begin{aligned}\Pi_s(p) &= \Pi_s^o(p) + \Pi_s^n(p) \\ &= p\Big[\theta(1-\theta)^2 + \theta^2(1-\theta)\left(2 - F_m^o(p) - F_l^{o,n}(p)\right) \\ &\quad + \theta^3\left(1 - F_m^n(p)\right)\left(1 - F_l^{o,n}(p)\right)\Big]\end{aligned}$$

- Similarly, the medium firm, by charging a price p, earns a profit of

$$\Pi_m^o(p) = p\theta(1-\theta)^2 + p\theta^2(1-\theta)\left[1 - F_s^n(p)\right]$$

where this firm serves its captive customers with a density mass of $\theta(1-\theta)^2$ and its local customers with a density mass of $\theta^2(1-\theta)$ when its price falls below that of the small firm with a probability of $1 - F_s^n(p)$.

Also, the medium firm can win new customers from the local market, with a density mass of $\theta^2(1-\theta)$, when its price falls below that of the large firm with a probability of $1 - F_l^{o,n}(p)$. In addition, it attracts customers from the global market, with a density mass of θ^3, when its price falls below both the small and the large firm with a joint probability of $\left(1 - F_s^n(p)\right)\left(1 - F_l^{o,n}(p)\right)$:

$$\Pi_m^n(p) = p\theta^2(1-\theta)\left[1 - F_l^{o,n}(p)\right] + p\theta^3\left(1 - F_s^n(p)\right)\left(1 - F_l^{o,n}(p)\right)$$

Summing up, the medium firm's total profits in the second period becomes

$$\begin{aligned}\Pi_m(p) &= \Pi_m^o(p) + \Pi_m^n(p) \\ &= p\Big[\theta(1-\theta)^2 + \theta^2(1-\theta)\left(2 - F_s^n(p) - F_l^{o,n}(p)\right) \\ &\quad + \theta^3\left(1 - F_s^n(p)\right)\left(1 - F_l^{o,n}(p)\right)\Big]\end{aligned}$$

- Lastly, the large firm, by charging a price p, earns total profits of

$$\begin{aligned}\Pi_l^{o,n}(p) = p\Big[&\theta(1-\theta)^2 + \theta^2(1-\theta)\left[2 - F_s^n(p) - F_m^n(p)\right] \\ &+ \theta^3\left(1 - F_s^n(p)\right)\left(1 - F_m^n(p)\right)\Big]\end{aligned}$$

where this firm serves its captive customers with a density mass of $\theta(1-\theta)^2$ and its local customers, each with a density mass of $\theta^2(1-\theta)$, when its price falls below that of the small (medium) firm with a probability of $1 - F_s^n(p)$ $(1 - F_m^n(p)$, respectively). Finally, it attracts customers from the global market, with a density mass of θ^3, when its price falls below both the small and the medium firm with a joint probability of $\left(1 - F_s^n(p)\right)\left(1 - F_m^n(p)\right)$.
- To support the medium firm's price randomization strategy to its old customers, $p \in \left[\underline{p}, v\right]$, where $\underline{p} = (1-\theta)v$ represents the reservation price of unreached customers with a density mass of $1-\theta$, we need $\Pi_m^o(p) = \Pi_m^o(v)$, solving

$$p\left[\theta(1-\theta)^2 + \theta^2(1-\theta)\left(1 - F_s^n(p)\right)\right] = v\left[\theta(1-\theta)^2 + \theta^2(1-\theta)\left(1 - F_s^n(v)\right)\right]$$

Substituting $F_s^n(v) = 1$ into the above expression, we obtain that

$$(1-\theta) + \theta\left(1 - F_s^n(p)\right) = \frac{v(1-\theta)}{p}$$

Rearranging and solving for $F_s^n(p)$ yields

$$F_s^n(p) = 1 - \frac{1-\theta}{\theta}\left(\frac{v}{p} - 1\right),$$

and it is straightforward to show that $F_s^n\left(\underline{p}\right) = 0$ and $F_s^n(v) = 1$.
- To support the large firm's price randomization strategy for all customers, $F_l^{o,n}(p)$ for $p \in [\underline{\underline{p}}, v]$, where $\underline{\underline{p}} = (1-\theta)^2 v$ represents the reservation price of unreached customers with a density mass of $(1-\theta)^2$, we need $\Pi_l^{o,n}(p) = \Pi_l^{o,n}(v)$. Inserting $F_s^n(p) = 0$ for all $p \in [\underline{\underline{p}}, \underline{p}]$ and $F_m^n(v) = F_s^n(v) = 1$, we find

$$p\left[\theta(1-\theta)^2 + \theta^2(1-\theta)\left(1 - F_m^n(p)\right) + \theta^2(1-\theta) + \theta^3\left(1 - F_m^n(p)\right)\right] = v\theta(1-\theta)^2$$

which can be rearranged into

$$(1-\theta)(1-\theta+\theta)+\theta(1-\theta+\theta)\left[1-F_m^n(p)\right]=\frac{v(1-\theta)^2}{p}$$

Simplifying, we obtain

$$1-F_m^n(p)=\frac{v(1-\theta)^2}{\theta p}-\frac{1-\theta}{\theta}$$

Solving for $F_m^n(p)$ yields

$$F_m^n(p)=1-\frac{1-\theta}{\theta}\left[\frac{(1-\theta)v}{p}-1\right]$$

and it is straightforward to show that $F_m^n(\underline{\underline{p}})=0$ and $F_m^n\left(\underline{p}\right)=1$.

- To support the medium firm's price randomization strategy to its new customers, $F_m^n(p)$ for $p\in[\underline{\underline{p}},\underline{p}]$, we need $\Pi_m^n(p)=\Pi_m^n(\underline{\underline{p}})$, solving

$$p\left[\theta^2(1-\theta)+\theta^3\right](1-F_l^{o,n}(p))=\underbrace{(1-\theta)^2 v}_{\underline{\underline{p}}}\left[\theta^2(1-\theta)+\theta^3\right]\left(1-F_l^{o,n}(\underline{\underline{p}})\right)$$

Simplifying the above expression, we obtain

$$p\left(1-F_l^{o,n}(p)\right)=(1-\theta)^2 v$$

Further rearranging yields

$$F_l^{o,n}(p)=1-\frac{(1-\theta)^2 v}{p},$$

and it is straightforward to show that $F_l^{o,n}(\underline{\underline{p}})=0$ and $F_l^{o,n}\left(\underline{p}\right)=\theta$, so that the large firm charges its customers at the reservation price v with a probability of $1-\theta$.

- To support the small firm's price randomization strategy to its new customers, $F_s^n(p)$ for $p\in\left[\underline{p},v\right]$, we need $\Pi_s^n(p)=\Pi_s^n\left(\underline{p}\right)$, solving

$$p\theta^2(1-\theta)\left[1-F_m^o(p)\right]=\theta^2(1-\theta)\left[1-F_m^o\left(\underline{p}\right)\right]\underline{p}$$

Substituting $F_m^o\left(\underline{p}\right)=0$ into the above expression yields

$$1-F_m^o(p)=\frac{(1-\theta)v}{p}$$

Rearranging and solving for $F_m^o(p)$, we obtain

$$F_m^o(p)=1-\frac{(1-\theta)v}{p},$$

and it is straightforward to show that $F_m^o\left(\underline{p}\right) = 0$ and $F_m^o(v) = \theta$, so the medium firm charges its old customers at the reservation price v with a probability of $1 - \theta$.

- Since the large firm randomizes between $[\underline{\underline{p}}, v]$, it must be the case that $\Pi_l^{o,n}(p) = \Pi_l^{o,n}(v)$. Substituting $F_s^n(v) = F_m^n(v) = 1$ into its profit function yields

$$\Pi_l^{2nd}(p) = v\left[\theta(1-\theta)^2 + \theta^2(1-\theta)[2-1-1] + \theta^3(1-1)(1-1)\right]$$
$$= \theta(1-\theta)^2 v$$

- Since the medium firm randomizes between $\left[\underline{p}, v\right]$ for its old and $[\underline{\underline{p}}, \underline{p}]$ for its new customers, we substitute $F_s^n(\underline{\underline{p}}) = F_l^{o,n}(\underline{\underline{p}}) = 0$ into its profit function to obtain

$$\Pi_m^{2nd}(p) = \Pi_m^o(v) + \Pi_m^n(\underline{\underline{p}})$$
$$= \theta(1-\theta)^2 v + \theta^2(1-\theta)\left[1 - F_s^n(v)\right] v$$
$$+ \underline{\underline{p}}\theta^2(1-\theta)\left[1 - F_l^{o,n}(\underline{\underline{p}})\right] + \underline{\underline{p}}\theta^3\left(1 - F_s^n(\underline{\underline{p}})\right)\left(1 - F_l^{o,n}\left(\underline{\underline{p}}\right)\right)$$
$$= \theta(1-\theta)^2 v + (1-\theta)^2\left[\theta^2(1-\theta) + \theta^3\right] v$$
$$= \theta(1-\theta)^2 v + \theta^2(1-\theta)^2 v$$
$$= \theta(1+\theta)(1-\theta)^2 v$$

- Since the small firm charges a price v for its old and randomizes between $\left[\underline{p}, v\right]$ for its new customers, substituting $F_m^o\left(\underline{p}\right) = 0$, $F_m^n\left(\underline{p}\right) = 1$, and $F_l^{o,n}\left(\underline{p}\right) = \theta$ into this firm's profit function, we obtain

$$\Pi_s^{2nd}(p)$$
$$= \Pi_s^o(v) + \Pi_s^n\left(\underline{p}\right)$$
$$= \theta(1-\theta)^2 v + \underline{p}\theta^2(1-\theta)\left[2 - F_m^o\left(\underline{p}\right) - F_l^{o,n}\left(\underline{p}\right)\right]$$
$$+ \underline{p}\theta^3\left(1 - F_m^n\left(\underline{p}\right)\right)\left(1 - F_l^{o,n}\left(\underline{p}\right)\right)$$
$$= \theta(1-\theta)^2 v + \theta^2(1-\theta)^2(2-\theta) v$$
$$= \theta\left(1 + 2\theta - \theta^2\right)(1-\theta)^2 v$$

- We next show that profit ranking satisfies $\Pi_s^{2nd}(p) > \Pi_m^{2nd}(p) > \Pi_l^{2nd}(p)$ where

$$\Pi_s^{2nd}(p) = \theta\left(1 + 2\theta - \theta^2\right)(1-\theta)^2 v > \theta(1-\theta)^2(1+\theta) v = \Pi_m^{2nd}(p)$$

reduces to

$$\theta(1-\theta) > 0$$

that holds, since the small firm charges its old customers the highest price v. Also,

$$\Pi_m^{2nd}(p) = \theta(1-\theta)^2(1+\theta)v > \theta(1-\theta)^2 v = \Pi_l^{2nd}(p)$$

simplifies to

$$1+\theta > 1$$

that holds, since the large firm faces the fiercest competition in the second period.

- Substituting $\theta = \frac{1}{2}$ and $v = 4$ into the above distribution functions, we obtain cutoffs $\underline{\underline{p}} = 4\left(1-\frac{1}{2}\right)^2 = 1$ and $\underline{p} = 4\left(1-\frac{1}{2}\right) = 2$, yielding the following pricing decisions:
  - The small firm charges the monopoly price $v = 4$ to its old customers, and a randomized price to its new customers in the interval [2, 4] with

$$F_s^n(p) = 1 - \frac{1-\frac{1}{2}}{\frac{1}{2}}\left(\frac{4}{p}-1\right) = 1 - \left(\frac{4}{p}-1\right)$$

    that satisfies the regularity conditions $F_s^n(2) = 0$ and $F_s^n(4) = 1$.
  - The medium firm never charges its old customers a price below $\underline{p} = 2$ since deviating to any $p < \underline{p}$ does not improve its profits. In this context, the medium firm charges a randomize price to its old customers in the interval [2, 4] with

$$F_m^o(p) = 1 - \frac{\left(1-\frac{1}{2}\right)\times 4}{p} = 1 - \frac{2}{p}$$

    that satisfies the conditions $F_m^o(2) = 0$ and $F_m^o(4) = \frac{1}{2}$, since there is a discrete mass of customers, $1-\theta = \frac{1}{2}$, buying from the medium firm at the price $v = 4$.
  - The medium firm charges a randomized price to its new customers in the interval [1, 2] with

$$F_m^n(p) = 1 - \frac{1-\frac{1}{2}}{\frac{1}{2}}\left(\frac{\left(1-\frac{1}{2}\right)\times 4}{p}-1\right) = 1 - \left(\frac{2}{p}-1\right)$$

    that satisfies the regularity conditions $F_m^n(1) = 0$ and $F_m^n(2) = 1$.
  - The large firm never charges its customers a price below $\underline{\underline{p}} = 1$ since deviating to any $p < \underline{\underline{p}}$ does not improve its profits. In this context, the large firm charges a uniform

price to all its customers randomized in the interval $[1, 4]$ with

$$F_l^{o,n}(p) = 1 - \frac{\left(1-\frac{1}{2}\right)^2 \times 4}{p} = 1 - \frac{1}{p}$$

that satisfies the conditions $F_l^{o,n}(1) = 0$ and $F_l^{o,n}(4) = \frac{3}{4}$, since there is a discrete mass of customers, $(1-\theta)^2 = \frac{1}{4}$, buying from this firm at a price of 4.

- Similarly, substituting $\theta = \frac{1}{2}$ and $v = 4$ into the above profit functions, we find
  - for the small firm, equilibrium second-period profits become

$$\Pi_s^{2nd} = \frac{1}{2}\left(1 + 2 \times \frac{1}{2} - \frac{1}{2^2}\right)\left(1 - \frac{1}{2}\right)^2 \times 4 = \frac{7}{8},$$

  - and for the medium firm, equilibrium second-period profits are

$$\Pi_m^{2nd} = \frac{1}{2}\left(1 - \frac{1}{2}\right)^2\left(1 + \frac{1}{2}\right) \times 4 = \frac{3}{4},$$

  - and for the large firm, equilibrium second-period profits are

$$\Pi_l^{2nd} = \frac{1}{2}\left(1 - \frac{1}{2}\right)^2 \times 4 = \frac{1}{2},$$

thus satisfying the ranking $\Pi_s^{2nd} > \Pi_m^{2nd} > \Pi_l^{2nd}$, as required.

(c) *First period, no data sharing.* Still in a context of no data sharing, find the equilibrium price that each firm sets in the first period. Evaluate the total profits that each firm earns across both periods, assuming no payoff discounting.

- Since firms are indifferent between all prices in the support $F_{pd}(\cdot)$ for all $p \in [\underline{p}_{pd}, v]$, the profit function of every firm across both periods is given by

$$\Pi_{pd}^{tot}(p) = \underbrace{p\theta\left[(1-\theta) + \theta\left(1 - F_{pd}(p)\right)\right]^2}_{\Pi_{pd}^{1st}(p)} + \left[1 - F_{pd}(p)\right]^2 \underbrace{\theta(1-\theta)^2 v}_{\Pi_l^{2nd}(p)}$$
$$+ 2\left[1 - F_{pd}(p)\right] F_{pd}(p) \underbrace{\theta(1+\theta)(1-\theta)^2 v}_{\Pi_m^{2nd}(p)}$$

$$+ \underbrace{F_{pd}^2(p)\,\theta\left(1+2\theta-\theta^2\right)(1-\theta)^2 v}_{\Pi_s^{2nd}(p)}$$

where subscript pd denotes price discrimination.

Evaluating at $p = \underline{p}_{pd}$, we have $F_{pd}(\underline{p}_{pd}) = 0$, yielding

$$\Pi_{pd}^{tot}\left(\underline{p}_{pd}\right) = \underline{p}_{pd}\theta + \theta(1-\theta)^2 v$$

Evaluating at $p = v$, we have $F_{pd}(v) = 1$, yielding

$$\begin{aligned}\Pi_{pd}^{tot}(v) &= \theta(1-\theta)^2 v + \theta\left(1+2\theta-\theta^2\right)(1-\theta)^2 v \\ &= \theta\left(2+2\theta-\theta^2\right)(1-\theta)^2 v\end{aligned}$$

where $2\theta(1-\theta)^2 v$ are the guaranteed profits from uniform pricing, as in part (a), and $\theta(1-\theta)^2\left(2-\theta^2\right)v$ are the additional profits from practicing BBPD.

Setting $\Pi_{pd}^{tot}(\underline{p}_{pd}) = \Pi_{pd}^{tot}(v)$, we obtain

$$\underline{p}_{pd}\theta + \theta(1-\theta)^2 v = \theta\left(2+2\theta-\theta^2\right)(1-\theta)^2 v$$

Rearranging and solving for $\underline{p}_{pd}$, the lower bound of first-period price becomes

$$\underline{p}_{pd} = \left(1+2\theta-\theta^2\right)(1-\theta)^2 v$$

Next, setting $\Pi_{pd}^{tot}(p) = \Pi_{pd}^{tot}(v)$, and using the fact that $F_{pd}(v) = 1$, we find

$$\begin{aligned}&p\theta\left[(1-\theta)+\theta\left(1-F_{pd}(p)\right)\right]^2 + \left[1-F_{pd}(p)\right]^2\theta(1-\theta)^2 v \\ &+ 2\left[1-F_{pd}(p)\right]F_{pd}(p)\,\theta(1+\theta)(1-\theta)^2 v + F_{pd}^2(p)\,\theta\left(1+2\theta-\theta^2\right)(1-\theta)^2 v \\ &= \theta\left(2+2\theta-\theta^2\right)(1-\theta)^2 v\end{aligned}$$

Simplifying the above expression, we obtain

$$\begin{aligned}&p\left[1-\theta F_{pd}(p)\right]^2 - (1-\theta)^2 v\left[1-\theta F_{pd}(p)\right]^2 + 2(1-\theta)^2 v \\ &= \left(2+2\theta-\theta^2\right)(1-\theta)^2 v\end{aligned}$$

which is equivalent to

$$\left[p-(1-\theta)^2 v\right]\left[1-\theta F_{pd}(p)\right]^2 = \theta(2-\theta)(1-\theta)^2 v$$

Rearranging yields

$$1-\theta F_{pd}(p)=\sqrt{\frac{\theta(2-\theta)(1-\theta)^2 v}{p-(1-\theta)^2 v}}$$

Solving for $F_{pd}(p)$, we obtain

$$F_{pd}(p)=\frac{1}{\theta}-\frac{1-\theta}{\theta}\sqrt{\frac{\theta(2-\theta)v}{p-(1-\theta)^2 v}}$$

Specifically, when $p=\underline{p}_{pd}$, we have

$$\begin{aligned}F_{pd}(\underline{p}_{pd})&=\frac{1}{\theta}-\frac{1-\theta}{\theta}\sqrt{\frac{\theta(2-\theta)v}{(1+2\theta-\theta^2)(1-\theta)^2 v-(1-\theta)^2 v}}\\&=\frac{1}{\theta}-\frac{1-\theta}{\theta(1-\theta)}\sqrt{\frac{\theta(2-\theta)v}{\theta(2-\theta)v}}\\&=\frac{1}{\theta}-\frac{1}{\theta}\\&=0\end{aligned}$$

and when $p=v$, we find

$$\begin{aligned}F_{pd}(v)&=\frac{1}{\theta}-\frac{1-\theta}{\theta}\sqrt{\frac{\theta(2-\theta)v}{v-(1-\theta)^2 v}}\\&=\frac{1}{\theta}-\frac{1-\theta}{\theta}\sqrt{\frac{\theta(2-\theta)v}{\theta(2-\theta)v}}\\&=\frac{1}{\theta}-\frac{1-\theta}{\theta}\\&=1\end{aligned}$$

that satisfy the regularity conditions.

- Substituting $\theta=\frac{1}{2}$ and $v=4$ into the above distribution function, we obtain cutoff $\underline{p}_{pd}=\left(1-\frac{1}{2}\right)^2\left(1+2\times\frac{1}{2}-\frac{1}{2^2}\right)\times 4=1\frac{3}{4}$, so that every firm randomizes its price in the first period according to the following cumulative distribution function:

$$\begin{aligned}F_{pd}(p)&=\frac{1}{\frac{1}{2}}-\frac{1-\frac{1}{2}}{\frac{1}{2}}\sqrt{\frac{\frac{1}{2}\left(2-\frac{1}{2}\right)\times 4}{p-\left(1-\frac{1}{2}\right)^2\times 4}}\\&=2-\left(\frac{3}{p-1}\right)^{\frac{1}{2}}\end{aligned}$$

on the support $[\underline{p}_{pd}, v]=\left[1\frac{3}{4},4\right]$, where $F_{pd}\left(1\frac{3}{4}\right)=0$ and $F_{pd}(4)=1$, as required.

In this setting, every firm earns expected total profits across both periods, as follows:

$$\Pi_{pd}^{tot} = \frac{1}{2}\left(2 + 2 \times \frac{1}{2} - \frac{1}{2^2}\right)\left(1 - \frac{1}{2}\right)^2 \times 4$$
$$= 1\frac{3}{8}$$

(d) *Second period, data sharing*. Let us now consider a setting where firms can share data about their customers in the first period. Find the second-period equilibrium prices and profits in this context. [*Hint*: Note that if the small firm learns the identities of its rivals' past customers, it can charge three different prices (one to its own past customers, one to those of firm 2, and one to those of firm 3. However, the medium firm has no incentives to acquire information from the small (large) firm's past customers because these customers do not value its product (because this information would not allow the medium firm to more finely price-discriminate its customers). Similarly, the large firm does not have any incentives to acquire information about the past customers from any of its rivals because these customers do not value its product. Overall, only the small firm would have incentives to acquire information about its rivals' past customers, and the medium and large firms compete in selling their customers' list at a price of zero.]

- Let $F_i^j(p)$ be the cumulative distribution function characterizing the price randomization strategy of firm i in market j in the second period, $\Pi_i^j(p)$ be its associated expected profits, and p_i^j be the resulting equilibrium price.
- The small firm's expected profits from the medium firm's past customers are

$$\Pi_s^m(p) = p\theta^2(1-\theta)\left[1 - F_m^m(p)\right],$$

while the medium firm's expected profits from its past customers become

$$\Pi_m^m(p) = p\left[\theta(1-\theta)^2 + \theta^2(1-\theta)\left(1 - F_s^m(p)\right)\right]$$

- Since both small and medium firms do not price below $\underline{p} \equiv (1-\theta)v$, we obtain that $p_s^m, p_m^m \in \left[\underline{p}, v\right]$. Setting $\Pi_s^m(p) = \Pi_s^m\left(\underline{p}\right)$ yields

$$p\theta^2(1-\theta)\left[1 - F_m^m(p)\right] = \underbrace{(1-\theta)v}_{\underline{p}} \times \theta^2(1-\theta)\left[1 - F_m^m\left(\underline{p}\right)\right]$$

Since $F_m^m\left(\underline{p}\right) = 0$, the above expression becomes

$$1 - F_m^m(p) = \frac{(1-\theta)v}{p}$$

Solving for $F_m^m(p)$, we obtain

$$F_m^m(p) = 1 - \frac{(1-\theta)v}{p}$$

and it is straightforward to show that $F_m^m\left(\underline{p}\right) = 0$ and $F_m^m(v) = \theta$, so the medium firm charges its customers at the reservation price v with a probability of $1 - \theta$.

- Similarly, we set $\Pi_m^m(p) = \Pi_m^m\left(\underline{p}\right)$ to obtain

$$p\left[\theta(1-\theta)^2 + \theta^2(1-\theta)\left(1 - F_s^m(p)\right)\right]$$
$$= \underline{p}\left[\theta(1-\theta)^2 + \theta^2(1-\theta)\left(1 - F_s^m\left(\underline{p}\right)\right)\right]$$

Since $F_s^m\left(\underline{p}\right) = 0$, the above expression becomes

$$p\left[(1-\theta) + \theta\left(1 - F_s^m(p)\right)\right] = (1-\theta)\left[(1-\theta) + \theta\right]v$$

which can be rearranged into

$$1 - F_s^m(p) = \frac{(1-\theta)v}{\theta p} - \frac{1-\theta}{\theta}$$

Solving for $F_m^m(p)$, we obtain

$$F_s^m(p) = 1 - \frac{1-\theta}{\theta}\left(\frac{v}{p} - 1\right)$$

and it is straightforward to show that $F_s^m\left(\underline{p}\right) = 0$ and $F_s^m(v) = 1$.

- The small firm's expected profits from the large firm's past customers are

$$\Pi_s^l(p) = p\left[\theta^2(1-\theta)\left(1 - F_l^l(p)\right) + \theta^3\left(1 - F_m^l(p)\right)\left(1 - F_l^l(p)\right)\right],$$

and the medium firm's expected profits from the large firm's past customers become

$$\Pi_m^l(p) = p\left[\theta^2(1-\theta)\left(1 - F_l^l(p)\right) + \theta^3\left(1 - F_s^l(p)\right)\left(1 - F_l^l(p)\right)\right]$$

while the large firm's expected profits from its own past customers entail that

$$\Pi_l^l(p) = p\left[\theta(1-\theta)^2 + \theta^2(1-\theta)\left(1 - F_s^l(p)\right)\right.$$
$$\left. + \theta^2(1-\theta)\left(1 - F_m^l(p)\right) + \theta^3\left(1 - F_s^l(p)\right)\left(1 - F_m^l(p)\right)\right]$$

- Since the small and medium firms are in a symmetric position facing competition of the large firm, we have $F_s^l(p) = F_m^l(p)$, which yields $\Pi_s^l(p) = \Pi_m^l(p)$. Besides, since the large firm does not price below $\underline{\underline{p}} \equiv (1-\theta)^2 v$, we obtain that $p_s^l, p_m^l, p_l^l \in [\underline{\underline{p}}, v]$. Setting $\Pi_l^l(p) = \Pi_l^l(\underline{\underline{p}})$, we find that

$$p\left[\theta(1-\theta)^2 + \theta^2(1-\theta)\left[2 - F_s^l(p) - F_m^l(p)\right] + \theta^3(1 - F_s^l(p))(1 - F_m^l(p))\right]$$

$$=\underline{\underline{p}}\left[\theta\,(1-\theta)^2+\theta^2\,(1-\theta)\,[2-F_s^l(\underline{\underline{p}})-F_m^l(\underline{\underline{p}})]+\theta^3(1-F_s^l(\underline{\underline{p}}))(1-F_m^l(\underline{\underline{p}}))\right]$$

Since $F_s^l(\underline{\underline{p}})=F_m^l(\underline{\underline{p}})=F_l^l(\underline{\underline{p}})=0$, the above expression becomes

$$(1-\theta)^2+2\theta\,(1-\theta)\left(1-F_s^l\,(p)\right)+\theta^2\left(1-F_s^l\,(p)\right)^2=\frac{(1-\theta)^2\,v}{p}$$

which can be more compactly expressed as

$$\left[(1-\theta)+\theta\left(1-F_s^l\,(p)\right)\right]^2=\frac{(1-\theta)^2\,v}{p}$$

Simplifying the above expression, we have

$$1-F_s^l\,(p)=\frac{1-\theta}{\theta}\sqrt{\frac{v}{p}}-\frac{1-\theta}{\theta}$$

Rearranging and solving for $F_s^l\,(p)$ yields

$$F_s^l\,(p)=F_m^l\,(p)=1-\frac{1-\theta}{\theta}\left(\sqrt{\frac{v}{p}}-1\right)$$

and it is straightforward to show that $F_s^l(\underline{\underline{p}})=F_m^l(\underline{\underline{p}})=0$ and $F_s^l\,(v)=F_m^l\,(v)=1$.

- Similarly, we set $\Pi_m^l\,(p)=\Pi_m^l(\underline{\underline{p}})$ to obtain

$$p\left[\theta^2\,(1-\theta)\left(1-F_l^l\,(p)\right)+\theta^3\left(1-F_s^l\,(p)\right)\left(1-F_l^l\,(p)\right)\right]$$
$$=\underline{\underline{p}}\left[\theta^2\,(1-\theta)\left(1-F_l^l(\underline{\underline{p}})\right)+\theta^3\left(1-F_s^l(\underline{\underline{p}})\right)\left(1-F_l^l(\underline{\underline{p}})\right)\right]$$

Substituting $F_s^l\,(p)$ and since $F_s^l(\underline{\underline{p}})=F_l^l(\underline{\underline{p}})=0$, the above expression becomes

$$\left(1-F_l^l\,(p)\right)+\left(\sqrt{\frac{v}{p}}-1\right)\left(1-F_l^l\,(p)\right)=\frac{(1-\theta)\,v}{p}$$

Simplifying the above expression, we have

$$1-F_l^l\,(p)=\frac{(1-\theta)\,v}{p}\sqrt{\frac{p}{v}}$$

Rearranging and solving for $F_l^l\,(p)$ yields

$$F_l^l\,(p)=1-(1-\theta)\sqrt{\frac{v}{p}}$$

and it is straightforward to show that $F_l^l(\underline{\underline{p}})=0$ and $F_l^l\,(v)=\theta$, so the large firm charges its customers at the reservation price v with a probability of $1-\theta$.

- Since the large firm randomizes between $[\underline{\underline{p}}, v]$, its profits become

$$\begin{aligned}\Pi_l^l(p) &= \Pi_l^l(v) \\ &= v[\theta(1-\theta)^2 + \theta^2(1-\theta)\left(1 - F_s^l(v)\right) \\ &\quad + \theta^2(1-\theta)(1 - \underbrace{1}_{F_m^l(v)}) + \theta^3(1 - \underbrace{1}_{F_s^l(v)})(1 - \underbrace{1}_{F_m^l(v)})] \\ &= \theta(1-\theta)^2 v\end{aligned}$$

- Since the medium firm randomizes between $\left[\underline{p}, v\right]$ in its own market and $[\underline{\underline{p}}, v]$ in the market previously served by the large firm, its profits become

$$\begin{aligned}\Pi_m^m(p) + \Pi_m^l(p) &= \Pi_m^m(v) + \Pi_m^l(v) \\ &= v[\theta(1-\theta)^2 + \theta^2(1-\theta)(1 - \underbrace{1}_{F_s^m(v)})] \\ &\quad + v[\theta^2(1-\theta)(1 - \underbrace{\theta}_{F_l^l(v)}) + \theta^3(1 - \underbrace{1}_{F_s^l(v)})(1 - \underbrace{\theta}_{F_l^l(v)})] \\ &= \theta(1+\theta)(1-\theta)^2 v\end{aligned}$$

- Since the small firm charges a price v for its past customers and randomizes between $\left[\underline{p}, v\right]$ ($[\underline{\underline{p}}, v]$) in the medium (large) firm's market, its profits become

$$\begin{aligned}&\Pi_s^s(p) + \Pi_s^m(p) + \Pi_s^l(p) \\ &= \Pi_s^s(v) + \Pi_s^m(v) + \Pi_s^l(v) \\ &= \theta(1-\theta)^2 v + v\theta^2(1-\theta)[1 - \underbrace{\theta}_{F_m^m(v)}] + v[\theta^2(1-\theta)(1 - \underbrace{\theta}_{F_l^l(v)}) \\ &\quad + \theta^3(1 - \underbrace{1}_{F_m^l(v)})(1 - \underbrace{\theta}_{F_l^l(v)})] \\ &= \theta(1+2\theta)(1-\theta)^2 v\end{aligned}$$

- Substituting $\theta = \frac{1}{2}$ and $v = 4$ into the above distribution functions, we find that
  - the small firm charges a monopoly price v to its past customers, and a randomized price to the customers of the medium firm in the interval [2, 4] with

$$\begin{aligned}F_s^m(p) &= 1 - \frac{1-\frac{1}{2}}{\frac{1}{2}}\left(\frac{4}{p} - 1\right) \\ &= 1 - \left(\frac{4}{p} - 1\right)\end{aligned}$$

  that satisfies the regularity conditions $F_s^m(2) = 0$ and $F_s^m(4) = 1$.

– the medium firm never charges its past customers a price below $\underline{p} = 2$ since deviating to any $p < \underline{p}$ does not improve its profits. In this context, the medium firm charges a randomize price to its past customers in the interval [2, 4] with

$$F_m^m(p) = 1 - \frac{\left(1 - \frac{1}{2}\right) \times 4}{p}$$
$$= 1 - \frac{2}{p}$$

that satisfies the conditions $F_m^m(2) = 0$ and $F_m^m(4) = \frac{1}{2}$, since there is a discrete mass of customers, $1 - \theta = \frac{1}{2}$, buying from the medium firm at the price $v = 4$.

– both the small and the medium firm charge a randomized price to the customers of the large firm in the interval [1, 4] with

$$F_s^l(p) = F_m^l(p) = 1 - \frac{1 - \frac{1}{2}}{\frac{1}{2}}\left(\sqrt{\frac{4}{p}} - 1\right)$$
$$= 1 - \left(\sqrt{\frac{4}{p}} - 1\right)$$

that satisfies the regularity conditions $F_s^l(1) = F_m^l(1) = 0$ and $F_s^l(4) = F_m^l(4) = 1$.

– the large firm never charges its customers a price below $\underline{\underline{p}} = 1$ since deviating to any $p < \underline{\underline{p}}$ does not improve its profits. In this context, the large firm charges a uniform price to all its customers randomized in the interval [1, 4] with

$$F_l^l(p) = 1 - \left(1 - \frac{1}{2}\right)\sqrt{\frac{4}{p}}$$
$$= 1 - \frac{1}{2}\sqrt{\frac{4}{p}}$$

that satisfies the conditions $F_l^l(1) = 0$ and $F_l^l(4) = \frac{1}{2}$, since there is a discrete mass of customers, $(1 - \theta) = \frac{1}{2}$, buying from the large firm at a price of 4.

- Similarly, substituting $\theta = \frac{1}{2}$ and $v = 4$ into the above profit functions, we find that
  – for the small firm, second-period profits under information sharing become

$$\Pi_{s,is}^{2nd} = \frac{1}{2}\left(1 + 2 \times \frac{1}{2}\right)\left(1 - \frac{1}{2}\right)^2 \times 4$$
$$= 1,$$

where subscript *is* denotes information sharing,

– and for the medium firm, second-period profits under information sharing are

$$\Pi^{2nd}_{m,is} = \frac{1}{2}\left(1+\frac{1}{2}\right)\left(1-\frac{1}{2}\right)^2 \times 4$$
$$= \frac{3}{4},$$

– and for the large firm, second-period profits under information sharing are

$$\Pi^{2nd}_{l,is} = \frac{1}{2}\left(1-\frac{1}{2}\right)^2 \times 4$$
$$= \frac{1}{2},$$

thus satisfying the ranking $\Pi^{2nd}_{s,is} > \Pi^{2nd}_{m,is} > \Pi^{2nd}_{l,is}$, as required.

(e) *First period, data sharing*. Anticipating the expected profits found in part (c), find the equilibrium prices that each firm sets in the first stage. Then, identify the equilibrium profits across both periods, assuming no payoff discounting.

- Since firms are indifferent between all prices in the support $F_{pd,is}(\cdot)$ for all $p \in [\underline{p}_{pd,is}, v]$, the profit function of every firm across both periods is given by

$$\Pi^{tot}_{pd,is}(p) = \underbrace{p\theta\left[(1-\theta)+\theta\left(1-F_{pd,is}(p)\right)\right]^2}_{\Pi^{1st}_{pd,is}(p)} + \left[1-F_{pd,is}(p)\right]^2 \underbrace{\theta(1-\theta)^2 v}_{\Pi^{2nd}_{l,is}(p)}$$
$$+2\left[1-F_{pd,is}(p)\right]F_{pd,is}(p)\underbrace{\theta(1+\theta)(1-\theta)^2 v}_{\Pi^{2nd}_{m,is}(p)}$$
$$+F^2_{pd,is}(p)\underbrace{\theta(1+2\theta)(1-\theta)^2 v}_{\Pi^{2nd}_{s,is}(p)}$$

where subscript pd, is denotes price discrimination under information sharing.

Evaluating at $p = \underline{p}_{pd,is}$, we have $F_{pd,is}(\underline{p}_{pd,is}) = 0$, yielding

$$\Pi^{tot}_{pd,is}(\underline{p}_{pd,is}) = \underline{p}_{pd,is}\theta + \theta(1-\theta)^2 v$$

Evaluating at $p = v$, we have $F_{pd,is}(v) = 1$, yielding

$$\Pi^{tot}_{pd,is}(v) = \theta(1-\theta)^2 v + \theta(1+2\theta)(1-\theta)^2 v$$
$$= 2\theta(1+\theta)(1-\theta)^2 v$$

where $2\theta(1-\theta)^2 v$ are the guaranteed profits from uniform pricing, as in part (a), and $2\theta^2(1-\theta)^2 v$ are the additional profits from practicing BBPD with information sharing.

Setting $\Pi^{tot}_{pd,is}(\underline{p}_{pd,is}) = \Pi^{tot}_{pd,is}(v)$, we obtain

$$\underline{p}_{pd,is}\theta + \theta(1-\theta)^2 v = 2\theta(1+\theta)(1-\theta)^2 v$$

Rearranging and solving for $\underline{p}_{pd,is}$, the lower bound of first-period price becomes

$$\underline{p}_{pd,is} = (1+2\theta)(1-\theta)^2 v$$

Next, setting $\Pi^{tot}_{pd,is}(p) = \Pi^{tot}_{pd,is}(v)$, and using the fact that $F_{pd,is}(v) = 1$, we find

$$\begin{aligned}
&p\theta\left[(1-\theta)+\theta\left(1-F_{pd,is}(p)\right)\right]^2 + \left[1-F_{pd,is}(p)\right]^2\theta(1-\theta)^2 v \\
&+2\left[1-F_{pd,is}(p)\right]F_{pd,is}(p)\,\theta(1+\theta)(1-\theta)^2 v + F^2_{pd,is}(p)\,\theta(1+2\theta)(1-\theta)^2 v \\
&= 2\theta(1+\theta)(1-\theta)^2 v
\end{aligned}$$

Simplifying the above expression, we obtain

$$p\left[1-\theta F_{pd,is}(p)\right]^2 + (1-\theta)^2 v\left[1+2\theta F_{pd,is}(p)\right] = 2(1+\theta)(1-\theta)^2 v$$

which is rearranged as

$$\theta^2 p F^2_{pd,is}(p) - 2\theta F_{pd,is}(p)\left[p-(1-\theta)^2 v\right] + \left[p-(1+2\theta)(1-\theta)^2 v\right] = 0$$

Solving for $F_{pd}(p)$, we obtain

$$F_{pd,is}(p) = \frac{p-(1-\theta)^2 v-(1-\theta)\sqrt{\left[(1-\theta)^2 v-(1-2\theta)\,p\right]v}}{\theta p}$$

Specifically, when $p = \underline{p}_{pd,is}$, we have

$$\begin{aligned}
F_{pd,is}(\underline{p}_{pd,is}) &= \frac{\underline{p}_{pd,is}-(1-\theta)^2 v-(1-\theta)\sqrt{\left[(1-\theta)^2 v-(1-2\theta)\,\underline{p}_{pd,is}\right]v}}{\theta\,\underline{p}_{pd,is}} \\
&= \frac{2\theta(1-\theta)^2 v - 2\theta(1-\theta)^2 v}{\theta(1+2\theta)(1-\theta)^2 v} \\
&= 0
\end{aligned}$$

and when $p = v$, we find

$$\begin{aligned}
F_{pd,is}(v) &= \frac{v-(1-\theta)^2 v-(1-\theta)\sqrt{\left[(1-\theta)^2 v-(1-2\theta)\,v\right]v}}{\theta v} \\
&= \frac{v-(1-\theta)^2 v-\theta(1-\theta)\,v}{\theta v} \\
&= 1
\end{aligned}$$

that satisfy the regularity conditions.

- Substituting $\theta = \frac{1}{2}$ and $v = 4$ into the above distribution function, we obtain cutoff $\underline{p}_{pd,is} = \left(1 + 2 \times \frac{1}{2}\right)\left(1 - \frac{1}{2}\right)^2 \times 4 = 2$, so that every firm randomizes its price in the first period according to the following cumulative distribution function:

$$F_{pd,is}(p) = \frac{p - \left(1 - \frac{1}{2}\right)^2 \times 4 - \left(1 - \frac{1}{2}\right)\sqrt{4\left[\left(1 - \frac{1}{2}\right)^2 \times 4 - \left(1 - 2 \times \frac{1}{2}\right) p\right]}}{\frac{1}{2}p}$$

$$\frac{2(p-2)}{p}$$

on the support $[\underline{p}_{pd}, v] = [2, 4]$, where $F_{pd,is}(2) = 0$ and $F_{pd,is}(4) = 1$, as required.

In this setting, every firm earns expected total profits across both periods, as follows:

$$\begin{aligned}\Pi^{tot}_{pd,is} &= 2 \times \frac{1}{2}\left(1 + \frac{1}{2}\right)\left(1 - \frac{1}{2}\right)^2 \times 4 \\ &= 1\frac{1}{2}\end{aligned}$$

(f) *Comparison.* Compare the total profits of every firm under uniform pricing (part a), under BBPD without information sharing (part c), and with information sharing (part e). Interpret your results.

- We demonstrate that the ranking of total profits satisfies

$$\Pi^{tot}_{pd,is}(p) \geq \Pi^{tot}_{pd}(p) \geq \Pi^{tot}_{u}(p)$$

where the part $\Pi^{tot}_{pd,is}(p) \geq \Pi^{tot}_{pd}(p)$ entails

$$\Pi^{tot}_{pd,is}(p) = 2\theta(1+\theta)(1-\theta)^2 v \geq \theta\left(2 + 2\theta - \theta^2\right)(1-\theta)^2 v = \Pi^{tot}_{pd}(p)$$

which simplifies to

$$2 + 2\theta \geq 2 + 2\theta - \theta^2$$

and is equivalent to

$$\theta^2 \geq 0$$

that holds for all $\theta \in [0, 1]$.

The second part, $\Pi^{tot}_{pd}(p) \geq \Pi^{tot}_{u}(p)$, implies

$$\Pi^{tot}_{pd}(p) = \theta\left(2 + 2\theta - \theta^2\right)(1-\theta)^2 v \geq 2\theta(1-\theta)^2 v = \Pi^{tot}_{u}(p)$$

which simplifies to

$$2 + 2\theta - \theta^2 \geq 2$$

and is equivalent to

$$\theta\,(2-\theta) \geq 0$$

that holds for all $\theta \in [0, 1]$.

- Intuitively, firms have incentives to share proprietary information about consumers' preferences, since every firm has symmetric probabilities to be the small firm on an ex ante basis, and can obtain higher profits from price-discriminating the consumers previously purchased from the medium firm against those previously brought from the large firm in the second period. Even when information sharing is not allowed, firms still profit from price-discriminating their customers.

Networks and Switching Costs

11

Introduction

This chapter studies the strategic interaction of firms in the presence of network effects, where the adoption of the same technology or standard by several firms allows each firm to benefit from larger market demand, lower production cost, or both. This can happen even when the technology that firms adopt is inferior to other technologies, such as the Blue-ray disk, often regarded as inferior to high-definition DVDs.

Exercise 11.1 examines a simultaneous-move game in which firms coordinate or miscoordinate the adoption of technologies in the presence of network effects. We show that excess momentum (inertia) may arise when firms coordinate on the superior (inferior) technology even though the inferior (superior) technology is socially preferred. Exercise 11.2 follows the same setting as in Exercise 11.1 but considers a sequential-move game where the leader adopts a technology and the follower, after observing the leader's choice, responds with its own technology adoption. Interestingly, equilibrium outcomes coincide in both the simultaneous and sequential version of the game. Exercise 11.3 studies firms' intertemporal pricing strategies when consumers face switching costs. We demonstrate that firms can reduce their first-period price to lock-in consumers and charge the full price in the second period, particularly when switching costs are higher than the price premium that consumers pay for the product.

Exercise 11.4 analyzes the welfare effects of entry in the presence of switching costs. We find that, when entry costs are relatively high, entry may still be profitable, but it decreases welfare because the switching costs that consumers suffer more than offset their gains in lower market prices. Exercise 11.5 investigates buyer coordination in procurement auctions. While buyers can all buy from the entrant at a lower price, they may continue to buy from the incumbent when the deviation of some buyers to the entrant does not generate enough demand to induce entry. Exercise 11.6 models the strategic interaction between an upstream manufacturer and two downstream retailers. As consumers' transportation costs to travel between the retailers increase, firms can set a higher price and offer lower quality of services, as expected. Surprisingly, the wholesaler sets the highest possible price to make consumers indifferent between purchasing the product from either retailer.

P.-S. Choi et al., *Industrial Organization*, Springer Texts in Business and Economics,
https://doi.org/10.1007/978-3-031-38635-0_11

Exercise #11.1: Network Effects with Simultaneous Moves, Based on Farrell and Saloner (1985)[B]

11.1 Consider an industry with two firms, 1 and 2, and two technologies, A and B. The profit that every firm earns from technology $i = \{A, B\}$ is

$$\pi_i = b_i + n_i k \text{ if } k \text{ firms adopt technology } i$$

where b_i denotes the baseline profit that the firm earns from technology i and n_i measures network effects. When $n_i = 0$, the number of firms adopting technology i does not affect the profits from adopting it, while when $n_i > 0$ technology i becomes more attractive as more firms adopt it. Profits are symmetric across firms. We assume that $\beta \equiv b_B - b_A > 0$, which indicates that, in the absence of network effects, technology B is regarded as superior by both firms.

Every firm simultaneously and independently chooses between technology A and B.

(a) Represent the game in matrix form. Find the best responses for each firm.

- *Matrix representation.* Representing the above game with network effects, we obtain the following payoff matrix where firm 1 chooses technologies in rows and firm 2 in columns.

		Firm 2	
		Tech. A	Tech. B
Firm 1	Tech. A	$b_A + 2n_A, b_A + 2n_A$	$b_A + n_A, b_B + n_B$
	Tech. B	$b_B + n_B, b_A + n_A$	$b_B + 2n_B, b_B + 2n_B$

For instance, when both firms adopt technology $i = \{A, B\}$, each firm earns a profit $b_i + 2n_i$ since $k = 2$ firms adopt the same technology. However, when firm 1 adopts i while firm 2 adopts $j \neq i$, firm 1 earns a payoff $b_i + n_i$ and firm 2 earns $b_j + n_j$ since only one firm adopts each technology.

- *Best responses.* Since firms' payoff is symmetric, we focus on firm 1's best responses. Firm 2's are analogous.
  - If firm 2 adopts technology A (in the left column), firm 1 responds with technology A if and only if $b_A + 2n_A > b_B + n_B$, which simplifies to

$$2n_A - n_B > b_B - b_A \equiv \beta.$$

  - If firm 2 adopts technology B (in the right column), firm 1 responds with technology A if and only if $b_A + n_A > b_B + 2n_B$, which simplifies to

$$n_A - 2n_B > b_B - b_A \equiv \beta.$$

  Finally, note that cutoff $2n_A - n_B$ satisfies

$$2n_A - n_B > n_A - 2n_B,$$

since this inequality simplifies to $n_A > -n_B$, which holds by assumption given that $n_i > 0$ for every technology $i = \{A, B\}$.

(b) Find all Nash equilibria in this game. Interpret.

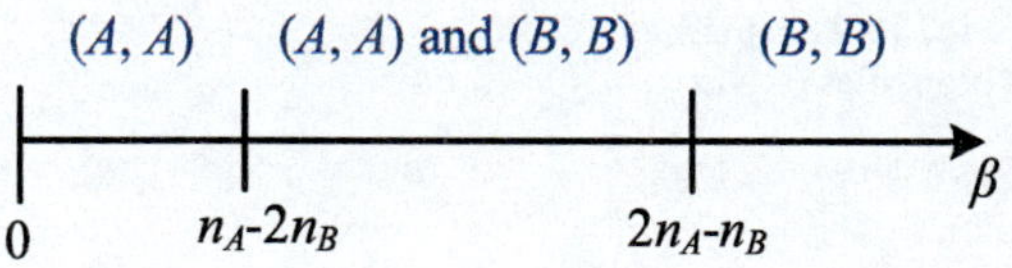

Fig. 11.1 Nash equilibria in the network effects game

- The above best responses generate three pure strategy Nash equilibria, depending on the precise value of β relative to cutoffs $2n_A - n_B$ and $n_A - 2n_B$, as depicted in Fig. 11.1:
  - If $\beta < n_A - 2n_B < 2n_A - n_B$, every firm responds with A to every technology of its opponent, i.e., adopting technology A is a strictly dominant strategy for every firm. Therefore, strategy profile (A, A) is the unique pure strategy Nash equilibrium of the game. Intuitively, the technological advantage of B relative to A is so low (low β) that both firms adopt A.
  - If $n_A - 2n_B \leq \beta < 2n_A - n_B$, every firm responds with A when its opponent chooses A, but with B when its opponent selects B. In summary, firms coordinate choosing the same technology in equilibrium, yielding strategy profiles (A, A) and (B, B) as pure strategy Nash equilibria of the game. In this case, the game resembles a Pareto coordination game with two Nash equilibria, but where (B, B) yielding a higher payoff for every firm than (A, A) does.
  - If $n_A - 2n_B < 2n_A - n_B \leq \beta$, every firm responds with B to every technology of its opponent, implying that technology B is strictly dominant for both firms. Therefore, strategy profile (B, B) is the unique pure strategy Nash equilibrium of the game. Intuitively, technology B is so superior to A (high β) that both firms adopt it.

(c) *Social optimum.* Under which parameter conditions is technology B Pareto optimal?

- For technology i to be Pareto optimal, we need that joint profits under this technology are larger than with technology j. That is, if both firms adopt technology i, joint profits are higher than when both firms adopt technology j if $\pi_i + \pi_i > \pi_j + \pi_j$, which entails

$$(b_i + 2n_i) + (b_i + 2n_i) > \left(b_j + 2n_j\right) + \left(b_j + 2n_j\right)$$

which simplifies to $b_i - b_j > 2(n_j - n_i)$. In the case of $i = B$, this condition can be expressed as

$$\beta \equiv b_B - b_A > 2(n_A - n_B).$$

(d) *Inefficient equilibria?* Are the Nash equilibria you found in part (b) Pareto optimal?

- Comparing cutoff $\beta > 2(n_A - n_B)$ against the two cutoffs that we found in part (b), $2n_A - n_B$ and $n_A - 2n_B$, we find that
  - $2(n_A - n_B) < 2n_A - n_B$, since this inequality simplifies to $2n_B > n_B$.
  - $2(n_A - n_B) > n_A - 2n_B$, since this inequality simplifies to $n_A > 0$.

  Therefore, cutoff $2(n_A - n_B)$ lies between cutoffs $n_A - 2n_B$ and $2n_A - n_B$, so that

$$n_A - 2n_B < 2(n_A - n_B) < 2n_A - n_B.$$

- We are now ready to evaluate the Pareto properties of each Nash equilibrium found in part (b) depicted, as a reference, in Fig. 11.2:
  - *Region I*: When β satisfies $\beta < n_A - 2n_B < 2n_A - n_B$, strategy profile (A, A) is sustained in equilibrium. In this setting, we have that $\beta < 2(n_A - n_B)$, which means that (A, A) is socially optimal.

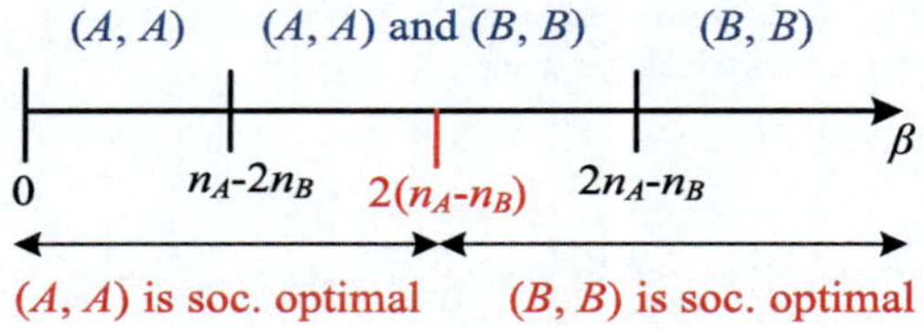

Fig. 11.2 Efficiency properties of NEs

– *Region II*: When β satisfies $n_A - 2n_B \leq \beta < 2n_A - n_B$, strategy profiles (A, A) and (B, B) emerge in equilibrium. In this setting, if $\beta < 2(n_A - n_B)$, strategy profile (A, A) Pareto dominates, while otherwise strategy profile (B, B) Pareto dominates.

  Farrell and Saloner (1985) say that, in the former case, there is "excess momentum" with firms adopting the superior technology B even if A is socially preferred. In the latter case, Farrell and Saloner (1985) say that there is "excess inertia" with firms not adopting technology B despite being socially preferred.

– *Region III*: When β satisfies $n_A - 2n_B < 2n_A - n_B \leq \beta$, strategy profile (B, B) arises in equilibrium. Since in this context, $\beta > 2(n_A - n_B)$ holds, the Nash equilibrium (B, B) is socially optimal.

(e) *Symmetric network effects.* Assume that firms experience symmetric network effects from both technologies, $n_A = n_B = n$. How are your equilibrium results affected?

- Cutoff $2n_A - n_B$ in this context becomes $2n - n = n$, while cutoff $n_A - 2n_B$ simplifies to $n - 2n = -n$, thus being not binding (as $\beta > 0$ by assumption). In terms of Fig. 11.1, this means that only two regions emerge:
  – If $\beta < n$, we can sustain two pure strategy Nash equilibria, (A, A) and (B, B).
  – If $\beta \geq n$, we can only sustain one Nash equilibrium, (B, B).
- Regarding social optimality, cutoff $2(n_A - n_B)$ in this context becomes $2(n - n) = 0$, so strategy profile (B, B) is socially optimal for all admissible values of β (see Fig. 11.2, considering that cutoff $2(n_A - n_B)$ in this symmetric setting shifts all the way to zero). Therefore, two cases emerge:
  – When $\beta < n$, we can have a socially inefficient arising in equilibrium, (A, A), or a socially optimal equilibrium, (B, B).
  – When $\beta \geq n$, the unique equilibrium that emerges in this case, (B, B), is socially optimal.

Exercise #11.2: Network Effects with Sequential Moves, Based on Farrell and Saloner (1985)^B

11.2 Consider the setting in Exercise 11.1. Assume now that firm 1 is the first mover, choosing between technology A and B. Observing firm 1's choice, firm 2 responds selecting either technology A or B. All information is common knowledge among firms. We solve the game by backward induction, starting with the second mover.

(a) *Second mover.* Find firm 2's best responses to each technology choice of firm 1.

- Firm 2 exhibits the same best response as in part (a) of the previous exercise. In particular:
  – If firm 1 adopts technology A, firm 2 responds with technology A if and only if $b_A + 2n_A > b_B + n_B$, which simplifies to

$$2n_A - n_B > b_B - b_A \equiv \beta.$$

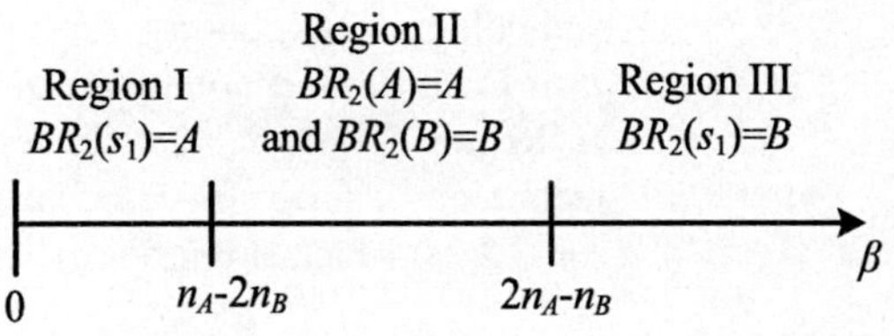

Fig. 11.3 Second mover's best responses

– If firm 1 adopts technology B, firm 2 responds with technology A if and only if $b_A + n_A > b_B + 2n_B$, which simplifies to

$$n_A - 2n_B > b_B - b_A \equiv \beta.$$

Finally, note that cutoff $2n_A - n_B$ satisfies

$$2n_A - n_B > n_A - 2n_B,$$

as this inequality simplifies to $n_A > -n_B$, which holds by assumption given that $n_i > 0$ for every technology $i = \{A, B\}$.

- This ranking in the cutoffs helps us divide the line of β in three regions, as depicted in Fig. 11.3:
  - *Region I*: $\beta < n_A - 2n_B$, so firm 2 responds with technology A regardless of firm 1's choice. In other words, technology A is strictly dominant for firm 2.
  - *Region II*: $n_A - 2n_B \leq \beta < 2n_A - n_B$, so firm 2 responds mimicking firm 1's choice (technology A after firm 1 selects A, and B after firm 1 chooses B).
  - *Region III*: $2n_A - n_B \leq \beta$, so firm 2 responds with technology B regardless of firm 1's choice, i.e., technology B becomes a strictly dominant strategy for firm 2.

(b) *First mover.* What is firm 1's technology in equilibrium?

- Anticipating the above best responses, firm 1's decision depends on the exact region where β lies:
  - In Region I, where firm 2 responds with technology A regardless of firm 1's choice, firm 1 anticipates that firm 2 responds with A. Therefore, firm 1 chooses A rather than B if $b_A + 2n_A > b_B + n_B$, which simplifies to

$$2n_A - n_B > b_B - b_A \equiv \beta.$$

which holds in Region I. Therefore, in Region I, the subgame perfect Nash equilibrium (SPNE) of this sequential-move game has firm 1 choosing A and firm 2 responding with A regardless of firm 1's choice.

  - In Region II, firm 1 anticipates firm 2 mimicking its decision. Therefore, firm 1 chooses A rather than B if $b_A + 2n_A > b_B + 2n_B$, or

$$2(n_A - n_B) > b_B - b_A \equiv \beta$$

where cutoff $2(n_A - n_B)$ satisfies $n_A - 2n_B < 2(n_A - n_B) < 2n_A - n_B$. Graphically, cutoff $2(n_A - n_B)$ then splits Region II into two areas. First, if $\beta < 2(n_A - n_B)$, we have a SPNE in which firm 1 chooses technology A and firm 2 follows by also choosing technology A. In contrast, when $\beta \geq 2(n_A - n_B)$, we have another SPNE in which firm 1 chooses technology B and firm 2 follows by choosing technology B as well.

– In Region III, firm 1 anticipates that firm 2 will respond with B regardless of its technology choice. In this context, firm 1 chooses A rather than B if $b_A + n_A > b_B + 2n_B$, which simplifies to

$$n_A - 2n_B > b_B - b_A \equiv \beta.$$

This condition is, however, incompatible with Region III, where $n_A - 2n_B < \beta$, and thus firm 1 chooses B in equilibrium. In summary, the SPNE that arises in this region has firm 1 choosing B and firm 2 responding with B regardless of firm 1's choice.

(c) Compare your equilibrium results with those in the simultaneous version of the game.

- In the previous exercise, we also characterized our equilibrium results as a function of Regions I–III, so we can compare the equilibria that arise in the simultaneous and sequential versions of the game for each region:
  - In Region I, (A, A) is the unique Nash equilibrium (NE) of the simultaneous game, and we found that it is also the equilibrium path in the SPNE of the sequential-move game.
  - In Region II, (A, A) and (B, B) are the NEs of the simultaneous-move game. We find that firms coordinate in both the sequential and simultaneous version of the game, since firms' profits are the same whether they move first, second, or at the same time. Specifically, firms adopt (A, A) in Region I and segments of Region II where $\beta < 2(n_A - n_B)$, but (B, B) in segments of Region II where $\beta \geq 2(n_A - n_B)$ and in Region III. Intuitively, only when technology B is sufficiently superior to technology A (as captured by the parameter β) to offset the network effects between the two technologies (as measured by cutoff $2(n_A - n_B)$) firms coordinate on technology B.
  - In Region III, (B, B) is the unique NE of the simultaneous-move game, and we found that it is also the equilibrium path in the SPNE of the sequential-move game.

Exercise #11.3: Switching Costs, Based on Klemperer (1995)^C

11.3 Consider an industry with N consumers distributed along the [0, 1] line, which measures consumers' linear cost of learning how to use a product. Firms A and B sell the same homogeneous product but are located at 0 and 1, respectively, and have the same unit cost $c = 0$ in each period. Therefore, a consumer located at point x has a learning cost tx of using firm A's product and $t(1 - x)$ of using B's product, where $t > 0$. Consumers do not have any physical transportation cost. Consumer utility is given by

$$u = r - p_i - t\,|l_i - x|$$

for every firm $i = \{A, B\}$, r denotes the consumer's reservation price, and l_i represents the location of firm i.

The time structure of the game is the following:

(i) At period 1, every firm sets prices and consumers, observing these prices, respond buying at most one unit of the good.

(ii) At period 2, each consumer has a reservation price r to buy the good.

Goods cannot be stored and consumers do not discount future payoffs. Goods are perceived as perfectly homogeneous, but there are switching costs: if a consumer changes provider, he has to pay a switching cost $s > 0$, which for simplicity is assumed to be independent of the distance that separates the consumer from the firm. Firms set simultaneously prices in each period. Assume that $r > c$, that $s \geq r - c$, and that $r - 2t > c$.

(a) *Second period.* Find equilibrium prices in the second stage for every firm i, $p_{i,S}^2$, where subscript S denotes that our setting considers switching costs. Find equilibrium profits in the second period, for given first-period sales q_A^1 by firm A and q_B^1 by firm B.

- A consumer who previously bought from firm $i = \{A, B\}$ in period 1 can be induced to buy from its rival, firm j, in period 2 if and only if price $p_{j,S}^2$ satisfies

$$p_{j,S}^2 + s < p_{i,S}^2.$$

That is, if firm j's price, plus the switching cost that the consumer suffers, is lower than firm i's price. Firm j has incentives to set this price if $p_{j,S}^2 > c$, or $p_{j,S}^2 + s > c + s$. Combining this condition with the one we obtained above yields

$$c + s < p_{j,S}^2 + s < p_{i,S}^2 < r$$

where the last inequality ensures that the market exists. Therefore, we must have that $c + s < r$, but this condition violates our initial assumption ($s \geq r - c$).

As a result, firm j does not have incentives to undercut firm i to steal its customers. In other words, in the second period, both firms keep all their customers. In this context, both firms set the highest price that consumers are willing to pay (reservation price r), that is,

$$p_{i,S}^2 = p_{j,S}^2 = r.$$

- Finally, equilibrium profits in the second period for firm i are given by

$$\pi_i^2 = (r - c)Nq_i^1,$$

where $r - c$ represents the margin that firm i earns per unit sold and Nq_i^1 denotes the units that this firm sells (which coincides with its customers in the first period).

(b) *First period.* Find the equilibrium prices in the first period of the game, $p_{i,S}^1$ for every firm i.

- *Finding demands.* We first need to find the indifferent consumer in order to obtain the demand for each firm. A consumer at position x (in the unit line representing his learning cost) buys from firm A, rather than from B, if

$$r - p_{A,S}^1 - tx > r - p_{B,S}^1 - t(1 - x)$$

or, after solving for x,

$$x \leq \frac{1}{2} + \frac{p_{B,S}^1 - p_{A,S}^1}{2t}$$

Intuitively, all consumers with learning costs in $[0, x]$ buy from firm A, while those in $(x, 1]$ buy from firm B. Therefore, firm A's demand in the first period is

$$q_A^1 = \frac{1}{2} + \frac{p_{B,S}^1 - p_{A,S}^1}{2t}$$

and firm B's is

$$q_B^1 = 1 - \left(\frac{1}{2} + \frac{p_{B,S}^1 - p_{A,S}^1}{2t}\right)$$
$$= \frac{1}{2} + \frac{p_{A,S}^1 - p_{B,S}^1}{2t}.$$

- *Overall profits.* Therefore, total profits (including first- and second-period profits) for firm A are

$$\pi_{A,S} = \overbrace{\underbrace{N\left(\frac{1}{2} + \frac{p_{B,S}^1 - p_{A,S}^1}{2t}\right)}_{q_A^1}(p_{A,S}^1 - c)}^{\text{First-period profits}} + \overbrace{\underbrace{N\left(\frac{1}{2} + \frac{p_{B,S}^1 - p_{A,S}^1}{2t}\right)}_{q_A^1}(r - c)}^{\text{Second-period profits}}$$
$$= N\left(\frac{1}{2} + \frac{p_{B,S}^1 - p_{A,S}^1}{2t}\right)(p_{A,S}^1 + r - 2c)$$

and, similarly, those of firm B are

$$\pi_{B,S} = \overbrace{\underbrace{N\left(\frac{1}{2} + \frac{p_{A,S}^1 - p_{B,S}^1}{2t}\right)}_{q_B^1}\left(p_{B,S}^1 - c\right)}^{\text{First-period profits}} + \overbrace{\underbrace{N\left(\frac{1}{2} + \frac{p_{A,S}^1 - p_{B,S}^1}{2t}\right)}_{q_B^1}(r - c)}^{\text{Second-period profits}}$$
$$= N\left(\frac{1}{2} + \frac{p_{A,S}^1 - p_{B,S}^1}{2t}\right)(p_{B,S}^1 + r - 2c).$$

- *First-order conditions.* Differentiating firm i's overall profit $\pi_{i,S}$ with respect to its first-period price, $p_{i,S}^1$, we obtain

$$N\left(\frac{1}{2} + \frac{p_{j,S}^1 - p_{i,S}^1}{2t} - \frac{p_{i,S}^1 + r - 2c}{2t}\right) = 0,$$

where $j \neq i$. In a symmetric equilibrium, both firms set the same first-period price, $p_{i,S}^1 = p_{j,S}^1 = p_S^1$, which simplifies the above first-order condition to

$$N\left(\frac{1}{2} - \frac{p_S^1 + r - 2c}{2t}\right) = 0.$$

Solving for p_S^1, we find the equilibrium price

$$p_S^1 = t - r + 2c,$$

which is increasing in learning cost t because it becomes more costly for consumers to switch in the second period, so that firms can command a higher price and still retain consumers. Interestingly, it is decreasing in reservation value r because this is the price that firms set in the second period, and therefore, lowering the first-period price boosts second-period demand for firm i's products.

(c) *No switching costs.* Find the equilibrium price for the game where the first period is like the above, but in the second period, there are no switching costs at all. Label prices $p^1_{i,NS}$ and $p^2_{i,NS}$ for every firm i, where subscript NS denotes that our setting assumes no switching costs.

- *Second-period prices.* In this setting, consumers regard products as completely homogeneous. Therefore, firms face a standard Bertrand duopoly game of price competition in the second stage, where each firm charges a price equal to marginal cost, that is,

$$p^2_{A,NS} = p^2_{B,NS} = c.$$

In this period, profits are then zero for both firms. Therefore, second-period profits are unaffected by first-period sales, as opposed to what happens when consumers face switching costs, where we showed that these profits increase in first-period sales. Intuitively, when consumers face switching costs, every firm has incentives to increase its pool of customers during the first period, as they become captive in the second period. When switching costs are absent, however, firms do not have this incentive, as we show next.

- *First-period prices.* In the first period, firms anticipate that consumers will buy the less expensive good in the second period. Therefore, they face a standard Hotelling game, where every firm i solves

$$\max_{p_{i,NS}\geq 0} \pi_{i,NS} = \underbrace{N\left(\frac{1}{2} + \frac{p^1_{j,NS} - p^1_{i,NS}}{2t}\right)(p^1_{i,NS} - c)}_{\text{First-period profits}} + \underbrace{0}_{\text{Second-period profits}}$$

which all originate from first-period profits since second-period profits are zero. Differentiating with respect to price $p^1_{i,NS}$ yields

$$N\left(\frac{1}{2} + \frac{p^1_{j,NS} - p^1_{i,NS}}{2t} - \frac{p^1_{i,NS} - c}{2t}\right) = 0.$$

At the symmetric equilibrium, both firms set the same first-period price, $p^1_{i,NS} = p^1_{j,NS} = p^1_{NS}$. Inserting this property in the above expression, we find

$$N\left(\frac{1}{2} - \frac{p^1_{NS} - c}{2t}\right) = 0.$$

Solving for price p^1_{NS}, we obtain the equilibrium first-period price:

$$p^1_{NS} = t + c,$$

which coincides with that in a standard Hotelling game, thus being increasing in learning cost t and in the firms' production cost c.

(d) *Price comparison.* Show that first-period prices are lower with switching costs, $p^1_{i,S} < p^1_{i,NS}$, but second-period prices are higher, $p^2_{iS} > p^2_{iNS}$, for every firm i.

- *First-period prices.* Comparing first-period prices with switching costs, p^1_S, and without them, p^1_{NS}, we find that

$$p^1_S = t - r + 2c < t + c = p^1_{NS}$$

since $r > c$ by definition.

- *Second-period prices.* Comparing second-period prices with switching costs, p^2_S, and without them, p^2_{NS}, we find that

$$p^2_S = r > c = p^2_{NS}$$

since $r > c$ by definition. Our results confirm our intuition described above: with switching costs, firms have incentives to set lower prices in the first period to expand their pool of customers before the second period, when consumers are captive and can be exploited by setting a monopoly price $p^2_S = r$.

Exercise #11.4: Welfare Effects of Entry with Switching Costs, Based on Klemperer (1988)^C

11.4 Consider an industry with an incumbent and a potential entrant, both facing marginal production cost c, where $1 > c > 0$, and inverse demand function $p(q) = 1 - q$. In the first stage, the potential entrant chooses whether to join the industry, at no cost. In the second stage, firms compete à la Cournot. If a consumer purchases from the incumbent, he does not suffer switching costs (since he bought units from that firm in the past), but if the consumer purchases from the entrant, he suffers a switching cost $s \geq 0$.

(a) *Second stage, no entry.* In the subgame that ensues after no entry, find the equilibrium output, price, profits, and welfare.

- Upon no entry, the incumbent is the only firm in the industry and solves the following monopoly problem:

$$\max_{q \geq 0} (1 - q)q - cq.$$

Differentiating with respect to q and solving for q yields

$$q^{NE} = \frac{1 - c}{2},$$

where the superscript NE denotes "no entry."

- In this setting, price is

$$\begin{aligned} p^{NE} &= 1 - q^{NE} \\ &= 1 - \frac{1 - c}{2} \\ &= \frac{1 + c}{2}, \end{aligned}$$

and profits are

$$\pi^{NE} = \left(1 - \frac{1-c}{2}\right)\frac{1-c}{2} - c\frac{1-c}{2}$$
$$= \frac{(1-c)^2}{4}$$

- Therefore, welfare is

$$W^{NE} = \frac{1}{2}\left(q^{NE}\right)^2 + \pi^{NE}$$
$$= \frac{1}{2}\frac{(1-c)^2}{4} + \frac{(1-c)^2}{4}$$
$$= \frac{3(1-c)^2}{8}.$$

(b) *Second stage, entry.* In the subgame that ensues after entry, find the equilibrium output, price, profits, and welfare.

- *Incumbent.* Upon entry, the incumbent (firm i) solves

$$\max_{q_i \geq 0} (1 - q_i - q_j)q_i - cq_i.$$

Differentiating with respect to q_i and solving for q_i yields the incumbent's best response function:

$$q_i(q_j) = \frac{1-c}{2} - \frac{1}{2}q_j$$

- *Entrant.* Upon entry, the entrant (firm j) solves

$$\max_{q_j \geq 0} (1 - q_j - q_i - s)q_j - cq_j$$

since $p_E(q) = p_I(q) - s$, or $p_E(q) = 1 - q_j - q_i - s$. Graphically, the entrant's inverse demand function shifts downward by s, in a parallel fashion, relative to the incumbent's inverse demand function, as consumers have switching costs when purchasing from the entrant.

Differentiating with respect to q_j and solving for q_j, we obtain the entrant's best response function:

$$q_j(q_i) = \frac{1-s-c}{2} - \frac{1}{2}q_i$$

which is symmetric to the incumbent's when switching costs are absent, $s = 0$, but otherwise originates at a lower vertical intercept.

- Combining the incumbent's and entrant's best response functions, and simultaneously solving, yields output levels:

$$q_i^E = \frac{1-c+s}{3} \text{ and } q_j^E = \frac{1-c-2s}{3},$$

where superscript E denotes "entry." Price for the incumbent in this setting is

$$\begin{aligned} p_i^E &= 1 - q_i - q_j \\ &= 1 - \frac{1-c+s}{3} - \frac{1-c-2s}{3} \\ &= \frac{1+2c+s}{3} \end{aligned}$$

and price for the entrant is

$$\begin{aligned} p_j^E &= 1 - q_j - q_i - s \\ &= 1 - \frac{1-c-2s}{3} - \frac{1-c+s}{3} - s \\ &= \frac{1+2c-2s}{3}. \end{aligned}$$

Incumbent profits are, therefore,

$$\begin{aligned} \pi_i^E &= \frac{1+2c+s}{3}\frac{1-c+s}{3} - c\frac{1-c+s}{3} \\ &= \frac{(1-c+s)^2}{9}, \end{aligned}$$

and entrant profits are

$$\begin{aligned} \pi_j^E &= \frac{1-c-2s}{3}\frac{1+2c-2s}{3} - c\frac{1-c-2s}{3} \\ &= \frac{(1-c-2s)^2}{9}. \end{aligned}$$

- Finally, welfare under entry is

$$\begin{aligned} W^E &= \frac{1}{2}(q_i^E + q_j^E)^2 + \pi_i^E + \pi_j^E \\ &= \frac{(2-2c-s)^2}{18} + \frac{(1-c+s)^2}{9} + \frac{(1-c-2s)^2}{9} \\ &= \frac{8(1-c)^2 + 11s^2 - 8s(1-c)}{18}. \end{aligned}$$

(c) *First stage.* Find under which parameter conditions the potential entrant joins the industry in the first stage of the game.

- At the first stage of the game, the potential entrant joins the industry if and only if it anticipates a positive profit, $\pi_j^E \geq 0$, which entails

$$p_j^E - c = \frac{1+2c-2s}{3} - c \geq 0,$$

which simplifies to $\frac{1-c-2s}{3} \geq 0$. Solving for switching cost s, we find that entry occurs if

$$s \leq \frac{1-c}{2} \equiv s_1.$$

Intuitively, switching costs, which capture the "demand disadvantage" that the potential entrant faces, must be sufficiently low, $s \leq s_1$, for the entrant to have incentives to join the industry. When switching costs are high enough, $s > s_1$, entry does not occur. In this case, we say that entry is blockaded since the incumbent does not need to take any strategic actions in the first stage to prevent entry.

(d) Does welfare increase or decrease when the entrant chooses to enter?

- Comparing welfare with entry, W^E from part (b), and without entry, W^{NE} from part (a), we find that $W^E > W^{NE}$ if and only if

$$\frac{8(1-c)^2 + 11s^2 - 8s(1-c)}{18} > \frac{3\,(1-c)^2}{8}$$

which, after rearranging, yields

$$[5(1-c) - 22s]\,[(1-c) - 2s] > 0$$

and, solving for s, we find that entry is welfare enhancing if

$$s < \frac{5(1-c)}{22} \equiv s_2.$$

- Comparing cutoff s_2 against s_1, we see that $s_2 < s_1$ since $\frac{5}{22} \simeq 0.22 < \frac{1}{2}$. Therefore, these cutoffs divide switching costs in three regions:
  - When $s < s_2$, entry occurs and it is welfare improving.
  - When $s_2 \leq s < s_1$, entry also occurs in equilibrium, but it is welfare reducing.
  - When $s_1 \leq s$, entry does not occur in equilibrium.

  Intuitively, entry produces two effects: a decrease in prices (positive effect) and consumers suffering from switching costs that they could have avoided if buying from the incumbent (negative effect). When switching costs are relatively high, the second (negative) effect dominates, and entry becomes welfare reducing.

(e) *No switching costs.* Evaluate your results at $s = 0$ and interpret.

- When switching costs are absent, then entry would occur, since condition $s = 0 < s_1$ holds, and it is welfare improving, since condition $s = 0 < s_2$ also holds. In this context, equilibrium output simplifies to

$$q_i^E = q_j^E = \frac{1-c}{3},$$

equilibrium price coincides for both firms and becomes

$$p_i^E = p_j^E = \frac{1+2c}{3},$$

profits are

$$\pi_i^E = \pi_j^E = \frac{(1-c)^2}{9},$$

and welfare is

$$W^E = \frac{8(1-c)^2}{18}.$$

(f) *No production costs.* Evaluate equilibrium outcomes at $c = 0$. What values of s make entry welfare improving?

- Substituting $c = 0$ into the results from part (b), we obtain

$$p_i^E = \frac{1+s}{3}$$

$$p_j^E = \frac{1-2s}{3}$$

$$\pi_i^E = \frac{(1+s)^2}{9}$$

$$\pi_j^E = \frac{(1-2s)^2}{9}$$

$$W^E = \frac{11s^2 - 8s + 64}{18}$$

The entrant finds it profitable to enter into the market when $p_j^E = \frac{1-2s}{3} > 0$, which entails $s < \frac{1}{2}$.

- Otherwise, when $s \geq \frac{1}{2}$, the entrant does not find it profitable to enter, so that the incumbent monopolizes the market. From part (a), we find that

$$q^{NE} = \frac{1}{2}$$

$$p^{NE} = \frac{1}{2}$$

$$\pi^{NE} = \frac{1}{4}$$

$$W^{NE} = \frac{3}{8}$$

which coincide with the equilibrium outcomes in Exercise 1.1 without fixed costs.

- From part (d), we know that entry is welfare improving when entry cost, s, satisfies

$$0 \leq s < \frac{5}{22}$$

since cutoff $s_1 = \frac{1-0}{2} = \frac{1}{2}$ in this context, and cutoff $s_2 = \frac{5(1-0)}{22} = \frac{5}{22}$.

Exercise #11.5: Buyer Power Coordination, Based on Fumagalli and Motta (2008)^C

11.5 Consider an industry with an incumbent firm, I, which already invested sunk cost, $F > 0$, and a potential entrant E which has not incurred this sunk cost yet. If it enters, the entrant produces the same homogeneous good as the incumbent. The potential entrant is more efficient than the incumbent, with marginal cost c_E satisfying $c_E < c_I$. Buyers have a unit demand for the good and their maximum willingness to pay is v.

The time structure of the game is the following:

(i) At $t = 0$, N buyers call a procurement auction for the good.
(ii) At $t = 1$, the incumbent and the potential entrant simultaneously make their (public) bids to all the buyers.
(iii) At $t = 2$, each buyer observes the bids and, independently of the other buyers, decides whether to accept the incumbent's or the entrant's offer.
(iv) At $t = 3$, the incumbent fulfills all the orders it has received. The entrant observes the number of buyers who addressed it and decides whether to actually enter the industry or not. If it does not enter, the entrant's payoff is zero.
(v) At $t = 4$, the buyers whose orders have not been fulfilled by the entrant can turn to the incumbent.

We assume that $F > v - c_E$, i.e., a single buyer is not enough to trigger entry, and $F < N(c_I - c_E)$, i.e., entry is viable if the entrant charges a price $p \geq c_I$ and is addressed by all buyers. For which parameter conditions can you support a subgame perfect equilibrium where the potential entrant does not enter the industry? Interpret.

(a) *Fourth stage*. Find the incumbent's selling price if the buyers are not addressed by the entrant.
   - If buyers are not addressed by the entrant, then the incumbent would charge $p_I = v$ to capture all buyer surplus while buyers will still purchase from this incumbent.

(b) *Third stage*. Find the entrant's profit and the minimal number of buyers that supports entry.
   - Let n_E (n_I) be the number of buyers who purchase from the entrant (incumbent, respectively). In order for the potential entrant to enter into the industry, its profit, π_E, must satisfy

$$\pi_E = n_E(p_E - c_E) \geq F$$

   which, after rearranging and solving for n_E, yields the minimum number of buyers $\underline{n}_E$ to induce entry:

$$n_E \geq \frac{F}{p_E - c_E} \equiv \underline{n}_E$$

   - We further verify that a single buyer is not enough to trigger entry, that is, cutoff $\underline{n}_E$ must satisfy $\underline{n}_E > 1$. To show this, notice that

$$F > v - c_E \geq p_E - c_E,$$

   because the entrant cannot set a price higher than the buyers' valuation, i.e., $p_E \leq v$.

Rearranging the above inequality, we obtain

$$\underline{n}_E = \frac{F}{p_E - c_E} > 1,$$

so that there must more than one buyer buying from the entrant to support its entry.

(c) *Second stage.* Will the buyers accept the incumbent or the entrant's offer?

- We consider two cases, (i) $c_I \leq p_I < p_E \leq v$ and (ii) $c_I \leq p_E \leq p_I \leq v$, ignoring cases in which either firm charges a price higher than v (which generates no sales) or lower than c_I (profit loss for the incumbent and lower profit for the entrant, since $c_E < c_I$).
  - In case (i), all buyers buy from the incumbent, so that $n_I = N$ and $n_E = 0$.
  - In case (ii), we have two equilibria, one in which all buyers buy from the entrant, so that $n_I = 0$ and $n_E = N$, as in typical Bertrand competition, and another in which all buyers buy from the incumbent, so that $n_I = N$ and $n_E = 0$, since deviation of any one buyer to the entrant will not be addressed.

(d) *First stage.* Characterize the equilibrium bids that the firms make and find their associated profits. Under which equilibrium is miscoordination among the buyers possible?

- In case (i), where firms bid prices such that $p_I < p_E$ cannot be supported as a Nash equilibrium because the entrant enjoys a cost advantage $c_E < c_I$. Specifically, if the incumbent sets a price p_I below the entrant's, the entrant can respond by setting a price lower than the incumbent's marginal cost, i.e., $p_E = c_I - \varepsilon$, where $\varepsilon \to 0$, and still make positive profit, leaving no sales to the incumbent. Therefore, a price pair satisfying $p_I < p_E$ cannot be an equilibrium in the first stage of the game.
- In case (ii), where firms bid prices such that $p_E \leq p_I$, we have two subgame perfect Nash equilibria:
  - The first one occurs at $p_E^* = p_I^* = c_I$ in which neither firm has incentives to further undercut its rival's price, and buyers accept the offer from the entrant. This happens because, in this price profile, entrant's profits become

$$\pi_E^* = n_E (p_E - c_E) = N (c_I - c_E) > F$$

that supports its entry into the industry to capture all sales, so that $n_E^* = N$ and $n_I^* = 0$, leaving the incumbent with zero profit, $\pi_I^* = 0$.

  - The second equilibrium happens at the incumbent charging a price equal to buyers' valuation, $p_I^* = v$, and the entrant setting its price in the interval $p_E^* \in [c_I, v]$. Interestingly, buyers will buy from the incumbent in this price profile, $n_I^* = N$ and $n_E^* = 0$, yielding equilibrium profits:

$$\pi_I^* = N (v - c_I) \text{ and } \pi_E^* = 0.$$

Intuitively, this happens because buyers do not coordinate their actions to buy from the entrant; instead, they "miscoordinate" to buy from the incumbent, making unilateral deviations of one buyer to the entrant not profitable to be addressed.

Exercise #11.6: Retail Price Maintenance, Based on Winter (1993)^C

11.6 Many industries are characterized by a manufacturer (upstream firm) supplying goods to retailers (downstream firms). In some cases, we observe that the manufacturer lets every retailer i set any price p_i for the product, but in other settings, the manufacturer sets a price floor that the retailer cannot exceed. In this exercise, we study the manufacturer's incentives to set this price floor.

Consider an industry with a manufacturer and two retailers, selling a homogeneous product. Consumers purchase at most one unit from either retailer, which is uniformly distributed in a line of unit length, and each retailer is exogenously located at one of the end points of the line. If a consumer located at x buys the product from retailer 1 at a price p_1 when the retailer offers a service s_1, he enjoys utility of

$$u(x, p_1, s_1) = r - p_1 - \underbrace{\theta\,[x + T(s_1)]}_{\text{Transportation cost}}$$

where the last term represents the transportation cost, which includes (i) the distance x to firm 1, and (ii) $T(s_1)$, which represents the time at the store, which is decreasing in the amount of service s_1 that this retailer provides, indicating that customers need to spend less time at the store when the retailer offers a better service. Parameter θ can be interpreted as the unit transportation cost of each consumer or, alternatively, as the opportunity cost of time.

A similar expression applies if this consumer purchases the good from firm 2, located at a distance $1 - x$ from this consumer, as follows:

$$u(x, p_2, s_2) = r - p_2 - \underbrace{\theta\,[(1 - x) + T(s_2)]}_{\text{Transportation cost}}.$$

The profit of retailer i is

$$\pi_i = (p_i - w)q_i - C(s_i),$$

where q_i denotes the units this retailer sells and $C(s_i) = s_i$ represents its cost from offering service level s_i. The manufacturer's profits are then

$$\pi = (w - c)(q_1 + q_2),$$

where c denotes the marginal cost of production.

In the first stage, the manufacturer sets the wholesale price w that every retailer must pay per unit of output; in the second stage, every retailer i simultaneously and independently chooses its price p_i and service quality s_i; and in the third stage, observing the profile of prices and services from both retailers 1 and 2, consumer purchase from either (or none) of them.

(a) *Third stage.* Find the location of the indifferent consumer, $\widehat{x}$, to obtain the demand function of retailers 1 and 2.

- The indifferent consumer enjoys the same utility purchasing the product from retailer 1 or 2, that is, $\widehat{x}$ solves

$$r - p_1 - \theta\,[\widehat{x} + T(s_1)] = r - p_2 - \theta\,[(1 - \widehat{x}) + T(s_2)]$$

Rearranging yields

$$\theta\left[(1-\widehat{x})-\widehat{x}-(T(s_1)-T(s_2))\right]=p_1-p_2, \text{ or}$$

$$1-2\widehat{x}-[T(s_1)-T(s_2)]=\frac{p_1-p_2}{\theta}.$$

Solving for $\widehat{x}$, we find

$$\widehat{x}=\frac{1}{2}-\frac{T(s_1)-T(s_2)}{2}-\frac{p_1-p_2}{2\theta}.$$

Intuitively, the indifferent consumer starts at the midpoint between the two retailers (as denoted in the first term of $\widehat{x}$), adjusts based upon the relative amount of time she must spend in each store (the second term of $\widehat{x}$), and then considers the relative price of each store in relation to her transportation cost. It is straightforward to show that $\widehat{x}$ is increasing in s_1 and decreasing in s_2. Intuitively, if retailer 1 spends more on retail services, its product is more attractive to consumers, which shifts the location of the indifferent consumer upward (a symmetric argument exists for retailer 2).

Along a similar reasoning, $\widehat{x}$ is decreasing in p_1 and increasing in p_2. As retailer 1 increases its price, consumers switch to retailer 2, as indicated by the median consumer shifting downward. Lastly, $\widehat{x}$ can be either increasing or decreasing in θ, depending on the relationship between p_1 and p_2. As θ increases, consumers must pay a higher cost to switch from one retailer to another. This favors a retailer with a price advantage in this market, as more consumers favor its product.

(b) *Second stage.* Write retailer i's profit maximization problem, and find its best response function for price $p_i(p_j)$ and for service $s_i(s_j)$. Interpret. For simplicity, assume for the remainder of the exercise that $T(s_i)=\gamma-\frac{1}{2}s_i^2$, where γ is sufficiently high so that $T(s_i)>0$.

- *Retailer 1.* This retailer solves

$$\max_{p_1,s_1\geq 0}\ \pi_1=(p_1-w)\overbrace{\left[\frac{1}{2}-\frac{\left(\gamma-\frac{1}{2}s_1^2\right)-\left(\gamma-\frac{1}{2}s_2^2\right)}{2}-\frac{p_1-p_2}{2\theta}\right]}^{\widehat{x}}-s_1$$

Differentiating with respect to p_1 and s_1 yields

$$\frac{\partial\pi_1}{\partial p_1}=\frac{2(w+\theta)-4p_1+2p_2+\theta\left(s_1^2-s_2^2\right)}{4\theta}=0, \text{ and}$$

$$\frac{\partial\pi_1}{\partial s_1}=(p_1-w)s_1-2=0,$$

respectively. Solving for p_1 in the top expression and for s_1 in the bottom equation, we find

$$p_1(p_2)=\frac{2(w+\theta)+\theta\left(s_1^2-s_2^2\right)}{4}+\frac{p_2}{2}, \text{ and}$$

$$s_1=\frac{2}{p_1-w}.$$

This retailer's best response function $p_1(p_2)$ increases in its rival's price p_2. Graphically, $p_1(p_2)$ is positively sloped, indicating that retailer's prices are strategic complements. However, firm 1's service, s_1, is not a function of its rival's, s_2, but is decreasing in its own price p_1.

Intuitively, this means that retailers do not directly compete in the service they offer, although they compete indirectly via prices. Specifically, an increase in retailer 2's service, s_2, leads to a downward shift in retailer 1's best response function $p_1(p_2)$, suggesting that retailer 1 sets more competitive prices to attract more customers, which, in turn, increases its level of services.

- *Retailer 2.* Similarly, retailer 2 solves

$$\max_{p_2, s_2 \geq 0} \pi_2 = (p_2 - w)\overbrace{\left[1 - \left(\frac{1}{2} - \frac{\left(\gamma - \frac{1}{2}s_1^2\right) - \left(\gamma - \frac{1}{2}s_2^2\right)}{2} - \frac{p_1 - p_2}{2\theta}\right)\right]}^{1-\widehat{x}} - s_2.$$

(Recall that retailer 2's profit maximization problem is not symmetric to that of firm 1 since it sells to $1 - \widehat{x}$, rather than $\widehat{x}$, customers.) Differentiating with respect to p_2 and s_2 yields

$$\frac{\partial \pi_2}{\partial p_2} = \frac{2\,(w + \theta) - 4p_2 + 2p_1 + \theta\left(s_2^2 - s_1^2\right)}{4\theta} = 0, \text{ and}$$

$$\frac{\partial \pi_2}{\partial s_2} = (p_2 - w)s_2 - 2 = 0,$$

respectively. Solving for p_2 and s_2, respectively, we find

$$p_2(p_1) = \frac{2\,(w + \theta) + \theta\left(s_2^2 - s_1^2\right)}{4} + \frac{p_1}{2}, \text{ and}$$

$$s_2 = \frac{2}{p_2 - w}$$

thus being symmetric to those of retailer 1.

(c) Find the equilibrium price, service, and time spent in the store.

- *Finding equilibrium prices.* Inserting $s_1 = \frac{2}{p_1 - w}$ and $s_2 = \frac{2}{p_2 - w}$ into retailer 1's best response function for its price, we obtain

$$p_1(p_2) = \frac{2\,(w + \theta) + \theta\left(\overbrace{\left(\frac{2}{p_1 - w}\right)}^{s_1}{}^2 - \overbrace{\left(\frac{2}{p_2 - w}\right)}^{s_2}{}^2\right)}{4} + \frac{p_2}{2}$$

which simplifies to

$$p_1(p_2) = \frac{w + \theta\left[1 + 2\left(\frac{1}{(p_1 - w)^2} - \frac{1}{(p_2 - w)^2}\right)\right]}{2} + \frac{p_2}{2}.$$

Since retailer 2's best response function, $p_2(p_1)$, is symmetric, we can claim that, in a symmetric equilibrium, both retailers set the same price, $p_1^* = p_2^* = p^*$, which further simplifies the above best response function to

$$p = \frac{w+\theta\left[1+2\overbrace{\left(\frac{1}{(p-w)^2}-\frac{1}{(p-w)^2}\right)}^{0}\right]}{2}+\frac{p}{2}$$
$$= \frac{w+\theta+p}{2}.$$

Solving for p, we obtain the equilibrium price:

$$p^* = w + \theta.$$

Therefore, the equilibrium price is increasing in the wholesale price, w, and in the transportation cost, θ.

- *Finding equilibrium services.* Inserting this equilibrium price into $s_1 = \frac{2}{p_1 - w}$ and $s_2 = \frac{2}{p_2 - w}$ yields equilibrium services:

$$s^* = s_1^* = s_2^* = \frac{2}{\underbrace{(w+\theta)}_{p^*} - w} = \frac{2}{\theta}.$$

Therefore, the equilibrium service is decreasing in the transportation cost, θ. Intuitively, when transportation costs are high, consumers are less willing to travel to a different store, so retailers do not need to offer much service at their stores. In contrast, when transportation costs are low, consumers can easily travel to another store, implying that retailers compete in the service they offer to attract more customers.

- *Finding equilibrium time in store.* Substituting s^* into $T(s)$, we obtain

$$T^* = \gamma - \frac{(s^*)^2}{2} = \gamma - \frac{1}{2}\left(\frac{2}{\theta}\right)^2 = \gamma - \frac{2}{\theta^2}$$

Intuitively, when traveling to the other store becomes more costly (θ increases), firms offer lower quality of services, and as a result, consumers have to spend more time in the store. Note that since consumers must spend positive amount of time in the store, the above expression also allows us to find the infimum of γ, where

$$\gamma > \underline{\gamma} \equiv \frac{2}{\theta^2}.$$

(d) Find the equilibrium profits of retailer i. How does it change with θ? Interpret.

- Substituting p^* and s^* into retailer i's profit function, we obtain

$$\pi^* = \frac{p^* - w}{2} - s^*$$

$$= \frac{w+\theta-w}{2} - \frac{2}{\theta}$$
$$= \frac{\theta^2-4}{2\theta}$$

which is positive if and only if $\theta \geq 2$.

- Differentiating π_i^* with respect to θ, we find

$$\frac{\partial \pi_i^*}{\partial \theta} = \frac{1}{2} + \frac{2}{\theta^2} > 0$$

so that the higher the transportation cost θ, the higher profits does every retailer i earn. Intuitively, consumers find it more costly to travel to the other retailer, so that they are willing to pay higher prices, and as a result, every retailer makes more profits.

(e) *First stage.* Anticipating equilibrium behavior in subsequent stages, find the equilibrium wholesale price w^* that the manufacturer sets in the first stage of the game. Given that consumers must spend positive amount of time in the store, what is the lowest wholesale price that the manufacturer can set? Interpret.

- From our above results, the equilibrium sales of retailer 1 are

$$q_1 = \hat{x} = \frac{1}{2} - \frac{(\gamma - \frac{1}{2}(s^*)^2) - (\gamma - \frac{1}{2}(s^*)^2)}{2} - \frac{p^* - p^*}{2}$$
$$= \frac{1}{2},$$

and those of retailer 2 are

$$q_2 = 1 - \hat{x} = 1 - \left[\frac{1}{2} - \frac{(\gamma - \frac{1}{2}(s^*)^2) - (\gamma - \frac{1}{2}(s^*)^2)}{2} - \frac{p^* - p^*}{2}\right]$$
$$= 1 - \frac{1}{2} = \frac{1}{2}.$$

- Substituting equilibrium price and service into the consumer's utility function, we obtain the utility of the indifferent consumer as follows:

$$u(\hat{x}, p^*, s^*) = r - p^* - \theta\left(\hat{x} + \gamma - \frac{1}{2}(s^*)^2\right)$$
$$= r - (w+\theta) - \theta\left(\frac{1}{2} + \gamma - \frac{2}{\theta^2}\right)$$
$$= r - w + \frac{4 - (3+2\gamma)\theta^2}{2\theta}.$$

- The manufacturer chooses the wholesale price w to solve

$$\max_{w \geq 0} \pi = (w-c)(q_1+q_2).$$

Inserting the equilibrium sales for each retailer, we can express the manufacturer's problem as a function of w alone as follows:

$$\max_{w \geq 0} (w - c)\left(\frac{1}{2} + \frac{1}{2}\right) = w - c$$

Since the above function is linear in w, we find a corner solution in which the manufacturer sets the highest possible price on both retailers that makes consumers indifferent between buying from either retailer or not, where $u(\widehat{x}, p^*, s^*) = 0$ yields

$$w^* = r + \frac{4 - (3 + 2\gamma)\,\theta^2}{2\theta}.$$

- Substituting the infimum of γ into the above expression, we find

$$\begin{aligned} w^* > \underline{w} &\equiv r + \frac{4 - \left(3 + 2\underline{\gamma}\right)\theta^2}{2\theta} \\ &= r + \frac{4 - \left(3 + 2 \times \frac{2}{\theta^2}\right)\theta^2}{2\theta} \\ &= r - \frac{3\theta}{2} \end{aligned}$$

so that the equilibrium wholesale price that the manufacturer sets increases in consumers' reservation utility r but decreases in the unit transportation cost θ.

Bibliography

Armstrong, M. (2007). Recent developments in the economics of price discrimination. In R. Blundell, W. Newey, & T. Persson (Eds.), *Advances in economics and econometrics 3: Theory and applications (Ninth World Congress)*. Cambridge: Cambridge University Press.

Belleflamme, P., & Peitz, M. (2015). *Industrial organization: Markets and strategies* (2nd ed.). Cambridge: Cambridge University Press.

Cabral, L. (2017). *Introduction to industrial organization* (2nd ed.). Cambridge: The MIT Press.

Calveras, A., & Ganuza, J. J. (2018). Corporate social responsibility and product quality. *Journal of Economics and Management Strategy, 27*, 804–829.

Carlton, D., & Perloff, J. (2004). *Industrial organization: Markets and strategies* (4th ed.). London: Pearson.

Che, Y. (1998). Consumer return policies for experience goods. *Rand Journal of Economics, 44*, 17–24.

Cho, I., & Kreps, D. (1987). Signaling games and stable equilibrium. *Quarterly Journal of Economics, 102*, 179–222.

De Nijs, R. (2017). Discrimination and customer information sharing. *International Journal of Industrial Organization, 50*, 319–334.

Delbono, F., & Denicolo, V. (1991). Incentives to innovate in a Cournot oligopoly. *Quarterly Journal of Economics, 106*, 951–961.

Deneckere, R. J., & McAfee, P. R. (1996). Damaged goods. *Journal of Economics and Management Strategy, 5*, 149–174.

Esteves, R.-B., & Cerqueira, S. (2017). Behavior-based pricing under imperfectly informed consumers. *Information Economics and Policy, 40*, 60–70.

Farrell, J., & Saloner, G. (1985). Standardization, compatibility, and innovation. *RAND Journal of Economics, 16*, 70–83.

Fershtman, C., & Judd, K. L. (1987). Equilibrium incentives in oligopoly. *American Economic Review, 77*, 927–940.

Fikru, M., & Gautier, L. (2016). Mergers in Cournot markets with environmental externality and product differentiation. *Resource and Energy Economics, 45*, 65–79.

Fudenberg, D., & Tirole, J. (1984). The fat cat effect, the puppy-dog ploy and the lean and angry look. *American Economic Review, 74*, 361–366.

Fudenberg, D., & Villas-Boas, J. M. (2007). Behavior-based price discrimination and customer recognition. In T. J. Hendershott (Ed.), *Economics and information systems*. Amsterdam: Elsevier.

Fumagalli, C., & Motta, M. (2008). Buyers' miscoordination, entry and downstream competition. *Economic Journal, 118*, 1196–1222.

Gelves, J. A. (2014). Differentiation and cost asymmetry: Solving the merger paradox. *International Journal of the Economics of Business, 21*(3), 321–340.

Harrington, J. E. (2014). Penalties and the deterrence of unlawful collusion. *Economic Letters, 124*, 33–36.

Harrison, M., & Kline, J. (2001). Quantity competition with access fees. *International Journal of Industrial Organization, 19*, 345–373.

Heller, M., & Eisenberg, R. (1998). Can patents deter innovation? The anticommons in biomedical research. *Science, 280*, 698–701.

Heyes, A., & Oestreich, A. (2018). A theory of social license when regulatory pressure is jointly produced by an EPA and an NGO. *Journal of Regulatory Economics, 54*, 219–243.

Huck, S., Konrad, K., & Müller, W. (2001). Big fish eat small fish: On merger in stackelberg markets. *Economics Letters, 73*, 213–217.

Irmean, A., & Thisse, J.-F. (1998). Competition in multi-characteristics spaces: Hotelling was almost right. *Journal of Economic Theory, 78*, 76–102.

Klemperer, P. (1988). Welfare effects of entry into markets with switching costs. *Journal of Industrial Economics, 37*, 159–165.

P.-S. Choi et al., *Industrial Organization*, Springer Texts in Business and Economics,
https://doi.org/10.1007/978-3-031-38635-0

Klemperer, P. (1995). Competition when consumers have switching costs: An overview. *Review of Economic Studies, 62*, 515–539.
Kreps, D., & Scheinkman, K. (1983). Quantity precommitment and Bertrand competition yield Cournot outcomes. *Bell Journal of Economics, 14*, 326–337.
Lambertini, L., Poyago-Theotoky, J., & Tampieri, A. (2017). Cournot competition and green innovation: An inverted-U relationship. *Energy Economics, 68*, 116–123.
Levin, D., Peck, J., & Ye, L. (2009). Quality disclosure and competition. *Journal of Industrial Economics, 57*(1), 167–196.
Liu, Z. (2005) Stackelberg leadership with demand uncertainty. *Managerial and Decision Economics, 26*(5), 345–350.
Matsumura, T. (1998). Partial privatization in mixed duopoly. *Journal of Public Economics, 70*(3), 473–483.
Martin, S. (2003). *Advanced industrial economics* (2nd ed.). Hoboken: Blackwell Publishers.
Maskin, E., & Riley, J. (1984). Monopoly with incomplete information. *Rand Journal of Economics, 15*, 171–196.
Milgrom, P., & Roberts, J. (1982). Limit pricing and entry under incomplete information. *Econometrica, 50*, 443–466.
Milgrom, P., & Roberts, J. (1986). Prices and advertising as signals of product quality. *Journal of Political Economy, 94*, 796–821.
Motta, M., & Polo, M. (2003). Leniency programs and cartel prosecution. *International Journal of Industrial Organization, 21*, 347–379.
Motta, M. (2004). *Competition policy, theory and practice*. Cambridge: Cambridge University Press.
Pepall, L., Richards, D., & Norman, G. (2008). *Industrial organization: Contemporary theory and empirical applications*. Hoboken: Wiley-Blackwell Publishing.
Reynolds, R. J., & Snapp, B. R. (1986). The competitive effects of partial equity interests and joint ventures. *International Journal of Industrial Organization, 4*(2), 141–153.
Salant, S. W., Switzer, S., & Reynolds, R. J. (1983). Losses from horizontal merger: The effects of an exogenous change in industry structure on Cournot-Nash equilibrium. *The Quarterly Journal of Economics, 98*(2), 185–199.
Shaffer, G., & Zhang, Z. (2000). Pay to switch or pay to stay: Preference-based price discrimination in markets with switching costs. *Journal of Economics and Management Strategy, 9*, 397–424.
Singh, N., & Vives, X. (1984). Price and quantity competition in a differentiated duopoly. *RAND Journal of Economics, 15*(4), 546–554.
Shy, O. (1996). *Industrial organization: Theory and practice*. Cambridge: The MIT Press.
Takalo, T. (2001). On the optimal patent policy. *Finnish Economic Papers, 14*, 33–40.
Tirole, J. (1988). *The theory of industrial organization*. Cambridge: The MIT Press.
Varian, H. (1980). A model of sales. *American Economic Review, 70*, 651–659.
Winter, R. (1993). Vertical control and price versus nonprice competition. *Quarterly Journal of Economics, 108*, 61–76.

Index

P.-S. Choi et al., *Industrial Organization*, Springer Texts in Business and Economics,
https://doi.org/10.1007/978-3-031-38635-0